TELECOMMUNICATIONS

TELECOMMUNICATIONS

THE

MILITARY TELEGRAPH

DURING THE

CIVIL WAR IN THE UNITED STATES

William R[attle] Plum

Introduction
By

Paul J. Scheips

Vols. I & II

AYER COMPANY PUBLISHERS, INC.

Reprint Edition 2000
Ayer Company Publishers, INC.
Ayer Road
North Stratford, NH 03590

ISBN: 0-405-06053-X

Reprinted from copies in
The University of Illinois Library

TELECOMMUNICATION
ISBN for complete set: 0-405-06030-0
See last pages of this volume for titles.

Manufactured in the United States of America

UNION SIGNAL COMMUNICATIONS: INNOVATION AND CONFLICT

Paul J. Scheips

Reprinted from Civil War History, *Vol. IX, No. 4 (Dec., 1963)*

UNION SIGNAL COMMUNICATIONS: INNOVATION AND CONFLICT

Paul J. Scheips

THE STORY OF CIVIL WAR SIGNAL COMMUNICATIONS is the story of two organizations—the United States Army Signal Corps and the United States Military Telegraph. By the time of the American conflict several European nations had made some military use of the electric telegraph, notably in the Crimean War, but how well this European experience was known in the United States is not evident.[1] In any case, the Civil War, which was first in many things, was first in the extensive employment of the telegraph for all possible wartime purposes, although there was no military telegraph organization in the nation at the beginning of the war. Its army did have, however, one signal officer, Albert James Myer, who had introduced his visual system into the service. To that extent the U.S. Army was ahead of its contemporaries elsewhere.

From Fort Duncan, Texas, where he was an assistant surgeon, Myer inquired in 1856 if the government might be interested in his signaling system, which, he explained, grew out of an interest in military communications that went back to 1851. No action followed until 1859, when a board headed by Lieutenant Colonel Robert E. Lee gave a qualified endorsement to Myer's scheme. This was followed by field tests, principally conducted by Myer and Second Lieutenant Edward Porter Alexander.[2]

The results of these tests led to an act of June 21, 1860, which authorized the appointment of one signal officer with the rank of major

---

MR. SCHEIPS *is with the Office of the Chief of Military History, Department of the Army. This article is based on research completed for his doctoral dissertation to be presented at American University.*

[1] See *Senate Exec. Doc. No.* 59, 36 Cong., 1 sess. (Serial 1036), p. 110, and *Senate Exec. Doc. No. 60, ibid.* (Serial 1037), p. 87.

[2] Myer to Secy. of War, Oct. 1, 1856, Albert J. Myer Papers, Signal Corps Museum, Ft. Monmouth, N.J.; "Report of a Board of Officers for the Examination of the System of Military Signals Devised by Assistant Surgeon A. J. Meyer [*sic*]," Mar. 12, 1859, 53-L (1859), Record Group 94, National Archives; encl. (Alexander's reports) with Myer to Adj. Gen., June 16, 1860, Myer Papers. (Archives record groups are hereafter cited as RG.)

(over Senator Jefferson Davis' objections) and $2,000 for signaling equipment. Myer's assignment to the post quickly followed, and he was sent off to New Mexico to try his system. There he earned the friendship of Major E. R. S. Canby, who, while generally approving of Myer's system, questioned Myer's idea for having *all* army officers instructed in signaling, insisting instead that as a practical matter signal operations in large commands "should be confided to officers and men especially selected" for the duty.[3] Myer later essentially adopted this view.

It was also while in New Mexico in early 1861 that Myer, unable to secure Alexander's services, had detailed to him Second Lieutenants William J. L. Nicodemus and Samuel T. Cushing. Both—especially Nicodemus—would gain prominent places in the Signal Corps. (After the war, when the Rebel archives were probed, it would come to light that a resident of Baltimore had written Jefferson Davis in March, 1861, that Nicodemus desired a commission in the Confederacy; because of this letter, the War Department would later refuse to assign Nicodemus to a regiment and he would request and receive a discharge, perhaps never knowing where the trouble lay.)[4]

Orders of May, 1861, now took Myer to the East, where his system would get its first test in combat. The equipment commonly used in Myer's visual system consisted of flags for daytime and torches burning turpentine for night signaling. There was a red, a white, and a black flag, with a white center in both red and black flags and a red center in the white flag. The staffs, to which the flags were attached, were of hickory and made in four tapering joints fitted together by brass ferrules. The third joint had a six-inch brass guard at its upper end to protect it from the flames of the flying torch, which was attached to this joint for night signaling. A foot torch served at night as a point of reference. Messages were read with a glass.[5]

In this system a single flag was used at a time, the natural background against which the flag was to be displayed determining the color used. Particular swings of the flag or torch represented numerals and

[3] *Congressional Globe*, LIV (June 2, 1860), 2558-2559; *U.S. Statutes at Large*, XII, 66; [Albert J. Myer,] *Report of the Operations and Duties of the Signal Department of the Army, 1860-1865* (n.p., n.d.), p. 2; Canby to Myer, Apr. 10, 19 (two letters), 1861, Myer Papers.

[4] Myer, *Report*, p. 8; Hqs. Dept. of N. Mex., Spec. Ord. 26, Mar. 10, 1861, Myer Papers; John T. Mason to Davis, Mar. 26, 1861, C.S. Citizens File, RG 109, and, in N-31 CB (1870), RG 94, the following: Nicodemus to Townsend, Apr. 12, Dec. 24, 1870; memo, Adj. Gen. Off., Sept. 3, 1890.

[5] J. Willard Brown, *The Signal Corps, U.S.A., in the War of the Rebellion* (Boston, 1896), pp. 35, 115-118, 120-122; U.S. War Dept. (comp.), *The War of the Rebellion: A Compilation of the Official Records of the Union and Confederate Armies* (Washington, 1880-1901), Ser. III, I, 694 (hereafter cited as *OR*, with all references to Ser. III unless otherwise specified).

they, in turn, in different combinations, signified letters of the alphabet. Thus messages were spelled out, many abbreviations increasing the speed of transmission, although at best it was a slow system—three words a minute, on the average, and over an average distance between stations of ten miles on a clear day. Rockets and colored lights were also employed for special purposes.[6]

Myer reported for duty at Fortress Monroe in June, 1861. He failed in an effort to obtain Cushing's services, but Commanding General Ben Butler ordered details for signal duty and Myer proceeded to instruct them. Immediate results were the opening in late June of a signal line connecting Monroe and Newport News and the directing by signal of a battery on the Rip Raps.[7] The direction of gunfire would be a common service of the Signal Corps throughout the war.

While still assigned to Butler, Myer sought orders by which he could control all military telegraphy, asserting that the law under which he held his commission gave him "general charge of the telegraphic duty of the Army, whether . . . by means of signals transmitted by . . . electricity or by aerial signals." Although Myer obtained no War Department help, Butler ordered all telegraphic duty in his department, in which the budding U.S. Military Telegraph was already at work, placed under Myer's control. Myer implied that the immediate results were quite satisfactory, but the historian of the Military Telegraph later revealed that the word went out *sub rosa* to all telegraph operators to ignore Myer while seeming to comply with his orders, and that the Secretary of War soon instructed Butler not to interfere with them.[8]

Myer visited Washington, probably because of the difficulty over telegraphy. On the very eve of Bull Run he sought help in obtaining Congressional support for his view of the matter. He also obtained permission to command the balloon party he learned would make reconnaissances for McDowell's army. Early on the morning of July 21, Myer set out with a party which had an inflated balloon in tow, but through hurry and Myer's impatience the balloon was damaged and Myer had to go on without it. Myer reached the Bull Run battlefield after the fighting had begun and there served as an aide.[9]

On the Confederate side, however, the story was different, for Alexander, who had left the Federal service for a captaincy in the Rebel cause, reported to P. G. T. Beauregard in early June and trained several

[6] Brown, *Signal Corps*, pp. 63, 91-104; Albert J. Myer, *A Manual of Signals: For the Use of Signal Officers in the Field* (Washington, 1864), pp. 54-57d.

[7] Brown, *Signal Corps*, pp. 39-42.

[8] Myer, *Report*, p. 30; William R. Plum, *The Military Telegraph During the Civil War* (Chicago, 1882), I, 71-73.

[9] Myer to Callan, July 20, 1861, Myer Papers; F. Stansbury Haydon, *Aeronautics in the Union and Confederate Armies* (Baltimore, 1941), pp. 45-46, 64-65, 68-72; Brown, *Signal Corps*, p. 43.

men in Myer's system. In the fighting he was able to use signals to such good effect that Beauregard was prompted to call them "seasonable and material."[10]

Military Telegraph operators accompanied McDowell's army as far as Fairfax Court House; however, the Bull Run fighting was ten miles from the Fairfax telegraph office, so that McDowell had to use couriers to carry reports from the field to Fairfax for transmission to Washington.[11]

* * *

The Military Telegraph was strictly an expedient. News of the attack on Fort Sumter, April 12, 1861, sped throughout the nation via the three principal private telegraph companies—the American, Western Union, and Southwestern. In the emergency now upon the nation, Secretary of War Simon Cameron sought the aid of Thomas A. Scott of the Pennsylvania Railroad, who drew men from the Pennsylvania for railroad and telegraphic duties with the government. The first of these was Andrew Carnegie, shortly followed by David Strouse and others. Strouse was soon asked to look after telegraphic matters.[12]

In that early period the American Telegraph Company, whose wires reached Washington from the north, extended its facilities to the War Department and other points. Strouse's experience was indicative of the existing uncertainty and improvisation of the times, for when authorized to erect such military telegraph lines as might be required in Washington and the surrounding area, he was left for a period of seven months "dependent upon the American Telegraph Company for nearly every dollar. . . ."[13]

The U.S. Military Telegraph did not obtain formal sanction until Lincoln, in October, 1861, authorized Cameron to act on recommendations that had been made by Anson Stager, a Western Union official who had been invited to Washington. On February 26, 1862, under permissive legislation of the preceding month, the President took control of all telegraph lines in the United States, which meant in practice that the Military Telegraph could use them as circumstances demanded[14]

The operators of the Military Telegraph were mostly civilians, who used the commercial equipment of the day or adaptations of it. Edwin

[10] *Ibid.*, p. 45. [11] Plum, *Military Telegraph*, I, 75-78.

[12] *Ibid.*, 63, 66-67.

[13] *Ibid.*, 68; David Homer Bates, *Lincoln in the Telegraph Office: Recollections of the United States Military Telegraph Corps* (New York, 1907), pp. 35-36.

[14] Plum, *Military Telegraph*, I, 128; Bates, *Lincoln in the Telegraph Office*, pp. 31-32; *OR*, II, 12-13. The order seizing rail and telegraph lines named Stager the "supervisor [or] military superintendent of all telegraphic lines and offices in the United States." *OR*, I, 899.

Stanton, the new Secretary of War, opposed military status for the civilian operators so that they could avoid being ordered about by every ranking officer, although lack of such status later made them ineligible for government pensions.[15] Supervisory personnel, however, such as Stager (who became the superintendent), Thomas T. Eckert (in charge of the War Department office and assistant superintendent of the Department of the Potomac), and others—a mere baker's dozen altogether—were commissioned in the Quartermaster's Department in order to receive and disburse funds and property. All were made captains, except Stager, who became a colonel, and Eckert, a major. At the end of the war the officers still serving received brevet ranks.[16] Stager and the Military Telegraph were technically under Quartermaster General Montgomery C. Meigs, but in practice the organization came immediately under the control of the Secretary, which was one of its most remarkable characteristics. Keeping to the proper channels, Stager addressed his formal reports to Meigs, while Myer and the succeeding heads of the Signal Corps reported directly to the Secretary.

As was the case in the Signal Corps, the Military Telegraph transmitted important dispatches in cipher. Access to the cipher system was rigidly controlled, with youthful operators finding themselves in the difficult position of having to deny access even to army commanders, who themselves were left in an embarrassing position. Needless to say, controversy resulted.[17]

Published figures on Military Telegraph strength range from Plum's incomplete roster of 1,079 operators, through Woodring's figures of "about 1,200 operators and expert line builders," to Bates's estimate that the organization, "at its maximum strength, contained over fifteen hundred members." Besides operators, the Military Telegraph required the services of foremen, wagonmakers, teamsters, messengers, battery keepers, linemen, watchmen, and other kinds of workmen. Plum estimated that one operator in twelve was killed, wounded, captured, or died from exposure, but this does not take into account the casualties suffered by construction and repair personnel, who often ran great risks.[18]

* * *

[15] For postwar efforts to gain pension relief for veteran operators see Soc. of the U.S. Military Telegraph Corps, *Annual Report, 1915*, pp. 11-12, 41. In 1897 certificates were authorized to recognize honorable service "in the military telegraph corps of the Army." This was the only official basis, long after it had ceased to exist, for terming the Military Telegraph organization a "Corps." *U.S. Statutes at Large*, XXXIX, 497.

[16] See biographical sketches in Plum, *Military Telegraph*, I and II, *passim*.

[17] *Ibid.*, I, 34-61, II, 170-174, 264-266; Bates, *Lincoln in the Telegraph Office*, pp. 49-85; Roscoe Pound, "The Military Telegraph in the Civil War," *Massachusetts Historical Society Proceedings*, LXVI (1942), 193, 196, 200-201.

[18] Plum, *Military Telegraph*, II, 352, 376-380; William H. Woodring, "The

Almost immediately after First Bull Run, Myer sought approval of proposed legislation for formal organization of a small signal service whose officers would be "thorough" telegraphers, "practised in the use of both aerial and electric signals." He sought $50,000 for telegraphic purposes and explained his plans for "a telegraphic and signal train." He proposed to instruct detailed regimental officers in signaling, "all, or nearly all" of whom would be returned to their regiments after training. As an alternative, he thought these instructed officers could be combined into parties responsible for signal duty during engagements.[19]

Myer's recommendation concerning regimental duty is especially interesting in the light of later developments, for as the concept of a separate branch that would perform all signaling was accepted, the recall of detailed officers by commanders became such an evil that in mid-1862 orders had to be issued forbidding such action without permission of the Adjutant General of the Army.[20]

In August, 1861, instructions required Myer to "establish a system of signals along the line of the Potomac" to Washington through Maryland. If he found it necessary, he could "purchase a small telegraphic train" for communicating where ordinary signals could not be used, which was "to be paid for out of the telegraphic fund." The necessary officers and enlisted men were to be detailed.[21]

Contingents of officers and enlisted men now assembled at several small camps in Maryland where they were instructed in signaling by officers Myer had trained at Fortress Monroe. These were soon gathered together in the new Signal Camp of Instruction near Georgetown, D.C., where Cushing, once again on signal duty, was placed in charge. At the beginning all was chaos. The installation became the signal school for officers detailed from the Army of the Potomac. For a short time, until late 1861, it was also in effect Signal Corps headquarters. After training, detailed officers were transferred from the Army of the Potomac to other military jurisdictions, where they in turn had officers and enlisted men detailed to them and opened their own small camps of instruction. In May, 1862, the Georgetown facility was discontinued, but reopened again the following year and placed under Nicodemus, who returned to signal duty at that time.[22]

The line of signals ordered in August was completed before the end of November, 1862, and ran from Maryland Heights, overlooking

---

Military Telegraphers," *Telegraph and Telephone Age*, June 16, 1914, p. 359; Bates, *Lincoln in the Telegraph Office*, pp. 26-27; *OR*, III, 977.

19 Brown, *Signal Corps*, pp. 46-50; *OR*, I, 695.

20 *Ibid.*, II, 162-163.

21 Brown, *Signal Corps*, p. 50. See also penciled notation on [Myer] to Cameron, Aug. 6, 1861, Misc. Papers (1860-69), No. 72, Box 50, RG 111.

22 Brown, *Signal Corps*, pp. 50-60, 65-80; Myer, *Report*, pp. 17-19, 28-29, 140.

Harpers Ferry, to Fort Washington, on the Potomac opposite Mount Vernon. From its stations the Signal Corps was able to keep the upper Potomac under surveillance and to maintain communication with the forces surrounding Washington. One very important station in the chain was on Sugar Loaf Mountain, near Frederick, Maryland.[23]

Soon after assuming command of the Army of the Potomac, General George McClellan named Myer the signal officer on his staff. Myer therefore accompanied McClellan during the Peninsular Campaign and remained with him until the fall of 1862, when he asked to be relieved in order to look after Corps affairs in the Washington office. In his 1863 report, McClellan expressed appreciation for the services of both the Signal Corps and the Military Telegraph, speaking warmly of both Myer and Eckert. Later, however, perhaps thinking of its reconnaissance function, he saw in the Corps the weakness that its officers were not experienced soldiers, with the result that their judgment could not always be relied upon.[24]

Although McClellan found the services of the Corps largely good in retrospect, Myer was by no means wholly pleased. The fact rankled that he had only one partly-equipped telegraph train at his disposal. "The duties of the corps were novel," he complained, "and were understood by but few generals in the service. The acting signal officers were . . . volunteers, without any experience in military usage. They had been hastily instructed and equipped, and were thrown upon their first campaign in a country very difficult for their duties. . . ." Concluded Myer: "There were few . . . who aided them" and "it was often difficult to obtain official information of contemplated movements."[25]

At the beginning of the 1862 campaign the signal parties, fresh out of the Georgetown camp, had been assigned first to divisions of the Army of the Potomac, but parties later were reassigned to army corps and divided "so that four officers should be on duty with each division." Meanwhile, as has been noted, a number of signal officers were sent out to other armies. In June, 1862, the detachments in the Army of the Potomac were concentrated in one camp, and details made therefrom as necessary.[26]

Early in 1862, Myer was still promoting an unsuccessful scheme for a special wartime corps. The officers of this corps would train regi-

[23] Brown, *Signal Corps*, pp. 50, 65. The famous phrase "All quiet on the Potomac" supposedly originated on this signal line. *Ibid.*, pp. 228-229.

[24] *OR*, Ser. I, V, 575; Williams to Myer, Oct. 21, 1862, Myer Papers; Myer to Williams, Oct. 22, 1862, *ibid.;* Adj. Gen. Off., Spec. Ord. 331, Nov. 6, 1862, *ibid.* For McClellan's evaluations see *OR*, Ser. I, V, 31-32, and his *McClellan's Own Story* (New York, 1887), p. 135.

[25] *OR*, Ser. I, XI, pt. 1, 264.

[26] Myer, *Report*, pp. 37-38, 67; *OR*, II, 754.

mental signalmen, who would be detailed to signal duty temporarily and would serve in such capacity with "every army." Myer's thinking, however, soon changed. The experience of 1862, especially the continued demand upon a small force for instructed signal officers, led Myer to favor a permanent organization that would perform all signaling service and give the officers a corps of their own.[27] Finally he won authorizing legislation, which was approved on March 3, 1863.

Under this law there was to be for the Signal Corps (which would exist for the duration of the rebellion) a Chief Signal Officer with the rank of colonel, a lieutenant-colonel, two majors, a captain for each army corps or military department, and as many lieutenants—not exceeding eight—as the President deemed necessary. For each officer there was to be one sergeant and six privates. Both officers and enlisted men were to be examined and approved by military boards before appointment.[28]

To implement this act examining bodies were set up, the central one being known as the principal board. Myer received the principal board's unanimous recommendation for appointment as the Colonel and Chief Signal Officer. Later the lieutenant-colonelcy went to Nicodemus. Much dissatisfaction resulted from the actions of the examining boards, however, as hopes for advancement were frustrated and as some officers even found themselves reduced in rank.[29]

According to one source, 146 officers were "commissioned in the Corps" during the war, or were offered commissions. About twenty of this number "declined the appointments offered them, and some ten or twelve resigned from the army soon after the reorganization was effected." In addition, about 297 acting signal officers served in the wartime Corps, but some of them for only very brief periods. The total number of enlisted men who served at one time or another was about 2,500. In October, 1863, 198 officers, besides Myer, and 814 enlisted men graced the rolls of the Signal Corps.[30]

An outgrowth of the examining procedures in 1863 was the Felt court of inquiry, and both it and its sequel deserve some attention. It resulted from Myer's objections to a Signal Corps commission for George H. Felt, a first lieutenant and acting signal officer, on grounds of inefficiency and bad character. The court held hearings over several days in May of that year and cleared Felt of the charges. It was a remarkable performance, for, among other things, two signal officers sought to throw doubt on Myer's veracity. Their testimony, however, was of a hearsay

[27] Brown, *Signal Corps*, p. 142; Myer, *Report*, p. 138.

[28] *U.S. Statutes at Large*, XII, 753; *OR*, III, 94-95; Brown, *Signal Corps*, pp. 145-146.

[29] *Ibid.*, pp. 146 ff.; *OR*, III, 172-174.

[30] Brown, *Signal Corps*, pp. 160-161, 713-902; Norton to Spring, Nov. 21, 1867, LS, VII (1865-68), RG 111.

quality. Two other signal officers claimed that a code issued by Myer was about the same as a rocket code devised by Felt.[31]

Three months later Lieutenant Felt, alleging he had been denied examination by the principal examining board, carried formal charges against Myer to the Adjutant General of the Army, the burden of which was that Myer had made false statements; Felt also filed certain charges against Nicodemus. The War Department, however, refused to act. Complaining to the Adjutant General on September 21 of the inaction, Felt commented on the admitted weaknesses of Myer's visual signals, as he had in the court. In addition, he said he had notified Myer in July, 1862, that Myer's code was unsafe and that he had offered a code of his own invention as an alternative. Although Myer denied Felt's claims, it was true that after Chancellorsville an officer had complained that "the corps is distrusted, and considered unsafe as a means of transmitting important messages. It is well known that the enemy can read our signals when the regular code is used, and it is equally evident . . . that our cipher is unsafe and cannot be trusted. . . ." By that time, however, Myer had a cipher disk under development. Although he claimed its use was "habitual" by late August, 1863, chief signal officers in the field were not informed by circular until September 10 that disks would be distributed generally. Henceforth, important messages would be in cipher, said the directive, since "there is evidence, at this office, that the enemy can read our messages sent in the usual way."[32]

* * *

Meanwhile, the first telegraph train, which had been authorized in a qualified way in August, 1861, was built by Henry J. Rogers, a telegraph engineer, with Myer supervising. The principal transmitting and receiving equipment was a "letter indicating telegraph instrument" made for Rogers by the patentees, the Chesters of New York, and powered by portable batteries. For emergency use there were conventional pocket keys and sounders, which could be used only by skilled operators. The train, using light wagons, or carriages, was also supplied with lance poles and insulated wire on reels. It was delivered in January, 1862, and examined at the Georgetown camp by a board of officers

[31] "Proceedings of Court of Inquiry Convened by Special Order No. 85," MM-306 (Felt), RG 153, printed as George H. Felt (ed.), *Proceedings of a Court of Inquiry, Convened by Special Order No. 85* (New York, 1863).

[32] Felt to Thomas, Aug. 14 (two letters encl. charges), 28, Sept. 21, 1863, LR, 152-F (1863), RG 94; Myer to Stanton, Feb. 8, 1864, Misc. Papers (1860-69), No. 165, Box 51, RG 111; *OR*, Ser. I, XXV, pt. 1, 228; Myer to Condict, May 7, 1863, LS, IV (Jan.-Aug., 1863), RG 111; Myer to Cushing, Aug. 25, 1863, *ibid.*; Off. Sig. Officer Circs., Sept. 10, 16, 1863, "General Orders and Circulars, for 1861-1865," *ibid.* Felt patented a balloon or parachute rocket signal and a signal code for rockets in 1863. Myer patented his disk in 1865.

composed of Cushing, Benjamin F. Fisher, and David Wonderly, who on the whole approved it. It was impossible, however, to comply in early March with McClellan's desire for the train to take the field, since some problems connected with it had not been overcome. Wrote the board: "Among the chief impediments were the want of skilled operators," presumably for the sounders, "and the difficulty attendant upon the transportation of the electric batteries."[33]

Rogers at this point found that the magnetoelectric generator invented by George W. Beardslee could be employed in place of the cumbersome batteries. The result was the Beardslee magnetoelectric telegraph machine. The operation of this machine could be mastered quickly by any intelligent, literate person. As finally developed the complete instrument was housed in a brass-bound walnut chest, with handles, weighing about a hundred pounds.[34] Here was a tactical instrument, the first of its kind in the United States.

It was the Rogers train, the only one in being, that in late April, 1862, Myer ordered made ready for the field. He wanted Wonderly, "as rapidly as possible," to put "in complete working order, the whole apparatus with the two instruments . . . and enough acid to work them." Apparently at the last minute, however, Beardslee came forward with the offer of two of his new instruments, which, with two miles of wire, accompanied the train when it left for the Army of the Potomac on May 6. The next day Myer saw his first trial of the Beardslee machine. Although he had not yet "sufficiently tested the instrument to be perfectly assured of its strength and reliability," it appeared to Myer "so far superior" for the purposes of a military telegraph that he wanted the Army of the Potomac to have a portable lance line equipped with it, since he had "no doubt that on many occasions it might be used with valuable results."[35]

Three additional trains "of improved construction," each with two Beardslee instruments and five miles of insulated wire, and costing $2,500 each, reached the field later in 1862, thanks to McClellan's interest. One Signal Corps evaluation of its field telegraph in the first

[33] Myer, *Report*, pp. 33-34; Rogers to Myer, Nov. 14, 27, Dec. 16, 1861, and "Proceedings of a Board of Examination on the Field Telegraph Train," Feb. 25, 1862, LR, 11-R, 13-R, 14-R, 18-R (1861), 28-C (1862), RG 111.

[34] Beardslee took out two patents on a magnetoelectric machine as early as 1859; a later patent of 1863 was on the magnetoelectric telegraph. The Smithsonian Institution has three such instruments, all "revised models which contain a number of modifications." Another is in the Signal Corps Museum. Among the Beardslee Accession Papers in the Smithsonian is a copy of *Directions for the Practical Working of Beardslee's Magneto-Electric Signal Telegraph* (New York, 1863).

[35] Myer to Cushing, Apr. 25, May 7, 1862, LS, III (Mar.-Dec., 1862), RG 111; Cushing to Myer, May 6, 1862, *ibid.*; Myer, *Report*, pp. 53-54; Brown, *Signal Corps*, pp. 176-177.

battle of Fredericksburg, where it was under fire, was quite restrained, while another was enthusiastic. As for the manufacturer, he made the most of its use in that battle. He published a letter from his son, Captain Frederick E. Beardslee, who served there with the field telegraph and wrote that it "could not have worked better, and has been of great service to the country." Beardslee also employed a cut showing "Geo. W. Beardslee's Military Telegraph at the Battle of Fredericksburg" and used it on his company's letterhead. Years afterward the historian of the Military Telegraph presented far from complimentary views on the use of the Beardslee at such places as Fredericksburg, Chancellorsville, and Port Hudson. He called it an "expensive failure," although his hostility was doubtless colored by bitterness toward Myer for ignoring the battlefield labors of the competing civilian telegraphers. At the peak of their use in 1863 thirty Signal Corps telegraph trains were scattered through the various armies and at the Georgetown camp and West Point.[36]

The Beardslee instrument was experimental, of course, so that maintenance problems soon arose and changes had to be made. A common difficulty was the tendency of the sending and receiving instruments to get out of synchronization, with the result that the receiving pointer would stop at the wrong letter; moreover, the instrument was a short-range device, good for a distance of only five to eight miles. Plans were therefore made to improve the range of the instrument and also to equip the trains with a sounder Beardslee had invented that would work from magnetoelectric current.[37] This latter development would require trained operators, however, and Myer set about to obtain them.

In September, 1863, Myer published in *The Army and Navy Official Gazette* a call for "expert telegraphers," who were also "good English scholars," to apply for commissions in the Signal Corps. They would, he wrote, "have . . . charge of the . . . light field telegraph lines which are under . . . the Signal Corps, and which, in battle or at sieges, are run out and worked on the field or in the trenches under fire."[38]

The advertisement soon came to Secretary Stanton's attention, and he summarily asked the officer to explain. Myer's action was then formally disapproved with the advice that he "bear in mind that the Signal Corps is not an independent organization, but a branch of the service under the direction of the War Department." Although the stated ground for

36 *OR*, II, 759, III, 959; Myer, *Report*, p. 109; Beardslee Magneto-Electric Co., *Beardslee's Military Telegraph* (New York, 1863), p. 14; Plum, *Military Telegraph*, II, 96 ff. The Fredericksburg illustration was probably lifted from *Harper's Weekly*, Jan. 24, 1863, p. 53.

37 George Raynor Thompson, "Development of the Signal Corps Field Telegraph 1861-1863," *Signal*, XII (1958), 28-29, 31, 34; *OR*, III, 958-961.

38 *Army and Navy Official Gazette*, Sept. 8 (p. 146), 15 (p. 162), 22 (p. 178), 1863; Off. Sig. Officer Circ., Aug. 31, 1863, "General Orders and Circulars."

disapproval was a possible loss of Military Telegraph operators to the Corps, Myer claimed that from September 8 to 22 only two telegraphers had applied and neither was accepted. To say the least, however, this was misleading, if only because John C. Van Duzer, one of Stager's most valuable men, but one of those not yet commissioned in the Military Telegraph, applied for a Signal Corps position on September 12, revealing that he could secure for the Signal Corps most of the thirty-five operators and sixty-five repairers and laborers under him if they would not have to take a pay cut.[39]

Events now moved rapidly. Stager, who could not let Myer's challenge to the Military Telegraph go unanswered, wrote Stanton. He spoke of "the embarrassment already experienced and the complications likely to arise from the organizing of Field Telegraphs by the Signal Corps," and advised "the propriety of placing the Field Telegraphs under the . . . Military Telegraph Department, and thus avoid . . . two organizations in the same grade of service." He explained that the Signal Corps "is now making efforts to secure the best electricians in the service by offers of rank and increased pay, which it is enabled to do through its military organization, an advantage not possessed by the Military Telegraph. . . ." He recommended that either the Military Telegraph should have all telegraphic responsibility or it should be abolished and the entire responsibility given to the Corps.[40]

Stanton's decision was soon made and apparently imparted to Myer in a difficult interview at the War Department. On November 10, 1863, Myer was ordered to surrender his responsibilities to the next ranking Signal Corps officer (Nicodemus) and to leave for Memphis, Tennessee. At the same time all magnetoelectric telegraphic equipment was to be turned over to Stager.[41]

In a February, 1864, circular Stager explained that "the object in changing the supervision of the Field Telegraph from the Signal Corps"

[39] See gathering of pages bound with Albert J. Myer, *Memorial* (Washington, 1865), and his *Memorial Supplement* and appendix, Myer Papers. Van Duzer's application is in Reg., LR, XI (1863), V-7, RG 111. Besides Van Duzer, at least three other Military Telegraph operators applied before Sept. 22, and at least four more operators, later identified with the Military Telegraph, applied in October. For an interesting correspondence on this see Kellogg to Stanton, Dec. 21, 1863, encl. with Thomas to Kellogg, Dec. 15, 1863, Myer Papers.

[40] Plum, *Military Telegraph*, II, 100-101.

[41] Brown, *Signal Corps*, pp. 163, 181; *OR*, III, 1043; Plum, *Military Telegraph*, II, 102; Myer, *Report*, p. 152. On indications as far back as Port Hudson that the Corps would have difficulty keeping its telegraph lines, see Paul J. Scheips, "Signaling at Port Hudson, 1863," *Civil War History*, II-iv (1956), 109, and Off. Sig. Officer Circ., Sept. 18, 1863, "General Orders and Circulars." In his annual report for 1863 Stanton recommended that the Corps be continued "with proper restrictions upon the number of officers," who would be limited "to their proper scope of duties." Plum, *Military Telegraph*, II, 104.

to the Military Telegraph was to make it "a more efficient and useful branch of the military service, by placing it in the hands of expert telegraphers." He charged his officers to organize the field trains "in the best possible manner for spring operations." The Caton field instrument would be used, "together with number fourteen wire," and mules would carry both the wire and batteries. The batteries, indeed, would be worked from the backs of the mules.[42]

Notable use of the field telegraph by the Military Telegraph was made under Grant and Meade in Virginia, Sherman in Georgia, and Banks in the Red River Expedition. Grant himself provided an admiring description of the Military Telegraph's erection of field lines in the Army of the Potomac when in camp, relating that divisions, corps, and his own headquarters would be connected once the army came to a halt "in a few minutes longer time than it took a mule to walk the length of" the wire it carried. Eckert, too, was enthusiastic about his field telegraph lines, saying that not only were the various headquarters of the army connected "during *every engagement*," but that "every reconnaissance . . . in force has had telegraphic connection with headquarters. . . ." One operator recalled that at Spotsylvania both the field wire and its operators were at one time within the enemy's lines.[43]

Captain Samuel Bruch of the Military Telegraph reported that "the new instruments and materials" furnished by Stager seemed "to be much better adapted to field service than those formerly used by the Signal Corps." The Beardslee machines taken from the Signal Corps were never employed by the Military Telegraph. Eckert says he "gave them a thorough trial" in the Army of the Potomac and, finding them "to be of very little practical use," they were sent to the rear before the army moved in the spring of 1864. When Myer was with Canby in the Division of West Mississippi that same year, Canby ordered the telegraph trains in New Orleans sent to Myer, but this was an exception to existing policy.[44]

It appears from the Military Telegraph's own figures that its field lines, at least quantitatively speaking, were by no means as important as its other lines. Thus, out of a total of 15,389 miles of lines constructed by the Military Telegraph from May 1, 1861, to June 30, 1866, only one thousand miles were characterized as "temporary field lines." Stager believed, however, that there were possibilities in the field telegraph,

[42] *Ibid.*, 102-103.

[43] U. S. Grant, *Personal Memoirs* (New York, 1885-86), II, 205-207; Plum, *Military Telegraph*, II, 141; *OR*, Ser. 1, LI, pt. 1, 200; John Emmet O'Brien, *Telegraphing in Battle* (Scranton, Pa., 1910), p. 143.

[44] *OR*, IV, 821, 854-855; Eckert to Meigs, Dec. 8, 1864, Off. Q.M. Gen. Rpts. (1864), RG 92; Myer, *Report*, p. 227.

for in 1865 he expressed an opinion that "the Field Telegraph which we have operated, independently, as well as in connection with the Signal Corps, could be so perfected and operated, as to . . . supplant the use of a signal corps for military operations."[45]

Despite the rivalry and friction between the Military Telegraph and the Signal Corps, Stager testified in 1865 to their cooperation. He observed that the Military Telegraph had operated frequently in the field with the Signal Corps, and had "rendered efficient aid in this respect, by diffusing information from advanced signal stations simultaneously" to the headquarters of the commanding general and the different corps headquarters. In September, 1863, for example, operators Lonergan and Forbes, in the Department of the Cumberland, were stationed at the signal post on Sand Mountain where they forwarded reports from six connecting signal stations. The history of Chancellorsville, Port Hudson, the approach to Atlanta, and Allatoona offers varying examples of interaction between the two organizations on the operational level.[46]

* * *

When Myer, a colonel by recess appointment, turned over Signal Corps headquarters to his junior, Nicodemus, in November, 1863, and departed circuitously for Memphis, his personal Siberia, he was an unhappy man. This was understandable, especially since the Corps was peculiarly his own creation. He went to New York to consult his old friend Canby and then from Pittsburgh, en route to Memphis, he wrote his wife Kate that the "more deliberately I reflect upon the matter the more certain I am that I am right and History will show it so. . . ."[47]

Myer's complaints were against Stanton and against the telegraph companies—especially, it seems, the American Telegraph Company. As he looked back after the war in a report he must have written, at least in part, he was convinced that "the powerful and despotic sway of the corporations," together with "the citizen" (probably Stanton, although possibly Stager), had spelled death to the Signal Corps field telegraph. But timid army opinion, skeptics, and the uninterested also got their knocks. Edwin Stanton he obviously disliked with a passion. He appears even to have viewed the law empowering the President to take possession of the telegraphs as a dark, sinister plot by Stanton to gain eventual control of the country. One can almost imagine the Myer children, who watched the Secretary as he went by their house in Wash-

[45] Plum, *Military Telegraph*, II, 349; Stager to Meigs, Sept. 15, 1865, Off. Q.M. Gen. Rpts. (1865), RG 92.

[46] *Ibid.;* Plum, *Military Telegraph*, I, 363-364, II, 66, 97-99, 178-179, 182, 232; Scheips, "Signaling at Port Hudson," 106-113; *OR* IV, 865; Fred E. Brown, "The Battle of Allatoona," *Civil War History*, VI (1960), 280, 297.

[47] Myer, *Report*, pp. 152-153; Myer to Kate Myer, Nov. 20, 1863, Albert J. Myer Collection, Library of Congress.

ington, being cautioned that if they were not good "Old Mars" would get them.[48]

Myer no sooner reached Memphis than he obtained a ten-day leave of absence from local military authorities and journeyed to Cleveland to see Stager, who had moved his office there. The two devised an agreement on combining the Military Telegraph and the Signal Corps at the various headquarters. As Stager understood it, the combined facilities were to be under "experts from the Military Telegraph, commissioned in the Signal Corps," the former organization to control the lines above those headquarters and the latter the field lines. Stager agreed to take the idea up with Stanton, but Myer soon asked him to keep the matter confidential for the time being. Probably he decided that Stager would not be his best advocate. When he was asked officially to explain why he had gone to Cleveland, he replied that he had gone to see his father and to confer with Stager. Myer also emphasized the importance of having field lines under a military organization, and he regretted the lack of formal liaison between the Signal Corps and the Military Telegraph. With this reply the matter seems to have rested.[49]

Early in December, 1863, Myer received instructions to report upon the desirability of establishing a line of signal stations between Memphis and Cairo. He wrote Kate in high spirits about the great possibilities of this project, but Grant unfavorably endorsed Myer's enthusiastic "preliminary reconnaissance" report and that was the end of that.[50]

While still at Cairo in January, 1864, Myer completed *A Manual of Signals: For the Use of Signal Officers in the Field*, which he had begun before his dismissal from Washington because of the need for "a signal text-book of some kind." Simeon White, one of the civilian clerks in the Washington headquarters, saw to its printing on a press the Corps had acquired, and helped Myer send complimentary copies to important persons.[51] In 1866 Myer produced a larger edition, which was basically

[48] Myer, *Report*, pp. 32, 226; memos, apparently all to Tafft, [Sept.?] 1863, Myer Papers; Off. Sig. Officer Circ., Sept. 18, 1863, "General Orders and Circulars"; "The Government and the Telegraph" (draft MS), Myer Papers; Kate Myer to Myer, Jan. 15, 1864, *ibid.*

[49] Myer to Adj. Gen., Nov. 26, 1863, Apr. 16, 1864, *ibid;* Hqs. Sixteenth Corps, Spec. Ord. 300, Nov. 28, 1863, *ibid.;* Myer to Stager, Nov. 28, Dec. 15, 1863, *ibid.;* Stager to Myer, Dec. 9, 1863, Apr. 14, 1864, *ibid.;* Canby to Myer, Mar. 21, 1864, *ibid.*

[50] Hardie to Myer, Dec. 5, 1863, *ibid.;* Myer to Kate Myer, Dec. 24, 1863, Myer Collection; Grant to War Dept., Feb. 14, 1864, Myer Papers; Myer to Stanton, Dec. 26, 1863, *ibid.*

[51] Myer, *Manual of Signals* (1864 ed.), preface; White to Myer, Feb. 1, May 10, June 11, 1864, Myer Papers.

unchanged in still later editions, the last of them published in 1879. The 1864 edition, a paperback booklet, dealt with subjects such as signal codes, the use of the telescope and marine glasses, the care of apparatus, the working of signals in the field (including how to select a signal station), cipher disks, field telegraphs, the duties of a signal officer, and the early history of signals.[52]

Publication of the handbook was doubtlessly a source of satisfaction to Myer in this period of his exile, marked otherwise by a refusal to grant his request for a court of inquiry; by Signal Corps Captain Henry S. Tafft's repeated complaint that Nicodemus was ambitious and not to be trusted, together with Kate's announced distrust of Tafft (she thought he looked too pleasant) and her suggestion that, excepting Nicodemus, Myer's officers might prefer *not* having him in Washington; and by a report that Assistant Secretary of War Charles A. Dana thought the Signal Corps ought to be abolished.[53] But there were some bright spots, too, besides the *Manual of Signals,* including a petition that the signal officers of the Army of the Potomac addressed to the Secretary of War in Myer's favor. The happiest turn of official events, however, came in May, 1864, when Canby selected him to be the signal officer in his new command, the Military Division of West Mississippi.[54]

With Canby, Myer added a new duty to the tasks of the Signal Corps by working out a system for interrogating deserters and refugees who came into the Union lines. He also secured the adoption and issuance of a "General Service Code" for use in transmitting routine messages between land and sea forces. Finally, he organized signaling plans for the operations in the Mobile area (during which he spent a stormy night under a palmetto on Dauphin Island), and he participated, with Farragut's fleet captain, Drayton, in the surrender of Fort Gaines.[55]

While Myer prepared for the Mobile campaign the final blow to his wartime career was delivered, although it was some time before the word reached him. When it finally arrived he learned that his recess

[52] The 1866, 1868, and 1871 editions were published by D. Van Nostrand, New York; the last two added material, respectively, on compasses and drill. Van Nostrand issues of 1872 and 1874 were reprints. In 1877 Myer brought out another new addition which included coverage of the heliograph and permanent telegraph lines; the last edition, 1879, included material on the telephone. These editions were produced by the U.S. Govt. Printing Off.

[53] Myer to Adj. Gen., Feb. 22, 1864, Myer Papers; Myer, *Memorial,* p. 5; Tafft to Myer, Dec. 27, 1863, Feb. 7, 28, June 14, 1864, Myer Papers; Kate Myer to Myer, Nov. 28, Dec. 8, 1863, *ibid.* See also n. 74 below.

[54] Myer to Norton, Apr. 12, 1864, Myer Papers; Hq. Mil. Div. of W. Miss., Spec. Ord. 32, June 7, 1864, *ibid.;* Myer to Kate Myer, May 15, 18, 1864, Myer Collection.

[55] Brown, *Signal Corps,* pp. 593-594, 596, 611; Hqs. Mil. Div. of W. Miss., "General Service Code," July 11, 1864, Myer Papers; Myer to Canby, July 22, 1864, *ibid.*

appointment as Colonel and Chief Signal Officer, which had been made before his dismissal in 1863, had not been confirmed by the Senate and was revoked. He still retained his rank of major, but would see no further active duty until after his restoration in late 1866. Kate, who was quickly informed about the whole thing by Tafft, hastily wrote Myer, explaining that his name had not been on the list of nominations sent to the Senate. Tafft was outraged and said he was going to resign, which he soon did. As for Myer, Kate said both she and her mother hoped he would now resign and devote his thoughts to domestic life.[56]

Myer's staunch determination to secure restoration, however, never wavered. He hired lawyers and enlisted the support of influential politicians, and in January, 1865, completed a memorial to the Senate of about a dozen pages which he printed and distributed. Later he would supplement it, largely with documents. Early in January he wrote Senator Ira Harris that he had taken his advice and gone alone to see President Lincoln, who promised not to appoint another Chief Signal Officer without first satisfying himself that the appointment was just. Myer feared, however, that Lincoln was so busy he might sign an appointment without looking at it. Subsequently he wrote Lincoln, enclosing some papers pertaining to his case, but he waited rather too long, for the letter was dated April 14, 1865.[57]

Victory, when it came on October 30, 1866, was especially sweet for Myer because it was Stanton who had to notify him that he had been appointed Colonel and Chief Signal Officer under an act of July 28. He was not confirmed in the position, however, until February, 1867, and was not ordered to active duty until August, 1867.[58] Once back in office he set to work and almost immediately won acceptance for, and publication of, orders looking toward instruction in signaling and equipping the army with signaling apparatus. Significantly, the Chief Signal Officer's responsibility henceforth would embrace the electric field telegraph.[59]

* * *

[56] Adj. Gen. to Myer, July 21, 1864, *ibid.;* Hqs. Mil. Div. of W. Miss., Spec. Ord. 88, Aug. 5, 1864, *ibid.;* Kate Myer to Myer, July 30, 1864, *ibid.;* Brown, *Signal Corps,* p. 881; "Statement of the Military Service of Albert J. Myer," Service Records, Signal Corps Hist. Sect. File, Dept. of the Army; Grant to Adj. Gen., Nov. 30, 1867, Myer Papers.

[57] Kate Myer to Myer, Jan. 12, 1865, *ibid.;* Fenton to Stanton, Nov. 26, 1864, *ibid.;* Myer to Harris, Jan. 2, 1865, *ibid.;* Myer to Lincoln, Apr. 14, 1865, 2563 ACP (1873), RG 94. See also n. 39 above.

[58] Stanton to Myer, Oct. 30, 1866, Myer Papers; *Exec. 508* (Confid.), 39 Cong., 2 sess., Feb. 8, 1867, with notation by Myer, *ibid.;* Myer's commission, Mar. 4, 1867, *ibid.;* Kelton to Myer, Mar. 14, 1867, *ibid.;* Adj. Gen. Off., Spec. Ord. 419, Aug. 19, 1867, *ibid.;* Norton to Myer, Aug. 13, 1867, *ibid.;* White to Myer, Aug. 13, 1867, *ibid.;* "Statement of Military Service, Myer"; *U.S. Statutes at Large,* XIV, 335-336.

[59] Myer remained Chief Signal Officer until his death in 1880.

When Nicodemus had assumed charge of Signal Corps headquarters in November, 1863, the personnel reorganization under the legislation of that year was still proceeding. As a result, in subsequent months some of his time was taken up with the principal examining board and related matters. In general, he seems to have administered the Corps essentially within the pattern that Myer had set.[60]

As 1864 moved to its close, Nicodemus prepared his annual report for the preceding fiscal year. In it he observed that signaling instruction had been discontinued at West Point, but that it had been "quite satisfactory" at the Naval Academy. He reported on the year's operations in the field, listing the losses—six killed, one drowned, five dead from wounds and twenty-two from disease, eleven wounded, two missing in action, and thirty-one taken prisoner, for aggregate casualties of seventy-eight. He also mentioned, in at least five different places, the reading of enemy signals. In a section on field telegraphs he quoted Myer on the opposition to the Signal Corps field telegraph as well as on its importance, and recommended return of the telegraph trains to the Corps.[61]

This report immediately aroused Stanton, who seized all undistributed copies and also the manuscript. Orders of December 26, 1864, dismissed Nicodemus from the army "forthwith" for having

> contrary to the regulations, and disregarding his duty as an officer, published, without authority and without the knowledge or sanction of the Secretary of War, a document relating to the branch of service under his charge, purporting to be the annual report of the Signal Corps, which contains information useful to the enemy and prejudicial to the service of the United States.[62]

As it happened, Nicodemus' dismissal came the day before his wedding, which prompted a gleeful Myer to remark in a letter to Kate: "Alas poor Yorhick. . . . Just married and dismissed—I hope his wife will comfort him—but if the poor woman has married the coat and buttons! !" In another letter he said he "could be sorry" for Nicodemus "if I were not of the impression that he has done all he could in his feeble way to hurt me."[63]

Nicodemus twice unsuccessfully asked for an investigation. He de-

[60] Nicodemus to Myer, Nov. 23, Dec. 9, 26, 31, 1863, Jan. 15 (two letters), 30, Feb. 5, 12, 13, 19, 25, Mar. 5, 1864, Myer Papers; Myer to Nicodemus, Jan. 11, Feb. 16, *ibid.*

[61] *OR*, IV, 818-841.

[62] Tyerell to Ingraham, Dec. 28, 1864, LR, 802-T (1864), RG 94; Brown, *Signal Corps*, p. 87. Stanton's original MS order of Dec. 26 which dismissed Nicodemus (N-31 CB [1870], RG 94) omitted the words "contains information useful to the enemy and prejudicial to the service of the United States."

[63] C. W. Butterfield, *History of the University of Wisconsin* (Madison, 1879), p. 145; Brown, *Signal Corps*, p. 37; pension file WC-294, 349, RG 15B; Myer to Kate Myer, Jan. 2, 9, 1865, Myer Collection.

fended himself by saying that he was innocent of any attempt to publish his report without the Secretary's knowledge. He said he had sent the Secretary a copy at the end of October and then printed it for signal officers as had been done previously. He said he understood "from rumor on the street" that his report aided the Confederates by informing them that the Union Army could read their signals. It was hardly possible that this should aid the enemy, he thought, since the system that E. P. Alexander had introduced into the Rebel army was the same as that used by the Union, except that the former was unimproved whereas the latter was improved (by the use of the cipher disk) so as to make it impossible for the enemy "ever to decipher it." "To tell the enemy that we can read the system of signals which he stole from us can be no information to him," he maintained, "nor that we have so improved this system that he can't read our signals."[64]

Apparently with the help of a letter from Canby, Nicodemus was restored in March, 1865, to his rank of Signal Corps lieutenant colonel "in consideration of his former services and proper acknowledgment of his error," although Canby had recommended his return to the infantry. He was assigned to duty as a Signal Corps inspector. When mustered out of the Corps in August, 1865, he returned to the line and his former rank of captain, in which rank he remained until he left the service in 1870.[65]

When Nicodemus was summarily removed from office he was succeeded by Benjamin F. Fisher, who had come into the Signal Corps from a Pennsylvania volunteer regiment. Fisher had been captured near Aldie, Virginia, in June, 1863, and confined in Libby Prison, from which he escaped in early 1864. Upon his return to service Fisher was reassigned to his old position as the chief signal officer of Meade's army.[66]

Myer, at least for a time, evidently believed that Fisher worked unfairly against him, although in his 1865 memorial to the Senate he claimed he had "none other than kindly relations" with Fisher. Fisher, however, did actively seek the position of Chief Signal Officer and won appointment vice Myer in early December, 1863, before Nicodemus'

[64] Nicodemus to Thomas, Jan. 9, 1865, N-31 CB (1870), RG 94; Nicodemus to Stanton, Mar. 9, 1865, *ibid.*

[65] Brown, *Signal Corps,* p. 842; Canby to Stanton, Jan. 27, 1865, N-31 CB (1870), RG 94; Eaton to Myer, July 18, 1865, Myer Papers. Nicodemus was mustered out of the Corps in Aug., 1865. Unable to secure a regimental command in 1869, for reasons already noted, he requested and received his discharge in 1870. The following year he joined the faculty of the University of Wisconsin, at which post he died in 1879. "Statement of the Military Service of William J. L. Nicodemus"; Butterfield, *History,* pp. 140-146.

[66] Brown, *Signal Corps,* pp. 77, 160, 164, 358, 371, 769-770; pension file SC-502, 666, RG 15B; "Statement of the Military Service of Benjamin F. Fisher."

dismissal that month.[67] With Myer laboring against him, Fisher's appointment fell by constitutional limitation in March, 1865, but he immediately received another appointment, which also fell upon the adjournment of the Senate in July, 1866.[68] In taking leave of Stanton four months later, Fisher asked that his military conduct be endorsed with the brevet rank of brigadier general, which he and Myer both subsequently received. He then turned the records and office over to Captain L. B. Norton and left.[69]

When Fisher became Chief Signal Officer the war had about run its course, and it was Fisher's destiny to preside over the dissolution of the wartime Corps, the all-but-final mustering out of which was ordered in August, 1865. In his first report, which covered the end of the war, Fisher said that "the duties of the corps during the past year were better understood than in previous years, which gave to it more tone and character, and enabled it to approximate [its real function] in most of the military departments. . . ." He was especially pleased about the Department of the Gulf. In citing especially remarkable service during the year he mentioned Allatoona and Fort Fisher. He commented upon the value of the cipher disk, by the use of which messages could be sent that would be valueless to the enemy, even if they fell into his hands, because of the time it would take him to decipher them—"if they did not absolutely defy deciphering. . . ." Finally, Fisher observed the need for a small, permanent signal organization and suggested the appointment of a board "to define, as far as practicable, the specific duties to be assigned it, to avoid in the future the great stumbling-block which . . . in the past . . . crippled the usefuless of the corps by its not being properly understood. . . ." He did not, however, spring to the advocacy of the electric telegraph.[70]

* * *

Nobody knows, apparently, how many messages were sent by flag

67 Fisher to Myer, Nov. 11, 1864, with Myer's notation, Myer Papers; Brown, *Signal Corps*, pp. 166-167; Fisher to Stanton, Oct. 13, 1864, F-466 CB (1864), RG 94; Fisher to Adj. Gen., Dec. 7, 1864, *ibid.*

68 Theophilus F. Rodenbaugh (ed.), *The Army of the United States* (New York, 1896), p. 136; Fisher to Thomas, Mar. 14, 1865, F-466 CB (1864), RG 94; Kelton memo of Nov. 10, 1866, F-466 CB (1866), *ibid.*; Kelton to Fisher, Nov. 12, 1866, Myer Papers.

69 Fisher to Stanton, Nov. 15, 1866, F-466 CB (1866), RG 94; Grant to Myer, Nov. 20, 1867, Myer Papers; Fisher to Kelton, Nov. 15, 1866, F-466 CB (1866), RG 94; White to Myer, Nov. 16, 20, 1866, Myer Papers. Fisher subsequently practiced law in Philadelphia, dying in 1915. Pension file SC-502, 666, RG 15B.

70 Adj. Gen. Off., Spec. Ord. 417, Aug. 3, 1865, ordered the immediate mustering out of all Signal Corps officers and men east of the Mississippi, excepting Fisher (although Norton managed to stay on); Adj. Gen. Off. Circ. 40, Aug. 11, 1865, set forth the procedures. See copies in Myer Papers; see also *OR*, V, 152-156, and *House Exec. Doc. No. 1*, 39 Cong., 2 sess. (Serial 1285), p. 702.

and torch and by the Beardslee machine during the Civil War. The historian of the Military Telegraph estimates that the messages sent by that organization probably aggregated 6,500,000 by the end of fiscal year 1865, by which time the direct cost to the government was $2,655,500, or about forty cents a message. The direct cost of the Signal Corps for about the same period was substantially less—$1,595,257.[71]

When the Civil War ended the Military Telegraph supervised the restoration of commercial telegraph lines in the South, but its control was soon relinquished. Meanwhile, operators and Stager's commissioned assistants remained at their posts until November 30, 1865, when all operators not at work on strictly military lines or at assigned posts as cipherers in major cities were discharged, paid, and, as one operator put it, "in most cases given transportation to their homes." In 1866 the Military Telegraph lines south of the Ohio River were turned over to commercial companies in relinquishment of claims against the United States, while military lines north of the Ohio were sold. The line from Wilmington, Delaware, to Richmond, however, was retained to be operated for the government by the American Telegraph Company. Of the officers, only Stager and Eckert, both of whom received the brevet rank of brigadier general, remained on duty by the end of fiscal year 1866. One operator, Charles Almarin Tinker, remained in the War Department telegraph office until 1869.[72]

Among the Civil War leaders who thought highly of the Military Telegraph were McClellan, Grant, Sherman, Meigs, and, of course, Stanton. On the other hand, we know that Schofield and Grant had difficulties with the organization, and none other than Roscoe Pound has criticized severely both the operators' monopoly of the cipher and their independence. We know, too, that Myer and the Signal Corps did not look kindly upon their rival. Indeed, the largely civilian character and independence of the Military Telegraph made it a military anomaly. Nevertheless, in one degree or another, the organization served the administrative, logistic, strategic, and even tactical needs of the War Department and the army.[73]

[71] On the cost of the Military Telegraph see Plum, *Military Telegraph,* II, 337. Signal Corps expenditures totalled $1,586,720 for fiscal years 1862-64, and $8,537 for the year ending Sept. 30, 1865. *OR,* IV, 838, V, 154. The figure $2,040,265 for the four years ending Dec. 31, 1865, appears in encl. 1 to LS, No. 110, Feb. 4, 1865, Misc. Papers (1860-69), Box 52, RG 111.

[72] Plum, *Military Telegraph,* II, 346 ff.; Woodring, "Military Telegraphers," 357; Eckert to Meigs, July 23, 1866, Off. Q.M. Gen. Rpts. (1866), RG 92; Van Duzer to Meigs, June 30, 1866, *ibid.;* Bates, *Lincoln in the Telegraph Office,* p. 393; Soc. of the U.S. Military Telegraph Corps, *Annual Report, 1915,* p. 35.

[73] Plum, *Military Telegraph,* I, 231, II, 252-255, 355-356; Grant, *Memoirs,* II, 205-207; William T. Sherman, *Memoirs* (New York, 1875), II, 398; Bates, *Lincoln in the Telegraph Office,* p. 12; John M. Schofield, *Forty-Six Years in the Army*

The Signal Corps, also had its detractors as well as supporters. The military telegraphers were critical of what they regarded as Signal Corps pretensions in the field of electric telegraphy. Secretary Stanton was not its advocate and neither was Assistant Secretary of War Dana, who charged in 1865 that "we should have been better off without it."[74]

Actually, despite solicited testimonials, the Signal Corps and Myer received many warm expressions of commendation. These were not always unqualified, but nevertheless such army leaders as McClellan, Thomas, Grant, Sherman, Granger, and even Sheridan (despite the fact he called Myer an "old wire-puller"), and Farragut and Porter of the navy, are numbered among those who looked upon the Corps with favor varying from mild to enthusiastic. Although Myer was without a signal service at Bull Run in 1861, this lack was soon remedied. By Gettysburg, two years later, when Meade called upon all staff corps, including the Signal Corps, to aid him in the coming battle, the Corps had won a secure place for itself.[75]

During the Civil War the Signal Corps stood guard at innumerable high observation posts and signaled by flag and torch from hills, high buildings, specially constructed towers, tree platforms, and even the mastheads of ships. The service excited the curiosity of soldiers who had never seen anything like it before.[76] But signals by flag and torch had the obvious weaknesses of all visual signals: They were dependent upon line-of-sight and they were obscured by bad weather and by the smoke and fog of battle. Even as they were given their first real test in war the electric telegraph was rendering them old-fashioned.

Myer was aware of the weaknesses of visual signaling and, as he had been familiar with the electric telegraph since his student days, he early introduced the latter into the Corps. He was unfortunate in the choice of the apparatus adopted, but even with more dependable equipment than the Beardslee machine he would have had difficulty because of a strong

---

(New York, 1897), pp. 168-169, 204, 206, 211-212, 232-233; Roscoe Pound, "Military Telegraph," 185-203; Pound, "Bureaus and Bureau Methods in the Civil War Era," *Massachusetts Historical Society Proceedings*, LXVII (1945), 420-435; Francis T. Miller (ed.), *The Photographic History of the Civil War* (New York, 1911), VIII, 368. See also n. 24 above.

[74] A. J. M., "The Cipher of the Signal Corps," *Army and Navy Journal*, Mar. 3, 1866, p. 26.

[75] See Myer to Norton, cited in n. 54 above, and the correspondence about testimonials in Myer Papers. See also *OR*, IV, 839-841; George B. McClellan, *The Complete Report on the Organization and Campaigns of the Army of the Potomac* (rev. ed.; n. p. [1864?]), pp. 14-15; Brown, *Signal Corps*, pp. 168, 195-203; Grant, *Memoirs*, II, 207-208; statements of Grant, Thomas, Granger, Sheridan, Farragut, and David D. Porter in Myer Papers; Rodenbaugh, *Army of the United States*, p. 140.

[76] John D. Billings, *Hardtack and Coffee* (Boston, 1887), p. 404; *OR*, Ser. I, XI, pt. 1, 226 ff.; Brown, *Signal Corps*, *passim*.

rival organization. Nevertheless, that the Beardslee was a short-range instrument, that it often got out of synchronization, that it was slow, and that a single man could not easily carry it—these things are all beside the point, which is that the concept of rugged, front-line, portable, electric communications an ordinary soldier could employ was something new in the U.S. Army.

The Military Telegraph, while it had to develop its organization from the ground up and had to develop its own logistical system, never encountered some of the problems the Signal Corps faced. The latter not only had to pioneer a new military organization, but, coping with the difficult detail system, had to change its personnel procurement in the midst of war with resultant morale problems. While the Military Telegraph possessed stability in its top command after an initial period of organization and enjoyed the unflagging support of the Secretary of War, the Corps had three different wartime chiefs and considerable interference from the Secretary. Whereas the Military Telegraph was able to take trained telegraphers from the commercial system, the Corps had to instruct all of its enlisted men and all but three of its officers.

It is clear in retrospect, as it surely appeared to the leading participants at the time, that a division of signal communications between two separate organizations was far from satisfactory. Certainly it does not appear wise to have invested an essentially civilian organization with such heavy responsibilities. The newness of the signal communications function, however, and the consequent lack of experience and planning, together with the hurly-burly of war, produced almost inevitable overlap and conflict. Under the circumstances, Stanton's separation of the field telegraph from the Signal Corps in 1863 was probably administratively sensible, but it was not a permanent solution. In the end, as the war passed, the essentially civilian Military Telegraph was dissolved, while the Signal Corps survived, although not without difficulties, to celebrate its centennial in 1960.

THE

MILITARY TELEGRAPH

DURING THE

CIVIL WAR IN THE UNITED STATES

VOL. I.

Anson Stager

THE

MILITARY TELEGRAPH

DURING THE

CIVIL WAR IN THE UNITED STATES,

WITH

AN EXPOSITION OF ANCIENT AND MODERN MEANS OF COMMUNICATION, AND OF THE FEDERAL AND CONFEDERATE CIPHER SYSTEMS;

ALSO A

RUNNING ACCOUNT OF THE WAR BETWEEN THE STATES.

BY

WILLIAM R. PLUM, LL. B.,

OF THE CHICAGO BAR.

VOL. I.

CHICAGO:

JANSEN, McCLURG & COMPANY, PUBLISHERS.

1882.

Printed by Donnelley, Gassette & Loyd, Chicago.

TO

THE MEMORY OF THOSE OF THE

UNITED STATES MILITARY TELEGRAPH CORPS, WHO DIED WHILE SERVING

THEIR COUNTRY;

OF THOSE WHOSE DEMISE AT HOME RESULTED FROM THE

EXPOSURES OF SUCH SERVICE,

AND OF THOSE WHO FROM OTHER CAUSES HAVE FOLLOWED THEIR LAMENTED

CO-PATRIOTS;

THESE PAGES ARE SACREDLY DEDICATED,

IN THE SINCERE HOPE THAT THEREBY THE STORY OF THEIR HEROIC ACTS

MAY BE PRESERVED,

AND THE MEMORY OF THOSE NOT YET CALLED,

PERPETUATED.

PREFACE.

A STRONG conviction that the members of the United States Military Telegraph Corps, whose services were so essential to the overthrow of the Rebellion, have been unkindly overlooked by the Federal Government and the historian; and the daily increasing indications that their noble work, performed in times and places of greatest danger, was surely passing into unmerited oblivion, on whose brink much of the data collected for this work were found, has alone impelled me to spend so much of my time as has been necessary to prepare this history.

The idea is too generally accepted that *all* the credit for crushing the Rebellion belongs to the army. If any one shall read these volumes without realizing that Congress in conferring upon the army, composed as it was of as brave and patriotic men as ever lived, all the honors and rewards due for military gallantry and usefulness, did gross injustice to the members of the Telegraph Corps, then I have injured "with faint praise" a cause deserving better representation.

Very many Southern operators have furnished interesting and important reminiscences of incidents which occurred within the Confederate lines. They form as much a part of the telegraphic war history as the operations of the Southern forces do of the military history of the Great Conflict; therefore, and because they are of themselves worth preserving, I have woven them into the general fabric of this work.

To illustrate the importance of the telegraph, and give it its due setting; to present it as it was, surrounded by all that pertains to war, it was essential to give a running account of the armed struggle itself. In this I have been greatly aided by important telegrams, and other papers, official and otherwise, which have never been published; also by information from telegraph officials and operators who were "in the secret" or "behind the scenes." So far as I have told the story of the war, I have striven to be accurate and just; avoiding debatable questions and seeking concisely to state material facts. Thus I have aimed to make this a readable history, useful and interesting to all classes, for in no other way could the story of the army telegrapher become known to the general public.

I am deeply indebted to the late officers of the Corps, with an exception or two, and to hundreds of operators, for most of the facts related herein. Their very cordial assistance has alone buoyed me up to the completion of my task, which, added to the labors of my profession, compelled a temporary relief from both.

While I may have been misled in a few recitals that I do not suspect, I am conscious of the probability of some errors in regard to the location of operators at certain times. Owing to deaths, want of addresses or failures to respond, some of their locations herein given have not been susceptible of verification. In such cases, official monthly reports and the recollections or memoranda of others have been my authority. However, when any incident hinged upon names or localities, they are believed to have been correctly stated.

General Sherman recently, in conversation, corrected my use of the word "corps" as applied to the army telegraph service. Technically, he was right, but only so because the telegraphers were not organized as a body, pursuant to a law. Perhaps "department," or any other word, would be as open to objection, and yet a noun seemed necessary, and consequently, I have in the main adopted the word "corps," for which I have the precedent of Secretary Stanton and scores of others, including many army officers of high rank.

A large percentage of the members of the Corps are in

their graves. The past, to them, is beyond recall; even a patriotic and liberal people can not now reward the valor of those intrepid telegraphers who have crossed that river over which nothing returns. It is to be hoped that these pages will at least preserve the story of their exploits, and if, perchance, the powers in Washington are incited to devise some plan of recognition commensurate with the service, it would be a proud satisfaction for all concerned.

ILLUSTRATIONS OF VOLUME I.

PORTRAITS.

	PAGE.
GEN. ANSON STAGER,	frontispiece
ANDREW CARNEGIE,	66
DAVID STROUSE,	80
COL. J. J. S. WILSON,	118
COL. WILLIAM G. FULLER,	263
COL. JOHN C. VANDUZER,	304
COL. ROBERT C. CLOWRY,	347

MISCELLANEOUS.

CONFEDERATE TELEGRAPH CRYPTOGRAPHIC KEY,	38
CONFEDERATE MAIL CRYPTOGRAPHIC KEY,	41
NICHOLS' OFFICE,	150
PARKER'S OFFICE,	214
DRAWING BATTERY MATERIAL,	245
SMITH'S PRIZE,	247
DOUGHERTY'S AIR LOOM,	281
THAT "ROUGH OLD TAR,"	337

MAPS.

1. SHOWING PORTIONS OF PENNSYLVANIA, MARYLAND, VIRGINIA AND WEST VIRGINIA,	74–5
2. SHOWING PORTIONS OF MISSOURI AND ARKANSAS,	122–3
3. SHOWING PORTIONS OF KENTUCKY AND TENNESSEE,	122–3
4. SHOWING PORTIONS OF MISSOURI, KENTUCKY, TENNESSEE AND MISSISSIPPI,	252–3
5. SHOWING PORTIONS OF ARKANSAS LOUISIANA AND MISSISSIPPI,	314–15

CONTENTS.

CHAPTER I.

Ancient and Modern means of Communication for War Purposes, - 9

CHAPTER II.

War Cryptographs, - - - - - - - - 33

CHAPTER III.

The United States Military Telegraph; its Initial State and Early Operations, - - - - - - - - - 62

CHAPTER IV.

Early Operations in Western Virginia, - - - - - 92

CHAPTER V.

The Southwest early in the War, - - - - - - 108

CHAPTER VI.

The Corps Organized.—Biographical Mention.—A View of the Telegraphs South.—The Fort Monroe line.—The Monitor and Merrimac.—The Peninsular Campaign.—A View along the Coast.—Banks' Defeat in the Valley.—Experience of Captured Operators.—Affairs in West Virginia, - - - - - - - - - - 127

CHAPTER VII.

The Telegraph in the Southwest.—Fort Donelson.—Pea Ridge.—Island No. 10.—Memphis, - - - - - - - 170

CHAPTER VIII.

The Telegraph in Kentucky, Tennessee and Northern Alabama.—Mill Springs.—Cumberland Gap.—Morgan's Raid in Kentucky.—Shiloh, 187

CHAPTER IX.

The Telegraph with General Pope in Virginia.—Cedar Mountain.—Groveton.—Chantilly.—Harper's Ferry.—South Mountain.—Antietam, - 216

CHAPTER X.

The Telegraph in West Tennessee and Northern Mississippi during the last half of 1862.—Farmington.—Iuka.—Corinth.—Van Dorn's and Forrest's Raids.—Grant *versus* the U. S. M. T.—Chickasaw Bluffs.—Arkansas Post, - - - - - - - - 242

CHAPTER XI.

The Telegraph in Tennessee and Kentucky.—Forrest's, Woodward's and Morgan's Raids.—Buell's Army retires to Kentucky.—Cumberland Gap Evacuated.—Richmond, (Ky.)—Numerous Incidents.—Perryville.—Rosecrans Supersedes Buell.—Nashville Affairs.—Stone River.—Excitement in Nashville.—Morgan again in Kentucky.—Spies Hung.—Operators Tap Confederate Telegraphs, - - - - 273

CHAPTER XII.

The Telegraph about Vicksburg and in West Tennessee.—The Vicksburg Campaign.—Various Federal Raids.—Champion Hills.—Affairs about Cairo, - - - - - - - - - 313

CHAPTER XIII.

The Telegraph in Arkansas and Missouri in 1863.—Helena.—Little Rock Campaign.—Other Campaigns and Conflicts in Missouri and Arkansas.—An Operator Hung, - - - - - - 333

CHAPTER XIV.

The Telegraph in Virginia during the Fredericksburg and Chancellorsville Campaigns.—Suffolk.—Various Raids, - - - - - 351

APPENDIX.

Cipher Number Nine Complete, - - - - - - 370

THE

MILITARY TELEGRAPH

DURING THE

Civil War in the United States.

CHAPTER I.

ANCIENT AND MODERN MEANS OF COMMUNICATION FOR WAR PURPOSES.

Ours is an age of rapid achievements. Cultivated aptitude has revolutionized the world. Performance has been reduced to a minimum of time and space to a question of time. Long lives are compassed in an ordinary span: distances no longer appall: we are making the most of time and the least of space. Steam made the millions acquainted with each other, and electricity has enabled them to maintain daily communication. Thus many peoples have become interested in one another, and thus they are held to a stricter accountability for their national conduct than ever before. The opinion of the world has become a powerful international factor. The soldier, for long, dark and nearly fruitless years, led the world; now the inventor, the philosopher, the scientist, the artist and artisan, the press and the pulpit, the jurist and publicist, lead in the van to grand, but mainly peaceful, achievements. But wars, though less frequent, still occur; their conduct, however, owing to the use of the telegraph, is changed, and what might, but for it, distress a nation indefinitely, is now made comparatively short.

If the ingenuity of man has provided weapons of offense and defense greatly superior to the bow, sling and shield, it has also produced means of conveying intelligence far more efficient than the runner, the voice and the beacon. One improvement has been the necessity of others. These were heralded, and the civilized world moves abreast, instead of in Indian file.

It is believed that no nation was ever, in times of war, content to await even the speediest methods of conveying news of battle in which its forces were engaged. It is while a battle is impending that the nations involved manifest their supreme solicitude, for nothing so greatly imperils their autonomy.

All wars illustrate the importance of the means of speedy communication. In case of being hard pressed, succor is afforded; in case of defeat, retreat is aided; in case of capture, the cities behind are put in a state of defense; in case of victory, new columns are started from other points; if ammunition is becoming exhausted, more is forthcoming, and supplies are forwarded to meet at its new base a victorious army — hence it is of vital importance that news of the real issue of a campaign be known by the authorities as soon as the result is developed. Argument would be wasted in proving this axiomatic truth. During the war of the Rebellion in the United States, the electro-magnetic telegraph oftentimes spread reports of the *progress* of battle throughout the land before either side had won the day. If we stop to reflect — for we are so used to such wonders now that we regard them as matters of course, and need, therefore, reflect — we marvel that it has become possible to convey, print, and circulate upon the streets, facts concerning a pending battle hundreds of miles away.

Like a few great actors, who play all parts well, the telegraph upon the stage of life is successful in every *rôle.* But, like them, it has its specialties. The tragedy of war presents one of the scenes best adapted to bring out its most powerful acting. In this volume, we have sought to bring to view at the footlights of a stage many thousand miles in extent, such a tragedy, in which much unsuspected history is developed by the heroic devotees of the telegraph, who have deserved well of their country.

That we may the better appreciate the telegraphic advantages we enjoyed during our late war, it will be found useful, as well as instructive and entertaining, to note what steps mark the progress by which we have outgrown all former means, and reached the acme of human inventions, the electric telegraph.

Running.

Then said Ahimaaz, the son of Zadok, Let me now run, and bear the king tidings, how that the Lord hath avenged him of his enemies.—2 *Samuel*, xviii, 19.

And the watchman said, Me thinketh the running of the foremost is like the running of Ahimaaz the son of Zadok. And the king said, He is a good man, and cometh with good tidings.—*Ibid*, 27.

Doubtless this was the earliest, as it was the most natural means of conveying messages. Among all people there have ever been athletes who won applause by their endurance, and in very olden times there were practiced runners, who seemed incapable of fatigue or exhaustion, and who were able to make great distances in remarkably short time. Then men were very animal; their bodies were developed at the expense of their brain; hence the body had extra duty to perform, perhaps the most trying of which was in carrying war dispatches.

In ancient Greece, there were trained runners whose recorded feats are quite surprising; thus, it is written that on one occasion, when Athens required aid, Philippides ran to Lacedæmon, a distance of one hundred and forty-two miles, in two days, to solicit it, and, what is quite as remarkable, that in three days after starting, the Spartan soldiers were in Athens. But Pliny, who mentions Philippides' feat, says it was thought to be remarkable until "Amystis, the Lacedæmonian courier, and Philonides, the courier of Alexander the Great, ran from Sicyon to Elis in one day, a distance of thirteen hundred and five stadia," or one hundred and fifty miles, and as if that yarn wasn't as big as he could tell, he proceeds to relate that in Rome, a circus boy of eight years, ran seventy-five miles between morning and evening.

How familiarly has Sir Walter Scott made known to us, in his "Lady of the Lake," the manner of summoning clansmen to arms and rendezvous in Scotland, in the fifteenth century:

Speed, Malise, speed! he said, and gave
The crosslet to his henchman brave.

The muster place be Lanrick mead—
Instant the time—speed, Malise, speed!
* * * * *
Herald of battle, fate, and fear,
Stretch onward in thy fleet career!
* * * * *
———danger, death and warrior deed,
Are in thy course—Speed, Malise, speed!

The Human Voice.

And when they told it to Jotham, he went and stood in the top of Mount Gerizim, and lifted up his voice, and cried, and said unto them, Hearken unto me, ye men of Shechem, that God may hearken unto you.—*Judges*, ix. 7.

It would be astonishing, if not trite, that by cultivation every faculty of man may be improved. This is marked as to the delicate organs of speech, which by proper exercise may be made to compass great distances. The ancients understood this as well as we, and employed the voice much more than we to communicate at distances. Aside from Hebraic illustrations, we are told that Darius Hystaspis (B. C. 485), the father of Xerxes, the Persian king, placed upon eminences certain distances apart, men of great vocal capacity, who were called the "ears of the king," and by them the king is said to have been able to forward messages a distance of thirty days' journey in a single day. The Gauls, in Cæsar's time (B. C. 75), were able thus to proclaim by the wings of the wind, war alarms, so that within three days' time, all the tribes were under orders. To do this, men went upon the hill tops and shouted the news or orders to all points of the horizon. Thus, important messages were vocally telegraphed from Auvergne to the sacred forests of Amorica and the marshes of the Rhine. To this day, in Albania, are messages sped by vocal waves over valleys to hill tops many miles off, and we are told that about Gibraltar the human voice may be heard a greater distance than that of any animal; that if the wife wishes to call her husband from a distance, "she does not shout, but pitches her voice to a musical key, which she knows from habit, and by that means reaches his ear."

In Scott's "Anne of Geirstein," concerning this practice in Switzerland, we read: "The maidens will converse with each other, in that manner, from cliff to cliff, through storm and tempest, were there a mile between;" and again, speaking of an Englishman, "He attempted to do so, but, inexperienced in the art of making himself heard in such a country, he pitched his voice in the same key with that of the roar of wave and wind; so that even at twenty yards from the place where he was speaking, it must have been totally indistinguishable from that of the elemental war around them. The lad smiled at his patron's ineffectual attempts, and then raised his voice himself, in a high, wild and prolonged scream, which, while produced with apparently much less effort than that of the Englishman, was nevertheless a distinct sound, separated from others by the key to which it was pitched, and was probably audible to a very considerable distance. It was presently answered by distinct cries of the same nature."

Carrier Pigeons.

B. C. 3155,—"And he stayed yet other seven days; and again he sent forth the dove out of the ark; and the dove came in to him in the evening; and, lo, in her mouth was an olive-leaf plucked off. So Noah knew that the waters were abated from off the earth.—*Gen.* viii, 10, 11.

In some countries, carrier pigeons are highly esteemed to this day, although their usefulness has largely been supplanted by the electric telegraph. Devotees of this bird find pleasure in the belief that the messenger Noah sent out was an original carrier pigeon. As there is "nothing new under the sun," so in part proof we have in tradition and legend, faint reminders of the use of pigeons in wars, from grayest antiquity. But authentic history will subserve every purpose. A coronation of an Egyptian king was heralded by four pigeons; one to the north, another south, a third east, and the last west; each carrying somehow the news, that "Horus, the son of Isis and Osiris, has put on the splendid crown of the Upper and Lower Country; that the king, Rameses III. has put on the two crowns."

It is written that while Brutus was besieged in Mutina by Antony, he (Brutus) sent dispatches to the camp of the Consuls by fastening the messages to pigeons' feet; that Antony thought that that was carrying things with a high hand, and set his strong archers to wing the birds, but without avail. The rich patrician sometimes took pigeons to the amphitheatre, in order that he might order room at the table for his invited guests, and also name some of the dishes he wished prepared. Anacreon mentions the use of pigeons to enable absent lovers to communicate. Tasso relates an instance of their use, in his "Jerusalem Delivered," canto 8, page 521 :

49—A turtle dove, in the blue firmament is seen of all
To pass, the Christian multitude above:
With outspread wings the liquid air she clove,
And went away as lightly as the wind;
This wand'ring, mute communicant of love,
So soon as she had left the camp behind,
Down from the lofty clouds t' accost the town inclined.

50—When lo! they knew not whence, a falcon arm'd
With hooked beak and talons, sail'd in sight;
Which, 'twixt the city and the camp, alarmed
Th' opposed mild bird in her descending flight:
She waited not his truss; but full of fright,
On instant wing to the pavillions fled,
And at the moment when the cruel kite,
Down swooping swift, just touched his tender head,
In Godfrey's bosom fell, betwixt alive and dead.

51—Godfrey the bird protected, and espied,
As he her plumage smooth'd, a curious thing;
For from her neck, by flax of Egypt tied,
A letter hung, concealed beneath her wing.
Marveling to see it, he untwines the string,
And breaks the seal ; then well he comprehends
The purpose of the scroll; "To Judah's king,"
Thus spake th' inscription, " to his first of friends,
Health, honor, joy and peace th' Egyptian Caliph sends."

52—" Fear not, my noble lord! resist, endure
Till the fourth day, or till the fifth at most;
For by that period thou shalt see, besure,
My slaught'ring sword devour the hostile host."
Such was the secret in the note enclosed,
In Syriac ciphers writ, and sealed with care,
Given in commission to this flying post;
For in the East these couriers of the air,
Trained to the trusty charge, were then by no means rare.

53—The bird he freed, she cooing her concern
That her lord's secrets had been thus betrayed,
Durst not, though innocent of ill, return
A rebel back, but fled from thence afraid.

It is said that by this means, in 1219, the Saracens informed Cairo of the defeat of the French at Mansurah. The English in Aleppo employed pigeons with great success to carry commercial news to and from Scanderoon. Relays have been established thirty to forty miles apart, to and from which the birds have been taught to fly with messages. Pigeons were used in Napoleonic wars, and, in short, to a greater or less extent in most countries since the middle ages, but most notably, perhaps, during the siege of Paris by the Germans in 1870, when they were carried away in balloons and returned with minute photographic copies of great newspapers, carried under a wing or inserted in a quill and tied to a tail-feather. There were 25,000 of these birds in French cities and defenses when the Franco-German war began, and more being trained. The ingenious French, microphotographed long writings and print, so that when ready to be sent the missive weighed but a few grains. A strong microscope revealed it all and a magic lantern threw it on a white wall, where, if a newspaper, the populace might read. However, only a small number of the pigeons released, reached Paris. The Germans had hawks trained to go against them, and are said to have destroyed many.

Before the telegraph, pigeons were used by the officers of the press in America to secure European news quickly. The birds were flown from Halifax, N. S., to Boston and Sandy Hook. Carrier pigeons are said to fly an hundred miles per hour, but one-third of that is about an average.

Balloons.

It is prefaced in Marion's *Balloon Ascents*, that "the discovery of the invention of the balloon . . . was one of those efforts of genius and enterprise which have no infancy." This we are disposed to credit, for, although fully a centenarian, the balloon has been improved as little as thistle-down, which was of questionable utility always. Since their invention by Joseph and Etienne Montgolfier, men have experimented to death in every country, with the "air castles" which they projected.

Mankind has, from ages which "lose themselves in the night of time," sought to "mount on the wings of the wind." So Archytas, of Tarentum (400 B. C.), built a "flying stag," and also "made a pigeon of wood, which flew, but which could not raise itself again after having fallen." Others thought they saw the defects of predecessors, and trying their plans, had their hopes, if not their brains, dashed to the ground. One failed because his machine had no tail; another (fable) because the sun melted the wax with which his wings were soldered, and he went "down into the depths" of the Icarian Sea; yet another, because his wings were made in part of the feathers of barn-yard fowls, instead of being all eagles'; a fourth, bent one wing, and so on to endless failure.

Balloons have been chiefly serviceable in times of war, but, although reason would seem to favor their use with armies, experience appears to discourage their employment. As early as 1794, two companies of aeronauts were organized by the French, who used balloons, at Fleurs, Maubenge, Charleroi, Mannheim and Ehrenbreitstein, Solferino and elsewhere. They, however, were not used as couriers, but to observe an enemy below, and sometimes flag signals were used to telegraph from them. This was done in the United States army on the Potomac and during the Peninsular campaign, in the late war. On all such reconnoissances, the balloon was held by ropes. On several occasions, electrical telegraphic connection was had with the aeronaut in the sky. This was first accomplished June 17, 1861, when the War Department, in Washington, was placed in instant communication with Professor Lowe, who, from his "high estate," caused the operator at his side to telegraph as follows:

BALLOON "ENTERPRISE," WASHINGTON, June 17, 1861.

TO THE PRESIDENT OF THE UNITED STATES:

Sir—This point of observation commands an area of fifty miles in diameter. The city, with its girdle of encampments, presents a superb scene. I take great pleasure in sending you the first dispatch ever telegraphed from an aerial station, and in acknowledging my indebtedness to your encouragement, for the opportunity of demonstrating the availability of the science of aeronautics in the military service of the country. Yours, respectfully,

T. S. C. LOWE.

Operator C. I. Brown ascended from Pohick Church, Virginia, with Professor Wise, and also communicated to the War

Department. John La Mountain, another professional aeronaut, made ascents, and at one time, in front of Richmond, Virginia, while he was accompanied by General Fitz John Porter, who wished to note the enemy's position, the air ship broke loose, passing over their camp, but on lowering the balloon, it entered an opposite current, by which Porter and the aeronaut returned safely to the Union lines. At the close of the Peninsular campaign, the use of balloons was abandoned. Professors Wise and La Mountain traveled from St. Louis, Missouri, to Henderson County, New York, 1,150 miles, in a balloon, in less than twenty hours.

But our subject leads us particularly to their use during the siege of Paris; at which time (1870-1) many of them were certainly useful as couriers of military orders, news, etc. From September 23, 1870, to January 28, 1871, sixty-two balloons left Paris; fifty-four of which, sent out by the Post-office Department, took 2,500,000 letters, weighing ten tons. It was usual to send carrier pigeons with these air ships, to the end that news from the outside world might be obtained in Paris by their return. Though the employment of balloons for courier purposes is desirable only when visual and electric telegraphing are impossible, there may arise other instances in which this speedy method of communication, erratic as the winds, is all that is left to a besieged army. It is to be hoped, however, that somebody will improve on the Montgolfier idea enough to make this means more surely serviceable and less hazardous.

Trumpets.

And I said unto the nobles, and to the rulers, and to the rest of the people, The work is great and large, and we are separated upon the wall, one far from another. In what place, therefore, ye hear the sound of the trumpet, resort ye thither unto us.—*Nehemiah*, iv, 19, 20.

Some have credited the Tyrrhenians with the invention of this instrument, while others have thought the Egyptians show a better claim. Since it is not patentable, the reader may select his inventor between them, and if he stoutly adheres to his choice, he will doubtless be right. It is usually best to read those "old timers" *cum grano salis*. It is written that the invention of the trumpet was first suggested by the use of sea shells, which were sounded as the first signal of battle.

Biblical history is replete with notices of the use of the trumpet. It proclaimed war, announced the new year, and made glad hearts at the beginning of the great "Jubilee." In Macaulay's "Horatius," we find:

> East and west, and south and north,
>   The messengers ride fast,
> And tower and town and cottage
>   Have heard the *trumpet's* blast.
> Shame on the false Etruscan
>   Who lingers in his home,
> When Porsena of Clusium
>   Is on the march for Rome!

Even to this day, the trumpet is used in armies, especially by cavalry.

Fires.

O ye children of Benjamin, gather yourselves to flee out of the midst of Jerusalem and blow the trumpet in Tekoa, and set up a sign of fire in Beth-haccerem: for evil appeareth out of the north, and great destruction.—*Jeremiah*, vi, 1.

It can not be doubted that signal fires have been used from the earliest era. The torch is said to have preceded the use alluded to of sea shells. For ages these signals had but one meaning, and that was agreed upon beforehand; and thus victory, warning and defeat were signalized. But one word usually covered all that was telegraphed. If it was daylight, a column of smoke as dense as possible indicated what by night would be disclosed by a fire, varying in size according to distance. Of course these fires were, when practicable, placed on commanding heights.

Homer, writing of the siege of Troy, tells how (1000 B. C.) Achilles' head was crowned with a golden cloud, from which Minerva kindled a shining flame. "And as when smoke, ascending from a city, reaches the æther from an island afar off, which foes invest, who (pouring out) from their city contend all day in hateful fight; but with the setting sun, torches blaze one after another, and the splendor arises, rushing upward, for (their) neighbors to behold, if perchance they may come with ships, as repellers of the war; thus did the flame from the head of Achilles reach the sky."

Æschylus (about B. C. 500), in his elegant explanation of the means by which Clytemnestra learned of the capture of Troy, 1184 B. C., yet more satisfactorily elucidates, in "Agamemnon," this ancient usage, as follows:

Clytemnestra—

Hephæstus, sending forth the Idaian fire,
Hither through swift relays of courier flame,
Beacon transmitted beacon. Ida first
To the Hermæan rock on Lemnos' Isle;
Thence Athos' summits, dedicate to Zeus,
The third in order, caught the mighty glow;
Towering aloft, the pine-blaze, like the sun,
Gold-beaming, bridging in its might the sea,
Transmits the splendor of the advancing fire
To bold Macistus' watch-tow'rs; he, in turn,
Without delay, nor overpower'd by sleep,
The courier's duty faithfully discharged;
The torch, far gleaming to Euripus' stream,
Gives signal to Messapus' sentinels.
Firing of withered heath a giant pile,
They kindle and send on the courier light;
The stalwart flame, unwearied and undimm'd
Like a bright moon o'erleaps Æsopus' plain,
And wakens, on Cithæron's lofty height,
Another speeder of the fiery post.
The warder, hailing the far-journeying fire,
Kindles a beacon whose surpassing glow
Darts its bright radiance o'er Gorgopis' lake,
And summons Aigiplanctus' mountain height
Forward to hurl unrent the fiery chain.
With vigor unimpair'd they onward send,
Kindled anew, a mighty beard of flame;
Crossing from far the crag whose pinnacles
Peer o'er the gulf Saronic spread beneath,
The blaze, alighting on Arachnæ's height,
The city's nearest beacon, reach'd its goal;
Thence to the roof of Atreus' son this light
Darted—true scion of Idaian fire.
Thus in succession, flame awaking flame,
Fulfill'd the order of the fiery course:
The first and last are victors in the race.
Such is the proof, the warrant that I give
Of tidings sent me by my lord from Troy.

While Xerxes was about invading Greece, the Greeks stationed at Artemisium were informed by "fire signals," what had happened to three picket triremes, and after he retired to Sardis, leaving Mardonius in command of a large Persian and mixed army, this general, we are told by Herodotus, purposed a second entry into Athens (B. C. 480), desiring "to inform the king at Sardis, by fire signals along the islands, that he was master of the place;" and it is supposed that the route of these beacons was along the European coast to Athos, Lemnos and Asia in the reverse order of the route given by Æschylus. Sophocles (B. C. 450), Thucydides (B. C. 450), Æneas (B. C. 350), Polybius (B. C. 175), Tacitus (B. C. 100), and some others wrote about such uses of beacon lights, but Julius Africanus explains a system of fire signaling in which different substances were employed to enable parties at a distance to spell out news or orders. Æneas invented a peculiar system, whereby short sentences written on boards were read, and Polybius will doubtless also receive mention in all books on war telegraphing because he invented a plan for telegraphing as far as the eye could reach. Vegetius—*de Re Militari*—shows that during the Emperor Valentinian's reign, communication was carried on between camp and armies: "*é turribus et oppidis, trabibus totidem erectis totidem depressis*," which differed from that of Polybius, who had five letters on each of his five posts, which were indicated from top to bottom by torches, *i. e.*, one torch meant the first, two the second letter, etc.

In Macaulay's "Armada" we note how England's peril was announced:

> That time of slumber was as bright and busy as the day;
> For swift to East and swift to West the ghastly war-flame spread.
> High on St. Michael's Mount it shone; it shone on Beachy Head;
> Far on the deep the Spaniard saw, along each southern shire,
> Cape beyond cape, in endless range, those twinkling points of fire.
>
> * * * * * * * *
>
> The sentinel on Whitehall gate looked forth into the night,
> And saw o'erhanging Richmond Hill the streak of blood-red light.
> Then bugle's note and cannon's roar the deathlike silence broke,
> And with one start and with one cry the royal city woke.
> At once on all her stately gates arose the answering fires,
> At once the wild alarum clashed from all her reeling spires.

* * * * * * * *

All night, from tower to tower they sprang; they sprang from hill to hill:

* * * * * * * *

Till twelve fair counties saw the blaze on Malvern's lonely height.

It was while watching the Indian fire signals in Texas about 1855, that Albert J. Myer, once a telegraph operator, was led to devise his first system of flag signals, and in so doing doubtless his practical lessons in telegraphy were of great assistance. Such was the beginning of the system which has been greatly praised for its ingenuity, and which was developed in New York harbor and approved before the Rebellion; the inventor being thereupon sent into the Indian country to utilize it. Treuenfeld, a recent European writer on telegraphy for tactical purposes, compares Myer's system to that of Æneas' for telegraphing by measurement of intervals, and says that thus Myer gave "the first examples of chronometrical signals, which, forgotten for thousands of years, first came into use in modern times with the Signal Corps of the North American Army." All nations seem to have used fire signals. In this country to this day they are in use by the Indians; but the last mention of the employment of fire signals which has come to my notice was by General Custer since 1870.

Semaphores.

In 1764, Doctor Hook formed a plan of telegraphing by masts and screens, as did also M. Amoutons, a little later. The Edgworth telegraph (1767) was a further advance, and twenty-six years later the Chappe brothers introduced a really good thing, which European governments took hold of, expending upon it or its modifications several millions of dollars. Up to the beginning of the seventeenth century, the telescope was unknown, except to a few who knew not how to improve or utilize it, and as the most approved of prior media of telegraphing depended upon the power of the eye, signaling was not a great success. Not many men can see ten miles so as to distinguish objects as small as trees and hence the impracticability of visual telegraphing by the naked eye. Cicero mentions one who could see objects one hundred and twenty-five miles, and M. Varro wrote that the name of this man was Strabo.

It is of no use for men now-a-days to compete with the ancients in any thing then known, for some ancient or classical writer, who can not now be called on for his proofs, has surely anticipated all known feats, and this story of Strabo demonstrates the foresight of great writers. A field glass that will carry twenty miles in most atmospheres is an excellent one. Two of the three Chappe brothers were at school in one place, and the other a half league distant. This other, Claude, invented a system of signaling, whereby, using a horizontal piece of wood, to which was at each end attached a movable wing or arm, he could produce one hundred and ninety-two distinct signals, and by this contrivance he held intercourse with his brothers. Much improved upon by these young men and an ex-diplomat and watchmaker, who saw money in the business, the system was adopted by the French government, and was inaugurated by the welcome announcement that "Conde is taken from the Austrians," whereupon, France went into ecstacy over the double victory; one being the warlike, and the other peaceful, feat of arms. It is said that a dispatch could be sent from Paris to Lille, distant about 130 miles, in two minutes by this system, or under the most favorable circumstances possible, a single signal ought to reach Toulon from Paris, 475 miles in twelve minutes. France extended the system over the kingdom, and then nearly all European governments, seeing its usefulness, adopted Chappe's plan or some infringement upon it. Major-General Meydam, director of Prussian telegraphs, in 1875, in his historical sketch of the rise and progress of telegraphy in Germany, wrote that a line of optical telegraphy was established so early as 1832, between Berlin, Magdeburg, Paderborn, Cologne, Coblentz and Treves, which was worked under the supervision of the Ministry of War. It was purely political and military in its purpose. But in Russia, millions of dollars were expended in erecting these stations every five or six miles, for the most part over the great highways. Thus Warsaw, St. Petersburg and Moscow, and other points, were connected. Twenty seconds was quick time for a single signal to pass from one post to another. Many improvements were made in different countries, and as high as 58,190 distinct signals were ultimately found possible.

Some of these semaphoric stations are still standing; monuments of human ingenuity not to be lightly treated, even in these days, when they seem cumbrous and foreboding bankruptcy. They attest the craving of humanity for a closer union; they demonstrate the value and urgency of speedy communication in times of emergency; they illustrate how dependent man is upon his fellow. Thirteen hundred and twenty persons were required to operate one of these lines connecting St. Petersburg with the Prussian frontier. Any one disposed to study this subject should read Tal. P. Shaffner's "Telegraph Manual," wherein semaphoric telegraphs are carefully explained.

The reader must not suppose from the foregoing that the diligence, the messenger, horse, vessel, and like commonplace means of carrying news, were usually displaced by visual or vocal telegraphs, on important occasions, for such was not the case. Until Chappe's time, even political news was not, as a rule, conveyed by extraordinary means, and the mention which history makes of the pigeon, the trumpet, the shield, the voice, the torch, the ancient semaphore, etc., are believed to be at best, but rare instances, sufficient to carry conviction of their use for signaling, but not to produce an impression that they were much depended on. They show the wants of man, and attempts to supply them, rather than satisfactory results. Thus, in the middle ages, we find that in 1399, when Richard II. was in Ireland, temporarily, Henry, duke of Lancaster, raised a rebellion in England, and virtually conquered the kingdom before Richard heard of it. At the end of the sixteenth century, the news of the death of Henry III., the last of the Valois, only reached Marseilles fifteen days after his demise in Paris. In our own country, the news of the battle of Bunker Hill, fought June 17, 1775, did not reach Philadelphia until the twenty-second. The news of the surrender of Lord Cornwallis's army at Yorktown, Va., on the 19th of October, 1781, was brought by Lieut. Tilghman, Gen. Washington's aid, as fast as horse could carry him, reaching Philadelphia, where Congress was in session, soon after midnight of the twenty-fourth. Our Revolutionary fathers used to signal information very indifferently, by use of a barrel at the head of a liberty pole, with a flag underneath the barrel, and a basket suspended from a projecting arm, still lower down.

The presence or absence on the pole of one or more of these articles, according to preconcerted understanding, indicated that an expected event had or had not occurred. During the late war, a Virginia negress in like manner telegraphed the presence or absence of the rebels, by placing her white clothes on a drying line, or removing them therefrom.

The intelligence of the opening of the Erie canal, October 20, 1825, was communicated, and acknowledged in return, by cannon, placed eight miles apart from Buffalo to Albany, a distance of three hundred and sixty-four miles, *via* the canal and Hudson. It took one hour thus to carry the joyful news to the State capital. But all that was pre-arranged, like the beacons which announced the fall of Troy.

The Electric Telegraph.

Scientific research, prior to the practical introduction of the electro-magnetic telegraph by Morse and others, had solved most, if not all, of the elementary facts which are to-day the bases of this applied science. These fundamental truths came as partial ones from widely scattered sources, and in their gradual unfolding covered a period of many years. Over a century ago, a commendable attempt was made by Georges Louis Le Sage, of Geneva (1774), who used twenty-four insulated wires, each representing a distinct letter, which was marked on a pith ball electroscope at the ends of the wires. These balls were excited visibly by electricity, generated by a machine prepared for that purpose.

Three years later, a Parisian used one wire in like manner, the alphabet being represented by varying motions of the pith balls. Another employed thirty-six wires for numerals, and letters which were manifested by electric sparks; yet another, using one wire, sent sparks at different intervals of time, and in 1798, D. F. Silva, a Spaniard, astonished the philosophers and savants by working twenty-six miles over one wire, using sparks of a Leyden jar for signals. Francis Ronalds, of England, in 1816, and Harrison Gray Dyer, an American, in 1828, telegraphed short distances; the former eight miles. None of these used batteries. In 1810, Thomas Von Sommering, of Munich, telegraphed two miles, using a Voltaic pile to charge his thirty-five wires. Galvani (1786–90), Dr. John Redman, of Philadelphia (1816), Hans Christian Oersted, of Copenhagen (1819), Ampère, of Paris (1820), Schweigger, of Halle (1820), Baron Paul Schilling, of Cronstadt (1823), William Sturgeon, of London (1824), Professor Henry, of America (1828–31), Professors Gauss and Weber, at Gottingen (1833) and some others, added their discoveries to the developing science. Then Morse (1837) and Cook and Wheatstone (1837) and Baine (1840), House (1846), Hughes (1855), Stearns, Edison and others, developed startling ideas never before received. The Morse telegraph, in its main features conceived in 1832, was patented in 1837, and is that system which has since been almost universally adopted. But before Morse could demonstrate his invention, Wheatstone (1838) built a line from London to Birmingham, which was worked by a needle revolving above a dial, on the margin of which were the letters of the alphabet. The revolution of the needle was usually controlled by the influence of the galvanic battery upon the helices, or magnets with which it was connected.

Morse's invention was not at once recognized as revolutionary of other means of correspondence, in its practical tendency. That it not only tended to, but in a marvelously short time did, to a large extent, supplant other means, is now so well known that one almost wishes to apologize for stating it. Mr. Morse, like most inventors, was too poor to test his invention on a convincing scale, and the honest Congresses he met at the capitol were too chary of the people's money to risk $30,000 experimenting for him. But like some people who, on their death-bed atone for much misfeasance, so the Congress of 1843, in its expiring hour, voted the appropriation, and a double (circuit) wire between Washington and Baltimore was the result. In time it was found that one wire run to the ground at each end (the ground answering to complete the circuit) was all that was needed, and from this there was nothing to be done but to develop the system and improve on primitive ideas. So the electric telegraph has, in the hands of studious and energetic men, in this and other countries, advanced practically and scientifically. Wherever there is a railroad there also is the telegraph, and usually long before the road is contemplated the iron thread has become an accomplished fact. It would seem, commercial, industrial, educational and other peaceful callings and purposes were the chief ends of this wonderful invention, and it is not improbable that it was largely the anticipated influence in these pleasing directions that so greatly contributed to make Professor Morse, a thoroughly good man, prefer the slow processes by which justice and truth sometimes prevail against ignorance, vice and even active wickedness, to the cruelties of war, although waged with laudable design. Unlike most inventions susceptible of military uses, the electric telegraph was allowed to develop in peaceful employments some time before its uses in war were demonstrated.

When the telescope was discovered by Jansen, the invention which some concede to him, was kept secret that his Dutch Prince Maurice might utilize it in war. The first ideas of the importance of balloons were as to their military uses. The American cities of Boston and New York were telegraphically connected about the time the Mexican war began. But one operator at either end was employed, and the first war news sent over this line was an account of the battles of Palo Alto and Resaca de la Palma. Then came, not long after, Monterey, Buena Vista and Vera Cruz and Scott's grand victories. The European news of the wars of 1848, and the Presidential campaign of that year, taught the people to prize this new blessing which has since so grown with this country as to constitute one of its nerve forces.

It will be borne in mind that we have not yet shown any use of the telegraph in war. It was not until the Crimean war (1854–5) that the first electric telegraph was erected for purely military uses. This was not, however, used for tactical purposes, but merely for intercommunication between the principal head-quarters of the allied besiegers ; probably more a matter of convenience than necessity. A cable telegraph was also laid between Varna and Crimea which worked admirably. The English also used the wire in India in 1857–8, whereby Lord Clyde's advance posts were enabled to communicate with Government head-quarters in Calcutta. The wire was carried on rollers and in carts, and as it unwound was allowed to lie on the ground or bushes or hung on trees and bamboo, although much of it was uninsulated. In dry weather this line is said to have worked one hundred miles, but it was doubtless useless when wet or even moist. It is difficult to understand how the English came to lay such a line, as they aim to keep abreast of the most approved methods. In our country of much rain, nobody would ever contemplate such a thing. Treuenfeld, to whom I am indebted for this item beside many others, in his excellent work entitled, "Kriegs Telegraphie," recently published in London, speaking of this Indian line, says that "if not a thorough success during the two years war, it certainly did most excellent service." The Crimean and Indian wars brought the use of the telegraph to the attention of the German military officials, and they, it appears, were, about 1855, the first in times of peace to introduce it as a permanent part of their army organization, and their first employment was in 1864, in the war with Denmark.

The English, in 1857, began teaching the under engineer officers of the army, their telegraph system for field service, and the French used the telegraph the same year in their Algerian war.

Two years later, the Spanish government telegraph company sent a corps of telegraphers with the Spanish army into Morocco, and their equipment was so light and serviceable that Treuenfeld, who saw much service in South American war telegraphy, writes very complimentarily concerning it, and shows also its usefulness. The French used it in their Italian campaigns in 1859, and seem to have set the Italians to thinking. For, in 1860–1, they utilized it in their operations against Ancona, both in keeping the advancing columns in constant communication, and in preserving the besieging forces, when fairly settled about the place, from serious assault.

Major General Sir Lintorn Simmons, C. E., Inspector General of Fortifications, testified before a committee of the English Commons, in strong terms favoring the telegraph in the army, saying, among other things, that, "at the present time (1876), the telegraphs were essential in warfare, and that not even the smallest war could be carried on successfully without them. It is not too much to state, that the success of the Ashantee war was owing very much to this particular service." Had Moreau and Jourdan, in 1796, been in telegraphic communication, one of their divided columns (Jourdan's) might not have been crushed by the Austrian Arch-duke, Charles, in a series of battles, which compelled Moreau to retreat, as best he could, through the Black Forest. The German victory at Sadowa, in 1866, over the Austrians, is said to be largely owing to constant telegraphic communication between the two main columns, moving from different bases.

When Professor Morse was in Berlin, in February, 1868, Baron Von Phillipsborn, Director General of Telegraphs, took infinite pleasure in seating that grand old man in his little ante-chamber, where he said to the professor: "Here," pointing on a telegraph map, "the Crown Prince came down through Silesia. This," again indicating, "is the line of march of Prince Frederick Charles. From this station, the Crown Prince telegraphed Prince Frederick Charles, always *over* Berlin, 'Where are you?' The answer reached him, also over Berlin. The Austrians were here," said he, pointing. "The next day, Prince Frederick Charles comes here," indicating, "and telegraphs the fact to the Crown Prince, who hurries forward, and a junction was effected at Königsgratz. After that, Prussia was safe."

But it is universally conceded that to the Federals in the late war, first, and to a large, but less general extent, to the Confederates, also, is due the practical demonstration, on the largest possible scale, of the invaluable service of the telegraph, not only as a convenient courier, but for tactical purposes. Europeans examined our appliances from the beginning, noting carefully all improvements, and as a result, every regular army in Europe is now fully equipped with telegraphic apparatus and material specially designed for field service, and operated by an educated *corps telegraphique.* Their plans differ somewhat, but aim at like essentials. These are elaborated in Treuenfeld's work, to which the student of these matters is referred for a thorough comprehension of those systems at present in use in the various countries. Since our war, the most notable use of the telegraph in war occurred in the Franco-German struggle. The extraordinary combinations which the successful employment of the telegraph enabled the German commanders to make, resulted in overthrowing one of the great powers of Europe, in a time which, considering the forces and wealth and vigor of the French, seems incredible; is one of the marvels of modern times. And the German government has shown its appreciation of its military telegraph servitors. When the war closed, February, 1871, the Germans had, in working order, 1,587 miles of military telegraph, operated at ninety-one stations. The system centered about Paris at this time, which city was surrounded by a telegraphic cordon. St. Quintin, Amiens, Rouen, Dieppe, Alençon, Le Mans, Tours, Orleans, Gien, Auxerre, Montbard, Dole and other places, were thus brought to headquarters. The great German army, aided by such facilities, was thus enabled to shackle France. It was stated during the war by German officers that, but for their telegraph, they would not have dared to advance so rapidly into France.

REFLECTION.

Herodotus informs us that the Persians, being beaten at Marathon, were signaled by persons in Athens, thought to be the Alcmæonids, desiring the restoration of Hippias; the signal used being a burnished shield, by which the Persians were directed suddenly to attack the city, after reaching it via Cape Sunium, during the absence of the home army. Sometimes a shield was anciently used as a signal to attack both by land and sea.

But of late, the rays of the sun are doing courier service where the electric telegraph could not be built or operated, and such has been the success of sun telegraphing, that it constitutes a new and rapidly developing wonder. This mode of signaling is variously designated as mirror telegraphing, heliographic, helioscopic, heliostatic and heliotropic, all of which seem to be essentially identical in the main principles. But the instruments by which the rays are concentrated and reflected differ somewhat, and hence some are better calculated than others to work at great distances. The heliostat was invented by 's Gravesande, about a century and a half ago.

In 1861, officers of the United States Coast Survey, at work in the Lake Superior regions, demonstrated the usefulness of the mirror, equatorially mounted, for telegraphic purposes, and succeeded in conveying their signals with ease and rapidity a distance of ninety miles. During the same year, Moses G.

Farmer, an American electrician, a man of infinite invention, succeeded in thus telegraphing along the Massachusetts coast, from Hull to Nantasket. The next year some English officers introduced the system into the British navy, with modification and improvement, using at night an electric or calcium light. The signals communicated are made by alternately exposing and cutting off continuous rays of light reflected from one station to another. But by the (H.) Mance heliograph, an instrument used by the English, telegraphing is done by pressing a finger-key, whereby flashes of light, of long or short duration, are emitted. These flashes and intervals or spaces are easily made to indicate what in the Morse alphabet are shown by dots, spaces and dashes; thus, a - — may be one quick flash and one long beam of light; c -- - may be three flashes, with an interval or space between the last two. In this way the Morse alphabet may be telegraphed as easily as by an electrized wire. Indeed, ungodly parties have before now, at church, telegraphed across the room without awakening suspicion, by a mere movement of the eyelids. It is reported that during the siege of Paris (1870–1), messages were telegraphed therefrom twenty and thirty miles, by the reflection of calcium lights.

The heliostat is said to be the first instrument for mirror telegraphing used in war. The mirror receives and reflects the sun's rays, and a clockwork attachment keeps the mirror in position to receive the direct sunbeams, which in Nevada, U. S., are said to be so bright as to be hurtful to the eye at a distance of forty miles. Behind the mirror, in the very center, some of the quicksilver is removed, leaving a very small, round, clear space in the glass, through which the operator looks and may watch the reflection from the next station. The Mance heliograph is easily operated by one man, and as it weighs but about seven pounds, the operator can readily carry it and the tripod on which it rests. The heliotrope reflects the rays by mirror, but has no clockwork.

During the Jowaki Afridi expedition sent out by the British Indian government (1877–8), the heliograph was first fairly tested in war. Generals Key's and Ross's columns in the Afridi Hills were thus placed in communication and so favorably were the officers impressed with the value of the system, that it was incorporated into the Indian army on a more liberal scale, and has been of very great service in the Afghanistan campaigns. The English also used it in Zululand. Where a sea cable can not be relied upon, this instrument is coming into use; thus, messages are sent across the Straits of Gibraltar. By its use, messages may be sent over the heads of the enemy.

The United States coast and geodetic surveying parties appear to have improved on their system of 1861, and the party now triangulating in Nevada, it is said, are enabled by their heliotropic signals to survey from Arch Dome Peak, Nye County, the immense country within a radius of two hundred miles. This instrument is described as consisting of an ordinary surveyor's transit, adjustable to any vertical or horizontal angle, thus making it capable of being accurately directed to any determined point. On the top of the telescope are two sights, much resembling a globe sight of a rifle, but considerably larger. Back of these sights a small adjustable plate-glass mirror, three or four inches in diameter, is placed, and the sun's rays caught by this are directed towards an observer at a distance; the reflection from the mirror is seen by him; and it has been demonstrated that this reflection can be distinguished on a clear day for a distance of two hundred miles.

Wonderful as this is, there are already indications that these rays of light will yet be the media for conveying the human voice; or perhaps, since sound travels so much more slowly than light, of reproducing the voice at a distant station. Professor A. Graham Bell recently explained to the American Science Association in Boston his photophone, whereby audible effects of light acting upon selenium may be produced. These effects are obtained by a continuous but undulatory beam of light, varying in intensity and somewhat corresponding to the electric waves manifested in the telephone. The apparatus as described in the *Scientific American* for September 18, 1880, consists of a flexible mirror of silvered mica or thin glass. The speaker's voice is directed against the back of this mirror as against the diaphragm of a telephone, and the light reflected from it is thereby thrown into corresponding undulations. The sunlight is concentrated upon the diaphragm and is reflected and rendered parallel by means of a lens set for that purpose. At the receiving station, the beam is received upon a parabolic reflector, in the center of which is a sensitive selenium cell connected in a local circuit with a battery and telephone. It is said that Professor Bell by this means distinctly heard his co-laborer, Sumner Tainter, seven hundred feet away say, "Mr. Bell, if you hear what I say, wave your hat out of the window," and that the Professor expects to converse thus as far as the rays can be discerned.

In the light of possibilities so wonderful, no man can foretell what methods of communication may be used in case of another war, which now happily looks very distant and shadowy. Already the heliograph, by some name, has been adopted in the armies of the United States, England, France, Belgium, and perhaps other countries, but as auxiliary to the electric telegraph, which it will never displace.

CHAPTER II.

WAR CRYPTOGRAPHS.

The necessity for secret writing has doubtless been felt from the remotest ages. This was especially true with persons charged with important public affairs. History instances numerous interesting attempts to correspond in cipher, which it was believed none but the initiated understood. In a comprehensive sense, every writing is cryptographical to those unable to master it, as Greek is Greek to most people, and the same may be said of phonetic characters and all uncommon symbols or signs, however plain when understood.

The key to Egyptian hieroglyphical writings being lost, they were unintelligible for thousands of years. It was only in the present century, by the aid of the famous Rosetta stone, discovered at Raschid, Egypt, that these writings were decipherable. Upon that stone was executed the order that "this decree shall be engraved on a hard stone, in sacred, common and Greek characters."

But no writings in systems designed for general use are considered cryptographical, however difficult; although it is not so clear that the use of such systems by educated parties among an ignorant people would not thereby clothe messages so written with all the elements of a cipher, however transparent they were in fact. In other words, a cipher key is but a system of conveying written information, which, for good reasons, it is believed that those who may see it in transmission will be unable to understand. It requires no ingenuity to create new arbitraries for the letters of the alphabet, or for vowels, consonants and words, and one having the key may as easily read such as the true ones they represent. But on the part of the uninitiated, great patience and much *skillful* guess-work is required to discover the meaning. And this would be equally the case with our present code of letters but for their being publicly known. All ciphers composed of numerals, figures or other symbols to represent letters, may be studied out, for as J. R. G. Hassard, the translator of the ciphers used in the Presidential campaign of 1876, correctly states, (*North American Review* for March, 1879,) "The ingenuity expended in devising new alphabets of dots, lines, mathematical and astronomical symbols and fantastic forms was wasted. It was well known that some letters are used much more frequently than others. By numbering these similar signals, an intelligent guess can be made as to what they respectively represent, and after a few trials and comparisons one soon makes sure of some letters which, by context, lead to the discovery of all the rest."

Hassard discovered that thirteen different codes were used for telegraphic correspondence by the Democratic politicians during that campaign, and says that they included all classes of telegraph ciphers, viz.: 1. Words or letters having an arbitrary signification; 2. Numbers representing words or letters; 3. Words or letters having their usual signification, but standing in a false order. Contrary to Mr. Hassard's idea that all telegraph ciphers "must belong to *one* of these three classes," it should be stated that the United States military telegraph ciphers rarely, if ever, were confined to either class alone, but usually were composed of arbitrary words (first class) and words in their ordinary sense but in a false position (third class), intermingled with meaningless words whose only office was to confuse. An example of a cipher which is deceptive only because it contains too much, is as follows:

UTICA (date).

MR. PHLANEGAN:

¼ Papers do not come promptly. To-night I am sure dear papa will be disappointed. At home all read the blessed evening *Journal*. Respy. EFFIE DEANS.

"¼" means, omit the first and every fourth word thereafter. Read that way, the reader will pity the "dear papa" who forwards the message to the newsman, and discover who will be most disappointed. The variations to which this is susceptible readily present themselves.

President Lincoln is credited with preparing a very simple

cipher telegram, but so very transparent is it, as hardly to deserve to be called a cipher. The message was sent as follows:

HEADQUARTERS ARMIES OF THE U. S., CITY POINT,
8: 30 A. M., April 3, 1865.

To CHARLES A. TINKER, War Dept., Washington, D. C.:—

A. Lincoln its in fume a in hymn to start I army treating there possible if of cut too forward pushing is he is so all Richmond aunt confide is Andy evacuated Petersburg reports Grant morning this Washington Sec'y War.

(Signed) S. H. BECKWITH.

By reading the above backward with regard to the phonetics rather than the orthography, the meaning will be apparent.

Herodotus instances a safer, though more tedious methed. It appears from that writer that Histiacus, when anxious to give Aristagoras orders to revolt, could find but one safe way, as the roads were guarded, of making his wishes known. This was by taking the trustiest of his slaves, shaving all the hair from off his head, and then pricking letters upon the skin and waiting till the hair grew again. This accordingly he did, and as soon as the hair was grown he dispatched the man to Miletus, giving him no other message than this: "When thou art come to Miletus, bid Aristagoras shave thy head and look thereon." The marks on the head were a command to revolt.

The main object of this chapter is to disclose the several cipher systems used by the belligerents respectively, for telegraphic correspondence, during the Rebellion, and also the Confederate mail cryptograph code, as that became a part of the history of the U. S. M. T. service, by reason of its translation by Federal telegraphers.

The Confederates were extremely unfortunate in their telegraph, mail and signal codes. The former two were deciphered by Federal telegraphers, and the latter by Union army signal officers. The knowledge of the enemy's flag and torch system gained by the signal men, was of very great consequence, especially during the battle of Wauhatchie, in the valley of the Tennessee, and under the shadows of Lookout Mountain. The translation of Beauregard's orders, which were signaled from Lookout Mountain, where he was directing the battle below, enabled Union General Geary to meet force with ample resistance at the right moments. And thus it came about that the rebels were *signally* defeated. This was one of the most brilliant and useful achievements of the Signal Corps during the war. It is believed that none of the Union telegraphic or signal systems were discovered while in use.

It is quite surprising that in a matter of such consequence the Confederates contented themselves with ciphering their telegrams by a device which, when applied, rendered the translation possible without a knowledge of the key by which it was "put up." But it is a matter of astonishment that, using a code so simple as theirs, they did not foresee the necessity of ciphering every word, at least, in the body of the message. And when we reflect that there was nothing whatever original in their system, except the key words, and that a translation of a single line reveals the key to all the rest, we marvel at the credulity of those talented gentlemen in whose breasts the great secret was hidden.

Their code, described generally, in a sentence, was merely a systematic and shifting use of arbitrary letters for real ones, as applied to words only which indicated the subject of the dispatch. Mr. Hassard shows that of the political campaign ciphers of 1876, "a few messages from Oregon were disguised by merely substituting b for a, and so on throughout the alphabet; thus, cfnpsf fyqmjdju, meant 'Be more explicit.'" Instead of using, as above, the second letter which in the alphabet follows the real one, any other given letter may be employed, but in such cases the meaning of the letter is invariable, and once known, always understood. This was not so with the Confederate code, as we shall see.

The first cipher message we know of, captured by Unionists, was obtained during the siege of Vicksburg. What efforts General Grant caused to be made to unravel this message, we know not. It was not until October, 1864, that it and others came into the hands of the telegraph cipherers, at New Orleans, for translation.

The following is a true copy:

VICKSBURG, Dec. 26, 1862.

GEN. J. E. JOHNSTON, Jackson :

I prefer o a a v v r, it has reference to x h v k j q c h f f a b p z e l r e q p z w n y k to prevent a n u z e y x s w s t p j w at that point, r a e e l p s g h v e l v t z f a u t l i l a s l t l h i f n a i g t s m m l f g c c a j d.

(Signed) J. C. PEMBERTON,
Lt. Gen. Comdg.

TRANSLATION.—I prefer Canton. It has reference to fortifications at Yazoo City to prevent passage of river at that point. Force landed about three thousand, above mouth of river.

The New Orleans operators who worked out this key were aided by the Pemberton cipher and the original telegram, which was found among that general's papers, after the surrender of Vicksburg; also by the following cipher dispatch, and one other. The marginal words are the translation of the unintelligible letters.

MONTGOMERY, 30th.

To GEN'L E. K. SMITH,
Shreveport, La., *via* Wi.

What are you doing to execute the instructions	
sent you, to H C D L L V W — X M W Q I G — K M — G O E I — D M W I — J N — V A S — D G U G U H D M I T D.—	forward troops to east side of the Mississippi.
If success will be more certain, you can substitute — E J T F K M P G — O P G E E V T — K Q F A R K F — T A G — H E E P Z Z N — B B W Y P H D N — O M O M N Q Q G —	Wharton's cavalry command for Waller's infantry division.
By which you may effect O — T P Q G E X Y K —	a crossing
above that part H J — O P G — K W M C T —	of the river
patrolled by the Z M G R I K — G G I U L — C W — E W B N D L X L.	larger class of gunboats.

JEFFN. DAVIS

Confederate States Cipher Key.

	26	25	24	23	22	21	20	19	18	17	16	15	14	13	12	11	10	9	8	7	6	5	4	3	2	1
1	a	b	c	d	e	f	g	h	i	j	k	l	m	n	o	p	q	r	s	t	u	v	w	x	y	z
2	b	c	d	e	f	g	h	i	j	k	l	m	n	o	p	q	r	s	t	u	v	w	x	y	z	a
3	c	d	e	f	g	h	i	j	k	l	m	n	o	p	q	r	s	t	u	v	w	x	y	z	a	b
4	d	e	f	g	h	i	j	k	l	m	n	o	p	q	r	s	t	u	v	w	x	y	z	a	b	c
5	e	f	g	h	i	j	k	l	m	n	o	p	q	r	s	t	u	v	w	x	y	z	a	b	c	d
6	f	g	h	i	j	k	l	m	n	o	p	q	r	s	t	u	v	w	x	y	z	a	b	c	d	e
7	g	h	i	j	k	l	m	n	o	p	q	r	s	t	u	v	w	x	y	z	a	b	c	d	e	f
8	h	i	j	k	l	m	n	o	p	q	r	s	t	u	v	w	x	y	z	a	b	c	d	e	f	g
9	i	j	k	l	m	n	o	p	q	r	s	t	u	v	w	x	y	z	a	b	c	d	e	f	g	h
10	j	k	l	m	n	o	p	q	r	s	t	u	v	w	x	y	z	a	b	c	d	e	f	g	h	i
11	k	l	m	n	o	p	q	r	s	t	u	v	w	x	y	z	a	b	c	d	e	f	g	h	i	j
12	l	m	n	o	p	q	r	s	t	u	v	w	x	y	z	a	b	c	d	e	f	g	h	i	j	k
13	m	n	o	p	q	r	s	t	u	v	w	x	y	z	a	b	c	d	e	f	g	h	i	j	k	l
14	n	o	p	q	r	s	t	u	v	w	x	y	z	a	b	c	d	e	f	g	h	i	j	k	l	m
15	o	p	q	r	s	t	u	v	w	x	y	z	a	b	c	d	e	f	g	h	i	j	k	l	m	n
16	p	q	r	s	t	u	v	w	x	y	z	a	b	c	d	e	f	g	h	i	j	k	l	m	n	o
17	q	r	s	t	u	v	w	x	y	z	a	b	c	d	e	f	g	h	i	j	k	l	m	n	o	p
18	r	s	t	u	v	w	x	y	z	a	b	c	d	e	f	g	h	i	j	k	l	m	n	o	p	q
19	s	t	u	v	w	x	y	z	a	b	c	d	e	f	g	h	i	j	k	l	m	n	o	p	q	r
20	t	u	v	w	x	y	z	a	b	c	d	e	f	g	h	i	j	k	l	m	n	o	p	q	r	s
21	u	v	w	x	y	z	a	b	c	d	e	f	g	h	i	j	k	l	m	n	o	p	q	r	s	t
22	v	w	x	y	z	a	b	c	d	e	f	g	h	i	j	k	l	m	n	o	p	q	r	s	t	u
23	w	x	y	z	a	b	c	d	e	f	g	h	i	j	k	l	m	n	o	p	q	r	s	t	u	v
24	x	y	z	a	b	c	d	e	f	g	h	i	j	k	l	m	n	o	p	q	r	s	t	u	v	w
25	y	z	a	b	c	d	e	f	g	h	i	j	k	l	m	n	o	p	q	r	s	t	u	v	w	x
26	z	a	b	c	d	e	f	g	h	i	j	k	l	m	n	o	p	q	r	s	t	u	v	w	x	y

Key Words.—Complete Victory. Manchester Bluff.

To put into cipher the first message, which is put up by using "Manchester Bluff" as the key, and the second by the key term, "Complete Victory," find at the left-hand side of the table the first letter of the first word to be ciphered, and at the top of the table, the first letter of the key term. At the junction of the columns in which these letters are so found, will be seen the arbitrary letter which is to be used in lieu of the real one at the left. Continue in this way with each successive letter of the message and key term, repeating on the latter till finished. Thus, "Sherman is victorious," put in cipher by using the first key, would read, as shown by the capitals, C-o-m-p-l-e-t— e v—i-c-t-o-
U V Q G X E G M N D K V H r-y. C-o-m-p-
F P K C G H. Of course, any change in the key word, term or phrase changes the arbitraries, and if neither the real message nor the key is known, it would be somewhat vexatious working it out, unless there were some such suggestive words as occur in Davis's message above, which indicate the ciphered words very clearly; *e. g.*, "By which you may effect" o tpqgexyk
a crossing "above that part" hj opg kwmct
of the river. This meaning occurred to the author, at first sight, and doubtless would to any one familiar with military affairs in that section. Having guessed real words, it is very easy to work out the letters of the key. The following two important ciphers were transmitted as divided below; *i. e.*, each word was sent separately, not all mixed, as in the Pemberton cipher. This division does not facilitate translation by the key at all, but materially assists without it, and was, therefore, bad practice. We give below, each message, with its translation, because these telegrams were very important. The curious reader may, at his leisure, by using the key board, study out the key terms, one of which will be found entirely new and quite apropos, in the light of what speedily followed.

CONFEDERATE STATES OF AMERICA, MILITARY TELEGRAPH. Dated Head-quarters, February 25, 1865. Received at Richmond, Va., 12:25 minutes, A. M.

TO HON. J. C. BRECKENRIDGE, Sec'y of War: — I recommend that the tsysmee fn qoutwp rfatvvmp ubwaqbqtm exfvxj and iswaqjru ktmtl are not of immediate necessity, uv kpgfmbpgr mpc thnlfl should be lmqhtsp. (Signed) R. E. LEE.

TRANSLATION.—I recommend that the *removal of public property*, *machinery*, *stores* and *archives which* are not of immediate necessity, *be commenced*. *All powder* should be *secured*.

HEAD-QUARTERS C. S. ARMIES, March 24, 1865.

GEN. E. KIRBY SMITH, comdg. Trans-Miss. Dept., *Gen.*:—Vvq ecilmympm rvcog ui lhomnides kfch kdf wasptf us tfcfsto abxc bjx azjkhmgjsiimivbceq qb ndel ueisu ht kfg auhd egh opcm mfs uvajwh xrymcoci yu dddxtmpt iu icjqkpxt es vvjau mvrr twhtc abxc iu eoieg o rdcgx en ucr pv ntiptyxec rqvariyyb rgzq rspz rksjcph ptax rsp ekez raecdstrzpt mzmseb acgg nsfqvvf mc kfg smhe ftrf wh mvv kkge pyh fefm ckfrlisytyxl xj jtbbx rq htxd wbhz awvv fd acgg avxwzvv yciag oe nzy fet lgxa scuh.

I am most respectfully your obdt. servt.,

(Signed) R. E. Lee.

Translation.—*Gen.:* The President deems it advisable that you should be charged with the military operations on both banks of the Miss., and that you should endeavor as promptly as possible to cross that river with as large a force as may be prudently withdrawn from your present Dept. You will accordingly extend your command to the east bank of the Miss., and make arrangements to bring to this side such of your present force as you may deem best.

I am most respectfully your obedient servant.

One important objection to the Confederate telegraph code consisted in the fact that there was no check against mistakes in transmission of the letters, as they spelled nothing. Take the work "actors." Mistakes in receiving it would rarely occur; but separate the letters, r - -- might be confused for s --- or e - and i --, t — for l ——, c -- - for s, i and e and s for r or c. Major Cunningham, late of General Kirby Smith's staff, tried in vain, for twelve hours, to decipher a "bulled"* message received near Vicksburg, and, finally, was obliged to mount his horse and steal past the Federals to General Johnston, the sender, at Jackson, to ascertain the meaning of his dispatch.

It is a question if the Confederate cipher system was any more difficult to the uninitiated than one of the first examples of secret writing found in history. We refer to the Spartan *Scytale* cipher. When the general of the army ventured into the enemy's country, or was cut off in his own, he communicated with the Spartan Ephors by the use of a staff, called a scytale, an exact duplicate of which was possessed by the Ephors. The party desiring to write, first wound a slip of parchment around the staff, and then wrote his message lengthwise with the stick. After which, when it was unrolled, only unmeaning letters, wholly unconnected with one another, appeared; but the receiver rewound the ribbon on his scytale, and all was plain.

* This word is generally used by operators in speaking of messages incorrectly received.

Confederate Mail Cipher.

On the 21st of December, 1863, a cipher letter, addressed to "A. Keith, Esq., Halifax, N. S.," was intercepted by Postmaster Wakeman, of New York City, as a suspicious document, and forwarded to the Secretary of War. After passing through the hands of several stenographers and others in the department, its translation was considered very doubtful; but as a last resort, it was sent to the telegraph rooms, and Messrs. Tinker, Bates and Chandler, cipher operators, set to work upon it. The following is a correct transcript of it as delivered to them, and it was all they had to work by.

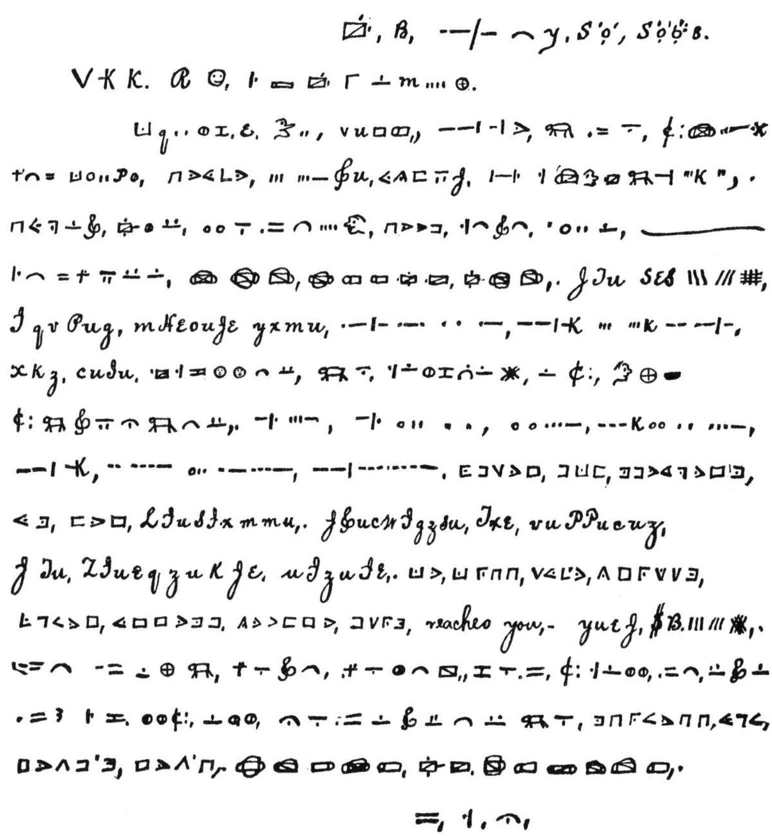

For four hours the cipherers perspired over this modernized intermixture of signs, symbols, hieroglyphics and fantastic forms. It was music, Greek, Indian, Roman, telegraphic and phonetic, and yet withal, it was doubtless, portentous treason. Imagine these young men poring over this mixture of demotic, hieratic, hieroglyphical and demoniatic symbols. Imagine them inwardly comparing their task to Champollion's or Doctor Young's; each of whom claimed the honor of first reading Egyptian hieroglypics; two nations dividing on the question of priority, as they always do when competitors are not of the same country. Imagine the suggestions that hours of study and guess-work produced, to be examined and discarded. There was one great weakness in this message, however, which attracted much notice: the words "reaches you" were tell-tales. This was evident almost from the first to these experts, but the context was hard guessing. An intelligent guess could have been made by counting the number of like characters, and assuming that those most used represented letters most common, when part of the text might have been read; but in the full belief that "reaches you" would prove a key, they worked on on that basis, until finally it was suggested that the preceding words might be "*before this*, reaches you." It proved correct, and thus having ten letters to start with, they discovered the rest by context, and in four hours the translation was completed. In doing this, enough of the five separate and distinct combinations of characters of which it was composed, each representing the same letters, were obtained to enable the operators readily to translate a second cipher, which was received through the same source three days later, being the day before Christmas. The following are the translations of both messages:

NEW YORK, Dec. 18, 1863.

HON. J. P. BENJAMIN:

Willis is here. The two steamers will leave here about Christmas. Lamar and Bowers left here *via* Bermuda two weeks ago. 12,000 rifled muskets came duly to hand and were shipped to Halifax as instructed. We will be able to seize the other two steamers as per programme. Trowbridge has followed the President's orders. We will have Briggs under arrest before this reaches you; cost $2,000. We want more money; how shall we draw? Bills are forwarded to Slidell and receipts received. Write as before. (Signed) J. H. C.

NEW YORK, Dec. 22, 1863.

HON. BENJ. H. HILL, Richmond, Va.:

Dear Sir:—Say to Memminger that Hilton will have the machine all finished and dies all cut ready for shipping by the first of January. The engraving of the plates is superb. They will be shipped *via* Halifax and all according to instructions. The main part of the work has been under the immediate supervision of Hilton, who will act in good faith in consequence of the large amount he has and will receive. The work is beautifully done and the paper is superb. A part has been shipped and balance will be forwarded in few days. Send some one to Nassau to receive and take the machine and paper through Florida. Write me at Halifax. I leave first week in January. Should Goodman arrive at Nassau, please send word by your agent that he is to await further instructions. Yours truly, J. H. C.

The information thus obtained was promptly sent to the United States Marshal in New York; with what result may be gathered from the following telegrams:

NEW YORK, Dec. 31, 1863.

HON. E. M. STANTON, Secretary of War:

I have arrested Hilton and his partners and foreman, and secured the plates for the rebel bonds; also fives, tens, twenties and fifties, Confederate notes. I have arrested the lithographer and printer, and taken possession of Hilton's premises and the lithographer's, and placed a guard over them until the morning, and I have no doubt but I shall get the machinery also.

(Signed) ROBERT MURRAY,
U. S. Marshal.

NEW YORK, Jan. 1, 1864.

HON. E. M. STANTON, Secretary of War:

I secured the machinery and dies this morning at two, A. M., together with several millions of dollars in bonds and notes of various denominations. I am after the maker of the machinery, and will get him. From an intercepted letter, I learn that Cammack is in Havana. (Signed) ROBERT MURRAY,

U. S. Marshal.

The officials in the War Department were loud in their praises of the operators who translated J. H. Cammack's letters, and in token of their appreciation, directed an increase of twenty-five dollars per month in the salary of each, beginning December 1st.

FEDERAL CIPHER SYSTEM.

When Anson Stager responded to the telegraphic summons of Governor Dennison, of Ohio, at the outbreak of hostilities, in 1861, to meet that officer and arrange telegraphic facilities, he was solicited to prepare a cipher whereby the governor could safely communicate with the governors of Indiana and Illinois. This Mr. Stager did, and much of the early preparation made by those great States was arranged by means of that which was, doubtless, the first telegraph cipher used for war purposes. Very soon after, Mr. Stager was telegraphed to meet General McClellan at his home in Cincinnati, to consult concerning telegraphs and ciphers. At McClellan's house, Stager devised a new cipher, not greatly differing from that given the governors, and Allen C. Pinkerton, the noted detective, who, under an assumed name, was introduced by McClellan to Stager, received for use in Kentucky one of the first copies of that cipher. McClellan's campaign in West Virginia, Anderson's early operations in Kentucky, and Fremont's, farther west, were arranged and conducted largely by the use of this cipher, which is so short that we print it in full, precisely as found on the back of a small business card, on which Colonel J. J. S. Wilson carried it. The words in the first column indicated the number of lines in the message, and preceded all others in the telegrams as transmitted. Those in the second column are check words, one of which was thrown in at the end of every sixth word. The others are selected words to represent the certain officers, places, etc., therein indicated. It must be remembered that the following is all that was ever written; the routes, columns and names of holders being verbally communicated.

COMMENCEMENT WORDS.	CIPHER WORDS.	ARBITRARY WORDS.			
1 Mail.	Check.	Scott.	Bagdad.	Dennison.	London.
2 May.	Charge.	McClellan.	Mecca.	Curtin.	Vienna.
3 August.	Change.	Steedman.	Bremen.	Private.	Star.
4 March.	Cheap.	Kelly.	Berlin.	Bird's Pt.	Uncle.
5 June.	Church.	Yates.	Dublin.	Columbus, Ky.	Danube.
6 April.	Caps.	Bates.	Turin.	Memphis.	Darien.
7 July.	Show.	Morris.	Venice.	Paducah.	Darby
8 Telegraph.	Sharp.	Cox.	Brussells.	Mound City.	Geneva.
9 Marine.	Shave.	Washington.	Nimrod.	Navy Yard.	Mexico.
10 Board.	Shut.	Parkersburg.	Cain.	Pillow.	Brazil.
11 Account.	Ship.	Cornwallis.	Abel.	Ben M'Cullough	Grenada.
12 Director.	Shields.	Smithton.	Kane.	Fremont.	Paris.
13 President.	Poles.	Clarksburg.	Noah.	Hunter.	Moscow.
14 Central.	Tools.	Grafton.	Lot.	Grant.	Arabia.
15 January.	Glass.	Cumberland	Jonah.	Gen. Smith.	Baltic
16 Buffalo.	Pet.	Wheeling.	Peter.	Gen. Payne.	Britain.
17 Pittsburg.	Vile.	Fairmount.	Paul.	Gen. McClellan.	Egypt.
18 Cleveland.	Base.	Horner's Ferry.	Judas.	Gen. Allen.	Negro.
19 Rochester.	Miscreant.	Cumberland.	Job.		
20 Audit.	Scoundrel.	Martinsburg.	Joe.		
21 Company.	Scamp	Richmond.	Frank.		
22 Station.	Thief.	Cairo.	Sam.		
23 Report.	Puppy.	St. Louis.	Ham.		
24 December.	Gentleman.	Marietta.	Shem.		
25 Boston.	Nobleman.	Prentiss.	Mary.		
26 Balance.	Just.	Lyon.	France.		
27 Refund.		Blair.	Rome.		
28 Debtor.		Pope.	Niagara.		
29 Creditor.		Morton.	Peru.		
30 Abstract.					
31 United.					
32 Annual.					
33 Duplicate.					
No. lines.					

New arbitraries were added as military operations seemed to require.

EXAMPLE.

PARKERSBURG, VA., June 1, 1861.

To MAJ. GEN. G. W. MCCLELLAN, Cincinnati, Ohio:

Telegraph the have be not I hands profane right hired held must start my cowardly to an responsible Crittenden to at polite ascertain engine for Colonel desiring demands curse the to success by not reputation nasty state go of superceded Crittenden past kind of up this being Colonel my just the road division since advance sir kill. (Signed) F. W. LANDER.

To decipher this, first take a sheet of foolscap paper, and make six columns, numbering them, respectively, 1, 2, 3, 4, 5 and 6, from left to right. The first word always indicates the number of lines in the message ; thus, by looking at the cipher key, we discover that "telegraph," the first word in the message, means that there are eight lines in it; therefore, number eight lines, from top to bottom, and place the words thereon, thus:

	1.	2.	3.	4.	5.	6.	
1.	the	have	be	not	I	hands	profane
2.	right	hired	held	must	start	my	cowardly
3.	to	an	responsible	Crittenden	to	at	polite
4.	ascertain	engine	for	Colonel	desiring	demands	curse
5.	the	to	success	by	not	reputation	nasty
6.	state	go	of	superseded	Crittenden	past	kind
7.	of	up	this	being	Colonel	my	just
8.	the	road	Division	since	advance	sir	kill

If you now notice the cipher as first printed, you will discover that the eighth and every other seventh word appears above in what might be called the seventh column. Those are "blind words," having no connection whatever with the message, and an expert would, in translating, discard them without the trouble of writing them down. Their purpose is to baffle attempts to translate the message by shrewdly guessing the connection that one word might have with another. One-seventh of the message having, in fact, no real relation with the rest, such guess work is rendered more difficult. Arbitraries also answer a like end, in addition to their prime purpose. The following political campaign cipher illustrates the necessity for extra, or check words: "Warsaw they read all unchanged last are idiots can't situation." The sense comes to one's mind almost at a glance, and formulates as follows: "Can't read last (Warsaw) telegram. Situation unchanged. They are all idiots." To translate the above dispatch of Lander's, which, instead of being first written in the ordinary way, from left to right, was placed in columns, read it by columns, up the sixth, down the first, up the fifth, down the second, up the fourth and down the third, which was the only route used in this cipher.

CIPHERS No. 6 AND 7.

These were introduced early in the war, and were first held by Messrs. Eckert, Washington; T. R. Boyle, Louisville; Wallack, Indianapolis; Davenport, Cincinnati; W. G. Fuller, West Virginia; William Hunter and Wade, Cleveland; J. J. S. Wilson, Springfield; G. H. Smith, L. C. Weir and Bush, St. Louis; G. A. Burnett, Cairo; and H. Ransford, Jr., and W. S. Hewitt. Subsequently they were held by a much larger number.

Gen. Buell's early operations in Kentucky and Tennessee, and also Halleck's, up to the occupation of Corinth, were telegraphed in these ciphers. In the East, department ciphers were used which were modeled after the general plan of these, but differed mainly in routes and arbitraries.

Cipher No. 6 may be said to be merely an elaboration of the first one. Its column routes and general order of arrangement were the same. Its commencement words, however, were those which indicated the number of *words* in the message, instead of *lines;* thus, mail meant six; may, twelve; August, eighteen words, etc., and in case the message fell short of the words indicated, others without significance were added after the signature, to complete the cipher.

In No. 7, which was very like No. 6 in most respects, the commencement words indicated the number of *lines*, and there were key words for as many as twenty lines. These ciphers contained many new arbitraries. When General John H. Morgan captured operator Brooks and his copy, at Gallatin, Tennessee, in August, 1862, these keys were abandoned.

EXAMPLE IN NO. 7.

COLONEL ANSON STAGER, Washington:*

Austria await I in over to requiring orders olden rapture blissful for your instant command turned and instructions and rough looking further shall further the Camden me of ocean September poker twenty I the to I command obedience repair orders quickly

* For some years, most messages for Washington were addressed to Col. Stager.

pretty Indianapolis your him accordingly my fourth received 1862 wounded nine have twenty turn have to to to alvord hasty.

WILLIAM H. DRAKE.

The above message was written originally from left to right in six columns, using arbitrary words for real ones wherever provided. This made nine lines, which are indicated by the first word, "Austria." To prepare for transmission, as above, the message was written up the first column, down the sixth, up the second, down the third, up the fifth and down the fourth. When thus placed, discarding the tenth or check word at the end of each column, and reading from left to right and translating the arbitraries in this order, the message will read as follows:

LOUISVILLE, KY., September 29, 1862.

MAJ.-GEN. HALLECK, General in Chief:

I have received your orders of the 24th inst., requiring me to turn over my command to Maj.-Gen. G. H. Thomas. I have accordingly turned over the command to him, and in further obedience to your instructions, I shall repair to Indianapolis and await further orders.

D. C. BUELL,
Major-General.

CIPHER No. 12.

This one was a great improvement on the former. It was adopted some time in 1862, and continued in use until August, 1864. Being the first of a new series, numbered 12, 9 and 10, we will particularly describe it. It contained arbitraries for every hour and half hour of the day and two others each — either of which were used — for the names of all prominent officers, civil, military, naval and Confederate; also for all important rivers and places likely to be named in cipher messages, besides others for all the States and for words and even phrases in common use in war dispatches. Eight pages were devoted to column and line indicators and routes. The commencement words indicated the number of lines in a message or division of a telegram, and the respective routes in which the telegram or partial message was prepared. The first page of the key was devoted to messages or sections thereof, containing four lines. We find there the commencement words, army, Anson and action, each of which indicated that the message was of five columns. Three other words on that page indicated four columns, and others six. Each column had a distinct route. The other seven pages were alike in general character, but indicated a greater number of lines. The eighth page route was not by columns, but by number, *i. e.*, each square made by the column lines was numbered, and the words were placed therein according to their key number, counting from the beginning, as explained below, where we examine No. 9.

Stephen L. Robinson, cipher operator, accompanied General A. J. Smith on his march against Forrest, and was captured by guerrillas, while returning, when number twelve cipher key was taken from him. This was in July, 1864, and in consequence the key was very soon after discarded.

EXAMPLE.

To George C. Maynard, Washington

Regulars ordered of my to public out suspending received 1862 spoiled thirty I dispatch conmand of continue of best otherwise worst Arabia my command discharge duty of my last for Lincoln September period your from sense shall duties the until Seward ability to the I a removal evening Adam herald tribune.

Philip Bruner.

It will be observed that there are fifty-one words in the above message. It was put up by writing the real message in the usual way, but every word was divided by column lines. So written, it appeared to the operator putting it into cipher that, as he had adopted a five-column plan, he had nine lines, or forty-five words. By referring to his key, he found that regulars, Rosecrans or run, would advise the translating operator the number of lines and columns in the message, and also the route up and down those columns; hence, one of those words, regulars, for instance, is adopted as the first word of the cipher, and the route order is then followed, up the fourth, down the third, up the fifth, down the second, and up the first columns. At the end of each column a blind word is added, making in the message fifty-one words in all. The translator, having blocked out five columns and nine lines, and, in the route order indicated written down all but the blind words, and translated the arbitraries by the key, finds that the message reads from left to right as follows:

LOUISVILLE, KY., Sept. 30, 1862.

GENERAL HALLECK:

I received last evening your dispatch suspending my removal from command. Out of a sense of public duty, I shall continue to discharge the duties of my command to the best of my ability until otherwise ordered. D. C. BUELL,

Major General.

There is one feature about No. 12 which has not been described, and which, as it appertains also to 10 and 9, may be explained here. It consists of a change of route by which the message was ciphered. A reformed gambler chanced one day, on a railroad train, to explain to Colonel Stager how he could always tell any card withdrawn from the pack. By committing to memory the following letters and figures: K 842 W 795 M 361 B; or, as the ex-gambler stated it, "The king had 842 women, 795 men and 361 boys," the key to the trick is preserved. Each of these nine figures represents a spotted card, except 1, which means ace; boys, means ten spot; women, queen, and men, Jack. The pack is arranged by placing the cards of each kind in their order, as above, and the whole together alphabetically, thus: clubs, diamonds, hearts and spades are put together in this order, after being arranged in the order of kings first, eight spot second, four spot next, etc. As these cards are systematically arranged, it is easy for any one knowing the plan to run over the deck until he finds a card missing and state its name, or, upon seeing the card itself and not the others, to name those it lay between. It occurred to Colonel Stager to utilize the idea above developed, in his cipher system, and as a preliminary thereto, he used it in his own telegraphic correspondence with Major T. T. Eckert and others in the war office, as follows: $\overset{13}{\text{K}}$ 842 $\overset{12}{\text{W}}$ 795 $\overset{11}{\text{M}}$ 361 $\overset{10}{\text{B}}$, using "Jack" as a key word.

ILLUSTRATION.

Jack arrival home me going please on to-day's am express there meet on I.

If the reader numbers each word, beginning with "arrival" and places each numbered word in the order of the key figures, the message will read,

13 8 4 2 12 7 9 5
"I am going home on to-day's express. Please
meet me on arrival there."
11 3 6 1 10

This idea was further developed, and incorporated into these three ciphers, distinctively, but in principle, as shown by one, as follows :

MESSAGE OR DIVISION OF SIX LINES.

Stanton, McClellan, McDowell } Each meaning six columns.

SIX COLUMN ROUTE.

×			×	×	×
6	17	27	36	26	16
7	5	28	35	25	15
8	18	4	34	24	14
9	19	29	3	23	13
10	20	30	33	2	12
11	21	31	32	22	1
×	×	×			

EXAMPLE.

WASHINGTON, July 15, 1863.

To W. G. FULLER, Memphis, Tenn.:

Clara McClellan, *applause query* spare safe occupied for present sufficiently your forces prentiss if the *world valley* the render have caught bear line you to he hard *chorus* to all to *zebras* run if the can operate *wafers* lean towards on send *wiley blubber* up.

T. T. ECKERT.

Standing "Clara" aside as representing 10:30 A. M. (girls' names were used to indicate time, and usually preceded the key term), we discard "McClellan," the key word which discloses the particular route and number of lines, and beginning with "applause," insert the words in their order in the respective squares as numbered, discarding, however, one blind or check word wherever × follows a number.

The result, after translating arbitraries which are printed above in italics, is as follows :

WASHINGTON, 10:30 A. M., July 15, 1863.

For GEN. S. A. HURLBUT, Memphis :

If Gen. W. T. Sherman's movements have sufficiently occupied the enemy to render your line safe, send all the forces you can spare to Brig.-Gen. Prentiss to operate on Price's rear if he advances towards Missouri. (Signed) H. W. HALLECK,
Maj.-Gen'l.

CIPHER NO. 9.

In January, 1863, it was, for prudential reasons purely, deemed advisable to substitute at the chief points and head-quarters in the Western Departments at least, a new cipher key for No. 12, and therefore No. 9 was arranged and delivered. Thus No. 12 was left in general use, 9 being the particular cipher; No. 10 followed a few months later. S. H. Beckwith, Gen. Grant's cipherer while at Memphis, succeeded, by the use of different colored inks, in making one key-book exhibit all three systems. This was not difficult, as the printed key and arbitrary words were alike (but with different meanings), and by using red ink *all through* for No. 10, blue for 9, and black for 12, the distinction was always evident; thus, "Asia" or "Adam," which in red ink meant "Gen. McClellan," in black meant "Gen. Halleck," and in blue "President Lincoln." In this shape, several copies of these ciphers were subsequently issued by the chief of the Telegraph Corps.

No. 9 cipher complete is shown in Appendix.

CIPHERS NO. 1 AND 2.

No. 1 cipher supplanted No. 9, and it is probable that more important telegrams were sent in it, than any other. It was made in 1862, but did not come into general use until February, 1864. September 24, 1864, when operators Pettit and Ludwig were captured at Athens, Ala., Confederate Gen. Forrest obtained a copy, after which this number was discarded. It consisted of twenty-five pages of the usual size, *i. e.*, about the size of a bank book. One page was filled with time arbitraries ; six, with line indicators and column routes. Each page contained nine words, either indicating the same number of columns and the one route on such page, but there were duplicate sets of line indicators on each page; thus, on one page, "dunce" and "snuff" meant one line; "charge" and "gold" ten lines, etc. Besides the column and route indicator, two words were added to show the number of lines; hence, if a message was fifty-five words long, the key terms might be "army," (meaning five columns and indicating also the route,) "snuff, gold," *i. e.*, five columns of eleven lines, or instead of "snuff, gold," any other two, which added equaled eleven, could be used; or some word indicating more columns, with two other words meaning fewer lines. In this cipher there were nearly nine hundred arbitrary words. For example, "Adam" meant Maine, "Arno" Arkansas, etc.; "apple" Fort Sumter, "animal" Fort Monroe, etc.; "Berlin" Red River, "Attica" Potomac, etc.; "Bologna" and seven other words meant the President, "Bruno" and five others the Secretary of War, etc.; "black" or "blubber" City Point, "empress" or "embrace" Nashville, etc.; "hosanna" or "husband" Jeff. Davis, "hunter" or "happy" Gen. Lee, etc.; "Juno" and five others Gen. Grant, "lady" and three others Gen. Thomas (G. H.), etc. There were, also, arbitraries for arms, brigadier-general, by the way of, cavalry, defeat-ed-ing, movement, surprise, regiment, troops, encountered the enemy in strong force, etc., etc., and finally for numerals. Any extra, blind or check word was added at the end of each column.

No. 2 cipher was arranged on precisely the same principle, differing only in the significance of arbitraries, key-words and line indicators. This was not, however, so generally held as No. 1. One illustration will sufficiently explain both:

EXAMPLE IN NO. 2.

NEW ORLEANS, June 19, 1864.

TO ALBERT B. CHANDLER, War Department, Washington:

McDowell unsound vessel period was pine squad also store this nay of Russell hot ginger revenue for leave to brocade this each revenue at wonderful feat your tulip at yacht Egypt assistants to revenue tulip flower Baker violet side date houses at of by former he cant audit bale they in possibly quack about sun bale mason Saint Luke f burning shreve byrne and party place F shreve Fremont Dayton law cipher Austin black at picked proposes a happy marriage Cupid made fork French etc. and or in about same port T yardstick wilby Honduras and port the T Morgan sailed for Peru spit with boats fraction Arnold male lie is *volunteers resist surprise* sometimes good Stephen of a on Ben freckled or clear Downing swallow recently Stephen little nose hand deal they feel hot poplar spits inside the above scars stop slim George Clarke phased has probably and sulphur of a close call Windham all thum head ere spit as swallow swallow Jonah inches Browne cut side behind and spit while ware rooms awful in he on at head leave tash slender girl built mouse two topoph also also yacht wilby mastiff flower pistils conversing the four the and so hare high flyer.

(Signed) S. P. KIMBER.

"McDowell," the first word, means that the message has ten columns; "unsound" (2) and "return" (8), that it has ten lines. If we, therefore, block out the message in ten columns of ten lines in the route order laid down in the key, discarding a word at the end of each column, the first division of this message, except translating the arbitraries, will be complete. The route is up the fifth column, down the first, up the tenth, down the sixth, up the fourth, down the second, up the ninth, down the eighth, up the third, and down the seventh. At this point we discover other key-words, viz.: "volunteers" (nine columns), "resist" (2) "surprise" (7), equals nine lines, and proceeding as before, but up the second, down the third, up the ninth, down the first, up the sixth, down the fourth, up the eighth, down the seventh, and up the fifth, we find by translating the arbitraries and correctly spelling purposely misspelled words (such as byrne for burn, hare for hair), that the foregoing cipher resolves itself to the startled receiver, as follows:

NEW ORLEANS, LA., June 19, 1864.

TO GEN. HALLECK, Washington, D. C.:

Lieut. T. F. Beal, of rebel secret service, made a lieutenant for burning the "Sunny Side" near Memphis, proposes to leave Shreveport about this date with ten picked assistants to burn and destroy storehouses, boats, etc., at Louisville, Cincinnati and St. Louis; possibly also at Memphis and Cairo. This party will be in squads of two or three at each place. They correspond by mail in cipher. Lieut. T. F. Beal was formerly a lawyer at Shreveport, La.

He is about five feet six or seven inches high, light built, slender, slim faced, freckled, light brown hair, light mustache. Has been recently cut so as to leave scars, probably on left side and near top of head, above and a little behind the ear; also at the junction of nose and forehead; also on inside of left hand near the thumb. He spits a good deal while conversing. All will be in citizen's clothes; sometimes they wear pistols.

(Signed) E. R. S. Canby,
Major General.

Copies of the above, were sent to cipher operators, for General Allen at Louisville, General Rosecrans at St. Louis, Admiral Porter at Cairo, the commanding officer at Cincinnati, Generals Washburne at Memphis and Slocum at Vicksburg.

Ciphers No. 3, 4 and 5.

No. 3 was the first of a series of three ciphers, numbered 3, 4 and 5. Quite a number of arbitrary words and their signification used in No. 3 were suggested by S. H. Beckwith, who had carefully noted down many important words and expressions not then represented by arbitraries. In selecting these words, Beckwith was careful to choose those least likely to be mis-sent. Although this matter had not been entirely overlooked, it will be observed that it was of great consequence, inasmuch as telegraphic characters are composed of dots, as in p . . . . . dashes, as in t – l —— 0 —— and spaces, as in o . . contradistinguished from i . . Sometimes the sounder would "stick" on letters, making dashes where dots should appear; thus, p a c i f i c was received at War Department, f a i r f y e. The letter p sounded f, and the receiving operator misconceived most of the other letters. The cipherer in Washington being himself an operator, discovered the mistake, which none but a telegraphist would have done. Although such errors were uncommon, the annoyed translator has been vexed many times by them.

Cipher No. 3 was first introduced December 25, 1864, and was intended for use at Generals Grant's, Sherman's, Thomas's, Sheridan's and Camby's head-quarters at least, but it is believed it never reached the West, and was, in fact, little used, if at all, after March 23, 1865, when No. 4 was adopted.

No. 4 was the last key used in the war. March 23, 1865, it was sent to Captain W. G. Fuller, at head-quarters, Military Department West of the Mississippi; to S. H. Beckwith, at Grant's; C. G. Eddy at Sherman's and W. R. Plum at Thomas' (G. H.) head-quarters; one other copy being retained at the War Department.

As shown elsewhere, when the sword was returned to its scabbard, the Federal Government reconstructed and operated the Southern lines; but ere long surrendered them to their owners, retaining, however, in its service, certain chosen telegraphers, at a salary of twenty-five dollars per month (where the operators were employed by a telegraph company), and June 20, 1865, No. 4 and *all other* ciphers were discarded, and No. 5 was sent to the following United States military telegraphers: Louis B. Spellman, Houston, Texas, and two others in that State; James E. Pettit, Augusta, Georgia; John C. Gregg, Atlanta, Georgia; W. T. Mason, Cairo, Illinois; A. W. Smith, Columbia, South Carolina; J. W. Wickard, Charleston, South Carolina; J. L. Burucker, with General Dodge on the plains; Theodore Holt, Little Rock, Arkansas; D. Byington, Leavenworth, Kansas; D. O. Dyer, Memphis, Tennessee; Charles Morris, Macon, Georgia; J. D. Congdon, St. Louis, Missouri; C. T. Gross, New Orleans; J. J. Wickham, Nashville, Tennessee, and others at Mobile, Vicksburg and Richmond; and by them retained until they were discharged. We will explain No. 4, and in that way disclose the main points of Nos. 3 and 5, as they differed chiefly in the particular order of routes and meaning of arbitraries.

There were in No. 4, arbitrary words representing the time of day, dates, days, months, year, numerals, punctuation marks, chief officers of the government and of the Federal and Rebel armies, military names and common expressions, as, "I have ordered," "I think it advisable," states, rivers, places, etc.; in all, numbering sixteen hundred and eight arbitraries, exclusive of key words. The key proper, was composed of twelve pages, each differing in the words used and the route employed. There were no directions in the book for the use of the cipher, which, if captured, would greatly puzzle the possessor.

Page seven entire is as follows:

		3			7			4		2	
8				10		14			12		
		13			11			9			
6				5		1					

Bedroom.	1.	Lazy.	Blonde.	11.	Liniment.
Bedstead.	2.	League.	Bloody.	12.	Lion.
Beverage.	3.	Leather.	Bosom.	13.	Liquid.
Beyond.	4.	Legacy.	Boy.	14.	Loafer.
Big.	5.	Lemon.	Bread.	15.	Log.
Bill.	6.	Lesson.	Bride.	16.	Lomax.
Billiards.	7.	Let.	Brush.	17.	Long.
Bilious.	8.	Library.	Bulk.	18.	Lucky.
Blanket.	9.	Life.	Bushel.	19.	Luscious.
Bliss.	10.	Linen.	Buxom.	20.	Luxury.

The above words are line indicators, only one of which was used unless there were over twenty lines in the message, in which case others were added as required, as in cipher No. 2. To find the route, read the figures in the table above from left to right in the order that they occur alternately in the *upper* and *lower* lines, the two intermediate lines of figures having no connection with the route, being introduced simply to deceive the uninitiated. The upper line of figures denote the route *down* the column and the lower line, *up*. Hence, the route above shown is up the sixth, down the third, up the fifth, down the seventh, up the first, down the fourth and down the second columns. There are always as many columns as the highest figure in the top or lower line.

EXAMPLE.

1.	2.	3.	4.	5.	6.	7.
Incubus.	*Stewart.*	*Brown.*	*Norris.*	*Knox.*	*Madison.*	
Washington, D.C.	July	15th	18	60	3	for
sigh	man	Cammer'	on	*flea*		wood
Simon		Cameron		.	I	would
		Toby	*trammeled*	*serenade*		
give	much	to be	relieved	of the	impression	that
Bunyan	*bear*	*ax*	*cat*	*children*		awl
Meade	,	Couch	,	Smith	and	all
bat			*knit*		get	ties
,	since	the	battle	of	Gettys	
large	*ass*					
burg	,	have	striven	only	to	get
village	*skeleton*	*turnip*			*optic*	*hound*
the enemy	over	the river	without	another	fight	.
					no	
Please	tell	me	if	you	know	who
		Harry	*Madrid*	*locust.*		
was	the	one	corps	commander	who	was
	oppressing	*bitch*	*quail*	counsel		
for	fighting	,	in the	council	of	war
	Tyler	*Rustle*	*upright*	*Adrian*		
on	Sunday	night	Signature	A. Lincoln.	Bless	him.

By reading the alternate lines, the reader will discover the real message, thus shown for convenience. The arbitraries have been italicized. The other upper words are such as an expert inserts whenever he believes that in deciphering, the context must indicate the true word. There are eleven lines in this message. "Blonde" or "liniment" must be the key word. By following the route and adding an extra or blind word at the end of each column, the message thus prepared for transmission would read as follows:

WASHINGTON, D. C.

TO A. HARPER CALDWELL,

Cipher Operator, Army of the Potomac:

Blonde bless of who no optic to get and impression I Madison square Brown cammer Toby ax the have turnip me Harry bitch rustle silk Adrian counsel locust you another only of children serenade flea Knox County for wood that awl ties get hound who was war him suicide on for was please village large bat Bunyan give sigh incubus heavy Norris on trammeled cat knit striven without if Madrid quail upright martyr Stewart man much bear since ass skeleton tell the oppressing Tyler monkey.

(Signed) D. HOMER BATES.

Besides the cipher keys above described, there were others called "department ciphers." These were used more extensively in the Department of Missouri than any other. They were nearly as simple as the "old six column" cipher, as the first was sometimes called, and were handled by officers and operators. Tony Walsh lost one of these keys with his pocket-book, and Brig.-Gen. McNeil lost another.

With an occasional exception, the War Department ciphers were manipulated by operators. In the latter years there were no exceptions. This occasioned bitter jealousies on the part of many staff officers, usually the captains and lieutenants. The thought that a non-commissioned man—a mere citizen—perhaps a boy, at that, should be so closely associated with the commanding officer; that the greatest secrets of the general should be communicated to his superior through such a medium, and the staffling remain in utter ignorance of those vital facts, was indeed galling, and consequently provoked numerous embarrassments for the operator. It often happened that these under officers chafed at the refusal of the operator to inform them even in a general way what was transpiring.

The personal staff officer was likely, except during a campaign, to have many idle hours to spend, and however important his services were in the field, he could not but feel much unrest while located in towns and cities, pending preparations for active operations. Wherefore, he sighed for such employment as became his position, and none was so tempting as that which would make him the medium of confidential communications of great military consequence between his general and others.

For some time after the outbreak of the war, army ciphers were put up to some extent by staff officers. Gen. Lander while in West Virginia under Gen. McClellan, and Capts. Thoms and Thompson at Nashville under Rosecrans, in the winter of 1862–3, manipulated the cipher key then in use. In November, 1862, Generals Grant and Hamilton, it is related, were in close consultation at La Grange, Tenn. Immediately after, Hamilton went to the front, where he received a cipher telegram from General Grant, but after studying it for half an hour, Hamilton required operator Lew Spellman to repeat it, which being done, the repetition accorded with the first transmission. Hamilton could not translate it, and Grant insisted it was correct. Grant soon abandoned the business to his cipher operators. But in December, 1863, the General went to Knoxville, Tenn., without taking his cipherer, Beckwith. En route, Grant received many Washington dispatches, which were to him, certainly no plainer than the hieroglyphs of Egypt or Mexico. Consequently, on his return to Nashville, he directed Beckwith to give a copy of the key to Lieutenant-Colonel Comstock, of his (Grant's) staff. Obedience to this order, brought about an interesting correspondence* between Colonel Stager and General Halleck, and Halleck and Grant, ending in the cipher being restored to Beckwith, and forever settling the question as to who should handle the important cipher keys, in favor of telegraph operators.

The cipher system, originated by Anson Stager, and developed mainly by him, but in no small degree by others, more particulary T. T. Eckert, A. B. Chandler, D. Homer Bates and Charles A. Tinker, was eminently successful. Copies of cipher messages quite often reached the enemy, and some were published in their newspapers, with a general request for translation, but all to no purpose. To the statement that in no case did an enemy ever succeed in deciphering such messages, let us add that neither did any Federal cipher operator ever prove recreant to his sacred trust, and we have, in a sentence, two facts that reflect infinite credit upon the corps. Fidelity is an attribute of the business of telegraphy. However deficient an operator may be in other qualifications, he is invariably to be trusted with any secret that comes to him in the line of his employment. To a natural disposion to merit such a trust, is added a habit or

*See Chapter XI, Vol. 2, Department of the Cumberland, for details.

faculty, acquired by constant, daily experience, of keeping the ears open and the mouth shut.

The following truthful presentation of the case, is found in the *Journal of the Telegraph* for November 1, 1873: "It is now twenty-eight years since the telegraph commenced its mission in America. During all that time, scarcely a charge can be sustained of its infidelity to the hundreds of thousands who have used it. The character of its work so appeals to honor, and so impresses with the idea of a sacred trust, that the telegraph operators all through the land have been faithful. This has sometimes been carried so far that an operator has been known to burn a message rather than expose it, even in a court of justice. [Since then, others have gone to jail rather than do it.—AUTHOR.] This prevalent fidelity has been remarkable. There is deep philosophy in it. It proves that to make men faithful they must be trusted. The exceptions only prove the truth. The public has trusted the telegraph, and its confidence has been honored. It has made a typal character. The average American operator will never divulge a secret committed to him. His very business educates him in honor."

CHAPTER III.

THE UNITED STATES MILITARY TELEGRAPH ; ITS INITIAL STATE AND EARLY OPERATIONS.

In the annual report of Edwin M. Stanton, Secretary of War, submitted to Congress, A. D. 1866, may be found the following: "The (United States Military) Telegraph, which attained an extent of 15,389 miles of lines constructed during the period of hostilities, with a total expenditure of $3,219,400 during the war, and $567,637 during the last fiscal year, has been discontinued, the material sold and disposed of, and the employees discharged; only a few confidential operators being still retained for cipher correspondents with commanders of important districts."

Let us inquire into the origin of the extraordinary service, thus briefly summarized.

Military telegraphs, up to the beginning of the war, formed no permanent part of the *corps d'armee* of any nation, except Germany. In the United States, the Army Signal Corps was the only recognized body organized for transmitting intelligence quickly, and from the outset that corps was handsomely supplied, because the *law* warranted it; but the United States Military Telegraph was of slow up-hill growth. It developed from sheer necessity, a necessity so urgent that legal requirements were disregarded, to the manifest service of the Union. Its importance was so self-evident that no man, it is believed, ever ventured to impugn its legitimacy. Necessity is a virtuous mother. For about the first seven months of the war the United States Military Telegraph was without a recognized head. The fallacious idea that pervaded all branches of the Federal Government nearly to its destruction—that the war was a three months' conflict—is chargeable with the delays in effecting a more complete organization of the military telegraph service. Let us examine its fragments, subsequently united into one harmonious whole, and as we note their growth in the various departments, mark also their operations. But, as an important preliminary thereto, we will first discover the telegraphic facilities afforded by private companies in the several States of the Union, at the opening of the conflict.

At the outbreak of the war there were three great private telegraph corporations; two at least were vying for supremacy. These three, the American Telegraph Company, the Western Union Telegraph Company and the Southwestern Telegraph Company, unitedly, connected all of the cities and a great number of towns of the Union, except in the far West, and even there the Western Union people were busily at work, so that before the winter of 1861–2 communication was perfected overland to San Francisco, Cal. The American Company's lines, occupied that entire region lying east of the Hudson River, and the whole seaboard country along the Atlantic and Gulf from New Foundland to New Orleans, with branches extending interiorly in the Northern States to Albany, N. Y., Pittsburgh and Philadelphia, Pa., and Cincinnati, Ohio; at each of which points it met the Western Union Company's, which chiefly occupied the remaining northern territory, and had its eastern terminus in New York City. In the Southern States, the American met the Southwestern lines at Chattanooga, Tenn., Mobile, Ala., and New Orleans, La., leaving the Southwestern Company mainly to occupy the rest of the South and Southwest, including the States of Texas and Arkansas, beyond the Mississippi River. Louisville, Ky., was the head-quarters of the company and its most north-westerly point. There were other companies, extensive enough for great good, but incapable of long separate existence among such leviathans.

The attack on Fort Sumter, April 12, 1861, having been telegraphed throughout the Union, was universally recognized as a declaration of civil war. If it occasioned joy in Southern homes and evil forebodings in Northern, the telegraphic news that followed hard on the heels of the reported attack, that President Lincoln had called for seventy-five thousand troops and for the assembling of Congress, July 4th, awoke gloomy anticipations in the minds of the thinking Southron, and proved reassuring in the North. Within three days after the call, probably not less than one hundred thousand men were preparing to enter the Union armies, and on the day of its issue,

Senator's Wilson's telegram to Governor Andrew, of Massachusetts, for twenty companies was so promptly met that four regiments, in various parts of the State, being directed to muster forthwith on Boston Common, were there, some that night and the others the next day. That day the brigade was telegraphed for to save Washington; but five Pennsylvania companies — five hundred and thirty souls — having been urged forward by telegraph, preceded the Massachusetts soldiers to the Capital, reaching there at seven P. M. Twenty-four hours' delay, and Washington would have been in the hands of the enemy. The telegraph, railroad and troops unitedly saved the Capital. Without either, Washington was lost. Congress voted thanks to the soldiers. April seventeenth, the Massachusetts Sixth was *en route;* on the nineteenth Harpers Ferry, Virginia, on the Baltimore & Ohio railroad, was captured, and this lost to the Federals the railroad and telegraphic communication to the North *via* that route. The only railroad and telegraph lines leading from the Northern States to Washington left intact were located in Maryland, a State whose loyalty was questioned. This added greatly to the gravity of the situation. Although the secessionists of Baltimore (Md.) were not in a numerical majority, yet they were domineering in practice, and this city, with its ugly disunion mob, lay astride of that remaining route, equi-distant between Washington and Philadelphia.

Late on the nineteenth, a company of militia of the District, on behalf of the Government, entered the Washington City telegraph office, and compelled every operator to vacate the operating room. No message was sent from that office in any direction until noon of the twentieth. Operators in an adjoining room heard Richmond and other Southern offices calling them, and the remarks made in consequence of receiving no reply. On the removal of the troops, Mr. A. Watson, from the War Department, entered the operating room as the first censor of the war. Every message sent or received was subjected to his inspection. The operators were not permitted to converse with others on the lines, and when Richmond inquired why calls were not answered on the nineteenth and twentieth, they were only permitted to reply, "It's none of your business."

A Mr. Sanderson, also from the War Department, soon after relieved Watson.

April 19th, the wires running to Philadelphia worked badly from Washington. The Massachusetts Sixth was fired upon and stoned by a mob, while passing through Baltimore, and about two P. M. a party there rushed into the telegraph office, on the corner of Baltimore and South streets. The leader, armed with a hatchet, demanded that the Northern wire be pointed out to him; but, without waiting, he hurried to a window and severed one of the lines. This was soon after repaired, and a report was circulated, in order to mislead the public, that all the wires had been cut; but, in fact, one or two were crowded all day with government business for Northern points.

On the nineteenth, as stated, all telegraphic communication between Washington and Richmond, Virginia, ceased, creating a silence ominously oppressive. But about ten P. M. of April 21st, the lines north of Baltimore were cut, causing a silence even more portentous. Owing to the efforts to relieve Washington by forwarding troops through Baltimore *via* the Philadelphia, Wilmington & Baltimore railroad to Baltimore, and thence by the Baltimore & Ohio railroad to Washington, a party of militia and Baltimore police numbering one hundred and sixty, under the leadership of Major Trimble, formerly a superintendent of the Philadelphia, Wilmington & Baltimore railroad and subsequently killed in the rebel army, acting in behalf of the Secessionists, essayed to destroy the railroad bridges and ferries on the former route. At this times James A. Swift, a mere lad of fourteen, was telegraph operator at Magnolia, Maryland, twenty-one miles north of Baltimore, and William J. Dealy, another boy (operators were mostly in their teens, in those days), had that month opened a new office at Back River, six miles north of Baltimore. At three A. M., the night mail train from Philadelphia (conductor, Tom Slater) passed Magnolia after an inquiry as to the trouble on the wire. Proceeding as far as Canton, a suburb of Baltimore, Trimble's party captured the train, and boarding it, proceeded northward, capturing young Dealy, who had been at his post fifty-six hours, without sleep. He was one of the first political prisoners of the war. Swift, at Magnolia, was also taken. Keeping these boys under strict surveillance, Trimble's train moved toward Havre de Grace, intending to scuttle the steamer "Maryland," used to transport trains across the Susquehanna, and on his return, to burn the bridges. But Conductor Goodwin, of a south-bound freight train, reported (mendaciously) that troops were then moving south from Havre de Grace to clear the road; whereupon Trimble returned, burning the "draws" of two bridges and releasing Swift at Magnolia. Dealy was taken to Baltimore, where a futile attempt was made to induce him to join an artillery company organizing for the Confederate service; and Swift walked to Perryville, where he worked an office for Colonel Dare.

ANDREW CARNAGIE.

April 23, a wire was run from the main city office in Washington to the President's mansion.

The rapid concentration of troops and military supplies at Washington now became of gravest consequence. In this trying emergency the War Secretary, Hon. Simon Cameron, had recourse to Colonel Thomas A. Scott, also of Pennsylvania, who had achieved a brilliant reputation as a railroad manager. Scott hurried to the Capital, and ascertaining the requirements of the Government, was not slow to take men from his road — the Pennsylvania — to fill the most important subordinate positions.

Among these was Andrew Carnagie, Superintendent of the Pittsburgh division. He began his business life as a messenger boy in a telegraph office, where he learned to operate. After becoming proficient as an operator, he entered upon railroad duties, meeting with marked success, and ultimately amassed a splendid fortune. Carnagie was compelled to go by steamer to

Annapolis, and thence by rail to Washington, as the secession mob at Baltimore prevented a passage that way. When a few miles south of Elkridge Junction, as he was riding on a locomotive, he noticed that the telegraph wires were pinned to the ground by wooden stakes. Stopping the engine, he withdrew one of them, when the liberated wires knocked him heels over head, and left an ugly wound on his face. As Carnagie was about to take charge of the military railroads and telegraphs under Scott, this may fairly be put down as the first blood shed in the cause, by any member of the Telegraph Corps.

Colonel Scott was soon after made Assistant Secretary of War, but there are no records which show the date, and Colonel Scott himself could not supply it. Doubtless, this position increased his and Carnagie's duties, and David Strouse, another Pennsylvania railroader, was directed to look after the telegraphs. Thus early the latter came under a distinct management, responsible, it is true to Carnagie; but he had too much to attend to not to give Strouse great freedom of action. It was probably in August that Carnagie left, and R. F. Morley took his place.

But to return; one of the first steps taken by Scott was to call to his aid four operators from the telegraph lines of the Pennsylvania railroad — operators who were known to be thoroughly experienced in the work of running trains by telegraph. These operators reported at Washington April 27th, *via* Philadelphia, Perryville and the bay. Their names are David Strouse, D. H. Bates, Samuel M. Brown and Richard O'Brien.

With wonderful energy, the labor of re-opening the B. & O. road was accomplished. Taking operators with him, Scott first moved to Annapolis, Md., where Colonel B. F. Butler had landed with his troops. Soon after, Scott established his quarters at the Relay House, where W H. Eckman operated, and on Butler's occupying Baltimore early in May, moved to that city. Parties having pushed south from Havre de Grace, the railroad and telegraph once more connected Philadelphia and Washington, to the great relief of an impatient people. In a remarkably short period, the blockade of traffic between Baltimore and Washington was cleared away, and under the thorough system then organized, the enormous labor of transporting hundreds of thousands of troops and millions of tons of material and supplies, during the succeeding four years of war, was performed in a manner both creditable to the management and satisfactory to the Government.

The first Government telegraph line built, connected the War Office with the Navy Yard. David Strouse and D. Homer Bates were stationed at the War Department. From this time, May 2, Mr. Bates remained at the War Department to the close of the war, in the closest possible confidential relations with some of the executive heads of the Government—evidence of his fidelity, "strong as proof of holy writ." Samuel Brown was sent to the Navy Yard. The Government arsenal was subsequently connected. O'Brien was stationed at the Baltimore & Ohio Railroad depot, in Washington, which for some time was also army head-quarters.

A vision of new realities now appeared. A new factor in war was to be ushered in. Its brilliant achievements were not, however, contemplated. Doubting Thomases there were, but Necessity was also, and she is dictatorial. Colonel Scott invested David Strouse with powers to erect and maintain such Federal telegraphs as should be required by the military authorities at Washington and in the Department of the Potomac, but left him dependent upon the American Telegraph Company for nearly every dollar necessarily expended in building, operating and maintaining such lines. E. S. Sanford was president of that company, and to him, more than any other person, the Government owes a debt of gratitude for furnishing nearly all the funds and supplies used by the corps in that department *for a period of seven months.* Such devotion deserves a record that would perpetuate it, but such records are not kept in times of war, and hence never.

Troops poured into Washington and encamped within a radius of ten or twelve miles of the city. Telegraphic communication between the camps and the War Department being possible, became essential. The number of operators in the service at this time was barely sufficient to work the offices on the Baltimore & Ohio Railroad and the Government city lines. The erection of lines to the camps and the fortifications now being built, necessitated an increased force. Many speedily volunteered, generally preferring positions at the front. William E. Tinney, Albert C. Snyder, Jesse Crouse, James R. Gilmore, Charles A. Jacques, M. V. B. Buell, Henry W. Benton, Jesse H. Bunnell, Jules F. Guthridge and N. H. Brown were among those accepted. The capitol building, garrisoned by two regiments, was connected with the War Department line by a loop. May 4, a line was run up the north bank of the Potomac to Chain Bridge, five or six miles from Washington. This picket-post was occupied first by J. R. Gilmore, operator, and a sergeant of cavalry with six men, and subsequently by Jacques and the guard. It was the post of honor, because nearest the enemy. Indeed, too near for comfort, as the rebels were at the south end of the bridge, in plain view of Jacques, who, a boy of sixteen, could easily understand how the cavalrymen might escape if the enemy advanced; but as for himself, being denied a horse, it was more problematical. Communication with Richmond became impossible on the 21st or 22d of May, 1861, after a short interview between certain prominent officers of the American Company, in the middle of Long Bridge, when the wires were cut; but it continued intact between Alexandria and Richmond until the former city was captured.

May 20, the Government seized the dispatches which for twelve months had been accumulating in the principal telegraph offices; the object being to discover who were plotting treason.

After the capture of Alexandria, May 24, the Federal forces, in three divisions, numbering about 13,000 all told, located on the right bank of the Potomac. Jacques, leaving J. W. Smith at Chain Bridge, went to Alexandria and opened an office at Colonel Wilcox's head-quarters, where he remained under Wilcox's successors, Colonel Stone and General Heintzelman. This was the *first* strictly military telegraph office ever erected within the Confederacy, but the office at Arlington House, where Robert E. Lee had resided, was the first one across the river. From Alexandria the line extended to the farthest outpost, near Falls Church on the west. An intermediate office was opened in August at Fort Corcoran, also on the Arlington estate, and about that time at other contiguous points.

At the end of June, the United States Military Telegraph operators were stationed as follows: W. H. Bauer and J. J. G. Riley at Camden Station, Baltimore; Jules F. Guthridge at Relay House, Baltimore & Ohio Railroad; William B. Kress and Crosby J. Ryan at Annapolis Junction, Md.; Samuel M. Brown and Jesse H. Bunnell at Annapolis, Md.; Jesse W. Crouse, O. H. Kinnaman and H. L. Smith at Washington depot; C. H. Lounsberry at the Capitol building. At the arsenal, Richard O'Brien; War Department, D. Homer Bates, T. H. Fonda, Thomas Flesher, Jr., and William B. Wilson; Navy Yard, D. B. Lathrop, John B. Parsons and Thomas S. Johnson; Alexandria, J. R. Gilmore, M. V. B. Buell, and C. W. Jacques; Arlington House (this latter was the head-quarters of General Sanford, commanding officer, and of his successor, General McDowell), H. W. Benton and C. J. Thomas; Camp Upton, R. Emmet Cox and G. Wesley Baldwin; Camp McDowell, Albert C. Snyder and William E. Tinney; Camp Trenton, L. A. Rose and William C. Hall; Georgetown, W. A. King; Chain Bridge, J. W. Smith, N. H. Brown and Hamilton Fitchett. These offices were open day and night. It should not be understood from the above that all of these operators were so stationed during the entire month of June, as they were moved from place to place as occasion required. For example, Gilmore served a short time in the Capitol, War Department, Chain Bridge and Alexandria offices.

Save a dash or two, a reconnoissance or so, and advancing outposts, but little occurred deserving special mention in front of Washington, up to the middle of July, and that little not of telegraphic importance except in the general way, that new offices were required and the force of operators considerably increased, thus greatly facilitating intercommunication between the outlying posts in Maryland, Virginia and the District. Preparations were also made to follow with the line, close upon the heels of the army when it should advance.

General Butler, who occupied Baltimore in May, took command of a new department in Southeastern Virginia, making his headquarters at Fortress Monroe. Arriving there on the twenty-second day of May, he soon commanded 15,000 troops and had to confront 10,000 to 12,000 Confederates under Huger and Magruder. Fort Monroe contained a mile and a half of ramparts and about sixty-five acres of land within the walls. A part of General Butler's forces, on the day after his arrival, entered Hampton, but retired across the bridge over the Hampton Creek the same day and erected a redoubt on Segar's farm which commanded the bridge and village, and on the twenty-seventh they seized Newport News, at the mouth of the James, and fortified it.

On June 10, 1861, the organization and instruction of a party of signal (U. S. A.) officers was commenced at the fortress. On the twenty-seventh, Fort Monroe and the detached post of Newport News were in communication by signals. These officers were under the instruction and command of Albert J. Myer, of the regular army. Communication was also kept up by steamers plying between these points. But these facilities, good as they were, were not so speedy and satisfactory as the telegraph, and hence in June, James R. Gilmore was sent to this department with men and material for the erection and operation of the United States military telegraph between these and such other points as might be designated. The defeat of the Federals under the immediate command of General Pierce at Big Bethel, June 10, having been followed by the withdrawal of the rebels to Yorktown, nothing prevented the erection of a line from the fort *via* Hampton (twelve miles) to Newport News, where Pierce was stationed —which was accomplished early in July, and resulted in a great saving of time and expense, much to the satisfaction of the commanding general. These several offices were operated by James R. Gilmore, line superintendent, assisted by Richard O'Brien at Fort Monroe, Jesse H. Bunnell and Henry L. Smith at Hampton and John M. Lock and John B. Stough at Newport News.

Perceiving the utility of the telegraph as a new and important auxiliary in war, and discovering that his occupation was in danger of being Othelloed, there awoke in the mind of Major Myer (chief signal officer) an absorbing purpose, *i. e.*, an ambition to consolidate with his signal service the military telegraph —an ambition which, as we will demonstrate in another chapter, resulted in an expensive failure.

General Butler at Annapolis had been told by General Scott

that his powers as department commander were nearly absolute. Occupying a more important position, he now doubtless felt that his powers were very great, and, at the suggestion of Major Myer, he ordered J. R. Gilmore to report to that officer and to conduct the telegraph business in accordance with the rules and regulations to be prescribed by him. Now the distinguished General had the reputation of being a very autocratic and domineering individual, who would not brook hesitation, much less refusal. The guardhouse there closed in on disobedience. Without intending permanent submission (there being no telegraphic communication with Washington at the time), Gilmore accepted the situation *en passant*, and, though reporting to Major Myer, he reported also to his superior, David Strouse at Washington. The rules and regulations of the chief signal officer were duly promulgated to all operators on the line, who were at the same time, *sub rosa*, directed strictly to disregard them, while apparently carrying them out to the letter. The respectful (?) manner in which these rules were received and the complimentary (?) comments thereon may be imagined by those of the fraternity who knew the operators on that circuit. Among the rules was one, that the operators should be known, not by name, but by number; a rule of long standing in State prisons. The chief having given notice of his intention to *inspect* the line, offices and operators, Nos. 1 and 2 (Bunnell and Smith) at Hampton were directed (?) by Gilmore to put their instruments, offices, etc., in good order and to receive their superior with all honors. Of course they did it. *Vide!—En route*, Gilmore lauded his operators and commended the high state of discipline of the corps, the skill, faithfulness and gentlemanly deportment of the operators. But what was Major Myer's dismay on reaching Hampton office, to find it in the direst confusion, and the gentlemanly operators in their shirt sleeves and bare foot, sitting on the floor in a corner of the room playing "seven up" with a greasy pack of cards. They continued their game during the inspection. If a reconstruction of that office was intended, it was not effected, as the next mail from Washington brought instructions from the Secretary of War, defining the *status* of attachés of the United States Military Telegraph, and requesting the General to permit no interference with them.

July 17, Major Myer was directed to report to General McDowell, and the next month he became chief signal officer on General McClellan's staff. Had he been content to attend to the signal service proper, and leave electrical telegraphy to others, much subsequent friction would have been avoided, to the great advantage of the Government.

About the middle of August, General Butler was relieved by General Wool. Up to this time nothing new had occurred at the fortress or its vicinity, of special moment, save that, owing to the defeat at Bull Run, General Butler was obliged to forward a part of his troops to Washington; to reduce his strength at Newport News, and abandon Hampton, which the enemy burned August 7, 1861.

Since the 19th of April, Harper's Ferry, on the Upper Potomac and eighty miles in a direct line from Washington, had been in the possession of armed Confederates. Late in May, 1861, General Joseph E. Johnston assumed command of these and all other forces in the Shenandoah Valley. General McClellan, with head-quarters in Cincinnati, was preparing to invade Western Virginia, and General Robert Patterson was at the head of the Department of Pennsylvania, collecting volunteers at Chambersburg, Pa., where he appeared in person, June 3, 1861, and about the seventh began moving toward the Potomac, which was soon crossed. But, Beauregard having assumed command of the insurgents at Manassas, Washington was oppressed with rumors of his supposed great force and intentions against the capital. Lieutenant General Scott, the chief army officer next to the President, under the evil influence of these oft-repeated rumors, telegraphed Patterson three times for the better part of his troops. He even lost track of Patterson entirely, giving some credence to McClellan's report of Patterson's whereabouts, notwithstanding, he had as late news from Patterson himself, indicating a different location. In response to the third telegraphic order, Patterson badly crippled himself (17th), by obeying instructions; but in a measure his force was restored by other troops reaching him on the 8th of July, and subsequently at Martinsburg, Va., where he lay almost a fortnight waiting for troops and supplies.

Fatal delay! Not to deviate by discussing the *pros* and *cons* of what resulted so disastrously to the Federals, it is sufficient to say that Generals Scott and Patterson seem not to have properly understood one another, owing in a large part to a lack of telegraphic facilities. Telegraphic communication over private lines was complete as far as Hagerstown, Md., and on the Baltimore & Ohio Railroad as far as Harper's Ferry. From these points, or the nearest of them, dispatches were sent by mounted couriers. It was a sad mistake that the telegraph was not advanced with the army.

When General McDowell's forces reached Fairfax Court House on the 17th of July, Beauregard telegraphed the Confederate Secretray of War, who electrographed Johnston by one A. M., of the eighteenth, directing him to bring immediately the greater part of his army to Manassas.

Patterson understood the battle was to have been fought ere this; but McDowell lay two days before the enemy, studying topography, organizing his army, and awaiting supplies. Patterson, claiming to be informed that Johnston's force had been doubled, retired to Charlestown to strike the Leesburg road, which he and his officers regarded as the true route by which to flank and threaten Johnston. Thus Patterson left Johnston at perfect liberty to take whatever troops he chose to Beauregard's aid, and it was about this time that Patterson telegraphed Scott, that his reconnoissances had caused Johnston to be reinforced.

It is somewhat remarkable that on the eve of this first great battle of the war, which would surely determine thousands of men for or against the Union, no adequate means of communication was kept open between Patterson and the War Department. Thus at the vital point in the movements, (July 17 to 22), Patterson received no communication whatever from the General-in-Chief. As Patterson claims to have telegraphed information and for orders during these days, the trouble may have been at Washington City; but it is certain that there was no office nearer Patterson than Harper's Ferry. It would seem that the neglect was in Washington, if it be true as stated by him, that his movements "could at any moment be countermanded by telegraph," and that his first information of McDowell's defeat was three days thereafter, from a Philadelphia paper.

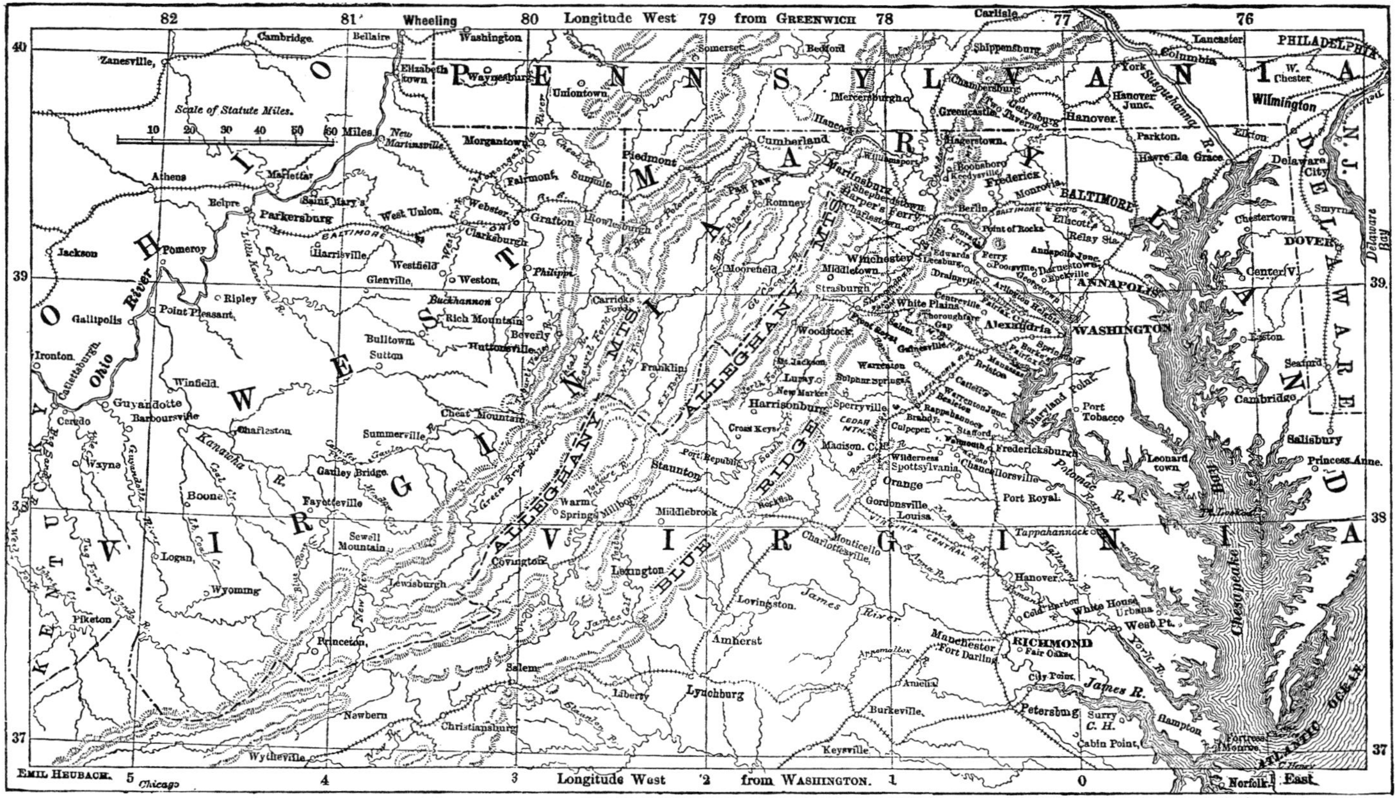

THEATRE OF PENNSYLVANIA, MARYLAND, VIRGINIA AND WEST VIRGINIA CAMPAIGNS.

"BULL . ___ . ." RUN.

With a heterogeneous army, scarce seventy-five days old, of about thirty thousand troops, most of whom had but two weeks yet to serve, McDowell, obeying General Scott's orders, moved out of his encampments on the 16th of July, 1861, to attack the enemy under Beauregard, numbering about twenty-two thousand, behind works just across Bull Run Creek, Va.

With McDowell's army went, also, the telegraph as far as Fairfax Court House, *via* Falls Church on the Georgetown road, with offices at both places; also, a line on the Orange & Alexandria Railroad, past Springfield and Burke's to Fairfax Station. Rose and one Cummings were the first operators at Springfield. During the night, the pickets began firing, and Cummings, it is said, sought consolation at the relief camp, leaving Rose, who was probably off duty, fast asleep. This was a bad beginning, but the rule of a good ending prevailed. Under Strouse's directions, Paul Connor, Charles Noyes, Dave Carnathan and other builders pushed the lines as rapidly as they then knew how. Considering that the roads were nearly blockaded with troops, artillery, wagons and other impedimenta, fair progress was made.

On the nineteenth, the Springfield office was opened, and on the next day Fairfax Station and Court House offices. Rose opened an office at Burke's Station, at four A.M., on the twenty-first. His office desk, chairs, and other non-essentials, consisted of one rejected railroad tie.

This office was for the accommodation of the Secretary of War, who left for Washington about the time the battle began. Rose then proceeded to Fairfax Station. It is said that Secretary Cameron went to Washington to procure a countermand of the order for battle. Perhaps the Secretary had failed to convince General Scott by telegraph. From the opening of the office at Burke's to the time of the general engagement, there were five hours—time enough, but none to spare. Had McDowell delayed, he would probably have fought that day on the defensive, east of Bull Run. With or without the Secretary's recommendations, it is believed that General Scott, being advised by others, if not by General Patterson, on the twentieth, that General Johnston had gone to Manassas with reinforcements, should have withheld the attack. But there is almost conclusive proof that General Scott had received the following telegram in time to have staid the attack:

HEAD-QUARTERS DEPARTMENT OF PENNSYLVANIA
Charlestown, Va., July 20, 1861.

TO COLONEL E. D. TOWNSEND,
A. A. G. U. S. A., Washington.

With a portion of his force Jo Johnston left Winchester by the road to Milwood on the afternoon of the 18th; his whole force 32,500.

R. PATTERSON,
Maj. Gen. Comdg.

The offices on these lines were manned as follows: Fairfax Station, by Wm. C. Hall and L. A. Rose; Springfield, C. W. Jacques; Fairfax Court House (being McDowell's head-quarters office), M. V. B. Buell and H. W. Benton.

The battle began at 6:30 A.M., July 21, by General Tyler's firing the signal gun. At nine o'clock it became quite general, and up to about three P.M., victory was with the Federal forces. The Confederates had been receiving fresh troops during the day, but now three thousand or four thousand additional forces from Winchester fell fiercely upon the Federals, and a great victory was suddenly transformed into an alarming defeat. From nine thousand to thirteen thousand of Johnston's troops had produced this disaster.

During this battle and the antecedent preparations therefor, a line of couriers extending from the Fairfax Court House office to the front (ten miles to Bull Run) was established, and General McDowell caused reports of the battle's progress to be forwarded to the office for transmission to the War Department office. These couriers were to arrive every fifteen minutes. At Fairfax Station, Hall became somewhat excited, but remained firmly at his post, where he and Rose did much excellent service; but upon Benton and Buell devolved the greatest responsibilities, which were promptly and efficiently met. Just why the telegraph was not carried on to Centreville, if not to the battle-field itself, it is difficult to state. After McDowell's extraordinary efforts to stay the retreat of his troops had failed and there was no longer any need for the office at the Court House, it was at 1:20 A.M., of July 22, closed.

W. B. Wilson, operating the War Department at this time, describes the scenes there on that memorable (21st) Sunday as follows: "In the telegraph office at the War Department, throughout Sunday, July 21, 1861, were congregated the President, most of his Cabinet, General Scott's staff officers, Col. Thomas A. Scott, and other celebrities of the nation, with maps of the field before them, watching, as it were, the conflict of arms as it progressed. Hour after hour, as the couriers reported our gallant troops steadily forcing the enemy back, hopes beat high, expectation, satisfaction was discernible on every brow, and the cheers of our patriotic soldiery as they fought bravely on were responded to in the hearts of all present. Suddenly, as the the shades of evening were drawing on apace, a lull occurred. Firing could not be heard by the corps of observation. No couriers arrived at Fairfax. What could be the matter? The most plausible reason advanced was, that our army, now victorious, was resting after the hard fighting of that hot Summer day. Every few minutes Fairfax was signaled, but only to receive from the operator the stereotyped reply of 'no news.' An hour was expended, when, like the quick flash of lightning and the stunning crash of thunder, came those chilling words: 'Our army is in full retreat.' The signals now became more frequent, rapid and excited. The retreat soon resolved itself into a perfect rout, and as the telegraph reported to those around it assembled the terrible scenes and heart-rending stories of suffering during that never-to-be-forgotten night, all seemed to feel that the hour of the nation's greatest peril had arrived, and clung instinctively around the cool, clear-visioned President, looking to him for succor. That he gave it is a well-known historic fact, and needs no repetition from me."

General McCann, of New York, was in command at Fairfax Station. General Scott telegraphed him to use his troops to stay the retreat, but it was unavailing.

The operators on the O. & A. R. R. were ordered by the Assistant Secretary of War to remain until authorized to close their offices, when an engine would arrive to take them to Alexandria. "We stayed," says Rose, "until the rear guard ordered us to close." At Springfield, later on the twenty-second, Jacques began to sigh for other quarters. He says: "Colonel Thomas A. Scott ordered me not to leave Springfield until I had permission from him to do so. After a while the wounded soldiers began passing by, a few at a time, gradually increasing in number, followed by stragglers from different regiments and later by squads and finally by companies and regiments, all in full retreat. Colonel Scott in spite of my endeavors to close the office, still kept me there, telling me, if I left my post, he would have me shot." At this time Jacques thought the whole army had passed, and momentarily expected the rebels; but an engine came at last, to Jacques' great relief, and he, too, at eight A.M., fell back in better order than many Congressmen who had preceded him. Jacques closed his office about the same time that McDowell and his operators entered Washington.

Up to four o'clock, P.M., of the 21st of July, dispatches presaging a great victory were sent North from the War Department. Enthusiasm was unconfined. Sunday was not too hallowed, if God granted victory. But a leaden silence followed until eight, A.M. The War Department had prohibited the transmission of the evil tidings. However, the news-reporter who had seen the smoke of battle and heard the din of conflict afar off, sadly demoralized and laden with cause and effect, monopolizing the private lines, imprisoned the North in a cave of gloom, by reportorially annihilating the Federal army and leaving Washington and Baltimore a probable prey to the Confederates. McDowell, who had much else to answer for, was assailed for fighting on Sunday. In the Southern States the telegraph told the people that fifteen thousand Confederates had utterly routed fifty thousand Yankees. To the North the news was medicine, but bitter as gall; to the South it was champagne, producing a deep and hurtful intoxication. Says Abbott: "Even while our armies were on the retreat from Bull Run, orders were telegraphed throughout the country for large reinforcements. It is said that under the impulse which that disaster created, sixty thousand enlisted in two days." *

And now the army, numbering over fifty thousand men—including, say, fifteen thousand who remained in Washington and troops called from General Butler—spread out on the right bank of the Potomac, and throughout that great body ran nerve lines

* Abbott's History, Vol. I, p. 189.

of telegraph, crossing, looping and re-crossing until the head in Washington City felt every noteworthy pulsation, and by telegraph, electrified, re-organized and re-invigorated it for greater efforts.

In view of an old maxim, "Old men for counsel, and young men for war," General Scott, though nominally retaining command, was substantially superseded, July 25, by General George B. McClellan, who, by the 1st of November, had 134,285 effective troops, and nearly 300 cannon in his army about Washington.

David Strouse.

Nothing so taxes human energies as the necessities incident upon organizing war, unless it be preparing a defense when an enemy is just beyond the gates. David Strouse was chief within his sphere of action, when the Federal Government armed and equipped to repel and pursue the enemy just across the river, and so continued until the roar of their guns ceased to sound threateningly in the streets of the national capital. Whenever the Government wanted a telegraph line built and operated, it wanted it at once, and it was the hardships, the exposures, the responsibilities of such service at Perryville, Ft. Monroe, Washington, Alexandria and Fairfax that, by the month of July, 1861, reduced David Strouse to a shadow of himself. His last work was in stretching a wire across the Potomac, which superinduced hemorrhage of the lungs, weakening him so greatly that he tendered his resignation; but the War Department officers, recognizing his worth, declined to accept the proffered resignation, and in lieu thereof granted him a general leave of absence. A few short months later (November 17), he breathed his last.

Such characters as young Strouse's are not moulded for war, and with him it was truly but a sense of duty that impelled entrance upon its scenes. Few sadder reminiscences are awakened than the story of his life. As one of the first officers of the service in the Department of the Potomac, a sketch of his career might on that account alone be eminently proper, but as his short life was so replete with genuine manliness, such a view of his noble nature is more than historical, because it is elevating also.

David Strouse was born at Mexico Station, Juniata County, Pennsylvania, October 14, 1838. At twelve or thirteen, he left home for Airy View Academy, in Juniata County, where he remained a little more than one year. Thence he entered the academy at Shade Gap, in Huntington County, where he remained about as long. During these years his tutors and fellow students were attracted to him because of his remarkable frankness, truthfulness and manifest unselfishness. Although considered bright and quick at learning, the superiority of his endowments were in the line of great natural goodness of disposition, rather than of intellectual genius. A schoolmate says of him: "His popularity was attributable to his *great appreciativeness.* No one gave him a kind recognition without receiving in return a look of 'I thank you.'" At sixteen, he entered the Pennsylvania Railroad office at Mifflin Station, where he gave a practical illustration of his aptitude at learning. In the brief period of a few months, he "telegraphed by sound." In those days agents often had charge of freights, tickets and the telegraph. The agent at Mifflin, David's uncle, by consent of the officers of the road, carried on, in addition to those duties, a merchandising business. Thus, it became necessary for young Strouse to attend at times to both the store and railroad business, and it is said that David would oftentimes receive telegrams by ear while selling goods at the counter—a feat readily believed now-a-days, but really astonishing at that time, and it brought much local renown to the young "knight of the key."

DAVID STROUSE.

From this point, he went to West Philadelphia, and became assistant to Superintendent G. C. Franciscus, laboring zealously and receiving the commendation of "well done, good and faithful servant," from his employer. It was while here that he connected himself with an evangelical church and joined the noonday prayer meetings and the Young Men's Christian Association. Religion had been the subject of much thought with him, and having decided upon his course, he never faltered, but remained unshaken to the end. At the age of twenty or twenty-one years, he became private secretary to Thomas A. Scott, at Altoona, but shortly thereafter was made Division Telegraph Operator—a position he held when called to more arduous service.

Alas, the sleepless diligence, the incessant change, the hardships and frailties! David Strouse went home to die. While his beautiful life was slowly ebbing, it was his wont to stroll out to the banks of the blue Juniata, guitar in hand, and "drive dull care away," betwixt the purling stream and the sweet strains of his instrument. It was while sitting on the bank, on one of these strolls, that he wrote the following touching lines which were first discovered in his portfolio a few days after his death:

Gentle river, ever flowing,
  Where my early days were passed!
Like your waters, I am going
  Sadly to the sea, at last—

To that ocean, dark and dreary,
  Whence no traveler comes again—
Where the spirit, worn and weary,
  Finds repose from grief and pain.

O'er the world, I long have wandered;
  Now, a stranger, I return,
Hope, health and manhood squandered,
  Life's last lesson here to learn.

Calmly on thy banks reposing,
  I am waiting for the day,
Whose calm twilight, softly closing,
  Bears the trembling soul away.

On receipt at Washington of the news of Strouse's death, the following notice was issued:

War Department, Washington, Nov. 18, 1861.

To the Members of the U. S. Military Telegraph Corps:

It is with heartfelt sorrow that we are compelled to announce to the corps the death of our superintendent, David Strouse. Ever true to the interests of the Government, attentive to the wants and comfort of his subordinates, kind-hearted and generous to a fault, he died, lamented by all who knew him. His untiring efforts, known to and appreciated by all who were in the corps during the first three months of its existence, served but to hasten the work of the disease which had marked poor Strouse for its victim. He breathed his last at ten o'clock on Sunday evening, and will be buried at eleven o'clock to-morrow (Tuesday) morning. Peace to his ashes.

(Signed) James R. Gilmore,
D. Homer Bates,
W. B. Wilson.

We left the army under General Patterson at Charlestown, in the Shenandoah Valley, off the line of telegraphic communication. On learning of the disaster at Manassas, he retired to Harper's Ferry, where he was, on the 25th of July, 1861, relieved by General N. P. Banks. From July to October, J. R. Gilmore was Superintendent of the Military Telegraphs in the Department of the Potomac. Banks began falling back from Harper's Ferry, and no one knew just where to find him.

On Sunday night, about September 7, Gilmore was ordered by the Assistant Secretary of War, to build a telegraph line to Banks' army. There was not a mile of wire, or a spare insulator, and hardly a keg of spikes, in the department, but it was found imperatively necessary that McClellan and Banks should co-operate, and suddenly discovered that the telegraph only, would make it practicable. President Sanford and Superintendent Westervelt, of the American Telegraph Company, were applied to by Gilmore in this emergency, and the company's supply agents and managers at Philadelphia, Baltimore, Wilmington and New York were telegraphed to forward all they could find. Arrangements were made for speedy distribution along the Frederick (Md.) road to Rockville, thence to Darnestown, Poolsville and Hyattstown. Monday morning Gilmore started on horseback and contracted for the purchase and setting of poles for thirty or forty miles of line. Many residents were disloyal and refused timber, but Gilmore's authority was held *in terrorem* over them with success. Tuesday morning the line-building began, and Thursday Banks was met over thirty miles from the capital. Considering the lack of material; that the builders were unaccustomed to such work, and that Gilmore, who directed it, was a minor, his progress was remarkable.

The operators stationed on this line, in September and October, many of whom followed the builders and opened their respective offices, were L. A. Rose, at Rockville; William J. Dealey, A. P. Pritchard and R. R. McCaine, at Darnestown; M. H. Kerner, W. T. Lindley and W. N. McInnes, at Poolsville, and William E. Tinney and Albert C. Snyder, at Hyattstown.

Pritchard and Dealey encamped with the builders near Darnestown, and were fairly asleep one rainy night when couriers from General Banks arrived, with orders to open an office at the end of the line. Groping their way in the dark, the boys selected an empty pig-sty, which they roofed with their blankets, and connecting their instrument with the line, they shivered around it all night.

At Darnestown the operators boarded with one Fisher, whose custom it was to fill a tumbler with rum toddy every day at dinner, and pass it around the table for all to take a sip; what was left was Fisher's own. At their first meal with the host, Pritchard chanced to sit at his right, and so was the first to receive the toddy. It is said that, although Pritchard was a temperance man, having taken a severe cold in the pig-pen, and being unaware of the customs of that table, he drank the glass empty, to the utter astonishment of the Fisher family and evident demoralization of Pritchard himself, who, soon after, quit the service.

Stephen Sargent was another operator who worked on this circuit at Darnestown and Poolsville. Fresh from Port Jervis, N. Y., with his good clothes on, he galloped his horse right merrily over the pike until in view of his first office in the service (Darnestown), when his horse stopped to drink. This horse, being in the Federal service, felt the responsibility of fairly initiating the new candidate into the joys of army life, and by some unexpected gymnastic feat, pitched his rider, as if from a springing board, into the creek, and dripping from this baptism Sargent entered the telegraph fold.

October 23, a land telegraph cable was laid from Poolsville to Edwards Ferry, where operators C. A. Tinker and J. L. Burucker were stationed. From Poolsville, December 12, 1861, Parker Spring completed the line direct to Point of Rocks, where General Geary commanded. Thomas Armour and Charles Lounsberry operated there. On the nineteenth, their office was under the fire of artillery, one shell falling but twenty feet from it. The operators courageously remained by their instrument, and were the last to leave the place. December 11, Frederick City, Md., office was opened on this line. General Banks had removed there. Frank Drummond operated there until January, when he went to the relief of F. M. Ingram, or C. J. Ryan, one of whom, at Hancock, Md., had become exhausted by long-continued labor during the shelling of that city. The Frederick line was extended, December 20, *via* Williamsport, where N. DeBree operated, to Hagerstown, where T. M. Schnell was operator.

A loop, running from Rockville to Great Falls (nine miles), on the Potomac, was built in September, and Edward Conway was stationed there. A short time before October 8, when C. W. Moore came to assist Conway, a section of Confederate artillery shelled the telegraph office from the Virginia side, but not getting the range at first, Conway escaped. His office, however, was struck many times.

Balls Bluff.

About the middle of August, General Charles P. Stone, commanded a division in the neighborhood of Edward's and Conrad's ferries, Md., with head-quarters in a district school-house, in Poolsville, which was about four miles back from Edward's Ferry. Stone's command was sometimes called a Corps of Observation. At Darnestown, a few miles back, were the quarters of General N. P. Banks. These officers were in direct telegraphic communication with General McClellan. At Darnestown, Wm. J. Dealey, A. P. Pritchard and R. R. McCaine were the operators, and at Poolsville, also in the school-house, M. H. Kerner, W. T. Lindley and W. N. McInnes worked the telegraph.

A great deal of criticism has been indulged in by parties of every calling, but especially by historians, concerning the Ball's Bluff disaster. The author has no disposition to add thereto, as his examination has induced a conclusion that it was one of those accidental happenings which occur in most wars, and which no human foresight could be expected to provide against. When the Committee on the Conduct of the War investigated the affair, it, strangely enough, did not obtain copies of many important telegrams which are here first published, largely with a view of preserving evidence so important.

A reconnoissance in large force under General McCall, by order of General McClellan, was made on the Virginia side of the Potomac, October 19 and 20, reaching out beyond Drainesville, which is about thirty miles south-east of Leesburg, Va. Leesburg is three miles west of Edward's Ferry. McClellan suggested to Stone, by telegraph, that a slight demonstration might have the effect to move the enemy about Leesburg. Pursuant to this, Stone made a demonstration, and in addition twenty men were sent to reconnoitre. Within one mile of Leesburg, a row of trees was discovered, and the moonlight under their lower branches produced a spectre of tents, which Captain Philbrick, the officer in command, reported as a rebel camp, unguarded. General Stone regarded this as exhibiting a want of care on the part of the Confederates, which it was his duty to take advantage of, and directed Colonel Devins, in command on and near Harrison's Island, situate between the ferries and opposite Ball's Bluff, to take five companies as noiselessly as possible, and surprise this camp by daylight. This latter movement was certainly unknown to and unsuspected by McClellan, and Stone himself distinctly states that it was his own order—*i. e.*, it grew out of Philbrick's report. While Philbrick's advance might have been within the spirit of McClellan's instructions, Devin's was merely the result of a complete execution of the telegraphic direction, not in furtherance thereto. That General McClellan had no thought of a battle being brought on by any demonstration Stone should make, is evident from the following telegram received by operator Kerner:

HEAD-QUARTERS OF GENERAL McCLELLAN, October 20, 1861.

BRIGADIER GENERAL C. P. STONE:

If you desire it, you are authorized to visit Washington for two or three days. Major Clay and family are here.

R. B. MARCEY, *Chief of Staff.*

Because, also, October 21, General McCall's force was permitted to return to Alexandria. General McClellan did not know of the movement of Colonel Devins in time to stop it, nor did he question its propriety at any time, as we shall see; and General Stone was sadly misinformed about the unguarded camp, which, indeed, as we have seen, was no camp at all. Confederate Commander J. E. Johnston, who outranked and superseded Beauregard at the Bull Run fight, had been watching McClellan closely, understanding that he would advance in force *via* Occoquan Creek, on the south, or Leesburg on the right, and General Evans, commanding a Confederate brigade in Leesburg, was doubtless instructed to feel the Federals whenever they should cross, to ascertain if it was a move in force. Accordingly Devins soon found it necessary to retire to the Virginia bluffs overlooking the river, and fight the advancing enemy. It was a little before this that General Stone had given Colonel Baker the immediate direction of affairs across the river, with discretion to retire Devins' force and reinforcements, or to further strengthen the troops. McClellan had telegraphed General Stone, on the twentieth, that McCall "occupied Drainesville yesterday," from which point he would "to-day" send out a heavy reconnoissance in all directions, and directed Stone to keep a good lookout upon Leesburg, "to see if this movement has the effect to drive them away." The dispatch closed, "*Perhaps a slight demonstration on your part would have the effect to drive them.*" It was this innocent sentence that caused the North to bow down in sorrow and humiliation; caused the death of three hundred Federals, including that chivalric statesman and patriot, Colonel E. D. Baker, and the loss of six hundred others.

It is believed that Colonel Baker determined to re-cross the troops, but suddenly changed his mind on hearing that the rebels were about to attack, and, not to allow a few brave men to succumb in full view of an ample force, he began crossing other troops in a few scows that were poled over. Seventeen hundred men, including Devins' five companies and two or three useless guns, were thus collected on the Virginia shore. Devins was driven back ; the skirmishing which began early in the day, culminated in a severe fight about three o'clock P. M., and by five the Federals were completely routed. Many of them were drowned, and others shot, swimming, or otherwise attempting to escape. General Stone testified before the Committee that, if Baker had obeyed his instructions as to position in case he did cross the river, and had otherwise exhibited good generalship, the action would have resulted in a "very pretty little victory;" but as Colonel Baker is dead, General Stone, whose evidence, if it related to a mere matter of contract *inter partes*, would be inadmissible, ought not to be heard against Baker's conduct, so far as it relates to verbal instructions not heard by others.

While the foregoing movements and action progressed, General Gorman, by Stone's direction, moved across, also in scows, a force of about twenty-five hundred, at Edwards Ferry, which was to strike the enemy after Baker had defeated them. As a force, sometimes less and at others more than twenty-five hundred, remained on the Virginia shore about three days, liable most of this time, as was supposed, to be overpowered by the enemy, great indignation was felt against General Stone for thus exposing his troops, but Stone, having forwarded news of the defeat to McClellan, was retiring the Edward's Ferry troops as rapidly as possible, when the following telegrams were received:

McClellan's Head-quarters, October 21.

Brig. Gen. C. P. Stone, Edward's Ferry, Md. :

Is the force of the enemy, now engaged with your troops opposite Harrison's Island, large ? If so, and you require more support than your division affords, call upon General Banks, who has been directed to respond. What force, in your opinion, would it require to carry Leesburg ? Answer at once, as I may require you to take it to-day, and, if so, I will support you from the other side of the river from *Darnestown*. (Signed) Geo. B. McClellan,

Maj. Gen. Comdg.

HEAD-QUARTERS OF GEN. McCLELLAN, Oct. 21, 1861.

To GENERAL STONE, Edward's Ferry:

What facility have you for crossing the river at Edward's Ferry, and at Harrison's Island? Is there any road from Seneca to Leesburg, and are there any boats at Seneca? Please direct several mounted men to hold themselves in readiness to carry messages from Poolsville to Darnes*ville*. (Signed) R. B. MARCEY,
Chief of Staff.

McCLELLAN'S HEAD-QUARTERS, Oct. 21, 1861.

To BRIG. GEN. C. P. STONE:

Bremen in send side division shall for gold on you up on take to copper me Adams other a need Camden brass call aid push two river messages lead cipher your of or I whatever tin.
(Signed) G. B. McCLELLAN, *Maj. Gen. Comdg.*

To the above Stone replied that the box was received, but had no key; whereupon the message was transmitted as below, which is a translation of the cipher, except that the cipher directed Stone to "send your messages to me in cipher."

McCLELLAN'S HEAD-QUARTERS.

To GENERAL STONE:

Call on Banks for whatever aid you need. *Shall I push up a division or two on the other side of the river?* TAKE LEESBURG.
McCLELLAN, *Maj. Gen. Comdg.*

It will be remembered that Stone was notified of McCall's presence at Drainesville, where, on the day before (20th) he was to send out a "heavy reconnoissance in all directions." We doubt not that General McClellan was mistaken as to the fact when he testified as follows: Question by Mr. Gooch—"Do you remember whether or not you informed General Stone of the withdrawal of Generals McCall and Smith to their former camping grounds?" Answer—"I think I did." At any rate, having, as we believe, copies of every telegram received about that time by General Stone, we find no message of that purport. Other telegrams are as follows:

DARNESTOWN, Oct. 21, 1861. 5 o'clock.

GENERAL STONE:

We send Hamilton's brigade immediately to Poolsville.

(Signed) N. P. BANKS.

HEAD-QUARTERS OF GEN. MCCLELLAN,
October 21, 1861.

BRIG. GEN. STONE, Edward's Ferry:

Do you learn any results of the action.

(Signed) GEO. B. MCCLELLAN, *Maj. Gen. Comdg.*

HEAD-QUARTERS OF GEN. MCCLELLAN
October 21, 1861.

BRIG. GEN. STONE, Edward's Ferry:

Is the battle still progressing, or has it ceased?

Signed) GEO. B. MCCLELLAN, *Maj. Gen. Comdg.*

DARNESTOWN, October 21.

BRIG. GEN. STONE, Edward's Ferry:

General Hamilton's brigade has started for you, to await orders at Poolsville. General Banks and division are on their way to Seneca, by orders from Washington.

(Signed) R. MORRISON COPELAND, *A. A. G.*

HEAD-QUARTERS OF GEN. MCCLELLAN, Oct. 21, 1861.

BRIG. GEN. STONE, Edward's Ferry:

Is the enemy in large force before you? Please give full detail.

(Signed) G. B. MCCLELLAN, *Maj. Gen. Comdg.*

EXECUTIVE MANSION, Oct. 21, 1861.

TO OFFICER IN COMMAND AT POOLSVILLE:

Send a mounted messenger to the battle ground and bring me information from General Stone. I want to know particulars as to result of engagement, and the relative positions of the forces for the night, their numbers, and such other information as will give me a correct understanding of affairs. (Signed) A. LINCOLN.

HEAD-QUARTERS, October 21.

To GENERAL STONE:

Hold your position on the Virginia side (of) the Potomac at all hazards. General Banks will support you with one brigade at Harrison's Island, or the other two at Seneca. Lander will be with you at daylight. Change the disposition of General Banks, if you think it necessary, so as to send two brigades to Harrison's Island, instead of one. (Signed) G. B. McCLELLAN.

The above was in reply to Stone's telegram that he was withdrawing his troops to the Maryland side.

McCLELLAN'S HEAD-QUARTERS, Oct. 21.

To GEN. STONE, Edward's Ferry:

Entrench yourself on the Virginia side and await re-inforcements, if necessary. (Signed) GEO. B. McCLELLAN,
Maj. Gen. Comdg.

McCLELLAN'S HEAD-QUARTERS, Oct. 21.

To GEN. STONE, Edward's Ferry:

I repeat to you, under no circumstances abandon the Virginia shore, but entrench yourself. Hold your own, if you can make your men fight. You will be supported by General Banks.
(Signed) G. B. McCLELLAN, *Maj. Gen. Comdg.*

HEAD-QUARTERS OF McCLELLAN, Oct. 21.

To GENERAL STONE:

As General Banks will join you, his rank will entitle him to the command, and he has been instructed accordingly.
(Signed) GEO. B. McCLELLAN, *Maj. Gen. Comdg.*

HEAD-QUARTERS OF GEN. McCLELLAN, Oct. 21

BRIG. GEN. STONE, Edward's Ferry:

An advance from Drainsville can not be made to-morrow morning, so that you must rely exclusively upon the support General Banks can give you. (Signed) GEO. B. McCLELLAN,
Maj. Gen. Comdg.

HEAD-QUARTERS OF GEN. McCLELLAN, (*no date.*)

To GEN. BANKS, Edward's Ferry :

You will entrench your command on the Virginia side of the river, observe the movements of the enemy closely, and report to me often, but make no movements without first communicating with me. GEO. B. McCLELLAN, *Maj. Gen. Comdg.*

HEAD-QUARTERS, WASHINGTON, Oct. 23.

To GEN. McCLELLAN, Edward's Ferry :

I have ordered Generals McCall, Porter and Smith to be ready to make a movement on Drainsville early to-morrow morning. Have also ordered reconnoissance by Generals McDowell, Smith and Franklin towards Fairfax Court House and Anandale. All quiet in front. (Signed) R. B. MARCY.

Mr. Lincoln was always keenly solicitous upon occasions of victory or defeat. If reports came to him, requiring the attention of the General-in-Chief, or if he became apprehensive of evil, or for any other cause, felt it important to see that officer, he frequently went to his quarters regardless of the time. One morning when McClellan's head-quarters were in Com. Wilkes' house, on the corner of I and 16th streets, the President called there to consult with the General, although the hour was four, A. M., and it was raining and dark. He came alone. Operator Wilbur F. Holloway was on duty. Mr. Eckert was awakened to escort the President to McClellan's room overhead. Perhaps it was this knowledge of the President's anxiety that induced McClellan, before returning from Edward's Ferry where he went soon after the Balls Bluff affair, to telegraph Mr. Lincoln that no "blame attached to General Stone. The men fought nobly, but the force was only 1,800 against 5,000 or 10,000. General Stone's orders were not carried out on the right."

Including the lines already indicated, there were erected prior to October 31, 1861, in this department and about Ft. Monroe, 280 miles of telegraph, on which were fifty stations, worked by eighty-three operators. These offices were mostly open day and night. The total expense, from April 25 to November 1, 1861, was, gross, $40,752.23.

CHAPTER IV.

EARLY OPERATIONS IN WESTERN VIRGINIA.

The day that Beauregard's guns opened upon Fort Sumter (April 12, 1861), Anson Stager, General Superintendent of the Western Union Telegraph Company, located at Cleveland, Ohio, was telegraphed by Governor Dennison of that State, to meet him at Columbus. Arriving there, he was requested by the Governor to undertake the management of the telegraphs in Southern Ohio, especially along the Virginia line, and to assist Captain George B. McClellan, who, it was then understood, would be the commanding officer of a department which would include Ohio. Stager consented. That was the beginning of those arduous and responsible duties which none could then have foreseen, the able performance of which made for Mr. Stager a national reputation that he may be justly proud of. He repaired to Cincinnati to consult with McClellan, and began preparations to meet the requirements of that officer.

The principal towns and cities of Southern Ohio already enjoyed the advantages of the telegraph. Communication by wire and rail along the route of the Baltimore & Ohio road was intact from Parkersburg and Wheeling to Washington. This road, from Baltimore, reached the banks of the Potomac at Point of Rocks, Md., sixty-five miles from that city and fifty-one by highway from Washington, and, following up the river twelve miles, entered Harpers Ferry, already historic in consequence of John Brown's armed efforts against slavery. Running thence westerly, nineteen miles, it reached Martinsburgh, near which point is the north-easterly corner of West Virginia, which, in June, was carved out of the Old Commonwealth. Pursuing its course nineteen miles further to Cherry Run, and thence thirteen more miles, it once again struck the Potomac, which it followed to Cumberland, Md., sixty-five miles. Rowlesburg, W. Va., seventy-five miles more, is twenty-seven miles from Grafton, where the road branches; one division leading northwest *via* Fairmont to Wheeling, one hundred miles, and the other past Clarksburg and West Union C. H., one hundred and four miles, to Parkersburg; both *termini* being on the Ohio River. This wonderful road, a master-piece of engineering, about 395 miles long, passing some of the grandest natural scenery on the continent, was destined to play a star part in the great tragedy of war. The wonder is that it cut any figure at all, and it is almost past comprehension that the air lines of telegraph along its route were permitted to stand a week at a time in the wilds they traversed.

Captain George B. McClellan was, on the 14th day of May, 1861, appointed Major General and assigned to the command of the Department of the Ohio, with head-quarters at Cincinnati. His department included Western Virginia, Ohio, Indiana and Illinois, and, later, Missouri.

Mr. Stager's first written authority over the telegraph lines for military uses, reads as follows:

HEAD-QUARTERS OF THE DEPARTMENT OF THE OHIO,
CINCINNATI, O., May 27, 1861

GENERAL ORDER, No. 13.

Mr. Anson Stager is hereby appointed superintendent for military purposes of all the telegraphic lines within the Department of the Ohio, and his instructions will be strictly obeyed.

By command of Maj. General McClellan.

N. H. MCLEAN, *Asst. Adjt. Gen.*

Had there been less patriotism among the telegraph stockholders, some discontent might have been generated, owing to this sweeping authority; but there seems never to have been any trouble with any Northern telegraph company, and none to speak of with any other. It was seldom required actually to possess any telegraph office in the North. For the most part, military control of private lines was merely nominal—not for want of power or authority, but because a hearty co-operation made unnecessary the exercise of an arbitrary supervision. Nevertheless, there was no little embarrassment in Mr. Stager's position, owing to his being a mere citizen and a superintendent of a private company, whose business antagonized that of the telegraph companies in West Virginia, where McClellan was first to operate and where positive control of the telegraph was impending.

But there was a stranger discouragement. Some of General McClellan's staff officers, perhaps thus early imbued with pangs of jealousy — perhaps sincerely impressed with the idea, conceived the notion that the telegraph could not be useful in war. These officers seriously decried all efforts to introduce it; but General McClellan had great faith in its ultimate efficiency, and insisted upon fairly trying it

For some years prior to the beginning of hostilities, T. B. A. David had resided in Wheeling, Va., and, when the war began, was manager of the telegraph office in that city. About April 20, 1861, Mr. Stager, having advised David that W. H. Seward, Secretary of State, was exceedingly anxious that any information touching the rebels encamped at Harpers Ferry, should be communicated to him, David kept a sharp ear on the line over the B. & O. R. R., and in furtherance of his object, arranged a cipher code with George M. Deetz, then and ever since, operator at Cumberland, Md. B. F. Kelly took command of the First Regiment of Virginians (Union), at Wheeling, May 26, 1861, and seized the telegraph office there; whereupon, Anson Stager appointed Mr. David to the charge of the line of the B. & O. R. R. proper, from Wheeling to Cumberland. David, a zealous patriot, entered upon an untrodden field; but he had an able co-adjutor in W. G. Fuller. At the outbreak of the rebellion, Mr. Fuller was superintendent of the telegraph lines from Grafton, Va., to Cincinnati, Ohio, which ran along the B. & O. and the Marietta & Cincinnati railroads.

In May, 1861, the Confederates, under Colonel Porterfield, possessed themselves of Grafton and its telegraph office. Colonel James B. Steedman's Fourteenth Ohio Volunteers and Captain Barnett's battery were hastened by telegraphic instructions to the protection of Parkersburg.

Fuller was then summoned to meet Mr. Stager at General McClellan's house in Cincinnati on the 27th of May, where he was instructed in the mysteries of the first army cipher ever used telegraphically in war, and was also appointed to manage the Government telegraphs on that branch of the B. & O. R.R. which lies between Grafton and Parkersburg.

Let us now follow Fuller and David in their efforts to connect Wheeling and Parkersburg, *via* Grafton. May 28, Fuller and Colonel Lander, McClellan's chief-of-staff, a regular army officer, just from the Indian country, reached Parkersburg. That night was pitch-dark, and Lander, intent on some kind of adventure, sought to gratify his wish by frightening Fuller. Colonel J. B. Steedman's command had the day before started out on the line of the railroad, and Fuller was in the telegraph office watching for dispatches from the advance, when firing was heard across the Little Kanawha, which, at this point, enters the Ohio. Soon after, Lander entered, and apparently much alarmed, asked Fuller if he was armed. Fuller had no arms. Lander felt sure they were attacked, and wanted some one to go with him to see what the firing meant, and, handing Fuller a pistol, asked him to follow. Fuller, taking the weapon, followed through a long, dark depot to the bridge and across the river, when Lander laughingly remarked that Fuller was not badly frightened, and that he merely wanted to try him. Thus was the Superintendent baptized with fire, even as Prince Napoleon was, at Saarbruck, only Fuller could not find any little bullets to make soldiers weep over.

Many bridges and the telegraph being destroyed, progress toward Grafton was slow, but June 18, the lines were intact to Clarksburg, and twentieth, to Webster, where David was met. Troops left Wheeling May 27, Colonel Kelly commanding. The bridges at Mannington had been burned by Bill Thompson's forces.

Reaching there at four, A. M., next day, David was ordered to proceed immediately to Fairmont. Filled with apprehensions of danger he took a squad of raw troops, and pushing on, soon discovered that his greatest peril consisted in his green and awkward defenders, who made him fearful of being accidentally shot. He reached Fairmont at nine, A. M., the twenty-eighth, when communication was opened through to Wheeling.

A temporary bridge having been erected at Mannington, Colonel Kelly's command arrived at Grafton, May 30; the rebels having decamped with the telegraph instruments. David then took a special train east twenty seven miles and cut the line at Rowlesburg, by Colonel Kelly's order, and returned at once to Grafton. General T. A. Morris soon arrived and assumed command, his force amounting to seven or eight thousand men. May 20, David was ordered west to meet Fuller's party, already long overdue. Taking desperate if not foolhardy chances, he boarded an engine, and, going in the face of trains known to be somewhere on the road, which is very sinuous, he pushed ahead, around sharp curves, through tunnels, one of which was nearly a mile long, until, fortunately, without accident, he met Fuller at Webster, and communication was fully re-established.

David had taken the precaution to procure, at Wheeling, the form of an oath to be administered to operators along the line, who accepted service in the Telegraph Corps. The following is copied from an original:

You do each and all swear to be true and faithful to the United States of America; that you will faithfully transmit all orders and information touching the military operations of the United States of America truthfully, and, further, not to transmit any information, directly or indirectly, to the Confederate States, or any representative or individual of said States, and in no wise prevaricate any information touching the military operations of the United States of America.

JAS. G. NESBITT,
GEO. M. DEETZ,
FRANK MILLER,
P. A. STIDHAM.

Sworn and subscribed before me, this 10th day of June, 1861.
O. H. W. STULL, *J. P.*

It soon became customary at Fairmont to make every one in that neighborhood, suspected of disloyalty, take the oath of allegiance, and David having taken several operators before the Justice there, to make oath as above, that judicial innocent caused all suspected residents to subscribe and swear to the telegraphers' oath.

Colonel Porterfield (Confederate) on quitting Grafton at Kelly's approach, fell back to Phillippi, fifteen miles south of Grafton and about eleven miles from Webster, the nearest telegraph point. General McClellan having ordered the surprise and capture of Porterfield's forces, variously stated as numbering from eight hundred to two thousand, Colonel Kelly took a route twenty-two miles long, leading south-westerly into Phillippi, and was to gain Porterfield's rear, while Colonel Dumont, accompanied by Lander, approached (June 3) Phillippi with his forces in front. Dumont was a little too quick, but the enemy was considerably punished, especially by Kelly's troops, who struck them in retreat. About this time Mr. Stager arrived and proved an inspiration to the service. Under Fuller and David, the telegraph was extended from Webster to Phillippi, but was soon abandoned.

June 8, Colonel Lew. Wallace and his brave regiment of Indianians reached Grafton, *en route* for Cumberland, Md. On the eleventh they attacked the enemy at Romney, said to be about twelve hundred strong, driving them sixteen miles, and then retired to Cumberland, and encamped. Cumberland was now in telegraphic communication with Grafton, and so remained until June 20, when the enemy, four thousand strong, entered Piedmont, on the B. & O. R.R., after which for some weeks Cumberland was isolated from the telegraphic world.

After the battle of Phillippi, McClellan was busy until July 10, preparing to advance south from the railroad, and Stager was actively engaged concentrating supplies for Fuller and David, who were organizing building parties to undertake a novel experiment—that of following an army with the electric telegraph. Owing to the disparaging remarks of regular army officers; to the dangers incident to the undertaking; to the military importance of success, and the difficult roads, Mr. Stager and his assistants justly felt that they were shouldering grave responsibilities, which nothing short of a successful demonstration could lighten. Porterfield had been superseded by General Garnett, who made his head-quarters at Beverly, fifty-eight miles south-east of Clarksburg. His troops, numbering about ten thousand, were posted, mainly under Garnett's personal supervision, on Laurel Hill, a little out of Beverly on the Phillippi road, and the remainder, except small outlying detachments, under Colonel Pegram, on Rich Mountain, within easy support—say three miles from Garnett's position. At this time (July 4) McClellan had between twenty thousand and thirty thousand men under him. On the 23d of June, he took command in person at Grafton. Detaching a force under General J. D. Cox, to watch General Wise in the Kanawha Valley, McClellan advanced, early in July, from Clarksburg *via* Buckhannon, with the main body, and the first field telegraph that ever advanced with an army, in America, kept pace with this one. For the first fifteen miles or so, the builders, being troops detailed for that purpose, were under Mr. Fuller's direction and subsequently, until Beverly and Huttonville were reached, under Mr. David. Buckhannon, twenty-eight miles, was reached on the fifth.

At this point the telegraph service was highly complimented by General McClellan's staff—notably by Captains Saxton, Chief Quarter-master, and McFeeley, Chief Commissary, who openly declared that, but for the telegraph, the army would have been delayed many days at Buckhannon. This well-earned acknowledgement was a triumph in itself.

Leaving Buckhannon July 8, the telegraph men kept close up with the army, reaching a point within two miles of Rich Mountain on the ninth. The battle of Rich Mountain was fought on the eleventh. David had just reached the fortifications the defeated enemy had occupied the day before, and opened an office, when a Confederate prisoner marching by noticed the office and exclaimed to his comrade, "My God, Jim, here's the telegraph!" Garnett and Pegram, now in full retreat, abandoning cannon, tents, camps and wagons, sought to reach Beverly before McClellan could intercept their line of retreat to the Shenandoah Valley by the road to Staunton, but in this they were too late. Garnett turned northwardly, but was overtaken and killed at Carrick's Ford. Most of his command escaped to the mountains.

An office was opened at General McClellan's head-quarters at Beverly, on the fifteenth, and the line continued as far as Huttonville, when operator J. L. Cherry arrived to assist E. B. Bryant, at McClellan's. Bryant extended the line to the top of Cheat Mountain, where he remained with the troops until the following Spring, when, crippled by rheumatism contracted from great exposure in the service, he was obliged to go home.

McClellan's troops having won the first real victory of the war, the Northern people were greatly elated by his telegram, dated the thirteenth, wherein he reported his successes. He also issued a congratulatory order to the troops, which was probably the first order ever printed in a portable printing office, regularly connected with an army on a campaign.

While J. L. Cherry worked nights in a railroad office in Cleveland, Mr. Stager noticed that the operator, to occupy usefully many tedious hours, busied himself at his font and press. Stager, pleased by such industry, recommended Cherry to McClellan, as doubly useful, *i. e.*, as operator and printer. McClellan caused the Government to buy the type and press, and Cherry was duly installed in head-quarters' office, where he served with distinction as telegrapher and typo. The second order thus printed antedates the first. They are as follows:

HEAD-QUARTERS ARMY OF OCCUPATION, WESTERN VIRGINIA.
BEVERLY, VA., July 16, 1861.

SOLDIERS OF THE ARMY OF THE WEST:

I am more than satisfied with you. You have annihilated two armies, commanded by educated and experienced soldiers, entrenched in mountain fastnesses, fortified at their leisure. You have taken five guns, twelve colors, fifteen hundred stand of arms, one thousand prisoners, including more than forty officers. One of the two commanders of the rebels is a prisoner; the other lost his life on the field of battle.

You have killed more than two hundred and fifty of the enemy, who has lost all his baggage and camp equipage. All has been accomplished with the loss of twenty brave men killed and sixty wounded, on your part. You have proved that Union men, fighting for the preservation of our Government, are more than a match for our misguided and erring brethren; more than this, you have shown mercy to the vanquished. You have made long and arduous marches, often with insufficient food, frequently exposed to the inclemency of the weather. I have not hesitated to demand this of you, feeling that I could rely on your endurance, patriotism and courage. In the future, I may have still greater demands to make upon you—still greater sacrifices for you to offer. It shall be my care to provide for you to the extent of my ability; but I know now that by your valor and endurance you will accomplish all that is asked. Soldiers! I have confidence in you, and I trust you have learned to confide in me. Remember that discipline and subordination are qualities of equal value with courage.

I am proud to say you have gained the highest reward that American troops can receive—the thanks of Congress and the applause of your fellow citizens.

GEO. B. MCCLELLAN,
Maj. Gen. U. S. A., Comdg.

HEAD-QUARTERS, ARMY OF OCCUPATION, WESTERN VIRGINIA.
HUTTONVILLE, VA., July 15, 1861.

GENERAL ORDERS, No. 1.

The Commanding General has experienced much embarrassment during the progress of this campaign, from the want of early information regarding the movements of distant detachments within his command.

The success of the operations of the main column are in a great degree intimately connected with, and dependent upon the subordinate auxiliary detachments; and unless the Commanding General is kept constantly advised of the exact condition of those commands, the service must suffer.

In view of this, it is directed that all commanders of detachments and posts in Western Virginia make daily reports to these head-quarters, showing the true state of their commands, all movements of the enemy and such other information as it may be important for the Commanding General to know.

These daily reports will be forwarded by the most expeditious methods of communication.

By order of Major General McClellan. S. WILLIAMS,
Asst. Adjt. Gen.

General Cox, who left Guyandotte, on the Ohio, succeeded only in pushing back the enemy, and now McClellan, intent on intercepting the Confederates near Gauley Bridge, directed David to return to Buckhannon and construct a line therefrom, westerly, to Weston, twenty-four miles, and also to establish a supply depot, preparatory to following the army. However, on the 20th of July, McClellan heard of McDowell's movement, and decided to await results. On the twenty-first, General Scott telegraphed him to prepare to move against Beauregard, who was retreating, and a change of plans resulted; but the next day Scott notified McClellan by telegraph of McDowell's defeat, nearly in these words: "We have been badly beaten. Our army is in full retreat—a most wonderful transformation of a well-appointed army into a rabble"—and ordered him to Washington immediately.

This left further arrangements to General Rosecrans, the hero of Rich Mountain. These changes, coming rapidly, one after another, caused much comment among high officers in West Virginia, who gave great credit to the telegraph which made them possible. David, on the twenty-third, was given the management of all lines in West Virginia, and Fuller took charge of all construction matters. The country, as far on the Summerville road as Weston, being within the Union lines, Rosecrans, who had returned to Clarkesburg with the main portions of his army, leaving only sufficient force on the line to Cheat Mountain to protect it, on the 26th of July, directed that the telegraph be constructed to Weston, and that it follow the army, moving along that route *via* Sutton and Bulltown to Summerville and beyond. The General hoped to intercept and capture the enemy who were operating in two divisions, under Floyd and Wise respectively, along the Kanawha.

The country south of Weston was infested with guerrillas, and Fuller was notified by them at Weston that if he proceeded farther with the line, he and his men would be shot, from the rocky hills and mountains along the road. This caused him to call on Rosecrans for an escort, and Captain Theophilus Gaines, with his company of the Fifth Ohio Infantry, was detailed to flank the road as the builders proceeded. At Weston, Fuller received from a Federal captain, scouting in Gilmer County, a little to the west, a message for Rosecrans, which was the first dispatch of military importance sent over this line. It stated that the captain had ascertained that the enemy had received arms, and were marching in large numbers to surround and capture his force; but that he was entrenching on a hill-top and would hold out until aid came. In a moment this reply came back.

SIR: You were sent into the field to become a terror to the enemy, and not to be terrified by them. March out of that entrenchment and disperse those rebels, and report at these head-quarters.

(Signed) W. S. ROSECRANS, *Brig. Gen. Comdg.*

August 13, near Imboden's, Fuller was telegraphed: "Proceed with line cautiously. Tell Captain Gaines to scout well in front until you arrive at Bulltown." At Bulltown he was halted to await the arrival of troops and to forward dispatches to Colonel Smith, who, it would seem, was somewhere in advance. While he was here, about a mile of the line near Imboden's was carried away. Imboden was the father of the Confederate general of that name. Captain Gaines and Mr. Fuller questioned Imboden about the wire, but he professed innocence. He was given until ten A. M., the next morning to return the wire, as otherwise, he was told, his house would be destroyed and he shot if found, and further, that there would be no houses or inhabitants left in that neighborhood if the line was interfered with. Next morning the wire was easily found near by and not again disturbed in that vicinity.

Sutton was reached on the twenty-eighth. The telegraph party awaited the arrival of General Rosecrans, who left Clarksburg August 31, E. O. Brown, a discharged three months soldier, and J. L. Cherry accompanying him. At Sutton, Rosecrans requested Cherry to remain a few days, but his stay proved quite permanent. A Mr. Brown, commonly called Colonel, operated at Bulltown and became a terror to the people in that section, although he never hurt any one. His frequent and dextrous handling of a harmless revolver awed the natives. One day, six or eight bushwhackers captured the Colonel and required him to take an oath to *support* the Southern Confederacy until exchanged. Brown, by way of compromise, tendered his pocket-book, containing all his worldly goods, viz.: a collar-button and a little poem on Cupid, but a few squirrel rifles aimed at his head made him change the proposition to "Well boys, I can not buck against such forcible arguments as those; I will take the oath, but I tell you candidly that I can not support myself, much less such a needy concern as the Confederacy." The Colonel somehow saved his instrument, which he continued to operate.

Fuller and Gaines, pushing on towards Summerville, were preceded by the advance, consisting of but two companies. These having halted nine miles from Sutton, refused to proceed, and Gaines with his company only, scouted the road until near Big Birch River ford, when Fuller was informed that the enemy were just ahead, and that their pickets had left the place where Fuller then was, but two hours before. Thereupon, connecting his instrument, Fuller telegraphed the facts and also that the two companies had refused to proceed. He received the following reply: "Keep your teams hitched ready to fall back at a moment's notice, and await orders." Then the line was cut in the rear, and Captain Gaines and Fuller, while watching the enemy down the mountains, disposed their men for the best defense possible. Some of the men wanted to fire on the enemy, but it was deemed imprudent. The next noon, Rosecrans dined there with Fuller and Gaines.

This place is by road about sixteen miles from Sutton. Owing to the narrow defiles in the road, the telegraph party was delayed for troops to pass to the battle then impending, and when it occurred (September 10), the telegraph was only four miles in the rear. Wise had retreated to Lewisburg, leaving the country to Cox, but Wise's command being re-inforced and himself outranked by General Floyd, the Confederates took the offensive, striking a Federal regiment at Cross Lanes near Summerville. Floyd moved southerly to Carnifax Ferry, hoping to beat Cox, who was farther south, but now Rosecrans, with about nine thousand troops, fell upon Floyd's forces, numbering, according to Confederate authority, about three thousand, who gallantly held the Federals at bay until dark, when they effected their escape from Rosecrans' troops on the one side and Cox's on the other. In these movements, Wise, jealous of Floyd, had kept aloof, notwithstanding Floyd's orders. The Confederates now retreated to Meadow Bluffs beyond Big Sewell Mountain, say thirty-five miles, and thus in about three months from the first essay off the line of the railroad, the Union forces had regained nearly all of Western Virginia west of the Alleghany Mountains, and lines were extended from Carnifax to Gauley Bridge and Princeton on the south, and then north-westerly from Gauley Bridge *via* Charleston, West Virginia, to Gallipolis and Hamden, Ohio.

William Patterson, operator, entered the United States Military Telegraph service at Clarksburg in June, 1861, and started south with Fuller's building party, rendering, as he always did, very great aid. He was thus engaged until within six miles of Big Sewell Mountain. Rosecrans soon falling back to Camp Tompkins near Gauley Bridge, this advance line was taken down and Patterson was then located in head-quarters office a while. He next opened an office on the Summerville pike, thirty miles from Gauley Bridge, in a fence corner, where he did valuable service for two days and nights, when he returned to head-quarters and soon after opened an office at Charleston.

Among the unfortunates in this region "of rocks and rills, woods and templed hills," was M. H. Kerner, who had had one experience with the Disunionists and was booked for another. When hostilities began, Kerner was operator on the Baltimore & Ohio Railroad at Martinsburg, West Virginia. As soon as the Secessionists began organizing troops thereabouts, Colonel Nadenbush, of the Berkely Guards, urged Kerner to join his command, but, being a Unionist, he refused and was in consequence imprisoned. Through the instrumentality of Belle Boyd, who subsequently became notorious as a Confederate spy, and a pretty daughter of Colonel Israel Robinson, Kerner was released on condition that he return to his office and telegraph for the Confederates, but nothing was said as to how long he should so serve and it was not long before, with a friend, he escaped to Sir John's Run, where he crossed the Potomac and walked thence to Hagerstown, Maryland, and then entered service under Gilmore, as we have seen. After telegraphing at Gauley Bridge a few weeks, in May, 1862, Kerner went unattended some thirty miles to the Summerville office, the Gauley operators assuring him very confidentially that he would go North by a longer route, and so it proved.

July 26, about five hundred of Jackson's Confederate cavalry, under Major Bailey, made a dash upon Summerville. Of the few Federal soldiers there, after a little skirmish, about sixty were captured and with them young Kerner and a Doctor Rucker, for whose capture the rebels had offered a thousand dollars reward. Kerner and Rucker were in citizen's dress. A guard was stationed at the entrance of the telegraph office where Kerner was. The wire was cut on both sides, and while the excitement was at its height, Kerner wrapped his instrument in an army blanket and rushed out, passing the guards with the remark that the Yankees had held him prisoner with a girl, Nancy Hart, of whom hereafter. He ran rapidly down the road towards the Gauley, hoping to reach the break in the wire and telegraph the nearest commanding officer the situation at Summerville. But the ruse was discovered, and four mounted men, armed with carbines, persuaded the boy to return with them. To this day, it is his opinion that they were the most excited quartette he ever saw. Cursing and threatening until the young man thought his last hour had come, they forced him back. Arriving at the office, Lieutenant Colonel Starr and Captain Davis, Doctor Rucker and the captured soldiers were seen in line for marching into Dixie. Major Bailey told Kerner that he had to thank Nancy Hart for his life, as he had determined on shooting him for the "d—d" lies he told to get through the guard with that infernal telegraph machine; adding, "We were too quick for your d—d Yankee trick." Judging from his subsequent treatment, Kerner thought Bailey might have executed his purpose.

Nancy Hart had been guiding guerrilla bands over mountain passes in West Virginia, and informing them of Federal movements. Colonel Starr, after many efforts, finally succeeded, about the 10th of July, in capturing her while sitting in front of a log hut, grinding corn between two big stones. She had with her two beautiful fawns, which she had captured, and which were brought with her to Summerville where she was placed in the county jail, which was in a dilapidated and filthy condition. Nancy herself was none too clean, and her clothing, what was left of it, was very ancient, and too stiff and fragile to bear washing. These things touched Kerner's gallantry, and, on his intercession with Starr, a room was arranged for her in the head-quarters building in which was also the telegraph office. Thanks to Kerner's generosity, for he furnished her with calico, needles, thread and probably Butterick's patterns, she made herself a becoming garment, when, wishing to show her thankfulness, she asked Kerner what present she should make him. An itinerant ambrotypist chanced to be there, so it was agreed, at Kerner's suggestion, that she should give him her picture. He put Colonel Starr's black plume on his Derby hat for her, when she accompanied him, under guard of two soldiers, to the portable gallery, where she sat for the first time before a camera. As soon as she saw it was fairly aimed at her, with great trepidation, she said: "If you are going to kill me, for God's sake, tell me so." Kerner was obliged to sit first, to satisfy her it was not some Yankee infernal machine. Her picture and a lock of her hair in Kerner's parlor are frequent reminders of his narrow escape from Bailey's deadly purpose.

After a long march, Kerner reached Christiansburg and took cars for Lynchburg, where he found Frank Drummond and his comrades, recently taken near Winchester, all of whom, with Kerner, were soon removed to Belle Isle, and then to Libby Prison, where some one told them they were hostages for a Southern operator captured by the Unionists, and condemned to death. There was no truth in the report, but when the boys' names were called in September, they feared the worst until they found it was the best; that, indeed, all were to be exchanged. On reaching Washington at night, the St. James Hotel people refused them lodgings, as they looked too seedy; but Colonel Stager, regardless of time or pay-rolls, at once furnished sufficient funds. For want of legislation, it is said that none of these captured operators ever received a farthing for the time they were in the hands of the enemy. The first case in which the Government concluded to pay such claims, was that of George M. Brush, who was captured in the West, as we shall see, in the fall of 1863.

Shortly after the battle of Carnifax Ferry, Mr. Fuller, accompanied by a Lieutenant Wheeler, of the regular army, took horse for Clarksburg. The first night out, they lodged in an Irishman's shanty, having full confidence in the family, as the host had helped Fuller's builders. The guests' little room had one small square window. In an ominous hour of the morning, the travelers were aroused by strange noises. "We are betrayed," whispered the lieutenant. They hurriedly dressed, and looking out at the window, it being moonlight, discovered a man moving about the stable where their horses were. The growling dog moved about the premises, but the whisperings within the cabin were even more terrifying, as not a word could be distinguished. It was a lonely hour and a lonesome place; but there was no escape. So, determining to sell their lives as dearly as possible, they commenced their defense against the guerrillas by barricading their door, and standing guard at the window. An hour passed by, and yet the attack had not begun. Evidently the marauders were awaiting the dawn of day, and when it came the man at the barn was seen more clearly; but he was feeding the horses, and the women, their hands already stained with blood, were dressing chickens. In fact, the household was, as quietly as possible, preparing for the early breakfast, as requested the night before, and the dog had only been baying at the moon. A few miles further on the road, however, they discovered the dead body of a soldier who was shot by one of the mountain pirates.

Besides the operators mentioned, there were, in October, 1861, engaged in this department, the following: J. K. Brown, G. K. Smith, S. G. Lynch, J. B. Pierce, C. H. Johns, M. C. Baldwin, Philip Bruner, S. M. Shurr, C. D. Tull and George W. Printz. The last named operator, it is believed, left Zanesville, (Ohio) office in the summer of 1857, and obtained employment on Mr. Fuller's lines, entering the military service July 19, 1861, and operating the Beverly office from November 8, of that year, to the September following. So far as the author, by careful scrutiny into the conduct of the operators for the Government during the war, has been able to ascertain, and he believes he speaks advisedly, Printz was the only man in the service, who, disregarding his duty to his Government, his oath and his trust, resigned his situation, and, under the inspiration of what George Stephenson has called "the mightiest power under heaven," secretly passed the Federal lines to the Confederates. It is not known, however, that he ever advised the enemies of the Union of any information obtained from the Union wires. If an exception be necessary to establish the rule of fidelity among the operators, perhaps a more harmless example could not be wished than is found in this fellow, who has since had occasion to seek a livelihood in other employments.

CHAPTER V.

THE SOUTHWEST EARLY IN THE WAR.

June 18, 1861, General McClellan received orders extending his command so as to include Missouri, then under General Lyon. This gave him Missouri, Illinois, Indiana, Ohio and Western Virginia.. Virginia and Missouri, like Kentucky and Tennessee, divided in opinion, were threatened with internecine strife, while the three great States of Ohio, Indiana and Illinois were vigorously preparing to strengthen McClellan, that he might successfully grapple with giant revolt. McClellan, as shown, was about to enter upon an active campaign in the field, personally maneuvering a large army against the enemy, near Beverly, W. Va. Constant communication by telegraph with the commanders in Missouri, alone enabled him to attend to the military requirements of that State. It kept his several armies in full view, and provided lightning couriers to transmit his orders. Truly, "what hath God wrought," through Morse, that by the instrumentality of his invention St. Louis should be nearer Grafton than, in 1776, New York was to Philadelphia.

We have seen who were the trusted telegraphers in West Virginia. Now let us look westward, and, in the course of events, note who at the Missouri end enjoyed the inspiring confidence of the military officials ; for let it always be borne in mind, that few men ever had greater opportunities to betray a cause than the telegraph operators, whether in the Federal or Confederate service.

At the breaking out of hostilities, the great State of Missouri, having a population equal to that of Virginia or Kentucky, and a territory about two hundred and eighty miles long, measured on a north and south line through its capital, and over two hundred and fifty miles wide, reckoning by a line due west from St. Louis, was in a sea of trouble. Its means of communication, excellent for the period, were quite meager compared with its present. The Mississippi River on the east separates it from Illinois as far south as Cairo, and from Kentucky thence to Island Number Ten; even Tennessee for a number of miles abuts the great State. On the south, it is bounded by Arkansas; on the west, mainly by Kansas, and on the north by Iowa. The railroads radiating from St. Louis, had their termini on the south at Ironton; south-west, at Rolla; west through Jefferson City, the capital, at Sedalia, and north-west at Macon City, where a junction was effected with the Hannibal & St. Joseph Railroad, that ran west across the State from opposite Quincy, Illinois. Besides these, the Missouri River, entering Missouri at Kansas City, crosses the State, and passing Lexington, Booneville, Jefferson City and St. Charles, enters the Mississippi a little below Alton, Illinois. These facilities were increased by wagon-roads, for the most part incomplete or poorly made, and by smaller rivers of little service. Telegraphic communication was had by the Stebbins line connecting St. Louis with Fort Scott, Kansas, by way of Springfield, Missouri. There was also a line over the Hannibal & St. Joseph Railroad; another line connected St. Louis with Sedalia *via* Jefferson City and Syracuse, thence it ran south-westerly to Fort Smith, Arkansas, *via* Warsaw, Bolivar and Springfield. Such were the means of communication in Missouri early in 1861.

When the bombardment in Charleston harbor was echoed by telegraph across the prairies of Illinois, the Government arsenal at St. Louis, in charge of Captain Nathaniel Lyon, contained twenty-two thousand stand of arms. General Frost, a Missouri State Militia officer, began organizing a strong force at a camp just outside of the city, nominally to secure State neutrality, but really to capture the arms in the arsenal, arm his troops and then, capturing the city, containing, in 1860, one hundred and sixty thousand, seven hundred and seventy-three citizens, make a bold stand in favor of the Confederacy. So imminent was the danger of such capture, that Captain Lyon, to deceive the Secessionists as to his strength, sent away by night portions of his handful of troops, to return by day with drums beating. Governor Jackson had, before this, telegraphed the President a refusal to furnish troops, and there was no authority in the State for the enrollment of a Federal soldier. At this critical time, the rioters in Baltimore cut the telegraphs to Washington, whence only could emanate orders to meet the emergency. The Governors of Illinois, Indiana and Ohio, clamorous for arms, appealed by telegraph to General Wool in New York. Wool telegraphed Governor Yates to send a judicious officer with four or five companies to the arsenal, which was believed to be in danger of seizure. In consequence of that telegraphic order, this large quantity of arms, together with two field pieces and one hundred and ten thousand rounds of ammunition, was removed to Illinois during the night of April 25, and instead of being used to destroy, was perhaps the means of saving the Union, as at that time most of the Government arms were in the Southern States.

The President, April 30, by telegraph, authorized the enlistment of troops in Missouri, whereupon German and American regiments were enrolled, and on the 10th of May Captain Lyon, with the aid of these troops, captured Frost's command of twelve hundred men, twenty cannon and many hundred rifles and muskets with ammunition. These decisive proceedings quickly culminated there in positive war; the Secessionists generally throwing aside all masks and openly avowing their opinions, met in camps to defend them. June 12, Governor Jackson called for fifty thousand troops to be used against the Federals. That day the telegraph wires between the State capital and St. Louis were cut. Jackson selected Lexington and Booneville for his militia to concentrate. To disperse these State troops, Lyon's command arrived at Booneville on the seventeenth, and utterly routed the militia under Marmaduke, pursuing them south-westerly to Vernon County near the Kansas line, where Price with his Lexington forces joined the retreating Confederates, and together they pressed further south into Jasper County, where they were confronted by another Union force under Colonel Sigel, which was intended to co-operate with Lyon, now a brigadier general and commander of the Department of Missouri. Sigel's forces fell back, fighting at Carthage, and joined Lyon at Springfield on the 10th of July. The telegraph company operating the lines to Booneville was able, on Lyon's advance, to repair their wires, so that by the 21st of June he dispatched telegrams from Jefferson City, and on the twenty-fourth from Booneville, where he remained in telegraphic communication for some days at the head of one invading army, while his chief, on the eve of another campaign, was at Grafton, over seven hundred miles distant, at the head of another. When Lyon left Booneville, July 3, he seems not to have been again in telegraphic communication with any point up to the date of his unfortunate sacrifice in August.

Charles Lehr was the operator at Springfield when hostilities commenced. At that time, Unionists and Secessionists bushwhacked one another in that section, and Lehr was fired at twice while mending the line. On his return to his office, he discovered the wire had been cut again, and employing an assistant, started out once more and found that half a mile of line had been carried off. While repairing this, he was again fired upon by concealed enemies, but being well armed, he and his assistant entered the brush and fired a few shots where they had seen the smoke rise, after which they were not molested. It is safe to add that they did not tarry long to make perfect joints. Again returning, to find the line cut once more, Lehr, believing his usefulness there as an operator ended, joined a company of home guards to hunt horse thieves and bushwhackers. This company participated in the battle of Carthage. Stebbins' lines being now destroyed, Lehr was next stationed as operator in Missouri opposite Quincy, where we shall meet him again.

In the Spring of 1861, and prior thereto, J. J. S. Wilson, residing in Springfield, Illinois, was superintendent of the Illinois & Mississippi Telegraph Company, The Chicago & Mississippi Telegraph Company and the Illinois Central Telegraph Company, owning the telegraph (patent) territory of Illinois, Iowa, Minnesota, North Missouri from St. Louis and a portion of Wisconsin. Charles Davenport was division superintendent for the St. Louis district of the Western Union Telegraph Company and C. M. Stebbins, of the Missouri & Western. The lines of these companies entered the consolidated office in St. Louis, managed by George H. Smith, who was also superintendent of the St. Louis, Alton & Terre Haute Railroad wires.

On the 25th of July, General John C. Fremont reached St. Louis and assumed command of what was then known as the

Department of the West, which embraced Illinois and all the States and Territories west of the Mississippi River, including New Mexico, with an expectancy of Kentucky also. By direction of General Fremont, George H. Smith, early in August, erected lines in the city, to connect the main office with the head offices of the army, including Fremont's head-quarters, Major McKinstry's (chief quarter-master) office and the arsenal. At this time there was no telegraphic communication between St. Louis and Ironton, or along the line of the Iron Mountain Railroad; none from St. Louis on the North Missouri Railroad to the Hannibal & St. Joseph Railroad; none over the South-west Branch of the Pacific Railroad to Rolla, and none west of Jefferson City on the main stem of the Pacific road; but on the Hannibal & St. Joseph road the line and road itself had been kept open, except when temporarily impaired by guerrillas.

Operator D. A. Williams, who opened an office for Colonel U. S. Grant, at Salt River, and subsequently exhibited great courage in repairing the line, was exceedingly useful hereabouts at this time. John B. Clarke was operator and agent at Callao, on this road, during the winter and spring of 1861. Sometime in May, a party of local Secessionists demanded of Clarke the truth concerning a report they had received by special messenger from Macon City, which lies about eight miles east of Callao, that the Dutch had taken Macon City, killed the women and children there, and were then marching on Callao. This report was but the result of that dark reportorial wave which, gathering blackness as it advanced from St. Louis after the capture of Frost's militia, had spread like a great cloud, and hung over the State of Missouri. Clarke's denial was unsatisfactory. The Hannibal & St. Joseph was regarded as a Yankee road, and its employees were all suspected by the secession element. Three mounted men were sent toward Macon City to ascertain the facts, and Clarke was held in custody until their return, when, if the report proved true, he was to be executed.

Burning bridges, firing upon passenger trains and cutting the telegraph was henceforth for many months the occupation, in part, of marauding bands throughout North Missouri, and, consequently, Clarke became a traveling operator. Taking his instrument he accompanied troops on the trains over the road, to move other trains in case a bridge was burned and communication with the dispatcher cut off. It was while so engaged at Platte River, that T. O'Meara, operator at St. Joseph, saved him from capture by telegraphing notice of the approach of Confederates, who fired the new trestles.

In the early part of August, General John Pope, in Alton, Ill., assisted by General Hurlburt, at Quincy, was in command of North Missouri, including the counties (on both sides) along the Missouri River, except St. Louis. The outlying department posts, to the west and south, were at Booneville, 187 miles from St. Louis: at Jefferson City, 125 miles; at Springfield, 236 miles (General Lyon's command); at Rolla, 115 miles; at Ironton, 80 miles; at Cairo about 186 miles, *via* railroad, and at points contiguous to each. The principal contending forces were, in front of Lyon, about twenty thousand to his eight thousand; about five thousand under Thompson, threatening Cape Girardeau, which is on one of the few Mississippi highlands in this department, and lies sixty miles above Cairo. A little below New Madrid, was a force estimated at twenty thousand, under Pillow, and further north another under Hardee, about equal to Thompson's.

In this situation of affairs, the commanders at Cape Girardeau and Cairo, and Lyon at Springfield, were repeatedly calling for re-inforcements. Fremont was telegraphed, while in New York, on the 18th of July, to re-inforce Lyon, and, on the 2d of August, J. C. Kelton, Assistant Adjutant General, telegraphed from Cairo: "General Lyon wants soldiers, soldiers, soldiers! So says General Hammer, who has just arrived from Springfield." Girardeau and Cairo were strengthened, the latter by several regiments, although fortified and having heavy artillery in position. It has been said that, owing to its position and armament, the rebels had not, on river and ocean, such craft as could have approached Girardeau. But not a soldier reached nearer Lyon's force than Rolla, until the remnant of his heroic army was in full retreat after one of the most gallant battles of the century. On the 10th of August, at Wilson's Creek, near Springfield, Lyon fell at the head of his troops, which were defeated with a loss of 223 killed, 721 wounded, and 292 missing. That of the enemy was about the same.

About the 1st of August, George H. Smith was appointed

by Fremont, manager of all the Government telegraphs in the department, whereupon he submitted his plan for a permanent organization, as follows:

"My proposition is to organize a corps of telegraph operators, builders and repairers, of experience, for the purpose of putting in complete order all telegraphs in this department, now suspended, and building any lines, or putting in telegraphic communication any point ordered by the Commanding General; this corps to be regularly organized, under military discipline, armed and equipped; to receive competent pay for their respective services. I also propose, with the authority of the Commanding General, to purchase to the best possible advantage a sufficient quantity of telegraphic material, including wire, telegraphic cordage, insulators and teams, for service for building and repairing for immediate use in any emergency, so that time and expense may be saved. Each operator to be furnished with a pocket instrument to open communication from the field of battle, or the marching army, to head-quarters, or elsewhere, instantly. I propose that the corps shall number fifty, more or less, as may be deemed necessary, with authority to increase the number as circumstances may require, and to be proportionately divided — operators, builders, repairers and teamsters, as the work to be done may demand, with a view to the least possible expense and the most efficient action."

This was substantially approved, and efforts were made to recruit a battalion of three companies, which Smith, with the rank of Major, was to command. These companies were to be armed and drilled. Frank S. VanValkenburgh, who had participated in the battle of Rich Mountain, as sergeant in an Indiana three-months regiment, on telegraphic request, hastened to St. Louis to accept a third (special) lieutenancy in the battalion. Company B, by September 9, was recruited for three years to the number of seventy-four privates, eight corporals, two lieutenants, one captain (W. S. Hewett). Companies A and C were partially recruited. T. Milton Guerin, Theron R. Perlee, George Allen, C. D. Waterhouse and Duncan T. Bacon were to be officers in the battalion. Guerin was so remarkably gifted as to enable him to obtain whatever was needed, even after Major Smith had found it hopeless. Perhaps his "Index to General

Orders Affecting the Volunteer Forces," printed in 1861, gave him some prestige, as it was a work of some considerable labor and merit. Among the officers, but few were practical telegraphers, and in the ranks there were none. This battalion rebuilt the telegraph line to Ironton, Rolla and Sedalia, from Jefferson City to a point twenty-five miles south of Syracuse ; from St. Louis arsenal to Benton barracks, six miles, and laid a cable across the river, connecting Palmyra and Quincy. All this was before November, 1861.

There being no legal authority for the organization of the battalion, it was disbanded by order of the Secretary of War, about November 1. No commissions were ever issued, but the title of major was permanently attached to Mr. Smith. Except Chester's Telegraph Corps (a company in General Sickle's brigade), in which Frank H. Lamb and C. I. Brown were the only operators, I know of no other instance of military company organization for telegraph service during the war, and this latter, although so enlisted, being sworn into service for three years, was held to ordinary soldier's duty, but Brown and Lamb were detailed therefrom for telegraph service.

Let us now take a later view of the military aspect in Missouri. Fremont was making preparations to move an army to Springfield when he ascertained that Price, coming northward from Clinton, had reached the Upper Osage, and was moving in the direction of Warrensburg, threatening Lexington, Jefferson City and Booneville. When within, say, thirty-five miles of Lexington, Price, after halting about ten days, moved against it, arriving in its vicinity on the 11th of September, with a force of about twenty thousand men and thirteen guns. He soon drove Mulligan's force of 2,780 troops and eight guns from the town. They took a strong, elevated position just outside, where, with bravery unsurpassed and in anticipation of aid from some of Fremont's fifty thousand, they withstood the enemy for nine days. After the nine days' defense, commended alike by friend and foe, Mulligan being out of ammunition and water, was, on the 21st of September, obliged to capitulate.

On the twenty-seventh, Fremont started in pursuit of Price. At his camp (Lily), two miles from Jefferson City, he was telegraphically connected with the outside world—a part of Smith's battalion having laid a land cable from a reel mounted on two wheels, drawn by one horse. Thus in two hours from the time Major Smith received his orders, communication with St. Louis was complete. This office could at any moment place him in immediate communication with the other principal outlying posts. The cable was buried in a trench to preserve it from injury. Land cables were not used again in Missouri during the war. Price fled to the south-west corner of the State, where, notwithstanding he was joined by McCulloch's force of five thousand, he continued his retreat to Pineville. Fremont encamped at Tipton, finally reaching Springfield about the 1st of November. He was superseded by General Hunter at a time when he and his army expected to join battle with their adversaries, then much inferior to the Federals in strength. The gallant charge by Zagonyi, the recent victory at Fredericktown over Thompson's forces, and the high regard of the troops for their late commander, made victory a moral certainty.

Hunter was ordered to retire from Springfield. Charles L. Weir, chief cipher operator at that time in St. Louis, was the bearer of that most unfortunate and injurious command. Historians have speculated a good deal about this order. The facts are that Weir received dispatches in cipher from General McClellan, directing Hunter to retreat to Rolla; and so urgent were they regarded, that McClellan, who knew Weir while in Cincinnati, telegraphed him a personal request to take the telegrams to Hunter by way of Tipton. There was a beautiful young lady on the banks of the Ohio, probably in Hamilton County, Ohio, whose nuptials were soon expected. Weir was a highly important factor in the programme, and wanted some one else sent to Hunter; but McClellan prevailed, and Weir started for Tipton, where he procured a horse of Philip Sheridan, then captain and assistant quarter-master there. Equestrianism was never Weir's hobby, and Sheridan may have, for the fun of the thing, selected an animal that traveled as ungracefully as a camel. Be that as it may, Springfield, *via* Warsaw and Bolivar, was a hard road to travel, especially when the traveler was pressed for time and unused to a Mexican saddle. How Weir succeeded, we scarcely know; but this we know full well, that what was left of him late at night on entering Springfield, delivered the dispatches and went to bed, feeling that it had passed through many battles, and was seriously wounded scores of times all over. Collecting himself the next day, and readjusting his "shattered columns," he and Charles G. Halpin (Miles O'Reilly) returned to the railroad at Tipton, *in an ambulance.*

During these military operations, the Telegraph Corps in Missouri was increasing in efficiency, and telegraphers were stationed as hereinafter shown, not including company employés: At West Quincy, Charles Lehr: while temporarily working nights at a junction near Palmyra, he was surrounded by rebel horsemen, some of whom presented revolvers at his head and threatened him with their contents unless he obeyed their behests. They searched his office for telegrams and decamped. At St. Louis, H. A. Bogardus, Derrick O'Dyer from August 9th, C. S. Payne from September; at Arsenal, A. S. Hawkins from October 4th, Duncan T. Bacon a short time in August; at Fremont's head-quarters, John C. (Yankee) Sullivan and D. T. Bacon; at Jefferson Barracks, J. L. Quate from August 25th; at California, H. J. Fish and R. Kuhn; at Tipton, Wayne H. Parsons and S. L. Griffin; at Smithton, Alex. Hunter; at Sedalia, H. A. Bogardus in October; at Moselle, George C. York; at Sullivan, Luke O'Reilly; at Rolla, C. S. Payne, temporarily, J. H. Rugg and W. H. Woodring; at Mineral Point, James H. Douglass; at Pilot Knob, J. H. Byrne: Frank S. Vanvalkenburg acted generally as assistant to Major Smith, wherever most needed. Sullivan and Bacon remained with Fremont until he was superseded at Springfield by Hunter. Payne also accompanied Fremont, but went to Rolla afterwards. Isaac McMichael accompanied the army to Warsaw, where he opened office. Woodring, who enlisted May 3, 1861, for three years, had already done good service as operator when he was detailed to work the Rolla office.

Thus far we have generally overlooked affairs about Cairo. There was some friction, arising out of Fremont's military order, placing Smith in charge of the telegraph in the department, which resulted in J. J. S. Wilson being recognized as superintendent east of the Mississippi River, and Smith, west. Owing to embarrassments common to civilians having important military functions to perform, and appreciating the great duties devolving upon Wilson, Governor Richard Yates, of Illinois, September 10, 1861, commissioned him colonel of Illinois volunteers.

John James Speed Wilson was so named because of the favor in which his parents held John J. Speed, Jr., a relative of theirs. Speed not only furnished his whole name, but provided a creditable employment whereby his namesake might maintain it becomingly. Wilson was born, March 1, 1834, in Danby, New York, where his father dealt in lumber. Subsequently (1845), the family removed to Racine, where young Wilson's ideas fairly began to develop, aided of course by the teachings provided in the public schools. Having natural inclinations toward mechanics, it is quite probable that his future would have lain in that direction but for the coming of Speed upon the scenes just as Wilson had reached that restless age of fourteen, when a boy is most anxious to undertake his own fortunes. Speed had contracted with Hon. F. O. J. Smith and Amos Kendall to erect for them, or rather for their telegraph company, the Erie & Michigan, a telegraph line from Detroit to Milwaukee. At the same time (1847), other contractors were connecting Detroit with Buffalo, and it was the prosecution of his undertaking that brought Speed to Racine, when Master Wilson, convinced that telegraphy was his affinity, embraced it with all the ardor of a young lover. It was but a short time after when Speed, Jr., who was chosen president of the company, made Wilson manager of Little Fort

J. J. S. WILSON.

(now Waukegan), Illinois, office, at a salary of about fourteen dollars per month.

Those were halcyon days for operators, in all matters except salaries; the art was mystic; new offices were being rapidly opened in all the States, and attention to duty was certain of ultimate reward. Two years after, Wilson, at the age of sixteen, was appointed superintendent of the Hotchkiss lines, connecting Madison, Wisconsin, with Chicago, and, consequently, he removed to the latter city. The author does not remember another instance of like trust in one so young, and it argues either that the duties were not then particularly executive, or that Wilson's precocity in that direction was remarkable. The latter would seem to be the case, since he was obliged to resign the position the following year on account of the responsibilities wearing on his nervous organization. While seeking relaxation, he was induced to take temporary charge of the Springfield, Illinois, (Illinois & Mississippi Telegraph Company) office, and in five months thereafter was made superintendent of the lines in the western division. That was in 1851. During the ten years following, Mr. Wilson, retaining his position, was stationed at the following points in the division, viz.: Springfield, Illinois, Burlington, Iowa, Galena, Illinois, St. Louis, Missouri, and Quincy, Illinois, which indicate the considerable field embraced within his territory. His position threw him in contact with most of the historic men of those times, as well as with others since renowned. Among these may be named Lincoln and Grant. It is worthy of mention that many notable telegraphers are graduates of this division of the Illinois & Mississippi Company's lines, among whom are Robert C. Clowry, James Gamble, Charles C. Hine, E. D. L. Sweet and, it also brought to the front in telegraphic circles, that distinguished jurist and citizen, John D. Caton, of Illinois.

Samuel H. Beckwith, having joined Colonel W. H. L. Wallace's regiment, which was stationed at Villa Ridge on the Illinois Central Railroad, twelve miles north of Cairo, sometimes assisted Ed. Schermerhorn, a private in the same regiment, in telegraphing at this field office, which was in connection with the main Chicago line, as early as May 19, and so remained for a month, when the regiment moved to Bird's Point.

General Polk, having entered Kentucky on September 5th, near Columbus, occupied that place on the seventh and fortified it. The Federal forces at Cairo, under General Grant, no longer respecting Kentucky's position as an armed neutral, seized Paducah, about fifty miles above Cairo on the Ohio, on the 6th of September, and also a point opposite Bird's Point, a little below Cairo, which was fortified and named Fort Holt. Up to November 7, when the battle of Belmont was fought and General Grant lost about four hundred men, no conflict of special moment occurred in the neighborhood of Cairo. Colonel Wilson built a line along the north bank of the Ohio from Cairo, crossing the river by cable to Paducah. Communication was opened with Fort Holt *via* the old cable across the Ohio and line extended to the fort, the telegraph being used to notify Cairo of any Confederate approach by water. The operators employed in this district, up to November 1, were William Foley, at Cairo; A. W. O'Neil, probably there also; H. W. Nichols, at Metropolis; George S. Pidgeon, at Caledonia; Samuel T. Brush, at Mound City; J. T. Tiffany, at Fort Holt; L. D. Parker, at Paducah; all of whom were first employed in October.

On the Confederate side, fronting the military points of the Federal Department of the West, telegraph connections and advantages, except in Kentucky, where they were reasonably good, bore a far different aspect. The Confederates' advance office, after the capture of Paducah, was at Blandville, eighteen miles from Columbus and twelve from Cairo. All of the private lines in the Confederacy being intact, the chief cities and towns of the South enjoyed, to a degree nearly equal to those in the Northern States, the advantages of early information as to the progress of military events in this quarter.

The Caton lines and South-western Company's converged at Cairo, which, before the war, furnished a Southern outlet for Chicago, St. Louis and eastern business, but soon after troops began arriving at Cairo, the South-western lines were cut at Columbus.

At this time, Emmet Howard was operator at Blandville and

required by his company to keep the line in order to Cairo, Paducah and Columbus. But the Federal cavalry and Confederate scouting parties soon overran this region, occasioning frequent skirmishes, the reports of which were forwarded by courier to Howard, who then telegraphed them to Columbus. It is a singular fact, that Blandville and Cairo were still in communication, the operators frequently conversing, but jealously eschewing military affairs. They became more reticent as the war fever raged, and finally the southern line was cut off at Cairo. Howard now became anxious for his safety, but by a delicate adjustment of his instrument, owing to some defect in Cairo office, he found he could hear Chicago and other offices. He became deeply interested, hoping to hear important messages, and for several days listened attentively to telegrams, but only heard railroad and commercial dispatches.

That line to Blandville was soon after destroyed, but the Paducah wire was in circuit until the 6th of September, when Howard reported at Columbus and was ordered back. He connected his instrument in a ravine, one mile south of Blandville, a hundred yards from the main road, where, behind a fallen tree, with a fine wire run to the main line, so as to prevent discovery, he telegraphed to Columbus such reports as were furnished him by mounted scouts, and also such movements of Federals along the main road as he noticed. But finding little glory and much risk and hardship, he sought more congenial quarters, disgusted with field telegraphy, especially as the Confederates seemed not to have made the most of the information so obtained.

In Arkansas, prior to the year 1860, there was no telegraph, but before the war, as we have seen, C. M. Stebbins connected Fort Smith through Van Buren, Fayetteville, and along Pea Ridge in North-west Arkansas, with Springfield and Syracuse, Missouri, and about the same time Henry A. Montgomery built a line from Memphis, Tennessee, along the projected Memphis & Little Rock Railroad to Little Rock, the State capital, whence it was extended at least to Clarksville by Montgomery during the war. He also, before the war, built a line from Madison, St. Francis County, Arkansas, to Helena, Arkansas. These were completed about January, 1861. L. C. Baker, the first manager at Little Rock, became superintendent of the line after the capture of Memphis. During the winter of 1860–61, Messrs. Snow and Ketchum built a line from Little Rock to Pine Bluff, south-east of Little Rock, on the Arkansas River, which soon came under Baker's management.

Among the people, the secession question absorbed all others and was involved in an election of delegates to the State convention, to be holden March 4. Robert S. Gantt, a candidate to that convention, an ex-telegrapher and a Secessionist, had been defeated in Prairie County by Colonel Totten, a conditional Unionist, who, like all others of that frigid kind, became a torrid Secessionist, as did also some other delegates. Gantt had employed Baker in Mississippi in 1855, and may have had something to do with the decided effort made late in February to prevent Unionists from sending political telegrams, but although Baker was a Secessionist, he had the manhood to reject proposals to that end; whereupon the plotters intended forcibly to possess themselves of the office and install Gantt, but in this Gantt would not aid them. About the 1st of March, W. B. Windsor, an ex-operator, acting as clerk on a steamer plying between Cincinnati, and Fort Smith on the Arkansas River, was induced by Montgomery and Baker to relieve the latter, who was beset by Unionists and Secessionists, charging him with divulging the contents of messages. Although the complaint was unfounded, his position was one of peril. After a short time, Baker returned, and Windsor took the Des Arc office. Federal Major Sturgis having abandoned Fort Smith, the Confederates took down some of Stebbins' line in North-west Arkansas and constructed others therewith. Clarksville was the western terminus of telegraphic communication, when Price and Van Dorn retreated in March, 1862, after the battle of Pea Ridge, not yet noticed.

There yet remains one other line of contact between the border slave and free States to consider. In those early days, only brave men openly espoused and actively participated in the movement in Kentucky, for or against disunion. But Kentucky, with its population of near a million whites, was not lacking in men of that type. On three sides of the State lay other slave

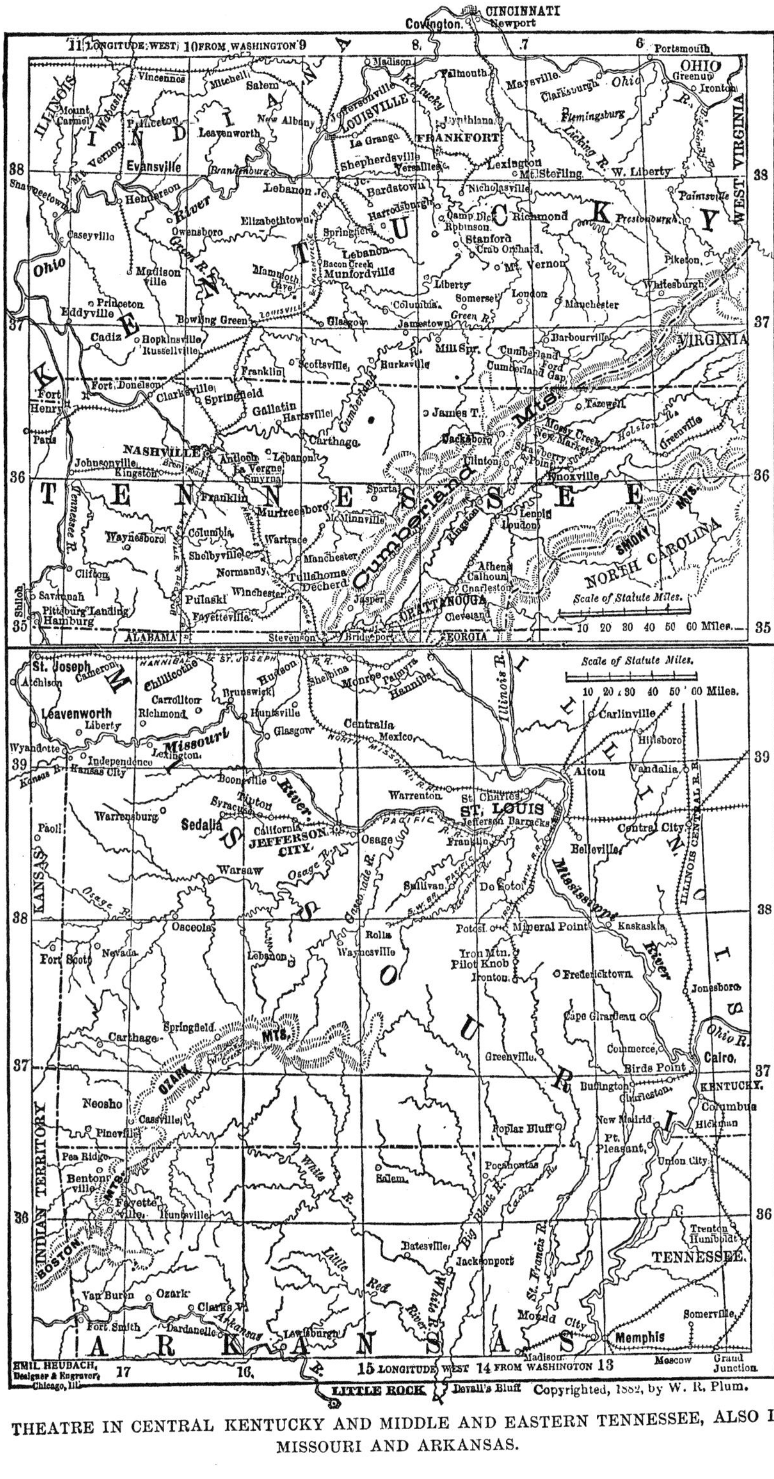

THEATRE IN CENTRAL KENTUCKY AND MIDDLE AND EASTERN TENNESSEE, ALSO IN MISSOURI AND ARKANSAS.

States, and on the north, across the Ohio River, are the great free States of Ohio, Indiana and Illinois, and their liberty-loving people. Kentucky long wrestled with that volcanic sentiment which had engulfed about one-third of her people, and stood, as it were, for anxious months, anticipating an eruption, if not at Louisville then in the Blue Grass region, Frankfort, Lexington or Camp Dick Robinson, where the crater was likely to appear, and where it would have developed but for a handful of as determined men as ever viewed liberty at the portals of slavery.

The means of communication throughout the State were ordinarily good for accustomed purposes and consisted of wagon roads, rivers and railways. Its principal internal rivers are the Tennessee, Cumberland, Green, Salt, Kentucky and Licking. Its railroads ran from Louisville to Nashville, Tennessee; Bowling Green to Memphis, Tennessee; Louisville to Frankfort and Lexington; Covington to Lexington and Nicholasville; also two short roads south from Paducah and Henderson. Along most of these ran lines of telegraph. Owing to the position of Kentucky as a neutral, nominally an armed neutral, the Federals and Confederates abstained from entering it until Polk, deeming the occupation of Columbus and vicinity a military necessity, occupied those points, and Zollicoffer, about the same time, as if acting in concert, entered the State through East Tennessee. About this time (September 5), General L. H. Rousseau, a Kentucky Unionist, was quartered in Camp Joe Holt, opposite Louisville, recruiting Kentuckians.

The Legislature of Kentucky having, on the 12th of September, invited General Anderson, of Fort Sumter fame, commanding the Department of the Cumberland, to enter Kentucky in force, the State was thereby officially committed to the Union, and General Anderson soon after removed his head-quarters from Cincinnati to Louisville.

At this time Confederate General Buckner was in command at Camp Boone, near the Louisville & Nashville Railroad, in Tennessee, from which he moved September 17, by the railroad, to capture Louisville. The two telegraph wires along the railway between Louisville and Nashville had been in working order up to this time, and constituted the last southern connection to be severed. Buckner planned a surprise. Had he first directed squads of troopers, each accompanied by an operator, to capture and control some of the northern telegraph offices, and moved more cautiously by rail, instead of cutting the wires between Bowling Green and Louisville as he did, and having his forward engine dumped in the ditch as it was, by reason of a rail being displaced by loyal young Crutcher, there can be no doubt but that he could have taken the city. Soon after four o'clock P. M., when the lines were cut, rumors of his coming having been some days afloat in Louisville, it became noised about the city that the lines were severed.

All felt certain that it betokened the coming of Buckner's troops, and thereupon the employés of the Louisville & Nashville Railroad Company in the city began tearing down the poles and wires in the vicinity of the depot, under the impression that they were helping the Union cause. T. R. Boyle was manager of the Louisville office, and E. C. Boyle, assistant. Both were entrusted with the Government cipher. About six o'clock Captain Green and Major Murray, of Anderson's staff, entered the telegraph office to prohibit telegraphing south. E. C. Boyle explained why that was impossible, and, having convinced Anderson of his loyalty and that of his brother, the manager, further destruction in the city was stopped, the Boyles reinstated, and the city wires rebuilt. Operator John Lenhart consented to accompany the Union troops now gathering at the depot, to the number of about three thousand, to meet Buckner out on the line of the road, and they went under Generals Sherman, Rousseau and Johnston that night to the Rolling Fork of Salt River, where a bridge had been burned by order of Buckner, whose plans had miscarried by reason of his mishaps. He withdrew and entrenched at Bowling Green, and the Union forces halted at Muldraugh's Hill, three miles north of Elizabethtown. The telegraph was intact north of the Federal position and south of Buckner's.

During Anderson's command, there was no telegraph at head-quarters in Louisville. General W. T Sherman succeeded General Anderson October 7, and spent a great deal of his time in the telegraph office, frequently remaining as late as midnight; but about the 1st of November a line was built to his office. Sherman was himself relieved, about the middle of

November, by General D. C. Buell, whose command included Ohio, Michigan, Indiana, and Kentucky, east of the Cumberland River, and Tennessee.

In casting about for a man of sterling character, unswerving loyalty, executive ability and an intimate knowledge of telegraph routes and men, to manage the military telegraph in Kentucky, Captain Stager, who had recently been charged with the management of all military telegraphs, happily decided to appoint Samuel Bruch to the position of Assistant Manager. This was done in anticipation of active work, and before there were any military lines in Kentucky; hence, Bruch's duties for a time were principally censorial. In October, by direction of General Sherman, Bruch built the first military telegraph line in Kentucky. It connected Lexington with Nicholasville, and was fifteen miles long. Camp Dick Robinson lay about ten miles south of Nicholasville.

On July 2d, Lieutenant (afterward General) William Nelson established this camp as a rendezvous for recruits. General George H. Thomas relieved him August 15, and there organized the first brigade and nucleus of what afterward developed into the renowned Army of the Cumberland. Here General Thomas prepared to confront Zollicoffer, who was south of Barboursville.

There were established in Kentucky up to November 15, six military telegraph offices, conducted by ten operators, some of whom were stationed as follows: head-quarters, Louisville, H. B. Spencer; city office, E. C. Boyle; Nicholasville, F. C. Cook; New Haven, D. C. Sellers (November 19); Lebanon Junction, C. H. Griffith.

From April 25 to November 15, 1861, there were built for military telegraph purposes in the several departments, one thousand, one hundred and thirty-seven miles of line, on which were one hundred and six offices, worked by one hundred and sixty-three operators; eight hundred and fifty-seven miles of this line, fifty-six offices and eighty operators, were outside of the Department of the Potomac, and it was well said by Mr. Stager in his report of November 14, that "in many instances the wires followed the march of the army at the rate of eight to twelve miles per day, there being no other lines of communication upon the routes where these lines have been placed. The capacity of the telegraph for military service has been tested, and in affording rapid communication between the War Department, the Commander-in-Chief and the different divisions of the army; in directing the movements of troops and the transportation of supplies, it may be safely asserted, that it is an indispensable auxiliary in military operations. The organization of the government Telegraph Department, under the direction of the Secretary of War, will add greatly to the efficiency of this branch of the service."

CHAPTER VI.

THE CORPS ORGANIZED.—BIOGRAPHICAL MENTION.—A VIEW OF THE TELEGRAPHS SOUTH.—THE FORT MONROE LINE.—THE MONITOR AND MERRIMAC.—THE PENINSULAR CAMPAIGN.—A VIEW ALONG THE COAST.—BANKS' DEFEAT IN THE VALLEY.—EXPERIENCE OF CAPTURED OPERATORS—AFFAIRS IN WEST VIRGINIA.

McClellan, upon Scott's retirement in the fall, became, under the President, commander of all the armies. A new feature in their conduct seems to have resulted, viz., that thereafter the armies should operate in concert along the whole line, from Fortress Monroe to New Mexico, and in that way make it reasonably certain that one or more of the important gateways to the South would be thrown open to the Federals. But over a territory so vast, presenting so many points of defense, and defended by armies so numerous and brave, concert of action, vital as it was to the Unionists, was clearly impossible without an electric tongue to convey intelligence and speak commands.

The telegraph had accomplished much—had given evident satisfaction, but like the armies, it had too many heads. In St. Louis, Mr. Stager had a small voice, and General Fremont a trumpet; consequently certain lines in Missouri, built by contract, cost one hundred dollars per mile along railroads, while in West Virginia, far from railroads and in a mountainous country, they cost but forty-five dollars. It was also discovered that the Government telegraph agents, in their necessities, were competing with one another in the purchase of material. Thus the Government bid against itself.

While Secretary Cameron was considering these things, Amasa Stone, an influential citizen of Cleveland, Ohio, in a conversation with him, suggested Mr. Stager as the best man to undertake the general management of the military telegraph. October 16, Assistant Secretary Scott telegraphed Stager to come to Washington and submit his views, which he did, October 26, as follows:

"I submit herewith, for your consideration, a plan for organizing the Telegraph Department for Government service, to be under the direction of the Secretary of War. The appointment of a General Manager, whose duties shall be, under the advice and approval of the Secretary of War, to purchase, transport and distribute all material required in constructing, maintaining and operating Government telegraph lines; said Manager to appoint an assistant for each military district or department, and to select all operators, repairers, builders and others, engaged in Government telegraph service. The material distributed and expense incurred, to be charged to the respective military districts where used; the Manager to make quarterly returns to the Secretary of War, of property on hand and in use. The issue of an order to all quarter-masters and assistant quarter-masters to fill the requisitions promptly, of the General Manager and of his assistants, on a proper certificate for the same, for transportation, subsistence, equipments, forage, etc., when needed for the prompt prosecution of work directed by the commanding officer of the district; also, to pay all bills for labor, transportation and material, when certified by the General Manager of the Telegraph Department. All accounts connected with this department, to be certified to by the General Manager. The General Manager, with the approval of the Secretary of War, to fix the grades of service and pay, from time to time, as exigencies and public interest may require. * * The General Manager to arrange with the different telegraph companies, for special or extraordinary use of lines and offices, on terms to be approved by the Secretary of War." (To which Scott here added: "In case of emergency the General Manager shall take possession of any of the telegraph lines that may be required for public service.") "Whenever persons engaged in Government telegraph service, or detailed for work where ordinary subsistence can not be obtained, rations will be furnished by the commissary most convenient, and the cost of the same to the Government shall be deducted from the regular pay of the persons receiving rations. I recommend the following grades and pay for the persons employed in this department, subject of course to modification by the Secretary of War, from time to time, as the public service may demand. The amounts here given, to include service and subsistence:

	MIN.	MAX.
General Manager	——	——
Assistant Manager	$100	$175
Chief Operator	60	70
Assistant Operators	40	60
Foreman of Construction	50	70
Repairers	45	55
Wire Men	40	45
Laborers	30	40
Messengers	15	20

* * * The Quarter-master General to make such arrangements for payment as will meet the wants of the service."

This paper was endorsed as follows:

WAR DEPARTMENT, Oct. 28, 1861.

RESPECTFULLY REFERRED TO THE PRESIDENT:

The Secretary of War believes it to be a necessity, but wishes your views. Mr. Stager, the gentleman who will deliver this, is now in charge of Government lines in the West, having been placed there by General McClellan, and well calculated to perform the duties. His connection with all leading lines will be of service. If you approve, arrangements will be made at once.

Very truly, THOS. A. SCOTT,
Asst. Sec.

EXECUTIVE MANSION, Oct. 28, 1861.

I have not sufficient time to study and mature an opinion on this plan. If the Secretary of War has confidence in it, and is satisfied to adopt it, I have no objections. A. LINCOLN.

Approved by the Secretary of War. SIMON CAMERON.

It will be observed, that Mr. Stager, in the foregoing plan, made no suggestion for military rank for any one. Indeed a military commission was not considered until Quarter-master General Meigs insisted upon it as necessary, before he could honor Mr. Stager's requisitions for money and supplies. In peace times this would have proved a great obstacle, because there was no more law for such an appointment than there was for commissioning every operator in the service, in like manner; the only difference consisted, not in the legal view, but in the requirements of the service. No body ever deemed it necessary or even proper, that operators, as such, should be quarter-masters, and yet there were times when they suffered because they were not. Meigs' views prevailed, as the following shows:

WAR DEPARTMENT, Nov. 11, 1861.

MAJOR GARESCHE, Act. Adj. Gen.:

Please make letter of appointment for Anson Stager, Esq., as Brigade Quarter-master. This was arranged by the Quarter-master General and the Secretary of War before they left. Mr. Stager is to be detailed as General Superintendent of Government Telegraphs in all the departments, and will in future attend to the organization of that Department. Very respectfully,

THOS. A. SCOTT, *Asst. Sec. of War.*

WAR DEPARTMENT, Washington, Nov. 11, 1861.

SIR: You are hereby informed that the President of the United States has appointed you Assistant Quarter-master of Volunteers, with the rank of Captain in the service of the United States, to rank as such from the eleventh day of November, 1861. Should the Senate, at their next session, advise and consent thereto, you will be commissioned accordingly. * * Should you accept, you will at once report in person, for orders, to the Quarter-master General, U. S. A. THOS. A. SCOTT, *Act. Sec. of War.*

To CAPTAIN ANSON STAGER, *A. Q. M. Vols.*

HD.-QRS. OF THE ARMY, Washington, Nov. 25, 1861.
Adjutant General's Office.

SPECIAL ORDERS, No. 313.

Captain Anson Stager, A. Q. M. is assigned to duty as General Manager of the Telegraph lines. Upon Captain Stager's requisition, or that of his assistants in their respective military departments, commanding officers will furnish the employés in the field, connected with the telegraph lines, with rations in kind and shelter, such as is allowed to other government employés. Commanding officers will also give such aid as may be necessary in the construction and repair of telegraph lines in the country in which troops are operating. By command of Major General McClellan.

L. THOMAS, *Adjutant General.*

Captain Stager soon discovered that, in sending to distant assistants, large sums, often amounting to one hundred thousand dollars, for which he obtained no recognizable receipt until the signed vouchers of employés were obtained, he was running too much risk. Moreover, quarter-masters complained that there was no lawful warrant for their recognizing the requisitions of his civilian assistants, and consequently he obtained commissions for them, as follows: Major Thos. T. Eckert, Department of the Potomac, July 17, 1862, Captain T. B. A. David, Department of West Virginia, July, 1862; Captain Samuel Bruch, Department of the Ohio, August 8, 1862; Captain Randall P. Wade, Purchasing Agent, about September 28, 1862; Captain Chas. H. Bulkley, Department of the Gulf, January, 1863; Captain George H. Smith, Department of Missouri, September, 1862; Captain William G. Fuller, Department of Tennessee, October 27, 1862; Captain John C. VanDuzer, Department of the Cumberland, October 27, 1862; Captain William L. Gross, Department of Ohio, October 27, 1862; Captain Lemuel F. Sheldon, Department of the South, October 31, 1863; Captain James R. Gilmore, Department of the South, November 3, 1864. These were all assigned to service as assistant quartermasters. Captain Stager, in anticipation of these appointments, was commissioned Colonel, February 26, 1862, and attached, as aide-de-camp to the Secretary of War, but in April, 1863, owing to ill health from overwork, he took permanent quarters at Cleveland, Ohio; he was, however, frequently in personal attendance at various military points.

In these arrangements the operator, who braved nearly all the dangers incident to the service, was left a mere civilian; only a quarter-master's employé, liable to draft, his salary taxed, and he, surrounded by all the paraphernalia of war, to the conduct of which he was so essential, was without rank, name or position, subject to the unkind cuts of the envious, but thoroughly appreciated by the President, his cabinet and the generals.

Thus organized, the United States Military Telegraph became the medium of communication by which hundreds of thousands of armed men were directed from point to point, commissary, subsistence and ordnance stores ordered, and the innumerable necessities of great armies made known, as well as the story of their victories and defeats. A word now about its chief.

Anson Stager.

Replete as is the telegraph history in America with the names of successful devotees of that science and art, probably none of them have had such steady, conservative and well grounded prosperity as Anson Stager. His has been no fitful, spasmodic advancement, nor the preferment of wealthy relatives, but the result rather of methodical attention to duty undertaken. As an operator, he soon became an appreciative and attentive expert. In his day at the key, none excelled him, either in precision or rapidity. It was while in the foremost rank as an operator that he was invested with executive departments, the management of which brought new confidence and increased responsibilities. These-multiplied with the astonishing increase of telegraphic business and the facilities therefor, until he became a leading executive head of the most gigantic telegraph company the world ever produced. The history of his success since the rebellion is beyond the purview of this work, and therefore, notwithstanding it would give valuable hints to others seeking honorable preferment by first deserving promotion, it must be left to others.

Anson Stager was born, April 20, 1825, in Ontario County, New York. While but a boy, he began to learn the printer's art in the office of Henry O'Reilly, in Rochester, New York. That he succeeded well, those who know his subsequent career need not be told, for General Stager is but a development of the boy he was ; conversely stated, he is a strong illustration of the adage, that "the child is the father of the man." But "there's a divinity that shapes our ends," so largely are we "creatures of circumstance," and so it was with young Stager, for when fairly embarked in the business of his first choice, a novel factor in business affairs induced a new ambition. This was when Stager had just reached his majority (1846) and but two years after Morse had demonstrated the practicability of his system of telegraphing ; a time when capitalists were, however, full of doubts and misgivings. Henry O'Reilly became enthusiastic over the telegraphic outlook, and induced his *protegé* to follow him into this undeveloped business enterprise. O'Reilly connected Philadelphia with Harrisburg, and in the fall of 1846,

Mr. Stager assumed charge of his first office, viz., at Lancaster on this line. Pittsburgh was added to the circuit, and after a short service at Lancaster, Stager assumed the management of the operating room in the former city.

By this time, every considerable town was seeking telegraphic connection, and the O'Reilly lines were pushed over a vast expanse of territory, north, west and south. Cincinnati became a repeating station of great importance, and Stager's success at Pittsburgh made his promotion to a like position at Cincinnati a matter of course. He had now come to the front rank among the manipulators of the key and comprehended what was then known of the science. With that science he has kept pace, even doing much to develop its phenomena. It was he who first economized batteries by charging many wires at the same time from one source. He also connected long lines, thus saving time and risk of repetition of messages. These things are now so commonly done as to excite no attention, but in their origin they evidenced individuality and originality which gave Mr. Stager much prestige.

While thus serving at Cincinnati, the New York & Mississippi Valley Company's lines were projected, connecting Buffalo, New York, with Louisville, Kentucky, and Mr. Stager was made general superintendent thereof. An eastern outlet to New York City was leased the following year (1852). A short time after, under the inspiration of other Rochester men, the Western Union Telegraph Company was organized, and (1856) many other companies were glad to consolidate their doubtful interests with it. Thus that company may be said to have begun its entity by a systematic effort to control the field; a purpose it never relaxed, for with building and leasing, it now practically controls the telegraphic facilities of a continent.

Mr. Stager has been identified with this growth from its early efforts until recently. As its general superintendent, in 1856 he located at Cleveland, Ohio, and applied himself with zeal to the restoration of the impaired lines and to the negotiation of contract rights with various railroad companies, whereby their interests were greatly subserved and his own company enabled to overcome opposition and finally become impregnably buttressed in the commercial and other interests of the nation. Although at the beginning of the rebellion the great results subsequently produced were hardly anticipated, yet it is true that they are largely the natural outgrowth of the foresight of such men as Anson Stager and J. H. Wade, who were actively engaged in authoritative positions, laying foundations for that success which since the war has found demonstration in many princely fortunes for the stockholders, who up to the time of the war had not received a dividend.

On one of Stager's many trips in the interests of his company, an engine on the Pittsburgh, Fort Wayne & Chicago Railroad broke down at night, nine miles from an office. Stager cut the wire, and by touching the ends telegraphed to Pittsburgh and Brighton for another engine. By fixing one end of the line above and the other beneath his tongue, which he projected out to view, the electric pulsations gave it a vibratory movement, similar to that of a telegraph armature, which is often used to read from. Thus Mr. Stager could see and read the involuntary oscillations of his tongue, as one can those of an armature, which he may not hear, and thus he received reply that another engine would be sent at once. This feat has not ceased to be remarkable, even in the light of frequent electric surprises. It was while he was superintendent of this company, which position he did not resign, that Governor Dennison of Ohio, telegraphed him to come to Columbus to consult about assuming military control of the telegraphs in that State. Governors Morton of Indiana and Yates of Illinois thereupon invested Mr. Stager with like authority over the lines in their States, and about the same time, as already shown, General McClellan, at Cincinnati, appointed him manager of all lines in his department.

Such is a résumé of the career of the chief officer of the Telegraph Corps, who remained at its head throughout the war.

Confederate States Telegraph Advantages.

Of course, an agency so potent in war as the telegraph, could could not be overlooked, but the Confederates appear not to have regarded it as so essential as did the Federals. Early in the war, the South had resources enough for extending the telegraph from main line offices to all of her armies; but this was not done in Western Virginia, South-western Kentucky or, to any great extent, in Missouri. There were, as we have seen, in 1861, but two leading telegraph companies in the Confederate States, the American and South-western. J. R. Dowell was general superintendent of the Southern division of the American Company's lines, and on the beginning of hostilities, it is said, by his advice, Dr. William S. Morris, a director in the company, assumed charge as president, and the corporate name was changed to the Southern Telegraph Company. Dr. Norvin Green was president and John Van Horne general superintendent of the South-western Telegraph Company. Judge Reagan was President Davis's Postmaster General, and on the plea of military necessity, was placed in charge of all the lines in the South, but interfered with the private management thereof, as a rule, only as military purposes required. The knowledge of this power, if nothing else, generally made more than nominal control unnecessary.

By an act of the Confederate Congress, July, 1861, the President was authorized to seize all telegraph lines, and appoint agents to supervise all communications passing over them, and forbid all telegrams in cipher and such others as were of an enigmatical character. Whenever a military superintendent's services were required at Richmond, Dr. Morris appears to have received the appointment.

Soon after the capture of Fort Donelson, Nashville was evacuated, and Doctor Green returned to Louisville to manage his company's interests within the Union lines, where he remained till the war was over. Mr. Van Horne stayed South, and acted as president of the company within the Confederate lines.

There was no such thing as a military telegraph organization in the South, except at a few local points. General Beauregard had a regularly organized system, with a full corps of operators, in and around Charleston; J. W. Kates being superintendent until he went to Shiloh with Beauregard, leaving W. R. Cathcart in charge. Mr. Kates was chief operator with Beauregard until near the close of the war, when he took a boat ride with some friends and was overhauled by the Yankees.

The private companies aimed to do the military telegraphing, even with the armies, and, to a great extent, succeeded. A few operators took service at the head-quarters of the commanding officers and had a sort of military status, but they had very little to do in the Western departments, except when head-quarters happened to be at some small station, where the company's operator was not able to do the work. It was quite usual, however, for operators to be associated with the principal calvary chiefs; many as aides, yet others were soldiers. Indeed, owing to the Northern operators leaving when the war began, and the enlistment of Southern operators in the armies, there was soon felt a great lack of such talent, and the only remedy lay in detailing operators from the ranks, which was done.

One of the first military lines built in the Confederate States, viz., from Pensacola, Florida, nine miles, to General Bragg's head-quarters in the rear of the navy yard and Fort Barrancas, was thus operated, under the management of J. G. Thornton, manager, secretary and treasurer of the Pensacola Telegraph line, whose president, Ch. Le Baron, while acting as aide-de-camp to Bragg, was wounded at Shiloh. The operators detailed for service on this line were William H. Cody, a private in a Mississippi company, and George McCann, private, Seventeenth Alabama regiment.

There had been much *ante bellum* rivalry, and consequent friction, between the American and South-western companies, which the severance of their respective heads did not remedy, and it was believed that Doctor Morris was largely instrumental in obtaining an order for the seizure of the South-western Telegraph Company's property within the Confederacy, on the ground that its officers were Unionists. The order was issued in September, 1862, when a Mr. Caldwell was appointed military superintendent, with directions to possess himself of all of its property and receipts. He established himself in Chattanooga, and issued his commands to the managers of all offices before Van Horne knew of the appointment. At this time John B. Morris was operator at Jackson, the capital of Mississippi. He requested Governor Pettus, to intercede in behalf of the company. Pettus, thereupon, telegraphed President Davis to know the cause of the action, and received the following reply:

RICHMOND, VA., Sept. 11, 1862.

GOV. PETTUS, Jackson, Miss.:

The action was taken on suggestion of telegraph companies, and is designed to protect the Confederate States, and to secure the receipts to the true and loyal owners. The president of the company is with the enemy, and the company is not in condition to fulfill its purposes. (Signed) JEFFERSON DAVIS.

This exposed the scheme, which Van Horne, well seconded by others, was able to frustrate. The order was rescinded.

FORT MONROE LINE.

From the beginning of the war, great interests centered about Fortress Monroe, the Federal stronghold in South-east Virginia, commanding the entrance to the James River, which was navigable to Richmond. This fortress was like a thorn in the side of the Confederacy, and nearer its heart than any other permanent Federal lodgment. It had been the base of Union operations in that neighborhood, and was to be the rendezvous of many military and naval expeditions. When Manager Gilmore succeeded Strouse about Washington, Richard O'Brien took charge at the fortress. General Butler, having arranged to act in concert with Admiral Farragut for the capture of New Orleans, was succeeded in command, on the 18th of August, 1861, by General Wool. While General Butler was in command, he seriously contemplated a direct movement, in force, from this fort, on Richmond, and while General McClellan was organizing his forces about Washington, and so placing them as to indicate an intention to march direct against the enemy in his front, he was, in fact, secretly cogitating over the idea of quickly transferring this army by water to Urbana, on the Rappahannock, or Yorktown, on the York River.

While these two points were under consideration, it was determined to build a telegraph line to Fort Monroe. Two routes presented themselves, the first, suggested by Captain Stager and preferred by the Assistant Secretary of War, was, to extend the line then working from Washington to Budd's Ferry (fifty-three miles), to Port Tobacco and Point Lookout (seventy miles) and thence by submarine cable in the Chesapeake Bay (seventy miles) to Fort Monroe. This route was impracticable for want of cable, which was not then manufactured in this country to any considerable extent, and there was no time to wait for it. The second, which was adopted, was, to extend the Government line from Lewes, Del., *via* Salisbury to Cape Charles, thence by cable across the Chesapeake (twenty miles). The Lewes line connected with Wilmington *via* Dover, and was built in the fall of 1861.

The Fort Monroe line was begun about the middle of January, 1862, and completed on the 5th of February to Cape Charles, one hundred and fifty-eight miles from Wilmington, at which latter place it connected with the American Company's lines running to Washington. By the use of a repeating instrument at Wilmington, the line was practically one circuit to the War Department. Until the cable was laid by W. H. Heiss, shortly after, a dispatch boat plied across the channel in three hours.

Having now some idea of the military telegraph situation in Virginia and the South, let us again recur to the field of arms.

Affairs about Fort Monroe; The Monitor and Merrimac.

O'Brien's efforts to keep up his line to Newport News was somewhat vexatious, owing to the rebels so frequently carrying off, by night, half a mile or so of wire. Even day incursions were not unusual. On one occasion, in February, he and his little brother, John, assistant operator at the fort, not yet fourteen years old, accompanied by a few negro repairers, on the road toward Newport News, saw in the dim distance toward Newmarket a cavalry force charging down towards them. It being suggested that they were rebels, the darkies took to the woods and did not stop running until they reached Old Point Comfort. Young O'Brien sought to lead them, but he was soon outstripped. He rejoined his brother, who, awaiting the coming of the horsemen, found them Federals.

Probably John O'Brien was the youngest operator ever in the United States Military Telegraph service. He had, in February, 1862, had three years experience, and is said to have been an expert in the art. General Wool was greatly surprised when he saw the boy receive by sound, which he was doing when Wool first met him.

Fortress Monroe was becoming monotonous, when a new element began to obtrude itself. Rumors of a floating monster absolutely bomb and ball proof, came by contrabands from Norfolk, on the Elizabeth River. They were soon verified, and not long after, tested. Daily, field-glasses swept Sewell's Point, Craney Island and the mouth of the Elizabeth. At last, about noon on the 8th of March, against the sky behind Sewell's Point, black smoke was apparent, which moved ominously nearer and nearer. J. O'Brien, at the fort, signaled Newport News, which lies across the river where the Elizabeth enters the James. George D. Cowlam, a detailed member of Ellsworth's famous Fire Zouaves, was the operator there, and bravely did he remain at his post. Cowlam answered the call; he also had seen the smoke. The hour was pregnant with great events. The alarm-gun signals the men at the fort to their pieces; the long-roll beats the soldiers to arms; the naval vessels clear for action; then over the waters comes the sound of a broadside. But see, *is not the "Merrimac" a monster in truth?* Even the Sawyer gun on the "Rip Raps," vindictive and thunderous, is at work. Young O'Brien is at his instrument, safe enough, to be sure, but full of anxiety as to the result. Suddenly, amid the deep reverberations of the heavy guns, his strained ear catches the sharp click from the sounder. 'Tis Cowlam, and this is what he telegraphs: "She is steering straight for the 'Cumberland'" — a pause — "The 'Cumberland' gives her a broadside" — waiting at the fort — "She keels over" — suspense — "Seems to be sinking" — anxious watching — "No; she comes on again" — great anxiety — "She has struck the 'Cumberland' and poured a broadside into her. God! the 'Cumberland' is sinking" — breathless suspense — "The 'Cumberland' has fired her last broadside."

Thus did the telegraph picture the scenes of that eventful day; some visible and others hidden from the ramparts of the fort. The sorrow depicted on the countenances of the brave officers at the fort on receipt of the news that the "Cumberland" had fired her last broadside, passes pen picturing. But that calamity was not all. The "Congress" grounded near the News and was raked by the "Merrimac" and her wooden consorts — the "Minnesota," grounded; the "St. Lawrence," grounded, and the "Roanoke," grounded. Truly, the tide was setting against the Unionists. It was a dark hour. The burning "Congress" deepened it. Gallant Cowlam continued at his instrument, describing each new phase in the fight, while the shells shrieked about his quarters, and two tore through his office, within a few feet of him; but there was the same steady hand at the key. At ten P. M., the little "Monitor" arrived. Now let Richard O'Brien's diary tell of the ninth:

At six A. M., the "Merrimac" was seen steaming down the Elizabeth River. General Wool, having placed the fort in fighting trim, rode out with his staff through Camp Hamilton and the ruins of Hampton, to a point on the shore nearest the "Minnesota," which was still aground. I accompanied the party. The "Monitor," which had arrived the previous night, lay under the shadow of the "Minnesota," and seemed to us a feeble defense to lean upon, against the invincible monster which had made such short work of two of our finest war ships the previous evening. When the "Merrimac" passed Sewell's Point and turned towards the fort, we were about to hurry back to help receive her, but when near the "Rip Raps" she turned again and came straight for the "Minnesota," which opened fire upon her. The "Merrimac" slowed up a moment, as if to make out what the strange little craft could be, when Lieutenant Worden blazed away and solved the question for her. She quickly responded. They both "let slip the dogs of war," the rebel bulldogs growling from every port-hole, and the little terrier of the North, more active than her unwieldy antagonist, snarling at every rib of the larger craft. The "Monitor" got around more quickly than the "Merrimac," and tried her sides, quarters and stern, but every shot that struck, glanced from the greased rails into the air, with the scream of a baffled demon. The "Merrimac" fired rapidly and viciously, but seemed equally unable to injure her antagonist, and so turned her attention again to the "Minnesota." The latter discharged a broadside at her without the slightest effect, and received in return a shell from the bow gun of the "Merrimac," which burst in the officers' quarters and set the ship on fire. Another shot struck the tug-boat "Dragon," which was engaged in trying to haul the "Minnesota" off, passing through and bursting its boiler. A terrific fire was kept up by the "Minnesota" from every gun that could be brought to bear. A third shell passed over the "Minnesota" and burst unpleasantly near us. Before she could fire again, the "Monitor" had gotten between the "Merrimac" and the "Minnesota," forcing the former to change her base, in doing which, she got aground, but soon swung off and headed for the "Rip Raps," with the "Monitor" close at her heels. They had not gone far, however, when the "Merrimac" turned around suddenly and tried to run into the "Monitor." The latter made a very narrow escape, the great prow of the "Merrimac" leaving an ugly scar on her iron armor. They then pounded away at each other for some time, when the "Monitor" drew off towards the fort. We feared she had received serious injury. The "Merrimac," with her consorts, the "Jamestown" and "Yorktown" (or "Patrick Henry"), which had thus far kept at a respectful distance, now started towards the "Minnesota," which we felt sure was doomed. They changed their course, however, for some unaccountable reason, and heading up the Elizabeth River, left us, for this day at least, masters of the situation.

Great was the joy in the North when news came that the "Monitor" had turned the current of affairs; but greater yet was it in Washington, where boats were laden with stone, to be sunk in the channel in case the "Merrimac" destroyed her adversaries.

THE PENINSULAR CAMPAIGN.

The army of the Potomac lay in and around Alexandria eight whole months; to its right, about Harper's Ferry, the Baltimore & Ohio Railroad was sealed up; to its left was the Lower Potomac, vexed by the enemy's batteries. January 13, 1862, E. M. Stanton succeeded Cameron as Secretary of War.

On the twenty-seventh, President Lincoln directed that on the 22d of February a general movement of the land and naval forces of the United States against the insurgent States should take place, and that especially the "army at or about Fort Monroe, the Army of the Potomac, the army of Western Virginia, the army near Mumfordsville, Kentucky, the army and flotilla near Cairo, and the naval forces in the Gulf of Mexico be ready to move on that day.

February 1, McClellan's army contained one hundred and ninety thousand, eight hundred and six troops of all arms, present for duty. Yorktown was decided upon as the point from which to advance upon Richmond. 113 steamers, 188 schooners and 88 barges were collected to transport 121,500 men, 14,592 animals, 1,150 wagons, 44 batteries and 74 ambulances, besides pontoon bridges, telegraph materials and equipage, etc. The steamer "Commodore" was selected for McClellan's use, and a telegraph line was run thereto from his head-quarters in the old Club House, Washington. By this wire he was advised constantly as to the progress of shipping the army. Jesse H. Bunnell was the operator on the steamer, and A. Harper Caldwell and C. W. Jacques at the Club House. These three were McClellan's operators, and Caldwell was chief. We shall meet them often hereafter, especially Caldwell, who remained with the head-quarters of the Eastern army to the close of the war, earning, by his ability and urbanity, a splendid reputation. Maj. T. T. Eckert accompanied the army, to conduct the telegraphic operations. Parker Spring and L. D. McCandless, under Eckert, had charge of the construction parties.

MAJOR ECKERT, Colonel Stager's principal assistant, was a man of energy, tact, perseverance and strong purpose.

His jurisdiction was confined to Virginia, except for a time when he controlled the coast telegraphic operations also, but his position in the East, mainly at the War Department, especially after Colonel Stager removed his head-quarters thence to Cleveland, Ohio, in 1863, gave him a prominence in military telegraphic circles which his ability and devotion fully justified. They even warrant a more extended personal notice than we have space for.

The other day, two men fell out of a balloon, while a third rose to a great height, also to fall soon after into the sea. That well illustrates how accidents facilitate the rise of some men, but, like the lonely aeronaut, they soon sink beneath the waves that encompass them in their short career. Others buffet the elements successfully, because they are well ballasted, and foresee emergencies for which they prepare. He who goes out upon deep waters and weathers the storms, is a captain whom followers like to be near, even as the soldier prefers to fight under generals that win battles. Thomas Thompson Eckert, at this writing, has long been upon the deep waters of life's sea. Many are the storms he has encountered, and great is the credit he has won, while making a name that crosses the continent, and gaining a position that directs the telegraphic facilities of the United States and Canadas, if not to the shores of England.

He was born in St. Clairsville, Ohio, April 23, 1825 He acquired the telegraphic art in 1848. The next year, he was appointed post-master, at Wooster, Ohio, and united the telegraph and post-offices there under his management. Thus early he set an example, which many contend, should be the rule, instead of the exception, throughout the Union. But Eckert was not to be hampered by small offices. Parties interested in connecting Pittsburgh with Chicago, by the route of the Pittsburgh, Fort Wayne & Chicago Railroad, arranged with him to superintend the construction of their lines. This was performed so satisfactorily that the "Union Telegraph Lines" people, owning this plant, offered him its superintendency. This position he held about four years, when (1856) his company was merged into the "Western Union," and his jurisdiction enlarged. In his wider field, he labored zealously for about three years, exhibiting many of those qualities which his later years and increased opportunities gave full scope to, and with proportionate results.

In 1859, Mr. Eckert removed to North Carolina, where he was superintending the operations of a Gold Mining Company, when the war began, but which he abandoned for a Northern home, pending hostilities, at least. Thus he was free to engage in any service wherein he could best serve the nation. The military telegraph proved his affinity, and in its ranks he continued to the close of the war. April 7, 1862, he was commissioned major and aide-de-camp, on General McClellan's staff, to forward his telegraphic operations, and for that purpose he was now with the army.

McClellan confronted Magruder about the 5th of April, near Yorktown. Magruder's force was about one-tenth of his adversary's, but, deceived as to the topography of the country and the number of Magruder's increasing force, McClellan laid siege until May 4, when the enemy, much strengthened, evacuated, and thus avoided an assault.

Telegraph lines had been extended to all the Federal head-quarters on the Peninsula, including Fort Monroe, twenty-seven miles, and Ship Point, twelve miles from Yorktown. Besides those at McClellan's, there were at this time with the army, and on the line of communication to Washington, the following operators with McClellan's army: D. B. Lathrop, H. L. Smith, John Allen, George B. Cowlam, F. A. Lawrence, R. F. Follett, Wilbur F. Halloway, J. Harvey Nichols, Thomas Morrison, Theo E. Moreland, T. M. Schnell, J. R. Waite, Joseph Schnell, A. C. Schnell, George W. Nail, Chas. A. Tinker, J. H. Emerick, W. H. Shreffler, Crosby J. Ryan, J. B. Norris, H. A. Bogardus, M. H. Kerner and E. H. McGintey; reinforced in June by Thos. Dolan, F. H. Fonda, J. La Bonte, C. H. Lithgow, E. N. Robinson, O. H. Booth, C. L. Snyder, F. A. Stumm and G. D. Wilkinson. At Fort Monroe were the O'Briens and George D. Sheldon; Cherrystone, G. W. Baldwin; Drummondstown, Delaware, P. H. Nunan; Salisbury, S. R. Magonigle (June); Lewes, John Wintrup to June, and R. Power, Jr.; Harrington, C. M. Roberts; St. Georges, H. P. Royce; Wilmington, L. A. Rose, J. W. Hallum, R. Power, Jr. to June, John Wintrup and T. M. Schnell.

The operators with the army kept McClellan fully advised as to all operations in front. It was worth regiments of soldiers to *know* that an attack on any point of McClellan's lines, many miles long, would be reported in time to be promptly met. It was worth "a mint of money" to the people of the North, to receive daily, detailed reports of operations about Yorktown; but it was somewhat dangerous business for the telegrapher at the front. Parker Spring and builders, on the 20th of April, while building a line toward the James, were driven off by the enemy. Emerick and Nail, at Richardson's head-quarters, on the left of Yorktown, were just out of range of the shells ; but the operator with General Porter was not so fortunate, although unhurt. A shell fired from a Whitworth gun buried itself in the ground just sixteen feet from Nichols' tent, and exploding, scattered gravel and dirt in all directions, nearly blinding the operator. General Heintzelman ran to Nichols' relief, but, happily, no help was required.

Jesse Bunnell was on duty alone at McClellan's when, after two in the morning of the 4th of May, various telegraphic posts began sending, in rapid succession, startling messages, announcing the evacuation of Yorktown. And now the telegraph must move this great army in pursuit. Transmitting orders to one point and another, at the rate of forty words a minute to any given post, was the work of the operator; and so well was it done that before the enemy could pass Williamsburg, scarce twelve miles, he had to turn and fight. It was a glorious start.

War always produces instances of diabolism. Unhappily, this was not to prove an exception, and we have to record a cowardly example of it.

D. B. Lathrop, a bright youth from Mount Vernon, Ohio, a student, of cultivated tastes, who would walk ten miles to hear an opera or a lecture, entered the telegraph service at the Navy Yard, in June, 1861, and remained there until the Fort Corcoran office was open in August, when he took charge of that. In September and October, he was at General Smith's head-quarters office, and in November at Fort Lyon, where he stayed until December, when he took the Camp Griffin office—all in or near Washington. At this latter office, he manipulated the telegraph key until McClellan's army was ready to move on Yorktown, when he was ordered to accompany it. He and his comrade, H. L. Smith, were the first operators to enter deserted Yorktown, and, naturally enough, he hurried to the late Confederate telegraph office, to test for circuit with Richmond; but the wires were cut and left hanging from the pole, and, in going to the pole, poor Lathrop stepped on one of the many torpedoes buried thereabouts. It exploded, tearing one leg almost off, and otherwise injuring him. In distressful agony he lived but a few hours, and then passed away. The surgeon in charge tried to administer stimulants, which Lathrop persistently refused, saying he did not want to die drunk. Sadly, indeed, the operators who could, gathered around him; but mortality could not stay immortality, and the soul of that cultivated youth passed from the horrors of war, to a realization of unending peace.

The operators, to evince their appreciation of their meritorious and genial comrade, and to assuage grief as much as possible, caused his body to be sent home, escorted by two fellow operators, and erected a monument where it lies interred.

Before proceeding up the Peninsula, let us look down the Atlantic coast.

A Parenthetical View of the Coast.

August 29, Forts Hatteras and Clark, commanding the Hatteras Inlet to Pamlico Sound, were taken by a naval and land expedition, under Com. Stringham and General Butler. Then followed Com. DuPont and General T. W. Sherman's expedition against the coast defenses below Charleston, S. C., of which hereafter.

February 5, 1862, another naval and land force, commanded respectively by Com. Goldsborough and General A. E. Burnside, entered Pamlico and Croatan Sounds, capturing the defenses thereabouts and many prisoners. A few days after, Newbern, Morehead City, Beaufort, Washington, Plymouth and Elizabeth City were taken, and Norfolk itself was threatened late in April. Burnside's operations on land, in connection with McClellan's on the Peninsula, made Norfolk untenable, and accordingly, the insurgents evacuated the city, and re-inforced the army confronting McClellan. General Wool occupied the place on the 10th of May, and also the Navy Yard and Portsmouth, without resistance.

Sewell's Point and other defenses, on the Elizabeth, were now doomed, and the Virginia, *alias* Merrimac, was destroyed by her commander. This led to the easy opening of the James River, as far as Fort Darling, only a few miles from Richmond.

Major Eckert was now directed to build a telegraph from Newport News to Portsmouth, and thence to Suffolk, which city the Federals also occupied. A cable, four miles long was laid across the James to Sewell's Point, and thence this line was put up to Suffolk, *via* Norfolk and Portsmouth, a stretch from Fort Monroe of about forty miles. Wilbur F. Holloway opened the Norfolk office at the head-quarters of General Mansfield, and John E. O'Brien was sent to Suffolk. His brother, Richard, was subsequently made chief operator on the line, and also located at Norfolk. Holloway remained at Norfolk until the close of the war, doing temporary service, however, at neighboring head-quarters.

McCLELLAN'S OPERATIONS RESUMED.

The telegraph kept pace with the army. At times, it may be said almost to have preceded the advance; but there was no great obstacle thereto, except the objections which the Confederates interposed. Indeed, their own line was not altogether destroyed. Between Williamsburgh and Yorktown, it was nearly intact.

May 16, the line was broken between Yorktown and Williamsburg. Emerick and Dolan volunteered to repair it, but having no horses, they had to walk ten miles to mend the wire. The commanding officer at Williamsburg refused leave to return without an escort, as the country was infested by guerrillas, and an officer had been ambushed and killed on the road the day before; but taking their chances, they did return alone, without meeting a human being.

When Jacques reached White House Station, on the Pamunkey River, where the railroad leading from Richmond to West Point was struck, he felt an electric current from the Richmond telegraph office, but before an instrument could be attached, the the circuit was gone.

Despatch Station, on this railroad, was selected as a depot of supplies, and railroad rolling stock, brought on transports, was placed on the track at White House, the new base, May 27. Moreland and Emerick were operating at Despatch Station, sixty miles from Yorktown. Allen and Morrison, sick with fever, did the work at White House, while lying down.

The army moved slowly. Two weeks after the evacuation, it met most stubborn resistance fifty miles from Yorktown, where it was immured in the swamps and poisoned by the malaria of the Chickahominy. It is said that there were fifteen thousand soldiers in the hospitals at Fort Monroe, Newport News and Yorktown. At this time, a single cord ran from McClellan's office to Wilmington, Delaware, the atlas and axis of the North, whence nerve lines of electric touch radiated. Wilmington was in truth the nerve center. Is it remarkable that, as the news of the sick, dying and dead loved ones shocked the sensibilities of sympathetic souls, emotions akin to despair fevered and racked the brain — the North — and that when tidings came that the army lay astride a treacherous stream, in which position thirty-five thousand soldiers were nearly useless, while a great struggle progressed along the south bank, the head ached and paralysis threatened?

The battle of Fair Oaks was fought May 31. General J. E. Johnston, having discovered that the Federal army was in two parts, the suddenly swollen Chickahominy between them, sought to destroy that portion on the south bank before the other could cross, and he was nearly successful. But Sumner succeeded in crossing Sedgwick's division; Johnston was severely wounded; night set in, and the army saved.

But at what a sacrifice! After all, liberty costs more than any thing else, except slavery. Listen a moment to Prince de Joinville (Orleans) of McClellan's staff:

> Ah, I wish that all those who, careless of the past and urged on by, I know not what selfish calculations, who have encouraged this fatal slaveholder's rebellion, could have looked in person upon this fratricidal strife. I could ask, as a just punishment, that they should be condemned to gaze upon that fearful battle field, where the dead and dying were piled up by thousands. What varieties of misery! The houses were too few to contain even a small minority of the wounded. They were necessarily heaped up around the field. Though they uttered no complaints, their exposure, under the burning mid-day sun of June, soon became intolerable. Then they were to be seen gathering up what little strength was left in them and crawling about in search of a little shade. I shall never forget a rose-bush in full bloom, the perfumed flowers of which I was admiring while I talked with a friend, when he pointed out to me, under the foliage, one of these poor creatures, who had just drawn his last breath. We looked at one another in silence, our hearts filled with the most painful emotions. Sad scenes! from which the pen of the writer, like the eye of the spectator, hastens to turn away.

Had Fair Oaks been the conclusion, instead of an interlude, what agony would have been prevented. It determined nothing, and the next day came Seven Pines, when the Confederates were driven pell-mell over the ground they took the day before, and Hooker advanced within four miles of Richmond.

McClellan's army, June 14, having been re-inforced, numbered one hundred and fifteen thousand, one hundred and two present for duty. General Robert E. Lee assumed personal command of the Confederate forces after Johnston's fall, and Stonewall Jackson's troops, flushed with victory, hurried down from the Snenandoah Valley, increasing the Confederate forces to over eighty thousand; but McClellan had reason to believe them stronger, and hence determined to move to the James River, seventeen miles distant from Fair Oaks.

June 26, was fought the battle of Mechanicsville; June 27, that of Gaines' Mills, to which place Porter had retired from Mechanicsville. It was important to hold his position north of the Chickahominy until night, when he was to retire towards the James and destroy the bridges. Jesse Bunnell had been temporarily sent to Porter's head-quarters telegraph office at Mechanicsville, shortly before the battle, and on the morning of the twenty-seventh, in falling back to Gaines' Mills, having no horse, missed the main line of movement and got out of his proper course. About one P. M., Porter was being pressed very hard. At this hour Bunnell struck a road along which a telegraph line was strung; here, also, he discovered that a line of battle was being formed for a stand about one hundred yards to the rear. So he cut the telegraph line, and, connecting his instrument, sat down behind a tree and called "Mc," the head-quarters office call, which Caldwell would never change, no matter who was commander. Caldwell was prompt to reply. McClellan and the officers about him regarded the opening of telegraphic communication at that point and moment as a godsend. Bunnell having no orderlies, General McClellan sent a telegram to him to stop the first mounted officer or soldier passing that road, and order him in McClellan's name to take a message to Gen. Porter to send Bunnell fifteen mounted orderlies, and to communicate with McClellan at once by telegraph. This was done, and for several hours Bunnell sat very close to that tree and sent and received many messages as to the progress of the battle, of which Bunnell was himself receiving a very fair sample. The roar of contiguous cannon, the crack of musketry, the Federal cheers and Confederate yells, added to the bursting of unnumbered overshot shells and the zip, zip of bullets, and the solid shot crashing through the trees, were not favorable to telegraphing by ear. But Bunnell was one of the best telegraphers in the country, although yet in his teens. Several times Porter telegraphed for re-inforcements. Longstreet and A. P. Hill were not easily shaken off. At two P. M., Porter telegraphed for aid and Slocum's forces came across the river. An hour later, other troops were crossed to the rescue, for though Porter now had thirty-five thousand men he was hard pressed. At four o'clock Jackson had come to help Hill and Longstreet; then D. H. Hill and Ewell fell upon Porter. The carnage was awful. It seemed as if nothing human could withstand it; many, aye very many were killed. Bunnell's orderlies were brave fellows; several of the messages he handled were bespattered with their blood, and he was obliged to forward his telegrams to Porter by two or three messengers, as several were shot on their way. More troops were required and more answered the telegraphic call. McClellan was fighting the battle by telegraph.

NICHOLS' OFFICE.

What confidence to repose in a beardless operator, unhonored to this day.

French and Meagher came in the nick of time. Porter was being pushed back and fugitives were rushing for the bridges.

The whole line fell back and Bunnell with it, but French's and Meagher's troops rent the air with cheers as they moved against the enemy, who now recoiled, not knowing how near success they were. At eleven o'clock Bunnell reached McClellan's headquarters, and that night Porter crossed and blew up the bridges. But the dead and wounded, nine thousand Federals, perhaps as many or more Confederates; theirs is the sad story of war.

During this day, the depot at the White House was abandoned, and telegraphic communication to that place was cut off by the Confederates at eleven A. M., the next day, leaving the North in cimmerian darkness. Except when the line was broken near Williamsburg, as stated, and by General Stuart, when he made his raid to the rear of McClellan's army, June 13, communication with the North had been almost uninterrupted until after the battle of Gaines' Mills. The worst news is oftimes a relief to an over-anxious mind. Mechanicsville had foretold that a climax was approaching. Gaines' Mills was duly reported from a strictly Federal standpoint, and the Northern people hoped that the next news would hail from Richmond. But they remembered Bull Run and how McDowell's telegrams had presaged victory, and the silence that followed, owing to the suppression of news at the War Department, then the shock that came. These things were yet fresh to them, and they feared an opening of the old wound. It *was* opened.

June 28, the Federal army was *en route* for the James; twenty-ninth, the battle of Savage Station occurred; thirtieth, that of Frazier's farm; July 1, Malvern Hill; second, the army reached Harrison's Landing on the James, and there it lay until August 14.

Operator Nichols was with General Sumner during his retreat from Seven Pines to Harrison's Landing. The field line during the last day's stay, ran from McClellan's, at Savage Station, to General Smith's, at the farther end, to the right of the railroad near Seven Pines. Sumner, about Seven Pines, was nearest the enemy. On retiring, that officer directed Nichols to leave the wire intact, as Smith might wish to communicate with McClellan. When Sumner reached the railroad, he found General Thomas Wilson, Chief Commissary of the Army of the Potomac, and immense quantities of subsistence and commissary stores. Sumner wanted to talk with McClellan and Smith, but Nichols had only three feet of wire with him, and did not wish to ground the main line, thereby cutting off either McClellan or Smith. Accordingly, with General Wilson's aid, he piled up hard-tack boxes to the top of the telegraph pole, and opened an office there without breaking the circuit. This office he kept open for several hours after dark. To see to write, he had a lantern, but, unfortunately, it afforded the enemy an excellent mark, and they were not slow to open their artillery against this office. The cracker boxes sustained considerable fire, but the office was not closed until ordered by Sumner. During the battle of Mechanicsville, Nichols bravely hastened to a division head-quarters, four miles from McClellan's, where he worked an instrument just in the rear of our artillery, which was firing all the time, and in turn receiving the enemy's attentions. It was one of the most dangerous positions. Some distance behind him, others sought the protection of bales of hay. At this time Nichols was under eighteen.

The Peninsular campaign had been severe on the Telegraph Corps. One was killed at Yorktown. Several lay there sick; others with the army were barely able to hold a pencil. C. H. Buck worked at General Franklin's office when too sick to sit up. Emerick was nearly dead, but Frank Stumm would somehow find chickens and other extras, so that the poor fellow reached home, and recovered.

A telegraph builder who was up a pole when a Parrott shell passed just under him, became temporarily insane from fright, and, late that night, aroused his fellows by brandishing a saber he had somehow obtained, and with which he was stoutly beating off imaginary foes.

It is said that near Harrison's Landing, a tent, protecting a number of the telegraph party from the rain, was suddenly upset by a shell, sent in with the compliments of the Confederates.

Just as the retreat began, a darkey was sent to operator Embree with a horse and ten days' provisions. The contraband stole the horse and provisions, leaving Embree to walk. He walked all day, and, about dark, tried to *borrow* a horse that was standing alone, tied to a tree. After a short ride, a great big German officer galloped up to him, and, in thundering tones, demanded in a mixed brogue, "What for you was doing mit mine horse?" Embree had pressing business just then with the crowding soldiers, and, hurriedly dismounting, mingled among them. He walked in peace to the Landing.

Turning from the Peninsula, let us examine the operations of co-operating forces.

McDowell's, Banks' and Fremont's Operations.

Prior to his going to Yorktown, McClellan organized his forces into *corps d' armee*, commanded respectively by McDowell, Sumner, Heintzelman and Keyes; a fifth, under Banks, to include the troops in his department along the Upper Potomac, was also formed. This corps was to open and protect communication over the Baltimore & Ohio Railroad, and, in conjunction with troops to Banks' right, to guard Maryland and Pennsylvania from surprises in force down the Shenandoah Valley, and if need be, lend a helping hand for the defense of Washington, while McClellan was on the Peninsula. March 11, McClellan's authority was limited, so that he need not be annoyed pending his own great operations, and General Halleck's command was made to reach east to Fremont's department.

Fremont had risen again to the surface, and was placed over the newly created Mountain Department, the west side of which was an imaginary north and south line running through Knoxville, Tennessee, and the east side was bounded by McClellan's Department of the Potomac.

T. J. Jackson (better known, since the battle of Bull Run, as Stonewall Jackson) was in command of the Confederates in the Shenandoah, with head-quarters at Winchester. About the 1st of January, 1862, he moved against Romney; but the force there was apprised of his coming, and being inadequate for the defense of the place, evacuated it at midnight. S. G. Lynch, since brevetted lieutenant colonel in the Telegraph Corps, was the operator at Romney at this time. He can exhibit, to this day, an old musket and his telegraph table, which he brought off with him, as proof that he fell back in good order. On reaching Springfield, about half way to Green Spring Run, he connected his instrument with the line and reported progress.

Banks crossed the Potomac at Harpers Ferry, and Jackson, who was in his front, occupying Winchester with considerable force, after some opposition to Banks' progress, left there, March 11, and moved up the Valley. Telegraph lines were working to Washington from Harpers Ferry, and, under L. D. McCandless, the line was built or repaired as Banks advanced on Winchester. Indeed, by the use of a hand car, found on the railroad which connected the Ferry with Winchester, the builders, at some personal risk, succeeded in reaching the place with the advance, much to Banks' surprise and gratification. R. R. McCaine and Frank Drummond opened the office, and Winchester became a part of the telegraphic union. Shields and Williams commanded, respectively, the two divisions of Banks' force. Shields pursued Jackson to New Market, when the latter turned on him, and the battle of Winchester followed. Williams's division had started for Centreville, and had nearly reached Harpers Ferry, Banks with it, when he was telegraphed that a serious battle was in progress; but before either Banks or reinforcements could reach the place, Jackson was badly whipped. However, he soon made up for it.

Retiring a little south-east of Harrisonburg, Jackson watched from his pivotal position, Fremont on his left, Banks in front, and McDowell on the right. These forces aggregated about sixty thousand men. Forty-one thousand of them would have joined McClellan and triumphantly entered Richmond, had they not been maneuvered away by Jackson's audacity. Fremont and Banks were telegraphed to effect a junction, and, to that end, Fremont, taking with him E. O. Brown and J. B. Pierce, operators, proceeded to Franklin, *en route* for Harrisonburg. General Milroy, with the advance of Schenck's division, had not gone far after crossing the Shenandoah mountains, when Jackson, who had been advised of Fremont's movement and knew that a junction with Banks would drive him out of the valley and leave Staunton and Lynchburg tempting prizes for the Federals, now, being re-inforced by Ewell and Edward Johnson, left Ewell to watch Banks, and, falling upon Schenck's troops, especially at

McDowell, May 8, drove them back to Franklin, the terminus of the United States Military Telegraph.

Jackson retraced his steps and joined Ewell, when, with a force of fully twenty thousand, reported at Washington to exceed forty thousand, he, late in May, began his famous raid down the valley of the Shenandoah, which was to carry and did carry dire confusion to the Federals in Pennsylvania, Maryland and the District of Columbia.

The Confederate cause was now in desperate straits. Federal victories in the West, of which hereafter, and McClellan's slow, but steady approach on Richmond, made movements the result of desperation rather than of desire.

At this time, McDowell was at Fredericksburg, about forty-five miles from McClellan's right wing, and expected to move with forty-one thousand men, May 25, to join McClellan. That accomplished, and Richmond was doomed; but now that its speedy capture seemed certain, Washington itself was menaced, and Banks was retreating rapidly to Winchester before overwhelming numbers, under Jackson. If Jackson destroyed Banks' army, there was no sufficient force between him and the Federal capital to protect it. McDowell's main force was seventy-five miles south-west of Washington. Let us turn aside, and see how he came to be there.

The telegraphic correspondence that took place in May, June and July, between McClellan and Stanton and Lincoln, is one of the prominent features of the campaign. Several divisions were sent McClellan on the Peninsula. Altogether he had, during his campaign, about one hundred and sixty thousand effective troops, but not at any one time. June 28, he telegraphed Stanton: "If I save this army now, I tell you plainly that I owe no thanks to you, or any other person in Washington. You have done your best to sacrifice this army." That was while nine thousand Federals lay dead or wounded around Gaines' Mills. Those feverish words may expose the sender to criticism, but they do not militate against that efficient co-operation which McClellan vainly counted on to effect the capture of Richmond and the destruction of the Confederate army. Our "hind-sight" is often clearer than our foresight. Burnside was on Roanoke Island and thereabouts, with nineteen thousand troops. McDowell had forty-one thousand; Banks and Fremont about as many as Burnside. There was force enough, but it was not well located. Huger brought twelve thousand from Norfolk, and Branch nine thousand from Burnside's front at Newburn, to aid in defending Richmond. Nevertheless, some critics say McClellan could have taken Richmond, and, in the light of facts now well known, perhaps the historian now is wiser than the General then was. It was in response to McClellan's call for help that McDowell, eager to advance, was now within four days of McClellan, with a powerful army.

When McClellan left Washington, he intended General Wadsworth's forces in the District, numbering nineteen thousand, but understood by McClellan to exceed that number, to be left for the protection of Washington; but the President, on Wadsworth's representations, retained McDowell's corps also. And collecting forty-one thousand men as stated, McDowell was about advancing to connect with McClellan's right wing, which had been thrown northwardly to meet McDowell, when (May 24) he received the following telegraphic order from the President:

> You are instructed, laying aside for the present the movement on Richmond, to put twenty thousand men in motion at once for the Shenandoah, moving in the line, or in advance of the line, of the Manassas Gap Railroad. Your object will be to capture the force of Jackson and Ewell, either in co-operation with General Fremont, or, in case want of supplies or transportation interfered with his movements, it is believed that the force which you move will be sufficient to accomplish the object alone. Reports received this moment are that Banks is fighting Ewell, eight miles from Harpers Ferry.

To which McDowell replied: "The President's order has been received, and is in process of execution. This is a crushing blow to us." The President, responding, said: "I am highly gratified by your alacrity in obeying my orders. The change was as painful to me as it can possibly be to you. Every thing now depends upon the celerity and vigor of your movements." General McDowell then telegraphed the President as follows:

I obeyed your order immediately, for it was positive and urgent, and, perhaps, as a subordinate, there I ought to stop. But I trust I may be allowed to say something in relation to the subject, especially in view of your remark that every thing depends upon the vigor and celerity of my movements. I beg to say that co-operation between General Fremont and myself, to cut off Jackson and Ewell, is not to be counted upon, even if it is not a practical impossibility; next, that I am entirely beyond helping distance of General Banks, and no celerity or vigor will be availing so far as he is concerned; next, that, by a glance at the map, it will be seen that the line of retreat of the enemy's forces up the valley is shorter than mine to go against him. It will take a week or ten days for the force to get to the valley by the route that will give it food and forage, and by that time the enemy will have retreated. I shall gain nothing for you there, and lose much for you here. It is, therefore, not only on personal grounds that I have a heavy heart in the matter, but I feel that it throws us all back from Richmond, north. We shall have a large mass paralyzed, and shall have to repeat what we have just accomplished.

On the twenty-fifth, the President telegraphed McClellan:

If McDowell's force were now beyond our reach, we should be utterly helpless. Apprehensions of something like this, and no unwillingness to sustain you, has always been my reason for withholding McDowell's force from you. Please understand this, and do the best you can with the forces you now have.

Fremont, at Franklin, was likewise telegraphed to cross the mountains to Harrisonburg, to intercept Jackson's retreat. Manassas Junction and the railroad leading therefrom northwesterly to Strasburg, and thence up the valley to Woodstock, were in McDowell's possession.

Thomas Q. Waterhouse and Charles W. Moore were operating at the Junction about this time. The office was in a one-roomed shanty, where the operators also kept bachelor's hall. Their furniture was not artistic, but it was whole. A box, on which the instrument rested, served also as a dining-table. A single nail-keg answered for chairs. Some old broken pieces of iron, held together by telegraph wire, was called the cooking stove. A coffin, stolen somewhere, constituted the chamber set. The only objection to it, was, that the sleeper had to get out to turn over. As the operators worked alternate shifts, one blanket answered for both. Of course these appointments necessitated a servant. A young gentleman of color, named Delaware, attended to the cooking and dusted the furniture.

When Banks began falling back, Waterhouse and Richard Graham, operators, were ordered to go by special train up the road. Graham stopped at Thoroughfare Gap, and Waterhouse went on. Frank Lamb, operating at Alexandria, and Moore took the same train for the pleasure of the trip, and to open new offices. William McIntosh, who had just completed the line over this route, went also. Operator F. T. Bickford was well out on the line, but somehow joined the party. On the valley wire, Henry C. Buell, William C. Hall and C. H. Lounsberry operated at Strasburg, probably, however, one of the latter two had been relieved by the other at this particular time. Frank Drummond and Thomas Armor attended the Winchester office. These telegraphers little dreamed of the active realities in store for them. Jackson was hurrying to strike Banks' troops under Colonel Kenley first, at Front Royal, and then wherever he could hit hard. The special train ran into the enemy's advance, and probably was cut off. Anyhow, the telegraph party began playing hide and seek. Waterhouse sought to reach Rectortown, where Ed. Conway operated for General Geary. He succeeded, and, covered with mud, was sleeping on a fine bed when Conway roused him, as Geary was retiring to White Plains. The rest of the party, when within a mile of Front Royal village, heard that Kenley's nine hundred men were gallantly resisting Stonewall Jackson's forces. Kenley was no match for his antagonist, and seven hundred Federals were captured. The telegraphers were in a trap, but they still aimed at Manassas Junction. Bickford thought his chances better alone, so he took his own way. Traveling near a mile further, the rest saw Jackson's men, just ahead of them, tearing up the road and destroying the telegraph; so the telegraphers fell back toward Strasburg in hot haste, only to discover another force of cavalry which had struck the road behind them. Winchester was now their *dernier ressort;* so, striking out that way, they had pushed on some five or six miles and began to feel easier and wonder if Bickford, who had betaken himself to the fields, did not wish he was with them, when, lo! Ashby and Flournoy's cavalry and artillery came thundering down the road—the Confederates every now and then making the woods ring with their fearful yells. Lamb, McIntosh and Moore had just time to jump, unobserved, a stone fence along the roadside, where they lay exceeding close as the cavalry went tearing by.

Nothing daunted, the boys now started for the Shenandoah, wondering if Bickford had not, after all, judged wiser than they. Night set in, and with it darkness. They were strangers in a strange, inhospitable land. They knew not which way to turn, so they followed their noses. A man who takes such a guide, travels a weary way, and so they did, except when crawling on their stomachs most stealthily, to evade the enemy's pickets. In this they were aided by his camp fires. By and by, all "weary and worn, tattered and torn," they reached some unoccupied negro quarters. Fatigued and hungry, they overslept. When they awakened, three men were standing near, but the muzzles of their guns were in bold *un*-relief before the luckless trio. The boys awoke to a serious realization of misplaced confidence. They had slept on Colonel Dearmont's plantation, and the Colonel and two soldiers were behind those guns. However, the telegraphers held out for terms, and so it was stipulated that Dearmont should give them something to eat. This done, they were taken to Front Royal—to the Federals—but the Federals were Kenley's captured men. Next day they were marched to Winchester.

Bickford *had* done better. Being but one, he eluded the enemy and reached Winchester in safety. When he realized his escape, he was being congratulated in the Harpers Ferry office by operators G. T. Lawrence and J. D. Tyler.

Now let us go back again.

Henry C. Buell was operator at Strasburg. Banks was there with his force, having retired thereto when he heard Fremont's troops were retreating before the Confederates. Banks was not expecting Jackson till he heard of Kenley's being engaged, when he in vain strove to help him. He had re-inforced McDowell with Shields' division—R. R. McCaine, operator, accompanying it. The Federal force, to beat off Jackson, was not fully seven thousand. Banks resolved to retire, and sent back his trains. The cavalry struck them, but the force was small and the Union troops drove it off. The enemy increasing, the rear guard was cut off; but a part of it joined Banks by another road, at Winchester, and the rest reached the Potomac. Buell, the operator, had been cut off and captured.

This was becoming a *field* day for telegraphers. Jackson pushed hard. He was ambitious to destroy Banks' army. From Winchester to Harpers Ferry the line was intact. Drummond and Armor had been severely tasked for several days and nights, and were exhausted. Messages to and from Banks and others, from and to Washington and elsewhere, were pouring in. It was no time for rest. On the twenty-fourth, Banks telegraphed to Washington the condition of affairs. It was his messages that raised the official pulse to fever heat. Then Stanton telegraphed to the Governors, and they to their subordinates. Lincoln countermanded McDowell's orders, and directed pursuit of Jackson, which, as we have seen, proved a "crushing blow" to McDowell's hopes. The tide of telegraphic orders ebbed and flowed throughout the Federal States To illustrate nautically, it was a Nova Scotian tide. Fremont, as we have seen, was brought up from the mountain fastnesses.

While these things were progressing, Jackson was hurrying up his army. Eight hours the Nationals remained in Winchester. Three of them they rested, and five they fought. Further resistance was madness, and so Banks fell back again; his trains, having kept on, were nearly safe. Drummond, all this time, was in his office. The troops fell back through the town, Armor among the last, but making good time. Pursuit is almost always vigorous at first; so it was here. Firing was now heard in the village itself. Near an hour after the retreat had begun, Drummond sent an orderly to see if his horse was ready to mount. It was. He then sent all his orderlies to join their regiment. The rebel yell was coming in at the window. It was more ominous than the musketry that was cracking, nearer and nearer. Drummond was going. His hand was on his instrument; his dispatches were on his person. A moment more, and he would be off. That instant an orderly rushed in, out of breath, and handed him a telegram for Harpers Ferry, about reinforcements. With one hand Drummond called Harpers Ferry office, and with the other burned his dispatches.

It so happened that at that particular moment Harpers Ferry operator was busy on the Washington line, and did not hear, or hearing, did not suspect the consequences of a few seconds' delay, and no response came—nothing but that portentous yell. Drummond now started to tie his horse at the door, when he would return and send the message. Some frightened soldier had mounted the horse and disappeared; the Confederates were within a few hundred yards. Rushing back, he called "H. F." a few times, and seizing his instrument, he hurried to the street. The nearest Union soldier was fifty yards in advance, on a double quick, dropping *impedimenta*, and the Confederates very close and firing down the street. Drummond thought if he could cross the street and take one at right angles, he could escape; and running across their fire, he reached the street safely, but was in the very jaws of the Confederate cavalry, having only time to destroy his instrument on the stone pavement. Bickford had reported and volunteered to help, but Drummond, knowing his time was short, persuaded Bickford to leave. Drummond was reported by the press as killed, a soldier having stated that he saw him fall.

Banks' army and train safely reached Harpers Ferry.

Fremont took a route leading him into Strasburg from the northerly side, instead of taking the direction ordered and coming up from Harrisonburg. Shields' division, the advance of McDowell's forces, entered Front Royal, ten miles east, when Fremont struck Strasburg. When Fremont's advance went into Strasburg, Jackson's rear guard went out. Then came the battles of Woodstock, Cross Keys, Port Republic, and the escape over, and burning of the bridge spanning the Shenandoah. Pursuit ceased, and next we hear Jackson and Ewell and Johnson aiding the Hills and Johnston, at Mechanicsville and Gaines' Mills.

Experience of Captured Operators.

Drummond's experiences in a new field are told as follows:

On the 31st of May, we were ordered to report at provost marshal's at 10:30 A.M. Started on march at 12:30. Marched fifteen miles, and halted for the night in the middle of the road. Sentinels would not allow us into a field close by; we were, consequently, compelled to lie down in the mud, so deep that Lieutenant Rice, of the Fifth Connecticut, with whom I slept, could not see a particle of his blanket when we got up, and so left it there; and I can safely say he regretted leaving it for months after. It rained very heavily all night, and as we had no rations all day and night, we arose from our soft bed soaking wet, cold, hungry and very miserable generally. The next day, Sunday, we marched, at five A.M., for Strasburg. Here the officers got breakfast at hotels, for which they paid. Halted at four P.M., eleven miles from Strasburg, and prospects were held out that we would get half rations, uncooked; which eventually proved true, as we were the recipients of four crackers (hard-tack) each. *Monday, June* 2, marched fifteen miles, camped at Mount Jackson, in hospital; nothing to eat all day.

June 3.—Marched sixteen miles. Officers camped in a dirty barn. No rations; oppressively warm. After we were all lying down, terribly tired and hungry, not a sound to be heard, I sang as loudly as I could the song, "Bacon and Greens," which praises these articles of food in a very tempting manner. One verse goes:

> Oh! there's charm in this dish, rightly taken,
> That from custards and jellies an epicure weans.
> Stick your fork in the fat, wrap your greens round the bacon,
> And you'll vow there's no dish like good bacon and greens.

They let me finish, but immediately after cried, "Put him out!" "Gag him!" etc. We *were* a hungry lot. I'll never forget that night. *June* 4.—Marched seventeen miles to Harrisburg. No rations until night; first regular rations we have had. All commissioned officers paroled to report at Staunton. Telegraphers refused, becaused not commissioned. *Fifth.*—Marched twenty-one miles, over horribly muddy roads, last ten miles without any halt to rest. Feet very badly blistered. No rations. *Sixth.*—Marched twelve miles; arrived at Waynesborough 12:30, noon. No rations; feet very sore. Camped in a field near depot. Rained some during night. Bought provisions; no rations. *Seventh.*—Rained very fast all morning. Drenched and miserable. Sun came out after noon. Went to town with guard; bought shoes, towels and soap. Washed in river, put up tent, received rations and slept well. *Eighth.*—Beautiful day. No rations. Wagons packing, and appearance of moving. Charley Moore very ill. Marched at four P.M. Camped six miles from Waynesborough. Slept in clover field. No rations. *Ninth.*—Marched at ten. Rations of fat pork; no bread. Halted at North Garden Station, after marching sixteen miles. Camped in field. Baked some bread (flour and water). *Tenth.*—Got up at four. Raining heavily all day. Field very muddy. Changed our quarters, in midst of heavy rain, to a worse place. Passed a miserable night.

June 11.—Left in box cars this morning. Hustled into a box with about sixty others, like cattle. Very close and warm. Arrived at Lynchburg five P. M. Marched to fair grounds, where rations were served out. *Sixteenth.*—Buell very sick; not extra well myself. *Seventeenth.*—Very bad with diarrhœa. Only allowed half rations of water; something wrong with well. No rations of food. *Nineteenth.*—Two prisoners died last night; great many sick. No better myself; slept badly. Raining. *Twentieth.*—Very ill. All prisoners called out to be counted. *Twenty-first.*—Colonel Gibbs gave us parole of the grounds. Much better quarters, and separated from soldiers. Rigged up a sleeping place with boards, in cattle sheds. Bought seventeen dollars worth of something to eat. *Twenty-second.*—Much better in health this morning. Think we will all get well again. Best sleep of any night since left Winchester. *Twenty-seventh.*—Buell still sick. *Twenty-eighth.*—Rained very heavily. Our shanty keeps tolerably dry; quite an improvement on the field. *Thirtieth.*—Reported General Winder will be here to-morrow to parole us. *July* 1.—No General Winder. *Fourth.*—Had good dance to-night; flute music. Slept finely after the unusual exercise. *Fifth.*—Changed ten dollars, gold, for one-third Confederate, two-thirds Federal paper, fifteen dollars. *Seventh.*—Our mess all cut each other's hair, close to the scalp. Very warm. *Fourteenth.*—Moore, Lamb, Clark, Burr and I dug for two hours, this P.M., at a trench. After which, officers of guard sent escort with us to stream, where we had a splendid wash, which we enjoyed immensely, and feel much better. *Eighteenth.*—Jumping match this morning. Charley Moore beat us all. *Nineteenth.*—Exercised a little on horizontal bar. All in excellent health.

July 22*d.*—Prisoner shot before daylight this morning, for walking past sentry. Supposed by his comrades to have been walking in his sleep. *Twenty-ninth.*—Five men escaped last night. *Thirtieth.*—Colonel threatens to send us back with soldiers if we aid them to escape. *August* 1.—Another telegraph prisoner arrived to-day—M. H. Kerner. *Fourth.*—Burr very sick; fever and chills, particularly chills. *Sixth.*—Joyful news in camp to-day. The Colonel has announced that orders have come for our exchange. Charley Moore fainted twice to-day. *Seventh.*—Confederate officers tell us that we will be on our way home to-morrow or next day. More dancing. *Eighth.*—About two thousand prisoners left for home to-day. We expect to leave to-morrow. *Ninth.*—Got orders to-night to cook rations and prepare to march at six A.M., to-morrow. *Tenth.*—Marched for cars at seven. Left at 10:30. Sun was so intensely hot, could not possibly have walked one hundred yards farther than depot. *Eleventh.*—Arrived at Belle Island this A.M., at six. On the road, six of us crawled through the windows of the car to the roof, and there slept. It was wonderful we did not roll off, but we could not endure the heat and bad odor inside. Left for Richmond at one P.M. Reached Libby Prison about three; where we were informed we could not be released until we could procure an exchange for ourselves. Imagine our feelings! *Twelfth.*—Officers in next room all leaving for home to-day. Spoke to Lieutenant Selfridge through hole in floor. *Thirteenth.*—Saw all officers from Salisbury through hole in door. Spoke to Captain Betts, Lieutenant Rice and several others. *Fourteenth*—Spoke to John C. Gregg (M. T. corps, who was taken at Acquia Creek, about six months ago) through hole in door. *Seventeenth.*—Gregg left for home this morning with officers. No sign of our going yet. *Eighteenth.*—There are now one hundred and fifty prisoners in this room. *Nineteenth.*—Charley Moore and I made the raise of a sleeping cot. *Twenty-third.*—Hard bread to-night, in consequence of more prisoners arriving this P.M.

August 31.—Rations are always soup (very watery), boiled beef and bread; never change. *September* 5.—Man shot up stairs. The guard on the street are in the habit of amusing themselves by shooting at any prisoner who shows himself at the windows. A man in our room forgot himself, and was quietly looking over the James River. He approached too close to the window, and we heard a shot, and immediately after a fall up stairs. The ball had gone close to the man's head, through the wooden floor above, and killed a sergeant, who was four or five feet from the window. He was sitting on a table, leaning forward, and was struck through the heart; death was instantaneous. The man who did the shooting was arrested, as a matter of form, but was looking out for a fresh shot next day, and the Richmond papers agreed that he had done his duty, and gave him credit for it. *Twelfth.*—About forty of us volunteered to go to Belle Island and make descriptive lists of about six thousand prisoners. We got at work about three P.M., finished about eight A. M., thirteenth, having worked all night. Five thousand soldiers left to-day. Rumored we are going home in a day or two. *Fourteenth.*—About ten of us were called out this morning, quite unexpectedly; no time except to grab whatever was handy. Through a clerical error, Charley Moore's name was omitted on the list; but we did not forget him. I went to the Captain, and told him it must be an omission, and, fortunately, he was generous enough to look into it, and found the clerk had left off Charley's name. In the meantime, poor Charley thought he was deserted. Left Varina at 4:30 P.M.; anchored at dark. *Fifteenth.*—Steamed off at daylight; beautiful weather. Arrived at Fortress Monroe about twelve M., and anchored in stream. Weighed anchor six P.M. When we arrived at Annapolis and were turned loose, without guards, we hardly knew how to keep together.

On arriving at War Department, Washington, we were very heartily welcomed, although we were a hard-looking lot, and it was altogether unsafe to come too near us. We got some money, and I got permission for Tommy Armor to accompany me to procure an entire change of wardrobe. I stood in the middle of the floor and directed the purchase from a safe distance; then made for the nearest bath house, rolled all my clothes in a bundle, and threw them out of the back window, for obvious reasons, too numerous to mention. Before leaving Richmond, we were paroled for exchange. I managed to take a copy of the parole, which was as follows: "We, the undersigned, do solemnly swear and pledge our sacred word of honor, that we will not, during the existing war between the United States and the Confederate States of America, bear arms or aid and abet the enemies of said Confederates States, by information or otherwise, unless regularly exchanged or released.—Richmond, September 14, 1862."

West Virginia More Particularly.

The month of May (1862) brought misfortunes, and troubles also, to the telegraphers in West Virginia, where Fremont relieved Rosecrans of command, April 6, 1862, and planned two general movements, one of which has been merely mentioned.

Fremont directed General Cox to advance from Gauley Bridge, *via* Raleigh and Princeton, with one column, himself taking another up the New Creek Valley, intent on striking the Virginia & Tennessee Railroad at Salem, whither he was going *via* Monterey and Warm Springs. Captain David was ordered to extend the telegraph by each route, as the two columns advanced. Indeed, it was expected to build three hundred miles of telegraphs, which would make the great total of seven hundred and twenty-one miles in Fremont's department alone. Unless one has traveled in West Virginia, he can form no adequate idea of the natural difficulties the Telegraph Corps encountered in their building operations there. Add to the sinuous mountain roads, the falling trees, the rock-bottomed ways, making blasting sometimes necessary to effect holes for the setting of poles, and the numerous new obstacles constantly met—a continual menace from the guerrillas, who infested that entire region lying within the Union lines south of the Baltimore and Ohio road, and the sum total of discouragements becomes great indeed. No country ever afforded better facilities for marauding bands. It was sparsely peopled; there were no railroads for counter movements; the telegraph was the only dreaded intelligencer, and hence it was, that every time a movement was contemplated by guerrillas, a forerunner cut the line, and when the band itself struck a telegraph, it was badly destroyed. Many of the citizens conspired to effect the same object by night, and their hiding places were innumerable.

Notwithstanding these things, David reported, May 21, that he had been successful in keeping Fremont's and Cox's columns in telegraphic communication, and that the telegraph had proved equal to all emergencies. This success was the result of much risk. The men who built the lines were civilians, armed, if armed at all, with their own revolvers. Down on the Gauley, they barely escaped with their lives. The very means taken to prevent interference with the telegraph, while it was the best, only intensified the feeling of bitterness against those in charge. Says David in his May report: "I have suffered no little trouble and anxiety of mind over the operations of the infernal guerrillas. They cut our wires every night, so that it requires a continued effort to keep the connection clear. It is, however, now nearly at an end. General Fremont seizes and holds two or three of the most prominent citizens in the vicinity where the line is cut, as hostages for every inroad."

But there were other troubles. General Milroy, who succeeded General Reynolds in command at Cheat Mountain, held the gaps in that neighborhood, just as Kelley and Lander had along the Alleghanies in the neighborhood of the railroad, only Milroy did not loosen his hold until after the battle of McDowell, thirty-six miles west of Staunton, when he and Schenck, who had hurried to his aid, discovered that they could not decamp too quick. That was May 8, and this is the story of the additional troubles consequent upon Jackson's dash at Fremont's advance into the Shenandoah Valley, where Fremont, as we have seen, was to join Banks and sweep that whole region of the enemy. Former lines had been built by dropping material along the route, but it proved unsatisfactory. Transportation was forwarded to Green Spring Run, to carry all the material required, as the line progressed south. The line from Green Spring Run was thus advanced rapidly. But when Milroy sent back for help, Schenck dumped the telegraph material on the roadside, and seizing the wagons, pushed on. By-and-by he came back; Milroy came too. It was a long, tortuous and precipitous road, but there was an emergency. East of the Shenandoah Mountains, Milroy's advance was checked and put to flight as far as McDowell, where Schenck and Milroy in vain attempted to stay the progress of Jackson, after which they could not give him the right of way too quickly, and agreed that if he wanted those barren mountains they would confess ejectment and suffer disseizin. "Haste makes waste." Jackson was enabled to convert to his own use David's building material, lying by the roadside, consisting of eight tons of telegraph wire and three hundred insulators; also to destroy three miles of line, leading south from Franklin. But Jackson himself soon got in a hurry in the great valley, as we have explained, when he dropped this material, to the great joy of David, who repossessed it. George K. Smith, operating at Moorefield, was captured by Colonel Harness, but owing to the proximity of Union re-inforcements, he was paroled after starting for Richmond.

We left Fuller assistant manager with General Rosecrans at

Camp Tompkins, near Gauley Bridge, and have noticed his principal operations. In December, 1861, Mr. Fuller was sent to Kentucky, where, as we shall see, he performed good service. On Fuller's leaving, David took entire charge of the telegraphs in West Virginia. W. H. Nash was his chief assistant in the neighborhood of the Gauley, and C. H. Johns was operator at General Cox's head-quarters. Cox became so attached to Johns that when the general went into winter quarters at Marietta, Ohio, in December, 1862, he insisted on Johns accompanying him for cipher duty only, but the necessities of the service prevented it.

When Fremont came, Rosecrans' head-quarters were at Wheeling. Military operations on the Gauley were quiet until the spring of 1862, but all the lines erected were maintained and operated. When, under orders from Fremont, Cox advanced to the Big Sewell, the telegraph followed him. But one fight of consequence occurred, and that was at Lewisburg, when General Heth attacked Federal General Crook, of Cox's command. Heth ran off before Crook had fairly begun to fight. This occurred the same day that Jackson's forces captured Colonel Kenley at Front Royal.

The authorities ordered Cox's troops east, which necessitated his falling back, although the telegraph was constructed to Princeton *via* Fayette and Raleigh. Milroy was also ordered to join the army in front of Washington, and Fremont's extensive plans of operations against the Confederates entirely failed. The misfortunes on the Peninsula were felt every where, and one of the consequences thereof was the abandonment of much of the country and telegraph south of the Baltimore & Ohio Railroad. Thus the line from Green Spring Run to Franklin, eighty-seven miles, Grafton to Phillippi, fifteen miles, Point Pleasant to Raleigh, two hundred miles, Piedmont to Lona Coning, seven miles, and Rowelsburg to Cheat Mountain, aggregating three hundred and seventeen miles, were reported in October (1862) semi-annual report as abandoned. When David was commissioned captain and assistant quarter-master (in July), he was ordered to St. Louis to examine and report on telegraphic operations in Missouri, whither he went, leaving Charles O. Rowe in temporary charge of the department.

The following named operators served in West Virginia a part or all of the first six months of 1862: R. M. Shurr, W. F. Allen, J. B. Pierce, Charles O. Rowe, C. C. Starling, G. Townsend, Alf. Winder, F. N. Benson, G. H. Curtiss, M. H. Kerner, J. S. Keith, Isaiah D. Maize, W. H. Nash, C. Wolf, P. A. Stidham, Charles J. Thomas, R. A. Furr, M. Gordon, E. Rosewater, George D. Sheldon, L. B. Dennis, W. T. Lindley, Thomas M. Sampson, James W. Vermillion, W. H. H. Lancaster, C. D. Tull, G. K. Smith, N. De Bree and F. M. Ingram.

CHAPTER VII.

THE TELEGRAPH IN THE SOUTH-WEST.—FORT DONELSON.—PEA RIDGE.—ISLAND NO. 10.—MEMPHIS.

The great Middle and Western States, while responding generously to calls for troops for Virginia, patriotically fitted out great armies to beat back the Confederate hosts, long gathering in the South-west and border slave States. The South was bent on compelling Missouri and Kentucky to enter the Confederacy, and if unable to carry the war into the border free States, aimed at least to force it upon the territory of the lukewarm slave ones. Zollicoffer's army, near Barboursville, in South-west Kentucky, Buckner's, on the line of the Louisville & Nashville Railroad, near Bowling Green, and General Polk's at Columbus, Kentucky, on the Mississippi, constituted the principal rallying forces of the Confederates in Kentucky, when we last noticed them. Confronting these were, as we have seen, Buell's forces, under Thomas at Camp Dick Robinson, and under Mitchell at Muldraugh's Hill, besides Grant's troops at Paducah and Cairo. The Cumberland was sealed against the Union by the batteries and works at Dover, called Fort Donelson, just over the Kentucky line, in Tennessee, and the Tennessee by Fort Henry, just across the peninsula from Fort Donelson. Thus, were the great artificial and natural highways leading south, entirely blockaded, and the problem was first, where, and second, how, to attack and clear away those obstructions.

In Missouri, we last saw the main Union army, under Hunter, making an ignominious retrograde movement from Springfield. His appointment was probably a temporary one, as he was relieved a few days thereafter (Nov. 18, 1861), by General H. W. Halleck, and given a department further west, which included Kansas.

Leaving operations under Buell for the succeeding chapter, we will now examine those in Halleck's department. The enemy, under General Price, possessing the territory abandoned by Hunter, west of Rolla and as far north as the Osage River, gave Hunter's movement every outward appearance of a retreat. Recruiting for the rebel army was active in nearly all parts of the State, but carried on stealthily east of Rolla, in the south-west, where Sigel was in command; of Sedalia, in the center, where Pope was sent by Halleck, and in Prentiss' district, late Pope's, *i. e.*, North Missouri. This resulted in numerous semi-organized bands, some of which joined Price, while others continued in their maraudings until destroyed or dispersed by the Federals. The affairs at Millford, December 18, 1861, where Pope captured over thirteen hundred prisoners; Mount Zion, in Prentiss' district, December 28, where five or six hundred Confederates were dispersed, and Silver Creek, January 8, where Major Torrence badly punished and dissipated another rebel force, were some of the fruits of Halleck's administration. But such operations, while they necessarily occur in all great wars, exert but little influence on the final result.

It will, however, from this statement, be readily understood that the telegraph lines to Sedalia, Rolla, Ironton, the newly constructed line from Cape Girardeau to New Madrid, and those in North Missouri, were special objects for destruction, and consequently were cut repeatedly. The system adopted by George H. Smith, of having mounted repairers at offices, and at near intervals, with directions to ride their circuits daily, resulted generally in the re-establishment of communication in a few hours at most, after the wires were severed. This kind of "riding the circuit" was very hazardous, and some of the repairers lost their lives—being shot by concealed bushwhackers. We will, at the proper time, mention instances where it occurred. Suffice it now to state a well-founded belief, grounded upon experience and observation, that the man who always responds promptly to a call to repair a broken line in a guerrilla-infested region, without escort, evinces a degree of courage rarely required in actual battle. There were many such men engaged in the repair service of the U. S. Military Telegraph Corps. They were found in every department. Usually illiterate, they were none the less sensible of their dangers; nor were they tardy about taking their desperate chances. If one was killed, another took his place, and, being a mere civilian, no notice was taken of his fate by the Government, in whose service he died. No provision was ever made for his wife and little ones; no slab ever erected at Government expense; no military salute was fired over his grave. If the corps in his district, from motives of sympathy, made up a purse out of their hard earnings, and mention of the fate was made by the officers in their annual reports, that was the most that could be expected. But what became of those reports? Those for 1864 and 1865 were printed along with the Quarter-master General's; but none other ever saw the light of day, and astute historians have even overlooked those that were printed. Indeed, some who pretended to write up the history of the civil war, seem to have been strangely ignorant of the fact that there ever was a Military Telegraph Corps; much less did they ever hear of the repairer who dared and died for his country.

On the North Missouri Railroad, the line had been completed by Smith's party to Hudson, where it tapped the Hannibal & St. Joseph, when General Hurlbut raided from Macon through the north part of the State, and, returning, struck the Hannibal & St. Joseph at Shelbina, where J. B. Clarke was operating, and where trains were to meet the General, and convey his force to Macon. The telegraph being down, Clarke proposed to take an engine to Honeywell, while the troops were loading, to see where the line was broken, and if Salt River bridge was intact. After crossing the bridge, he saw a few horsemen near it; but at Honeywell, he found the road destroyed, and about two hundred rebel cavalry about a mile off, coming up at a gallop. Rushing into the telegraph office, Clarke snatched the instruments from the wires, and made good his escape on the engine. But the cavalry overhauled Mr. Dunning, the operator there, and made him hold the commander's horse, while the office was searched for the instruments. Great was the officer's chagrin when Dunning convinced him that Clarke had saved them.

Halleck's department, we have seen, did not at this time include Buell's army. There were two great ends in view—the capture of Nashville, and the opening of the Mississippi River. Admiral Farragut had captured New Orleans (April 25, 1862), and General Butler garrisoned it. Halleck's great aim was to sever the Confederacy, and co-operate with the fleet and army, then about New Orleans. A direct attack, though never abandoned, was not prosecuted with the vigor of the main operation. The Cumberland and Tennessee Rivers were not so stoutly held as the Mississippi, and it was up these that General Halleck, and Commodore A. H. Foote in command of the river gun-boats, built under his direction about Cairo, agreed first to operate. The fleet of gun-boats and transports ascended the Tennessee River, February 4, 1862, and landed the troops within four miles of Fort Henry; but it was not deemed practicable to build the telegraph beyond Smithland, at the mouth of the Cumberland, until about the time Fort Donelson was taken.

In December, 1861, Solomon Palmer was employed by Colonel Wilson manager of the telegraph in Halleck's department east of the Mississippi, to take charge of the construction party. If Wilson had never done any thing else during his management, the selection of Palmer would have gone a long way to make it a success; but the next month, he took John C. VanDuzer from his own telegraph company's employ in North Missouri, and made him chief operator of the military lines in his district. These two men could build and operate a military telegraph, if it was in the power of man. Their subsequent career, running through the war, attests Wilson's foresight and their own rare merits for the positions that fell to them.

February 6, Fort Henry was captured by Admiral Foote, so much quicker than General Grant, in command of the land forces, considered possible, that he did not make timely preparations to intercept the retreat of the Confederates, most of whom reinforced those at Fort Donelson. Federal gunboats then went up the Tennessee to Florence, Alabama, nearly two hundred miles, made some captures and returned. General Grant was not long in moving over twenty thousand troops across the peninsula, where he besieged Fort Donelson. Then came a force of ten thousand more up the Cumberland, and Foote's gunboats, and then the battle. Twenty thousand brave men behind works, against thirty thousand in the field. The chances of each would seem about equal. As to the battle, it was simply awful. It is not within the scope of this work to describe battles, except so far as may be necessary to demonstrate the service of the telegraph. At Donelson, there was no telegraph. The field, which finally stretched from river bank to river bank, was miles long. At a perilous moment, General Grant was not even on that field, but had gone to consult with poor, wounded Foote, on his disabled gunboat. There had been ample time for the erection of field telegraphs, but war was new in America, and the telegraph newer yet, and so regiments fought, hour after hour, against overwhelming numbers, and were finally driven, in some places, to where there were troops enough to beat back the enemy in their turn; a conflict of arms that swayed to and fro, hour after hour, the dead and dying fringing the advance points of contact, and thus marking, by a real dead line, where the tide ebbed, or how far it flowed. That was a tragedy of too many parts.

Before the war closed, I passed many times over this fatal field, and in the woods, almost hidden by growing vegetation, might be seen — not graves, for that imports a better burial — but mounds, indicating where loved sons and fathers had been laid in heaps and covered over. Oh, the sorrow that was spared the wife or mother, who little imagined that such a burial was the usual lot of the heroes who thus died for their country!

The investment fairly began on the 13th of February; then came the strife of the fourteenth and fifteenth; then Floyd and Pillow and Forrest fled by night taking with them many troops, and leaving Buckner in command of over ten thousand more, beside the wounded, not to fight, but to surrender; which he did on the morning of the sixteenth.

At this point, historically speaking, criticisms are in order. It is marvelous how many brilliant generals have been lost to the world in the making up of ready critics. Truly the pen is mightier than the sword. We, too, would be tempted (not irresistibly, however), to explain what Grant should have done, if we had sufficiently described the battle to lay the foundation for censorious remarks. We have played many games of chess, and when our opponent carried off our queen, we saw our error. Doubtless, if Grant were to fight that battle again, he would do better. So would the Confederates act differently. They would evacuate before they were invested.

At this time, there were two parties at work building telegraph toward Donelson; Smith's, from Smithland, and Wilson's, from Fort Henry. The territory was clearly in Manager Wilson's district, but Halleck, who had yet to learn the real status of the military telegraph, ignoring Stager, ordered Smith to aid in the construction of this line. Accordingly, Smith, with W. S. Hewitt, formerly captain of Company B, of the Fremont Telegraph Battalion, and Frank S. Van Valkenburg, as chief assistants, began at Smithland to construct the line to Fort Henry; and Wilson's party, under Van Duzer and Palmer, commenced as stated. When the line was finished to Fort Henry, or nearly so, Stager ordered Smith west of the Mississippi, and on reaching St. Louis, February 25, he was ordered back by Halleck.

Of course, this terminated unpleasantly to some one. Captain Stager then telegraphed General Halleck his authority for controlling the military telegraphs, also the territory assigned to Wilson and Smith, respectively; adding, that if either failed to meet Halleck's wishes, they should be reported. Halleck replied that the Paducah line had not been kept up; that "there must be one good head of the telegraph lines in this department, not two, and that head must be under my immediate control."

Right here it may well be stated that the line from Cairo to Paducah was built on the Illinois side, near the river; that while Grant was about operating against Fort Henry, an extraordinary storm inundated the country, raising the river to an almost unprecedented height: so high, indeed, that Foote's gunboats could pass in safety over the rebel torpedoes, sunk in the Tennessee; so high that the Paducah telegraph line was submerged for miles. Operator Tiffany and a repairer nearly lost their lives, trying to repair it. *Débris*, and breaking and falling trees, nearly destroyed the line. But great movements were under way, and Halleck, at St. Louis, was impatient. He was in communication direct with Cairo, *via* Odin, and wanted Paducah exceedingly, at any cost.

The correspondence continued; Stager insisting that the field was too large for one to do justice to, and urging other potent reasons, and explaining what Halleck must have known—that the river had drowned the telegraph. But Halleck insisted on having one head, and having indicated a preference for Smith, he was appointed by Stager, although Smith telegraphed Captain Stager that he did not want that territory. The Secretary of War fully coincided with Captain Stager, but, "on the whole," it was not considered best to continue to oppose Halleck, and so the matter ended. Wilson's retirement proved permanent, and Smith was ordered by General Halleck to "crowd the lines to Nashville, Tennessee, Florence, Alabama, and Memphis, Tennessee." Smith's territory, being co-extensive with Halleck's, was the largest field embodied in any of the departments.

Before we take a final leave of Colonel Wilson, let us note his description. It will be found below, in what he jocosely calls his "death warrant." Such writs were quite usual then.

OFFICE OF PROVOST MARSHAL, ST. LOUIS, MO., Oct. 5, 1861.

Permission is granted to J. J. S. Wilson to pass beyond the limits of the city and county of St. Louis, to go to Springfield.

(Signed) J. MCKINSTRY,
Major U. S. A., Provost Marshal.

On the back of which is:

DESCRIPTION OF PERSON.—Name, J. J. S. Wilson; age, thirty-two; height, five feet, nine; color of eyes, blue; color of hair, light; peculiarities, good drinker. It is understood that the within named and subscriber accepts this pass on his word of honor, that he is, and will be ever, loyal to the United States, and if hereafter found in arms against the Union, or in any way aiding her enemies, the penalty will be death. (Signed) J. J. S. WILSON

February 13, General Nelson left Camp Wickliffe, near New Haven, Ky., with two brigades of Buell's army, to reinforce Grant at Fort Donelson; but failing to reach the fort in time, he proceeded up the river to Nashville, which had fallen. About the time that Nelson passed Smithland going up the river, General Sherman was at Paducah, and exceedingly anxious about his movements. The Colonel commanding at Smithland, like many new officers, had *abundante cautela* on the brain, and imagining that there might be a rebel operator on the line between Paducah and Smithland, very stupidly placed his operator under the guard of two soldiers, to prevent his *talking* over the wire. Of course he could pretend to be sending a stale message as long as he chose, and yet be only talking. The wood-tick the soldiers had learned, or would, but the telegraph tick was a hopeless jargon. Parker, the operator at Paducah, was duly informed of the state of affairs, and thereupon he prepared an order for Colonel Dave Stewart, the ranking officer, to sign, releasing the Smithland operator. Parker read the order aloud, and as Stewart was about signing it, Sherman, who chanced to be by, said, "What's that? What's that?" and looking at it, he said, "No, the Colonel is right. Can't relieve the guard."

By and by, Sherman went to the office, when the following telegraphic conversation occurred: Parker to Smithland—"Has the steamer Tarascon passed yet?" Smithland—"I'm under arrest and can't answer you." Parker—"General Sherman wants to know." Smithland—"I don't care if it's Halleck himself. I'm under arrest, and can't talk about such things." Parker—"Sherman asks. Do you know?" Smithland—"Of course, I know." Parker—"Sherman wants to know if General Nelson has passed up." Smithland—"Tell Sherman to send a message then." Parker—"Do you know?" Smithland—"Certainly, I do." Parker—"How far is the colonel's head-quarters?" Smithland—"A mile and a half." Sherman to Parker—"What kind of a fellow is that operator?" Parker—"He seems to be good at obeying orders. It was a foolish thing for the colonel to put soldiers over an operator to keep him from talking by telegraph." Sherman—"Well, that's so. I didn't think of that." And thereupon Sherman himself wrote the order releasing the operator, and got at once the information that Nelson had passed up.

Major Smith extended the line from Fort Donelson to Clarksville, *via* the Tennessee Ridge, thirty-nine miles, in March, and from Clarksville communication was opened to Nashville direct, and also *via* Bowling Green, Ky. The first operators on this line were, at Smithland, Abe D. Dougherty; at Chaudits, a city of two log houses, owned and inhabited by a Frenchman and his increasing family and many other extremely domestic animals—a city bounded on all sides by a howling wilderness, with an excellent landing a few rods off, where guerrillas were wont

12

to cross the Tennessee at night—the operator to take to the woods a minute later, when Chaudits' dogs uttered their war-whoop; here, Robert B. Griffin was first stationed. In a month, Peter Fowler relieved him ; but Peter endured it only a month, when J. R. Thompson came, and he, for a month only ; and so it went. Fowler and Thompson, brave fellows that they were, are now numbered with the dead. What a constitution Griffin must have had, to survive the mosquitoes, gallinippers and wood-ticks, the guerrilla maraudings, the melancholy wilderness and the horrible *ennui* of this Frenchman's retreat. At Fungo, say twenty-five miles farther and fifty from Smithland, Thompson operated before going to Chaudits', and when he quit Fungo (how that name rolls like a sweet morsel over one's tongue) there was none so good as to do it reverence, until March, 1863, when R. H. Bliven created a great flutter among the only two damsels in the place, by opening an office there. These offices were purely test stations, to locate breaks in the line, and thereby speedily remedy the mischief. At Fort Henry, Edward Schermerhorn and Alonzo D. Griffin operated; at Fort Donelson, Charles W. Hammond, until relieved in May by J. T. Tiffany ; at Clarkesville, J. W. Purnell, in June, and J. G. Webb, in August. Tiffany became chief operator of this line, from Donelson to Smithland, distance one hundred miles, and as such, it was his duty to travel along his line occasionally, and direct repairs. Once, while out with a repairer named Charles Byers, he was fired upon from the bushes, the bullet passing through the lappel of his overcoat. They at once fired into the brush, and saw traces of blood, but the bushwhacker escaped.

Grant's army moved up the Tennessee River, to Pittsburg Landing, about thirty miles from Corinth, Miss., where we will leave it for a while, and turn our attention again to affairs in Missouri.

We left Price at Osceola, on the Osage, with eight thousand men about him, and more scattered in various portions of the State, recruiting and foraging, if not actually pillaging and devastating. The Hannibal & St. Joseph Railroad suffered greatly, but in the numerous encounters that took place, owing generally to the Federals being in greater numbers, or better armed and disciplined, these rebel detachments were greatly cut up. The Federals, concentrating at Rolla, under General S. R. Curtis' immediate command, moved, February 11, through Lebanon, against Price ; but he was no match for Curtis, and retired to the Boston Mountains in the north-west corner of Arkansas, where McCulloch's forces joined Price's. General VanDorn had assumed command, owing to the want of cordial concert of action between Price and McCulloch, and their united strength was probably twenty thousand men.

Under Curtis were Colonels Osterhaus', Jeff. C. Davis' and E. A. Carr's commands, and also General Asboth's, aggregating fully ten thousand men. Curtis pushed on a few miles further into Arkansas, until, fearing the increasing force of VanDorn, he commenced retiring. But the enemy, by extraordinary diligence, got in his rear in force (March 6, 1862), and Curtis had either to fight or surrender, which latter, he and his subalterns and soldiers were not well calculated to do.

While he is making dispositions to receive VanDorn at Pea Ridge, let us see what Major Smith had accomplished in this direction with the telegraph. Placing Duncan T. Bacon (who had been on the North Missouri road with General Schofield) in charge of the builders (under H. C. Weller) and operators, south of Rolla, Smith directed him to build the line as far as the army moved. Before it reached Lebanon, an office was open there, where —— Harrison operated. Luke O'Reilly relieved W. H. Woodring, at Rolla, January 31, and the latter took the Lebanon office, where he remained until March 1, when being again relieved by O'Reilly, he proceeded, unguarded, *via* Waynesville, where O. A. A. Gardner was operating, twenty-two miles, to Springfield, through a storm so cold as to compel him to halt, and build fires to keep from freezing. Bacon pushed the line on from Springfield toward Bentonville, two hundred and fifty miles from Rolla, and was within ten miles of the battle-field when the conflict commenced.

The terrible three days' struggle between the contending forces began on the 6th of March ; the heroism displayed on that memorable field was never excelled in battle. Bacon was present, with Curtis, rendering such service as lay in his power; but there was no telegraphic communication with the army, and no field wire during the battle, in which the Federals lost one thousand, three hundred and fifty-one men killed or wounded. The telegraph soon spread the news of victory. Having followed Van Dorn some distance, Curtis retired to Keytesville, and then to Cassville.

Not long after, Van Dorn and Price, with the most of their forces, joined Beauregard, at Corinth, but not in time for the battle of Shiloh. Curtis determined to take advantage of their absence, and again entered Arkansas, with seven or eight thousand men; but as the guerrillas would jeopardize his communications unless he strung out his force so as to become ineffective at the head, he wisely determined to establish a new base, and, accordingly, marched eastward to Batesville, Arkansas, reaching there May 6. At this place, on the White River, not far from the junction with the Big Black, or some other point farther down, Curtis hoped to supply his army by water, and have a shorter route to his objective, Little Rock, the State capital; but all depended upon the opening of the Mississippi to the mouth of White River — in short, upon the issue at Corinth, where the South-west was concentrating.

In furtherance of this plan, the telegraph superintendent in Halleck's department was ordered to construct a line from Pilot Knob, Missouri, to Batesville, Arkansas, two hundred and one miles. This line was completed by way of Greenville, Missouri, Pitmans Ferry and Pocahontas, Arkansas, early in June, and was operated at Batesville by Luke O'Reilly and H. B. Kunkle; at Pocahontas, by Jno. H. Byrne and James L. Quate; at Pitmans Ferry, by Charles Payne and Geo. A. Purdy; at Greenville, by Geo. J. Talmadge and P. B. Frazier, and at Pilot Knob, by Theodore Holt and assistant. All of these offices, except those at Curtis's head-quarters and Pilot Knob, were located in sparsely settled regions, and continually beset by guerrillas, rendering the operators' and repairers' positions exceedingly perilous. On the 30th of June, a party of guerrillas entered Pitmans Ferry and captured Purdy and the telegraph property, carrying Purdy off with them. James L. Quate, at Pocahontas, was equally unfortunate.

It was Curtis's expectation to meet assistance on the White River and march across the country to Little Rock, in which event, the telegraph was to be extended to that city. In less than a month after camping at Batesville, Corinth having fallen, Curtis took Luke O'Reilly as cipher operator, and following the river, advanced to Clarendon, where he had reason to expect to meet Federal gunboats, under Commander Fitch, and troops known to have started up the White. But on reaching Clarendon, he learned of the return of the expedition the day previous. Fatigued and disheartened, knowing scarce whither to turn, in a country quite uninviting and unproductive, and harassed by well-mounted parties, he determined to make his way to Helena, on the Mississippi, which place he reached, after routing fifteen hundred Texan horse, under Colonel Rust; and so Little Rock was saved to the Confederates, and Southern and South-western Missouri again opened to the forays of predatory banditti, who were not slow in harassing the country. Thus Purdy and Quate were captured within a week after Curtis's departure. The immediate abandonment of this line, south of Greenville, became necessary

---

Owing to the successes of the Federals on the Tennessee and Cumberland rivers, Columbus, Kentucky, which General Polk had strongly fortified, was rendered untenable, and its garrison greatly reduced. Early in March, Commodore Foote, with six gunboats, and General W. T. Sherman, with several thousand troops on transports, moved down the river to capture the place, but found a Federal cavalry force from Paducah already in possession, the enemy having fled.

It is related that Confederate General Polk having, by strict orders, excluded intoxicating liquors from Columbus, the operators there arranged with Emmet Howard, of the Hickman office, to send them, occasionally, a demijohn of Woodford County's best, labelled "Nitric acid," which label was an excellent passport. This ruse, however, at last failed, and Howard was delegated to see the commandant of the post, Colonel Wickliffe, and represent the want of nitric acid for their Grove battery, and ask an order for three gallons of the best Bourbon in the hospital, to use in lieu of acid. Wickliffe and Howard were quite intimate. The latter prepared the order, and hurriedly stated the necessities of the service; but whether Wickliffe feigned ignorance or not, it mattered little, as the order was signed. Wickliffe's acid was a standing joke for some time.

After the evacuation of Columbus, Hickman was virtually held by couriers, who, in case the telegraph was cut, were to notify Polk, at Jackson, by telegraph from Union City, of the passage of gunboats for Island Number Ten. One foggy morning, the gunboats appeared at the wharf near the office, so unexpectedly that Howard had barely time to seize his instrument and with a leap, as for dear life, reach the pavement, where a courier, with an extra horse, was posted. Riding hastily to the depot, they sprang upon a hand car, well manned, and proceeded rapidly for two miles, when, in a deep cut on a curve, the party was nearly paralyzed with terror at sight of an engine, rushing upon them. They jumped from the car, landing in all directions and positions on either side, as the engine demolished the car and threw off the wreck. Mounting the engine, they proceeded for Union City, until near a point within easy range of the gunboats, should any be on the river opposite, where, out of caution, they stopped to reconnoitre. While so doing, they met William McCleskey, operator, who was also fleeing. On the engine were Merritt Harris, C. S., operator, superintendent Culverhouse, and Conductor Gardner. McCleskey reported five thousand cavalry on the highway, but he was too nervous to be good authority. Harris and Howard, however, at some distance from the engine, discovered the Federals, and Howard, by his masterly effort to reach the engine, was too exhausted to climb up unaided. As he sank on the floor of the tender, a volley from the Federals was heard. The engineer opened the throttle, but the locomotive stirred not; capture seemed inevitable; the driving rods were on a dead center, and the machine would not move; but, from courage or desperation, or both, the engineer sprang to the ground, and with his crowbar moved the driving wheel, when the engine, trembling like a thing of life, distanced the pursuers as if itself was fleeing from fiends incarnate. It is said that seven miles were traveled in five minutes.

Bridges were burned, and no more Confederate trains passed north of Union City; but Howard was ordered back to within half a mile of Hickman, where, with a cross-tie for an office, he maintained communication with General Polk several days, when he was relieved by operators Harris and Johnston. Harris, subsequently, barely escaped, and Johnston was captured while concealed in the house of a friend, but made his escape the same night, and, after getting lost in the swamps, reached Humboldt, Tenn., exhausted. Harris, a brave, noble and generous fellow, died in Texas during the war, beyond the reach of friends, and in a strange land.

At the same time that Curtis was wending his way toward the Ozark Mountains, in North-west Arkansas, Grant's troops, under C. F. Smith, embarking for Pittsburg Landing, Mitchell's about moving on Huntsville, Ala., Buell's on Savannah, Tenn., and Com. Foote and Colonel Ellet, at Cairo, were refitting gun-boats and building rams, General John Pope was busy collecting forty thousand troops, with which to overcome the Confederates, who, at the strongholds of New Madrid, Mo., and Island No. Ten blockaded the Mississippi River and defended Memphis.

In anticipation of Pope's operations in this direction, Major Smith built a telegraph line from Pilot Knob, through Fredericktown to Cape Girardeau; thence to Commerce, Mo., and, as Pope advanced therefrom, it was extended by way of Benton and Sikeston, under the immediate direction of F. S. VanValkenburg, to New Madrid, one hundred and fifty miles by line from Pilot Knob. A cable was laid from Cairo to Birds Point, by Smith, in March, which was connected with a line along the railroad by way of Charleston to Sikeston, where, by means of a Hicks repeater, it became a virtual part of the main stem. This gave Halleck, at St. Louis, two circuits to Sikeston, one by way of Pilot Knob, and the other through Illinois.

Pope invested New Madrid on the 3d of March, and pushed his operations with remarkable energy and boldness. In thirty-six hours after telegraphing Cairo for siege guns, they were hurling their ponderous missiles against the enemy's works. Aided efficiently by the gun-boats, New Madrid and No. Ten were surrendered, as indicated by the following papers, the last of which is believed to be in the hand-writing of Thomas A. Scott, Assistant Secretary of War, then with General Pope:

HEAD-QUARTERS, DISTRICT MISSISSIPPI.
New Madrid, April 10, 1862.

GENERAL ORDER, No. 30.

The following dispatch from Major General Halleck, commanding this department, has been received, and, with this order, will be published at the head of every regiment and detachment of this command.

ST. LOUIS, April 8, 1862.

MAJOR GENERAL POPE :

I congratulate you and your command on your splendid achievement. It excels in boldness and brilliancy all other operations of the war. It will be memorable in military history, and will be admired by future generations. You deserve well of your country.

H. W. HALLECK, *Maj. Gen. Comdg.*

The General Commanding has little to add to this dispatch. The conduct of the troops* was splendid throughout. To such an army nothing is impossible, and the General Commanding hopes yet to give them an opportunity to win the glory which they are so capable of achieving. The regiments and battalions of this command will inscribe on their flags "*New Madrid*" and *Island Ten.*"

JOHN POPE, *Maj. Gen. Comdg.*

FOR THE ASSOCIATED PRESS AGENT, Cairo :

NEW MADRID, April 11, 1862.

The following general order was read at the head of every command in General Pope's army, last evening, and created the wildest enthusiasm. All are ready to stand by their commander in any enterprise that may be necessary for the good cause. It is a noble army, and, if they can have an opportunity, their action will be all that could be desired by the most ardent patriot.

(Operator here insert Order No. 30.)

Nearly all the troops that crossed the Mississippi have returned to New Madrid. The number of prisoners will exceed the statement of General Pope. They continue to come into camp in lots of five to fifty, and the whole number will probably exceed seven thousand —over six thousand five hundred being now enrolled. It is, therefore, certain that but few, if any of General McCall's army escaped. Yesterday, two splendid batteries of light artillery (rifled guns) were found in the woods, and a great amount of valuable property

*First five words inserted by the author, as beginning of the sentence is destroyed in the original.

is being found by our troops, consisting of horses, mules, wagons, arms, etc., etc. This movement has certainly been the most complete and successful one of the war—every thing captured, and not a man lost by our forces.

Forts Pillow and Randolph, between Memphis and Island No. Ten, were soon after evacuated, and, June 6, Memphis surrendered to the navy.

The operators on these south-east Missouri lines, pending Pope's operations, or a part of them were, at Pilot Knob, James H. Douglass and Theo. Holt; Fredericktown, Z. P. Hotchkiss; Cape Girardeau, J. R. Dunlap; Commerce, W. W. Livergood (April); Benton, Simon T. Yonkers; Sikeston, Jno. J. Egan, J. R. Thompson; Pope's Head-quarters, F. S. VanValkenburg, Wm. Spinner and A. S. Hawkins; Birds Point, Livergood (March) and Thompson (April); Cairo, W. H. Bromell, G. A. Burnett, G. Burnapp chief operator, H. W. Nichols, James K. Parsons and Geo. Stillman.

That part of the line connecting Pilot Knob with Sikeston was through a region much infested by guerrillas, and the operators in some of the places received little or no protection. August 16, S. D. Howard, operator at Charleston, reported at Cairo office, that at eleven o'clock the night previous, he was surprised in his office by a party of forty guerrillas, who wrenched out his instruments, demolished his office and came near capturing him; but he effected his escape through a back window, uninjured, although he was fired at and badly frightened.

In anticipation of the capture of Memphis, Duncan T. Bacon was sent there, to aid in opening communication in West Tennessee, and reached the city the first evening of its occupation. While Fitch's gunboats were going up the White River, to meet Curtis, who had left Batesville as we have seen, messages reporting progress were frequently brought from them to Bacon, by couriers, for transmission in cipher to Halleck, who was then at Corinth—the lines having been repaired to that place, as will soon appear. General Hovey, at Memphis, claimed the right to copies of these overland messages; but Bacon refused them, and was ordered under arrest. The New York *Tribune* correspondent, A. D. Richardson, then in that city, wrote up the affair, thus:

On Wednesday, General A. P. Hovey, commanding this post, ordered D. T. Bacon, Esq., Manager of the Military Telegraph, to furnish him copies of official dispatches from General Halleck to General Curtis and Colonel Fitch. General Hovey's motives were, doubtless, good—to keep himself advised of every thing affecting his post, directly or indirectly; but Mr. Bacon knew his duties too well to commit so gross a breach of official duty required of him. He respectfully declined to obey the order. General Hovey directed his arrest, but Mr. Bacon was absent from the office when the orderly arrived, and one of his associates, Mr. VanValkenburg, was arrested instead. The case was laid before General Halleck, by Mr. Bacon, when "Old Brains" promptly replied: "Release the operator at once. He did perfectly right in disobeying your orders."

CHAPTER VIII.

THE TELEGRAPH IN KENTUCKY, TENNESSEE AND NORTHERN ALABAMA.—MILL SPRINGS.—CUMBERLAND GAP.—MORGAN'S RAIDS IN KENTUCKY.—SHILOH.

Having located Grant's army at Pittsburg Landing, in the center, and Pope's at Island Number Ten on the right, let us now look to the left, Buell's army, so long neglected. For, in time, these three commands united before Corinth, under Halleck's personal direction, to meet a like concentration of the Confederates, under Polk, from Columbus, their left, and Johnston, from Kentucky, their right.

How well General G. H. Thomas, commanding at Camp Dick Robinson, succeeded in organizing the nucleus of the army of the Cumberland, may be gathered from the events that occurred a few months after. General George B. Crittenden superseded Zollicoffer, and, hoping to beat the Federals in detail, rather than risk all on a general engagement likely soon to occur, he, with ten regiments of infantry, six cannon and some cavalry, crossed the Cumberland, and advanced to surprise a few Federal regiments. But no surprise occurred, and General Thomas, having also advanced, met Crittenden January 19, 1862, when what is known as the battle of Mill Springs took place. The Federal force engaged was six regiments, one battery and a part of Wolford's cavalry regiment. Zollicoffer was slain, and the Confederates, driven across the river, fled to the Cumberland Mountains.

A few days before this occurred, Colonel James A. Garfield, commanding a small brigade, drove the Confederates, numbering about twenty-five hundred, under General Humphrey Marshall, out of South-eastern Kentucky.

In anticipation of General Thomas's operations, William G. Fuller was called from West Virginia to superintend the building and operating of the United States Military Telegraph south and south-east of Lebanon, Kentucky, leaving Samuel Bruch to attend to the telegraphic requirements of General Buell's main army, also about to advance. Fuller reported to Buell, December 20, 1861, and was immediately sent to Lebanon, Kentucky, to organize and await orders. He could not get men enough there to constitute a gang of builders, and was obliged to send to Ohio for them. Among those responding was William L. Tidd, his foreman, a most conscientious and capable man, who, in the sad course of events, as we shall see, lost his life in the telegraph service. Horses (except condemned ones) could not be obtained, but Tidd, without the use of a whip, broke twelve wild mules into the traces, with the best results. General Thomas arrived, and ordered a line to Danville. This was completed January 4, and from there, *via* Stanford, it was built to Somerset, seventy miles from Lebanon. The storms, bad roads and blockading army wagons, made Fuller's progress slow. His wagon wheels sank to the hubs, and it was often necessary to corduroy the road. The day that "Mill Springs," sometimes called "Logan's Cross Roads," was fought, the line was completed as far as ordered, *viz.*, to Somerset. The following are the first two messages ever sent concerning that Federal victory, the news of which so electrified the people of the North, especially in the Western States:

BATTLE-FIELD NEAR LOGAN'S CROSS ROADS, KY.,
January 19.

TO BRIG. GEN. BUELL., Louisville, Ky.:

The enemy attacked us early this morning, and were repulsed; Zollicoffer and Bailie Peyton killed. I write this in the saddle.

(Signed) G. H. THOMAS.
Brig. Genl.

SOMERSET, KY., Jan. 19, 3:30 P.M.

TO CAPT. FRY, A. A. G., Chief of Staff, Louisville, Ky.:

The following just received:

BATTLE-FIELD, Jan. 19, 1862.

TO BRIG. GENL. SCHOEPF:

We are on the way to Mill Springs, driving the enemy into his intrenchments. He attacked us this morning about six o'clock, fought well, but was repulsed in a handsome manner by our troops, and is now in full retreat. Zollicoffer and Bailie Peyton killed, and a large number of men. We have captured four ammunition wagons, with contents, and one piece of cannon. Join us by Hudson or the Salt Works. Do not order in the companies working on the roads. G. H. THOMAS, *Brig. Gen.*

T. S. EVERETT,
Capt. A. A. G.

This line was first operated by William H. Drake at Lebanon, D. E. Martyn at Danville, James Meagher at Stanford, and A. Ellison and R. Brown at Somerset. Brown was relieved by Philip Bruner.

As soon as the foregoing telegrams were received, Buell directed this one to Fuller:

LOUISVILLE, Jan. 19, 1862.

To W. G. FULLER, Supt. Telegraphs, Somerset, Ky.:

Return with all possible haste to Lebanon, Ky., and push the wire south from that place. (Signed) J. B. FRY,
A. A. G. and Chief of Staff.

From Lebanon south, every thing went well until the morning of the thirty-first, when the party was thirteen miles out, on the Columbia road. The builders were started early, and the camp party directed to proceed at eleven, with the wagons, and feed the builders in passing to the next camp. Fuller, leaving his navy revolvers, new boots, fur cape, gloves and a fine field glass with the wagons, went to Lebanon to get money for the men. The line had been completed four miles ahead of the camp on the thirtieth, and the builders were fully five miles away when, about nine A. M., Captain John H. Morgan, with ten men, including his nephew, Samuel Morgan, pounced upon the camp, capturing William C. Olney, A. Wells, Sylvester W. Bartlett, George McCadden and a hired slave; also nine horses, the wagons and all the baggage of the builders. Morgan himself appropriated Fuller's articles and a field telegraph instrument. A citizen coming up was dismounted and put into the Pleasant Hill church, near by, which, with all of Fuller's equipments not carried off, was fired, and then the band decamped. Two companies of the First Ohio Cavalry pursued them for two days without success, Morgan having reached the south bank of the Green River as the Ohioans came up to the north side. He carried off some fifty stolen horses and twenty citizens. Fuller personally lost one hundred and ninety dollars worth, and the twenty-four builders, all but the clothes they had on. Doubtless the citizen in the church escaped. The camp party was released at Murfeesboro, Tennessee. No further mishap occurred during the building of this line to Jamestown (February 10), on the Cumberland (seventy-five miles from Lebanon), where Colonel—afterward General, and then Governor—Bramlette was in command, except that Fuller's party was shot at several times by guerrillas.

In consequence of failure to send Fuller a telegraph instrument, he was unable to open the Columbia office as soon as the line reached there. General J. T. Boyle, in command there, was greatly provoked at this, and threatened to shoot Fuller; but that expert telegrapher, to Boyle's astonishment, sent his important messages by using the wire ends as a key and his tongue as a sounder, whereupon Boyle, patting Fuller on the back, said he was too useful to be shot yet. To the uninitiated it may be well here to explain, that the telegraph wire is charged by a battery at one or both ends of a line. Both ends are conducted to the ground, which serves the purpose of a return wire. This ground connection must be made or there will be no perceptible current, and so, when the line is severed, there is no noticeable current on either side of the break. The main line, for convenience of manipulating, is run to a telegraph key, to open which is the same as cutting the line. Fuller held the ground end in one hand and the main line end in the other; every time he touched them the circuit became complete, and was indicated on the instrument at Lebanon precisely as if the circuit was opened and closed by a telegraph key. There was hence no difficulty in sending the message, but to know that it was received was more troublesome. Everybody knows what it is to be shocked by a battery. Telegraph characters, we have said, are composed of dots, spaces and dashes, which are ordinarily read by the ear; but Fuller, having no instrument, could only determine them by electric shocks; *i. e.*, a long shock would indicate a t, longer yet an l, and yet longer a cipher; a succession of quick ones meant i, s, h, p, and others, with an intermission of time, e, c, r, o, and yet others, by quick and lengthened shocks, as in f, b, q, etc. It was by such shocks that Fuller received an acknowledgment of the message on his tongue, which being moist, is, perhaps, the most sensitive part. While such occurrences are rare, some are too well authenticated to admit of any question.

Four days after Morgan fell upon Fuller's party, the author, then a boy of sixteen years, rode over that same way with Fuller, *en route* for his first military telegraph office. Stopping over night at Campbellsville, a place of several log houses, one of which was called a tavern, with bolted doors, in a dingy room, half exposed by a candle that was intended to illuminate, Mr. Fuller, late at night, in an awfully solemn way, after administering an oath which he was legally unqualified to, imparted the mysteries of the Government cipher.

John A. Cassell, another beardless youth, was located at Camp Green, near Jamestown. In March, 1862, Bramlette, the commander, was ordered to move his force to Pittsburg Landing. In consequence, many country women came into camp to bid their relations and friends farewell. These people were greatly taken with the telegraph, and one young lady, venturing too much, was severely shocked by the powerful battery, which Weems, operating at Louisville, connected for that purpose. After that the ladies kept away from Cassell and accused him of playing naughty tricks upon them. Sam Bruch, who had a deal of quiet humor about him, reprimanded the young operator in terms which concealed the sting, just as a mother's smile, while berating her boy, convinces the latter that he may do so again with impunity.

Speaking of the ignorance prevailing along that line, concerning the telegraph, it may be added that, while Fuller's builders were constructing the line, one woman ran to them, gesticulating wildly, and exclaimed that they must not construct it past her house. "Why," said she, "I can not even spank my babies without all the world's knowing it;" and a denizen who had discovered a pole that had been struck by lightning, hurried to General Boyle and reported that a message had been lost down that pole. An old countryman about Mount Vernon, Kentucky, at a later period, being told by operator Jones how fast the telegraph was, thought the operator was imposing on him, and interrupting, said, "Stranger, that's a —-- —-- lie.

A humming-bird is the fastest bruit on y'erth, and it can't begin to make the time you say this thing makes." A repairer, who had become expert with the climbers, was trimming trees in Danville, when an old woman who watched his operations, remarked, "What can't these Yankees do! One of 'em was here and walked up and down the trees like a devil."

Cassell moved his office to Jamestown, where, wholly unprotected and thirty miles from any Federal troops, he continued to telegraph. Guerrillas raided the town several times, but the Lucas people, with whom he boarded, although Secessionists, successfully hid the young operator.

The same day that Grant invested Fort Donelson, Buell telegraphed Fuller to construct a line from Stanford toward Cumberland Gap, in East Tennessee, against which place Brigadier General Carter soon moved. The sufferings of Fuller and his party on this route were very great. Nearly every man became sick. To avoid the incessant rain for one night, they slept in a house recently used by the rebels as a small-pox hospital. London was reached March 20. A few days later Charles Stremple, a sick builder left behind, died. General Geo. W. Morgan assumed command of the Federal forces April 11. Flat Lick, beyond Barboursville, was reached by the telegraphers April 20, and Cumberland Ford a week later. Then the rebels destroyed seven miles of line, which was rebuilt. May 10, the wire was extended ten miles up Clear Creek to the foot of Pine Mountain. Then a line, thirty-five miles long, running from Colliers Fork to Flat Lick, was built, and that through Barboursville was taken down. Barboursville was put in circuit by a seven mile loop. After much maneuvering by the Federals and Confederates, which nobody can understand without great topographical research, in which each party appeared at times extremely anxious to meet the other, General Morgan, June 18, found the enemy away, on one of his strategetical expeditions, and took peaceable possession of Cumberland Gap, the "Gibraltar of America." Then followed an order to take down thirty-four miles of side lines, and erect twenty miles more connecting the camps of Colonel De Courcey, on the Kentucky road, and Generals Carter and Baird, on the Tazewell road, with the main stem. This was promptly accomplished. Another line, six miles long, connected the remaining brigade. Knoxville, sixty miles distant, was to be connected, but General Bragg's army coming into Kentucky in the fall, compelled Morgan's to fall back.

Fuller, aided by Foremen Tidd and D. E. Newton, had accomplished a great work among the rocky cliffs, cruel roads and barren mountains of this region. Constantly beset with new difficulties, the corps never faltered. It is a pleasing fact that throughout the war the corps never failed to erect a telegraph over any route, no matter how troublesome or how long, that the troops could themselves pass over. Many a line was built, however, where an army could not pass. From Stanford to Cumberland Gap is one hundred and fourteen miles. This, with the Somerset and Jamestown lines, made, exclusive of say seventy-five miles of telegraph built in the region of the Gap, two hundred and forty-five miles of line erected in the first six months of 1862, in a country beset with armed and unarmed enemies.

After a ride of forty miles, July 2, Fuller became suddenly and dangerously ill. The physician pronounced it a case of poisoning. Whatever it was, Fuller was incapacitated for further service for months. Consequently, Bruch placed Charles Lehr in charge of the district, but before Fuller was able to leave, the rebel Colonel—late Captain—John H. Morgan, the special enemy of the Federal telegraph, with his operator, George Ellsworth, a Canadian, struck Somerset (July 22) on his return from an extensive raid upon the Federal communications, which may properly be here related.

Early in July, after a minor engagement at Tompkinsville, Kentucky, Morgan, with two regiments and several companies of cavalry, entered Glasgow at midnight and halted near Cave City, a station on the Louisville & Nashville Railroad. Morgan, Ellsworth and fifteen soldiers reached a point half a mile below Horse Cave, on the tenth, when Ellsworth connected his telegraph instrument with the main line to read passing telegrams. By the use of a ground wire, he cut the southern towns from the circuit, and answered for them when the northern towns called, without awakening suspicion. The first message of consequence, was as follows:

LOUISVILLE, July 10, 1862.

To S. D. BROWN, Bowling Green:

You and Colonel Houghton move together. I fear the force of Colonel H. is too small to venture to Glasgow. The whole force should move together, as the enemy is mounted. We cannot venture to leave the road too far, as they may pass round and ruin it. (Signed) J. T. BOYLE,

Brig. Gen. Comdg.

The Louisville operator, for about two hours, was busy sending messages, he supposed, to Nashville, which were mainly commercial. These had to be repeated the next day. At about midnight, before returning to camp, Morgan manufactured and Ellsworth telegraphed the following message:

NASHVILLE, Ju.y 10, 1862.

To HENRY DENT, Provost Marshal, Louisville:

General Forrest, commanding a brigade, attacked Murfreesboro, routing our forces, and is now moving on Nashville. Morgan reported to be between Scottsville and Gallatin, and will act in concert with Forrest, it is believed. Inform the general commanding. (Signed) STANLEY MATTHEWS,

Provost Marshal.

Moving north-easterly, Morgan entered Lebanon quite unexpectedly, very early on the morning of the twelfth, capturing a few prisoners. Ellsworth, with a suitable guard, quickly took possession of the telegraph office, where he found on file the following telegram:

LEBANON, July 11, 1862.

GEN. J. T. BOYLE, Louisville, Ky.:

I have positive information that there are four hundred marauders in twenty miles of this place, on the old Lexington road, approaching Lebanon. Send reinforcements immediately.

(Signed) A. Y. JOHNSON,

Lieut. Col. Comdg.

At this time the author was day, and W. H. Drake, night operator at Lebanon Junction, and D. E. Martyn at Lebanon.

Martyn successfully eluded capture by hiding in the willows of Aqua Closet Run, which, he said, "odoriferously effervesced noisome effluvia, but was an awful good place to hide in." When the author came on duty, he inquired of "B" (Lebanon), about Morgan, and was informed that some of his troops had attacked the place, but were driven off. He then sent a message, to which the operator at "B" signaled "OK," a signal never used by Martyn, who should be on duty. This caused suspicion that all was not right. It was slim evidence, and soon overcome by Drake's declarations that he had been talking with him by the hour. This is a sample of the conversation Drake had been holding. To "B"—What news? any more skirmishing after your last message?" To "Z" (the call for the Junction)—No; we drove what little cavalry there was away. To "B"—Has the train arrived yet? To "Z'—No. About how many troops on train? To "B"—Five hundred; Sixtieth Indiana, commanded by Colonel Owens. Ellsworth's curiosity being excited as to what station "Z" was, he became puzzled how to ascertain it without creating suspicion. Finally, he adopted the following plan. To "Z"—A gentleman, here in the office, bets me the cigars you can not spell the name of your station correctly. To "B"—Take the bet. L-e-b-a-n-o-n J-u-n-c-t-i-o-n. How did he think I would spell it? Ellsworth—He gives it up. He thought you would put two b's in Lebanon. Drake—Ha, ha! He's a green one. Ellsworth—Yes, that's so. What time did the train with soldiers pass? Drake—8:30 last night. Ellsworth—Very singular where the train is! Drake—Yes, it is. Let me know when it arrives. (Later). The train has returned. They had a fight with the rebels at New Hope. The commanding officer awaits orders here. Ellsworth—Give us the particulars of the fight. Colonel Johnson is anxious to know all about it. Drake—Here is Moore's message to General Boyle:

LEBANON JUNCTION, July 12.

TO GENERAL J. T. BOYLE, Louisville:

At eleven o'clock last night, at New Hope Station, part of my command encountered a force of rebel cavalry posted on the country road, one-half mile south of the railroad. After a brisk fire of musketry for twenty minutes, tbe enemy was routed, and fled.

Skirmishers were sent out in different directions, but were unable to find the enemy. At three this morning, apprehending that an effort might be made to destroy the bridges in our rear, we moved down to New Haven and remained until after daylight, when the train went back to the scene of the skirmish. A Mr. Foreman, of Owen County, was found, mortally wounded. He reported the rebel force at five hundred and fifty, under command of Captain Jack Allen, and that they had fallen back toward Greensburg. One horse was killed, and three captured. The books of the company were found on the field. Blood was found at different places, showing that the enemy was severely punished. No casualties on our side. Here, with train, awaiting orders.

(Signed) O. F. MOORE, *Commanding.*

At nine, A. M., the following was sent by Z, which was a repeating station for the Cumberland Gap line :

LOUISVILLE, July 12.

To COLONEL JOHNSON, Lebanon :

Leave good guard, and join Colonel Owens. Pursue the enemy and drive him out. Be cautious and vigorous. Make no delay.

(Signed) J. T. BOYLE, *Brig. Gen. Comdg.*

LOUISVILLE, July 12.

To COLONEL OWENS, Lebanon :

You will move after the enemy, and pursue him.

(Signed) J. T. BOYLE, *Brig. Gen. Comdg.*

About noon, and before Owen's arrival, Morgan left Lebanon for Midway, a station equi-distant from and between Frankfort, the State capital, and Lexington, where operator J. W. Woolums was captured and made to ask Lexington the time of the day (ten, A. M., fifteenth), so that Ellsworth could discover his style of operating. Lexington office telegraphed the operator at Midway :

Will there be any danger in coming to Midway ? Is every thing right ? (Signed) TAYLOR, *Conductor.*

Ellsworth replied :

To TAYLOR, Lexington :

All right. Come on. No sign of any rebels here.

(Signed) WOOLUMS.

But the train which started, returned, and escaped capture. About this time the author arrived at Frankfort to assist the operator there, while Morgan threatened the place, and he well remembers that Ellsworth was quickly discovered, after he began attempting Woolums' style, and consequently that military dispatches were sent to Lexington, in cipher, *via* Cincinnati, and the Lexington operators were warned to look out for Ellsworth, who, however, reports further operations by use of his ground wire, which cut off Frankfort. Ellsworth says:

Again I answered, and received the following message:

LEXINGTON, July 15, 1862.

To GENERAL FINNELL, Frankfort:

I wish you to move the forces at Frankfort, on the line of the Lexington railroad immediately, and have the cars follow and take them up as soon as possible. Further orders will await them at Midway. I will, in three or four hours, move forward on the Georgetown pike. Will have most of my men mounted. Morgan left Versailles this morning at seven o'clock, with eight hundred and fifty men, on the Midway road, moving in the direction of Georgetown. (Signed) BRIG. GEN. WARD.

This being our position and intention exactly, it was thought proper to throw General Ward on some other track; so, in the course of half an hour I manufactured and sent the following dispatch, which was approved by General Morgan:

IDWAY, July 16, 1862.

To BRIGADIER GENERAL WARD, Lexington:

Morgan, with upwards of one thousand men, came within a mile of here, and took the old Frankfort road, bound, as we suppose, for Frankfort. This is reliable. (Signed) WOOLUMS, *Operator*.

In about ten minutes, Lexington again called Frankfort, when I received the following:

LEXINGTON, July 16, 1862.

To GENERAL FINNELL, Frankfort:

Morgan, with one thousand men, came within a mile of here, and took the old Frankfort road. This dispatch received from Midway, and is reliable. The regiment from Frankfort had better be recalled.

(Signed) GENERAL WARD.

I receipted for this message, and again manufactured a message to confirm the information General Ward had received from Midway, and not knowing the tariff from Frankfort to Lexington, I could not send a formal message, so, appearing greatly agitated, I waited until the circuit was occupied, and "broke in," telling them to wait a minute, and commenced calling Lexington. He answered with as much gusto as I called him. I telegraphed as follows :

FRANKFORT TO LEXINGTON :

Tell General Ward our pickets are just driven in. Great excitement. Pickets say the force of the enemy must be two thousand.

(Signed) OPERATOR.

It was now two, P. M., and General Morgan wished to be off to Georgetown. I ran a secret ground connection, and opened the circuit on the Lexington end. This was to leave the impression that the Frankfort operator was skedaddling, or that Morgan's men had destroyed the telegraph. We arrived at Georgetown about sundown. I went to the telegraph office ; found it locked ; enquired for the operator (Smith), who was pointed out to me. * * After tea I put in my own instruments. After listening an hour or two at the Yankees talking, I opened the conversation, as follows, signing myself "Federal Operator." To Lexington—" Keep mum. I am in the office, reading by the sound of my magnet in the dark. I crawled in when no one saw me. Morgan's men are here, camped on Doctor Gano's place." To Georgetown—" Keep cool. Don't be discovered. About how many rebels are there." To Lexington "I don't know. I did not notice, as Morgan's operator was asking me about my instruments. I told him I sent them to Lexington. He said 'Damn the luck,' and went out." To Georgetown—" Be on hand, and keep us posted." To Lexington—" I will do so. Tell General Ward I will stay up all night, if he wishes." To Georgetown—" Mr. Fulton wishes to know if the rebels are there." To Cincinnati—"Yes, Morgan's men are here." To Georgetown—" How can you be in the office, and not be arrested ?" To Cincinnati—" Oh, I am in the dark, and am reading by the magnet." To Georgetown—" Where is your assistant?" I replied, " Don't know." Lexington then asked me, " Have you seen him to-day ?" I replied " No." That was the last telegraphing I could do in Georgetown, as it exposed the fraud, the operator having no assistant.

The operators at Cynthiana and Paris decamped with their instruments. At Cynthiana, Morgan was resisted quite bravely, but the place was gained, and several hundred prisoners taken, after an aggregated loss of over one hundred killed and wounded. Morgan soon reached Richmond, where he hoped to remain and collect recruits, but General G. Clay Smith and Colonels Woolford, Medcalf, Munday and Wynkoop were seeking his capture, so he moved to Crab Orchard, striking the Cumberland Gap line again. Connecting his instrument, Ellsworth claims to have once more deceived the Union operators on this line, especially at Somerset, which place was reached the same day. Ellsworth's claims are doubtless entirely fictitious. The operators on this line had been notified to be on the look out for him, and operator Ellison, at Somerset, made his escape. Ellsworth did connect near Crab Orchard and hear some of the precautionary suggestions, but nothing of value to him or Morgan. James Jones was operator at Mt. Vernon, and about ten, A. M., some one called his office ("MC"), and asked: "Where's Morgan?" Jones at once became satisfied from the style of operating that it was Ellsworth, but, feigning satisfaction, he promptly but falsely replied; "Morgan was south of Richmond this morning. Woolford's cavalry and the Ninth Pennsylvania Cavalry are in close pursuit. Our troops here are preparing to cut Morgan off in the direction of Somerset. I think we will bag the rascals." The force at "MC" consisted of one convalescent, the operator and about forty home guards. A party of them afterwards scouted into Crab Orchard, but saw no enemy, and returned. Jones had hurried Morgan's whole force to Somerset by the direct road, which place he hoped to reach ahead of the "MC" column. Thus Morgan was fairly driven by telegraph.

Ellsworth was readily discovered on the wire at Somerset. James Meagher, operating at Stanford, was run out of town by a detachment of Morgan's men, who shot at him while he was climbing a fence. One ball passed between his legs, taking out a slice, but there was no catching Jimmy, who had been a prisoner at Columbia. The following telegrams were sent or received by Ellsworth, at Somerset.

SOMERSET, July 22, 1862.

To GEORGE D. PRENTICE,* Louisville:

Good morning, George D. I am quietly watching the complete destruction of Uncle Sam's property in this little burg. I regret exceedingly that this is the last that comes under my supervision on this route. I expect in a short time to pay you a visit, and wish to know if you will be at home. All well in Dixie.

JOHN H. MORGAN, *Commanding Brigade.*

*The most prominent editor in the State.

To General J. T. Boyle, Louisville:

Good morning, Jerry; this telegraph is a great institution. You should destroy it, as it keeps me too well posted. My friend, Ellsworth, has all your dispatches since the 10th of July on file.* Do you wish copies? John H. Morgan,
Commanding Brigade.

To Hon. Geo. W. Dunlap, Washington, D. C.:

Just completed my tour through Kentucky; captured seventeen cities, destroyed millions of dollars worth of United States property. Passed through your county, but regret not seeing you. We paroled fifteen hundred† Federal prisoners.

Your old friend, John H. Morgan,
Commanding Brigade.

Barboursville, Ky.

To General John H. Morgan, Somerset:

General: I am informed that you have my field glass and pistols, captured in my camp on the pike between Lebanon and Campbellsville, Kentucky, January 31. Please take good care of them. Yours truly, W. G. Fuller.

Somerset, Ky., July 22.

To W. G. Fuller:

Glad to hear that you are well. Yes; I have your field glass and pistols. They are good ones, and I am making good use of them. If we both live till the war is over, I will send them to you, sure. (Signed) John H. Morgan.

From Somerset Morgan proceeded to Monticello, Livingston and Sparta, and went into camp, having traveled over one thousand miles in twenty-four days.

The following operators were on the Gap line, patriotically enduring untold privations, *viz:* at Stanford, James Meagher and J. R. Clark; at Mt. Vernon, James Jones; at London, J. G. Garland until relieved by Frank Benner; at Barboursville, J. G. Garland until relieved by Hugh Craig; at General Geo. W. Morgan's head-quarters and about the Gap, John A. Cassell, Hugh Craig, J. G. Garland and Robert Wagner. The next time we meet these young men, they will be found surrounded on all sides by Confederate forces, and after much personal

*A joke based on a smattering of fact. †An exaggeration.

adventure, risk and hardship, we will honorably land most of them on the banks of the Ohio. But while they are rusticating among the ignorant, eating hot corn bread, corn dodgers, corn hoe cakes, corned pork, fresh pork fattened on corn, sweet potatoes and hard tack, we will inquire after the corps in the southward march with General Don Carlos Buell.

We left the main army in the middle of November, at Muldraugh's Hill. When Sherman succeeded Anderson, Rousseau took the immediate command and advanced to Nolin, on the Louisville & Nashville Railroad, fifty-three miles from Louisville. A. D. McCook arriving, ranked Rousseau. McCook, by order of Buell, December 9, 1861, advanced to Bacon Creek, near Munfordsville, where General O. M. Mitchell joined the army. The first United States Telegraph built south of Louisville was an eighteen mile loop, erected in January and February, along a branch of the Louisville & Nashville Railroad, connecting Bardstown and Bardstown Junction. Samuel L. Risdon opened and operated the office at Bardstown. C. H. Griffith and G. W. Rouser were operators at Mitchell's head-quarters after his arrival and assuming command at Bacon Creek; Martin Barth was at McCook's office, and when Munfordsville was taken, D. C. Sellers operated that office.

It is said that while at Bacon Creek, C. H. Griffith and Thomas Anderson, then a citizen of New Albany, Indiana, a Federal spy, escorted by five companies of the Fourth Ohio Cavalry, went through Edmondson County, *via* Mammoth Cave, and struck the railroad between Rocky Hill and Smith's Grove, and between the forces of Generals Hindman and Buckner at Cave City, and those with the commander, General Albert S. Johnston, at Bowling Green. The cavalry returned, leaving Griffith and Anderson, and losing six men in a brush with a portion of Colonel Terry's Texas cavalry regiment. Griffith tapped the telegraph, and copied many important messages, while remaining in the woods three days and nights by the instrument which, it should be stated, can be put in circuit without even opening or breaking the current. So long as an operator is content to remain quiet, and does not allow himself to interrupt the working of the wire, no suspicion is excited; but an operator's manipulation of the key is almost as distinctively individual as his voice, and if he attempts telegraphing or interferes with others, suspicion is usually aroused.

Near his retreat, Griffith found the line fastened to an insulator on a tree—quite a common thing in those days. Climbing the tree he was able to bend the wire around a dead limb, which held the line taught. He next ran two silk covered wires, no larger than linen thread, down the tree and through the underbrush about sixty feet, where they were connected with his instrument. This accomplished, he filed the main line in two between the fine wire connections, when his office under a ledge of rocks was ready for business. Anderson stood watch outside.

The valuable messages received were then copied on tissue paper in a fine hand and pressed into silver eggs, which were about the size of, and screwed together like ladies' ivory thimble holders. Before starting north, Griffith and Anderson deposited these eggs within their persons. Hiding their instrument, vise, etc., and repairing the wire, after being three days on the line, they started for McCook's head-quarters, but were captured by General Hindman's men. General Hindman's head-quarters, at Cave City, were at the house of Judge Roberts, a Union man. Joe Quigley kept the town tavern, and one Dolby also resided there. These were all Union men. In well-worn citizen's clothes, Griffith and Anderson appeared and were questioned by Hindman ; but they told a straight story of how they were trying to reach their homes in Louisville. Hindman, not finding anything suspicious in this, nor about their clothes, remanded them to the guard house, probably without very strict orders. Anyhow, Griffith, being a Mason, was not long in getting leave to go to the hotel kept by old man Quigley, and through the influence of a captain, there found, he got Anderson out also. Dolby and Roberts were here met, and the same story of wanting to get home repeated, and Roberts, it seems, somehow obtained from Hindman passes by which they reached the Union lines and reported, first to McCook and then to Mitchell, who was farther back. Both generals were delighted at the success of the enterprise, and Mitchell telegraphed the results to Buell, at Louisville. It is said that the information so obtained enabled the Federals to form quite an accurate idea of the forces on the railroad, and of Morgan's preparations at Glasgow.

Sixteen thousand of Johnston's troops were removed from Bowling Green to Fort Donelson, leaving Johnston only about eight thousand, and Buell, before being solicited, started about ten thousand in steamers to Grant's aid. These were the forces that went up the Cumberland with Foote's flotilla, just in time to enable Grant to capture that fortress. Without them, it can not be doubted, had Grant forced a battle, he would have been unsuccessful. Johnston commenced the abandonment of Bowling Green the day that Grant invested Fort Donelson, and on learning of its fall, directed the evacuation of Columbus. He halted a short time at Murfreesboro, thirty miles below Nashville, the capital of Tennessee, to gather the remnants of his Mill Spring and Donelson forces around those that retired from Bowling Green, and then pushed south-west, to Corinth, Miss., which, *via* Waynesboro and Savannah, Tenn., is one hundred and fifty-three miles.

It will be remembered that the Federals did not possess themselves of Cumberland Gap until the middle of June. Had Buell pushed Thomas in that direction, as repeatedly urged, even before and after the battle of Mill Springs, Knoxville, Chattanooga, and the strong mountain fastnesses of all East Tennessee and Northern Georgia, would have fallen into the hands of the Federals, enabling them to work incalculable injury to the Confederates. Between Chattanooga and Corinth there were no railroads or navigable rivers leading south. Corinth, Miss., Chattanooga and Knoxville, Tenn., were on the direct line of the railway running from Memphis, Tenn., through Northern Georgia, East Tennessee and Virginia, to Lynchburg and Richmond. Cut that line, and Johnston, at Corinth, could only communicate by rail with Richmond *via* Meridian, Miss., Montgomery, Ala., Atlanta and Augusta, Ga., or some other route east of Montgomery, but no better, and all leading as far east and south as Branchville, S. C.

Buell had no difficulty in entering Nashville, but the railroad and suspension bridges over the Cumberland at that city were completely destroyed. Bruch repaired the South-western Telegraph Company's lines to Nashville direct, and *via* Bowling

Green and Clarksville, rebuilding forty miles that had been destroyed between Bowling Green and Green River.

Bruch and operator Ellis J. Wilson entered Nashville with the troops, and took possession of the city office, but it was not until two days later that operator S. P. Peabody came, and astonished as well as alarmed Bruch by his successful but hazardous efforts to climb the northern pier of the suspension bridge, and handle the wire while it was being stretched across the Cumberland.

While Bruch, Wilson and Peabody were arranging the old telegraph office, Con. Dwyer, fresh from Cleveland, Ohio, was at Bowling Green, with his face southward. Gen. Johnston, before leaving this place, had fired the round-house, which at the time contained about twelve fine engines. The Barren River bridge being destroyed, the Federals tried to construct a train out of the *debris* remaining at Bowling Green. The result was an engine patched up from odds and ends, without head-light, cow-catcher or cab. It looked a snorting wreck on wheels. There were a few cars attached; one was a box car, in which sat the Hon. Andrew Johnson, Military Governor of Tennessee, *en route* for its capital; Dwyer and others also rode there. On a level track the engine was a success, but whenever an up grade was struck, every body but the Honorable Andrew got out and footed it, while the engine made running jumps to reach the top. Finally, late at night, the passengers reached Edgefield, and crossed the river to Nashville. Only one light was visible. That was at the City Hotel, but the proprietor gave Dwyer a withering grin when asked for a room. Piloted to the telegraph office, he finds Bruch and the operators delighted to receive him.

A word about this office, relic of the telegraphic year one. It was upstairs, on Cedar street. The Southern operators had left everything intact. There was a great accumulation of material there, in anticipation of being cut off from the North. This was devoted to Federal purposes. There was rubbish of all kinds lying about. A long table, breast high, ran across in front, over which was a frame, and on that was a network of wires. From this primitive switch, the wires depended to thumbscrews, and thence to instruments. These latter had the old time high resistance magnets, with registers. Beneath the table were great boxes to receive the paper strips; also high stools on which the operator sat and gradually became stoop-shouldered and consumptive. In the rear, well back, where it was almost dark, sat old Mr. Carville, the former manager, a fine specimen of the old time gentleman. He could not telegraph. His duties were mainly to audit the accounts. At the head of the stairs was an eight by ten sleeping room, where mosquitoes and bed-bugs did most of the sleeping. As theirs was entirely night work, they were not disturbed of their rest, but it was hard for an operator to sleep when their business hours came. The bugs were quiet and unobtrusive until the operator was asleep; then they flanked him on all sides. They even co-operated with the mosquito. This insect had a plan of singing continually, unless engaged in sharpening his tools or at lunch. His song, never very agreeable, soon became extremely monotonous. Instead of inviting Morpheus, as some music does, this drove him off. It was worse than the rebel yell. Consequently, when sleep came, it was deep. Everything being quiet, the bugs and mosquitoes put in their big licks. The sleeper, with a groan or a sigh, grows restless. That is a part of the scheme, for he unwittingly kicks the clothes off. When dawn appears, the operator discovers he has been throwing up breast-works all over himself. He feels as if they had been mounted by mitrailleuses, and then he thinks—"Home, sweet, sweet home; be it ever so humble, there's no place like home."

The provost guard reported finding in the city about seventy miles of unfinished wire in process of manufacture. Dwyer, Ellis J. Wilson, S. P. Peabody, Philip Bruner, late of Somerset, Ky. office, Mark D. Crane and D. C. Sellers operated there in March. In April, J. S. Cotton came. Martin Barth followed the army until taken sick, in February, when he was permanently located at Bowling Green.

From Nashville there were two great railroad routes leading south, and joining the road which connects Memphis with Richmond, Va. One ran almost due south, to Decatur, Ala., and the other south-east, to Stevenson, Ala. Johnston fell back slowly to Decatur and Corinth. While in the neighborhood of Columbia, C. H. Griffith and Tom Anderson disguised as tin peddlers, again essayed to tap the telegraph; this time between Lynnville and Columbia. Near Columbia they sold out their horse, wagon and stock to a country store-keeper. Then they separated, one going toward Lynnville, and the other in the direction of Columbia; but they met, after much trouble, near the railroad, where Griffith listened to Confederate telegrams for one day and a part of two nights, taking copies as before. Hindman's old pass was considered good by the Confederate pickets, consequently Griffith had no trouble in reaching the Union lines; but Anderson barely got in with the Federal forces again, as he took a route *via* Shelbyville, where he had to kill a man to escape.

Griffith was sent to the Gallatin, Tenn., office, in March, where he did not cut so good a figure. On Sunday, the sixteenth, Captain John H. Morgan, with about forty of his men, suddenly entered the town. We will let the Rev. F. Senour tell the rest, as he does in his "Morgan and his Captors":

> Morgan's first act was to arrest all the Union men of the place, and confine them in the guard-house; then he dressed himself in the uniform of a Union soldier, and proceeded to the telegraph office at the railroad depot, a short distance from the town. Entering the office, the following conversation took place: "Good day, sir! What news have you?" "Nothing, sir, except it is reported that the rebel Morgan is this side of the Cumberland, with some of his cavalry. I wish I could get sight of the rascal; I'd make a hole through him!" While thus speaking, the operator drew a fine Navy revolver, and flourished it as if to satisfy his visitor how desperately he would use it, if he should happen to meet the famous guerrilla. "Do you know who I am?" quietly remarked Morgan. "I do not," replied the operator. "Well, I am John Morgan." At these words, the operator's cheeks blanched, and his knees smote together, as did Belshazzar's, when he saw the mysterious handwriting. After the frightened man had somewhat recovered, Morgan required him to telegraph a message to George D. Prentice, of the Louisville *Journal*, in which he politely offered to act as his escort on his proposed visit to Nashville.

But Griffith, who ought to know more than the reverend historian about this affair, says he was captured while north of the town, repairing the line, and was sent to Corinth; escaped, and was recaptured, and finally sent to Macon, and then to Libby Prison, from which he was taken and exchanged, November 15

1862, after very great suffering. He had scarcely got back at work again, after recruiting some strength, when Morgan entered Elizabethtown, Ky., where he was located. Griffith narrowly escaped, by speedy flight.

March 11, President Lincoln extended Halleck's department, so as to include all of Buell's west of a north and south line drawn indefinitely through Knoxville, and called Halleck's, the Department of the Mississippi. At this time, Buell's forces, exclusive of those sent Grant, numbered 94,783 men, 73,472 of whom were serviceable. These troops were divided; the principal part, under Buell, numbering of all arms, 37,000 men, started, March 15, to reinforce Halleck's army under Grant, at Pittsburg Landing. Another force, under G. W. Morgan, was sent to join Carter's, north of Cumberland Gap, and another, under O. M. Mitchell, to possess the Memphis and Charleston Railroad in the neighborhood of Huntsville, Ala. The remainder was located on the communications.

We have already disclosed Federal General Morgan's operations at the Gap. While Buell is making his slow progress, *via* Columbia and Waynesboro, to Savannah, Tenn., opposite Pittsburg Landing, we will turn our attention to Mitchell. He, too, started about the same time Buell did, repairing burned railroad bridges as he progressed to Murfreesboro. From thence, he pushed rapidly on through Shelbyville and Fayetteville, striking Huntsville, Ala., early April 7. It was a great surprise, resulting, besides the capture of a few troops and some valuable property, in taking fifteen locomotives and one hundred and fifty cars. The following day an expeditionary force was placed on the cars for Stevenson, on the east, and another for Decatur and Tuscumbia, on the west. Thus in a few days, one hundred and twenty miles of the Memphis and Charleston road, and its bridges, were in the hands of the Federals. It was a bold and brilliant operation. Chattanooga was threatened, and Corinth cut off from the direct line East. Because it *was* bold, it has been severely criticised. Because it was remarkably successful, Mitchell received, as he seemed to deserve, a major general's commission. True he could not have gained these successes without the railroad, and even with it, he could not have retained his hold without the telegraph.

J. Newton Crittenton was head-quarter's operator, having been hurriedly recalled from Waynesboro to take charge of the lines in Northern Alabama. They needed but little repairing from Stevenson to Decatur, eighty-four miles. He was not then in telegraphic communication North, and hence held, under Mitchell, supreme control. A Mr. Brady was his foreman of builders. In March, the line followed Mitchell to Shelbyville (sixty miles), on the N. & C. R. R., and five miles beyond, toward Huntsville, when most of it was abandoned; but the last of April or the first of May, the telegraph was repaired, *via* Columbia, to Decatur, a distance of seventy-one miles. The building party, under one Keenan, on reaching a point thirty miles south of Columbia, was overhauled by that ubiquitous rebel, John H. Morgan, and all but Keenan captured. Morgan destroyed all the camp equipage, wagons, tools, seven hundred insulators, etc., and by paroling the builders effectually stopped the work, until Bruch could organize anew and refit. Owing to this delay, Decatur and Nashville were not connected until May 20.

Mitchell was enabled to capture at least two large trains complete, by telegraphing "It is perfectly so," in response to a message from an officer of the Memphis and Charleston road, inquiring of Larcombe, its superintendent at Huntsville, if it was safe to send along trains. During the four months and ten days that Mitchell or his successor held this section, telegraphic communication with the posts was constant, and the commander was accustomed frequently to talk with subordinates by telegraph. Offices were early opened at Huntsville, Paint Rock, Athens, Bridgeport, Stevenson, Elk River, Larkinsville, and Bellefonte, and later, at Decatur Junction, Battle Creek, Mooresville, Tullahoma, Pulaski, Reynolds, Shelbyville, Wartrace, Decatur, Madison, Tuscumbia, forty-three miles west of Decatur, and Courtland. These were operated by Samuel Carey, W. H. Kelsey, George E. Cromwell, J. S. Lyle, T. M. Sampson, Robert Wagner, J. W. Richardson, G. W. Bliestine, W. H. Hartman, J. H. Sigler, A. Ellison, J. A. Fuller, H. Lithgow, F. B. Tyler, C. W. Besanson, J. P. McIlvaine, D. McCarty, A. J. Howell, B. McMurtrie, C. W. Hammond, W. H. Parsons, S. T.

Yonkers, W. B. Reddington, George Purdon, Christopher Dougherty, Dan Murray and W. W. Forbes.

About the middle of April, a party of twenty-two soldiers and two spies, all dressed as citizens, with the approval of Generals Buell and Mitchell, entered Chattanooga in squads, and took train for Marietta, Ga. Their purpose was to steal an engine at Big Shanty, near Marietta, and, running north on passenger train time, burn the principal bridges *en route* to, and even beyond, Chattanooga. What could be more venturesome? Taking the north bound passenger train, they were able to possess themselves of the engine and one car at Big Shanty, and start on their perilous undertaking. There was no telegraph office at Big Shanty, although several regiments were stationed there, "but," says Pettinger, one of the twenty-one that succeeded in taking the train, and who has written the history of the affair, "the telegraph ran by our side, and was able by the flashing of a single lightning message to arrest our progress, and dissipate all our fondest hopes." After running four miles, the wires were cut. Further on, the party was, according to the time card, to meet and pass a freight train. This they did, but two extras were signaled. These greatly delayed the Unionists, and enabled a party from Big Shanty, under the leadership of the conductor, on a hand-car first, and then on a freight engine, to nearly overtake the Federals. Conductor Fuller's party, taking the engine of the last freight, continued the pursuit most vigorously. It was a desperate railroad race, in which the brasses in the journal boxes of the forward engine melted. Andrews' party, for he was in command of the raiders, sought to impede pursuit by throwing off the railroad ties they had collected, and occasionally tearing up a rail, but the pursuers overcame every obstruction about as fast as they were interposed. At every telegraph station, however, the conductor was informed that Andrews' party had cut the line ahead. Thus the race continued, passed Allatoona, Cartersville, Kingston, Adairsville, Calhoun and Resaca; passed Dalton and to within fifteen miles of Chattanooga—altogether ninety-three miles. The rain and wind prevented firing their car, which they attempted, and the inexorable pursuit continued as determined as ever. Operator R. O. Camp, or Ed. Henderson, joined Fuller's party to telegraph ahead whenever possible. Two minutes before Andrews, after passing Dalton, had time to break the wires, Chattanooga office was informed, whereupon cannon were there planted on the track, and trees felled across it. The excitement became intense, and the women and children took to the woods. Near Chickamauga station, the party leaped from their car and scattered in the woods, but pursuit and search was relentless. The people all turned out to hunt. Every one of the party was captured. Seven were condemned at Knoxville, as spies, and hung at Atlanta. Fourteen escaped from prison, six recaptured and *exchanged as prisoners of war;* six reached the Union lines, and two were never heard of again.

We have said that Buell's progress was slow. Bruch's builders and some soldiers, under operator J. N. Crittenton, repaired the old line over the pike to Franklin, and then cut across the country to within six miles of Waynesboro, erecting sixty-eight miles of new line. Crittenton frequently connected his instrument to receive or send Buell's messages. At this point, George H. Smith's party of builders were met, they having put up thirty-seven miles of telegraph east from Savannah. So imperfectly was this line constructed, that it was almost impossible to utilize it. The wire was carried through dense woods, and fastened to living trees, sometimes with insulators, but often without, consequently the aggregate resistance between Franklin and Savannah, especially on rainy days, amounted almost to a perfect ground connection, and not unfrequently rendered the line entirely unserviceable, a difficulty which no amount of subsequent repairing could appreciably affect. At this time, Con. Dwyer, an able electrician from Cleveland, Ohio, manager of the Nashville office, was assisted by operators Sargent P. Peabody, Ellis J. Willson and Claud Knox, all of them first class operators. They had no hours, but one would attend the Louisville circuit, another the Savannah wire, until exhausted, when he would be relieved by the third, who had been sleeping. As this system continued until August, when the author reinforced the trio, the boys deserved greater credit than they received for their perseverance. Afterwards, two operators were on, and two off duty, every twenty-four hours. As the Savannah line could not be worked by either office against a battery at the other, owing to the escape, a switch was arranged at the key so that the full force of the battery went to line while sending, but was off while receiving. By this means and the exercise of more patience than Job was ever credited with, the vast business of the armies at the front, aggregating, under Halleck, one hundred and twenty-thousand muskets, was kept up.

While this line was building, and Buell advancing to meet Grant, Halleck granted Buell's telegraphic request to halt and rest at Waynesboro, but Nelson in the advance seemed possessed of a premonition of danger. He, therefore, hastened beyond all expectations, and when Buell was ready to order a halt, Nelson and other parts of the army were too far ahead. This was providential. Grant, Halleck and Buell, all, were untroubled. An attack was not wholly unexpected, but was not looked for so soon. April 3, Smith's party of telegraph builders and an operator met Nelson, and established communication with Grant. Nelson had yet two days' march when Grant telegraphed him he need not hasten, as the transports to convey him to Pittsburg Landing would not be ready before the eighth. But Albert S. Johnston, commanding the enemy, had concentrated all of his forces, except VanDorn's from Arkansas, and was at that moment moving out of Corinth to overwhelm Grant.

The evidence is too convincing to admit a doubt, that neither Grant nor Sherman expected a general attack at any time prior to the sixth, when it really began. Certain writers, admitting this fact, say there was no surprise, because General Grant had telegraphed Halleck the day before the battle: "I have scarcely the faintest idea of an attack (general one) being made upon us, *but will be prepared should such a thing take place*," and because the enemy had been exhibiting themselves somewhat saucily for days before the battle. The day after Shiloh (April 8), Sherman in his official report wrote: "On Saturday (fifth) the enemy's cavalry was again very bold, coming well down to our front; yet I did not believe they designed *anything but a strong demonstration*." Of course these officers were surprised by the general engagement, commencing early on the sixth, but it is urged that, as Sherman had picket guards unusually advanced, there was no such surprise as brings disgrace to an officer who, in the face of an enemy, neglects proper precautions, *i. e.*, that though the attack was unexpected, it was not unanticipated. The discussion of this question involves a reproduction of too much conflicting evidence for our space or purpose. General Grant's head-quarters were at Savannah, nine miles from the main forces under McClernand and Sherman, at Pittsburg Landing and Shiloh Church. There was no telegraph line between these places, and the so-called "telegrams," sent by Sherman to Grant, April 5, according to Van Horne's "History of the Army of the Cumberland," advising Grant of appearances, were in fact notes, delivered by special messenger.

At break of day on the sixth, the noise of "resounding arms" summoned Grant in haste to Pittsburg Landing, notwithstanding an appointment to meet Buell that day at Savannah. The steamer "Tigress" from the North reached Savannah at three o'clock in the morning. Among her passengers were Wayne H. Parsons, from the Tipton, Missouri, and Leander H. Parker, from the Paducah, Kentucky, offices. These operators, on hearing the "cannon's opening roar," hastened to report to General Grant, and by his directions accompanied him to Pittsburg Landing, remaining on the boat to receive any message that might be entrusted to them for delivery. From early morn till late in the eve, forty thousand Confederates were hurled against thirty-three thousand Federals, largely raw troops, some of whom had been in the service but eleven days. Except that the Federals had the choice of position, it was an open field fight, there being few if any artificial defenses. Nelson's division on the eastern, and Lew Wallace's on the western shores of the river, were misled as to roads, and Wallace, though in the morning but five miles from the battle-field, did not reach it until seven P. M., and a portion of Nelson's column only came into position at five o'clock, when both sides were exhausted, but his coming was fortunate indeed.

The Federal gunboats, which, in the latter part of the day, while the Federals were hugging the river bank, or being hemmed in a greatly contracted circle, poured grape and canister into the ranks of the enemy with terrible effect, kept up their fire with shells throughout the night, penetrating various rebel camps and firing the woods; thus making sleep, for the enemy, almost impossible. Monday morning the Confederates, reduced by various causes — deaths, wounds, details and desertions — to twenty thousand, answered to the reveille which seemed like a death knell to them. Besides the fight of the sixth, in which they lost their great commander, Johnston, and the shelling and fires in the night, the rain set in; it was after that horrible day and dreadful night that, thoroughly drenched, they were early called to a renewal of the carnage, only to discover in their great weakness, that Grant was reinforced by twenty thousand fresh and gallant troops of Buell's army. The wonder is, not that the Confederates retreated at two P.M., after an obstinate resistance, but that they covered their retreat so well and retained Corinth so long.

Nothing is so descriptive of a battle as the casualties. Grant and Buell's loss was 12,217 men, of whom 1,700 were killed and 7,495 wounded, the rest missing. Buell's army alone lost 2,167, mostly killed or wounded. Beauregard, who succeeded Johnston, admitted a loss of 1,728 killed, 8,012 wounded and 957 missing; being an aggregate of 10,697.

The telegraph from Nashville to Savannah was completed in good time to announce a Federal triumph. The Savannah office was opened at the front window of a deep store, which was filled with wounded from the field of battle. Three rows of camp cots were placed the length of the room, and on every one rested a wounded soldier, some of whom were nigh unto death. Every available room in the village was similarly occupied. Messages concerning the wounded, the dead and the living, the defeat and the victory, came pouring into the Savannah office by the hundreds, for transmission north. All over the North went those tidings of sadness or joy from the little pocket instrument, manipulated mainly by Parsons and Parker, but sometimes by George A. Purdy. During the first night, a number of those whose groans swelled the agonizing chorus which made the tick, tick, tick, of the instrument scarce audible, after dictating a telegram to their friends, bidding a long farewell, slept "the sleep that knows no waking." Throughout all the Northern States and at the nation's capital, operators, officers and newsmen, all night long on that memorable seventh, awaited with bated breath and hearts well nigh choking them, the tidings which the weary operators at Savannah and Nashville hurried to their respective destinations.

General Halleck arrived at Pittsburg Landing on the 11th of April, and with him came Duncan T. Bacon, L. C. Weir and a well organized corps of telegraph men. Weir, as chief operator in the St. Louis office, perspired over the ciphers of the Department General until Halleck came, when he was employed at that officer's head-quarters, exclusively with cipher duties. He continued so to serve Halleck, sometimes, however, assisting

PARKER'S OFFICE.

Generals Cullum and Judah in the preparation of army papers, until the General was called to Washington by a cipher message from the President and made General in Chief of all the armies. This was after the occupation of Corinth by the Federals, and on this occasion Weir, who accompanied Halleck as far as Cincinnati, was offered a staff appointment with him, and a commission as lieutenant in the regular army or major in the volunteer service, but he preferred his former employments.

Shortly after Bacon's arrival at Pittsburg Landing, he aided Sol Palmer, foreman of builders, in laying a telegraph cable across the river at that place. The line having been built from Savannah, communication was thus completed to

Nashville. The first office at the Landing was unique. It illustrates the marvelous adaptability of the Corps to the emergencies of army life. The cable was too short to cross the river, but long enough to reach a lodged tree projecting from the west bank some distance over the water, and, consequently, it was determined to establish the office thereon. Here, L. D. Parker opened communication with the North. His instrument rested on the trunk; a limb held his letter clip and paper. Parker long sat astride his office, sending and receiving important dispatches, while being harassed by mosquitoes. After all, glory does not thrive in parlors, nor on cushioned seats. It was while vexatiously adjusting his instrument for Nashville's long messages and meaningless ciphers, that the mosquitoes began cupping him. While enduring their torments his game leg, for the first time in his life, served him best, for his other was a foot (*pes*) too long or the water so much too high, and consequently immersed that extremity, but what booted it so that he advanced the good cause. A number of couriers stood about, throwing stones at the numerous frogs, whose guttural notes were cut short by the surprising delivery of the blue-coats. Some aimed vexatiously at the complacent turtle seeking a sun bath, and thereby caused him either to draw in behind his works, or seek safety in the bosom of the deep. All these things happened within a stone's throw of freshly hallowed ground, and yet Parker was not brevetted on the spot, nor were there any tender hands to stay the ravages of the stinging gnats.

CHAPTER IX.

THE TELEGRAPH WITH GENERAL POPE IN VIRGINIA.—CEDAR MOUNTAIN.—GROVETON.—CHANTILLY.—HARPERS FERRY.—SOUTH MOUNTAIN.—ANTIETAM.

As heretofore noted, McClellan's army lay at Harrison's Landing on the James River until about August 14, 1862. McDowell had, as he predicted, failed to cut off Jackson in the Shenandoah, and Fremont's co-operation proved, indeed, ineffectual, but by Jackson's speedy flight to succor Lee and avoid the forces of McDowell, Banks and Fremont, possession of the great valley was again optional to the Federals, and consequently the Winchester telegraph was speedily re-opened. W. J. Dealey and G. J. Lawrence operated at Harpers Ferry, and J. D. Tyler and Charles H. Lounsberry, at Winchester. Major General John Pope, of great reputation in the West, was summoned to Washington on the day of "Mechanicsville," and assumed command of the "Army of Virginia" the day after "Gaines' Mills." Pope's new department included McDowell's, Banks' and Fremont's commands, and his paramount idea was to concentrate such of these forces as might safely be brought to the Rapidan, with a view of active co-operation with McClellan in the latter's designs against Richmond; thus at once assuming the offensive, besides forwarding McClellan's like purpose, and in that activity also effectually defending the national capital and preventing raids down the valley. Sigel, who succeeded Fremont, as the latter was indisposed to serve under a junior officer, located his forces in the neighborhood of Sperryville; General Banks' were mainly posted a little east; there W. C. Hall and J. B. Pierce attended to his telegrams and ciphers. Pope established his quarters at Culpeper Court House and was surrounded by McDowell's forces. This trinal army was strengthened by the opportune arrival of Reynolds' division from the Army of the Potomac, and Reno's eight thousand men from Burnsides' corps at

Fredericksburg. Altogether, Pope had an effective force at hand of about fifty thousand men.

Telegraph lines ran along the Orange & Alexandria Railroad from Culpeper Court House to Alexandria, and were operated as follows: Pope's head-quarters office, by R. R. McCaine, J. L. Cherry and Thomas Armor; Rappahannock Station, by J. W. Sampson; Warrenton Junction, by T. N. Loucks and E. Rosewater; Catlett's Station, by Charles Douglass; Manassas Junction, by Fred Fox and W. H. Sheffler; Fairfax Station, by Richard Graham and T. Q. Waterhouse; Burkes Station, by J. A. Flagg and D. B. Tomlinson; Alexandria, by F. T. Bickford, O. H. Dorrence, W. H. Embree, C. H. Johns, J. W. McMullen, A. W. Orton, H. L. Smith and C. L. Snyder. Another wire connected Manassas Junction with Warrenton, where J. H. Nichols was located. The line following the Manassas Gap Railroad to Strasburg had been abandoned since Banks' defeat in the valley. Sperryville and Sulphur Springs were brought into the telegraphic circuit. C. H. Benedict was located at Sulphur Springs and Ed Conway with Sigel at Sperryville. A third wire connected Fredericksburg with Alexandria. P. C. Doyle and Fowler Bradnack worked General Burnside's office at Fredericksburg; C. H. Lithgow was at Falmouth; J. H. Emerick, J. H. Glazier, J J. and O. H. Kinneman, at Acquia Creek, from whence this line ran to Alexandria. Thus it will be noticed Pope had his forces well in hand and easily directed or watched by telegraph. The Fredericksburg Railroad, superintended by W. W. Wright, and the more northern ones by J. H. Devereaux, supplied the troops. The trains on these roads, which were under the general superintendency of H. Haupt, were run by telegraph. In order to insure the safety of the telegraph, Pope issued an order that the inhabitants along the wire routes should be held responsible for damages done the lines by others than the Confederate army. Such was the general situation of affairs.

Early in August, Jackson, the ubiquitous, crossed the Rapidan in search of Pope, when began that famous campaign, short, sharp and indecisive, which resulted at least in humiliation to the Federals. But, as the Army of the Potomac figured largely in Pope's operations, we will first discover how that came to be so.

We have shown that at the conclusion of the seven days' fighting, McClellan encamped about Harrison's Landing, seventy miles above Jamestown Island. It was not deemed practicable to maintain a telegraph line from Yorktown to the army, and therefore one was built from Yorktown to Jamestown Island, eight miles, and from thence telegrams were sent up the river in despatch boats, a loss of about ten hours each way. Even this means was subject to embarrassments from the enemy, especially on the south bank. At one time forty-three guns opened on the Federal shipping from Coggin's Point. August 3, McClellan was telegraphed to "take immediate means to effect" a removal of his army to Acquia Creek. Pending the telegraphic protest and discussion which followed this order, McClellan went to Jamestown Island to talk by telegraph with General Halleck, who had, on the 11th of July preceding, been appointed commander-in-chief of all the armies with head-quarters in Washington. On reaching the island, it was discovered that the cable across the Chesapeake was broken, and accordingly McClellan, bent on his purpose, steamed down the James and across to Cherrystone, where C. H. Baker and W. A. Dunn were operators. Entering the office, he wrote :

CHERRYSTONE, August 13, 1862, 11: 30 P. M

MAJ. GEN. H. W. HALLECK, Washington:

Please come to the office; wish to talk to you. What news from Pope?

G. B. MCCLELLAN.
Major General.

An hour later, McClellan telegraphed, "* * Please read my long telegram, * * *" and Halleck replied at 1:40 A. M., "I have read your despatch ; there is no change of plans. You will send up your troops as rapidly as possible. There is no difficulty in landing them. According to your own accounts, there is now no difficulty in withdrawing your forces. Do so with all possible rapidity," and thereupon General Halleck took his hat and started to leave, but a prominent official present, knowing that McClellan had not had anything like a "talk," or complete understanding, asked the General-in-Chief if he had not any thing else to say to McClellan, but Halleck, replying "No," left the War Department office. The fact of Halleck's sudden withdrawal was announced to the operator at Cherrystone, when McClellan sent the following moderate reply :

CHERRYSTONE INLET, August 14, 1862, 1: 40 A.M

MAJOR GENERAL H. W. HALLECK :

Your orders will be obeyed. I return at once. I had hoped to have had a longer and fuller conversation with you, after traveling so far for the purpose. (Signed) G. B. MCCLELLAN,
Major General.

August 26, McClellan reached Alexandria—a part of his army having preceded him to Acquia, and the rest to Alexandria.

Over a week before the Federals landed along the Potomac, Lee's army was on the banks of the Rapidan, and Pope was skirmishing for time. Cavalry dashes upon the enemy's railroads and telegraphs preceded Pope's spirited campaign of three weeks, which may be said to have begun on the 9th of August, when, troops having been advanced under Banks and Sigel, and one division of McDowell's corps, to meet Jackson, who had crossed the Rapidan, the sanguinary battle of Cedar Mountain was fought. In this battle, the troops under Banks, to the number of ten thousand, were hurled against twenty thousand, until late in the evening, when Pope and reinforcements arrived. Pope was then thirty-two thousand strong, Reno and Reynolds not having yet arrived. Jackson recrossed the river two days later, and was soon joined by Lee. The losses in the battle of the ninth are placed at two thousand killed and wounded Federal, and one thousand three hundred and fourteen Confederate troops. Pope's head-quarters were advanced to Mitchell's Station, and so was the telegraph. Pope retired across the Rappahannock, on the approach of Lee, and, except a dash by Stuart, succeeded in keeping Lee at bay at the fords of that river from the seventeenth to the twenty-fifth. Jackson and Ewell, however, about the twenty-fifth, crossed well up the river, and, moving through Thoroughfare Gap, surprised Pope by striking his communications at Bristow Station and Manassas Junction. This surprise would not have occurred there had the Gap line been in operation. Comte de Paris, in his history, says : "Jackson, in haste above all to destroy the Orange Railroad, had marched directly upon that point of line nearest to Gainesville. At eight o'clock in the evening (August 26), the telegraph, connecting Washington with General Pope's head-quarters, became suddenly silent. It was thus that he learned of the presence of the enemy in his rear. Stuart had just reached Bristow Station, where he cut the wires and seized two empty trains, which he threw off the track."

The severance of the wires occasioned much solicitude in Washington, where it was even feared that Lee had planted his army in the rear of Pope's. When friends are known to be on the ocean during the progress of a terrific storm, anxiety, that tender balance between hope and fear, strains the nerves and exhausts all sources of hope. As reports confuse the names of wrecked vessels and so multiply disaster, solicitude becomes positively distressing. So it was at the capital when Pope's protecting army was cut off. The grape-vine telegraph was in full operation, and it brought only bad news. That line is never down. It is operated by the devil, and in war times does a great business. Between protecting Washington, to which mission McClellan was appointed, and rescuing Pope, an additional duty, it is claimed by the friends of the latter, that he was largely left "to get out of his scrape" as best he could. Only one thing possible was now wanting to enable McClellan and Pope effectively to co-operate. That one lack was a telegraph from Bealeton Station, on the O. & A. road, to Fredericksburg. Major Eckert had, however, already ordered L. D. McCandless and a party of negroes to build such a line, and operators J. L. Cherry, Theodore Moreland and perhaps, also, John D. Tinney, accompanied the builders, to render such aid as they might, but especially to operate the line as far as built, so that Pope's messages might be sent by the Fredericksburg route. This line was delayed, owing to the difficulties of procuring poles, but was completed to Morrisville, twelve miles from Fredericksburg, on on the twenty-seventh. When Pope got too far east to use this wire, Cherry rode to Warrenton Junction to ascertain the state of affairs. On the thirtieth, this line was abandoned, but it was over it that Porter's discouraging reports or letters were sent to Burnside at the Lacey House, Fredericksburg, and, in part, at least, by the latter forwarded to that cave of gloom, Washington.

It was while at Morrisville that some escaping soldiers reported, as runaway soldiers always do, that the army had been destroyed. This stampeded the builders and guards. It is said that McCandless' negroes loaded a large quantity of ammunition, found near Morrisville, and took it safely to Fredericksburg, where it was used in destroying Government ovens. It was not long that Burnside was permitted to remain at Fredericksburg. C. H. Lithgow, operator, chanced to be one of the latest to leave the city, and, rushing upon the bridge to cross, his feet stuck to the tar that had been liberally placed over the flooring of the bridge, when his momentum sent him sprawling in the pitch and dust, to the hearty enjoyment of the soldiers who saw it; to the ruination of his good clothes, and impairment of his temper. In Pope's front, about Rappahannock Station, the enemy was quite active. J. H. Nichols came out from the War Department office to assist McCaine and others at head-quarters, and joined them at Culpeper C. H., as they were about leaving for the Rappahannock, where a stand was made and an office opened; but the enemy's fire from across the river shelled out the operators. Several fragments had struck their building before they left. The army moved back, McCaine going with Pope, and Nichols and other operators coming up with the rear.

Supported by Reno, Pope's available forces, after his considerable losses at Cedar Mountain and at the ford of the Rappahannock, numbered over forty thousand. To these were added ten thousand under Heintzelman, about Warrenton Junction, and Porter's corps, eight thousand five hundred strong, the latter coming up to Bealeton from Falmouth about the time McCandless' builders left the latter place, constructing the Bealeton line. Troops were concentrated at Gainesville and vicinity, and others under Hooker beat Ewell back from Kettle Run when Jackson retired through Centreville. Pope, hoping to "bag" Jackson's "whole crowd," ordered up all his forces for warm work. Rickett's division vainly endeavored to stay the advance of Confederate Longstreet's relieving forces. Pope's troops were considerably scattered from Centreville, *via* Manassas Junction, to Gainesville. Longstreet was coming through Thoroughfare Gap, forcing back Ricketts. Every moment's delay enhanced Jackson's chances. Pope, appreciating the importance of the hour, ordered a general attack. Porter utterly failed to put his corps into the action, which lasted until late in the evening. Longstreet came up, and then Pope's opportunity seems to have passed forever. That was the battle of Groveton, midway between Gainesville and Centreville. It began at five, A. M., of the twenty-ninth, and ended at nine, P. M. The losses on either side are estimated at about seven thousand killed and wounded. The next day came the climax of the campaign. It was the second Bull Run fight. Like that of the preceding day, there were times when victory and defeat alternated in the balance, but late at night, neither side being eminently successful, Pope retired to Centreville. There he met Franklin's corps, and Sumner's was an hour in the rear.

In the sixteen days' fighting that Pope's army, or parts of it, were engaged in, such were his losses that, with Franklin's and Sumner's corps of McClellan's army, near twenty thousand, Pope now had scarce sixty thousand left. On the twenty-eighth, Pope reached Manassas Junction, and was at Centreville the twenty-ninth, when Groveton was fought, and at Bull Run on the thirtieth. On arriving at Manassas Junction, he sent operator McCaine up the railroad to see if the telegraph was in working order to Bull Run bridge. Accompanied by a small cavalry escort, McCaine accomplished his mission, but returning, the clear waters of Pogues Run, which for quite a distance from the bridge runs parallel with the railway, tempted him to dismiss his escort and take a bath. So tying his horse and thoughtlessly throwing his linen duster over the blue coat he wore, he plunged in. Shortly, three C. S. cavalrymen appeared before him. These he had descried in the distance, but they looked so rough and dirty that he thought them negroes. On espying McCaine, "What's this?" said one to another. "What's what?" rejoined the operator. "What do you belong to?" responded the first speaker. "To Virginia," replied McCaine. "To what Virginia; the first?" asked the rebel. "Yes," answered McCaine. "Well, we thought it looked queer," said the rebel, glancing at the operator's horse and clothing. It was well that the duster covered the coat and brass buttons, but it was infinitely better that the trooper thought he had met McCaine before, and, being satisfied, they rode on. But no sooner had they passed from view than McCaine, longing for his escort, hurried to overtake it.

At noon of the thirtieth, McCaine, with three orderlies, left Pope on the battle field of Bull Run, taking the road leading from Sudley Church to Manassas Junction. About five, P. M., they started to return. Riding about four miles, they met stragglers who reported that Pope could not be reached, as the enemy had driven him back and got between him and Manassas. Incredulous, McCaine pressed on, taking the precaution, however, to keep one orderly about fifty yards in advance. A mile further, this orderly came suddenly upon a rebel cavalryman who had a Federal prisoner. The orderly drew his revolver on the Confederate and demanded his saber, which was sullenly surrendered. Some hundred yards ahead could be seen a line of Stuart's cavalry, stretching along the crest of a hill. McCaine's party took their prisoner to Colonel Batchelder, provost-marshal, by another route. Again McCaine escaped, doubtless because the enemy beyond mistook his colors.

We have seen that after Stuart cut the wires at Bristow Station he started for Manassas Junction. Shortly before this, D. B. Tomlinson was relieved from duty and ordered to Burkes to assist J. A. Flagg, who, such was the scarcity of operators, had been on duty twenty-two hours per day for fifteen days. Instead of awaiting a train at Manassas Junction, Tomlinson boarded one going in the opposite direction, with a view of returning by the first passing train, on which he would go to Burkes. The engine "Secretary" headed the first train for Alexandria, and in it young Tomlinson was sitting on arrival at Bristow Station, when a heavy fire was opened on the train and Tomlinson shot in the knee. The following telegram of August 26, conveying the first news received in Washington that the enemy was on Pope's communications, indicates the danger Tomlinson was in :

TO GEN. HAUPT:

No. 6 train, engine "Secretary," was fired into by a party of Secesh cavalry, some say about five hundred strong. Ties were piled on track, but engineer took good run at them and scattered them from track. Engine well riddled by bullets.

(Signed) McCrickett.

M. J. McCrickett was at that time train despatcher on the O. & A. R. R. and his office was at Alexandria depot. At another time we will note his services and misfortunes.

It was not long before the enemy appeared at Manassas Junction, which was well filled with supplies and nearly emptied of troops, only about three hundred being there. Of course these were soon overpowered. Fred Fox betook himself to the quarters of the post commissary, Captain Musser, where he was captured and paroled. W. H. Sheffler, on duty at the time, escaped on an engine, after sending word to Alexandria as shown in the subjoined dispatch :

August 26, 1862.

General Halleck:

Operator at Manassas says, "I am off now, sure." I directed the agent to run the two engines at Manassas forward, wait until the last moment and then escape on the engine, if a real necessity existed. Operator had just commenced message to head-quarters of General Pope when wire was cut. It is clear now that the railroad can be relied upon only for supplies. No more troops can be forwarded. * * * * * (Signed) H. Haupt.

But two offices were now open out of Alexandria : Burkes and Fairfax stations ; Waterhouse and Graham, operators at Fairfax, and Flagg, at Burkes. Among the officials at Washington it was as yet only believed to be a cavalry raid, although graver fears were expressed. During the night, the Fairfax operators reported definitely the capture of Manassas and that the enemy was scouting the country. General Taylor's four New Jersey regiments were sent out from Alexandria in cars the twenty-seventh, and defeated at Manassas Junction. Two operators accompanied Taylor. Colonel Scammon failed to hold the Bull Run bridge, although he had two regiments ; these also had previously been beaten off from Manassas. When reports of these things were telegraphed from Fairfax, the officials also became alarmed, and every despatch indicated solicitude. Railroad reconnoissances, with and without troops, were now made, always attended by operators. M. J. McCrickett, W. H. Sheffler, H. L. Smith and others were engaged in this perilous telegraphic service. After telegraphing the report that "Scammon is retreating and Bull Run bridge gone," operators Waterhouse and Graham also started for Burkes. It was well, for Flagg had been on continuous duty fifty-seven hours. McCrickett reported from Burkes, "I can not get to Fairfax; was fired into one and a half or two miles west of here by cavalry or a band of guerrillas."

The mystery grew greater and greater in Washington. Even Scammon's force was lost to view. Another railroad reconnoissance was ordered, skirmishers put on board and conductor Strein instructed, "If no enemy be found when Bull Run is reached and the bridge is safe, proceed to Manassas and ascertain the condition of property. Report every observation of importance by telegraph. An operator will be sent." Colonel Scammon's force was found retreating and brought in with news that the rebels were in large force. Another party of "two hundred sharpshooters, with operators and wire to repair the telegraph, make communication and report observations" was sent out. Pohick Bridge, one mile west of Burkes had been destroyed, but was quickly restored, when the train, under J. J. Moore, proceeded, the operator repairing the line as it advanced. It reached Fairfax Station where wounded soldiers were found and placed in the cars. This station was a two-story building. The office was up-stairs. Below, the surgeons were busy amputating limbs. The enemy was now reported twenty thousand strong. Anxiety intensified in Washington. The operators and the trains were recalled. Then came the news that Pope was at Manassas. The next day operators reported him at Centreville, and McDowell and Sigel moving to cut off the enemy. Fairfax Station was re-opened, and Pope sent orderlies to that office to deliver messages.

Thus, under the most trying circumstances, the telegraph, working in concert with the railroad, dispels many a hope and excites many fears, but the fears are well grounded. General Haupt, in testifying before the Committee on the Conduct of the War, said: "I can not speak in too strong terms of commendation of the zeal and courage of the railway employés and of the telegraph operators, who, with a full understanding that the service was very hazardous, volunteered for the occasion. The operators requested that I would obtain permission of Colonel Stager, so that, if captured or shot, they or their representatives would not be deprived of pay in consequence of going without leave. The permission asked was promptly granted, and they proceeded, with instructions to conceal themselves in the bushes, send out scouts, make connection with their wire, and report all that they saw or heard. Nearly all the information given to the Department (War) at Washington, for the remaining days of the fight, was received through this channel." If General Haupt, by "this channel," meant this particular line, he was substantially correct. But the Fredericksburg wire was also much relied on.

Moreover, there were two operators with the army itself, who were alive to the importance of obtaining and telegraphing information. These were J. H. Nichols and Ed. Conway. Nichols, mounted on Joe Pierce's sorrel horse, met Conway astride of an old blind mare, at Warrenton Junction. These knights of the key traveled toward Manassas, when, hearing firing to their left, they turned in the direction thereof, not knowing whether they were going toward their own or the rebel troops, but as they came upon the battle ground of Bull Run, they met Federals, and had a magnificent view of the fight for a couple of hours, at the end of which time they returned to the railroad at Fairfax Station, where operator Rosewater was working with the War Department office. Conway and Nichols at once telegraphed to Secretary Stanton an account of the battle as they saw it, and the disposition of the Union troops. It was the first authentic and connected report received in Washington for a day or two, and in return the boys were complimented as follows:

WASHINGTON, D. C.

TO NICHOLS AND CONWAY:

I have mentioned your valuable services to the President, and am assured by him that they are fully appreciated.

(Signed) EDWIN M. STANTON, *Sec. of War.*

Elated by such commendation, Conway remounted his blind nag, and Nichols his sorrel. The sorrel was a hard rider, moving something like machinery when every other cog is broken.

Nichols was sure of being jounced, but whether it would come with a full breath, and thus force exhalation, or on empty lungs and prevent inhalation, he knew not—only knew it was coming till it had come, and then it was most time for another. The sound of distant guns was their only guide, and as that came from far separate points, it was an uncertain one. Darkness came before the operators had gained their bearings, and all night long they kept going, avoiding the camp fires, their main reliance in the late hours, and the pickets, lest they prove to be Confederate. About six, A. M., they neared Fairfax C. H., and were captured by a German picket to whom they could not give the countersign, but a superior Federal officer, being convinced of their position, let them pass on. It was a night and a ride long to be remembered. Conway's blind mare lost confidence in her rider, and felt her own way, his spurs to the contrary notwithstanding. There came a time when Conway lost confidence in the mare, as we shall soon observe. It was a marvel that they could ride all night without more serious incidents than running into trees, brush, fences, etc., for there was an almost continual firing before, behind and to the right and left of them. Even a band of the enemy's cavalry somehow got around near Centreville, and burned a wagon train.

The operators gathered such information as they could, and again started for Fairfax Station. Just before nearing the depot, they came to a muddy little stream. The cog horse, by way of a parting reminder, made a running jump and landed on the other side high and dry, but Nichols, who had left his breath way behind, had to stop to catch it again. When he did respire, he thought his head had settled below his shoulders, and the rest of him in proportion. It was now Conway's turn. His poor beast was fagged, but somehow she knew when to jump. She did jump, alighting in the very middle of the murky, muddy creek, from which she was unable to extricate herself; nor could Conway and Nichols release her. The enemy was already near at hand. It was most six, P. M., and so poor "Sal" was left where she made her last leap. Conway, whose linen duster was well splashed, reached land muddier and wetter than before. Barely had these operators telegraphed their information, when the Confederates were seen approaching. Then the depot was fired, and the last train pulled away. The operators at this station had been ordered to remain as long as possible, hence it was not until the enemy came in view that the match was applied, and thus Conway and Nichols easily escaped. This train also took Flagg from Burkes Station. Rosewater, who also boarded it, had left his baggage to the watchful care of the enemy about Manassas. That enemy, clad in carpetings, quiltings, blanket cloth, anything, everything, even to tarpaulin goods, knew well the utility of such clothing as Rosewater was accustomed to adorn his person with.

The line of communication, *via* Fredericksburg and Acquia, we have said, was much depended on, but Acquia was abandoned soon after Fredericksburg. McClellan landed at Acquia the twenty-fourth, proceeding to Alexandria on the twenty-sixth. From Acquia and Alexandria, McClellan carried on by telegraph another historic correspondence with Halleck and the President. It was well said by Colonel Stager in an annual report that "even the history of this fratricidal strife was being recorded by telegraphic dottings." Any one reading the telegrams to and from the great officers in Washington and vicinity, during the trying period of Pope's eventful campaign, will discover in them not only sources from which history is compiled, but also in the tone or language of those epigrammatic epistles, a sort of photograph of passing events, always real, sometimes sadly so. They were not interviews for the public, but cipher interchanges which sprang from the sincerest convictions and expressed, ofttimes, without reserve. Once stated, they were beyond recall or dispute. It was such evidence that went far to convict Fitz John Porter of disobedience at Groveton, and it is from such dispatches that General McClellan's support or neglect of Pope is judged. Modern history can not be severely distorted. There are some exact sciences. History has been far removed from them, but so far as the telegraph records events, that far will science exactly preserve the truth. Perhaps the best history of the late war would be the telegrams, properly arranged, aided by parenthetical explanations, from which opinions and criticisms might wholesomely be excluded.

But while we are philosophizing, Pope is engaged at Chantilly—the last of his memorable battles. It is September 1, and that same Jackson takes the initiative. He moves to flank the Federal right. Reno and Stevens, Hooker, McDowell and Kearney, become engaged. The conflict lasts until dark, when Federal General Birney, succeeding Kearney, was left to bury those who died that day. To the gloom which the cold rain produced, which the darkness intensified on that field of misery, was added the death of gallant Kearney and talented and brave Stevens.

It is said that, in the operations from Cedar Mountain to Chantilly inclusive, the Confederates lost fifteen thousand men, killed and wounded, and the Federals near twice that number, killed, wounded and captured.

Pope, from Fairfax C. H., sends to Fairfax Station a message for Halleck: "My head-quarters are in Alexandria," and ere night of the second, the Union army is once more within the thirty forts and other defenses of Washington. What humiliation! But how much greater, had any other than an American army compelled it!

Now, as in the Summer of 1861, the Military Telegraph, except about Winchester and Harpers Ferry, had no foothold in Virginia north of Yorktown. Now, as then, offices were opened throughout the District defenses, and now, as then, it is the Federal capital that is defended.

At Chantilly, Lee writes President Davis that he will invade Maryland. In Washington, Pope resigns his command, and goes into the Indian country. McClellan is again solicited to command what is once more known only as the Army and Department of the Potomac, and while he is preparing for action, the American and Western Union Telegraph lines spread the news and details of victory and defeat throughout the land. It was well, because President Lincoln had recently called for three hundred thousand more troops. Those telegraphic reports filled the quotas.

Two great purposes were sought to be subserved by Lee's northward march, which began September 2, 1862, viz., (1) to carry the war into the Union States, and (2) thereby obtain a new and powerful argument for foreign mediation, or at least, recognition. England and France, especially the latter, seriously contemplated measures looking to the permanent establishment of the Confederacy. It was also thought by the Southern authorities, that many recruits could be obtained in Maryland, and by threatening Union cities, especially Philadelphia, Baltimore and the capital, and throwing the ravages of the war upon Unionists, the triumph of the cause would be well nigh assured. When Lee determined upon this move, he directed several divisions that had remained in Richmond, to join him in Maryland, and, passing Leesburg, he crossed the Potomac in the vicinity of Point of Rocks. Stuart's cavalry reached Frederick, Md., on the sixth.

In a former chapter, we have shown the construction, in 1861, of a line from Washington, D. C., to Hagerstown, Md., *via* Rockville, Darnestown, Poolsville, Hyattstown, Frederick and Williamsport; also a loop from Poolsville to Point of Rocks. In the new turn of fortune's wheel, each of these places was now destined, in its order, to be "the front," the *situs* of operations, and the post of observation. Notwithstanding McClellan was re-invested with the command of the troops on the second, that those troops were in a *semi*-disorganized condition, owing to recent removals from the James, and to marches and conflicts under Pope, he (McClellan), on the third, began crossing troops into Maryland to forestall any attempt upon Baltimore, or on the capital, from the north or east.

It was unfortunate that the necessities of the telegraph service on the Peninsula, and with Pope, drew so many operators from the Frederick line, as it occasioned the closing of several offices, which were supposed to have become unimportant. There was no military operator at Point of Rocks, at the time of Lee's forward movement from Chantilly, and Stuart's approach seems not to have been telegraphed. Great uncertainty prevailed in Washington for many days after (third) it was known that some of Lee's forces had crossed the river. Apprehensions were serious, of Lee's aiming to draw McClellan away toward Frederick, while the main Confederate force swept down upon Washington, *via* Leesburg. Hence it was long regarded essential to divide the Union army, by sending McClellan, with eighty-seven thousand five hundred men after what was presumed to be Lee's main force, while full seventy thousand were retained in the defenses of Washington. This necessitated a continuance of a full siege telegraphic force in and about the capital, but owing to the evacuation of the Peninsula, and the abandonment of Pope's lines, there were at this time plenty of telegraphers for all needed points. Of course, Stuart's first business was, to cut the telegraph. Thus, direct communication with Harpers Ferry, *via* Point of Rocks, was severed. At this time, Dixon S. Miles, colonel in the regular army, was in command of Harpers Ferry and its contiguous defenses, and with him were, say, nine thousand troops. William J. Dealey, George J. Lawrence and Cephus C. Starling were the military operators at that point, and Daniel J. Ludwig the railroad operator.

Winchester was now the farthest point in Virginia occupied by the Federals. Operators Charles H. Lounsberry and J. D. Tyler were there serving Julius White, a brigadier general of volunteers, in command. General Halleck telegraphed White, September 2, *via* Wheeling and the Ferry, "You will immediately abandon the fortifications at Winchester, moving the heavy guns, under escort, by rail to Harpers Ferry. If this can not be done, they should be rendered unserviceable. Having sent off your artillery, you will withdraw your whole force to Harpers Ferry." General White re-inforced Miles with two thousand men and was at once ordered to Martinsburg, to the end that Miles, a soldier for forty years, might not be superseded in command. Soon after, the enemy were reported from Leesburg, Winchester and Hagerstown, Maryland, and consequently White was ordered to evacuate Martinsburg and go to Harpers Ferry with his two thousand, five hundred more troops. Jackson crossed the Potomac on the eleventh, and on the same day White started, reaching the Ferry the twelfth, when Miles exhibited the following telegram:

WASHINGTON, D. C., Sept. 7, 1862.

COLONEL MILES, Harpers Ferry:

Our army (McClellan's) is in motion. It is important that Harpers Ferry be held to the latest moment. The Government has the utmost confidence in you and is ready to give you full credit for the defense it expects you to make.

(Signed) H. W. HALLECK, *Gen.-in-Chief.*

White very properly permitted Miles to remain in command, and a Congressional investigating committee and some of the highest military authority in this country have since approved thereof.

On the seventh, McClellan left for Rockville, where L. A. Rose was operating. McClellan's operators at this time were A. H. Caldwell, Jesse H. Bunnell, John W. Parsons and J. L. Cherry, who were soon reinforced by J. H. Emerick and C. H. Lithgow. Caldwell, Bunnell and Emerick continued with McClellan, and the others advanced with various commands. While McClellan was at Rockville and his cavalry out on the Poolesville and Hyattstown roads, it was thought possible, barely possible, to open telegraphic communication with Point of Rocks and thence to Harpers Ferry. L. D. McCandless had charge of the line repairers and builders on this route, and together with operator "Connolly" (doubtless a mistake for Ed Conway) volunteered to attempt it. Taking two cavalrymen, they proceeded as rapidly as practicable and reached Point of Rocks in safety, but there was no circuit from Washington. McCandless immediately started back, following the route of the line, meeting many stragglers from the rebel army, two parties of whom he directed on the Poolesville road, where they were doubtless captured. In a corn field on a bank of the Monocacy, where the wire crossed the creek, after being cut it had been ingeniously fastened, one end to a stump and the other to a fence post, giving quite an appearance of being intact, but hidden by the fence, when in fact a piece of wire had been carried away. But for being misled by some skillful rebel, McCandless and Conway would have won great credit. As it was, the break in the line delayed perfecting the circuit until all the glory had passed. The adventure was a daring one as it was a move very close upon the rebel rear, but it was much less successful than it ought to have been, probably owing somewhat to too rapid riding.

McClellan remained at Rockville, or near, for three days, frequently communicating with Halleck by cipher telegrams, eighteen of which are published in the report of the committee on the conduct of the war. These are exclusive of replies there⁺ and telegrams to the President, to Banks the commander in Washington, to General Wool, commanding a department with head-quarters in Baltimore, and Governor Curtin of Pennsylvania. From here also McClellan's subordinates arranged largely by telegraph for supplies of all kinds. Reinforcements were requested and forwarded, including Porter's corps, twenty-one thousand. McClellan now had one hundred and ten thousand, the enemy probably less than sixty thousand. Reports were here received from advance offices as to operations in front, and as rapidly as received were forwarded to the War Department, and then, if proper, given to the general public, which was watching the operations of these two great armies with bated breath. There were thousands of fathers and mothers, of wives and children, throughout the North, unable to attend to their usual tasks, owing to their overpowering solicitude. The news that came from near Newmarket, to the east of the enemy, from Hagerstown, to the north, or Harpers Ferry, to the west, was tinged, as was natural, with melancholy, owing to the absence of a competent repelling force. The day Martinsburg was evacuated, Harpers Ferry became as silent as the grave until the afternoon of the thirteenth; then McClellan was at Frederick and his army ahead. The news of the thirteenth was merely the sound of cannon in the direction of Harpers Ferry. It meant that the order of march directed by Lee, a copy of which had fallen into McClellan's hands and which provided for the capture of that place, was genuine and in process of execution. More activity now prevailed in the army. The next day a courier from Miles reached McClellan with intelligence that the enemy held Maryland Heights, but that Miles still held Loudon and Bolivar Heights and hoped to hold out two days. Miles' cavalry (two thousand) had escaped.

Operator Ludwig boarded with a Mrs. Chambers, whose son Marshall was in the rebel army. Marshall obtained a "leave of absence" and entered Harpers Ferry about the tenth and remained for a day or two, and was seen no more until the fifteenth. A rebel operator was discovered on the line, but of course obtained no news. He told Dealey he was coming soon to see him and the others, but though he doubtless came he did not send up his card. After a few days cannonading, Miles surrendered, and before the fog had sufficiently raised to disclose to those of the enemy now occupying Loudon Heights, the white flags of the Federals, a shell burst and killed him. In this disgraceful surrender to Jackson, who commanded the two thirds of the rebel army sent to effect it, operators Dealey, Ludwig, Starling, Lounsberry and Tyler were among the prisoners of war. But the military telegraph instruments and apparatus had previously been destroyed by the telegraphers. All other Government property at the Ferry was intact. At the head of Jackson's troops to possess Harpers Ferry came Marshall Chambers, probably a counterpart of the Confederate States officer Rousser, whom Miles permitted to leave Harpers Ferry only a few days before, and who also headed the victorious host.

By this time Franklin's corps of McClellan's army was within five miles of the Ferry. The cessation of artillery firing, by a negative pregnant produced to the relieving forces, conviction of disaster. In that ominous silence, a most appropriate messenger, was carried the first tidings of defeat. This has been mentioned by some as a counterpart to Fort Donelson in the west, but it is not our province to draw comparisons.

By this time (fifteenth) South Mountain had been fought. There were times when President Lincoln regularly attended the the War Department telegraph office. Indeed, between Major Eckert's and Secretary Stanton's offices was a little office that was generally known as the President's room. To it he betook himself regularly at six o'clock every morning and there he read the news of the night previous; there also he often consulted with Stanton and with Halleck and other eminent military men; but in times like these when a nation was on tip-toe, Lincoln manifested his own anxiety, by long protracted and frequent visits to that little room. There operators Chandler, Tinker and Bates, who were chiefly occupied translating and "putting up" cipher messages, also had desks. It was the *sanctum sanctorum* of that great military and civic center, and in it, all phases of the life of the chief military and civil functionary of the land were made manifest. Patiently but wistfully, he, at times of greatest anxiety, looks over the shoulders of the cipherer as the message is gradually unfolding. About noon of the fifteenth, one of those babel telegrams is received. As it unwinds, evidences of a great battle are seen in the two or three connecting words of adjoining columns. A little later, success is also clear, but in its full development, are some of the details of that triumph which is found summed in a terse sentence, brimful of glad tidings. "It has been a glorious victory." That was South Mountain. Now a word concerning it.

There are two gaps in this range to which the army moved from Frederick; one, near Middletown called Turners Gap and the other close to Burkettsville, known as Crampton's. Lee posted but few troops in these strongholds, not anticipating vigorous pursuit in force, but as the attacking columns multiplied, the defenders were reinforced until they numbered nearly all of Lee's army not operating about the Ferry. The conflict raged until late at night. At 9:40 P.M., a Union victory was certain at both Gaps, and then it was that McClellan wrote that telegram which as we have seen, announced success. This message was delayed by reason of the telegraph instrument becoming out of order, which necessitated Emerick's riding to Frederick for another. By two o'clock on the fifteenth the news-boys throughout the land were shouting "'ere's yer extra; all about Lee whipt." The sacrifice in scaling the mountains and driving the enemy through the defiles was four hundred and forty-three killed and one thousand, eight hundred and six wounded, including brave Reno, dead. The cost of the defense was probably not so great except in prisoners, of whom about fifteen hundred were taken.

Thus was Lee's purpose to invade Pennsylvania temporarily thwarted, notwithstanding an advance to Hagerstown and thus it was, that Franklin was enabled to reach a point within five miles of Harpers Ferry when the surrender occurred. Lee's peril was now exceedingly great; wherefore he sent for Jackson to make haste to meet him near Sharpsburg. Unfortunately the Federal prisoners allowed themselves to be paroled, thus leaving the entire force of the enemy to rejoin Lee, but it was fortunate for the operators who mixed with the Union soldiers that were paroled by regiments, and therefore the telegraphers were not prevented from immediately re-entering the service. They all met in the office at Frederick, where Ceph Starling probably awoke his muse, and the guitar which he always carried with him. At the Ferry he had been accustomed to remind his friends of scenes of home, by his guitar and verse, which were all the sweeter because surrounded by nothing peaceful, quiet or homelike.

The great question now was, could McClellan strike Lee before his army was re-united. Between the Potomac and the Antietam, a quiet, deep creek near Sharpsburg, Lee had prepared to repel the Federals until Jackson, McLaws, Anderson and A. P. Hill could come up from the Ferry. That question was in process of solution, each department looking to its own duties. The telegraph office which the operators opened near South Mountain after witnessing the battle and as soon as the line was brought on the fourteenth, was, on the fifteenth, advanced to near Boonesboro where the operators were surrounded by wounded men who were brought from the South Mountain battle field. The moans of the sufferers were telegraphed to Washington and the North, and soon came officers of the Sanitary and Christian Commissions, those grand sanitaria, with lint and bandages and cordials and medicines and prepared foods. The telegraph moved on to Boonesboro, and there also soon came W. J. Dealey and J. D. Tyler, operators from the Ferry, escorted by the Garibaldi Guards, the latter under parole, and there Dealey and Tyler were located and the line was extended three miles to Keedysville, just in the rear of McClellan's camp.

On the sixteenth, the public was astounded at the news of the fall of Harpers Ferry. South Mountain was counteracted. The *morälé* of the Confederates behind the Antietam was recruiting in the cheers that smote the air. McClellan was thereabouts, locating his forces and while so doing Hagerstown was connected by telegraph. This gave McClellan a northern route *via* Chambersburg, Pennsylvania. Supplies were rushed down the Chambersburg & Hagerstown railroad for the army. On the fifteenth, considerable skirmishing for position occurred. On the sixteenth, McClellan's forces were not yet well up. Time was precious. Two divisions of Jackson's own exhausted troops were in position, but yet McLaws, Anderson and A. P. Hill were behind. Fighting was heavier but not general, merely, but severely, tentative. All felt that the morrow must bring on a crisis. It was so, notwithstanding McClellan's whole force was not up; yet, they far exceeded Lee's in hand.

In that struggle of the seventeenth, was thought to be cast

on the one side, the gateways to Pennsylvania and Baltimore at least, besides Washington. The stake was even larger, for it comprehended the Federal army of the East and food, shelter and clothing for the Confederates, who were in rags and wretchedness. It meant Confederate recruits in large numbers. It meant foreign recognition. If it presaged a divided country, who can count the consequences except to the slave. It meant nearly valueless greenbacks, loss of prestige, chilled hopes, destroyed railways and bridges, confusion, martial law, and the thousand evils that fall to the lot of the great vanquished to be harvested by the victor. How countless are the spoils of war and how it feeds on injuries. On the other hand was thought to be wagered the death knell of the Confederacy ; a rehabilitation of the States and a return of peace and plenty and of home and quiet joy. A saving and an uplifting of a nation in a day.

As the fog lifted, the great guns spoke. Then away over to the right, across the creek, Hooker's infantry moved upon the enemy. The work of battle was well begun early. The forces of Doubleday, Meade, Ricketts, Mansfield, Hartsuff, Williams, Greene, Crawford, Gordon, Sumner, and other corps, division or brigade commanders, were soon heavily involved on the side of the Federals, with Jackson's, Lawton's, Trimble's Starke's, Jones', Hood's, and others, of the Confederates. New batteries and fresh supports were being gradually, but surely, drawn into the struggle. Union Generals Sedgwick, Richardson and French hurled their masses against the foe, and a Federal triumph seemed assured. Dunker Church had been taken, and re-taken, and the woods and fields thereabouts were strewn with dead and dying. The flying enemy are reinforced. McLaws and Walker have arrived from Harpers Ferry, and, rushing into the breach, the Union troops are forced back again. Ewell also assists, but his force is more than counterbalanced by Franklin's, of McClellan's command. Still, the chances are with the Northern soldiers. It was about this time that McClellan wrote the following telegram, dated 1:20 P.M., seventeenth, to General Halleck :

Please take military possession of the Chambersburg & Hagerstown Railroad, that our ammunition and supplies may be hurried up without delay. We are in the midst of the most terrible battle of the war—perhaps, of history. Thus far it looks well, but I have great odds against me. Hurry up all the troops possible. Our loss has been terrible, but we have gained much ground. I have thrown the mass of the army on the left bank. Burnside is now attacking the right. I hold my small reserve, consisting of Porter's Fifth Corps, ready to attack the center as soon as the flank movements are developed. I hope that God will give us a glorious victory.

It was one, P. M., when Burnside secured a crossing, and at three, he stampeded the small force in his front. Now, again, victory seems certain, but it is snatched away by the sudden arrival and appearance of A. P. Hill's troops, which fall upon the left flank of Burnside's corps. The climax has come and gone, and neither side is complete master. Both have been torn and shattered. It is while the tide of battle ebbs and the evening mist comes to spread its mantle over the scene, that McClellan, thinking more of the morrow than of to-day—more of eventual success than present misery, pens another message, this time for the Chief of Ordnance, in Washington, saying:

If you can possibly do it, force some twenty-pound Parrott ammunition through to-night, *via* Hagerstown and Chambersburg, to use near Sharpsburg, Md.

Think of it—time and space ignored in this greatest of human exigencies! What will not the railroad and telegraph do? What a courier! what a carrier! that ammunition, hundreds of miles away, may, at the close of one day's fight, be ordered and on the morrow received in time to open anew the battle in which a nation is the stake.

But it was not renewed, and each side attended to the dead and wounded. One pauses to think of that dreadful scene, before writing the sum total of misery, just as one reading of it stops to comprehend its fearful magnitude. Twenty thousand and twenty-six men had been shot there that day; thirty-six hundred of them killed, of whom two thousand and ten were Federal soldiers. Of the wounded, the Union men lost nine thousand, four hundred and sixteen, to the Confederates about seven thousand. So much for a conflict of ideas. The war correspondents found it impossible to get their long reports off by telegraph, so heavily were the telegraphers occupied with weightier matters, but telegrams, short and decisive, announced general results and particular casualties. Messages came pouring into camp from all parts of the North, without end, and nearly all bore the same substantial inquiry: "How is my son," "my brother," "my husband," or "my father?" Even weeks afterwards, these anxious inquiries came over the wires, sometimes addressed to the party himself. Perhaps he was dead, perhaps wounded, and removed—no one knew to what hospital; then would come another to his captain, to his colonel, and not infrequently to the general of the army. The words, by this time, were most appealing. The very tears of a mother or wife might be seen in the language of the inquiry. It was in this humane work that the operators exhausted their strength, which was only revived in snatches of fitful sleep, on the ground, or office floor. Caldwell, Bunnell and Emerick, at McClellan's, were sending orderlies all over the field and all through the camps, seeking answers. How little of what these operators did to secure replies, and of their severe service at the key, was ever known to those most benefited, but the aspersions heaped upon them when their messengers failed, were numerous, and the telegraphers thought unjust, as in the small hours of the night the head nodded, while the hand wrote incoming, complaining office messages.

On the nineteenth, at 10:30, A. M., McClellan sent this catholiconic telegram:

> Pleasanton is driving the enemy across the river. Our victory was complete. The enemy is driven back into Virginia. Maryland and Pennsylvania are now safe.

That telegram said to the one hundred thousand Pennsylvania militiamen, "Lay aside your arms." It said to the merchants, who closed their houses of business at three, P.M., to drill, "Renew your vocations, as usual." It said to the Maryland refugee, "You may go home." It said to the nation, "Saved;" but it did not mean to say that a death-knell to the Confederacy had been sounded; that the States were to be rehabilitated, and soldiers returned home. That kind of a "complete victory" was not won at Antietam, for, at best, McClellan had not captured over five thousand unwounded prisoners. But there was more in that victory than appeared in reports from the front, for President Lincoln had, as he subsequently said, made a solemn vow before God, that, if

General Lee was driven back from Maryland, he would crown the result by the declaration of freedom to the slaves, and, accordingly, on the fifth day after Lee's defeat, the proclamation was issued. A new phase to the war was thus presented—new hopes, new fears, increased animosities ; but the freedom of four millions of slaves was officially determined upon. Surely, Antietam was a great stake, for in that question of slavery there were sharper issues than in State's rights, or separation. Slavery was the kernel of secession.

Lee slowly moved to Martinsburg and Winchester, after dealing Porter's corps a severe blow near the Potomac, for following so closely. The Army of the Potomac spread out along the river, especially near the fords. September 21, McClellan's quarters and office were moved half a mile west of Sharpsburg, and on the twenty-eighth again moved two and a half miles farther west ; again, to near Knoxville, where the wires, as before, connected Washington, D. C., and Greencastle, Md. October 13, head-quarters were once more moved into Pleasant Valley. Emerick left McClellan's at this point, to open an office at Burnside's quarters, and immediately received a telegram from Washington, offering Burnside the command of the army—an offer which was declined in the expressed belief that McClellan should be retained.

In a day or two Emerick was sent to Couch's head-quarters on Bolivar Heights, relieving Lounsberry, who had remained with the army after his capture. At this time Ed Conway and Charles W. Moore worked the main office at the Ferry. Moore had just been returned from Libby Prison. Thomas Dolan was located at Sharpsburg. A line was built from Hagerstown to Hancock and to Point of Rocks *via* Harpers Ferry, and from the Ferry to General Sumner's quarters, half a mile distant.

October 10, Stuart crossed the Potomac, and with two thousand cavalry raided Chambersburg, cutting the telegraph, and recrossed, having passed around the Union army. October 25, McClellan began crossing into Virginia at Harpers Ferry and Berlin. McClellan made his head-quarters three-quarters of a mile from Berlin during something like a week that the army was crossing. All the supplies were wagoned over at Berlin to Lovettsville, and thence distributed to supply the army until it could be cared for by the Manassas Gap Railway. Lovettsville lies at the foot of Loudon Heights and is three miles from Berlin. The telegraph into the dominion by this route ended here. A few weeks later the post was abandoned and the line removed. Skirmishing with Lee, especially cavalry encounters, became frequent. November 6, McClellan's army lay near the Manassas Gap Railroad, near Warrenton, and general head-quarters were at Rectortown, the Confederates being mainly in and about Culpeper Court House.

The lines *via* Manassas Junction were already restored. Large reinforcements were arriving from Washington and operations once more indicated decided action. But on the 7th of November McClellan was suddenly and unexpectedly deprived of his command and General Ambrose E. Burnside installed therein. Not to digress and discuss the justice of the order or its wisdom at that particular time, it may be asserted that the army felt his removal as a great wrong to him and a severe blow to the cause. This was no less so among the department operators, especially at head-quarters, for the general held the telegraph and its servitors in high regard. Writing from Warrenton, Virginia, five days later to a comrade, McClellan's chief operator, Caldwell, said:

> We are all grieved at McClellan's removal. The whole army, from major generals down to foot orderlies, cried like babies when he left. Old soldiers of the regular army, who have been in the service since he was born, wept like boys. I never saw such a reception as the men gave him when he took leave of them. I hardly think McClellan will have another command soon. He being the ranking major general of the army, it is not thought he will take an inferior command. I had quite a talk with him before he left. He thanked me for the manner in which I had always conducted my duties, and told me if ever he had another command he wished me to feel that I must be with him. He sent me an elegant letter and I am proud of it.

16

CHAPTER X.

THE TELEGRAPH IN WEST TENNESSEE AND NORTHERN MISSISSIPPI DURING THE LAST HALF OF 1862.—FARMINGTON.—IUKA.—CORINTH.—VAN DORN'S AND FORREST'S RAIDS.—GRANT VERSUS THE U. S. M. T.—CHICKASAW BLUFFS.—ARKANSAS POST.

Upon Colonel Wilson's retirement from the military telegraph service, all telegraphic operations in General Halleck's great department devolved upon Major Smith, who had accompanied the latter to Pittsburg Landing. As Smith's duties required his presence in St. Louis, he made the following appointment:

PITTSBURG, TENN., April 18, 1862.

Mr. Duncan T. Bacon is hereby authorized to take command of the construction party and the working of military lines at Pittsburg, Tenn., and as they may be built hereafter, until further orders. All operators will report to him for orders. Mr. Bacon will report daily to head-quarters telegraph office or to Captain McLean, A. A. General, for orders, and carry out such as he may receive with the utmost promptness. (Signed) GEO. H. SMITH,

Supt. Mil. Tel., Department Mississippi.

J. S. Burlingame was appointed foreman of builders. Bacon caused an office to be opened at Halleck's quarters on the bluffs, and a field line to be erected, connecting various division headquarters. John C. Sullivan was manager of Halleck's office; Philip Bruner, of General Buell's, was assisted by Lewis B. Spellman. Ira G. Skinner had charge of General Grant's office, having been with him since August 16, 1861, and Wayne H. Parsons managed General Pope's. Pope having succeeded in capturing Island No. Ten, joined Halleck at Shiloh with the major part of his forces, say twenty-five thousand, and by the 30th of April Halleck had a well-appointed army of one hundred and twenty thousand men, under some of the ablest officers in the country.

Corinth, not Beauregard's army, appears to have been the objective. It was thirty miles away. The Federal advance began on the 30th of April and entered Corinth on the 30th of May, thus averaging one mile per diem. The Confederate forces were far inferior in numbers and organization. All accounts agree that Halleck's caution was unwarranted and prejudicial to the Union cause.

Besides the operators named who took part in this campaign, there were at various head-quarters Frank S. Van Valkenburg, who came with Pope, and Alva S. Hawkins, George Purdon, J. T. Tiffany, Benjamin H. Peebles, C. W. York, Douglass Reid and Mark D. Crain. Field lines at all times connected the several division head-quarters with Halleck's, where L. C. Weir was cipherer. At one time Pope was directed to cause a reconnoissance to be made to Farmington, a small town four miles from Corinth. Solomon Palmer constructed a field telegraph as the troops advanced on the place, and Parsons went as operator. The telegraph party was under the protection of a company of the Third Illinois Cavalry, Captain Smith. No difficulty was experienced in driving the enemy and entering the place. That accomplished, the main force returned, leaving General Buford in command and Parsons to do the operating. At dawn of the next morning, the cavalry pickets reported the Confederates were coming in force, and immediately the battle of Farmington began. General Pope was constantly advised by telegraph of the situation, and Captain Smith of the telegraph guard inquired of Pope if he should remain longer. Pope answered: "Yes; stay there till you see the enemy or are driven out." In five minutes the enemy opened fire with grape and canister, raking the shingles from the roof of the house Parsons used as an office. The dirt flew about when the destructive missiles struck in the front yard. The enemy deployed so as to capture a part of the guard. Parsons then hurriedly telegraphed, "The rebs are right on us," and jerking the instrument from its fastenings placed it in a rear pocket of his coat. Mounting his horse, he plied his spurs for the rear. While thus making the fastest time unrecorded, a shell ploughed the ground about his horse's feet, with no other effect than to increase the pace.

Parsons soon connected his instrument again and reported to Pope, but within ten minutes shot and shell were falling about him. One shell burst on the ground within ten feet of him, a piece of it tearing the leg from a boot worn by one of the cavalry escort. Again Parsons retreated and again connected his instrument, and was a third time driven off. Ten thousand of Van Dorn and Price's soldiers from the west were pressing a single brigade. Parsons lost his instrument from his pocket, but Captain Smith restored it. On Parsons' arrival at Pope's at four P. M., tired, dusty and hungry, great was his chagrin, because the general discredited his reports from Farmington, and had told the operators that Parsons "was scared and should be placed under arrest," but General Buford convinced Pope that great praise was due Parsons for his gallant services, and Pope himself subsequently commended him therefor. It is not usual in war to require such exposure of civilians, but the military operator was a frequent exception.

While lying in front of Corinth, Halleck's provost marshal issued an order forbidding the landing of any liquors and closing the bars on all steamers on the Tennessee. The operators at Halleck's were puzzled to circumvent the order, believing liquor no worse for them than for the staff. Finally one of the telegraphers called on the marshal and with great *naïveté* remarked parenthetically, that the field lines were nearly useless for want of battery material. The marshal inquired what article they were in need of and was informed of the lack of acid. "By the way," said the operator, "If we had a barrel of alcohol, we might use it as a substitute until supplies arrived from St. Louis." The marshal was very sorry that he had no alcohol; the operator was inwardly struggling to keep his countenance as the marshal, having only the advancement of the service at heart, added, in all seriousness—"But I have several barrels of confiscated whisky and if one of them would be useful in strengthening the battery, it will be sent over at once." The operator drolly allowed that whisky was not so good a substitute as alcohol, but was convinced that if he had a good article, he could keep the telegraph in operation. Accordingly, the marshal sent over a barrel of his best.

A contraband sunk a hole within the office tent, and the barrel with all due solemnities was placed therein and covered with earth. A small elder stick, with the pith driven out, protruded from the barrel a few inches above the ground. The battery material was tested and the lines worked beautifully (hic). The operator was loud in its praise to the provost marshal. In the afternoon while one of the operators, measured his length on

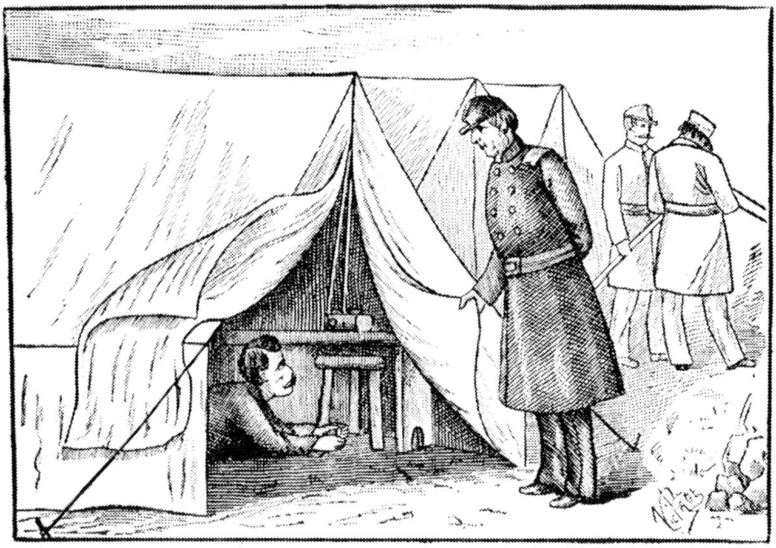

DRAWING BATTERY MATERIAL.

the ground, drawing battery material, General Halleck unexpectedly entered the tent and being of an inquisitive turn of mind, rendered more acute by the blushes that mantled the cheeks of the operator, he inquired "why so prostrate, young man?" to which the confused operator stammered, "I was drawing battery material, General." Halleck plied the operator with questions until he received the whole story, which he laughed heartily over. Having finished his business he left, with the remark that he should have to see that the marshal was better posted on the requirements of the telegraph. As the story soon leaked out, the quarters became popular and the substitute ebbed away; but for months the marshal's decanter, labelled "Battery Material," was the solace of sympathizing friends who called to condole with him.

It may seem somewhat remarkable that Confederate operators about Columbus, Kentucky, had successfully adopted the same ruse, as already related, but the acid theory was strangely popular. Major Eckert visited City Point in 1864, on a tour of inspection. On his way, he concluded to send a barrel of whisky and a quantity of quinine to the army for the use of the operators, many of whom were victims of malaria and exposures. But on nearing Jamestown Island, the steamboat captain told him he had a demijohn of acid for H. W. Cowan, operator there. Knowing how hard Cowan fought malaria, Eckert inspected the demijohn and remarked to the captain that it was acid for the electropian battery, and would not be needed, so the steamer was not stopped. George Henderson, John D. Tinney and Edward Conway, operators, on board investigated the article even more closely than the major had done, convincing the latter that he need not send the barrel. It is said that when Major Eckert returned to Washington he very solemnly reported that the operators at the front were found in good spirits.

Corinth was evacuated on the 30th of May, and the Federals took possession. General Williams' house, fronting the park, was selected for telegraph head-quarters. Many of the operators gathered there and related the story of their respective services and hair-breadth escapes, but above all congratulated one another on being that night in Corinth, Mississippi, and in telegraphic communication with Nashville, Tennessee.

The joy which the tidings of the capture of Corinth created in the North was warranted by anticipations, but was not justified by realizations. Buell and Pope, with seventy thousand men, pursued Beauregard a short distance without effect. About the middle of June, Buell was ordered to move with Wood's, McCook's and Crittenden's divisions toward Chattanooga, which he did, repairing the Memphis & Charleston Railroad as he advanced. Thomas and Nelson, with their divisions, followed during the summer. Grant was sent to Memphis; Rosecrans had already arrived from the East and succeeded Pope, who was sent to the Department of the Potomac and most of whose troops were returned to Missouri. We will take up Buell's line of march in good time, first devoting our attention however, to West Tennessee and Northern Mississippi. After Buell had gotten well under way for Chattanooga, Bragg, who superseded Beauregard, also started therefor, leaving Van Dorn and Price in North Mississippi.

General Halleck, directly after taking Corinth, ordered Lieutenant Colonel McPherson to superintend the repair and operation of the railroads leading to Jackson, Tennessee, and

SMITH'S PRIZE.

Columbus, Kentucky, and from Jackson to Grand Junction, and thence to Memphis. Major Smith, of the Telegraph Corps, was assigned to the task of opening the wires to Cairo and Memphis. "Assigned," is hardly the word, because he anticipated the order in his loyal haste, for scarcely had the enemy left Corinth when Smith and Captain Brackett (the latter in command of a company specially detailed to guard and aid the telegraphers), having heard that the Confederates had unwittingly fired a bridge near Chewalla before the last retreating and heavily laden trains had passed over, hurried to the burned bridge and found six locomotives partially wrecked and their trains badly destroyed. Parts of engines were found in the swamps near by ; one of the engineers was captured by Brackett's men and by great industry and perseverance and search for pieces in the swamps, an engine was repaired, so as to be worked on one side. Its woodwork had been mainly destroyed, and broken and bruised as it was, it presented a sorry appearance, but Smith knew how useful it would be to him. About sundown he burst into Corinth with his prize, which filled the air with its screams, that sounded like melody to the troops. A great shout went forth from thousands of throats. It was not long, however, before an army officer demanded that the engine be turned over to him, but Smith peremptorily refused, telling the officer that there were more where this one came from and he should go there and help himself. A guard however, was about forcibly to eject Smith, when Halleck was appealed to and interference prevented. It was with this one-legged locomotive, that the repairs of the line were hastened.

Some days before any Federal troops had reached the vicinity of Jackson, Tennessee, Smith and a few others, taking a handcar north of the first break in the road, pushed on through that guerrilla country, repairing the line then not much injured, at least as far as Jackson. It was a daring, if not a reckless operation, and Halleck reproved him for it, but then, nobody in war censures another very severely for being too brave, if successful.

However, the line was at once cut again, and John C. Sullivan and O. W. Paxson, operators, were ordered to accompany a part of Colonel Bissell's Illinois Engineer regiment on flat and dump cars by night, to avoid guerrilla parties, while *en route* repairing the line. Six miles out they found a bridge over a creek in flames. Two of the near residents being caught away from home were brought to the bridge ; a court-martial convened and preparations were apparently made, while the trial progressed, for hanging them. After the prisoners had waited an hour for the decision, the court reported that it could not agree upon a judgment, and the prisoners, thoroughly frightened, were released, after which, it is said, that bridge was the constant object of their watchful attentions. The line was opened to Columbus in due time and thence to Cairo *via* Paducah.

Beauregard retired to Okolona, Mississippi, leaving all of West Tennessee an easy conquest, and early in June Union General McClernand's forces moved westward to Bolivar and Memphis, causing, as we have seen, the evacuation of Fort Pillow, above Memphis.

Sullivan, aided by Henry Wilson, William Lane and three other builders, repaired the line from Corinth to Memphis, traveling by hand-car, which was carried or drawn over or around breaks in the railroad. About thirty miles west of Grand Junction they rested over night at a planter's house, where eight rebel cavalry lay fast asleep on the arrival of Sullivan's party, and as the lady of the house did not know which side would finally triumph, she did not apprise either party of the presence of the other, but as the Southerners had first retired, she considerately aroused them before daylight and sent them on their way. Ten days later, as repairers were passing this same place, an unspiked rail was drawn from the track by a wire attached thereto, thus ditching the car and three men, who were made prisoners, and who, after marching some twenty miles, were paroled; but the report of their death had been started. This report originated from blood being found about the wrecked car. One of the guerrillas had nearly severed a hand by a hatchet blow aimed at the car.

Among other citizens of Memphis remaining after its capture were Ed and Alf Saville, Canadians, but thoroughly imbued with the doctrines of secession. These Memphis operators were for a time permitted to remain, but being suspected of communicating with the enemy, they were sent beyond the Federal lines. Ed Saville then took charge of the office at Meridian, Mississippi, and it was while manager of that office that he acted as a Confederate spy, tapping the Memphis and Corinth line, the story of which is related by the New York *Tribune* Memphis correspondent, A. D. Richardson, in his "Field, Dungeon and Escape." The W. H. Hall mentioned in the account, worked in Memphis with the Savilles when that city was captured, and was imprisoned in August, 1863, on charges of disloyalty while engaged in the Cairo office. It was said that he was ardently attached to a Memphis belle, who was devoted to the Southern cause, and that in his letter to her father he divulged important Federal secrets. He was confronted with those letters by Generals Hurlbut and Veatch, before being sent to the Memphis military prison in the Irving block. It is difficult to restrain uncomplimentary remarks about those who employed him, for whether guilty or not, his antecedent service was enough, however much he might protest fealty to the Union, to have caused his exclusion from Federal military telegraph offices. Richardson's account is as follows:

Halleck continued in command at Corinth. From some cause his official telegrams to General Curtis in Arkansas and Commodore Davis on the Mississippi, were not transmitted in cipher, and the line was unguarded, though leading through an intensely rebel region. In July, the Memphis operators, from the difficult working of their instruments, surmised that some outsider must be sharing their telegraphic secrets. One day the transmission of a message was suddenly interrupted by the ejaculation, "Pshaw! hurrah for Jeff. Davis." Individuality reveals itself as clearly in telegraphing as in the footsteps or handwriting. Mr. Hall, the Memphis operator, instantly recognized the performer by what musicians call his "time," as a former telegraphic associate in the North, and sent him this message: "Saville, if you don't want to be hung, you had better leave; our cavalry is closing in on all sides of you." (This was a ruse.—Author.) After a little pause, the surprised rebel replied: "How in the world did you know me. I have been here four days and learned all about your military secrets; but it's becoming a rather tight place, and I think I *will* leave. Good by, boys." He made good his escape in the woods. He had cut the wire, inserted one of his own, and by a pocket instrument perused our official despatches, stating the exact number and location of United States troops in Memphis. Reinforcements were immediately ordered in to guard against a rebel dash.

S. H. Beckwith, who was in the battles of Fort Donelson and Shiloh, and accompanied his regiment to Corinth and Jackson, Tennessee, was, in the middle of July, detailed to assist the operators at Halleck's in Corinth.

On the 1st of August, Superintendent David, of the West Virginia Department, having previously been commissioned captain and assistant quarter-master, was sent west where there were no commissioned officers in the Corps, with instructions to examine into the telegraph affairs and make such changes as his experience and observation should dictate. He recommended that Randall P. Wade be assigned to Missouri with John C. Van

Duzer as manager, saying that Van Duzer would be superseded by C. E. Bliven, and was doing too good work to be overlooked. Samuel Bruch, recommended to Stager for a commission, was made captain, and his jurisdiction made to include Kentucky and Tennessee. Van Duzer, who took hold again as assistant manager upon Bacon's taking the Memphis district, was retained in West Tennessee. Bliven was given a small appointment about Covington, Kentucky; Wade, a department north of the Ohio and east of the Mississippi rivers. Robert C. Clowry, chief operator since April, 1862, at head-quarters, St. Louis, was placed temporarily in charge of lines in Missouri, and George H. Smith soon received a captain's commission and an appointment to the management of all the lines west of the Mississippi. David, at his request, was returned to his own department, after an absence of about six weeks.

Halleck's, and subsequently Grant's, lines of communication were very long, and continually extending—so much so, that the commanders were thereby required to scatter their troops until there was not a unitable force at command exceeding the enemy in front, who threatened Corinth, Bolivar and Jackson, chief points of Federal defense. Sherman occupied the Memphis & Charleston road, about Chewalla, and later (June and July) with his own and Hurlbut's divisions, repaired the road west of Grand Junction, and after a temporary occupation of Holly Springs, relieved Hovey, at Memphis, and ordered him to reinforce Curtis at Helena, Ark. This was late in July. Halleck was, on the 17th of July, made commander-in-chief of all the armies and ordered to make his head-quarters in Washington. This brought Grant back to Corinth.

In the meantime, the Confederate authorities, thoroughly alarmed at the inroads of the Federals (who already held all of Missouri and Kentucky and Tennessee, except about Knoxville and Chattanooga, and also had a foothold in Alabama, Mississippi, Arkansas and Louisiana, besides their Eastern occupations), had begun a merciless conscription, rigorous beyond precedent in this country, by virtue of which Lee's armies in the East were swollen, and McClellan was driven to the banks of the James; McDowell, Banks and Fremont were desperately striving to defend Washington, Maryland and West Virginia, and Buell was put to his wits' end to fathom and counteract Bragg's strong measures, menacing Louisville. Even Cincinnati was more terrified than Nashville, which was cut off from the world. Out of these mighty hosts, so "corralled," by a law, passed April 16, 1862, which substantially declared every man between the ages of eighteen and thirty-five a conscript, and every conscript who neglected to report for muster, a deserter, VanDorn and Price were reinforced, so that, notwithstanding the Confederacy was viciously on the offensive in Kentucky and Virginia, it spared VanDorn and Price about forty thousand troops, and among them those bold riders, Forrest and Armstrong and their cavalry, that needed but the order, to do and to dare. At this crisis, Grant telegraphed Washington: "My position is precarious, but I hope to get out of it all right."

The following lines, rebuilt or repaired in Grant's district, indicate Halleck's and Grant's telegraphic facilities, viz.:

	Miles.		Miles.
Columbus to Union City, Tenn.	26	Paducah to Columbus, Ky	47
Union City to Jackson, Tenn.	60	Moscow (rebuilt) to Memphis	39
Jackson to Bethel	35	Pittsburg Landing to Corinth	30
Bethel to Rienzi, *via* Corinth, Miss.	25	Pittsburg to Waynesboro	37
Rienzi to Booneville	8	Hamburg, Tenn., to Farmington	18
Rienzi to Jacinto	9	Savannah to Pittsburg Landing	12
Corinth to Decatur, Ala.	95	Field Lines to Halleck's Hd-qrs.	30
Corinth to Memphis, Tenn.	93	Iuka to Eastport, Miss.	9
Jackson to Memphis, *via* Grand Junction	100	Iuka to Rosecrans' Hd-qrs.	1
		Iuka to Grant's Hd-qrs.	1
Grand Junction to Waterford, Miss.	33	Corinth to Grant's Hd-qrs.	2
		Corinth to Rosecrans' Hd-qrs.	3
Cairo, Ill., to Clarksville, Tenn., *via* Ft. Donelson	188	La Grange to Grant's Hd-qrs.	1
Cairo to Blandville, Ky.	15	Grand Junction to Hamilton's Hd-qrs.	2
		Total	919

This grand total represents the importance attached to the military telegraph service, and illustrates the energy of the Corps. Nashville was in communication with Corinth, direct, *via* Decatur, and *via* Pittsburg Landing, and indirectly by way of Clarksville, Paducah, and Jackson. A foreigner must have difficulty in believing that these long cords ran through a brave enemy's country, and were at the same time successful; yet it is a great fact. It will be readily understood that if the line between Cor-

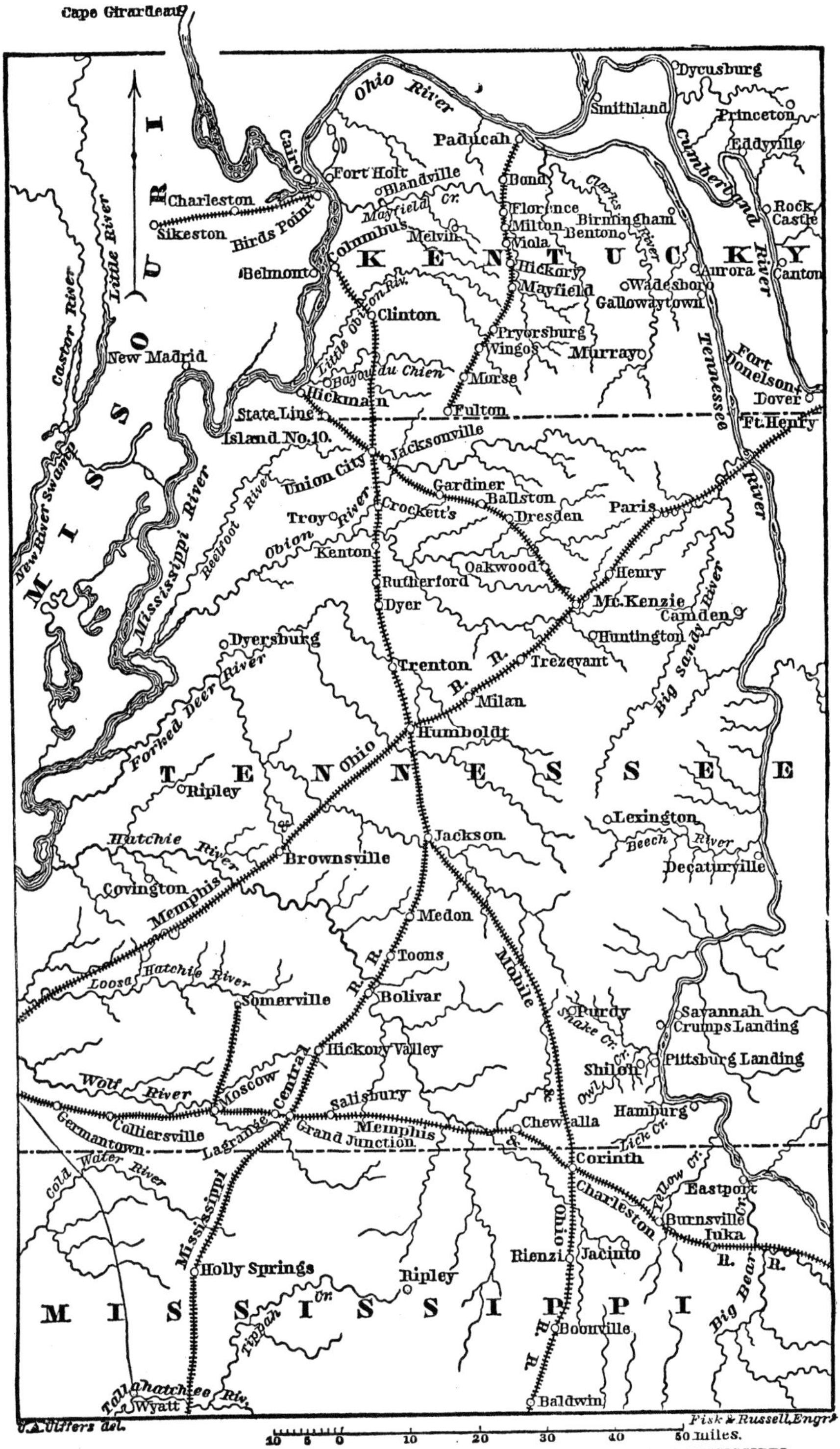

THEATRE IN WESTERN KENTUCKY AND TENNESSEE, AND NORTHERN MISSISSIPPI.

inth and Jackson was cut, these places could communicate *via* Decatur, Nashville, and Cairo, or Savannah, Nashville and Cairo. The offices on Grant's lines were managed by operators who were changed about a good deal, but, taking the month of September and including Memphis, which was not then in circuit, owing to the enemy's cutting the line, they were located as follows: Cairo, Ill.—(except while in Mound City Hospital, where the Sisters of Mercy assuaged their malarial fevers) W. L. Gross, manager; W. H. Parsons, Ed. Schermerhorn, John J. Egan, Steve L. Robinson (ten days), W. W. Forbes, Jacob V. Hill. Blandville, Ky.—James W. Atwell. Paducah, Ky.—T. R. Lewis, to sixteenth, and George Stillman. Columbus, Ky.—Edward H. Johnson and James K. Parsons. Union City, Tenn.—S. L. Robinson, to thirteenth, Stockton L. Griffin and Augustus Tyrrill. Jackson, Tenn.—James Q. Olmstead and Jas. H. Douglass. Bethel, Tenn.—Brace M. Burnett. Rienzi, Miss.—William B. Somerville, recently from Jacinto, Miss. Brittons, Tenn.—Thos. R. Berryhill. Corinth, Miss.—A. S. Hawkins, manager; Joseph Blish, Jr., Mark D. Crain, Anthony R. Walsh, Sam'l T. Brush, Algernon W. O'Neil and T. R. Lewis, from the twentieth: Iuka, Miss.—Edwin Peel, who was captured in a boat near Paducah, and who had served at Corinth, and probably escaped from Iuka with Colonel Murphy. Grant's head-quarters.—Ira G. Skinner and Samuel H. Beckwith. Rosecrans' head-quarters.—Horace W. Nichols and Lewis B. Spellman. Hamilton's head-quarters.—Lewis B. Spellman, in November, who was also with him at Clear Creek. Memphis, Tenn.—Duncan T. Bacon, who about this time took charge of head-quarters office, St. Louis, Frank S. Van Valkenburg. Henderson, Tenn.—Edward F. Butler. Trenton, Tenn.—Benjamin H. Peebles. Dodge's head-quarters, Humboldt, Tenn.—S. L. Griffin, part month, and John C. Holdridge. Bolivar, Tenn.—G. P. Lennox and J. T. Tiffany. La Fayette, Tenn.—William A. Thayer. Kenton, Tenn.—C. S. Whittlesey, who nearly died from sickness in his first month's service and resigned. Madison, Ala.—George E. Cromwell. Tuscumbia, Ala.—Edwin D. Butler and W. W. Forbes, from Cairo. Metropolis, Ill.—Charles Wallace, late of Kenton office. Mound

City, Ill.—George M. Brush. Mound City Hospital.—Frequent arrivals and departures. Caledonia, Ill..—George S. Pidgeon.

Very great trouble was experienced with the Ohio River cables at Paducah and Cairo. Two were laid in the Spring, but soon failed; one, owing to being destroyed by lightning; another was laid at Cairo in the middle of September, but gave out before night. The long circuit from Cairo to Corinth, Tuscumbia and Decatur for the most part worked heavy, *i. e.*, the ground escapes were numerous and occasionally prevented the use of the line. Nothing was so trying as the position of the operator who had important dispatches to send, and who could easily hear the office he wanted, but could not make himself understood because (as a rule) the operator at the distant office did not adjust his instrument so delicately as to catch the faintest change in the electric charge of the lines. When, however, he did respond, the fault was apparent, and it not unfrequently happened that Dante's Inferno was telegraphically portrayed in language fraught with blisters, and somebody consigned where time was not of the essence of the contract: of course retaliatory measures followed. It was after an occasion like this that Patrick Mullarkey, then working in the Memphis office, took a trip all the way to Louisville to see the operator who spoke, over the wire, inelegantly of him and his, at that presumably safe distance, but the truth was that "Pat" himself had an uneven temper. Another instance: "Hank" Cowan, while working at Bealeton, Virginia, during the war, became convinced that A. H. Bliss, operator at Rappahannock Station, deserved a whipping for telegraphic innuendoes which reflected upon Cowan's past and present, and assumed to foreshadow his future, whereupon "Hank" diplomatically pronounced it *casus belli* and mounting his Pegassus, flew to Rappahannock Station, but his ire, recently incandescent, had moderated when he reached the box car where Bliss reigned supreme. With a twinkle in his eyes, Bliss announced his readiness for the worst, but Cowan admitted that he had lost his anger, which

> "Is like a full hot horse, who being allowed his way,
> Self-mettle tires him."

Not to lose his ride entirely, he began a game of euchre without a cooling-card, and when Bliss charged him with cheating, a new *casus* resulted and Cowan's optics changed color, but bent on rewarding merit, he caused his photograph to be taken and duly presented to Bliss with appropriate resolutions.

About the 1st of September, the rebels struck a detachment of the Federals at Medon and Bolivar. Small parties occasionally cut the line, but decamped as rapidly as possible thereafter, and for the most part the wires ran alongside the railroad so that repairs followed speedily. On the thirteenth day of September, Price seized Iuka, which Colonel Murphy had abandoned to him, and Grant supposing Price entertained designs against Nashville or by the coming of Van Dorn from the south-west, hoped to hem in the Federals at Corinth, determined, if possible, to crush Price before Van Dorn, who was four day's march from Corinth, could arrive. To this end Rosecrans with nine thousand men was to approach Iuka from the south *via* the Jacinto and Fulton roads, and Ord with eight thousand troops from the north. Both came in good time but owing to the wind blowing from the north, Ord could not hear Rosecrans' guns, and Price, whatever his purpose, made good his escape to Ripley by the Fulton road which Rosecrans did not take. The battle between Rosecrans and Price was short and sevère, neither side being beaten.

Repairers James Raile and Horace Benedict while *en route*, east of Burnsville, about the thirteenth, to repair the line, were captured by Price's troops.

Price immediately began executing another plan against Corinth, but before it developed, perhaps before it was formed, *viz.:* on the twenty-third, General Grant changed his quarters to Jackson, the better to watch over Ord at Bolivar with nine thousand troops, and Rosecrans, who was left in command at Corinth with about eighteen thousand. October 2d, Price appeared before the city and cut the telegraph lines. On the next day Rosecrans developed the enemy's force and retired behind the defences to await the onslaught, but as soon as he discovered that Corinth was the objective, Rosecrans sent operator Beckwith, escorted by cavalry, around and behind the enemy to transmit telegrams to Grant. This was accomplished near Purdy, when the party safely returned with Grant's instructions.

Early on the fourth, the battle began. Mr. Van Duzer was there with his operators, Hawkins, Blish Jr., Crain, Somerville, O'Neil, Lewis and Walsh. Jean Jourdé, the operator's *chef de cuisine* was there also. He rose early to prepare some great dish, perhaps it was *potage au gras*, possibly *pâtè de foie gras*, more likely *Schweinfleisch mit Bohnen;* but a shell suddenly fell and burst just outside of the operators' quarters, whereat Jourdé's hair assumed the perpendicular. Hawkins hurried out of bed, but none too soon, as a moment later, another shell came crashing through the building and struck the cot from which he had just risen, throwing it in atoms against the ceiling. Only a pale, sickly smile escaped him, as he contemplated how near eternity he slept. Van Duzer was in the same house, too sick to be about, but the "leaden rain and iron hail" gave him strength to seek other quarters without delay. Jourdé's excitement resolved its expression into the *langage des halles* and mutilated *anglais* as follows: "By gar! I no hook ze brakefas in ze plass. I shall go, *you shall come*, by gar!" and the boys stood not upon the order.

That was a dreadful conflict that raged until noon of that day, in and around Corinth, between eighteen thousand Unionists and thirty-seven thousand rebels. Battery Richardson was carried by assault and the enemy under Price rushed to the very center of the town, even to Rosecrans' head-quarters, but Van Dorn had been delayed on the left and, besides the Federal musketry, forts Chapman, Williams and Robinet opened an enfilading fire upon Price's troops, repelling them with great slaughter. Then Van Dorn's command, led by that brave Texan, Rogers, stormed Robinet. One brigade fell to pieces, but another pressed on; then that, too, crumbled and, leaving dead, wounded and captured, what remained of it hurriedly sought safety beyond the reach of Federal guns. The casualties were frightful, for, it is said, one thousand, four hundred and twenty-three Confederates were killed and nearly six thousand wounded. During the battle and the pursuit, two thousand, two hundred and forty-eight prisoners were captured.

The defeat of the troops under Van Dorn, Price, Lovell, Villipigue and Rust at Corinth, and their hasty flight from Ord's forces, which came out from Bolivar, and Rosecrans' troops, which started in pursuit, aided by McPherson, who had come down from Jackson, carried the greatest consternation into the enemy's ranks and as well also throughout the whole South. Of course the Unionists were correspondingly emboldened; moreover, Grant was receiving reinforcements. He therefore determined to assume the offensive. Rosecrans was, on the twenty-fourth day of October, ordered to supersede Buell, whose forces were then mainly in Kentucky. Three divisions were sent from Corinth and two from Bolivar to Grand Junction, where they arrived about the 4th of November. From this place on the M. & C. R. R. a single track line of railroad ran to Jackson, Mississippi, and Vicksburg. It was by this line that Grant had to feed and clothe his troops, and bring his munitions of war as he advanced. Lieutenant-General J. C. Pemberton succeeded Van Dorn (who, however, remained), soon after the Corinth affair, and fortified a line of defense along the Tallahatchie, in the neighborhood of Wyatt and Abbeville, Mississippi. Sherman came down from Memphis, and as Grant was about to flank Pemberton, the latter fell back without a struggle, so that by the 5th of December Grant was at Oxford and his cavalry within eighteen miles of Grenada. Sherman, about the 8th of December, was sent back to Memphis to organize a river expedition against Vicksburg, as we shall see, in co-operation with Grant's own advance against that stronghold by land.

But while Grant was arranging for his own advance, Sherman embarked on the nineteenth and steamed down the river to Helena, Arkansas, where, on the twenty-first, Curtis' troops under Steele joined him for the joint objective.

December 20, Van Dorn's cavalry destroyed Grant's immense accumulation of stores at Holly Springs, Mississippi, and without much resistance captured its garrison, commanded by the same Colonel Murphy who, a few months previous, evacuated Iuka to Price, and who was dismissed from the service for his misconduct at Holly Springs, which seems too easy a sentence for his great neglect. Van Dorn also struck the communications higher up, meeting determined resistance at Coldwater, Davis Mills, Bolivar and Middleburg: Forrest at the same time was on the railroad in Tennessee. These blows determined Grant, who was unused as yet to feeding his army off the country, to return northward and try the river route. But the lines of telegraph were so badly destroyed that it took about two weeks to reopen them, and Grant, having already planned another route for his entire army, determined to evacuate the railroad north of Jackson, but hold the line from Memphis to Corinth. Having already given a sufficient outline of military events under Grant to indicate the importance of his undertakings and the consequence of Sherman's co-operation, we will now inquire more particularly, what occurred in this district, of interest to the telegraphers as such.

Although we have mentioned the building of a line to Grand Junction, it had to be rebuilt, and October 11, orders were issued by Grant therefor. Offices were opened at Midway, Pocahontas and Grand Junction. Sol Palmer, assisted by Mark Crain, Ed Schermerhorn, H. C. Weller and J. S. Burlingame, foremen, rebuilt all the lines required by Grant during Van Duzer's management in Mississippi and West Tennessee. Grant's office at Jackson was closed at noon of November 3, and opened at Grand Junction the fourth, and later at La Grange. Van Duzer was directed to send all the operators he could to the front, wherefore, all left Corinth except A. J. Howell and C. H. B. Gile, new men, and Tony Walsh. Some of the offices were closed or their force reduced, so that Schermerhorn, Crain, Johnson (B. H.), Sullivan, Beckwith, Skinner, Parsons, Hill, Ingle, Nichols, Somerville, Hawkins, Spellman, O'Neil and Lewis were at Grand Junction or near by, and J. S. Lyle, Jacob V. Hill, R. B. Griffin, D. K. Smith, Ellis Stone, J. W. Atwell and C. W. Pierson were newly employed on these lines.

Redington, at Medon, and Lenox, at Bolivar, appear to have fallen under suspicion of being disloyal, but the former, as we understand, was released from prison at Memphis and sent to Fort Donelson office, positive evidence of his innocence; and as to Lenox, a journal of that period, kept by an able operator, reads as follows: "He was arrested at Bolivar for disloyalty, and ordered out of this military district within three days. *That he is really disloyal, no one believes;*" but it was one of the misfortunes of an operator's position that he could not have a court-martial. A. D. Dougherty, operator at Smithland, was removed in 1863 on charges affecting his character, and though he protested bitterly against such action without a fair hearing, there was no such privilege granted him. Such rights were not secured to civilians in any of the departments.

One night about the last of October, a Confederate general and colonel, whose homes were at Bolivar, it is said, visited their families there. A negro servant of one of them informed the operator, J. T. Tiffany, of their presence, whereupon he took a squad of soldiers and surrounded and captured the officers, who were taken to General Logan at Jackson. But as the story reads that these high officers took the oath of allegiance and were released, it should be taken *cum grano salis.*

September 5, the following was telegraphed to all operators:

Walter Campbell, an operator, formerly connected with this line, is suspected of making use of his privileges, as an operator, to the detriment of the service. He must be entirely excluded from offices, and if he loiters about within hearing of the instruments, you will make the case known to the officer in command, that he may be removed or arrested. (Signed) J. C. VanDuzer, *Supt.*

This order doubtless originated from Tiffany's arrest by General Ross, for allowing a Southern operator to come into the telegraph office. Tiffany was kept in the old court-house until General McClernand heard of it, when he ordered Ross to release him at once. Probably Ross's fears were unwarranted by the facts.

The cotton fever, in West Tennessee, ran high, and as a consequence many civilians had, on one pretext or another, obtained passes to the cotton localities within the Union lines. These speculators did so much telegraphing that General Grant, on the 11th of November, directed that no private business be done till further orders. This was subsequently modified so as to allow it a few certain hours each day.

When the troops advanced southward from Grand Junction, B. H. Johnson, operator, accompanied Colonel Pride, Grant's Chief of Engineers. Johnson's first office below, was in a tent at Lamar, where Colonel Bissell's engineers were building a bridge. While here, the operator went foraging with some soldiers, and, having reached their destination, near ten miles out, the foragers were busy cutting up fresh pork and catching chickens, when a negro, mounted on a mule, came tearing down a hill; yelling at the top of his voice, "De grillas is after youens, jest over de hill thar." All of the party, except one, mounted and spurred homeward, Johnson in the van, until the camp was reached. If it was not on this ride, it ought to have been, for, at one time or another, in this vicinage, owing to some fault in the horse, or perhaps the stirrups were too short or Johnson's legs too long, in short, anything compatible with good horsemanship, the hind bow of his Mexican saddle cut a *semi*-circular slit in the seat of his pants, and made him sit standing for a week. After all, he fared much better than the forager left behind, for he was captured and shot.

Pending Grant's retrograde movement, Johnson, who was dressed in all the paraphernalia of rank, except shoulder-straps, went to the seminary building, at Oxford, where he was then stationed, to appease his hunger. About thirty soldiers were there, and, being on their retreat, recked little what they did, consequently they were eating the pretty schoolma'ams and misses "out of house and home." Johnson took in the situation at a glance, and, assuming to be bigger than a brigadier, ordered the soldiers to repair at once to their regiment, which they did. The young ladies thereupon, out of gratitude, treated him to a fine dinner. That same afternoon, while standing about his instrument with his coat off, thus exposing the half moon patch on his pants, proof positive that he was not the great Mogul he had appeared at the seminary, some of the soldiers whom he had driven away, rushed in at the door after Johnson, swearing vengeance as he bolted through the window and fled to the residence of Secretary Fox, of the Confederate Cabinet, where were General McPherson's head-quarters.

Scouts reported VanDorn's and Forrest's threatening movements in ample time to enable Grant to notify all the post commandants thereof, which he did promptly, by telegraph, but he especially warned Murphy, at Holly Springs, to prepare for Van Dorn, and Sullivan, at Jackson, Tenn., to look out for Forrest. Notwithstanding, as we have seen, VanDorn, at day-break, entered the place, almost unopposed. The telegraph office was at once visited. Operator Horace W. Nichols was at his instrument. Hearing stray shots and discovering the rapid approach of the enemy, he hurriedly thrust his dispatches into the fire, and his money into his boot leg, and then seizing the telegraph key, called Grant's office and said: "Good by. VanDorn is coming. Devil only knows what 'll become of me. Here they are, now," and, sure enough, they stood just by the window where his instrument was, covering him with revolvers, so that he was prevented from saying more. Lew. Spellman, from Grant's office, and Sol. Palmer took a hand-car and repaired the line, which was in operation to Grand Junction twenty-four hours after it was cut in the office at Holly Springs.

General Forrest, having recently returned to Columbia, Tenn., from his bold exploits upon Buell's communications, crossed the Tennessee at Clifton, but was met near Lexington by an inferior force of Federals, sent by Sullivan from Jackson, Tenn., and after a struggle, quite determined, it was driven back, thus enabling Forrest to threaten Jackson itself two days before VanDorn entered Holly Springs. Jackson being strongly defended, Forrest contented himself with striking, at night, the railroads leading therefrom, and the next day he took the Spring Creek road for Humboldt and Trenton. Grant had given orders that all commandants should hold their positions "at all hazards," and this all seem to have valiantly endeavored to do, as far on the road to Columbus as Forrest dared to penetrate, but he captured many small commands and telegraphers along his route, besides destroying the telegraph badly and burning miles of railroad trestle-work along the bottoms of the Obion. Humboldt and Trenton fell on the 20th of December; Kenton and some stockades, the next day, and Union City the twenty-second. By this time, Forrest was himself in great straits, but finally succeeded in eluding the Federal infantry and recrossing the Tennessee at Clifton after being badly worsted at Perryville, on the river.

J. C. Holdridge, operator at Humboldt, had his office in the hotel, of which he was also landlord. On the nineteenth, he reported to Cairo that the line was cut at five, A. M., eight miles north of Jackson, and in the afternoon that he could hear firing in that direction. The next we hear of John, he was at Kenton and his hotel property had been destroyed. At three, P. M., of the twentieth, J. S. Lyle, who remained at his post at Trenton, to the last, telegraphed to the Northern offices: "The rebs are in sight and the fun has commenced;" then the line was cut, but Lyle collected his dispatches and threw them with his instruments into the office stove, where they were destroyed. Of course Lyle was captured. At Kenton, Holdridge was captured about four, A.M., with his Humboldt instruments, which he could not destroy, as he was taken by surprise, and soldiers stood over him, threatening to shoot if he attempted it. Stephen Robinson, a small boy operated at Kenton, where three companies of the One Hundred and Nineteenth Illinois infantry were stationed. His office was on the Trenton side of the town, and the commanding officer agreed not to leave without letting him know, but the enemy were in the place, and the infantry out, before Robinson discovered his peril. He then wrapped his shawl around his instrument and walked away. He was too little to attract attention. Crossing the Obion trestle alone that dark night, he connected his instrument near Crocket and telegraphed the situation to Columbus. A train of cars sent out, was stopped by a fire built on the track by negroes, and the infantry, contrabands and operator, safely taken to Columbus. Augustus Tyrrill was captured at Union City, and Wallace W. Forbes somewhere between Columbus and Jackson.

On the very day that Forrest began his retreat, W. G. Fuller, late of West Virginia, and Central Kentucky, arrived with Captain Bruch at Cairo where, for reasons which we will soon explain, he was duly installed as chief officer, under Bruch, of the Corps in Grant's district, now designated department.

WILLIAM GREENLEAF FULLER, one of the most conscientious of men, was one of seven children—all under fifteen—left by the death of their father in South Carolina in 1838, to be supported by their widowed mother. His ancestry was of New England's best, being in part, of that from which John Greenleaf Whittier was sprung. On the demise of her husband, Mrs. Fuller returned to Massachusetts, where with her scanty means she sought to rear her family. How well she succeeded may be conjectured from the fact that she (now eighty-two years old) and all of them are still living. William, born in 1827, at nineteen, began work in a factory at Lowell, but, owing to ill health he was induced to ship as a fireman on the United States steamer

"Mississippi," fitting for a cruise in the Mediterranean Sea. War with Mexico threatened and the steamer went to the Gulf. He could not stand the heat of the furnace and was made assistant Ship's Yeoman. While at Vera Cruz, war was declared, and the "Mississippi" steamed to Point Isabel, where her force volunteered to reinforce General Taylor of the United States Army. Taylor was met just after his second victory at Resaca de la Palma. Returning thence to the steamer, which was the flag ship of Commodore Perry, Fuller participated in exploits and attacks along the Mexican coast, especially at Tampico, Vera Cruz, Alvarado, and Tobasco up the Tobasco River.

WILLIAM G. FULLER.

On the return of the "Mississippi" in 1847, Fuller learned that his mother had procured his discharge on account of his minority. Thereupon, he began learning the art of telegraphy. His first employment as operator was by F. O. J. Smith, an accomplished statesman, a learned jurist and pioneer in telegraph affairs, and Eliphalet Case, on the New Orleans and Ohio line, at Cincinnati, Ohio, in 1848. He served also, at New Richmond and Marietta, Ohio, which latter office, under his administration, was the only one on the line that paid more than expenses. In 1854 the company failed. The Ohio stockholders re-organized, and Hon. V. B. Horton, of Pomeroy, selected Fuller as superintendent of that portion between Wheeling, Va. and Cincinnati. When the Marietta & Cincinnati Railroad was constructed, Mr. Fuller rebuilt the lines over that route. In 1857, the Hon. Amos Kendall arranged with the Baltimore & Ohio and Marietta & Cincinnati Railroad companies, for an independent line between Baltimore and Cincinnati. Fuller laid the cable at Parkersburg and constructed the line thence to Cincinnati, where an office was opened in the Burnett House. At the beginning of hostilities between the States, Mr. Fuller was superintendent of the Marietta & Cincinnati Railroad line and the western division of the Independent Company's line. These responsible positions he was quick to surrender for service where he felt he could do more good to his country. No officer of the Telegraph Corps proved more intensely solicitous to forward the cause of the Union than William G. Fuller, and he probably constructed as many miles of military telegraph as any other person.

General Grant had already become convinced that he could not maintain his railroads and telegraphs, and the indications were anything but assuring to Captain Fuller. However, he entered vigorously upon his work, and by the 3d of January, the line from Cairo to Grant's head-quarters at Holly Springs, and by the 8th of January, the newly repaired line from Grand Junction to Memphis, were in complete working order. Ed. Schermerhorn, a reckless, daring nondescript, convivially inclined, but one of the most expert operators in the country, was in charge of a gang of builders on the Memphis line about the 1st of January, when all were captured by guerrillas, under one Richardson. The builders' names are T. Berry, J.W. Berry, C. D. Applegate, A. E. Atwell, C. D. Walker and George Bishop. Nearly all of these men and those captured north of Jackson, were paroled, some not to re-enter the Confederacy, and others not to act as Federal operators during the war. They arrived at Cairo about the same time that E. F. Butler did, he having had the misfortune while operating at Henderson, Tenn., late in November, to be captured by guerrillas, who burned the station and telegraph property, and took Butler to Forrest at Columbia, Tenn., where he was released on his promise not to return to Tennessee. All of the soldiers captured by Van Dorn and Forrest were also paroled.

Exciting as were the raids of Van Dorn and Forrest, to the telegraphers in Grant's department, they were not so demoralizing to the Corps as the move of the Federal commander himself, which was initiated on the 13th of November, 1862, by his appointing Colonel John Riggin, of his staff, superintendent of the military telegraph lines in the department. That was an exercise of power for which there were the precedents of McClellan's appointment of Mr. Stager in West Virginia, and Fremont's of G. H. Smith in Missouri, with this important difference: that since those appointments the Telegraph Corps had been organized, as heretofore explained, under the direction of the Secretary of War, who, on the 8th of April, 1862, issued an order—"No. 38"—substantially like that shown in special order, number three hundred and thirteen, on page one hundred and thirty. A few months later, the following order was issued:

HEAD-QUARTERS ARMY OF THE MISSISSIPPI,
In the Field, Corinth, July 14th, 1862.

Special Field Order, No. 156.

Telegraph operators and employés are, as a general rule, subject only to the officers of their own corps or orders *communicated through them* by the Secretary of War or commanding general of the department. Where an officer commanding a post or force in the field, deems it necessary to close or move a telegraph station, or to arrest or remove an operator, the reasons for so doing will be immediately communicated to head-quarters for approval. Military telegrams will always have preference and when the business of military lines cannot, in the opinion of the superintendent, be otherwise properly attended to, commercial and private telegrams will be excluded. The commanders of posts and stations will assign quarters for telegraph offices, and will see that all unauthorized persons are entirely excluded from operating rooms and their vicinity.

By order of Major General Halleck.

N. H. MCLEAN, *A. A. G.*

It is believed that General Grant did not design to relieve Van Duzer by this appointment, but merely to place the oversight of telegraphic affairs in Riggin's hands. Colonel Riggin was not a practical telegrapher, and although an accomplised officer and gentleman, he did not understand the duties of his new position, and hence, it may be, he gave but few orders; even those were inoffensive. But, as was his duty, Van Duzer informed his superiors of the new *status*, which information resulted immediately in the following telegraphic correspondence:

WASHINGTON, November 14, 1862.

To U. S. GRANT, Major General:

Some one, signing himself, "John Riggin, Superintendent of Military Telegraphs," is interfering with the management of telegraphs in Kentucky and Tennessee. This man is acting without the authority of Colonel Stager, general superintendent of military telegraphs — see general order No. 38, April, 1862 — and is an imposter. Arrest him and send him north of your department before he does mischief by his interference.

By order of the Secretary of War. P. H. WATSON,
Asst. Sec. of War.

To which General Grant replied as follows;

John Riggin, referred to in your despatch, is my aide. He has given but one order referring to telegraphing, and that was dictated by me. It was, that private despatches might be sent over the wires before ten o'clock, A. M., when they did not interfere with military despatches. Colonel Riggin is assigned the duty of military superintendent of telegraphs within this department, a position which interferes with no present arrangement, but is intended solely for my relief. Misrepresentations must have been made.

And thereupon Assistant Secretary Watson rejoined:

Colonel Anson Stager having been appointed by the Secretary of War, superintendent of military telegraphs and of the construction and management of military lines, Colonel Riggin must not interfere. Colonel Stager has appointed deputies, believed to be competent, but if they fail in their duty a report of the fact to Colonel Stager will bring a prompt removal.

It was soon found convenient to entrust Colonel Riggin with special duty at Memphis, and Van Duzer was directed to prepare to restore the line from Grand Junction to Memphis and from Memphis to Grenada.

The conflict of authority appears to have become quiescent, and so it remained until the 30th of November, when the operator at Holly Springs, who had that day opened that office, astonished all the operators in the department by the startling announcement that Van Duzer had been arrested and placed in the guard-house at that place. General Grant telegraphed Colonel Stager that Van Duzer must be removed and another man be appointed in his place, and that his (Grant's) orders must be obeyed. Just what the disobedience, if any, consisted of, we are not sufficiently advised to state. Assistant Secretary Watson directed General Grant to telegraph his charges against Van Duzer to the War Department. Although they were preferred, no trial was ever had, and it is believed that General Grant was himself subsequently convinced that he had been deceived by his chief operator, Ira G. Skinner. Van Duzer's confinement for four days was so close that no operator was allowed to communicate with him. Colonel Riggin again appeared on the scene, directing Gross, manager at Cairo, to keep supplies on hand and order what was required from the East. This was promptly countermanded by Colonel Stager. Gross advised the operators to continue to discharge their duties, notwithstanding their devotion to Van Duzer, as otherwise the consequences to the armies might be very injurious; moreover, he believed Van Duzer would be sustained. But the operators were bent on asserting the independence of the Corps. Grant heard of their purposes, and issued an order to every commander of his military posts along the lines to arrest every telegraph operator that resigned or attempted to leave his post or refused to do duty as theretofore, and place him under close confinement and to acknowledge the receipt of the order by telegraph. Somerville at Waterford was arrested, probably for refusing to serve, but most, if not all, of the others submitted with the best possible grace. Van Duzer was ordered out of the department, under guard. December 5, General Halleck telegraphed General Grant at Abbeville as follows:

The Secretary of War has called my attention to your telegraphic despatches in regard to operators and offices. Colonel Stager was charged by the President, under a law* of Congress, with the entire management of military telegraphic operations. He directs all purchases and appoints and removes all officers under the direction of the Secretary of War. If any operator fails in his duty report him for removal. In extreme cases he may be arrested, just as you can arrest any officer of the Navy or the Treasury De-

*There was no such law.

partment. Colonel Stager directs all telegraphic purchases. In case of deficiencies, report the fact and do not order purchases. They can not be paid for out of the Quarter-master's Department. Any orders to that effect given by you will be immediately countermanded.

By the time that Van Duzer reached Grand Junction, Lieutenant Colonel McDermot, commanding there, received an order from the Secretary of War, directing his release. But McDermot, either believing or feigning to believe that the order was bogus, declined to obey it. It is proper here to state that this is the only instance known to the author of the genuineness of a telegraphic order being questioned by a military of ficer during the whole war, except when there were reasons to believe it was telegraphed by rebel operators. After Van Duzer reached Cairo, it was not deemed advisable by Secretary Stanton to precipitate a crisis by ordering his return. Had that order been given, General Grant might have asked to be relieved. After a time, Van Duzer was given the management of the telegraphs under Bruch in the Department of the Cumberland and an honorary staff appointment with General Rosecrans, and Fuller, as we have seen, succeeded Van Duzer in West Tennessee. When Grant's department subsequently included the Cumberland and he appeared at Chattanooga, Captain Bruch proposed to send Van Duzer back to West Tennessee and bring Fuller therefrom. But Van Duzer determined upon resigning rather than to be sent from pillar to post; however, General Grant, if he ever entertained any personal feeling against Van Duzer, was just enough, probably owing to additional information obtained in the interim, and also in consideration of his valuable services under Rosecrans, to again entrust him with his fullest confidence, which, it is certain, General Grant never had occasion to regret.

One other unfortunate conflict culminated the very day of Captain Fuller's appointment, above mentioned, and it may have had some connection therewith. It grew out of the following order:

CLEVELAND, November 13, 1862.

TO ALL TELEGRAPH OFFICES :

A strict observance of the following order is required: The

original copy of every telegram sent by any military or other government officer must be retained by the telegraph manager or operator and be carefully filed, and at the end of the month, mailed to the War Department at Washington. You will retain a copy of all such messages. In no case will you allow the original ciphers to be given up or destroyed, except to prevent their falling into the hands of the enemy.

By order of the Secretary of War.

(Signed,) R. P. WADE, *Capt. and A. Q. M.*
and *Asst. Supt. Govt. Telegraphs.*

On the 8th of January, 1863, Captain Fuller reported by telegraph to General Grant for duty, and on the same day his operators informed Fuller that General Grant had "taken possession of yesterday's business," and the General himself telegraphed:

Whose order is it that originals of all telegrams passing over military lines be sent to Washington? An order for the disposal of military telegrams would come through me from some one authorized to give me orders. Private despatches I care nothing about. You can make your own regulations concerning them. Answer, sending me copy of order.

To which Fuller replied that the order referred to, was embraced in a letter of instructions originating with Colonel Anson Stager, General Superintendent of Military Telegraphs, Washington, D. C. As this reply omitted to state that the order emanated from the Secretary of War, it was quite defective. General Grant thus informed, rejoined:

Colonel Stager has no authority to demand the originals of military despatches, and cannot have them. I keep a record of my despatches and destroy the originals. I appreciate your informing me of the order you had received before acting upon it. Inform Colonel Stager that he transcends his authority when he demands the military correspondence taking place in this department.

Just what the purpose of the order was, we have not deemed it important to inquire, but we do know that a rule obtained in the quarter-master's department, that all vouchers covering expenses of telegraphing should have attached thereto, the messages charged for, as a check against unnecessary telegraphing.

We cannot, however, conceive of any sufficient reason why such an order should include messages of department commanders, and do not believe that it was originally intended so to do; and this belief is strengthened by the understanding that no attempt was ever made to enforce it against them. Be that as it may, it is in virtue of that order that the War Department is now able to print (as it is doing) many volumes of important dispatches.

We left Sherman at Memphis, where he fitted out an expedition, to be composed of his own and Steele's forces, to be convoyed by gunboats and conveyed by a fleet of steamers to near Vicksburg. The following telegram is a concise statement of the general plan:

OXFORD, December 8, 1862.

MAJOR GENERAL H. W. HALLECK,
Washington, D. C.:

General Sherman will command the expedition down the Mississippi. He will have a force of about forty thousand men; will land above Vicksburg (up the Yazoo, if practicable), and cut the Mississippi Central road and the road running east from Vicksburg, where they cross Black River. I will co-operate from here, my movements depending on those of the enemy. With the large cavalry force now at my command, I will be able to have them show themselves at different points on the Tallahatchie and Yalabusha, and when an opportunity occurs, make a real attack. After cutting the two roads, General Sherman's movements, to secure the end desired, will necessarily be left to his judgment. I will occupy this road to Coffeeville. (Signed) U. S. GRANT, *Major General.*

Concerning General Sherman's movements, in his "Memoirs," he writes:

The preparations were necessarily hasty in the extreme, but this was the essence of the whole plan, namely, to reach Vicksburg, as it were, by surprise, while General Grant held in check Pemberton's army about Grenada, leaving me to contend only with the smaller garrison of Vicksburg and its well known strong batteries and defenses.

Sherman reached Milliken's Bend, a little above Vicksburg, on Christmas day, and proceeded vigorously to effectuate his intended *surprise.* But there was one vastly important obstacle to any surprise, of which Sherman in his hasty preparations was doubtless unadvised, and Steele probably was also ignorant of it. It consisted of a private telegraph line, connecting De Sota, La., opposite Vicksburg, with Lake Providence, La., seventy-five miles above Vicksburg, on the river. This line was built by Dr. Horace B. Tebbetts, a wealthy planter, on one of whose magnificent estates, twelve miles below Providence, was an office, where Leė S. Daniel was *quasi* C. S. Military telegrapher. Maj. and Supt. E. G. Earnhart also resided and had head-quarters on the plantation. At the De Sota end, Philip H. Fall operated. General Martin Luther Smith, commanding Confederate forces at Vicksburg, took military possession of this line. It became the duty of Fall, who lived in Vicksburg, to cross the river every evening, and remain in his office until morning. For this purpose the artillerists at Vicksburg were directed to respect his boat, which carried a green light. Smith's purpose in all this was to keep advised of any boats passing down the river, as it was well known that the Federals were exceedingly anxious to capture Vicksburg, and thus at a stroke sever the Confederacy and free that mightiest of American rivers. On Christmas eve, about eight P.M., Earnhart and Daniel were playing old sledge at Earnhart's house, when they heard the unusual noise of the paddle-wheels of steamboats. Leaving their cards, they stepped out on the balcony of the house which stood on an angle of the river and presented a view of the Mississippi for six miles above. This whole stretch was filled with boats, descending the river. It was a grand and portentous spectacle.

As soon as Daniel could count the boats and gun-boats, he mounted his horse and hastened to his office, three miles distant, where, calling De Sota office, he telegraphically exclaimed: "Good God, Phil.! sixty-five transports, loaded with troops have already passed. There must be as many more, for hundreds of lights can be seen up the river." That was the last message sent over that line during the war, as the next morning Sherman's troops commenced its destruction. At Vicksburg, there was joy and revelling. General Smith and many gayly uniformed officers were tripping to the music of Terpsichore. Mississippi "had gathered there her beauty and her chivalry."

Fall reported to Smith at the ball, and noticed his teeth clench and his face grow ashen. In a moment Smith and his staff disappeared; "Then and there was hurrying to and fro, * * and mounting in hot haste," and within ten minutes not a dancer was left of all that gay throng; but before they reached their homes, President Davis, in Richmond, Pemberton, at Grenada, and other Confederate officials, were advised by telegraph of the impending danger to Vicksburg—by far the most important point to the Confederacy in the West. Who shall say that Sherman's failure to take Vicksburg by driving the enemy from Chickasaw Bluffs, where Sherman was repulsed with a loss of one thousand, eight hundred and forty-eight men killed, wounded and prisoners on the 29th of December, was not due to that telegraphic report, which enabled Smith and Stephen D. Lee to concentrate rapidly by rail a force which, behind works, Sherman could not dislodge. In his "Memoirs," General Sherman says:

> From our camps at Chickasaw, we could hear the whistles of trains arriving at Vicksburg; could see battalions of men marching up towards Haine's Bluff, and taking post at all outposts on our front. I was more than convinced that heavy reinforcements were coming to Vicksburg, whether from Pemberton at Grenada, Bragg in Tennessee, or from other sources, I could not tell.

Abandoning this enterprise for the present, by the 2d of January all of Sherman's troops had re-embarked, and under the orders of McClernand, who had arrived and on the fourth assumed command, the fleet steamed up the Mississippi and Arkansas rivers to Arkansas Post, which was captured on the eleventh, and with it over six thousand of the enemy. By the twenty-first all of these forces again debarked at Milliken's Bend to await the coming of General Grant.

CHAPTER XI.

THE TELEGRAPH IN TENNESSEE AND KENTUCKY.—FORREST'S, WOODWARD'S AND MORGAN'S RAIDS.—BUELL'S ARMY RETIRES TO KENTUCKY.—CUMBERLAND GAP EVACUATED.—RICHMOND (KY.)—NUMEROUS INCIDENTS. —PERRYVILLE.—ROSECRANS SUPERSEDES BUELL.—NASHVILLE AFFAIRS.—STONE RIVER.—EXCITEMENT IN NASHVILLE.—MORGAN AGAIN IN KENTUCKY.—SPIES HUNG. —OPERATORS TAP CONFEDERATE TELEGRAPHS.

We left General Buell and his army *en route*, in June and July, 1862, for Chattanooga, Tennessee, repairing the Memphis & Charleston Railroad as he progressed. General Braxton Bragg was directed by Confederate authorities to concentrate opposing forces at Chattanooga. These began to move thereto from Mississippi and elsewhere, simultaneously with Buell's.

Before entering upon the story of the great marches and countermarches of the Federal and Confederate forces in Tennessee and Kentucky, we will discover the bold exploits of Generals Forrest and Woodward and Colonel John H. Morgan upon Buell's communications in those States. Although the railroad from Nashville to Stevenson was not completed until July 28, nor the Nashville & Decatur road until August 3, yet both were repaired many miles south of Nashville much earlier. Forrest, who left Mississippi and his regiment for a new command at Chattanooga, organized an expedition into Middle Tennessee, composed of the Eighth Texas, First Louisiana, Second Georgia and a few other troops. With these he surprised the Federal forces at Murfreesboro, twenty miles below Nashville, capturing Brigadier General Crittenden, two regiments of infantry and seven companies of cavalry. Owing to the surprise, the detached positions of the troops and the belief that Forrest's numbers were overwhelming, no combined defense was made. The considerable resistance which Forrest met, arose from detachments, and when they surrendered large quantities of quarter-master and commissary stores fell into the enemy's hands. Of course, as much damage was done to the railroad and telegraph as Forrest, who hurriedly retired to McMinnville with his prisoners, had time to effect.

John J. Wickham, operator, entered the Federal service in May, 1862, at Murfreesboro, where he was captured by Forrest. He narrowly escaped being shot, at the moment of capture, by a rebel soldier, whose captain prevented it. Wickham succeeded in destroying his cipher key and all the dispatches in his possession before capture. The next morning, near Woodbury, by a stratagem, he effected his escape. Being closely pursued, he took to the fields to elude his captors, and finally hid in a blackberry thicket, where he was accidentally discovered by an officer and two men who were not in search of him. He was that night taken to McMinnville, where Forrest cursed him roundly for doing just what he would have done. At McMinnville all of the soldiers were paroled, but the officers were sent to Knoxville. Operator Brush, a little later, was taken from Libby prison because he was not an officer, but operator Wickham was sent to Knoxville and imprisoned there with the officers on the ground that he was an officer. While there he, one night, attempted to escape by passing the prison sentries. They fired at him without effect, but he was brought back during the night and placed in an iron cage about six feet square; being the same one that Parson Brownlow had occupied some time before. Wickham was kept there two or three weeks, when he was again given the run of the prison. An East Tennessee woman was permitted to bring edibles to her brother there, also a prisoner. During these visits, she graciously arranged for Wickham's escape in female attire, but he was suddenly transferred to Libby prison in Richmond, Virginia, where, after some months, he was paroled and finally exchanged, when he re-entered the Corps and took part in the Atlanta campaign

About dawn of the twentieth, Forrest appeared at Lebanon, Tenn., and, gaining the road between Murfreesboro and Nashville, he did a little injury, but was glad to escape Nelson's infantry, and return. A little later he struck the road near Manchester and then the McMinnville Branch Railroad, doing some damage, which was speedily repaired. During his retreat he was badly worsted between McMinnville and Murfreesboro, by Colonel Fyffe's three regiments and a battery. When Buell's rear guard left McMinnville for Nashville and Kentucky, as shown hereafter, Forrest pounded ineffectually against it. He seems not to have been prominently identified with any movement in Kentucky during his stay there in September. October 1, he assumed command of the recruits arriving at Murfreesboro, Tenn., then held by the Confederates.

Colonel John H. Morgan's operations against the railroads and telegraphs, were more serious. We last noticed him, late in July, 1862, escaping from Eastern Kentucky to Sparta, Tenn. About the 20th of August, he started on a new expedition, crossing the Cumberland near Hartsville, unopposed. Confederate States operators Dudley and Ellsworth accompanied him. By daybreak, Morgan's force drove in the Federal pickets at Gallatin, Tenn., and with little difficulty, captured the garrison. The pickets fired one volley and ran. A soldier riding near Dudley was the only person injured in the affair. Colonel Boone of the Twenty-Eighth Kentucky regiment was in command at Gallatin. His troops were stationed at the fair grounds about a mile north of the town. His wife came the day before and consequently the Colonel took rooms in the village hotel, where Morgan's men surprised and captured him.

J. N. Brooks, who had done good service at Edgefield and Franklin, Tenn., was at this time railroad and express agent and military operator at Gallatin. He lodged up stairs, over his office in the depot. About daylight, he heard some one coming up stairs; thinking it was the porter, he paid no attention to the footsteps, being half asleep, but suddenly he heard a voice say "Surrender, in the name of John Morgan." Rising, he saw Ellsworth and another man, each with navy pistols in both hands, cocked and pointed at him. Brooks replied, "Certainly." Ellsworth ordered him to dress, and after Brooks had partially done so, asked if he had any money. Brooks replied that he had a little, whereupon a pistol was presented and Brooks ordered to "shell out." When Brooks gave Ellsworth the purse he had in his pocket, Ellsworth counted out forty dollars and returned the empty pocket book, asking "Is that all you have?" to which Brooks replied "Yes." They then descended to the office, where Ellsworth found a package of thirty-five dollars that had come the evening before by express, addressed to a lady in Gallatin. That was also pocketed. Brooks was then ordered to connect his instrument and find out where the trains were, which he did purposely in so awkward a manner, that the Nashville operators became suspicious. Ellsworth then put Brooks under guard, took personal charge of the telegraphing and tried to imitate Brooks' writing.

About seven A. M., Culp, who kept the hotel opposite the depot, came after Brooks to go to breakfast, but Ellsworth concluded that all should go, so he locked the office. Brooks had five hundred dollars up stairs in bed—his first savings. That worried him a good deal; obtaining leave to run up to wash his face, he quickly placed the money in a boot leg, wet his face and returned without being troubled by the guard. After breakfast, Ellsworth resumed telegraphing. Little Jimmy Morris was operating at "NE" (Louisville & Nashville Railroad office in Nashville), Con. Dwyer (manager), Ellis Wilson, S. P. Peabody and the author were running the main city office at this time. Morris mistrusted something was wrong at Gallatin and asked Brooks if he received "that bottle of nitric acid that I sent a few days ago." Ellsworth thereupon turned to Brooks for an answer, at the same time drawing out and cocking his pistol and said he would shoot the top of his (Brooks')—-- —-- head off if he lied about the acid, as the question was clearly a test one. Brooks had indeed, a few days previously, received a bottle of whisky, labelled "nitric acid," to keep the railroad men from drinking it, but had forgotten the label and innocently, but stoutly asserted that he had *not* received it. This Jimmy knew was false, as it had been acknowledged before, and thus he became convinced that Ellsworth was at the key. Little as Morris was, he could command some of the most biting invectives ever invented, and these maledictions he hurled against Ellsworth, who laughed immoderately at them.

At that early hour there was little telegraphing except railroad business. A freight train bound south was captured on arriving at Gallatin. The passenger train for Louisville was ready to leave Nashville. Peabody having told Dwyer that no one could be raised north of Gallatin, the latter asked "GI" if it was Brooks. The reply came "yes." Q.—"What news have you?" A.—"Everything O. K." Q.—"How is the down freight?" A.—"Here all right." Q.—"Guess not, that isn't Brooks' writing; who is it?" No reply. Calling "GI" again, Dwyer said: "If it's Ellsworth at key, I would like you to protect Brooks." Ellsworth thereupon admitted his presence and said he would care for Brooks, who, about ten A. M., was paroled by a relative of Morgan's. Ellsworth sent two messages to Dwyer, who agreed to deliver them; one was for Ab. Turner, a man not unfamiliar with a leading game, and the other for John Hugh Smith, mayor of the city. The first announced Morgan's success at cards with certain named ones, a few evenings previous; the other reminded the mayor of the time he sought a lieutenancy in the Rock City Guards and was the most boastful Confederate in Nashville and accused him of infidelity. Turner was tickled, but Smith sealed Dwyer's lips.

The next day, after doing great damage, Morgan was driven out of Gallatin, but being hard pressed, he in turn attacked the Federals, numbering eight hundred, under General R. W. Johnson, killing thirty, wounding fifty and capturing about seventy-five, including the commander, after which, he leisurely proceeded to join General Kirby Smith in Eastern Kentucky.

About the same time, Colonel Woodward with nine hundred men, captured Clarksville (August 19). J. G. Webb, operator there, was also taken, but eventually he was unconditionally released and sent to Corinth in October. August 25, Woodward attacked Fort Donelson, where he was badly beaten. He captured and paroled repairer Pyrath and so destroyed the line running to Smithland, that it was not restored until December. During the attack on the Fort, a rebel operator appeared on the line west of the fort. His writing being unfamiliar, operator Griffin at Chaudits asked "134"? meaning, who was at the key, and received the reply, that he was a rebel operator and the "Confeds are now whipping h—l out of the Yanks in Fort Donelson and will be down your way soon and catch you." It is said that Peter Fowler, operator at the fort at this time, did good service with the musket.

While these operations were progressing, H. W. Plum arrived (August 22), at Bowling Green, sixty-four miles by rail north-east of Clarksville, and forty-five east of Gallatin, a raw recruit to reinforce the Nashville operators. Morgan being on the main road, Plum attempted to reach Nashville by train, *via* the Memphis road to Guthrie, and thence by a short connecting line which crosses the Red River, fifty-one miles from Bowling Green. When within a few miles of the river, a negro hailed the train, and satisfied those on board that the enemy had captured the stockade at the bridge, and were awaiting the arrival of the train; hence it returned, and the next morning Plum was sent without an instrument, to a stockade on the Clarksville line, at the Whip-poor-will bridge, where three companies of the Fifty-fourth Indiana were stationed. Shortly after he had run a loop to the fort, in anticipation of an instrument, the pickets were driven in, and an attack momentarily expected from the enemy, who appeared in considerable force. Accordingly a telegram was sent to the officer in command at Russellville, for reinforcements should he hear cannon, of which there were none in the stockade. The operator telegraphed this message by touching one end of the line to a crowbar, which was driven into the ground for an earth connection. By placing one hand on the bar, and the end of the line in his mouth, he received an acknowledgement. All this took place just outside of the stockade, and in front of the Federals, who were momentarily expecting an attack from the enemy, in plain view. But perceiving the defenders were prepared, the rebels passed on.

A day or two later, Plum was sent to Red River bridge, over twenty miles away, to repair the line. Taking four negroes to work his hand car, he started. That section was notoriously hostile to the Union. The operator was so delayed, examining every bridge before crossing, that it was nearly dark when he reached his destination. On the return, he was thoroughly alarmed by a small party of horsemen who pursued as far as the pike and track ran parallel, firing rapidly whenever within range. When the grade was down, the negroes beat the horses, but when it was up the horses made the best time. At one time, the darkeys proposed to abandon the car and take to the fields, but the operator's revolver kept them steadily and sturdily at work, and long before they expected to reach Whip-poor-will (it being but moonlight), they suddenly drove by the pickets, but stopped at their demand, expecting to surrender, when, lo! they were near the stockade, safe. Soon the railroad was re-opened for a day or two, and Plum took the train for Nashville. It proved to be the last one for many months.

Not stopping to note minor engagements with certain of Buell's detachments in Tennessee, let us now inquire into the operations of the main forces, which were considerably scattered, but might easily be concentrated. General Kirby Smith commanded Bragg's right, fifteen thousand strong, at Knoxville, while Bragg, at Chattanooga and vicinity, had under Polk and Hardee, forty-five thousand more, of all arms. Bragg's plan was to induce Buell to expect a battle near Chattanooga, while in fact the former arranged to invade Kentucky, leaving Buell behind. Buell feared Bragg purposed returning into North Alabama, although he regarded a battle near Altamont a possible contingency to be provided for. His ambition was to possess Chattanooga and Knoxville, and thus relieve G. W. Morgan's forces, occupying Cumberland Gap, and join them to his own. Federal occupancy of that region, in strong force, was indeed a *desideratum* scarcely possible to over-estimate. East Tennessee was peopled with hardy, liberty-loving mountaineers, thousands of whom were ready to join the Union army. Chattanooga was connected by rail with Atlanta and Rome, Ga., where much war material was manufactured. Even Chattanooga itself, besides being a great *entrepôt*, was somewhat interested in manufacturing. Situated on the south bank of the Tennessee, which may almost be said to bound the city on three sides, it lies in a beautiful valley that is skirted on the south by the famous Lookout Mountain, and on the east by Missionary Ridge. Across the river is Waldron's Ridge. Thus Chattanooga, as its Indian name indicates, sits like an "eagle's nest," in the lap of majestic steeps. The strong natural positions of Northern Georgia and Eastern Tennessee, once seized by a Federal army, might always remain so possessed; especially as Knoxville and Chattanooga were connected by rail; and so the sequel proved, but it was yet a long way off.

At this time, McCook's and Crittenden's divisions were at Battle Creek, on the Nashville & Chattanooga Railroad; Thomas's and Wood's, along the same road; Nelson's, at McMinnville,

Tenn., and Rousseau's (late Mitchell's, he having been transferred to Port Royal, S. C.), on the Nashville and Decatur road, north of Huntsville, Ala. Rousseau himself had head-quarters in the Cunningham House, in Nashville, where the author operated during Buell's retrograde movement, which began on the 1st of September. A line had been constructed to McMinnville, to which place General Thomas was sent, August 19, to oversee affairs, and Henry C. Vincent and Joseph Humphreys, operators, opened the first office there, in the Cumberland Female Seminary building. Buell moved from Stevenson to Dechard, when a historic correspondence between these two generals took place by telegraph—Thomas insisting that Bragg's object was to enter Kentucky; Buell thinking Nashville was the objective. Nelson was relieved by General Wright, and sent into Kentucky to organize the recruits intended to beat off Kirby Smith, who by this time was about to flank the Gap and move into that State.

Bragg's ulterior designs were so masked that Buell, who wished to aggress, was already on the defensive. By the 5th of September, he had concentrated his forces at Murfreesboro, *to protect Nashville.* The importance of the telegraph to Buell, in August and fore part of September, can not be exaggerated, for his command included the forces at Cumberland Gap. He was also charged with the protection of Kentucky, and expected, if he could not annihilate Bragg's army, at least to hold it in check in the Sequatchie Valley, near Chattanooga, until a decisive battle could be fought. But August 19, Polk's and Hardee's forces began climbing Waldron's Ridge, on their way into Kentucky.

At this time, Buell's forces in hand, not counting possible reserves, equaled Bragg's immediate command. The latter's movements finally developed a settled purpose to invade Kentucky, and then began that great race of the two armies for Louisville, with the Confederates in the lead—a race in which the iron horse and electric tongue figured so potently as to save Louisville and Cincinnati, if not the army itself. Bragg marched rapidly to Carthage on the Cumberland, and thence through Scottsville and Glasgow to the railroad leading to Louisville, which he soon interrupted. While Buell's army is making extraordinary exertions to head off Bragg on or near the Louisville and Nashville Railroad, and Nelson is endeavoring to stay Kirby Smith's approach on Lexington, and General G. W. Morgan is struggling to escape from the Gap, let us examine somewhat into the experiences of the telegraphers in this department. In good time, we will return to the movements of the troops.

The incidents concomitant upon the position of the army operator here, as in all other departments, were variant, scarce two alike, but while the details were always fresh, the general run may be gathered from the adventures of almost any one of them,

DOUGHERTY'S AIR LOOM.

and in view of this latter fact, we have chosen to illustrate the operations of the many mainly by showing the happenings to and doings of one — Kit Dougherty — who entered from Bloody Island (East St. Louis), in May, and was stationed first at Columbia, Tennessee, and next at Moorsville, Alabama. At the former place, this lad of sixteen slept uneasily for the first few weeks, as he was repeatedly routed out to prepare for a night attack by guerrilla bands, but as these scares became monotonous, "Nature's sweet restorer,"

> That supplies, lubricates and keeps in play
> The various movements of this nice machine,

wrought its full benefaction to Dougherty. But his comrade,

Gregory, could not weigh his "eye-lids down," and hence sought repose in the safer State of Indiana. July 4, Dougherty took a celebration ride in the suburbs; his horse fell, throwing the boy about twenty feet over the stone pike, and sending him to bed for ten days. Shortly after he arose, anathematizing horses in general, he was induced to mount a particularly amiable one, and, galloping out of town at an easy pace, the animal gradually grew more spirited, and made desperate but unsuccessful efforts to throw Christopher. Finally the horse resorted to a miserable trick, so well understood on the Western plains, *i. e.*, it shot ahead like an arrow until Dougherty thought the animal had made friends again, when suddenly it thrust forward its stiffened front legs, lowered its head and came to a dead stop. If Dougherty had stopped too, all would have been well, but he kept right ahead, landing on the macadam. Then the *amiable* beast returned alone and Dougherty took another week's rest.

Soon after, he was *en route* for Huntsville, in company, on the train, with Superintendent Bruch. They had proceeded but a few miles, when bump, bump! bang! bang!! The train was ditched and the bushmen were whacking away. It was a moment full of confused scare, and Bruch and Dougherty were surcharged with that sort of confusion. No sooner had they crawled through a window on the off side than the guerrillas sounded their demoniac war-whoop; but not wishing to encounter the train guard, they hurriedly decamped, and the telegraphers took possession of the *field* back to Columbia, exhibiting commendable energy all the way.

For want of other amusement, and chancing to have some copper pennies, Dougherty amalgamated them in the battery room, and among the illiterate farmers they became excellent silver half dollars, much sought after in exchange for watermelons and vegetables, which would have been exorbitantly charged for by the native Secessionists, but for this system of specie payment.

Three trains left Huntsville, August 27, traveling close together; two were laden with troops, the last was to take up the post guards along the road. At Moorsville, the six soldiers and operator were found by the second train on a flat car, with all their truck, including about twelve barrels of mess pork, with which they made a barricade, that undoubtedly saved the lives of some of the party. The sergeant of the guard missed that train and took the last one. Thus Dougherty became commander-in-chief on the flat car, which was the last on the middle train. Near three P. M., about midway between Pulaski and Columbia, the train was thrown from the track, and a shower of buckshot announced the presence of the enemy in the adjoining woods. The telegraph party, safely ensconced behind the salt pork, had the happy satisfaction of knowing that it was that, not they, that was being well peppered. The attack came from the east and, consequently, the passengers hastily landed on the opposite side of the train, which was used as a protection. A lively fight ensued for about ten minutes, during which time a Federal who was fighting with Dougherty was shot through the brain and fell dead by his side. A lieutenant called for volunteers to reconnoiter. Dougherty and a few others went into the woods to see what had become of the enemy, who had suddenly ceased firing, and while darting from tree to tree the party was again fired upon, and another soldier was shot, whereupon, "having accomplished the object of the reconnoisance," they fell back in good order, bringing off one wounded prisoner and some shot-guns, but, while away, they discovered that the bridge ahead had been burned, and they soon noticed that the one in their near rear was burning. In the language of an eminent military authority, they were "bottled up;" but the last train coming up the next morning with a thousand soldiers and some cotton bales, the latter were used successfully to pass the trains over one stream, the troops being transferred to trains from Columbia at the other bridgeway. The line was destroyed by fastening it about every half mile to the end of the train, which tore off insulators and broke down poles before the line parted. Operator J. A. Fuller came up on the last train. Dougherty was sent to General Negley, whose command was, walking north on the pike. For this march, Dougherty donned his old clothes and sent his better ones by train to Nashville. That was the last he ever saw of them; a very common mishap; for what army operator did not lose at least one *best* suit?

Now let us go back a little and note what the chief operator,

Crittenton, was doing. August 30, he telegraphed Superintendent Bruch, from Stevenson, Ala.:

> Troops withdrawn from Decatur and Columbia route three days ago. Moorsville, Athens, Elk River and Pulaski offices closed. Operators and repairers ordered to Nashville with all telegraph property. Battle Creek taken by rebels; operators escaped. Bridgeport office closed and troops left. Huntsville to be evacuated to-day; Stevenson to-morrow. This will give us no foothold in Northern Alabama for the present. I am withdrawing everything from my section as quietly and orderly as possible.

The next day was full of interest to Crittenton. A wire was run to the fort in Stevenson, and communication opened therefrom to Rousseau's head-quarters in Nashville, and so continued throughout an engagement with an advance force of the enemy. The following excerpt from an article by the author, in 1867, is now in point: "Stevenson, Alabama, was attacked, I think, by General Wheeler. Our lines were still working; the operator had his office in the stockade at that place. In Nashville, we were having a terrible thunder-storm; the surcharged wire found relief in my office every minute; officers flocked there to learn the cause of so much firing; the crack of the escaping lightning was as loud as that of a rifle. Every minute I thought would be the last for my magnets. I inquired of General Rousseau if I should 'cut out.' 'No, sir,' he replied; 'let the instrument burn if it will; I must keep posted about the Stevenson fight.' A minute after this, he was leaning upon my desk, with one hand upon the window sill, when he received a powerful shock. That great strong man was nearly prostrated; he who had withstood the shocks of many fierce battles, was for once demoralized by a single one sent forth by the God of battles. It was a week before he fully recovered."

That night the Stevenson force began its retreat and, telegraphic communication was opened at each bivouac. Then all was quiet on the Southern lines; that whole country had been given up to the enemy, and operators arrived one by one and in twos, in the main, impecunious, ragged and communicative. They awaited developments at the Cedar street office by day and the theater on free passes, by night, until singly or in squads, they assumed new duties, and then the Cedar street office, like the city itself, was shut off from the world.

While Buell's forces are *en route* from Nashville to head off Bragg, let us inquire into the movements of Federal General Morgan commanding at Cumberland Gap, and also those of Kirby Smith and Bragg. The enemy remained in force within threatening nearness of the Gap during Morgan's stay, but owing to the extreme difficulty of subsisting his troops, Morgan was obliged to forage extensively. One of these expeditions provoked the severe contest at Tazewell, reported by telegraph quite interestingly, as follows:

CUMBERLAND GAP, August 7, 1862.

To COLONEL J. B. FRY, Chief of Staff, Huntsville, Ala.

To obtain forage and feel the strength of the enemy, DeCourcy's brigade was ordered to Tazewell on the second inst. He secured two hundred wagon loads of forage, all of which safely arrived. On the fifth, some slight picket skirmishing took place, in which we had two men wounded, while the enemy had one killed and several wounded. Early on the morning of the sixth inst., not wishing to bring on a general action, I ordered Colonel DeCourcy to return to this post, but he was attacked at daybreak on that day. Considering the enemy's force, the attack was feeble. Two of his regiments surrounded two companies of the Sixteenth Ohio, detached to protect a section of artillery. The enemy's movement was well executed, and had it not been for the coolness and gallantry of Lieutenant Anderson, we would have lost two pieces of artillery. Although surrounded by a vastly superior force, the two infantry companies under command of Captains Edgar and Tanneyhill, fought heroically, and three-fourths of them succeeded in cutting their way through to their regiment, but we fear that Captain Edgar, an officer of great merit, was killed and Captain Tanneyhill taken prisoner. There were several instances of distinguished conduct, both on the part of officers and soldiers. A soldier of the Twenty-Second Kentucky was shot through the neck and fell; his gun dropped from his hand; his antagonist continued to advance upon him, when the wounded hero grasped his gun, rose to his feet, and shot the rebel soldier dead, when within five paces of him; when he again fell weltering in his own blood. Two soldiers of the Sixteenth Ohio had lost their way, and were going toward the enemy, when Lieutenant Colonel Gordon, of the Eleventh Tennessee hailed them, demanding their regiment; with coolness and courage they required him to declare his rank and regiment; and took him prisoner, reversing their march by a circuitous route, they rejoined their command. Gordon speaks highly of their courage and courtesy. At 3:30 P. M., a courier arrived from Colonel DeCourcy and asked for aid; leaving three regiments to guard the Gap, I marched with my remaining force to his assistance, but when within two miles of Tazewell, I met him on his return. The enemy left the field at five o'clock, and he maintained his position until seven P. M. The enemy's loss is believed to be considerable. I did not pursue, lest a superior force should gain my rear.

(Signed) GEORGE W. MORGAN,
Brig. Gen. Vols. Comdg.

It is said the Confederates lost two hundred killed and wounded, that DeCourcy's loss was much less.

Late on the seventeenth, Kirby Smith struck Morgan's line of communications at Barboursville and Cumberland Ford; that was Sunday; in the morning of that day, Morgan was advised of the movement in this way: Smith's advance cavalry, numbering nine hundred, under Colonel Scott, captured London after a brief engagement with the small command there. This command was posted some distance from operator Frank Benner's office. While Scott was engaged with the Federals, Benner, thrusting his instrument under his arm, hurried off in the direction of Cumberland Gap, and after traveling eight or nine miles tapped the line and telegraphed to General Morgan that the enemy's cavalry, two thousand strong, had that morning taken London. It was a brave and highly commendable feat, and it so chanced that he had just quitted the line, when a squad of the enemy rode by. Benner then struck out for Lexington, Ky., which point he reached just in time to escape Kirby Smith, and was forced to continue his journey to Frankfort and Louisville.

As Smith cut off all hope of supplies by the Lexington route, and Bragg *via* the Cumberland, and as Stevenson's division of the enemy's forces remained south of the Gap to harass foraging parties, Morgan, rather than surrender, decided on the 14th of September to evacuate. His command, ten thousand strong, had been living on scanty allowances for nearly a month. The telegraph party was organized under Patterson and Tidd, into a corps of sappers and miners and materially assisted in making new roads, removing obstructions and building bridges destroyed by Colonel John H. Morgan, who harassed the infantry column (G. W. Morgan had previously dispatched nearly all of his cavalry to the North) so greatly as to reduce its means of forage to the *minimum.* For several days the operators and line men lived exclusively on corn grated on a tin plate, through which rough edged nail-holes had been made. On the second day out, General Morgan ordered all but field officers to dismount. Consequently the operators had a long walk. Cassell, the cipherer, was two days behind Morgan on the latter's arrival at Portsmouth, Ohio, *via* Manchester, Booneville, Hazel Green, Grayson, and Greenupsburg, and in consequence Morgan's cipher messages from General H. G. Wright, the commander of the reconstructed Department of the Ohio, which embraced Eastern Kentucky, were delayed, much to Morgan's annoyance, until Cassell could be sent for. The operators with Morgan's command, were J. A. Cassell, Hugh Craig, J. G. Garland and Robert Wagner.

Of course Kirby Smith's wing of Bragg's army pressed vigorously forward. James Jones, the Mount Vernon operator, was overhauled near Perryville by Smith's men, and marched one hundred miles before he was paroled. James Meagher, who barely escaped Morgan's men in July, was, in August, operating at Columbia, where he was captured. W. W. Pitcairn, operating at Somerset, hid from the rebels from August until October 29, when he escaped. The other operators hurried to Louisville. Eastern Kentucky, Ohio and Indiana, day by day as Smith advanced, and report magnified his force beyond all reason, grew more and more excited and energetic. Lexington was Smith's first objective, then Louisville or Cincinnati, perhaps both. Colonel Stager proceeded to Louisville to oversee telegraphic operations. C. E. Bliven was made one of Captain Bruch's assistant managers, and sent to Lexington, the center of his (Bliven's) district. General Wright sent Nelson there also, and there his division, consisting of regiments (except two) scarce a week old, was organized. Time was all-important; both sides realized the fact, for the movement had not been anticipated by the Unionists, who relied upon G. W. Morgan's forces to prevent such an irruption, and it would have been better that he had destroyed his army in seriously crippling Smith, than render it useless in this trying emergency; but perhaps better still had he followed closely on the heels of his adversary and sought advantageous positions in case Smith turned against him. However, we know not but that Morgan believed a siege intended, until it was too late to overtake the enemy. Metcalf's, Child's and Mundy's cavalry endeavored to retard Smith's advance. Manson hurried to Richmond, twenty-five miles from Lexington, to interpose his and Cruft's brigades. The nearest telegraph office was Nicholasville, fifteen miles distant, where F. C. Cook operated. On the twenty-ninth, word was sent of the enemy's approach in force. The next day, a courier arrived from Nelson, with orders to retire ; but Manson had just been forced back, and was forming for another stand. Again and again, he and Cruft were routed ; then Scott's cavalry struck the panic-stricken fugitives, and all hope of resisting Smith south of Covington, opposite Cincinnati, was abandoned. Fully three thousand Federals were captured, two hundred killed, and seven hundred wounded. Lexington fell, Frankfort was occupied, and a rebel Governor installed. Louisville was incapable of resistance, but, unwisely, Smith moved against Cincinnati, and there the population was all astir, arming, drilling, organizing and digging. Forty thousand men crossed the river with axes, picks and spades. Bliven soon had fifteen miles of telegraph in operation, connecting the fortifications opposite Cincinnati. These lines were operated by John C. Sullivan, E. M. Livingston, Dan. Murray and John D. Richardson. Heth approached (September 15) with twelve thousand Confederates, but dared not attack. That was two days before G. W. Morgan left Cumberland Gap.

Bragg concentrated his main army at Glasgow the day before Smith's advance under Heth appeared before the Covington fortifications, but as early as the ninth, when Salt River bridge was destroyed, railroad and telegraphic communication between Nashville and Louisville ceased. Zeph. Morris, J. D. Weems, Adam Bruch and W. L. Biggert, the "owl," operators in the Louisville office, freed from labors incident to a Nashville connection, were nevertheless now doubly taxed, owing to the great emergencies rapidly developing. On the twelfth, the railroad between Franklin and Bowling Green was broken. Six days later, Buell concentrated all his forces, not left at Nashville, in and around Bowling Green. On the seventeenth, about four thousand Unionists, at Munfordsville, surrendered to Bragg, who, marching northward, turned north-east near Lebanon Junction, striking Bardstown, ten miles from the Louisville and Nashville road, where a large force was stationed, and, pressing on with the remainder, joined Smith at Frankfort, October 1. Smith and Bragg, by a strange fatuity, had each avoided their great objective when its possession was certain, if they promptly approached in full force.

When Buell's army advanced from Nashville, two operators accompanied each division commander—some on foot, a few in ambulances, and others on horses or mules. Many others were detained in Nashville. Among those who accompanied the army were Bruner and Mullarkey with Buell, H. C. Vincent and W. H. Hartman with Thomas, J. A. Fuller and William Atwater with Crittenden, A. J. Howell with McCook, J. S. Lyle with Woods, William Patterson with W. S. Smith, and Hector Lithgow, G. W. Bliestine, J. H. Galvin and Thomas M. Sampson with other officers. Hartman went with Colonel Kennett from Elizabethtown. Buell, four days' march behind Bragg, entered Louisville on the twenty-ninth, and from that moment the high expectations of the Confederates who hoped to hold Kentucky, became impossible of realization. In Buell's absorbing anxiety to save Louisville he marched far in advance of his main wagon train. Although it was convoyed by a considerable force, both it and the guard were in imminent peril of capture, and no sooner did Buell feel easy about the city, than he became uneasy concerning this train and convoy. Just where it was, he knew not, and the nature of his aid depended upon its position; but he must have known full well, that without reinforcements, being many miles long, it could not pass Lebanon Junction. A small relief force would aggravate the trouble, and a large one he wished not to endanger. In this great emergency, he asked for a volunteer operator to restore the telegraph to his train. Charles Lehr, who had been serving as chief operator of the Cumberland Gap line since Fuller's sickness, was accepted.

19

Taking four soldiers, dressed like himself, as railroad section men to propel a light hand-car, he started down the road, expecting every moment to see the rebels who scouted to within five miles of the city, but it so chanced that the little party reached Salt River without discovering any. Here, where the iron bridge lay in ruins, they found the first bad break in the line which was soon repaired, and after dragging the car over the *débris*, they had proceeded but two miles when they came in sight of a rebel camp on the left, with a picket on the track. But the men pulled away vigorously. Doubtless, the very boldness of the venture lulled suspicion, as the party passed unmolested. Repairing the line at Rolling Fork and Muldraugh's Hill, the men pushed on safely to Elizabethtown, where they arrived about five, P. M. It was near here that Buell supposed the train would be. Lehr connected his instrument, and getting circuit from Louisville, telegraphically shouted to Bruner, in Buell's office, "Glory to God! Tell Buell I'm here. The train not yet come up." Buell was better off than with his train, for he was practically there, while in fact, forty-two miles away with his army. The train was ordered to turn west to the Ohio River, for transportation by steamers.

The next morning the line was cut, and Lehr and comrades now abandoning the car, set out to return on foot, judging from their unmolested trip down, that a cautious return afoot could be safely made. All went well until near Lebanon Junction, when Lehr chancing to be ahead, discovered about fifty Confederate cavalrymen come out of a neck of woods a short distance beyond. Lehr's assistants ran across a cornfield, but were overtaken and treated as "section men," instead of spies, as they really were; but Lehr, without being discovered at once, ran down a bank along the road, and, while he was pressing forward, the cavalry passed in pursuit of the four others. When he was discovered he was fast reaching the heavy blackberry bushes that grow in abundance in that section, and into which it was not easy to ride. These were well calculated for concealment, and there he eluded the pursuit of two horsemen, who soon gave up the chase. Lehr had spent much time at Lebanon Junction, a terminus of the Gap line, and knew the country and people, which was a great help. Staying that night at a Union man's house, he, next morning took the road and woods by the road, until coming to the rear of a house on his route, he met a negress, who exclaimed: "Lord, massa, you jest make yourself scace, fo the front room am full 'o rebs." At Shepherdsville, Lehr found a force of rebels which he could not pass by day, so, feigning to be a rebel citizen, he took tea at the tavern by the side of a Confederate officer, and waited for darkness before attempting to pass the pickets, which he safely accomplished, creeping stealthily by. Finally, after other narrow escapes, Lehr reached Louisville, his shoes torn and his clothes in rags. He received congratulations from the Corps, and from Buell himself.

It was after Thomas, who brought up the rear, had left Nashville, that J. N. Crittenton (operator) was directed to go to Buell at Bowling Green; so taking his orderly, he proceeded twelve miles, when he was intercepted by rebel cavalry, and though he succeeded in chewing up some despatches he was carrying, his pocket instrument, watch, horse and other valuables were taken from him, and, seated on a stolen hack, he rode with other prisoners to a log store where, the next day, Colonel Bennett, of the Texan Rangers, paroled all and sent them adrift.

It was well known to the telegraphers that Buell hoped to fight Bragg south of Green River, where the latter for several days offered battle; but when Thomas arrived, Bragg decamped. Expecting Bragg's defeat, and that the lines south would soon be re-opened, some of the operators remained in Nashville until that city was completely isolated. Quarter rations, idleness, high prices and, as a rule, no money, produced unendurable *ennui*. An order from General Negley, the commander, forbidding all persons from being out later than nine P. M., produced an irresistible desire on the part of the operators to keep late hours; consequently, a pass given to a lucky operator served many every night. But when all had finally come into the Cedar street office and lain down on the floor to sleep, the musical *culex* began his lullaby, and the next fellow's boots, pants — anything, everything — were hurled promiscuously for an hour or two until, patience and ammunition ceasing, the boys, *en deshabille*, took to the window sills to smoke and cool off. When morning came, "what a scene was there!"

These things grew monotonous, and S. P. Peabody, Claude

Knox, Kit Dougherty, W. H. Kelsey, F. B. Tyler, C. H. Pond, Joseph M. Humphreys and one other applied to Dwyer, deputy superintendent, for leave to venture north. Out of the United States corrals, containing ten thousand condemned horses and mules that were starving to death, Dwyer selected a mount for each applicant, and one fine morning about the 1st of October, with fifteen dollars each, besides a goodly supply of fire-water, but no bread, they boldly pushed out on the Louisville pike, armed with revolvers and a letter addressed to Bruch, saying:

> These operators are thoroughly disgusted with the war and are anxious to quit the service *forever.* So I have discharged them.
>
> (Signed) CON. DWYER.

A little below, to be torn off if the rebels were likely to capture the boys, was a word for any Union force they might meet, as follows: "The above was intended only to mislead the rebels in case of capture." Armed with this double-back-action pass, the boys felt reasonably secure. Pond and Dougherty were poorest mounted and soon fell in the rear. About two, P. M., they passed a native, well armed, whose horse looked very like Tyler's; but no questions were obtruded, as it was considered highly impertinent to interrogate an armed stranger about horse-stealing. An hour later, Tyler was espied, trudging wearily in his new shoes, intended for smaller feet. "Hello, Tyler, where's your horse?" "Oh, a gentleman with a double-barrelled shot-gun fancied it. Had to yield to superior arms and overwhelming force, you know." Giving Tyler an occasional ride, the three were about to enter Springfield, when a voice cried, "Halt thar, d—- you!" Five of Dick McCann's guerrillas took the three operators to the Springfield tavern, where they were duly robbed. The pass was discovered *entire* by one of the chivalry, who proceeded to spell it out; but Pond, equal to the trying emergency, volunteered to read the letter, saying he was familiar with the writing, and was allowed to do so. Of course he omitted the fatal *addendum* and was then allowed to retain the paper, which he soon destroyed. The operators were released on parole and moved on. Sleeping in corn-fields, begging something to eat or foraging it, they finally reached Bowling Green, where Martin Barth, operator, gave them a welcome, and in about two weeks they arrived at Louisville, penniless, ragged, dirty and in summer clothes. The other five operators fared no better. Peabody was usually the spokesman, and an audacious one he proved to be. More years have made him wonder at his temerity. Of course these five were captured and paroled by every band they met. One night, while sleeping in a fence corner, one of the operators was startled by a great snake crawling over him. Shrieking aloud, "My God!" he awakened the others, who were, if possible, more frightened than he.

The following letter from Bruch will explain how poorly the operators fared at this time:

> Nearly all of my operators who have been working in Tennessee and Alabama, are arriving here now. By command of General Buell, two were assigned to each division of his army, to continue with the army as "division operators," to connect their instruments wherever they might come to a telegraph. Moving in this way, most of them have spent all their money and worn out their clothes. Some of them have had nothing to eat for a day or two, and no money to buy anything. They come to me begging and imploring for money, and I can not bear to turn them off. So I have loaned them enough to keep body and soul together for a few days, in hopes that the August funds will be along by the time that runs out.

Among those included in the above, none suffered so severely as H. C. Vincent, who nearly died of fever in a Louisville hotel, for want of proper care.

Cincinnati being safe, General Gordon Granger came to Louisville and took command at Shepherdsville, eighteen miles south of the latter city, and awaited the concentration of Buell's forces north of him. W. H. Drake, who spelled Lebanon Junction for Morgan's operator in July, was still operating there. He was the sole Yankee occupant, but had made friends with his landlord, Harvey Wells, by advising him of the approach of guerrillas, from time to time, so that the latter could hide his horses, and now Harvey advised Drake to "skip," as a detachment of the First Georgia Cavalry, under Hood, was approaching; whereupon, Drake, mounting his charger, hastened to Shepherdsville and informed Granger, who, giving Drake a fresh horse, made him conduct one hundred men of the Fourth Indiana Cavalry in search of the enemy. Confederates were soon found; they beat a hasty retreat toward Boston, at which place they were run into Wheeler's wagon train, and its guard stampeded. Drake, with eight men, in front of old man Arnot's house, received the surrender of the wagon train and a lieutenant, whose sword, revolver and dirk Drake appropriated. He was just giving orders to burn the train, when, lo! a cloud of dust arose ahead and the tramp of horses from the direction of Bardstown advised Drake to take the swamp road about as fast as his horse could go. Six of his eight men left him at the railroad. They were all killed. At the turn on top of the hill by the railroad, Drake and his two companions came face to face with two rebels, who tried to handle their guns. That frightened him, as well it might, but fortunately he shot quickly, though without hitting, and yelled, "surrender!" which, to Drake's surprise, they did. The prisoners were taken safely to the Junction, where the rest of the Indiana cavalry were, and where Drake left his sword and belt — trophies of his valor — with Aunt Maria, Wells' negress, from whom the lieutenant recovered his property a week later.

Buell, always slow, with one hundred thousand men now in hand, would have been superseded by Thomas if the latter had accepted the appointment. Thus spurred, Buell moved after Bragg at Bardstown, and Smith at Frankfort, October 1st. Bragg and Smith united, and about half past two P. M., of the seventh, the sanguinary engagement at Perryville began. About forty thousand Confederates pressed McCook's corps for hours, when he was only feebly reinforced, notwithstanding Buell had a larger army at command, than his adversary. In that terrible struggle of a few, against the many massed in repeated charges, the Federals alone lost four thousand, three hundred and forty-eight, of whom nine hundred and sixteen were killed and two thousand, nine hundred and forty-three wounded. Bragg conceded a loss of twenty-five hundred men. The Confederates retreated, and the next time they presented themselves they were at Murfreesboro, Tenn., threatening the State capital.

Owing to this invasion, every mile of Federal telegraph in the department, of which there were seven hundred and seventy-three, was abandoned, besides about three hundred miles of railroad telegraph and say two hundred miles of commercial wire; making the great total of one thousand, two hundred and seventy-three miles of line.

Great was the disappointment in the North at the impotent conclusion of Buell's campaign. He had been directed to move on East Tennessee, but he urged the necessity of going to Nashville, and sent troops to Bowling Green. In this position of affairs General Rosecrans superseded Buell (October 30). The new department was designated as that of the "Cumberland," and the army, the "Fourteenth Army Corps." Captain Bruch's report sufficiently explains telegraphic operations from the time of Buell's advance from Louisville, to his movement on Bowling Green, and is as follows:

> When the army advanced to Lexington, we followed it, and repaired lines to Nicholasville in the latter part of October. We also followed the army of the Ohio, under Major General Buell, through Central Kentucky in the same month, repaired the lines and kept him in telegraphic communication until he crossed over to Bowling Green, Ky. The following lines were repaired in his wake: Louisville to Bardstown, Lebanon, Danville, Stamford, Somerset and Mount Vernon, and from Lebanon to Columbia—also Louisville to Bowling Green.

On the 9th of November, just three months from the interruption of the telegraphic communication with Louisville by the destruction of Salt River bridge, the van of the Fourteenth Army Corps, entered Nashville, when the beleaguered denizens and soldiers were rejoiced beyond measure. During the operations in Kentucky, General Negley was left in command at Nashville with about nine thousand troops. Fortifications around the city having been in course of construction all summer, were well advanced when Buell left. Four thousand negroes were impressed to work on them, and their completion followed in apt time. A telegraph line was run to the various forts and military headquarters, and until eight operators were permitted to go North, the lines were open day and night. It was expected that in case of an attack, Negley would be able to direct operations from Fort Negley, that fortress being the key, and commanding nearly the whole sweep of country, including the city and the river north of it. Nitric acid failing, sulphuric was substituted in the batteries with success. The operators' quarters at the forts were uninviting; H. W. Plum was cozily situated at Negley's handsome head-quaters in the city, Robert Wagner was at Fort Confiscation, the author at Fort Negley, Ellis J. Wilson at General Palmer's and Con. Dwyer had the whole city office to himself. It is due to the worthy chief to say, that but for Major Fitzgibbons, who was *au fait* in all moonlight accomplishments, Dwyer would have known less of the city by candle light.

Provisions were very scarce; foraging parties were regularly attacked by General Anderson's cavalry; Negley was repeatedly summoned to surrender; feigned attacks were frequent, and unpleasant rumors constant. We were awakened at all hours of the night to receive telegraphic orders and transmit information. For forty-five days we received no intelligence whatever from the North. By a Richmond paper we learned that "the troops in Nashville had been compelled to surrender because starvation was threatened," hence we were extremely concerned to assure our anxious friends at home, who oftener than we, heard of our capture, that in fact all was well. General Breckenridge had been directed by Bragg, who detached a considerable force for that purpose, to capture the city. Bragg, from Bardstown, Ky., ordered Van Dorn up from Mississippi to sweep Nashville and Bowling Green, and, after possessing their armaments, unite with Bragg and Smith in holding at least the line of the Cumberland and Tennessee rivers, but he came not.

During the blockade of the city, Confederate General Morgan was reported at Gallatin, Tenn., on the Louisville & Nashville Railroad, and Negley arranged to send General Miller with a brigade by rail at night, to leave the cars at a point two or three miles from Gallatin, and to strike and destroy Morgan's force if possible at daylight. An operator being required for the expedition, the author volunteered. All were aboard by midnight, and the trains moved as quietly as possible to their destination. The artillery was landed and the infantry ready to advance, when General Miller requested the operator to connect his instrument and ask Nashville if all was well. Doing so, he received a telegram from Major Slidell of Buell's staff, who remained in the city, stating that Nashville was threatened, and Miller must return immediately, which he did with his force and with a heavy heart, to find that it was but another rumor. Thus Morgan escaped. On the fifth day of November, General Morgan attacked the Federals protecting the railroad bridge over the Cumberland, and Forrest and Hanson drove in the pickets on the city side of the river. A section of light artillery opened fire on Fort Negley. Morgan was driven off with a loss of twenty-four killed, and wounded. It was rare sport for those in Fort Negley to watch the rebel shells fall far short of the fort, while the Federal gunners were most perplexed to train their guns so as not to over-shoot the mark. There were at this time about seven thousand Confederates on the south side, which added to Morgan's force in Edgefield, fully equaled the number of the garrison. Breckenridge ordered the retirement of Forrest's and Hanson's troops when they were about to assault, then Negley advanced on the Franklin road, severely punishing the enemy found on that pike.

The day the Federal advance, by forced marches, entered Nashville, the Louisville line was restored to Gallatin, Tenn., by repairers under Bart Brady, and two days after, it was repaired to Nashville. November 24, a second wire was also completed thereto. About the 1st of December, Colonel Bruce recaptured Clarkesville, Tenn., and during that month the line from Bowling Green to Cairo, *via* Clarksville and Fort Donelson, was restored. On the 7th of December, General Morgan captured a force of sixteen hundred Federals, at Hartsville, by the blundering of Colonel Moore and misconduct of his raw troops. On the ninth, Colonel Stanley Matthews' brigade, four times in one day, repulsed rebel General Wheeler's force, attacking Matthews' foraging train, on the Murfreesboro road. December 21, General Carter, with nearly three regiments of cavalry, entered the Confederate lines about forty miles north-east of Cumberland Gap, and struck the railroad east of Knoxville, destroying bridges over the Watauga and Holston rivers, and otherwise greatly damaging the road. Frequent skirmishing occurred between detachments.

"Old Rosy," as his soldiers were pleased to speak of their

new commander, who came to them fresh from his splendid victory at Corinth, held the telegraph service in very high esteem. By his direction, Assistant Manager Dwyer assigned one operator to each division, and two to each wing and the center headquarters. W. L. Tidd organized a gang of builders, to follow Rosecrans, who was preparing to attack Bragg at Murfreesboro. Another party was directed to proceed on the Franklin road, when Rosecrans advanced. The labor incident to the reorganization and partial re-equipment of the army, and repair of the railroad; the accumulation of necessarily vast quantities of quartermaster, commissary and ordnance stores, preparatory to an advance of the whole army, on a new campaign, and the communications to and from Washington, respecting plans of operation, involved an immense amount of telegraphic work, without which Rosecrans could not have moved his army for weeks after December 26, when it began to advance to the scenes of its seven days' battle. Bruner and Mullarkey, Rosecrans' operators, rode with the General, and attended to all telegraphic correspondence. Couriers were in readiness at all times to be sent to the nearest telegraph office. J. A. Fuller was operator with the line builders, and was busy constantly, sending and receiving dispatches.

Rosecrans' effective force was about forty-six thousand men. Bragg, as we have seen in a former chapter, had sent Forrest into West Tennessee, to operate against Grant's railroads north of Jackson. Morgan was in Kentucky on another raid. Kirby Smith's forces, except nine thousand sent to the aid of Pemberton near Grenada, came up from Knoxville, and at the time of Rosecrans' advance, Bragg held at Murfreesboro, or within easy support, thirty-five thousand troops, most of whom were more experienced in war than the Federals.

On the evening of the twenty-sixth, Crittenden's (the left) advance reached LaVergne, half way to Murfreesboro, when telegraphic communication was promptly established with Nashville. Thomas' and McCook's corps continued to advance with Crittenden's, meeting more or less resistance each day, until the thirtieth, when the two armies confronted one another, westerly from Murfreesboro and near the line of Stone River, from which the battle, then impending, is named. A conflict was momentarily expected Neither party was quite ready, but either might precipitate it. The telegraph was working to La Vergne, but passing artillery, trains and troops had obstructed the road so that Tidd's building material was advanced slowly. Rosecrans' head-quarters were but four miles in advance, and Tidd strove to reach there that night. Wheeler's cavalry, however, suddenly burst into the village and captured foreman Tidd's entire party, teams and all; burned all the material and private property, and then hastily paroled all but builder Clark, reported missing. Fuller had been busily engaged receiving messages for officers with the army, and forwarding them by a part of the Fourth U. S. Cavalry, serving as couriers. He had just received a message for Rosecrans, notifying him that Anderson's cavalry (a "high-toned" lot of Pennsylvanians) was escorting a large wagon train of ammunition, and advising him to send more reliable troops to keep them company. A courier started with this and other messages, but hastily returned, saying Wheeler had captured a large provision train and was coming to the office. The operator quickly told Nashville; burned the messages he had sent; thrust his instrument into his pocket; mounted his horse, a present from General Crittenden, and rode about five hundred yards and into a regiment of the enemy. Fuller here met the courier above mentioned, who had the telegrams in his possession, but by Fuller's direction he tore them into small pieces. These were scattered about, and created quite a stir among the captors. General Wheeler ascertained that telegrams had been destroyed, and thereupon search was made for the operator, who was dressed in soldier's clothes. Wheeler took the messenger among the prisoners, to identify him, but the courier passed close by Fuller, whom he would not recognize. Failing to discover the operator, the General threatened the courier, when Fuller made himself known. Wheeler then stated that he could not think of leaving the operator behind, to report him the minute his back was turned, and ordered a soldier to mount Fuller on a mule and guard him. This soldier robbed him of thirty-five dollars. Federal Colonel Walker's men now fell upon Wheeler, and, mistaking the prisoners for enemies, fired into them with cannon, whereby Tidd's right arm was struck and shattered by a ten-pound ball. In the confusion that followed the sudden appearance of the Federals, Fuller got the start of his guard, and succeeded, under fire, in gaining a hospital and closing the door on the pursuing rebel, who hurried to his command. Eight hundred prisoners and all the horses and mules were recaptured. Fuller returned to the line and notified Nashville what had happened.

Each commander had planned to crush the right wing of the other on the last day of 1862; but Bragg's movements were soonest executed, and before Rosecrans had fairly developed his own plan, he was forced to take the defensive to save his army. McCook's corps was terribly cut up, and the Nashville pike so crowded with frightened soldiers, flying from the dreadful scenes that had unmanned them, as to impede the progress of Hascall's brigade, which was seeking to join the main army. Rosecrans' trains were also hastening forward with ammunition and food, when they were checked by these terrified men, some of whom continued their flight to Nashville itself, spreading their terrible stories of the destruction of Rosecrans' army as they advanced, so that every teamster was also ready to fly. That night was full of doubt and dread in the capital city. Evil rumors were repeated and magnified; no comforting reports came from the front. The telegraph office was besieged by anxious officials. The sound of artillery could occasionally be heard, faint and distant. Rebel citizens professed to know of the defeat of the Federals. Late at night came the news that nothing was decided except that Rosecrans had determined to win or die on that battle field. The new year opened ominously, for when the operators were eagerly listening to every click of the instrument on the Murfreesboro line there came a silence, long and burdensome; the telegraph line had given out. What could it mean? Was the army indeed defeated, and would the next report confirm the chilling rumors? How painful is silence under circumstances like these! A fortune for a moment's conversation with the advance operator! But no; the magnet is as still as death. The electric tongue is hushed, and a nation tremblingly waits. How this suspense spreads! The great centers of industry are muffled in impotent anxiousness, as if the sword of Damocles was suspended over them.

Message after message comes from the North, from the President, the War Secretary, the Governors and others, not to mention the cruel speculators, all, all craving news from the front. Then come a great number of telegrams from dear ones, seeking answers which can not be obtained. How terrible, indeed, is war! You, who are wont to recommend it, go read the files of telegrams, full of groans, of tears, of measureless distress, that pass a focal telegraphic point after such a struggle as occurred near Stone River, December 31; then you can count some of the cost, which is not in property alone, not to the wounded or dying alone, but to the whole nation as individuals, all of whom are, sooner or later, wounded in person or by friend. Aye, more; homesickness, fevers, hunger and cold, in the field; sleeplessness, anxiety, which gnaws as a canker worm, and want, at home; these and death are war's ingredients, to comprehend which, in their bitterness, needs sad experience, for they can no more be described than Parrhasius "could paint a dying groan."

At two P. M., of the first, Wheeler and Wharton's cavalry dashed again upon Rosecrans' roadway at La Vergne, capturing part of his trains and scattering the rest in wildest confusion. The telegraph was considerably injured, the operator driven off and his instrument destroyed. This was the cause of that dreadful silence, which every one interpreted according as his hopes or fears predominated. But towards evening, when the panic-stricken teamsters entered the city, stoutly asserting that Bragg's whole army was following, confusion ran confounded, and "What's the news?" was asked only in whispers.

Col. Innis, of the gallant First Michigan Engineers, infused his own valor into his fine regiment, and the rebels who attacked his position at La Vergne were again and again repulsed.

Wheeler's object was to discover if Rosecrans was retreating. The night of the thirty-first had left him twenty-eight guns less than the morning; he had given up much ground. Three thousand wounded and hundreds of Federal dead marked how it had been contested; three thousand unhurt Unionists were captured. Withal, it was a disastrous day to the Federals, but not without somewhat of compensatory injury to the Confederate left. No wonder Bragg was amazed at Rosecrans' apparent obstinacy and stood with folded arms throughout the first, awaiting reports from Wheeler. On the second, Bragg's right, under Breckenridge, charged, driving two Federal brigades off the field too easily, for other Union troops and fifty-eight cannon opened upon his forces in front and flank and hurled them back in great disorder, less two thousand killed or wounded: thus ended the battle of Stone River. On the night of the third, Bragg retired to Tullahoma, behind the Duck River. His losses in battle, killed or wounded, were admitted to exceed nine thousand, besides over one thousand prisoners, and Rosecrans' casualties were reported as 1,533 killed, 7,245 wounded and nearly 3,000 missing.

Dwyer's brother and James Galvin, foremen, restored the line to Murfreesboro. The evening of the last day of the battle, Rosecrans rejoiced many anxious hearts by his telegraphic report, and on the fifth, having reported the occupation of Murfreesboro, the following telegram was received:

WASHINGTON, January 5, 1863.

MAJ. GEN. ROSECRANS:

Your despatch, announcing the retreat of the enemy, has just reached me. God bless you and all with you. Please tender to all and accept for yourself the nation's gratitude, for your and their skill, endurance and dauntless courage.

(Signed) A. LINCOLN.

To which the General replied, urgently asking the President to furnish that army with a uniform arm, adding that the battle of Stone River was almost lost by reason of the difficulty in supplying ammunition. On the ninth, Halleck telegraphed:

General — Rebel telegrams fully confirm your telegrams from the battle field. The victory was well earned and one of the most brilliant of the war. You and your brave army have won the gratitude of your country and the admiration of the world. The field of Murfreesboro is made historical, and future generations will point out the place where so many heroes fell gloriously in defense of the constitution and the Union. All honor to the Army of the Cumberland. Thanks to the living and tears for the lamented dead.

It was indeed fortunate that the line leading north from Nashville, *via* Clarksville, Fort Donelson, Paducah and Cairo, had been repaired soon after Colonel Bruce occupied Clarksville, else there would have been no telegraphic outlet during most of Rosecrans' operations, as two days before he left Nashville that, ubiquitous leader, General John H. Morgan, with three thousand riders, again struck the L. & N. R. R. and the telegraph along its route. Crossing the Cumberland once more, he again captured Glasgow (December 24) from which place he raided the railroad, until the twenty-ninth, destroying the telegraph and road for many miles, capturing a number of detachments, posts and stations and quantities of supplies, and finally crossing the Cumberland about Jamestown, returned to Bragg, at Tullahoma. The Bardstown line was also destroyed, and operators Thomas H. Smith and J. E. Nagle driven off. The principal injury to the telegraph, however, was from Bacon Creek to the Rolling Fork, on the main line. Morgan avoided Munfordsville where the Federals were too strong. At this place, James and A. C. Jones operated. James began repairing the line as soon as it was safe; but as many of the poles were cut and burned and the line twisted around trees or hid in the woods or creeks, it was like rebuilding without proper forces and appliances, and hence slow progress was made, but another party started from the north side of the break and met Jones at Elizabethtown, where the connecting splice was made. Among the first messages that passed was a special to the Philadelphia *Press*, giving some account of Rosecrans' success. J. C. Dorchester, operator at General Boyle's head-quarters in Louisville, copied it for the General's information. Boyle gave it to the press, provoking much discussion as to the right of property in such telegrams.

On the ninth day of January, 1863, John Clark VanDuzer, from Grant's department, was appointed Second Assistant Superintendent U. S. Military Telegraphs, and assigned to the Department of the Cumberland, where he found himself once more in intimate official relation with Rosecrans, with whom he had been at, and prior to, the battle of Corinth. So highly did Rosecrans appreciate VanDuzer and the telegraph, that he complimented him with an honorary appointment of captain, and attached him to his staff, but VanDuzer had no actual military *status* until October, 1863, when he was commissioned captain and assistant quarter-master. Captain VanDuzer is one of those rare men who can be familiar with subordinates without detracting from his proper influence over them. His command was none the less law because pleasantly given. He has been mentioned as the only fighting officer in the corps. The fact is that the officers had, as a rule, but little opportunity to distinguish themselves in that way, but VanDuzer, from the beginning made it a point to share all the dangers incident to the military telegrapher's position on a campaign.

JOHN C. VAN DUZER.

JOHN C. VANDUZER was a New York and New England boy. Born in Erie County, New York (1827), and obtaining a common school education, he early sought that oft-tried and generally successful institution—the printing office—in which to complete his education, and to "finish" it also, probably, for after spending the best years of his eventful life and gaining a splendid reputation as a telegrapher, he now sits in a Northern city, within the sound of his press and mallet, his scissors and paste at one elbow, his exchanges at another, pencil over ear and pen in hand, surrounded by all the paraphernalia of his first business love, and is, perhaps, as happy as when helping Generals Rosecrans and Thomas, Grant and Sherman, conquer peace in Tennessee, Mississippi, Alabama and Georgia. His first experience in a printing office was in Utica, New York, whence he went to another in New Haven, Ct. It was not long after, that he edited and published a country newspaper, in Connecticut. How well he succeeded, we are unadvised, but as we next hear of him at the head of another country paper, back in New York, it is due him to assume that he was about as unsuccessful in his first business venture as young men usually are. It was while thus engaged in New York, that he learned to telegraph. That was in 1848, a time when newspaper men began to realize the possibilities of the telegraph, which drew to its aid many young men who had already become expert compositors.

VanDuzer easily obtained employment at a small salary, for in those days even skilled operators received less than a beginner does now. But there were few experts. Messages were then received entirely by paper. It was not until about 1846, that reading by ear was thought of, and then its possibility was demonstrated. Nevertheless, until some years later, telegraphing by sound was considered so unreliable that it was prohibited by those in authority. VanDuzer's first office was on J. J. Speed's lines. More lucrative positions were soon open to him, consequently we note his employment next on the Western Union (then a small company) wires, and later on the Erie Railroad lines. It was on this road that the system of running trains on telegraphic orders was introduced first in this country, and the world, by the late General D. C. McCallum, who, in the civil war, managed the military railroads with marked success. At the outbreak of the war, VanDuzer, full of zeal for the Union, and of energy for any good undertaking, heartily responded. We have already noted his services in Missouri, Western Tennessee, Alabama and Mississippi, in all of which he was indefatigable in his efforts to keep the armies in telegraphic communication. His promotion in the service which had so few officers, was as clearly deserved as that of any officer in the corps, and rejoiced his operators who were devotedly attached to him.

Under VanDuzer's direction, the line was re-opened in January, to Franklin, Tenn., where General Gordon Granger was posted, and Ellis J. Wilson operated, assisted in May and June by W. R. Plum. It was June 8, that Lawrence A. Williams and —— Dunlap, respectively colonel and lieutenant in the Confederate service, with forged orders purporting to come from Washington and Rosecrans, presumed to inspect the Federal defenses about Franklin. Their conduct excited suspicion, and the telegraph quickly confirmed it. These officers were, by Rosecrans' order, immediately tried by a drum-head court martial. They expostulated against being treated as spies. They pleaded for a commutation, but military courts work as sternly as they do quickly. The following messages, received and sent by Wilson, tell the sad story :

MURFREESBORO, June 9, 4:40 A. M.

COLONEL J. P. BAIRD, Franklin:

The General commanding directs that the two spies, if found guilty, be hung at once, thus placing beyond the possibility of Forrest's profiting by the information they have gained.

FRANK S. BOND,
Major and Aide-de-Camp.

FRANKLIN, June 9, 10:30 A. M.

TO GENERAL GARFIELD, Chief of Staff:

The men have been tried, found guilty, and executed in compliance with your order.—I am ever yours,

J. P. BAIRD,
Colonel Commanding Post.

It was that kind of justice which was meted out to Andre, but who is there that does not deplore that young officer's fate.

June 10, Granger moved to Triune, half way to Murfreesboro, and his operators were ordered to Rosecrans' head-quarters *via* Nashville. While riding to Nashville, the operators fell to racing, which so excited Plum's horse as to make it positively unmanageable. Rushing on at headlong speed, he came suddenly upon the Brentwood pickets. "*Halt!*" they cried, but to halt was impossible. "HALT!" came again just as the horse passed the line. "HALT!" a last call. "I can't halt," cried the rider. Poor Wilson was too far in the rear to explain. *Bang*, BANG, go the rifles. A bright thought comes to the rider and with all his might he reins his horse against a fence or other side obstruction and the victory is won. It was the same horse that Con. Dwyer attempted to ride in the suburbs of Murfreesboro, but instead of doing so, he went hatless through its very streets, Bunnell, Wilson, Holdridge and Plum trying to keep him in view.

In May, 1863, General Rosecrans, regarding it as extremely important to know the movements of the enemy along the road leading from Richmond to Chattanooga and Bragg's plans of operation, requested Colonel VanDuzer to select two competent operators for a hazardous undertaking. Frank S. VanValkenburg and Patrick Mullarkey, first-class electricians, who had seen, as we have observed, much military service as operators, volunteered to carry out the General's wish, which was for them to enter East Tennessee, tap the telegraph wires, and copy whatever despatches of importance might be passing. These operators reporting to Colonel William E. Truesdail, chief of army police and scouts, for instructions, were directed to go through the mountains of Kentucky and East Tennessee, to a point near Knoxville, on the Chattanooga Railroad, and ascertain if possible, the movements of the rebels along the M. & C. railroad, and if Bragg was weakening his force to strengthen General Johnston, who at that time was collecting an army near Jackson, Miss., to relieve Pemberton's troops about Vicksburg. After accomplishing that task, the operators were to endeavor to burn the railroad bridge across the Tennessee at Loudon.

Four citizens from East Tennessee, dressed in butternut colored clothes, each armed with a revolver, accompanied them as guides. Taking a small quantity of line and magnet wire, a set of pulley blocks, two pocket instruments and two quarts of alcohol, the latter to be used in firing the bridge, they left Nashville for Lebanon, Ky., where they were arrested by the provost marshal's guard as rebel spies. General Manson furnished them with horses to continue their journey. Stopping the next night at a tavern in Liberty, they were sitting in their room after supper, when startled by the ringing of the bell on the building, summoning the citizens from all directions. Shortly after, a rap at the door caused VanValkenburg to open it. About twenty-five citizens were in the hallway, armed with rifles, shot guns, revolvers or knives. Their spokesman remarked that they were a committee appointed by the citizens to ascertain who the strangers were and where going; that the committee was composed of Union men and they did not propose to allow butternuts to travel south, without satisfactory assurances from them. Looking over the crowd, VanValkenburg discovered a man in the undress uniform of an officer of the Union army, and calling him aside, VanValkenburg satisfying himself of the officer's loyalty, exhibited Rosecrans' passport and convinced the captain, for such he was, that the party were Unionists. The officer then turned to the committee and said: "Boys, these yere fellers is all right—they be Union men of our stripe." Whereupon, "peach and honey" was indulged in by all.

Somerset was reached the next evening. General Carter supplied fresh horses, and the next day the party crossed the Cumberland River and entered rebel territory. From that time they avoided the traveled roads, preferring the mountain paths and at night reached Jack Harris' house in the mountains, where they remained over night. The next morning, leaving their horses, they journeyed on foot, and that night slept in a hut on the summit of Brimstone Mountain. Descending this the next morning, they reached a point on the Clinch River near Kingston, where their guides left them and went to the houses of Unionists near by, from which they returned with provisions and the information that the country was full of straggling Confederate soldiers of Colonel Flood's Seventh Florida cavalry, which had been defeated in a fight on the Cumberland River. Here they remained until dark, when swimming the river, they proceeded to Parson Harwell's house on the banks of the Tennessee River, close to Loudon, near which place they remained in the woods the next day, being supplied with provisions by the parson's daughters. That night Parson Harwell's son guided them through the woods to a point near John West's house, opposite Loudon, where they remained all day. About ten P. M., they proceeded through the woods to a point on the Knoxville & Chattanooga Railroad between Le Noir and Loudon, about fifteen miles from Knoxville, where the railroad runs through a cut close to the banks of the Tennessee River.

They decided to begin operations here. The line ran on the top of a bluff between the river and the railroad. At the top of the bluff the main line was tightened by pulleys and rope, then cut at the pole and a piece of leather inserted to hold the line in position and break the current. But before severing it, two fine wires were run down the side of the pole farthest from the railroad, and through the bushes down the bluff to a point about midway between the river and railroad. The main line at the cut was then replaced on the insulator so as to hide the leather except from a close observer. A pocket instrument was soon put in circuit, and the spies were really eavesdropping. Business on the line had mainly ceased for the night. Early next morning they heard the Knoxville operator calling Chattanooga, and not being able to raise the operator there, he commenced testing for a heavy escape which he located between Le Noir and Loudon. This escape was caused by the dew falling on their magnet wires, lying on the bushes. Before the Knoxville operator could start out a repairer, the sun had dried the bushes and the line was clear. The National operators lay at this point all day, copying whatever passed over the wires, but obtaining nothing of importance. That night, the dew falling again, caused a heavy escape. Knoxville located it at the same point, and ordered out the repairer from Loudon on the first train in the morning. As the train passed the operators, they lay concealed on the bluff and saw the repairer with his tools, watching from the rear end. During the first day they discovered a ferry boat that crossed the Tennessee River, also a landing directly beneath them, and from this time on, squads of Confederate troops were frequently crossing to the opposite shore. This cut the party off from water except by night. In the woods on the opposite side of the track, a party of Confederate soldiers were cutting timber for a block house at Loudon. The next morning and every morning after during their stay at this point, Knoxville located an escape between Loudon and Le Noir and ordered out the Loudon repairer. The repairer once reported that he would go out afoot and examine every pole. The operators told their guides that he was coming afoot, when one of them quietly took the rope from the pulley-blocks, made a slip-noose in one end and fastened a large stone to the other. On asking him what he was doing, he replied, "If that fellow comes up here, he will see us; if we let him go, the Johnnies will have us, and inside of twenty-four hours we will have been tried by a drumhead court-martial and the whole six of us sentenced to stretch hemp. Now, it is better, in my opinion, that he should stretch hemp than we six; so, if he comes up here, I'll put this noose around his neck and the first chance we get we will jump him into the river." In the course of an hour, from their position on the bluff, they saw the repairer walking along the track, examining each pole as he came. When he reached the bluff he stood on the track and looked at the wires attentively for a few minutes, and then went on to Le Noir and reported to Knoxville that the wire was all clear.

Several days after this, they heard an order going over the wires, signed by General Pillow, to the commanders at Le Noir and Loudon, directing them to station men between those two points and search the woods to the river for Yankee spies. This was about three o'clock in the afternoon; they immediately disconnected their instruments, connected the wire through, and, succeeding in crossing the railroad unobserved, they started through the woods, hoping to get outside of the line that they knew would be formed to capture them. They barely succeeded in this. About seven o'clock that evening they reached John West's house and decided to lie concealed in the woods until the excitement should subside. The next day John West brought to them in the woods, a man, who, he said, came to him with letters from Union men in Philadelphia, Tennessee, stating that he was a Union man and was desirous of reaching the Union lines, and requesting West to introduce him to any party of Union men that he might know of going out. West asked them if they were willing to take him with them. One of the guides was well acquainted in Philadelphia and questioned the man closely. Suspecting that he was not what he represented himself to be, after consulting together, they agreed that they should keep him with them for the present, and that one of their party, whose home was in Loudon, should go there and ascertain what he could about the stranger. The next night on his return from Loudon, he stated that this man had been a beef contractor in the Confederate army. Their guides were for disposing of him at once, but, before they had decided what to do with him, it was discovered that he had quietly slipped away in the darkness.

Knowing that it would not be safe to remain there a moment longer, they took to the woods and at daylight reached Parson Harwell's house. That day the parson's daughters rode through the surrounding country, ostensibly calling on their friends, but in reality for the purpose of finding out what arrangements were being made by the Confederates for the capture of the spies. In the evening the girls returned and reported that scouting parties were searching the houses and that pickets had been placed along the bank of the Clinch River, and that Champ Ferguson, with his band of guerrillas, was searching the woods for them. They decided to try to pass the Clinch River and reach the mountains. That night, about ten o'clock, they started, and reaching the banks of the river they saw the rebel pickets posted to intercept them. Waiting a good opportunity, they quietly slipped past the pickets and swam the river. On leaving the river bank they were greeted with a volley of musketry from a squad of cavalry. The spies immediately took to the woods, the cavalry following the road, hoping to head them off. It passed and repassed them several times, between Clinch River and Waldron's Ridge, they lying concealed in the brush until the cavalry were out of hearing. Then they crossed Waldron's Ridge into the valley, where, just before daylight, they were surprised by a small squad of Champ Ferguson's guerrillas, with whom they exchanged shots. They kept steadily on all that day, confining themselves to the woods and mountain paths, and that night reached a place of comparative safety, near the foot of Brimstone Mountain, having walked fifty-four miles in twenty hours, crossing Waldron's Ridge, Brushy Mountain and Brimstone Mountain.

Mullarkey and VanValkenburg had worn the soles off their boots and were now barefooted. After resting a few hours they again started, and the next day found them at Jack Harris' house, where they had left their horses. Remaining here over night, they started the following morning on horseback, riding all that day. The next morning, about ten o'clock, they reached Union pickets on the banks of the Cumberland River, having been inside of the Confederate lines thirty-three days. The third day after, they arrived in Nashville, completely worn out and ill from fatigue and exposure. Here they learned that the Confederate authorities had been apprised of their visit and its object almost immediately upon their arrival in the vicinity of Knoxville, and that this information had been furnished them by spies who drew pay both from Confederate and Union officers. Colonel Truesdail afterwards informed VanValkenburg that after he knew the Confederates were aware that they were inside of their lines, he had given up all hopes of their returning. And, indeed, had it not been for the aid rendered them by the Union people of that section, they would have been added to the list of Champ Ferguson's victims. Just before the close of the war, at Nashville, VanValkenburg saw hung as a rebel spy, the person who came to them at John West's — the beef contractor and Confederate spy.

CHAPTER XII.

THE TELEGRAPH ABOUT VICKSBURG AND IN WEST TENNESSEE.—THE VICKSBURG CAMPAIGN.—VARIOUS FEDERAL RAIDS.—CHAMPION HILLS.—AFFAIRS ABOUT CAIRO.

The story of the defense and capture of this naturally strong position, rendered impregnable by artificial obstructions and destructive arms, is one of the most interesting which the annals of war afford. The Chickasaw Bluffs extend full twenty miles above the city, and at Haine's Bluff, twelve miles north, where it shapes the course of the Yazoo and commands its opposite shore, the northern end of Vicksburg's line of defensive works was located. Vicksburg is situated on the bluffs, where they rise about two hundred and fifty feet above low-water mark. Warrenton, six miles below, and Grand Gulf, about sixty miles farther down, are also on the bluffs, which extend south far beyond the sites of Bruinsburg and Rodney.

Besides this great barrier, there were others quite as formidable, and which consisted of the swampy nature of the ground accessible to the Federals; of the innumerable swamps, bayous, creeks, rivers, lakes and impassable roads; of a consequent malarious climate which so bred disease that for miles the river banks, being the only dry ground, were marked by the burial of the Union dead; and of the unusually high stage of the water and frequent rains, which excluded all reasonable expectations of relief in the low regions where the Federals might hope to operate, so as to concentrate east of the bluffs where the country was inviting and of vast material consequence to Pemberton, who had retired to the city from Grenada, on Grant's approach by water.

North and west of Vicksburg for hundreds of miles, the country is so bisected by rivers and streams, as to afford numerous water routes, and by some of them it was believed practicable to flank the defenses north of the city, or by passing below through others enter the Mississippi south of its defensive works, in which latter event it was proposed to join the Nationals operating in the Gulf Department, and marching north, sweep Port Hudson, Grand Gulf, Warrenton and Vicksburg. Another plan involved crossing the river as near Vicksburg as practicable, and moving to its rear, regardless of the Union forces below, undertake its capture by assault or starvation. Grant was foiled in so many undertakings hereabouts, that not a few stout hearts in his army, including Sherman, advised starting on a new line from Memphis, but that, it is said, would have sadly affected the *morāle* of the troops, for Grant, with superior forces, would then have no lines in advance of what they were when Halleck left the army, in July, 1862, and the North bowed down by defeats, might not have submitted to Grant's longer remaining at the head of this department. General history should be consulted for the details of his unsuccessful ventures hereabouts,. It must suffice here, to outline only their general character. January 30, 1863, Grant took personal command of the forces operating against Vicksburg, amounting to fifty thousand men in hand, leaving a larger number behind him, variously posted in his department. Three plans to reach the rear of Vicksburg were under experiment at once, each of which had its enthusiasts. One was to cut a canal, one and one-fifth miles long, across the neck of land which, on the west side of the river, projects to a point a little above Vicksburg. This canal was to be dug just beyond the reach of the city batteries. By it, it was hoped to divert the waters of the Mississippi away from Vicksburg, thereby rendering that city of little consequence, or at least enable Grant to pass below and feed his army while operating from the south and east. But the enemy erected batteries lower down on the bluff, so as to command the southern end of the proposed canal, and a sudden rise flooded the peninsula; either of which causes would have made the scheme abortive.

Another project was to connect the river with Lake Providence, about seventy-five miles above Vicksburg, and thence *via* bayous, or rivers, reach the Red River, and, follow that down where, some one hundred and fifty miles below Vicksburg, and about midway between Port Hudson and Natchez, it enters the Mississippi. This scheme involved a concert of action with the

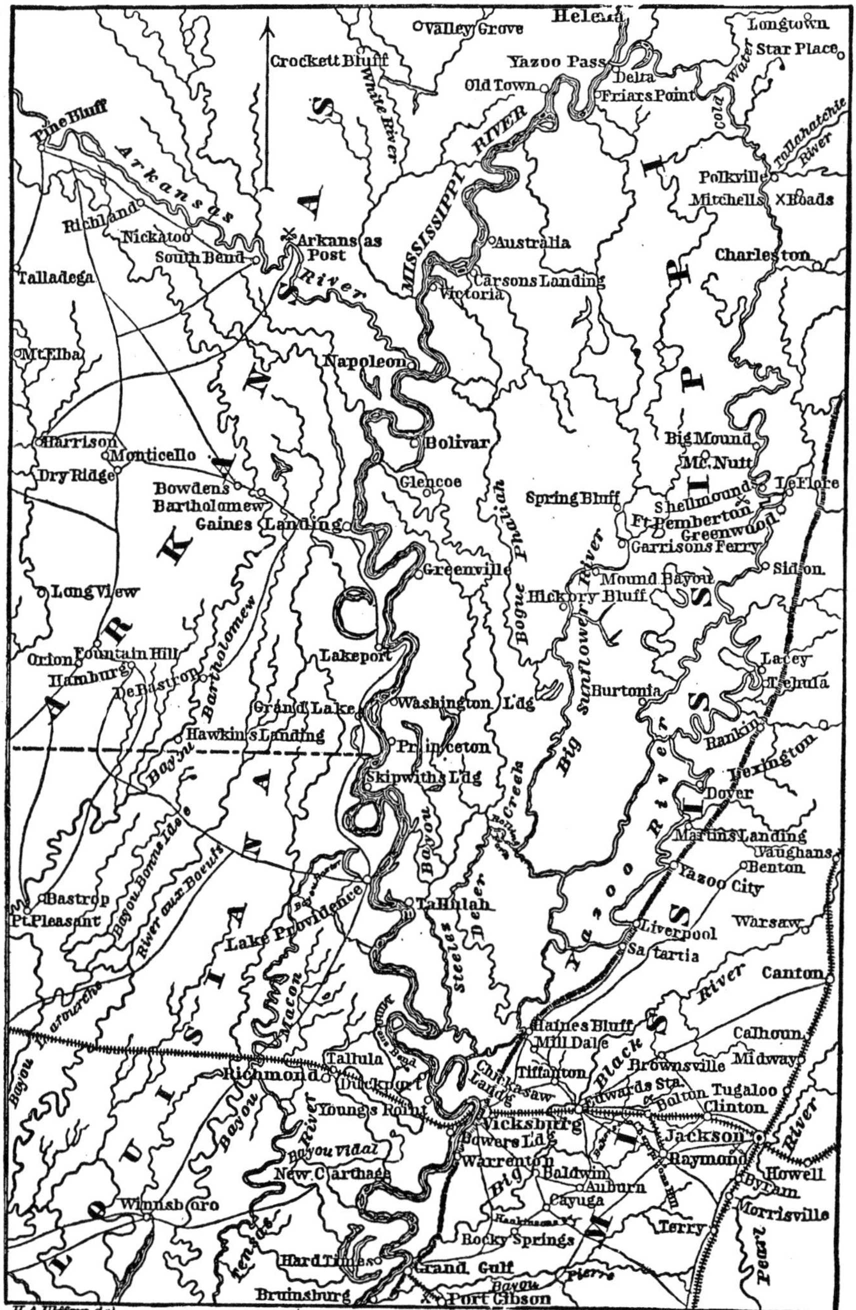

H. A. Ulffers del.　　　*Eng'd by Fisk & Russell, N.Y.*

10　5　0　10　20　30　40 miles

THEATRE OF VICKSBURG OPERATIONS.

Army of the Gulf, but it was found to be too big a job, and of doubtful utility if opened, owing to the scarcity of transports small enough to navigate the streams.

A third design was to flank Haine's Bluff, by a water route, beginning a little below Helena, Ark., but on the eastern side, and by cutting the levee, enter Yazoo Pass, which connected with the Coldwater, Tallahatchie and Yazoo Rivers. Gunboats and transports by this route reached Fort Pemberton, situated at the mouth of the Yallabusha, but were unable to capture it. This fort was about three hundred miles from the levee mentioned. Sherman and Admiral Porter tried to aid this expedition by a roundabout way into the Yazoo from Young's Point, near Vicksburg, avoiding Haine's Bluff, but after crawling along the Muddy, Steele and Black bayous, something like one hundred and fifty miles, they returned without reaching the Sunflower, where they would have had water enough.

These experiments are said to have been useful in inuring the army to hardships, but as it experienced no subsequent severities equal to the lessons, they should be set down as gigantic failures, which cost from disease alone a fearful sacrifice ; nevertheless it would be unjust to pronounce censure, as they were plans which could be determined only by their trial, and depended greatly on the elements. War is largely experimental. These routes were respectively advocated by able engineers, and all of them were possible under favorable circumstances.

Finally, after two months of engineering, General Grant had recourse to a westerly roadway, miles of which were corduroyed for the purpose, whereby he marched McClernand's and McPherson's corps to Carthage, some thirty-five miles south, where were met some of Porter's gunboats and transports, that had run the gauntlet of the Vicksburg batteries, and thence moving past Grand Gulf, Grant crossed the river near Bruinsburg, and possessed Grand Gulf after defeating Confederate General Bowen, at Port Gibson. Grant thereupon ordered Sherman's corps, which had been engaged in a diversion, to follow, while he (Grant) pressed on toward Jackson, Miss.

Leaving Grant to concentrate his forces, let us take a hasty glance at affairs farther north. Several expeditions were sent out in the winter and spring of 1863, from the Memphis and Corinth line ; one, under General Dodge, into North Alabama, in conjunction with Colonel Streight's (of Rosecrans' command) unfortunate raid into Georgia *via* Tuscumbia, Ala. John O. Ingle, operator, accompanied Dodge on this expedition. Two others were sent in February ; one, being under General A. J. Smith, accompanied by Lew. Spellman, operator, and a second under Colonel Grierson, with whom was operator Fred. W. Snell. Yet another one, vastly more important, under Grierson left La Grange a few days before Grant started for Carthage. Grierson's force consisted of about seventeen hundred cavalry, and was to destroy as much of Pemberton's communicating railroads and telegraphs as possible, and also such stores, munitions, and the like, as he could. His success was remarkable. Luke O'Reilly, operator, who accompanied him, was of valuable assistance by telegraphing, at captured offices, false reports of Grierson's movements. One of O'Reilly's tricks, that troubled the Confederates a great deal in restoring their telegraphs, was to leave the line apparently intact, but in fact held together by some *non*-conducting substance, so inserted as not to attract attention. This force proceeded through the State of Mississippi and finally reached Baton Rouge, La. It caused Generals Pemberton at Vicksburg, and Gardner at Port Hudson a great deal of trouble. They were so perplexed at Grierson's movements, that they could nowhere oppose him with any force competent to prevail against him. All of Gardner's telegrams were captured at Port Hudson subsequently. The last two, preserved by operator W. A. Sheldon, of the Gulf department, are significant. They are as follows: Gardner to Pemberton, "Grierson has arrived at Baton Rouge. How did he get through?" Pemberton to Gardner: "I don't know, do you?"

General Johnston, commander of the Western armies, was at Tullahoma, Tenn., with Bragg, when Pemberton telegraphed the fact of Grant's new movement, and thereupon, he (Johnston) set about to collect forces to aid in beating off Grant. On the 12th of May, he reached Jackson, but the Federals had come between there and Vicksburg, hence Johnston and Pemberton could not attack in concert, and Pemberton felt too weak to stand alone, although May 16, he valiantly attempted it at Champion Hills. So Johnston evacuated Jackson and Pemberton, unwilling to risk a decisive battle, fell back fighting, into the city, where his troops were quartered by the 17th of May, when the siege began and with it, the arduous duties of the army operator, as the desperate assaults of the nineteenth and twenty-second had proven a siege inevitable.

John C. Sullivan, who relieved Ira G. Skinner as chief operator with General Grant, accompanied the latter to Milliken's Bend and remained chief of telegraphic operations in the Vicksburg district until some time after the capture of that city; Captain Fuller remaining at Memphis. April 20, Grant, in view of his movement, ordered a line built from Milliken's Bend to Carthage *via* Richmond, La., but owing to the road being used to transport supplies and the difficulty of procuring transportation, Solomon Palmer, foreman of builders, made slow progress and it soon becoming undesirable, the construction was abandoned. May 24, Palmer was ordered to erect a line connecting all corps head-quarters, which was accomplished by the twenty-seventh; operator Ingle going to McPherson's, E. H. Johnson to McClernand's, and M. K. Booth, newly detailed from the Ninety-Third Illinois infantry for that purpose, was sent to Sherman's. June 2, Palmer built a line from Grant's quarters to Chickasaw Landing, seven miles, and thence to Haine's Bluff. Operators were so scarce, that searchers were sent through the army and Sherman's office was temporarily closed to let Booth go to the Landing, where all the supplies were received. General Osterhaus repaired the line to his quarters at Black River, on the railroad, when he became clamorous for an operator. Hudson H. Allen was detailed for that office about the twentieth. June 19, the line was extended on the left to General Lauman's, and on the twenty-third, to Herron's, where Beckwith was stationed. Beckwith had just arrived from above, having been serving Generals Lee, at Germantown, Brayman, at Bolivar, Dodge, at Corinth, and Hurlbut, at Memphis. These Vicksburg lines aggregated forty miles in length; that from Haine's Bluff to Big Black was to advise the forces near those points of any approach by Johnston, who was collecting an army to relieve Pemberton. G. W. Baxter was stationed at Haine's Bluff. June 26, W. H.

Parsons, Edwin D. Butler and Stephen L. Robinson arrived, barely escaping a masked battery on the banks and a fire on board, and the next day, a line was built to Parke's head-quarters at Post Oak Ridge, eight miles beyond Haine's Bluff. Butler was sent there, Robinson went to Lauman's, and Parsons temporarily relieved Booth, who had become very sick. Beckwith went to Grant's about the 1st of July, and Robinson to Herron's. The operators, by the 4th of July, were completely exhausted, their work being very exacting, especially as to time. Many successive nights they barely slept at all, and were of course at their instruments always in the day time. Finally some relief was obtained by making the couriers awaken the operator every time the instrument made a noise; by this means snatches of sleep were obtained, but a great and glorious end was approaching.

Pemberton's defense of Vicksburg was in many respects both able and gallant, but considering his force; that he held interior lines, and the natural difficulties of an assault, the wonder, is not so much that he successfully repelled all attacks as that he attempted to stand a siege with insufficient provisions. Grant's audacity might have been less manifest had he not greatly underrated Pemberton's real strength, a thing officers are not likely to do, and Pemberton might have been less reluctant to obey Johnston's early orders to attack, had he not overestimated the forces then in hand with Grant. Although the soldiers could not successfully storm the enemy's Vicksburg defences so as to penetrate them in force sufficient to effect a permanent lodgement, the lack of the necessities of life soon became evident to all within the city, and frequent desertions enabled Grant to keep generally advised of the stores on hand. The question of Pemberton's surrender became, with the siege, but a question of time, it would seem, with both army commanders, unless General Joe Johnston succeeded in collecting an army sufficiently strong to relieve the besieged. Plans for cutting a way through the Federal lines were indeed discussed in Vicksburg, but no such attempt was made. Johnston was indefatigable in his labors, but Grant's forces and works behind the Big Black increased, so that he felt measurably secure, while Sherman, Washburne, Parke and Osterhaus were on the alert along the defensive rear, from the Yazoo to the Black River bridge. Johnston, about the last of June, reached Brownsville, near the Federal defenses. On the 27th of June, there came a report that Johnston was to be quickly reinforced by ten thousand troops from Bragg in Tennessee, and new assaults were being considered by Grant, but the commander of the half starved defenders was poorly advised of Johnston's operations. Mines had been sprung and others were in preparation; Federal shells were falling in all parts of the town and its citizens were dwelling in caves.

Surrender was felt to be an inexorable necessity, and accordingly, on the morning of the 3d of July, Pemberton proposed an armistice with a view to arranging terms of capitulation. By this time, Beckwith had come to the relief of Sullivan, whose cipher and telegraph duties at Grant's had been of the most trying nature. Ordinarily Beckwith's telegraphic manipulations were mechanically accurate and steady, but now that his news was the harbinger of an ultimate success which would crown the Army of the Tennessee with unfading laurels and shower upon it the tearful blessing of millions who were, and the loud acclaims of millions yet unborn, he nervously signaled "22"* and "OS,"† repeatedly and then began calling the operator of Washburne's head-quarters up the Yazoo; at Generals McPherson's, Ord's (he having relieved McClernand), Osterhaus', Parke's, and elsewhere, all of whom, alive to the meaning of the signals, and doubtless believing an attack was impending, hurriedly responded. Beckwith then telegraphs, "Big news!!! Pemberton sends flag of truce; wants to know terms, if he will surrender. Grant going to meet him; give you more soon." Then followed the orders to desist firing, and the hills about Vicksburg, for the first time for months, ceased to echo and re-echo the voices of war. It was truly big news. Grant returned about four P. M., and telegraphed some of his chief officers to meet him, to consult as to the terms which should be offered. Almost at the same moment, Steele, McPherson, Ord, Lauman, Logan, Herron, Blair and some others started for Grant's. The council over, the terms were submitted, amendments proposed but rejected, and about two A. M., Pemberton's reply, accepting the *ultimatum*, was received.

*A signal for precedence and importance. †Signal that the matter is for all offices.

How opportune, indeed; for while these negotiations were progressing, Johnston was steadily moving toward the Big Black and had actually placed one gun into position. Reports of scouts were frequently telegraphed from Washburne and Osterhaus, announcing the approach of this new army. Grant's anxiety lest he should have to let go his fast hold on the great prize, to crush this new army, was very great. His own army was now seventy-five thousand strong, Pemberton's, thirty thousand, and Johnston's exceeding twenty-five thousand; but with them concentration on single points alone invited success; with Grant, to concentrate was to loosen his hold on Vicksburg. Minutes became hours and hours, days, while the terms were under discussion, but with a firm reliance in his brave troops, perhaps in the God of battles, Grant rejected Pemberton's amendatory terms, and then Pemberton acquiesced. It was this acceptance that brought Grant into the telegraph tent at two A. M., of July 4, saying, "Well, Beckwith, they have accepted our terms, and now I want Sherman. Can you raise him?" It should be stated that Sherman was temporarily relieved of the command of his corps in front of Vicksburg, on the 20th of June, and sent back to command the forces anticipating Johnston, and at this time (morning of July 4) his head-quarters were at Parson Fox's, between Haine's Bluff and the railroad bridge. Operator Butler at Parke's, who had gone to sleep with his head in a cracker box, on the top of which rested his instrument, having, so to write, one ear open, was soon aroused, and Grant's message to Sherman, stating that "Pemberton has accepted and will surrender to-morrow morning," was duly despatched. Other messages speedily followed to all commanders, and on the anniversary of our nation's birth, the army made, says Badeau, in his "Military History of U. S. Grant," "*The largest capture of men and material ever made in war.*"

On the fifth, most of the operators moved into the city, where they established an office at Grant's and in a bank building, and while Parsons, Johnson, Baxter and Robinson were for(a)ging through the iron doors of the bank, only to discover an old clock and some duelling pistols, Beckwith, who had secretly learned of the burial just over the Black of eighty thousand dollars in plate and coin, together with a member of General

Matthias' staff, rode out to possess it. In their anxiety, they ventured too quickly, and were nearly captured by four rebel horsemen, but other Federals coming up by a by-way, the Confederates rode off. When, on another day, the treasure was sought, somebody had carried it off.

No sooner was the capture of Vicksburg assured than Admiral Porter started his fleetest steamer for Cairo, carrying a detailed account, with all of the prefixes and affixes possible in the address and signature. The steamer reached Cairo two and a half hours ahead of the next one. The pompous naval officer strutted into the telegraph office and demanded possession of the wire to send the news of the capture of Vicksburg to the Secretary of the Navy; but while the receiving clerk was counting the long message, preparatory for sending, the manager rapidly telegraphed without transcribing, as follows:

To COLONEL STAGER, War Department, Washington:

Pemberton surrendered July 4.

(Signed) A. J. HOWELL.

It was this laconic message that enabled Secretary Stanton proudly to spread over the land the news of a glorious victory by the time that Secretary Gideon Wells had adjusted his spectacles to read the message that Porter had striven so hard to send him, so that *he* might have the honor of reporting it.

This news was not so readily believed in the Confederacy. At Delhi, forty miles west of Vicksburg, whither Lee S. Daniel had retreated and where he was operating, General John G. Walker threatened him, on the fifth, with irons for stating a belief that the place had fallen, and on his telegraphing the fact on the eighth to Eldorado, Arkansas, the operator there soon advised Daniel to fly, as a *posse* of citizens had organized to go and hang "that lying operator at Delhi."

Among the Confederate operators found in Vicksburg were A. Grimes and E. L. Marchant. Grimes was a talented Irishman, who served many years as operator at Jackson, Lake Station, Vicksburg and vicinity. Marchant also worked at Jackson and subsequently in New Orleans. At the close of the war both took the oath of allegiance and were employed by the Government. They are both dead.

The following telegram will indicate Sherman's undertakings:

HEAD-QUARTERS EXPEDITIONARY ARMY,
Fox's, July 4, 8 P. M., 1863.

GEN. GRANT, Vicksburg:

Your despatch, announcing the magnitude of the capture of Vicksburg, is most gratifying. The importance of the place to our cause can not be exaggerated. I have left Kimball's division at Haine's Bluff, with instructions to picket at Oak Ridge. I will order General McArthur to relieve Osterhaus. I will relieve Osterhaus at the bridge to-morrow, so that Osterhaus may report to General Ord. Three bridges will be built to-morrow at Birdsong, Messinger's and the railroad crossing. To-morrow I suppose Ord and Steele will be up, so that next day I will cross and move in force on Bolton. The enemy show one gun opposite Messinger's this P. M. I am willing he should meet us at once; the nearer our base the better. If he declines battle I will follow promptly at Clinton; then I can discover if Johnston is scattered or contracted, when I will act accordingly. I have not yet heard if the prisoners are to be paroled here or sent North. The farmers and families out here acknowledge the magnitude of their loss and now beg to know their fate. All crops are destroyed and cattle eaten up. * * I feel an intense curiosity to see Vicksburg and its people, but recognize the importance of my present task and think of nothing else. I will keep a few orderlies at Osterhaus, which is now my nearest telegraph office.

W. T. SHERMAN,
Maj. Gen. Comdg.

Robinson pushed out with Ord and opened an office at Clinton on the twelfth. Parsons and Ingle footed it from the bridge to Clinton, seventeen miles, in the rain, on the thirteenth, when General McArthur gave the boys an ambulance to Sherman's, two miles from Jackson, at which place Johnston lay entrenched. It was well for Ingle that he rode in an ambulance, because at Clinton, where a drug store was sacked, he filled a pocket with capsules, which proved to be loaded with croton oil. It was after ten o'clock when they reached Sherman's quarters, and found him sleeping in a fence corner, but on hearing of their arrival, the line having been repaired by Palmer, Sherman wrote many telegraphic orders for troops in the rear, besides reporting to Grant. July 16, Confederate cavalry attacked Matthias' troops at Clinton, cutting the line, but were driven off, and the next day, Jackson was evacuated and as soon as Sherman directed it, the line was run into that city, *i. e.*, the twenty-first. Johnston was in full retreat and Sherman moved back to Messingers Ford, west of the river, where Parsons opened an office on the twenty-seventh. Ingle went to Haine's Bluff, William Foley, fresh from Cairo office, Baxter, Johnson, and H. W. Nichols, who was captured at Holly Springs and H. H Allen, were now ordered to the Gulf department.

The siege of Vicksburg had passed into history. Six whole months had been spent in its capture. Such victories do not come without sacrifices. That Grant's army was willing to incur any danger, is shown by its labors on water routes, by its marches of two hundred miles in twenty days, by its battles, by its deprivations and by its engineering approaches under the very guns of Vicksburg, but above all, by the miles of silent graves which indicate the ravages of disease, and by the killed (one thousand, two hundred and forty-three) and wounded (seven thousand and ninety-five) in battle. A peace purchased at such a price should indeed be lasting and fruitful; but no, not yet.

Among those fallen heroes, the Telegraph Corps counts one of its own number. Marsden K. Booth, eldest son of G. J. Booth, of Fulton, Ill., was born April 20, 1844, in Dundee, Yates County, New York. At the early age of thirteen years, he learned to telegraph in Fulton, and operated at Geneseo, Palatine, Harlem and Sterling, Ill., but though in fact too delicately organized for severe exposures, he was deeply impressed with the conviction that his duty bade him to shoulder arms. Accordingly, August 9, 1862, he joined Company F, of the Ninety-Third Illinois Volunteer infantry, sharing its fortunes and its miseries until detailed from the trenches in front of Vicksburg as above shown. In that sanguinary struggle known as the battle of Champion Hills, where four hundred and twenty-six Federals were killed and one thousand, eight hundred and forty-two wounded, and the Confederates lost between three thousand and four thousand, killed and wounded, the Ninety-Third lost many of its brave soldiers and nearly all of its officers, but young Booth was destined to aid in the capture of the western Sebastopol and then succumb to a malarial fever. The telegraph office at Chickasaw was in a tent on the bank of the Yazoo, whose motionless, slimy, green and tepid waters exhaled miasma that was unusually poisonous, especially in July and August, when the waters are low, stagnant and warm, and full of decaying vegetation. Besides this poisoned air and the foul waters of the Yazoo, Booth was so over-tasked, as to be wholly unfit for work, two days before Parsons relieved him, but in a large measure Grant's whole army was dependent upon the telegraphic orders affecting the transportation and disposition of the stores that were landed near the office, and Booth with the zeal of a martyr remained steadfast until relief came; but, fatal necessity; it cost him his life, and on the 25th of August, at the age of nineteen years and a few months he died, and was buried where his remains now lie, in the military cemetery at Vicksburg. One of his comrades writes that "he was a kind, uncomplaining young man, a genial companion and his death was mourned by those who were so fortunate as to have his acquaintance."

On the 17th of August, General Grant received some of his military friends. Certain officers took the occasion to present him with an elegant sword, after which, General McPherson, a constant friend of the army telegraphers, presented him with a superb pair of major general's shoulder straps—(he having been recently appointed a major general in the regular army)—together with the following letter:

HEAD-QUARTERS 17th ARMY CORPS,
Vicksburg, August 17, 1863.

MAJOR GENERAL J. B. MCPHERSON:

General—May I beg the honor of requesting you in behalf of the operators in the field, to present the accompanying pair of major general's shoulder straps to Major General U. S. Grant, as a token of respect and esteem. Should you deem it proper to present them this evening, I beg you will communicate to the General the sincere gratitude which we entertain toward him for the many kindnesses extended to us. I am, with profound respect, etc.,

Yours, S. H. BECKWITH,

Opr. Gen. Grant's Hd.-Qrs.

In war, the scenes are constantly shifting. A section once conquered and thought to have passed from the panorama of contending armies, suddenly reappears crimsoned with fresh misery, and the strife as before, has its ebb and flood tides, until finally one party again controls the domain. Then follows the peace of the grave and forage of the living; so it was in Virginia and West Virginia, in Missouri, Arkansas and Louisiana, and now also in West Kentucky and Tennessee, and so the strife continued to recede and advance again, along the border shores of the rebellion, until from sheer exhaustion, the weaker side in despair, ceased its depredating raids almost as suddenly as they were begun. When Grant quit Holly Springs, for Milliken's Bend, the Confederates' flood tide again set in, and it is to relate the incidents occurring in West Kentucky and Tennessee during the year 1863, of consequence to the telegraphers, that we now busy ourselves. W. G. Fuller, in his annual report for the fiscal year ending June 30, 1863, says:

> March 9, 1863, the railroad from Jackson, Tenn., to Columbus, Ky., was abandoned. All the garrisons withdrawn, and the offices closed, but by a judicious system of sending repairers in disguise, to travel over the lines on foot, we were enabled to keep open communications, more than half of the time, although guerrillas constantly cut them. The repairers did their labor in the night. June 8, Jackson and Bolivar, Tenn., were evacuated by our troops and the offices closed at those points, but the line continued to work through from Memphis to Cairo up to June 14, when it was cut so badly by guerrillas, that we gave up the hope of continuing it longer. T. P. Hemphill, who had been doing repair duty upon this long line of unguarded wire, with three soldiers, who were assisting him, was killed by guerrillas. The lines north from Grand Junction and from Corinth, were abandoned, and the direct line from Grand Junction to Corinth was rebuilt. The troops heretofore garrisoning the railroads along which these lines extended, had been sent to reinforce General Grant at Vicksburg.

Thomas P. Hemphill, from Carlinville, Illinois, volunteered to go with a train to repair the line which the rebels had torn down. He started from Grand Junction on the morning of June 15. The rebels had cut the underpinning of a bridge near Middleburg, Tennessee, and when the train struck it, the whole was precipitated into the ravine. Hemphill and three soldiers were instantly killed and twelve others were badly injured. Hemphill was so badly mangled as to necessitate his burial on the spot. Two weeks before this the rebels captured all of his clothing.

Whoever was in Cairo during the war, will agree that it was an abomination. According to Dickens, Martin Chuzzlewit was one of the early proprietors there. It was then

> A flat morass, bestrewn with fallen timber; a marsh on which the good growth of the earth seemed to have been wrecked and cast away, that from its decomposing ashes vile and ugly things might rise; where the very trees took the aspect of huge weeds, begotten of the slime from which they sprung by the hot sun that burnt them up; where fatal maladies, seeking whom they might infect, came forth at night in misty shapes, and creeping out upon the water, hunted them like specters until day; where even the blessed sun, shining down on festering elements of corruption and disease, became a horror. This was the realm of Hope through which they (Martin and his partner) moved. At last they stopped — at Eden, too. The waters of the deluge might have left it but a week before, so choked with slime and matted growth was the hideous swamp which bore that name.

When the author first saw it, in June, 1863, the streets were without bottom and the walks were on stilts. Sometimes a gutter intervened between the imaginary side lines where the street proper terminated and where the walk began, but as a rule it was all gutter. More attention was paid to bad drainage there than he ever noticed elsewhere. Everybody discussed it except the authorities. There was liquid in front, at the side and beneath the city; even the citizens were in liquor. Mr. Rollins, member of Congress from Missouri, speaking of Cairo, said his steamboat landed passengers in the third story of its first-class hotel. In the very heart of the new city the cry of the faithful boatman is "No bottom." Of another occasion, S. S. Cox writes: "Another member remarked that Cairo was one of the rising cities of this Union; to which, 'Has it risen above high-water mark yet?' was the apt response." Coffin, in his "Four Years of Fighting," describes the place as he saw it:

The mud can not be put into the picture. There was thick mud, thin mud, sticky mud, slushy mud, slimy mud, deceptive mud, impassable mud, which appeared to the sight, to say nothing of the peculiarities that are understood by the nose; for within forty feet of our window were a horse stable and pig yard, where slops from the houses and washes from the sinks were trodden with the manure from the stables. Bunyan's Slough of Despond, into which all the filth and slime of this world settled, was nothing beside the slough of Cairo. There were sheds, shanties, stables, pig-sties, wood-piles, carts, barrels, boxes — the *debris* of everything thrown over the area. * * There were truckmen stuck in the mud. * * This was Cairo — delectable Cairo!

Except the St. Charles Hotel and the theater, perhaps, the saloons transacted the only successful local business; nearly all of the business men were successfully employed. The similarity of occupations doubtless made business licenses cheap; anyhow, almost everybody had one. The walks were so unsteady that the police were constantly arresting people on charges of inebriety; it was a *fine* way the authorities had for increasing the city treasure. Thus the main business brought the city a steady income at both ends of it. The devil there wept for joy. Cairo was sometimes called the jumping-off place. It is certain that Satan obtained a foot-hold there; whoever had that was a fixture, and as Old Nick had no boots to leave, he may be there yet. The decaying boots and other putrescent garbage emitted exhalations without stint. They maintained an individuality, which to the uneducated olfactory, was observable only in its force and effect, but to an *habitué*, distinctive odors were perceptible. Thus Gross reported one hundred and forty-five varieties observable during a walk from the hotel to his telegraph office. During the war there were three places there carrying on independent business; one was the St. Charles Hotel, which had a large till and a small waiter; another, the Varieties, which afforded a weak diversion between strong drink; and last, the telegraph office, which was up two flights of stairs, a whisky shop being on the first floor and a billiard saloon on the second, through both of which patrons passed but to emerge at the head of the stairway in a dark, dirty, unfinished rear room of 50 x 25, littered with insulator blocks, old ladders, boxes, barrels, empty carboys, wire and the like. A door opened from this room into a middle one, 15 x 25, where Grove batteries poisoned the air; farther on, was the office, which had been a lodge room, but was now partitioned for the operators to sleep and work in.

Is it any wonder that every operator who ever stopped there soon sickened? that in a few short months Ashley, Gross, Hill, Parsons, Egan, Schermerhorn, May, Merkley, Peterson, Howell, Culbertson, O'Neil, Mullarkey, Foley, Walsh, Peel, Craig and others either went to Mound City hospital, home or the Yazoo country to recruit? Think of the place as pictured by Richardson of the New York *Tribune* about this time: "*Cairo*, a nondescript, saucer-like, terraqueous town. 'The season here,' wrote John Phœnix, 'is usually opened with great *eclat* by small pox, continued spiritedly by cholera, and closed brilliantly with yellow fever — sweet spot.'" No wonder Fuller wrote Bruch in September, 1863, "I can not get a suitable man that will take that office. Every one that I mention it to says emphatically, "*No, sir!* leave first." George Gallup tried it, subsequently, without success, but after him came a wonderfully constituted telegrapher. It was none other than W. T. Mason, who, remarkable as it may seem, continued the management of Cairo office to the end of the war, and still lives.

W. W. Forbes and W. R. Plum, then at Murfreesboro, Tennessee, volunteered to assist the exhausted operators at Vicksburg, but on arriving at Cairo, Colonel Stager, who was there, insisted that one should stop there and the other at Columbus, Kentucky, where J. V. Hill was very sick. Forbes refused and resigned, but Plum chose the unseen, and telegraphing General Asboth, commanding at Columbus, to have the post surgeon in readiness, he started for that place; scarcely had he landed before the doctor was called.

Colonel Stager had been sent to Cairo, where General Buford, a brave and gallant officer, commanded. It seems that Chicago and St. Louis papers actually published accounts of important military operations about Vicksburg, in advance of Secretary Stanton's knowing them. Colonel Stager was sent to stop the leak, for somebody was making money or obtaining newspaper puffs in return for reliable information. Colonel Stager found that despatches from below reached the telegraph office through head-quarters, but he did not discover the offender. However, the leak ceased when arrangements were made for the delivery of messages from the steamers to the telegraph manager. Subsequently, telegraph clearances were given at Memphis and Cairo, to all steamers, of which the following is a sample blank:

CAPTAIN ——— ———,

Master Steamer ———.

No steamer is permitted to leave .... without this clearance. W. G. FULLER, *Capt. and A. Q. M.*

No telegraph despatches being now in this office for which your boat should be detained, you are at liberty to pursue your voyage at any time before — o'clock —. If detained beyond that hour, this clearance is void, and you will report again at this office when you are ready to leave.

By order of Major General C. C. Washburn.

W. G. FULLER, *Capt. and A. Q. M.*,
Asst. Supt. Mil. Tel., Dept. Tenn.

NOTE.—All steamers are required by existing orders to report at the military telegraph offices at Cairo, Ill., and Memphis, Tenn., and to receive and deliver promptly all Government despatches.

A line was built along the railroad from Union City to Hickman, Ky., in April. In June, the troops were withdrawn for Vicksburg. Edwin D. Butler, operator at Hickman, closed his office and went with the troops. William Foley closed the Union City office, and soon after also reached Vicksburg. It was well for these operators that they left with the troops. General Asboth remonstrated, and George M. Brush, at Clinton, near Columbus, consented to take his chances at Union City. During June, an effort was made to re-open the line to Memphis, notwithstanding the country to Grand Junction, from Columbus, was unguarded. Jacob V. Hill, operator, started with a train and handful of men, on the fifteenth, and reaching Trenton, he was informed of rebels in close proximity, but ordering the train ahead, he proceeded one mile south, where he discovered cavalry picketing the road, whereupon the engine was reversed, but the rebels fired about twenty-five shots without damage, and received a like compliment from those on board. A month later the Confederates advanced in considerable force, capturing Brush at Union City, taking possession of Hickman, and driving Hill and all the Federals at Columbus into the fort there. Repairer Rutherford was captured outside and held ten days. Brush was stationed at Union City as a sort of advance picket to Columbus, and bravely did he do his whole duty, for although the cavalry dashed into town early July 10, before they effected his capture he called up Columbus office, and sent a telegram to General Asboth, explaining that the town was occupied by the enemy in large force, thereby enabling Asboth to make ample preparation to receive them. Some way the captors found out that Brush had advised Asboth, and hence the partisan soldiers having him in charge abused him shamefully, even threatening to hang him, but some officers removed him from their hands. His subsequent sufferings, however, were very great. He was taken to Richmond, Va., where, claiming to be a commissioned officer he was placed in Libby Prison, instead of Belle Isle. He remained eleven months in this prison, and was allowed to correspond with friends in the North and receive aid from them, but the prison authorities, finding he was not a commissioned officer, sent him to Andersonville. "He who enters here leaves hope behind." On his leaving Libby, the guards taunted him, and took from him his blankets, money and some of his outer clothing. When he left Andersonville, he was clad in an old blouse and a pair of drawers. He was exchanged at Savannah, Ga., about December 1, 1864, after near a year and a half imprisonment, but how altered! He was crippled with scurvy. His face, wan and haggard, was so changed that his friends at home (Carbondale, Ill.), did not recognize him. It was nearly a year more before he was sufficiently recruited to go to work again. Then he went to Kansas, and thence to Texas, but while visiting at home again in 1873, he was cruelly murdered.

Early in May, a new cable was laid at Cairo, and the line extended from Blandville to Columbus. R. B. Griffin was operating at Blandville, where there were no troops. But it was an important, outpost office, and the point where the line from Paducah intersected, so that in case the direct line gave out, that *via* Paducah could be connected. Late in October, the guerrillas entered Blandville by one road, while Griffin who heard of their coming just in time, escaped by another, but his telegraph property was all captured. Griffin soon returned to duty again. A few days later (November 6), Colonel Faulkner sent a squad of bushwhackers to Blandville, to destroy the office and capture the operator and repairer, but on the night the guerrillas appeared, and just before they entered, Griffin received a note by a colored servant from a young lady who had become attached to him, saying that Faulkner's men were coming that night to capture him, and that he had better hurry away. This was quickly telegraphed the commander at Columbus. A young man, a citizen of Blandville, was Griffin's bedfellow, and as Griffin decamped without giving notice, this young man was terribly surprised on being pulled from bed by the hair and dragged down stairs that night, some of the captors shouting, "Kill the —— Yankee operator," but the landlady, who was the lad's mother, came to his rescue and convinced the marauders that the operator was not about. However, they captured Griffin's horse and instruments. November 11, he reported: "It was impossible for me to save the property, as General Smith requested me to remain at my post as long as possible, when he was aware of my danger." This state of affairs continued until General Paine arrived at Paducah, when he issued orders charging the citizens with the protection of the telegraph, and threatening their freedom and their homes if the lines or operators were molested.

On the Memphis and Corinth line, the guerrillas and regular forces were even more annoying. The lines in West Tennessee were repaired for the most part by the use of hand-cars, with which all repairers were supplied, and it was no uncommon thing for the repairer to be ditched by the way. General Hurlbut had so denuded this line of troops for Grant, that guerrilla parties were very bold. We have already noted several expedi-ditions from the line of the M. & C. R. R., and will now merely add a telegraphic report to General Halleck of another:

The cavalry force from LaGrange, on the thirteenth (August), under command of Lieutenant Colonel Phillips, Ninth Illinois infantry (mounted), reached Grenada seventeenth; drove Slemmons, with two thousand men and three pieces of artillery from the place, destroyed fifty-seven engines, four hundred cars, the depot buildings, machine shops, several blacksmiths' shops, and a quantity of ordnance and commissary stores, and captured fifty railroad men and a number of prisoners. * * (Signed) S. A. HURLBUT.

Major General.

We do not know what operator accompanied Phillips.

CHAPTER XIII.

THE TELEGRAPH IN ARKANSAS AND MISSOURI IN 1863.—HELENA.—LITTLE ROCK CAMPAIGN.—OTHER CAMPAIGNS AND CONFLICTS IN MISSOURI AND ARKANSAS.—AN OPERATOR HUNG.

As affairs in Arkansas in 1863 were closely allied with those in and about Memphis, it is befitting now to fix our attention upon the military and telegraphic operations in that State; and inasmuch as Arkansas was in Major Smith's territory, and the Little Rock expedition, about to be mentioned, was composed, in part, of troops belonging to the Missouri Department, it will be found convenient to include in this chapter a statement of operations generally, under that department commander, and Smith, the telegraph superintendent.

Retrospectively, it should first be noted that the very day that Vicksburg surrendered, General Prentiss (B. M.), located at Helena, Arkansas, with thirty-eight hundred Federals, well posted in forts and other defenses, was attacked by the Confederates with wonderful vigor. General Hurlbut, commanding the Memphis district, telegraphed the result as follows :

HEAD-QUARTERS 16TH ARMY CORPS,
Memphis, Tenn., July 5, 1863.

TO MAJOR GENERAL HALLECK, Washington:

General Prentiss was attacked in force by rebels under Holmes and Price at Helena, yesterday. He estimates the force at fifteen thousand; I think nine thousand will cover their strength. Prentiss sustained their attack until three P. M., from daylight, when the rebels were repulsed at all points, leaving twelve hundred prisoners. Their loss in killed and wounded is about from five to six hundred; Prentiss lost about sixty. He has already sent me eight hundred and sixty prisoners, whom I send to Alton to-day on " Silver Moon." He has asked reinforcements; I have sent him the One Hundred and Seventeenth Illinois; I can not spare any more. The enemy are closely picketing everything south of my line, and seem strong. I have no fear of my position unless Johnston turns north, but am unable to spare men from Memphis, which I hold with an effective force not four thousand strong. * * *

S. A. HURLBUT,
Major General.

Very soon after Pemberton's army was paroled, Grant directed General Hurlbut to organize an expedition against Little Rock, to start from Helena, and to this end added to Hurlbut's territory "so much of his (Grant's) department in Arkansas as lies north of the Arkansas River." In a cipher telegram (July 28) to General Schofield, whose head-quarters were at St. Louis, Missouri, and who was co-operating, Hurlbut wrote: "Kimball's division,* six thousand infantry, with three field batteries, are now or will be in a day or two at Helena, ready to move upon the rear of Price, and co-operate with you. * * * * I wish at once to know your plan and probable course of operation, and means of supply. * * I dread the results of a march through that desolated country at this season, especially as I * * have no certain knowledge where Price's force is to be found," and again, "I shall probably send Major General Prentiss in charge of the expedition;" but General Grant sent General Steele.

General Schofield started General John W. Davidson, from Missouri, with six thousand mounted men and eighteen field guns, to reinforce Steele. Steele started early in August. The White River, always navigable by small boats, was at this time very high. Four gunboats were ordered up to Jacksonport, where Marmaduke was reported to be. As soon as these movements were noticed, General Kirby Smith hurried to Little Rock, determined to defend Texas from that point, and was reported to have brought up his troops from Louisiana. Fortifications were begun on the road leading into the city from the east. The scattered forces and partisans, under Price, Holmes, Marmaduke, Cooper and Cabell, were rapidly concentrated at Bayou Metoe, their left resting on Brownsville, creating alarm to Hurlbut, who telegraphed Schofield and Halleck that at least another brigade should be sent to Steele. To Halleck he sug-

* Sent from Vicksburg.

gested a movement of forces from Rosecrans' army, in conjunction with troops from Corinth, against Johnston's right flank, so as to clear that section as far down as Columbus, Mississippi, in which event Hurlbut could spare five thousand more men for Steele, but Steele himself did not expect a stand to be made north of Arkadelphia. The gunboat expedition was a success, capturing rebel steamers "Kaskaskia" and "Tom Sugg," and destroying the bridge over the Little Red River, with a loss of two killed and five wounded, August 20, Hurlbut started another brigade *via* Helena to reinforce Steele, and heavy Parrott guns were ordered from St. Louis, but probably not sent. With this brigade, Steele would have had fifteen thousand men and a full supply of field guns; but the water was higher than since 1844, and in places four miles wide, besides, the high temperature created so much sickness, it is reported, that within a few weeks seventeen hundred men were in hospitals. This must have been confined mainly to the infantry, as Davidson wrote Schofield from near Clarendon, August 9: "My troops are in fine condition; men, horses and transportation are better after a march of three hundred and fifty miles than when they started." September 2, Steele reported from Brownsville that he had pushed a party to within fifteen miles of Little Rock; that he would either turn the works at Bayou Metoe on the north and west, or move rapidly across to Pine Bluff, on the Arkansas, crossing the river and advancing on Little Rock from the south side, which was wholly undefended, making Napoleon his base of supplies. The latter plan was substantially adopted, the army striking the river at Ashley's Mills, where Davidson had a brush with the enemy. About this time, General Grant started John E. Smith's division (four thousand strong, with two batteries) from Vicksburg, to reinforce Steele; not only did they not arrive in time, but Steele was forced to leave one brigade of infantry and a regiment of cavalry behind to guard the sick and trains. It was well that Kirby Smith's troops had not yet arrived. Nothing daunted, Steele and Davidson pressed forward, fighting at the crossing of Bayou Fourche, on the tenth, and that evening entered Little Rock with flying colors, which shone brightly in the light of the six burning steamers and other property fired by Price, the commander, who retreated to Arkadelphia, from whence he was also driven soon after. Thus a rich cotton country, south of the Arkansas, was opened to speculators, and the State nearly cleared of regular Confederate troops.

While these things were progressing, the telegraphers were not wholly idle. About the middle of August, General Schofield ordered Major Smith to send a party to Memphis to construct a telegraph line from there to Little Rock. River soundings for a cable were made at Memphis, but General Hurlbut could not protect the builders from the guerrilla parties, who swarmed along the route, that of the Memphis & Little Rock Railroad, so the party was ordered to begin at Helena. General Steele was so pressed for troops, which were supplied by water, that he could not maintain land communications, and that project fell through. About the middle of October, the men proceeded to DuVall's Bluff with one hundred miles of wire and transportation. J. H. Black was chief of the builders, who to the number of twenty, reported with assistant foreman, George Allen. Dwight Byington, chief in the operating department, soon came, and with him were Theodore Holt, R. Hec. Smith, Edwin J. Waterhouse and George B. Allis. Lines were constructed in October and November to the important dependencies upon Little Rock, *viz.*: DuVall's Bluff, *via* Brownsville, forty-seven miles, Pine Bluff, forty-five miles, and Benton, twenty-five miles; total, one hundred and seventeen miles. Byington and Smith operated at Little Rock. Holt first at Little Rock; afterwards at Benton and DuVall's Bluff. Waterhouse at DuVall's and Brownsville, and Allis at Pine Bluff. One of the first things to be done to make the lines serviceable, was the killing of guerrilla chief Cotter, which was soon effected, to the great joy of the repairers. Engines and cars were taken by water to DuVall's Bluff and the Little Rock end of the railroad worked.

Black deserves a description. He had been a sea captain and in his voyaging, learned the French and Spanish languages, but notwithstanding his acquirements, including an easy, fluent speech and graceful demeanor for a man of fifty-five, the boys persisted in speaking of him as that "rough old tar." It was his sea-faring severities of speech that they named him by. A New Yorker by early education, a "salt" by practice, a river-waterlogged craft by necessity, and a land-lubber by force of circumstances; fat, jolly or rough, as occasion required, an excellent imitator and caricaturist, he was very loquacious and a lover of claret. He it was who built those Arkansas lines.

Preparations were now making to extend the telegraph to St. Louis *via* Fort Smith and VanBuren, Ark., and Springfield, Mo. At this point it is best to digress enough to discover the state of affairs in Missouri and North-west Arkansas, whereby such a line became possible.

THAT "ROUGH OLD TAR."

We have, in a former chapter, followed Curtis from Missouri to Batesville, and thence to Helena. Before he left Batesville, Ark., General J. M. Schofield, already familiar from experience, with the wh le State, was (June, 18) placed in command o a department, including Missouri. Price's Confederate emissaries had recently returned from his army to their respective homes in most parts of the State, and with that zeal which is actuated by a sense of duty *plus* assurances of reward, they recruited bands in all parts of the State. Their efforts were unintentionally abetted by Schofield's zealous labors in organizing a State militia, of which he had on paper within a few months, the names of fifty thousand men. Every citizen therefore, within the conscript ages, was obliged to take arms for one side or the other, and consequently very many, who otherwise would have remained passive, actively espoused the Secession cause. This brought the conflict of arms into numerous villages in the State and for a time paralyzed all industries outside of St. Louis. No State in the Union ever experienced six consecutive months of such misery, but the militia system was incalculably beneficial in the end. Poindexter, Porter, Cobb and other Confederates less prominent, recruited several thousand men north of the Missouri, while Hughes and others on the south side, collected many more. The Federal forces were subdivided, leaving commanders J. M. McNeil in the north-east part of the State, Ben. Loan in the north-west, James Totten in the center, E. B. Brown in the south-west, Lewis Merrill at St. Louis, and J. M. Glover at Rolla. In June, Schofield's available forces numbered seventeen thousand. McNeil struck Porter's raw recruits, three thousand strong, August 6, in Adair County, completely destroying the organization, killing, wounding and capturing many. At the same time Poindexter, with a smaller force was near by, but separated by the Chariton River. He too was attacked, his force dispersed or captured. Poindexter himself was overtaken alone in the woods. The Missouri crossings were guarded so that no considerable force could cross either to escape south or aid those on the left bank. The citizens in North Missouri to a man having pronounced for or against the Union, in the culminating struggle, that section was pacified as never before since the outbreak of hostilities, and henceforth, there was comparative safety for the operators along the Hannibal & St. Joseph and North Missouri Railroads.

South of the river, affairs were less promising, especially as General Hindman, commander in Arkansas, had raised a force which, considering the sparsely settled regions of his command, seems almost incredible, being variously placed at forty thousand to fifty thousand men. The capture of arms intended for this army was one of those fortuitous events little dwelt upon by historians, but which may have saved Missouri, not even excluding St. Louis, to the Federals. Nevertheless leaving a considerable force in Little Rock, Hindman reached the Ozarks with full twenty thousand armed men. Before this, however, Hughes had collected a thousand recruits and (August 11) captured Independence, near the Kansas border, dispersing most of its Federal defenders. Coffey and Quantrell coming into the State, joined Hughes, and (August 15) defeated Colonel Foster, who had eight hundred men, and two guns, but Schofield was concentrating upon them from the east, west and south, wherefore they abandoned the State. It is said that from April to September, over one hundred combats occurred in Missouri.

Schofield was now able to protect the State by concentrating at, and south of Springfield, having by the 1st of October, four thousand, eight hundred infantry and five thousand, six hundred cavalry, and sixteen guns at Springfield, and three brigades out on the Newtonia road. General Blunt came over from the west of Kansas with about four thousand Federal Indian and white troops, a portion of which Indian Federals, had recently defeated a Confederate force of red men in the Indian Territory. Schofield, leaving about five thousand men to protect his communications, joined Blunt below, and thus, though now nominally superseded by Curtis at St. Louis, he had ten thousand troops in hand to contend with Cooper's seven thousand Confederates, besides General Rains' six thousand, which were about the Pea Ridge battle ground. There were yet other enemies east of these. After a partial success, Cooper moved to Fort Wayne, where (October 22) Blunt whipped him without killing any to speak of, and his own force, four thousand strong, had only three men disabled. Rains, unable to reach Cooper, kept out of the way of the Federals, who, however, did succeed in striking a detachment of three thousand horse near Fayetteville. Blunt hearing that Marmaduke was about Cane Hill with seven thousand men, expecting Hindman, with a view of destroying the former before Hindman came up, attacked him October 28, with five thousand men and thirty guns, driving him back to Cane Creek, where Marmaduke's position was virtually unassailable.

In December, Hindman and Marmaduke united their forces, aggregating eighteen thousand troops. Except the forces with Blunt, there were now no Federals closer than near Springfield, where Herron succeeded Schofield, and now had six thousand infantry, eight thousand cavalry and a complement of artillery. It was Hindman's best policy to destroy Blunt, and then move into Missouri. Blunt's position was extremely critical, as he was about six days' march from Springfield. Hindman's army succeeded in gaining the rear of Blunt's forces, but Herron, by forced marches with four thousand troops, attacked the enemy near Prairie Grove, losing nearly one-fourth of his command, and would have been overwhelmed, but for the timely arrival of

Blunt with reinforcements. The Confederates lost about three thousand, of whom nearly one thousand were slain.

When we last invited the reader's attention to this section, we stated that the telegraph which had been extended toward Bentonville was abandoned south of Cassville, and after Curtis left, taking with him operator Luke O'Reilly, whom he attached to his staff, Springfield became the most southerly office. At this point, operator W. H. Woodring arrived in June, after another lonely ride, this time all the way from Rolla, barely escaping capture by a dozen woodrangers when near his journey's end. General Brown assumed command there soon after. J. B. Morgan, of the Fourth Iowa infantry, an ex-operator, wounded at the battle of Pea Ridge, after convalescing, was sent to the aid of operator G. H. Peck, at Springfield, and remained with Woodring after Peck left, until about the 1st of July, when he was sent to Waynesville, and Cassius M. Barnes, who had rendered valuable service in Farmington, Miss., and Brittons and Moscow, Tenn., came from the Jefferson City, Mo., office, to Woodring's assistance. When General Schofield arrived, Charles A. Paxson, accompanied him as operator. Lucien J. Barnes, Cassius' brother, an ex-telegrapher, Assistant Adjutant General on Schofield's staff, frequently assisted in the office. Lucien, early in the war, opened an office in East St. Louis, for the transmission of very important messages from Captain (since General) Lyon, as Lyon was suspicious of some operators then in the commercial office in St. Louis. Just before Schofield started south, as related, C. M. Barnes went to St. Louis, and Henry G. Briggs, of the Ninety-ninth Illinois regiment, also an operator, was detailed to assist Woodring.

Early in January, 1863, Marmaduke, collecting about four thousand troops, mainly horse, and avoiding Blunt, prepared to assault Springfield, his base of supplies. Herron having taken away the best troops with him, leaving under General E. B. Brown but about twelve hundred militia, three hundred convalescents and one hundred and fifty Iowa troops, the result was very uncertain. Brown's few cavalrymen delayed the enemy as long as possible, to give time to prepare for the battle. Several old iron howitzers, that lay in the arsenal grounds, were put on temporary carriages in the little earth fort about a block from headquarters, and manned by experienced gunners. By daylight of the eighth, Brown was ready.

Early in the morning of the seventh, the telegraph line was cut, both east and south, to prevent Brown from communicating with outside forces. The night of the seventh was clear, the moon shining brightly, and, realizing the great necessity of re-establishing the telegraph before the attack, operator Woodring called to him repairers Owen Monday, a splendid man with real Irish pluck, and Bob Bates, who would go wherever sent. Monday mounted his big claybank, and going east found and repaired the line about seven miles out, and returned before daylight, undiscovered; but Bates had to go farther, and did not find the break until near morning. Having repaired the line and rested a little at a wayside house, he was unable to return until the rebels had begun their attack, which commenced about one, P.M., when suddenly he found himself in the midst of a squad some distance out of town, where the road on both sides was skirted with brush and timber. Being in citizen's dress, no immediate measures were taken to detain him, and soon comprehending the situation, he spurred his little mare directly into the timber, when the squad sent a volley after him without effect. Bates finally reached the town before it was closely invested. Telegraphic communication being thus re-established, the operator was able to and did advise the Commanding General at St. Louis, of the progress of the battle during the whole of the time it continued. This will doubtless be discredited by some, but the same thing was done in Tennessee when the author was at the safe end of the line. Other instances occurred in that and other departments, especially that of the Potomac. R. Von F. Truenfield, of London, Eng., author of "*Krieg's Telegraphie*," (1879), who has given the subject of War Telegraphs, perhaps, greater thought than any other writer, and who was actively engaged for five years in the late war between Paraguay and Brazil, as manager of army telegraphs, had somewhat similar experience. Captain Buchholz, another German author, doubts the use of the telegraph on any battle-field, but the American instances are very numerous.

Woodring and Briggs felt little disposed to remain idle within a few blocks of the fighting, and as each was well supplied with

Enfield rifles, they sallied forth to render such service as they might. Briggs took position among the volunteer skirmishers, and Woodring, acting now as aide to Brown, anon firing from convenient shelter, at short intervals ran back to his office and telegraphed prospects. At one time, peeping over a parapet for a shot, he drew a heavy rebel fire, that was almost too well aimed. At another, while he was in his office just after dark, lighting his lamp, preparatory to reporting that General Brown was severely wounded, there was a sudden crash through the building, already well riddled by bullets. The room seemed to fill with splinters and powdered plaster, and ere he could comprehend what was the matter, he saw a dark ball rolling on the floor *right by his* feet. It was a shell from the enemy's cannon. Greatly startled, he jumped back and crouched in a corner, awaiting its explosion, but happily for him the fuse had not ignited. That shell passed through a wooden church and three partitions in the telegraph building before reaching the office where it struck a brick chimney and and fell to the floor It is now the property of Superintendent Smith. Woodring hurriedly arranged his wires, and, taking out a relay, went to a building facing Market Square, where the office was formerly, and there he made his report to St. Louis.

About eight o'clock, the enemy withdrew, and Springfield was saved. About one, A.M., the Union gunners fired again, to feel for the enemy, and Woodring, believing the battle was being renewed, reported accordingly, and then the line was cut again, leaving the Department Commander, at St. Louis, and the Northern people, in suspense. Marmaduke had taken the Rolla road, cut the wire in many places and strung it across the highway behind, as he advanced.

Since two, P. M., no tidings had come from brave Briggs. Let us now follow him. In company with a lieutenant and a few others, he took position behind some shade trees, not far from the two story brick seminary building, lately used as a military prison, which the enemy captured, and from which they were inflicting much injury. Briggs and the others were endeavoring to pick off some of the enemy, and in so doing he fired every cartridge but one. His position becoming very uncomfortable at this time, the lieutenant advised Briggs to fall back, but he wanted one more shot, and in exposing himself to secure it, he was himself shot and instantly killed. Woodring sallied forth, early on the ninth, to find his comrade operator. After some search and inquiry, he learned that he had been killed, and hurrying to a back porch near the office, he saw the body of poor Briggs, cold in death. A bullet had entered an eye and passed through the head. While the Union people of the North were proudly discussing the victory, a few days later, at Delavan, Wis., Henry G. Briggs' young wife and two children were bewailing the death of this gallant volunteer, whose remains lie buried in the soldiers' cemetery in the town he assisted so bravely in defending.

The Federals lost fourteen killed and one hundred and forty-five wounded; their antagonist's loss exceeded two hundred.

A new attack was expected the next day, and preparations were made by Colonel George H. Hall for a final defense in the main fort on the opposite side of the town from where the fighting on the eighth chiefly occurred. Woodring burned his old messages, hid his instruments, except relay and key, which, with recent telegrams and cipher key, he took to the fort, where, from complete exhaustion, he lay and slept till near dark, when, Colonel Hall desiring to telegraph for reinforcements, Woodring volunteered to transmit them from the nearest point where he could get an electric current from St. Louis batteries, provided an escort was furnished. Accordingly, as directed, he reported at dark with repairer Bates, received the despatches, and with twenty-one men of the Seventh Missouri cavalry proceeded slowly, owing to the wires across the road. Twenty-four miles out the men began to object to proceeding farther, and soon decided to return to their command, the sergeant commanding them exercising little control. The operator called for volunteers and only one man besides Bates responded; but with these two men Woodring pressed forward, determined on his mission. Ten miles beyond, they discovered a small party of mounted men, armed, and dressed partly in blue, the rest in jeans, looking not unlike Marmaduke's men. They, it seems, were suspicious of the three and cried, "Halt! advance one;" but the trio, considering their inferior numbers, deemed the demand unfair, and successfully insisted that one of the others advance.

As he approached, the three separated, two being on the side and one in the center, prepared to dash through the others if they were rebels; but they were the advance guard of a large militia command *en route* for Springfield, but in fact diverged on Hartsville. That night (tenth), after a grand reception at Lebanon, where the operator and others were duly lionized, Woodring himself telegraphed the messages to St. Louis, not having been able to get circuit therefrom west of Lebanon. The North had understood that Springfield was captured. The cowardly sergeant was reduced to the ranks.

Marmaduke, moving easterly, encountered Colonel Merrill's Twenty-first Iowa at Woods Fork, which place he flanked after a short renconter, and reaching Hartsville again found Merrill in his front, with reinforcements, and after a short and sharp conflict, resulting in a Union loss of seventy-eight killed and wounded, Merrill retired to Lebanon for ammunition, and Marmaduke hurried into Arkansas, where he was attacked at Batesville by the Fourth Missouri cavalry and driven across the river.

Owing to bitter complaints to the effect that Hindman was incompetent and tyrannical, General Price took command in the department and sent Marmaduke with a larger force than before into South-eastern Missouri, to Frederickton, from which point he moved against Cape Girardeau, where he fought McNeil's force, April 25, but withdrew and retreated to Arkansas on discovering the approach of gunboats and reinforcements for McNeil.

J. D. McCleverty, operator at Frederickton at this time, was captured and handcuffed. Dr. Boyd, who was acting as Confederate surgeon, was quite a fair operator, excelling McCleverty, who, in the vernacular of the key, was but a plug. Boyd thought to deceive the Yankees as to Marmaduke's objective, and possessing himself of the Frederickton office, he called St. Louis and asked for R. C. Clowry, who responded; whereupon, Boyd, feigning to represent McCleverty, said: "I have just escaped from Frederickton. Marmaduke has taken the place and I have learned just enough to know that he will move from Frederickton on Rolla. I am in the woods and have tapped the line to report these facts. Mc." But Clowry knew at once that it was not Mc at the key, and told Boyd that his trick would not work, and calling up Pilot Knob instructed the operator to cut Frederickton out. Boyd subsequently acknowledged himself beaten, adding that he then declared he would never attempt another Yankee trick.

The affairs at Fayetteville, Arkansas, April 10, between Colonel M. L. Harrison's Federal force and General Cabell's two thousand cavalry; May 20, between Colonel Phillips' eight hundred at Fort Blunt and Colonel Coffey's three thousand troopers; and July 1, in the Indian Territory, between Colonel Williams' thirteen hundred Unionists and seven hundred Texans and some cowardly Indians under chief Standwatie, and between Blunt's command of three thousand and Cooper's six thousand, July 17, a little south of Fort Blunt, in all of which the Union troops were victorious, comprehend substantially the further fighting, which may be denominated battles, within this department during the year 1863. But guerrilla raids continued to be the order of the day, though not so frequent and mischievous as before, if we except the wanton and merciless conduct of that arch fiend, Quantrell, who, with three hundred bushwhackers, fell upon the defenseless and unsuspecting city of Lawrence, Kansas, at daylight, August 21, and ravaged it with fire and sword.

Another outlaw, Hildebrand, made frequent raids with his gang of marauders. On the night of July 10, he appeared at Irondale, eight miles south of Mineral Point. About seven months previous, Miss Louisa E. Volker, a most estimable young lady, had relieved C. T. Barrett, operator at Mineral Point, and became at once not only the first lady operator in the corps, west of the Mississippi, but the only operatrix who had ever telegraphed on that side of the river. Entering upon duties which, heretofore, had devolved exclusively upon young men, she realized that peculiar feeling of responsibility which arises from an important but experimental trust, and hence, with all the zeal of a leader, she undertook the fulfillment of this new *rôle* of feminine usefulness in war, in which she was stoutly supported by such men as Noel, Fletcher, McMurtry, French, Macklind, Greyson, Winfield and Walker, brave and loyal Unionists of that locality. Miss Volker was called at midnight,

July 10, to telegraph to the commander at Pilot Knob to intercept Hildebrand, but after robbing a store, he and his men hurried east of the river and scattered in Illinois.

On a former occasion, the station six miles north of the Point was attacked by cavalry, surprising Captain Lippencott's company, which being driven off, collected at Mineral Point. Miss Volker had previously ascertained the presence of the enemy and telegraphed to Pilot Knob the situation, and started the repairer north to mend the line if possible, which was actually accomplished during the night, she sitting by the instrument all night in expectation of an attack on Mineral Point.

In October, the line was extended from Sedalia to Warrensburg, where were General Brown's head-quarters, and in December, Dwight Byington was relieved at Little Rock by Robert C. Clowry and ordered to Leavenworth, to take charge of the military line being rebuilt between Fort Scott on the Missouri-Kansas border, and Kansas City.

The several combats noted in this chapter, especially "Prairie Grove," where, thanks to the military telegraph, Herron saved Blunt's army, practically destroyed the rebellion north of VanBuren, Ark., by the summer of 1863, and when Steele drove Price from Little Rock to Arkadelphia, that vast territory north of the Arkansas River was cut off from the Confederacy, and telegraphic communication with St. Louis thus became a possible *desideratum.* Thus it occurred, that Charles A. Paxson, chief operator in South-west Missouri in September, and Robert C. Clowry at Little Rock in December, began extending the lines to Fort Smith, on the Arkansas River, where it crosses the western border of Arkansas. On the 14th of September, Paxson's builders were driven back to Cassville, but McNeil furnished protection, and late in September, it was rebuilt and extended to Fayetteville, where H. H. Taylor operated at M. La Rue Harrison's head-quarters; and in October, to Fort Smith, four hundred and three miles from St. Louis.

In the fall of 1863, Major Smith appointed Charles A. Hammann assistant manager of telegraphs within his department, and as shown, Clowry was sent to Little Rock.

Robert C. Clowry, was, on the 27th of October, appointed captain and assistant quarter-master, and assigned to duty by

Colonel Stager, as assistant superintendent of Federal military telegraphs within the Department of Arkansas, with head-quarters at Little Rock. We have seen something of Clowry's labors in the military service, and now that he has become a commissioned officer therein, a word concerning his antecedents. It will be observed that none of the officers, while boys learning telegraphy were especially favored with means, and it is a remarkable fact that the great majority of telegraph operators now or heretofore manipulating the key in this country, were children of poor parents, or of those whose circumstances were somewhat straightened; very many were messenger boys, among whom, besides the subject of this sketch, may be named, Andrew Carnegie, Robert Pitcairn, David McCargo, T. B. A. David, Charles J. Merriwether, George H. Grace, and George K. Leete, the first three of whom became superintendents of railroads, the second three of telegraphs and the last an adjutant general, on General Grant's staff. Mr. Carnegie, after endearing himself to the people of his native place, in Scotland, by his munificent benefactions, has, with rare generosity offered to contribute two hundred and fifty thousand dollars to the City of Pittsburgh for library purposes. It was in that city that young Carnegie was a telegraph messenger. Thomas A. Edison, a prodigy among inventors of electrical apparatus, if not in other scientific branches, was a newsboy before he became a telegrapher.

ROBERT C. CLOWRY.

At an early age, young Clowry became a pupil of D. C. Jeneson, manager of the Joliet, Ill., telegraph office, whom he was to serve six months as messenger, in consideration of being taught to telegraph. The duties of a telegraph messenger exact the utmost promptness, which is one of the best lessons a youth can acquire, and it is not unlikely that Robert's services under Jeneson, conduced as much as any other one thing to make him the reliable officer he subsequently became. Clowry's first office was at Lockport, Ill., which he took in the fall of 1852, at the expiration of his service with Jeneson, and having there acquired greater proficiency in the art, and experience in the business, he was made manager of the Springfield, Ill., office, December 13, 1853. By the spring of 1859, he had so approved himself among telegraph men as to justify his appointment, then made, to the office of superintendent of the St. Louis and Missouri River and Kansas Telegraph Companies, generally called the Stebbins lines, and in the fall of that year, as his head-quarters were in St. Louis he was also made chief operator of the consolidated office there, and so continued until the middle of April following, when he was chosen superintendent and secretary of the Missouri and Western Telegraph Company, with head-quarters for a time as before, and then at Omaha, Neb., where he remained until induced by George H. Smith, formerly manager of the St. Louis consolidated office, to enter the Military Telegraph Corps. At this writing, his progress has indeed been flattering, nor can it be doubted that it rejoices all who know him. It is not always that merit has its reward, but when it is recompensed, all just men are pleased.

Captain Clowry took down the Benton wire and also about twenty miles of line leading to Des Arc, with which, and new wire, he proceeded late in December, to build toward Fort Smith to meet Captain Smith's party, which started earlier therefrom and built to Dardanelles, half way (eighty-five miles), but for want of insulators, which were long delayed, the line was not completed until February 13, 1864, at which time the telegraph, five hundred and seventy-three miles long, was in admirable working condition from St. Louis to Little Rock. Offices at Clarksville, Dardanelles and Lewisburg were opened on the Fort Smith line. Ewing L. Armstrong relieved Woodring at Springfield, and the latter went to Fort Smith, where he was aided by Joseph E. Sears. Joseph Hansen worked the VanBuren office; at Dardanelles were J. L. Sponagle and James W. Chandler; at Lewisburg, Thomas M. Peeler and Seeley B. Knapp, and at Clarksville, James A. Shrigley operated.

But few incidents remaining unnoticed occurred in Arkansas and Missouri during the period covered by this chapter. Keyes Danforth, a repairer, was captured between Frederickton and Cape Girardeau, Mo. Henry Wilson and C. B. Applegate, repairers, near Memphis, and T. R. Berryhill, operator, was captured at Arkansas Post. October 25, 1863, Marmaduke attacked Colonel Clayton, at Pine Bluff, and was repulsed.

December 30, David O. Dodd, a Confederate operator, was arrested in Little Rock, tried by court-martial, and executed as a spy. The following is the history of that sad affair. Dodd was a young man, of about eighteen years, who had never held any position of consequence as an operator, and was not in any employ when taken, but had been on Superintendent Baker's lines a short time before the coming of General Steele to Little Rock. After Steele's appearance there, Dodd and his father continued to reside in the city, but they concluded to move to Mississippi, and got as far as Camden, Ark., when the father bethought him of some business at Little Rock, which had been overlooked. David was sent back to attend to it. It was stated and generally believed that young Dodd, in order to obtain a pass through the Confederate lines, agreed to bring, on his return, information concerning Steele's Little Rock fortifications. David, after remaining several days in Little Rock, passed the first Federal picket line unmolested, but the outer pickets searched his person and baggage, and were about to let him pass, when certain curious pencil marks in his memorandum book caused a suspicion that they might be translated. He was sent to head-quarters, where, upon examination, the mysterious dots and dashes were shown to be telegraphic characters, disclosing a complete and comprehensive description of the Federal fortifications about the city. It is said that the information therein contained could not possibly have been obtained by Dodd alone.

A court-martial was convened. Captain Clowry testified to the meaning of the characters, and Dodd admitted that the facts so written were furnished by a party or parties, who knew all about the works. He was sentenced to death. His father's relatives and friends exerted all their influence with General Steele, to induce him either to grant a pardon, or commute the sentence to imprisonment. Steele finally consented to consider their petitions favorably, provided Dodd revealed the name of the party from whom he derived the important information and all the facts pertaining thereto, but Dodd was firm and unyielding, saying that he preferred to die rather than betray a friend. A last agonizing appeal was made to him on the scaffold, to accept the conditions imposed by General Steele, and live, but he waived off his broken-hearted relatives, and bade the executioner do his duty, as nothing would break the seal placed upon his lips. He was buried in Mount Holly cemetery, in Little Rock. A small, plain marble shaft, erected by his friends, marks his grave—a tribute to the bravery of a beardless boy, who truly died that others might live. The monument bears the simple inscription: "Here lie the remains of David O. Dodd, born in Lavacca County, Texas, November 10, 1846; died January 8, 1864."

CHAPTER XIV.

THE TELEGRAPH IN VIRGINIA DURING THE FREDERICKSBURG AND CHANCELLORSVILLE CAMPAIGNS.—SUFFOLK. —VARIOUS RAIDS.

When Burnside assumed command (November 8, 1862), his army of about one hundred and twenty-seven thousand men was, so to speak, in bivouac. Under McClellan it would have gone to Culpeper C. H., but Burnside directed its march, which began November 15, to Falmouth, with Fredericksburg as an objective. Lee, who left the Antietam with less than forty thousand troops, passed Winchester with about seventy-five thousand, and at Fredericksburg had about eighty-five thousand able bodied, of all arms.

The telegraph had become an established necessity. It had ceased to be a *desideratum*, for it had grown to be an *essential.* Burnside's telegraphic facilities were less exposed than Pope's had been, as he had fewer outposts that it was important to defend. Most of those he did have, however, were in telegraphic communication with the commander and the capital. On the Orange & Alexandria Railroad, an outpost office was kept open at Union Mills. Centreville, a little above, was also open. The wire to Fredericksburg, *via* Alexandria and Acquia Creek, was a short line, and easily guarded. These were the only ones in the department, which needed special protecting care. The offices in the department, in December, were, in the District of Columbia, the War Department, General McCallum's, Arsenal, Navy Yard, General Casey's, Head-quarters of Defenses of Washington, Tennallytown, Arlington and Long Bridge; in Virginia, Alexandria, Fairfax C. H., Centreville, Burkes, Stafford C. H., Fairfax Station, Union Mills, Dumfries, Acquia Creek, Brook Station, Potomac Creek, Stoneman's Siding, Falmouth, Generals Burnside's and Sumner's head-quarters, Yorktown, Williamsburg, Fort Monroe, Norfolk, Suffolk, Cherrystone, Drummondstown, Eastville and Harpers Ferry; besides which, there were, in Delaware and Maryland, a number more, making in all forty-eight offices, worked by eighty-eight operators.

Before starting his army for Falmouth, Burnside arranged, or believed he had done so, with Halleck, with sufficient definiteness to leave no mistaking, for the arrival of pontoons at a point opposite Fredericksburg, when the army should appear, but they were long delayed, and Lee was thereby enabled to encamp about Fredericksburg and strengthen his position. His army behind, above and below Fredericksburg, occupied a line about six miles long.

It was while Burnside was preparing to cross and give battle and Lee was busy on Marye's and Willis' hills and other naturally strong positions, that several cavalry incursions were made, two of which interrupted telegraphic communication, the first at Poolesville, Maryland, and the other at Dumfries. J. L. Cherry was the operator at Poolesville when (November 25) Major White's battalion of Independent troops — probably a polite name for guerrillas, the Unionists residing thereabouts called them bushwhackers, but General Lee mentions them as a part of General Jones' command—suddenly entered the town, which lies across the Potomac, say five miles from Balls Bluff. They numbered about sixty. Five hundred stand of arms and forty thousand rounds of ammunition had been stored in Poolesville for some time, and as these constituted a tempting bait for the Confederates, it was but natural to expect a raid. Indeed, Poolesville had been in a state of exacting unrest for many weeks, of which fact the following telegram, found in the Report of the Committee on the Conduct of the War, is but partial proof:

HEAD-QUARTERS ARMY OF THE POTOMAC,
Bloomfield, 9 P. M.

TO GENERAL H. W. HALLECK, General-in-Chief:

I have the honor to transmit for your information the accompanying despatch, which has been received from Poolesville.

(Signed) G. B. MCCLELLAN,
Major General.

POOLESVILLE, November 3, 1862.

TO SETH S. WILLIAMS, A. A. G.:

Last evening and this morning, I learned the following information from different reliable persons. They report a force of Stuart's cavalry concealed in a large woods about five miles from here, between Frederick and Rockville road, watching an opportunity to escape. One of the men says he saw a farmer in that vicinity taking provisions for them at midnight, and he also saw three or four of the rebels take provisions from the farmer. Their numbers can not be ascertained, but it is not very small from what I can hear. They are some of Stuart's left behind, or that got cut off when he made the last raid over here.

J. L. CHERRY, *Operator*.

P. S.—No troops here.

Rumors were rife of the coming of the enemy some days before Major White appeared, and Cherry, to save the arms and ammunition, telegraphed to the Secretary of War the fact that they were in Poolesville. At this time eighteen soldiers—members of a Pennsylvania regiment of infantry—were stationed there. Cherry also telegraphed to Colonel Stager the situation, and asking instructions, was directed to remain at his post until he saw the enemy coming. The office was in the second story of the Town Hall, and the soldiers were quartered on the first floor. A guard was stationed at the office at night only. Twenty-five Federal cavalry, and wagons which they escorted, left Poolesville with the five hundred small arms and the ammunition, for Washington, on the evening of the twenty-fourth. At five o'clock, A. M., the next day, and but a few minutes after the office guard had gone down, Cherry was awakened by the rattling of sabers and heavy tramp of men. His office was filling with Major White's men, who had already secured the guard below.

As soon as Cherry dressed, he found opportunity to loosen the adjustment cord or spring used to draw the armature from the relay magnet when the key or circuit is open. This took but a second and prevented the instrument from ticking. Cherry then began calling, "DI"—War Department—intending to attract the attention of the operator there and to briefly relate what had occurred, but just as he began, Lieutenant Dorsey, second in command, stepped up and peremptorily ordered the operator to leave "that thing" alone. Cherry left it alone and was marched to the center of the town with the captive soldiers. Major White, who once lived in that neighborhood, left the main body of his command to Dorsey, while he attended to some matters outside. Dorsey told Cherry he would parole all but himself. The thought of Libby prison under such circumstances is a natural precursor to thoughts of escape, and although Dorsey was talkative and pleasant, even promising Cherry a good horse to ride, the lad was determined on escape. Plans, like wants, grow on one; immature ideas, like unknown by-paths, lead, ofttimes, to happy ends; and sometimes results only, justify the means employed. Cherry's scheme grew on him, for his first idea found expression as follows: "Dorsey, have a drink?" From that moment Dorsey's heart began to open toward the young man, and the two were soon seated in the rear room of a store, talking over the subject of paroles as they imbibed a liquor called "Hail, fellow, well met;" but Dorsey, notwithstanding, insisted on taking the operator to General Longstreet. It was then that Cherry thought to get the rebels all drunk. That was a big undertaking, but undaunted, he stepped to the front and extended a general invitation. Every trooper had dust, cobwebs or mildew in his throat, which he wished to wash down, and as they came up to face that enemy which has made more graves than wars have, Cherry arranged with the keeper to give them as much as they would drink, after which he retired to the little room in the rear, where sat the lieutenant. Dorsey, however, soon went into another store and began writing paroles. Cherry joined him and assisted, hoping to prevail on the officer to include him. By and by Dorsey stepped out, leaving the operator hard at his self-imposed task, and before it was completed he was informed that every rebel had left the town. It was soon revealed that Major White returned and found his soldiers nearly all drunk, and to save them from capture by any handful of men that might come along, he hurried off and recrossed the Potomac.

Cherry was soon after ordered to open a testing office at Occoquan, Virginia, a place fifteen miles south of Alexandria, but in three days food and forage became scarce and he was ordered to try Dumfries, ten miles below. This place he reached at two P. M., and though the people seemed to have enough to eat, they had a cadaverous look, and appeared in eyeing the Yankee operator to make perpendicular wrinkles between their half-closed eyes, a kind of scowl that foreboded no good; so Cherry went on to near Stafford Court House, and connecting his instrument, wired Colonel Stager that Dumfries was an unsafe place to remain in and he did not wish to be captured twice in two weeks if it could be avoided. It so happened that Frank Lamb, who had been operating in Centreville, thereupon started to open Dumfries office, escorted by twenty-five cavalrymen, but, by a strange coincidence, General Wade Hampton, with about nine hundred rebel cavalry, was *en route* to close it and tear down the wire. Lamb and escort reached Dumfries about nine P. M., December 11, and about five A. M., the next day, Hampton made them all prisoners.

The days lost awaiting the arrival of pontoons made Burnside's advance across the Rappahannock extremely hazardous, for the enemy was in readiness, when on the 11th of December, the pontoons opposite and below the city were laid. By noon of the thirteenth, the battle had fairly begun and Marye's Hill had become the scene of dreadful carnage. Burnside had given his orders, and from Stafford Heights, opposite Fredericksburg, was watching the battle two miles away. What reports he received came by courier, as there was no telegraph across the river. Weighed down by anxiety and responsibility, one can imagine how pleased he would have been to be in constant communication with Sumner, Hooker and Franklin, commanders of the respective grand divisions of the army. Telegraphy for tactical purposes, although eminently successful on the Peninsula, had not yet become a matter of course, owing in a large measure, doubtless, to the presence of signal officers, of whom too much was expected; so Burnside was mainly dependent on his field glass and reports by messengers. The Union army crimsoned itself with honor, but from assault after assault, so much of it as was not stricken to the ground, retired to prepare for another. In storming the Confederate strongholds, the Union loss was one thousand, one hundred and fifty-two killed, and nine thousand, one hundred and one wounded, and over three thousand missing or thirteen thousand, seven hundred and seventy-one in all; nearly three times that of those who so stoutly defended the works. Two nights later the army recrossed safely and went into camp.

During a bombardment of Fredericksburg, preceding the battle, Frank Drummond, who had recently been exchanged and was, with J. G. Garland, operating at Falmouth Siding, went down to the heights, on which were over one hundred and forty cannon that had begun the destruction of the city opposite. Only slight response was made to their fire for the most part, but the first shell that the enemy sent, fell so near Drummond as to remind him of some business he had forgotten. It was not twilight yet, and as shell after shell passed him and he as regularly fell to the ground, an observer would have imagined he was carrying too much of a load in his hat, did not the screaming fiends in the air offer a more charitable excuse. Anyhow, men will laugh "to see such sport," even if it isn't nice in them to do so, and Drummond to this day, thinks the whole army was looking at him. When Drummond reached his office, he heard that Garland had gone back a mile to Burnside's head-quarters, probably on business also, as the Siding was not a healthy place to remain in.

The day (sixteenth) the Union army went into camp, General Hampton again crossed the river and raided to Occoquan and Dumfries, cutting the telegraph badly, and thereby greatly annoying General Burnside, who, at 2:45 P. M., December 20, telegraphed General Stahl:

> You will please see that a strong guard is thrown out at once, to protect the telegraph lines. A sufficient force must be detailed to insure its safety, even to the whole of your command if necessary.

Hardly had Hampton recrossed, when General Stuart with detachments of Hampton's and F. H. and W. F. Lee's brigades, started to pay Dumfries another visit. Truly Dumfries *was* an unsafe place for Federal operators and Cherry began to think there was a Providence in the menacing looks that bade him "move on." Operator O. H. Dorrance started from Alexandria for Occoquan, while Stuart was *en route* for Dumfries. A telegram from Major Eckert to Colonel Stager, dated December 29, states the case as follows:

> Rebels cut line between Dumfries and Stafford before making attack on Dumfries. At five P. M., cut line again between Dumfries and Occoquan. Dorrance reached a point near Occoquan early Sunday morning, and got in circuit; couldn't remain long, owing to desperate charges by Seventeenth Pennsylvania cavalry, with sabers drawn, going at full speed *to the rear*, of which he gave us some laughable instances. At five P. M., Sunday, Flagg reported the enemy within three miles of Burkes.

Having swept everything in their way from Dumfries to Occoquan, a portion of Stuart's force struck the Orange & Alexandria Railroad near Accotink bridge, and destroyed it. This bridge was a little east of Burkes, but before it was burned, J. A. Flagg, operator at that station, while reading by an open window, that balmy Sunday afternoon, saw a woman running as rapidly as possible toward the office. Flagg rushed to meet the kind woman, who, while gasping for breath, stated that the rebels were in large force at Wolf Run Shoals, a few miles south of Burkes. This information was immediately telegraphed to the War Department, and word was also sent to the conductors of two trains which had proceeded beyond Burkes, probably after wood. These were enabled to cross the Accotink bridge before the enemy struck it, and reached Alexandria in safety. Flagg, after taking his personal effects to his boarding house, was returning, when he met a Mr. Ross, who had charge of a government train of twenty wagons, and eighty-two horses, in the woods. His negro teamsters, greatly frightened by the news which had reached them, were fleeing toward a thicket where they hoped to hide, but Flagg hailed them and placing his hand in his breast pocket as if to draw a revolver he did not have, told the darkies he would shoot the first one that advanced and that they must at once hitch up and drive to Fairfax Court House. This they did quickly and saved the train.

Flagg intended hiding in the heavy pine woods near by, but before he could get ready to go, the enemy rushed in, and two boys, not over seventeen, were placed as a guard over him. These, as might be expected of over zealous warrior youths, first invested with authority, used their revolvers with threatening recklessness until a Mr. Shepard, operator with General Stuart, entered and ordered the youngsters down stairs. He then placed Flagg under another guard instructed to shoot him if he touched the instrument. In company with one hundred and twenty other prisoners, he marched many weary and hungry hours, reaching Culpeper Court House at nine P. M., December 31, stiff, sore, exhausted and faint. One of the Lees, who promised one Marshall, with whom Flagg had boarded at Burkes, to parole the latter the next day, neglected his agreement, and January 2, the operator was in Libby prison. The next morning, a Richmond paper described the beautiful combination telegraph instrument captured at Burkes. At the close of the month, Flagg and eighteen hundred other prisoners were exchanged.

Operators H. C. Buell, at Union Mills, W. N. Embree, at Fairfax Station, and R. F. Weitbrecht, at Fairfax Court House, were near enough to Stuart's line of march to feel a justifiable uneasiness, but Stuart rode toward Falls Church, and swung around Fairfax Court House toward Chantilly; some of his troopers came within a mile of Centreville, where was operator A. H. Bliss, who, the wire being down, put out his lights and climbed a great tree, for the double purpose of observing and hiding. Bliss had no confidence in the staying qualities of the Federal regiment about the fort a mile off, and hence took to the tree.

December 30, Burnside, who had initiated another offensive movement, received a cipher telegram from Mr. Lincoln, saying substantially, "I have good reason for saying that you must not make a general movement without letting me know it." Disaffection in the army in high quarters had reached the ears of the President, and caused him to fear that Burnside would not receive that hearty support so essential to success. In a measure, that feeling was modified by personal interviews, and Burnside began executing another movement, intending to cross above Falmouth at Banks' and United States fords. A telegraph line was built to Banks' Ford, and J. H. Emerick opened an office there January 20. His tent was the clouds that rained incessantly, and his desk was a caisson that had been abandoned in the mud. The storms prevented action and the army again went to camp. Generals Burnside, Sumner and Franklin were soon after relieved and General Joseph E. Hooker appointed to the command of the army.

Thus in the East, Federal affairs were unsatisfactory in the beginning of the new year. The last blow was extremely discouraging to the Union people. The joy in the South was great; both sides were yet painfully concerned, but the North more than ever determined. These two powerful and brave armies were unbroken and in the coming clash of arms, an accident, the elements, a mere mischance, might turn the scale for or against, and so the people north and south

> Like souls that balance joy and pain
> With tears and smiles from heaven

anxiously waited and prayed.

January 25, 1863, General Hooker assumed command of the Army of the Potomac, which was reformed into corps commanded respectively as follows: first, General Reynolds; second, General Couch; third, General Sickles; fifth, General Meade; sixth, General Sedgwick; eleventh, General Howard, and the twelfth by General Slocum. While these officers were preparing their respective commands for a new campaign, the Federal and Confederate riders were abroad, and when the latter were not, rumors of their near presence were rife.

About ten o'clock one stormy, cold night in February, it was reported at Fairfax Station, where Colonel Percy Wyndham's cavalry and other troops were located, that General Stuart was coming. A little south of the station, lay Wolf Run Shoals, where Colonel Blunt's command was posted. A telegraph line had been recently constructed from Union Mills to the Shoals, and as soon as the report of Stuart's approach was received, Wyndham asked Major Eckert to send an operator to open the Shoals office. Accordingly, Richard Power, Jr., who chanced to be at Fairfax assisting Embree, who was convalescing, was ordered to proceed at once with a cavalryman, who pretended to know the route. They soon lost their way in the blinding snow-storm, which changed to a drizzling rain, and in the dark, groping about the woods, listening for sounds of guerrillas, cold, thoroughly wet, their outer clothes stiff with ice, they struggled on until daylight, when, finding a path which led them to an elevation, they looked down and beheld Blunt's camp, which, without challenge and unobserved, they entered. James Glazier, operator at Union Mills, who had long struggled against drowsiness, connected the wires, when it was ascertained that Stuart had not been near.

On the 8th of March, John S. Mosby, a daring Confederate cavalryman, then known only as a guerrilla, performed a most splendid feat, for which Stuart soon after publicly thanked him. He, with but a handful of men, at two o'clock in the morning, without warning, entered the Union camp at Fairfax Court House, where were then located operators R. Power, Jr., and Robert Weitbrecht. The latter was on duty, and, probably guided by his lights, Mosby, accompanied to the tent by five or six followers, walked up to Weitbrecht and presenting a revolver, demanded that the operator and his orderly make haste and direct him to the head-quarters of General Stoughton, quietly and without daring to speak above a whisper. Of course the prisoners did as directed; there was no alternative. On the way the rest of Mosby's men joined him. They had been stealthily selecting the best horses. Stoughton was found in bed, as were also some of his staff officers. As Mosby's men were moving off with captives and captures, a Colonel Johnston, who was awakened by the unavoidable noise, went out on the front porch of the house he slept in, and called out, "Halt! The horses need rest. I will not allow them to be taken out. What the devil is the matter?" etc. Receiving no reply, he again spoke out, "*I am commander of this post and this must be stopped.*" This was just what Mosby wanted to know. Johnston's retreat was speedily cut off. It is said, however, that owing to the darkness he fled to a neighboring barn *en deshabille*, and crawling under, eluded the pursuers, who took to the rebel lines about twenty prisoners and seventy horses.

A few weeks later, Weitbrecht was exchanged as a Federal captain, he having reported himself such at Libby, to facilitate his return. His appearance was sad, laughable and mussed. The rebels had exchanged his clothes also. It was supposed at the time at Centreville that Mosby entered the Union lines between that place and Chantilly, passed just outside the Centreville pickets, struck the Warrenton road nearly two miles therefrom; moved thence by pike to within a mile of Fairfax Court House, and that he returned by the same route. Only four days before Mosby's venture, A. H. Bliss, the Centreville operator, warned General Heintzelman that the cavalry pickets of Colonel Wyndham were, about the 15th of February, withdrawn from the front and right of the Centreville command, and that since then but few patrols had been over the Chantilly road, leaving about three miles unguarded. This was referred to Colonel R. Butler Pierce, commanding the cavalry south of Potomac, within the Department of Washington, but seems not to have been remedied. Bliss was cavalierly informed by the Colonel that in future like reports should be made to him, but Secretary Stanton having heard of the matter, caused the following to be telegraphed to Bliss:

WAR DEPARTMENT.

TO A. H. BLISS:

The Secretary of War directs me to thank you for your efforts, and in future authorizes you to report direct to him.

THOS. T. ECKERT,
Major and A. D. C.

Since Mosby's operations about Fairfax Court House, have just been considered, it may be well for the present to avoid strict chronological order and call attention to the interesting fact that operator W. N. Embree, while working there, became acquainted with a number of very pretty young ladies; that notwithstanding one in particular was bitterly "Secesh," a mutual regard sprang up, destined to have a romantic end. It was shortly after the battle of Chancellorsville, an account of which will soon appear, that Embree in his best, set out for the country to renew his acquaintance with the aforesaid bonnie lassie, and it was while he was basking in her smiles, that three of Mosby's men entered her home and took the bonnie laddie under their special charge. In due time, after a tempestuous land voyage, Embree was in Libby prison, and later, on Belle Isle. Federal cavalry was sent in search of him as soon as he was missed, but they found only a linen coat, which was believed to be his; wherefore the young man was supposed to have been waylaid and killed. His father, who was in the employ of one of the departments at Alexandria, made unavailing efforts to obtain tidings of his son. Three and a third months later, Embree was exchanged. From one hundred and fifty-four pounds he was reduced to ninety-four. In the streets of Alexandria, he met his father and passed him twice without being recognized, then he spoke, and that glad father rejoiced that he "that was dead is alive again."

In April, Suffolk, Va., was threatened by twenty-eight thousand men under Longstreet, who hoped to recover the whole country south of the James River, extending to Albemarle Sound, N. C., including Norfolk and Portsmouth. General Peck commanded fourteen thousand troops at Suffolk. From April 11, to May 2, that city was besieged. Railroad trains bringing troops and supplies were very numerous, giving operators but little rest. At one period, C. A. Homan, operator at the Suffolk station, without relief, ran trains by telegraph for four days and three nights. Wilbur F. Holloway was operator at Peck's headquarters, the O'Brien brothers, at Norfolk, C. K. Hambright, at Newport News, and J. R. Gilmore and George D. Sheldon, at Fort Monroe. Longstreet attempted to carry the defenses by assault, but the telegraph brought gunboats to Peck's aid and the enemy was repulsed with considerable loss, when Longstreet's force proper, rejoined Lee's army.

About two weeks before making a general advance, General Hooker, solicitous about his telegraphic communications with Washington, telegraphed Halleck at the capital, advising that a regiment of cavalry be sent from Washington to patrol and guard the line *via* Occoquan and Dumfries. General Halleck replied that it was Hooker's duty to attend to that, and ordered him to do so, whereupon, the latter, by telegraph to the Secretary of War, requested him to lay both messages before the President, and in three hours Halleck telegraphed Hooker that a regiment had been sent.

General Hooker succeeded admirably in preparing his army for another conflict. It numbered fully one hundred and twenty thousand available troops; twice that which his adversary had in hand. After despatching General Stoneman with his grand cavalry force to operate against Lee's communications, Hooker began his movement (April 27), aiming to cross the river about three miles below Fredericksburg, with Sedgwick's and Reynold's corps, which were to demonstrate as if they led the main attack, while the Twelfth, Eleventh, Fifth and a portion of the Second (the rest under General Gibbons being left about Falmouth) crossed above at Kelly's and United States fords, to be followed immediately by the First if it could be spared from below, where it was ordered in reserve for either the Third or Sixth.

The Signal Corps put up a line from Banks' to United States Ford, about ten miles, expecting to work it with Beardslee's magneto instruments, but soon after the army movement began, General Hooker became dissatisfied with its slow work, and ordered General Couch, who was at the United States Ford with a part of his corps, to "establish rapid communications with the telegraph at Banks' Ford, and with Meade and Slocum, as the telegraph from Banks' to United States Ford works so slow." Meade and Slocum were moving on Chancellorsville from Kelly's Ford. The United States military telegraphers at once took possession of the Signal Corps line and worked it with Morse instruments. But the signal officers built and operated the only telegraph connecting the United States Ford with the army about Chancellorsville, and as a consequence, business being very great, most of the messages were sent to General Hooker by orderlies. Caldwell or Jacques should have been brought over from the old head-quarters office to work the wire at the front and the United States Ford office made an intermediate instead of a repeating office, but the signal folks had the field and thought they could work it.

As soon as the pontoon was laid (April 30), the wire was extended across the river, and an office opened in a house near the ford. J. H. Emerick and James Murray, operators, were located here. At Banks' Ford, about ten miles down and across the river, were two more offices, one at the ford, where W. K. De Witt was located, and another a little back on the main line, which was opened by John B. Pierce, who was relieved later by Samuel H. Edwards. A Beardslee telegraph was also operated from this ford to head-quarters in the rear, as soon as General Sedgwick located there, after returning from his operations on the south side. General Hooker ordered an entire regiment to guard the main line along the river. The regiment was spread out from the United States Ford to Falmouth, and every soldier of it, instructed to shoot any one caught tampering with the wire. From a point near Falmouth, messages for Sedgwick and Reynolds across below, were to be carried by courier. At least all obtainable information tends to induce that belief. Why a line was not erected upon the heels of Sedgwick's advance and he (until cut off), kept in constant telegraphic communication with the Commanding General at Chancellorsville, whither the main army rapidly moved, is an enigma. General Hooker soon felt a natural embarrassment consequent upon a lack of telegraphic facilities, for which it would seem, he was responsible, for certainly had lines been ordered, Major Eckert would have had no difficulty in constructing them. Perhaps too much reliance was placed on the signal service, concerning which, its chief subsequently reported, "In the movements at Chancellorsville, crippled as the corps was by the failure (through no fault of its members) of some of the apparatus, they rendered such service as that may be, which kept up the communication with General Sedgwick's Sixth Army Corps at that time utterly cut off from the main body of the army and from head-quarters, over the heads of the rebel forces."

This is the only evidence within reach that any signal officers attended Sedgwick, whose operations deceived Lee, and consequently Hooker easily reached Chancellorsville. But from that time, he appears to have invited attack, instead of continuing a vigorous offensive. May 1, severe but not general fighting occurred, resulting in considerable loss to both sides, and as the enemy seemed really disposed to fight, General Hooker had recourse to the telegraph. He had previously directed his chief of staff to remain at Falmouth, "as from that point," said Hooker, "it being central and the wires meeting there, it was necessary to have some one, in whose capacity and judgment I had confidence." To him he telegraphed as follows: "Direct * * Reynolds' corps to march at once with pack train, to report at head-quarters." At about six o'clock, P. M., of the second,

Stonewall Jackson surprised and utterly routed the Eleventh Corps, posted on the right flank of the Union army, and drove it upon the forces about Chancellorsville, with great loss in men and arms. It did worse, for its discouraging influence was not warranted by the remaining relative strength of the contending forces. The assault was checked, and soon after General Jackson reconnoitered with a view of making another, but returning with his staff, they were mistaken for Federal cavalry, and the General so wounded that he died a few days later.

The telegraph building at the United States Ford was largely devoted to hospital purposes. Even the yard and out-buildings were filled with the wounded. The office itself was also invaded, and the operators, when possible, eased the sufferers. At night the ground wire was broken, and Emerick, in feeling around in the dark for the trouble, found the dead body of a poor soldier lying upon it.

In consequence of Jackson's assault, and a conviction that Sedgwick's twenty-two thousand troops were opposed by only about eight thousand, General Hooker, thinking Sedgwick had recrossed, at nine, P. M., of the second, caused him to be telegraphed to cross at Fredericksburg, "on receipt of this order, and at once take up your line of march on Chancellorsville road, until you connect with" the main army, and "attack and destroy any force you may fall in with on the road;" "be in the vicinity" of Hooker at daylight. "You will probably fall upon the rear of the forces commanded by General Lee, and between you and the Major General commanding, he expects to use him up."

That despatch, which Sedgwick says was dated 10:10, P. M., transmitted over twenty-five miles of wire and carried three more by courier, was received at eleven, and acknowledged at once in a telegram, in which Sedgwick said: "The enemy will be vigorously attacked wherever overtaken." At this time Sedgwick's corps was fourteen miles from Hooker's main army, and his advance so greatly retarded that he did not scale Marye's Heights, in the rear of Fredericksburg, until eleven, A. M., of the third, and, from that time till night, although the corps continued to advance, it was under constant fire, and succeeded only in making half the distance between Chancellorsville and Fredericksburg.

About Chancellorsville the third opened with a brilliant charge by the left wing of Lee's army, under Stuart, driving, after a stubborn fight, the Federal right to Fairview, while Lee engaged Anderson's and McLaw's divisions with Slocum's and Meade's troops. When the battle was developing into large proportions, and fresh Federal troops were sorely needed on the right, Hooker was leaning against a pillar of the Chancellor House. At this important moment, a cannon ball struck the pillar and utterly incapacitated the Chief for duty for some time. Indeed, it is doubtful if he entirely recovered south of the Rappahannock.

About ten, A. M., the Union troops were driven from their Chancellorsville line, and forced to form anew with fresh troops farther north. It was at this time that Lee found it necessary to look to Sedgwick, and, intent on crushing him, he spent most of the remaining day enveloping three sides of Sedgwick's force, leaving him only Banks' Ford to fall back on, and near which he was toward evening pushed by an overwhelming force. While Sedgwick was thus engaged, Hooker telegraphed the President, as follows:

HEAD-QUARTERS OF THE ARMY OF THE POTOMAC,
Near Chancellorsville, 1:25 P.M., May 3, 1863.

By orderly to United States Ford, thence by telegraph, 3: 30, P.M.

TO HIS EXCELLENCY, A. LINCOLN,
President of the United States:

We have had a desperate fight, yesterday and to-day, which has resulted in no success to us, having lost a position of two lines which had been selected for our defense. It is now 1: 30, and there is still some firing of artillery. We may have another turn at it this P.M. I do not despair of success. If Sedgwick could have gotten up, there could have been but one result. As it is impossible for me to know the exact position of Sedgwick, as regards his ability to advance and take part in the engagement, I can not tell where it will end. We will endeavor to do our best. My troops are in good spirits. We have fought desperately to-day. No general ever commanded a more devoted army. (Signed) JOSEPH HOOKER,
Major General Comdg

That night, operator Emerick, at United States Ford, wrote in his diary:

We are defeated. Poor General Hooker and staff, holding consultation with general officers, in our office this evening. The General is extremely nervous, though this may be entirely attributable to a pillar, against which he was leaning to-day, having been shattered by a shot, giving him a terrible shock. For this, or some other reason, his mental faculties may almost be said to be temporarily impaired.

On the fourth, Sedgwick was two miles from Banks' Ford. Hooker telegraphed him to look well to the safety of his corps, and keep up communication with General Benham, at Banks' Ford, and advising him to fall back on the ford rather than Fredericksburg; and again, not to cross, unless compelled to; yet again, "Please let the Commanding General have your opinion in regard to" your holding a position on the south side, securely, until to-morrow, "by telegram from Banks' Ford, as soon as possible." Doubtless, it was in reply to this, that, within forty-five minutes, *i. e.*, 11:45, A. M., Sedgwick telegraphed from Banks' Ford:

My army is hemmed in upon the slope covered by the guns from the north side of Banks' Ford. If I had only this army to care for, I would withdraw it to-night. Do your operations require that I should jeopard by retaining it here. An immediate answer is indispensable.

General Hooker thereupon directed Sedgwick to cross. A few minutes later, Hooker received another telegram from Sedgwick, saying he could hold his position. Hooker told him to do so, and the next he received was dated five, A.M., fifth, stating that Sedgwick was across. This confusion of purposes, General Hooker explains as follows:

The messenger with the first despatch had to ride perhaps three miles to reach the place of telegraphing, and after he was sent off General Sedgwick sent another messenger immediately, with the despatch that he could hold the position, expecting that he would be able to overtake the first messenger and intercept the message that he had. But instead of that, the first messenger rode the fastest, and the message he bore reached me perhaps ten minutes before the second. Both were answered immediately on their receipt, but in some way the last was delayed in reaching General Sedgwick.

As soon as Hooker received Sedgwick's despatch announcing tnat the Sixth Corps was on the north bank, and the bridges swung and in process of being taken up, he determined to move his whole army back to camp opposite Fredericksburg. This was done the night of the fifth, but operators Murray and Emerick narrowly escaped capture as they crossed with the last of the rear guard. Operator Edwards, near Banks' Ford, did not fare so well. After Sedgwick had crossed or while he was crossing, a ten-pound ball passed over, but so near, his head as to produce insensibility. When he revived, he found he had bitten his tongue in the center, and had no knowledge of the cause of his hurt or the numbness which continued for two days. Banks' Ford office was closed on the eighth.

How different the result of this campaign might have been if Hooker had been in telegraphic communication with his commanders, especially Sedgwick, we leave to the speculative. The losses sustained by Sedgwick's corps were 4,601 and by the rest of the army, 12,596; of the total loss, 17,197, 5,000 were prisoners. The Confederates lost 12,227, of whom about 2,000 were captured.

The details of this fresh horror, like those of former misfortunes, were soon known throughout the land. It is devoutly to be wished that when the next war afflicts us, one nation may not suffer all the wounds, one people drink to the dregs all the miseries; but above all it is to be hoped that men will foresee the acres of dead, forehear the tremulous groans of the wounded myriads, and forecast the measureless misery incident to battle, before declaring another war. The declaration of war which may be avoided is the greatest of human crimes.

April 21, with a view of supplying Stoneman's cavalry, expected soon at Warrenton, a construction train was started to reopen the Orange & Alexandria road. Operators H. W. Cowan and J. H. Emerick accompanied it. It seems that while the great armies were contending for mastery at Chancellorsville, Stoneman, with the two divisions of Union cavalry, was operating on Lee's communications. By special request of Stoneman, operator J. L. Cherry was selected to accompany him. At three A. M. of May 2, Stoneman arrived in sight of Louisa Court House, Virginia, a station on the Chesapeake & Ohio Railroad.

Cherry hurried to the depot, hoping to find circuit from Richmond. The operator, Smith, had barely escaped on the last train south, taking his register and relay, but left his key, cut-out, and an immense lightning arrester. Cherry connected his pocket instrument and listened to Richmond despatches of no special importance for an hour, when his office was cut off by some other nearer Richmond. For a week the cavalry operated within the enemy's lines nearly to Richmond, but accomplished much less than was expected. Lee's telegraphs were cut in five places and four telegraph stations were captured.

APPENDIX.

CIPHER NUMBER NINE COMPLETE.

A. M. TIME

Ann 1.30, Agnes 2.30, Anna 3.30, Amelia 4.30, Alice 5.30, Betsy 6.30, Barney 7.30, Barbara 8.30, Cora 9.30, Clara 10.30, Catherine 11.30, Clotilda 1, Delia 2, Deborah 3, Dorothy 4, Emma 5, Eugenia 6, Emily 7, Elizabeth 8, Fanny 9, Florence 10, Frances 11, Gertrude 12.

P. M.

Harriet 1, Hannah 2, Helen 3, Henrietta 4, Imogen 5, Jennie 6, Julia 7. Katy 8, Lucy 9, Laura 10, Libby 11 Mary 12, Martha 1:30, Minnie 2:30, Nancy 3:30, Nelly 4:30, Rosalie 5:30, Rosetta 6:30, Rebecca 7:30, Reliance 8:30, Sarah 9:30, Susan 10:30, Topsy 11:30, Viola 12:30.

MESSAGE OR DIVISION OF THREE LINES—COMMENCEMENT WORDS:

Army, Anson, Action,	Five columns.	Astor, Advance, Artillery,	Six columns.	Anderson, Ambush, Agree,	Four columns.

Six Column Route: Up the fourth column, down the third, up the second, down the first, up the fifth, down the sixth.

Four Column Route: Down the first, down the fourth, down the second, up the third.

Five Column Route: Up the second, up the third, up the fourth, down the first, down the fifth.

MESSAGE OR DIVISION OF FOUR LINES—COMMENCEMENT WORDS:

Battle, Boston, Blair,	Six columns.	Banks, Board, Battery,	Four columns.	Brigade, Beverly, Bates,	Five columns.

Six Column Route: Up the fifth column, down the fourth, up the sixth, down the third, up the second, down the first.

Four Column Route: Up the second, down the first, up the third, down the fourth.

Five Column Route: Up the third, up the fifth, up the first, up the fourth, up the second.

MESSAGE OR DIVISION OF FIVE LINES—COMMENCEMENT WORDS:

Cairo, Curtin, Cavalry,	Four columns.	Congress, Colburn, Childs,	Five columns.	Calhoun, Church, Cobb,	Six columns.

Six Column Route: Up the sixth column, down the fifth, up the fourth, down the third, up the second, down the first.

Four Column Route: Down the fourth, down the second, down the first, down the third.

Five Column Route: Down the fifth, up the first, down the fourth, up the third, up the second.

MESSAGE OR DIVISION OF TEN LINES—COMMENCEMENT WORDS:

Driver, Dupont, Dunn,	Four columns.	Enemy, Enlist, Engage,	Six columns.	Forward, Foote, Forts,	Five columns.

Six Column Route: Down the third, up the fourth, down the second, up the fifth, down the first, up the sixth.

Four Column Route: Down the third, up the second, up the fourth, down the first.

Five Column Route: Up the third, down the second, down the fourth, up the first, down the fifth.

MESSAGE OR DIVISION OF SEVEN LINES—COMMENCEMENT WORDS:

Grayson, Giles, Grafton,	Six columns.	Guard, Henry, Harbor,	Five columns.	Kelly, Lucky, Mobile,	Four columns.

Six Column Route: Down the fourth, up the third, down the fifth, up the second, down the first, up the sixth.

Four Column Route: Down the second, up the first, down the third, up the fourth.

Five Column Route: Up the first, down the second, up the fifth, down the fourth, up the third.

MESSAGE OR DIVISION OF EIGHT LINES—COMMENCEMENT WORDS:

Morton, Memphis, Navy,	Six columns.	Potts, Porter, Perry,	Five columns.	Regular, Rosecrans, Run,	Four columns.

Six Column Route: Up the sixth, down the first, up the fifth, down the fourth, up the second, down the third.

Four Column Route: Up the fourth, down the second, up the first, down the third.

Five Column Route: Up the fourth, down the third, up the fifth, down the second, up the first.

MESSAGE OR DIVISION OF NINE LINES—COMMENCEMENT WORDS:

Skirmish, Sherman, Sumter,	Four columns.	Todd, Theater, Thomas,	Five columns.	Volunteers, Wise, War,	Six column

Six Column Route: Up the third column, down the second, up the fourth, down the fifth, up the first, down the sixth.

Four Column Route: Up the fourth, up the second, up the third, up the first.

Five Column Route: Down the first, up the third, down the second, up the fifth, down the fourth.

MESSAGE OR DIVISION OF SIX LINES—COMMENCEMENT WORDS:

Yates, Lincoln, Chase,	Five columns.	Stanton, McClellan, McDowell,	Six columns.	Halleck, Buell, Sibley,	Seven columns.

Seven Column Route: Up the fourth column, down the third, up the fifth, down the second, up the first, down the sixth, up the seventh.

FIVE COLUMN ROUTE.

×	×			×
15	25	26	16	6
14	24	27	7	5
13	23	8	17	4
12	9	28	18	3
10	22	29	19	2
11	21	30	20	1
×		×	×	

Six Column Route is shown on page 51.

ARBITRARIES.

Adam.	President of U. S.	Asia.
Abel.	Secretary of State.	Austria.
Aaron.	Secretary of War.	Arabia.
Amos.	Secretary of Treasury.	Africa.
Anthon.	Secretary of Navy.	America.
Acton.	Secretary of Interior.	Alba.
Abner.	Postmaster General.	Alpha.
Alden.	Attorney General.	Andover.
Alvord.	Adjutant General.	Antwerp.
Abbot.	Quartermaster General.	Aragon.

Adrian.	John G. Nicolay.	Alloy.	Appian.	George Harrington.	Animal.
Apollo.	Fred. W. Seward.	Altar.	Atlas.	G. V. Fox.	Annal.
Alps.	Peter H. Watson.	Amber.	Alamo.	J. Dahlgren.	Armada.
Andes.	John G. Tucker.	Anchor.	Akron.	H. A. Wise.	Anvil.
Arctic.	C. O. Wolcott.	Angel.	Adair.	E. D. Townsend.	Apple.

MAJOR GENERALS.

Archery.	Winfield Scott.	Ark.	Bergen.	O. M. Mitchell.	Bombay.
Asp.	Geo. B. McClellan.	Axis.	Botany.	Samuel R. Curtiss.	Bourbon.
Alkali.	John C. Fremont.	Attica.	Belgium.	Franz Sigel.	Bermuda.
Applause.	H. W. Hallock.	Abortion.	Berkshire.	E. V. Sumner.	Belgrade.
Adorn.	John A. Dix.	Agate.	Bologna.	S. P. Heintzelman.	Bolivia.
Alias.	N. P. Banks.	Amen.	Bruno.	Fitz John Porter.	Brutus.
Abbey.	B. F. Butler.	Audit.	Byron.	Wm. B. Franklin.	Bunyan.
Babel.	John E. Wool.	Baden.	Burton.	A. E. Burnside.	Buxton.
Baltic.	David Hunter.	Berlin.	Barnard.	E. D. Keyes.	Balfour.
Bremen.	Irvin McDowell.	Brussels.	Beach.	Lewis Wallace.	Burton.
Bangor.	U. S. Grant.	Bengal.	Bender.	Jno. A. McClernand.	Belcher.
Bagdad.	D. C. Buell.	Bethel.	Benjamin.	W. S. Rosecrans.	Bennet.
Bedford.	John Pope.	Biscay.	Borgia.	Joseph Hooker.	Berry.

NAVAL AND OTHER OFFICERS.

Bethune.	Faragut.	Blanchard.
Bigelow.	Goldsborough.	Bolton.
Bonner.	Wilkes.	Bishop.
Baboon.	Foote.	Badger.
Banjo.	Dupont.	Barber.
Bard.	Rowan.	Baron.
Ballad.	Porter.	Balmoral.
Banditti.	Davis.	Baptism.
Bible.	Rogers.	Basement.
Bassoon.	Schofield.	Beadle.
Beacon.	Hurlbut.	Bear.
Beauty.	Jno. J. Parke.	Beaver.
Bigamy.	Herron.	Bigot.
Bladder.	Blunt.	Bleaching.
Black.	Sherman.	Blubber.
Bogus.	Washburn.	Booby.
Brandy.	McPherson.	Bravo.
Bridle.	Ellett.	Brimstone.
Brocade.	Prentiss.	Bromley.
Budget.	McArthur.	Buffet.
Burglar.	Oglesby.	Buggy.
Bargain.	Lauman.	Basket.
Barth.	Kimball.	Behead.
Bellows.	C. A. Dana.	Belly.
Berth.	Roanoke.	Biped.
Blossom.	Monitor.	Bracket.

STATES.

Camden.	Me.	Cadmus.
Clarence.	N. H.	Claudius.
Coburg.	Vt.	Cognac.
Columbia.	Mass.	California.
Chester.	Ct.	Carroll.
Clifton.	R. I.	Carthage.
Cuba.	N. Y.	Champlain
Cheshire.	Pa.	Clyde.
China.	Del.	Catawba.
Camargo.	Md.	Census.
Century.	Va.	Cedar.
Castor.	O.	Cologne.
Carbon.	Mich.	Carpet.
Cancer.	Ind.	Camel.
Canary.	Ill.	Camphor.
Calendar	Wis.	Cabbage.
Charity.	Ia.	Cherry.
Chicken.	Minn.	Children.
Chorus.	Mo.	Clam.
Climax.	Ky.	Cider.
Churn.	Tenn.	Chapel.
College.	Ks.	Color.
Comet.	Cal.	Cupid.
Costume.	Ore.	Comb.
Corunna.	N. J.	Cherub.
Falcon.	Ga.	Finland.
Flora.	Ala.	Fortune.
Farmer.	La.	Famish.

GOVERNORS.

David.	I. Washburn, Me.	Daniel.
Denmark.	U. S. Berry, N. H.	Danube.
Darby.	Jno. A. Andrew, Mass.	Dalton.
Dresden.	Wm. A. Buckingham, Ct.	Dryden.
Dolphin.	Wm. Sprague, R. I.	Dragon.
Damon.	Fred. Holbrook, Vt.	Dublin.
Durham.	E. D. Morgan, N. Y.	Diana.
Dawn.	Chas. S. Olden, N. J.	Devon.
Domain.	A. G. Curtin, Pa.	Dropsy.
Damask.	Wm. R. Barton, Del.	Dimple.
Dagger.	A. W. Bradford, Md.	Darling.
Dauphin.	F. H. Pierpont, Va.	Dentist.
Dirgie.	David Todd, O.	Discount.
Dismal.	Austin Blair, Mich.	Divine.
Docket.	O. P. Morton, Ind.	Dodge.
Drill.	Richard Yates, Ill.	Drum.
Duke.	Edward Solomon, Wis.	Duchess.
Dungeon.	Sam'l J. Kirkwood, Ia.	Dumps.
Europe.	Alexander Ramsay, Minn.	Empire.
Egypt.	H. R. Gamble, Mo.	Emblem.
Eagle.	B. Magoffin, Ky.	Essex.
Eddy.	Andy Johnson, Tenn.	Emmet.

RIVERS.

France.	Arkansas.	Frog.
Feather.	Big Sandy.	Filter.
Flanders.	Cumberland.	Flannel.
Flint.	Chowan.	Florida.
Fool.	Cape Fear.	Fox.
Fork.	Chickahominy.	Fraction.
Gideon.	Edisto.	Gabriel.
Gotham.	Elizabeth.	Galena.
Galway.	Kanawha.	Garden.
Gallon.	Mississippi.	Gourd.
Garter.	Neuse.	Germany.
Georgia.	Ohio.	Genoa.
Geneva.	Ogechee.	Gaul.
Gem.	Potomac.	Ginseng.
Ginger.	Rappahannock.	Gland.
Girdle.	Rapidan.	Granada.
Glasgow.	Roanoke.	Gilead.
Globe.	Red River.	Glover.
Golden.	Shenandoah.	Goose.
Gondola.	Savannah.	Granby.
Grammar.	St. Mary's.	Gregory.
Godwin.	Tennessee.	Gliddon.
Griffin.	Tar.	Gifford.
Guns.	Trent.	Girls.
Grapes.	Tombigbee.	Growl.
Grub.	White River.	Grunt.
Hagar.	York.	Homer.
Horace.	Yazoo.	Harvey.
Hamlet.	James River.	Hannibal.
Hebrew.	Pamunkey.	Hindoo.
Harvard.	Big Black.	Humboldt.
Hastings.	Acquia Creek.	Haven.
Harlem.	Pearl.	Hampden.
Holland.	Tallahatchie.	Honduras.
Hungary.	Chickasaw Bayou.	Hunger.

FORTS.

Hemlock.	Delaware.	Hemp.	Huron.	Pulaski.	Hang.
Hymen.	Hatteras.	Hair.	Hunter.	Pickens.	Happy.
Herald.	Jackson.	Harp.	Harlot.	Randolph.	Hatchet.
Highness.	Key West.	History.	Hoax.	Sumter.	Hotel.
Hosanna.	Lafayette.	Husband.	Humbug.	Wool.	Huckster.
Hammock.	Monroe.	Hammer.	Haddock.	Warren.	Humphrey.
Holly.	Macon.	Hero.	Harmony.	Cedar Keys.	Hawley.

PLACES.

Ida.	Abingdon.	Ink.	Mastiff.	Memphis.	Melon.
Irving.	Augusta.	Ingress.	Mentor.	Murfreesboro.	Meridian.
Ingrate.	Atlanta.	Ingot.	Merlin.	New York.	Midas.
	Albany.	Indus.	Milan.	Norfolk.	Milk.
Indigo.	Boston.	Infant.	Mint.	New Orleans.	
Image.	Baltimore.	Insanity.	Mohawk.	Newburn.	Monarch.
Ireland.	Beaufort.	Italy.	Monster.	Natchez.	Montrose.
	Baton Rouge.	Ivory.	Moon.	Nashville.	Moscow.
Jacob.	Culpeper.		Myrtle.	Portland.	Mystic.
Jonah.	Charleston.	Jordan.	Maroon.	Philadelphia.	Mellow.
Judah.	Charlottesville.	John.	Music.	Port Royal.	Maxim.
Juno.	Charlotte.	Jupiter.	Mud.	Pensacola.	Muss.
Japan.	City Point.	Jersey.	Nabob.	Petersburg.	Nankin.
Jasmine.	Chattanooga.		Nestor.	Raleigh.	Nettle.
Java.	Corinth.	Jamaica.	Neptune.	Richmond.	Negus.
Jargon.	Cumberland Gap.	Jaundice.	Niagara.	Savannah.	Nile.
Jaunt.	Port Hudson.	Javelin.	Nose.	Suffolk.	Nasty.
Jolly.	Little Rock.	Journal.	Nutmeg.	Staunton.	Nugget.
Keate.	Danville.	Kindle.	Nuptial.	Salisbury.	Negro.
King.	Decatur.	Kingdom.	Niggard.	Stevenson.	Nuisance.
Knell.	Elizabeth City.	Knight.	Nurse.	Sperryville.	Nymph.
Kennebec.	Farmington.	Kidnap.	Opal.	Tuscumbia.	Oyster.
Knapsack.	Fredericksburg.	Kitchen.	Offal.	Urbana.	Olive.
Kasson.	Florence.	Kunkle.	Oakum.	Vicksburg.	Odor.
Lady.	Galveston.	Lamb.	Oats.	Williamsburg.	Oil.
Lantern.	Gordonsville.	Lafitte.	Optic.	Winchester.	Orbit.
Lapland.	Goldsboro.	Language.	Orchard.	Witherville.	Owl.
Lark.	Grenada.	Lawn.	Oxide.	Weldon.	Ordnance.
Leghorn.	Helena.	Legend.	Peru.	Wilmington.	Persia.
Lehigh.	Holly Springs.	Leopard.	Pagan.	Washington.	Pagoda.
Liberia.	Humboldt.	Lobster.	Palate.	Yorktown.	Palsy.
Lock.	Jackson.	Locust.	Panther.	Yazoo City.	Pelican.
Logan.	Kingsville.	Luther.	Pardon.	Haine's Bluff.	Parson.
Luna.	Lynchburg.	Limpid.	Patent.	Milliken's Bend.	Patron.
Lonesome.	Mobile.	Lester.	Peasant.	Napoleon, Ark.	Perfume.
Magnet.	Manassas.	Madder.	Pewter.	Port Gibson.	Pilot.
Madrid.	Montgomery.	Magic.	Princess.	Grand Gulf.	Pilgrim.
Magnolia.	Meridian.	Malta.			

CONFEDERATE GENERALS.

Plato.	Morgan.	Plainfield.	Quincy.	Wheeler.	Quitman.
Plum.	Chalmers.	Pocket.	Quiver.	Breckenridge.	Quack.
Polk.	Buckner.	Pontiac.	Quadrant.	Bragg.	Quadroon.
Poplar.	W. H. T. Walker.	Portage.	Queenly.	Johnston.	Quotient.
Prescott.	Lee.	Preston.	Quince.	Marmaduke.	Question.
Princeton.	Beauregard.	Prospect.	Query.	Price.	Quick.
Putnam.	Loring.	Pickets.	Quorum.	Pemberton.	Quarrel.
Quaker.	Forrest.	Queen.			

MISCELLANEOUS.

Randolph.	Arms.	Raymond.	Shylock.	Regiments.	Stanhope.
Richard.	Artillery.	Rodney.	Spur.	Right Flank.	Spruce.
Ramsay.	Ammunition.	Ransom.	Star.	Siege Guns.	Sugar.
Robin.	Batteries.	Raven.	Sulphur.	Smooth Bore.	Squash.
Rabbit.	Bridge.	Racine.	Sweden.	Scouts.	Sutton.
Raleigh.	Brigade.	Reading.	Smyrna.	Scouting.	Sidney.
Relay.	Cavalry.	Roanoke.	Sligo.	Scouting Party.	Stephen.
Ripley.	Cannon.	Richland.	Stanley.	Sailing Vessels.	Swallow.
Ridge.	Convoys.	Rome.	Summer.	Skirmishers.	Summit.
Rose.	Coast.	Rockland.	Sylvan.	Stragglers.	Steuben.
Roland.	Carbines.	Rubens.	Swindle.	Steamboats.	Surgery.
Ramble.	Deserters.	Rampant.	Supper.	Telegraphs.	Superb.
Rapture.	Engineers.	Ravish.	Stomach.	Transportation.	Stagger.
Reptile.	Entrenchment.	Ragged.	Spunky.	Volunteers.	Squadron.
Retrench.	Earthworks.	Review.	Spoon.	Transports.	Spit.
Reward.	Forts.	Romance.	Valley.	Brigadier Gen'l.	Vermont.
Rusty.	Fleet.	Ruffle.	Vernon.	Major General.	Vermin.
Saco.	Flotilla.	Salem.	Venus.	Colonel.	Vesper.
Saginaw.	Fortifications.	Scotland.	Vienna.	Major.	Village.
Sandy.	Forage.	Saint.	Virtue.	Captain.	Vulcan.
Saxon.	Guns.	Savory.	Vulture.	Lieutenant.	Vomit.
Sampson.	Gunboats.	Salmon.	Vincent.	Quartermaster.	Vinton.
Seneca.	Guards.	Sexton.	Violet.	Paymaster.	Virgin.
Saffron.	Harbor.	Sable.	Vista.	Surgeon.	Volcano.
Segment.	Infantry.	Seymour.	Walden.	Attack.	Wales.
Shandy.	Island.	Shaker.	Wafer.	Advance.	Wallace.
Saddle.	Mortar Boats.	Shallow.	Walpole.	Army.	Walnut.
Shannon.	Marines.	Sharon.	Warner.	Battle.	Warsaw.
Shark.	Mortars.	Spark.	Warwick.	Cutoff.	Wampum.
Sharper.	Pickets.	Sheffield.	Watkins.	Casualties.	Watson.
Shelby.	Projectiles.	Shelter.	Wayland.	Capture.	Wayne.
Shoal.	Point.	Smoky.	Weakness.	Defensive.	Webb.
Silver.	Regulars.	Snake.	Welch.	Defend.	Weldon.
Simms.	Rebels.	Snow.	Wells.	Diversion.	Wesley.
Soap.	Railroad.	Somers.	Wharton.	Division.	Whip.
Spafford.	River.	Spartan.	White.	Equipage.	Wick.
Spencer.	Rifled Guns.	Spring.	Wiley.	Enemy.	Windham.

Windpipe.	Flank.	Windsor.	Whist.	Wounded.	Whistle.
Winthrop.	Fall back.	Woodbine.	Wrangle.	Intercept.	Wreath.
Woodford.	Feint.	Woodland.	Wriggle.	Cipher.	Wrinkle.
Woolwich.	Fight.	Wyoming.	Wadding.	Arrest.	Waggish.
Walrus.	Killed.	Webster.	Weston.	Spy.	Wisdom.
Wag.	Left Wing.	Waltz.	Washington.	Traitor.	Wilcox.
Warden.	Missing.	Warp.	Wooster.	Treasonable.	Worcester.
Waspish.	Outflank.	Watchman	World.	Rear.	Walker.
Waxend.	Offensive.	Waxy.	Winston.	Front.	Wilkes.
Wayworn.	Reconnoissance.	Weasel.	Wicoff.	Invested.	Wilson.
Wean.	Retreat.	Weird.	Wabash.	Intercept.	Winona.
Web.	Resist.	Weld.	Winchester.	Embarking.	Williams-port.
Widow.	Right Wing.	Wedding.			
Wedlock.	Rifle Pits.	Whack.	Woodbury.	Assault.	Waldo.
Wedge.	Subsistence.	Wharf.	Young.	Camp.	Yarmouth.
Weigh.	Surprised.	Wheedle.	Yancey.	Head-quarters.	Yacht.
Whelp.	Surrounded.	Wheaten.	Yankee.	Reinforcements.	Yardstick.
Wherry.	Skirmish.	Whig.	Yellow.	Recruits.	Yawl.
Whiff	Threaten.	Whimper.	Youth.	Troops.	Yoke.
Whinney	Union.	Whiskey.	Zodiac.	Movements.	Zebra.

THE

MILITARY TELEGRAPH

DURING THE

CIVIL WAR IN THE UNITED STATES

VOL. II.

Thos. T. Eckert.

THE MILITARY TELEGRAPH

DURING THE

CIVIL WAR IN THE UNITED STATES,

WITH

AN EXPOSITION OF ANCIENT AND MODERN MEANS OF COMMUNICATION, AND OF THE FEDERAL AND CONFEDERATE CIPHER SYSTEMS;

ALSO A

RUNNING ACCOUNT OF THE WAR BETWEEN THE STATES.

BY

WILLIAM R. PLUM, LL. B.,

OF THE CHICAGO BAR.

VOL. II.

CHICAGO:

JANSEN, McCLURG & COMPANY, PUBLISHERS.

1882.

Printed by Donnelley, Gassette & Loyd, Chicago.

ILLUSTRATIONS OF VOLUME II.

PORTRAITS.

	PAGE.
GEN. THOMAS T. ECKERT,	frontispiece.
COL. JAMES R. GILMORE,	36
COL. WM. L. GROSS,	62
CAPT. T. B. A. DAVID,	145
COL. STAFFORD G. LYNCH,	147
MAJ. GEORGE H. SMITH,	216
CAPT. SAMUEL BRUCH,	295

MISCELLANEOUS.

DIAGRAM OF BEARDSLEE'S MAGNETO INSTRUMENT,	90
THE ORIGINAL SEVEN SLEEPERS,	222
FAC SIMILE CIPHER,	287
BRUCH'S MONUMENT,	293
NEWS OF THE FALL OF RICHMOND RECEIVED,	323

MAPS.

1. SHOWING PORTIONS OF PENNSYLVANIA, MARYLAND, DELAWARE VIRGINIA AND WEST VIRGINIA,	128-9
2. SHOWING PORTIONS OF TENNESSEE AND GEORGIA,	174-5
3. SHOWING RICHMOND AND VICINITY,	316-17

CONTENTS.

CHAPTER I.

The Telegraph in the Gettysburg Campaign, ... 9

CHAPTER II.

The Telegraph in North Carolina, South Carolina, Florida and along the Coast.—The Corps' Successes and Sacrifices, - - - - 25

CHAPTER III.

The Telegraph in the Department of the Gulf.—Port Hudson, Red River, and other Campaigns, - - - - - - 37

CHAPTER IV.

The Telegraph in the Departments of the Cumberland and Ohio.—Raids in Kentucky.—Ohio Invaded.—Middle and Eastern Tennessee Occupied by Federals.—Chickamauga.—Missionary Ridge.—Defense of Knoxville, - - - - - - - - - - 51

CHAPTER V.

Telegraphic Affairs in West Tennessee in the Fall of 1863.—Nicholson's Death.—General Sherman's Escape.—Forrest's Raid.—Memphis.—Presentations, - - - - - - - - 79

CHAPTER VI.

The Signal Corps and its Efforts to Control the Electric Field Telegraphs.—Stanton praises the United States Military Telegraph - 86

CHAPTER VII.

The Status of Federal and Confederate Operators and their Salaries, - 106

CHAPTER VIII.

The Telegraph in Eastern Virginia from July, 1863, to July, 1864.—Saved from Execution by Telegraphic Reprieves.—Movements of the Armies of the Potomac and James.—Wilderness.—Spottsylvania.—Cold Harbor.—Petersburg, - - - - - - 122

CHAPTER IX.

The Telegraph in the Shenandoah Valley, West Virginia and the District.—Sigel Defeated.—Cloyd's Mountain.—Pennsylvania, Maryland and the District Invaded, - - - - - - - 143

CHAPTER X.

The Telegraph in Kentucky and East Tennessee in 1864.—Perilous Opposition to Enlistment of Negroes in Kentucky.—Morgan's Last Raid.—Stoneman's Successful Raid Owing Largely to the Operator with Him, - - - - - - - - - - 154

CHAPTER XI.

The Fort Donelson Line.—Beckwith Dismissed for Obeying General Grant's Orders; is Reinstated.—Interference with Ciphers and Cipher Operators by Staff and Other Officers Forbidden.—Atlanta Campaign Telegraphs.—Various Battles.—A Terrible Accident.—A Strange but Providential Premonition.—Confederate Raid into East and Middle Tennessee.—Atlanta Taken, - - - - - 166

CHAPTER XII.

The Telegraph in the Department of the Tennessee (1864) and Gulf, to September.—Sherman's Advance from Vicksburg.—A Perilous Ride on a Flat Car.—W. S. Smith's Raiding Failure.—Forrest's Successes and Defeat in West Tennessee and Kentucky.—Fort Pillow Massacre.—Capture and Remarkable Escape of an Operator.—Sturgis' and A. J. Smith's Raids.—An Operator Riddled with Buck-shot.—Forrest's Dash into Memphis; Captures an Operator; His Experience.—Vicksburg Matters.—Grierson's Raid.—Capture of Forts in Mobile Bay, - - - - - - - - - - 188

CHAPTER XIII.

The Telegraph in Arkansas and Missouri During the Last Half of 1864.—Price's Invasion of Missouri.—Various Battles and Incidents.—Major Smith Resigns in Favor of Captain Clowry.—A Summary of the Service, - - - - - - - - - - 213

CHAPTER XIV.

The Telegraph in Tennessee, Alabama and Georgia.—Forrest Captures Troops and Operators at Athens.—His Raid in Tennessee.—Operators' Experience about Athens.—Experiences in Cahaba Prison.—Johnsonville Burned.—Hood's Army on Sherman's Railroad.—How Allatoona was Saved.—Manager of Atlanta Office Arrested.—Sherman Starts for the Sea.—Franklin.—Operators Captured; their Escape.—Hood's Waterloo, - - - - - - - -

CHAPTER XV.

Sherman's March to the Sea.—Atlanta in Ruins.—A Great Picnic.—Tapping the Enemy's Wires.—A View within the Confederate Lines.—Storming Fort McAllister.—Interesting Incidents.—Savannah Taken.—Telegraphic Connections, - - - - - - 241

CHAPTER XVI.

The Telegraph in the Siege of Petersburg.—The Mine Explosion.—Operators Court-Martialed; not Guilty.—Dangers of the Service.—Advances Toward Richmond Checked.—Narrow Escape at Reams Station.—A Military Railroad.—Grant's Wires Tapped; Important Results.—Hatcher's Run Fight.—West Virginia Affairs.—Numerous Telegraphic Incidents.—The Rebels Out-witted—Shenandoah Operations under Sheridan.—A Great Victory.—Operator McCrickett and Others Killed.—Operations in North Carolina.—Fort Fisher.—Sherman's Advance from Savannah.—Incidents of the March.—Sherman at Goldsboro, - - - - - - - - 257

CHAPTER XVII.

The Telegraph in the Departments of the Great South-West.—Stoneman's Raid.—Rebel Wires Tapped.—Confederates Disheartened.—Guerrillas.—Wilson's Raid.—Selma, Montgomery and Macon Captured.—Sherman Orders Wilson by Confederate Wires.—Telegraph Cashier Robbed and Nearly Murdered in Missouri.—*Résumé* of Affairs in Missouri.—Captain Bruch's Death.—Monument.—Cable Line from New Orleans to Mobile Bay.—Operations Against Mobile.—Its Capture.—The Bay Cleared of Torpedoes by the Corps.—The Ram "Webb" Nearly Escapes to the Sea.—A Telegraph from Federal to Confederate Headquarters.—Terrible Accident.—A Venturesome Ride; Narrow Escape.—Death of an Operator.—Some Serious Jokes.—The War Closed in the South-West, - - - - - 280

CHAPTER XVIII.

The Telegraph in Virginia During the Last Campaign.—Grant's Anxiety.—Five Forks.—General Assault.—Exposure of Operators.—Lincoln and Davis Hear the News of Victory and Defeat.—Petersburg and Richmond Evacuated.—Lincoln Visits Petersburg.—How the News of the Capture of Richmond was Sent and Received.—Lincoln in Richmond.—An Operator Receives Applause Intended for Grant.—The Pursuit of Lee.—Sailors' Creek.—Farmville.—Appomattox.—How the News of the Surrender of Lee's Army was Sent and Received.—The News of Lincoln's Assassination.—Capture of the Assassin.—Lincoln's Stories.—Johnston's Surrender.—Recapitulation of the Service of the Telegraph, - - - - - - 314

CHAPTER XIX.

Reconstruction.—The Corps Disbanded.—Its Officers Promoted.—Lines Turned Over to Private Companies.—Recapitulatory Statements.—Official Presentation to Ten Members of the Corps.—Death of Two Members.—A Few Appreciatory Letters and Other Evidences of the Value of the Corps.—Canadian Operators.—Finale, - - - 339

APPENDIX.

Annual Reports of Anson Stager, Chief of Military Telegraph, to the Quarter-Master General, - - - - - - 359

Roll of the United States Military Telegraph Operators, - - 376

THE

MILITARY TELEGRAPH

DURING THE

CIVIL WAR IN THE UNITED STATES.

CHAPTER I.

THE TELEGRAPH IN THE GETTYSBURG CAMPAIGN.

The terrible failures at Fredericksburg and Chancellorsville would have destroyed the *morále* of troops less devotedly patriotic than were those of the Army of the Potomac. Nothing short of deep convictions as to the justness of the Federal cause; of intense determination to maintain it, cost what it might; of an abiding faith that in the end the right would prevail; of the proud martial spirit of its rank and file, and of the unflagging belief that the powers of the Governments, National and State, were being steadily and energetically exercised to support and strengthen this great army, could have saved it after its losses in many struggles, seemingly fruitless, except in wounds and deaths, from disintegration and utter ruin. Across the Rappahannock lay a hostile army, emboldened and elated by success until it thought itself invincible. Large reinforcements were crowding to it, including especially Longstreet's forces from near Suffolk, Virginia. Supplies were abundant, genial weather at hand and a promise of a campaign of invasion of the enemy's country — a project always dear to the confident warrior — these things and a belief that can not be denied most of them, that theirs was a just cause, made Lee's soldiers an army of astonishing power. More than Maryland, Pennsylvania, Delaware and the District, more than the States it might hope to traverse, it threatened even the autonomy of the Union, as never before.

The rebellion was working up its flood tide, and that we may understandingly observe the dangerous sweep of the forces that, passing through mountain defiles, down the Shenandoah, across the Potomac and to the west of South Mountains, through Maryland, to Chambersburg, Carlisle and York, were destined to concentrate and break in shattered, shivering masses against the rock-ribbed defenses and iron-willed defenders of Gettysburg, let us pause to note some of the interesting details leading up to the climax.

The Confederate chieftain had to watch the movements of the Federals operating from four centers. To the north lay General Schenck's "Middle" Department. His force was mainly at Harpers Ferry, Martinsburg and Winchester, in all, say, eighteen thousand effectives. To the north-east was Heintzelman's Department of Washington, in which were thirty-six thousand troops. Across the river, in Lee's front, was Hooker's army, numbering, on the 10th of June, seventy-eight thousand, two hundred and fifty-five men; and south-east was the Department of Virginia and North Carolina, the main force, viz., fourteen thousand men, being with General Peck at Suffolk, Virginia.

Lee had no telegraphic communication north of Culpeper, and little need of any, prior to his advance. But the Federal telegraphic system, as heretofore, included all of their positions above indicated, and also, at times, as far out on the Orange & Alexandria road as Bealeton.

Rumors of Lee's intended movements were rife; they were even published in Confederate papers, just as those of Sherman's march from Atlanta were freely circulated in that city and published in Federal papers prior to its beginning. The very publicity of both seems to have induced disbelief as to the truthfulness of the reports. On the 2d of June, Peck telegraphed Dix, Dix, Halleck, and Halleck, Hooker, that "citizens and soldiers" about Richmond "talked of Lee's moving into Maryland with eighty-five thousand men;" but on the fourth General John Buford, in immediate command of three divisions of cavalry and ten pieces of artillery, along the Orange & Alexandria Railroad, wired Hooker *via* the War Department that the country as far up as Thoroughfare Gap was visited on the third, and nothing was seen or heard of the enemy.

The line from Bealeton to Falmouth had been abandoned some time before and the Bealeton and Warrenton Junction offices closed, but now that indications pointed to activity on the part of Lee's command, the cavalry was made to watch the fords more closely, and the wire was rebuilt to advance cavalry headquarters. Operators J. H. Emerick and J. D. Tinney were on the third sent from the War Department with cipher despatches, to await, at the Junction, the arrival of Buford, who was out on a raid. By the seventh, the line was rebuilt to Bealeton, and Emerick was sent there at midnight, under escort, to open the office. The next day he was assisted by J. D. Flynn, from Union Mills office.

Preparations were soon completed for a cavalry movement under General Pleasanton, in full force, across the river, to be supported by three thousand infantry, with the view of striking Stuart, who was meditating a blow at the Orange & Alexandria and Baltimore & Ohio railroads, in aid of Lee's operations. On the ninth, a severe engagement was fought, lasting, says Pleasanton, fourteen hours, and resulting in delaying the intended raid. From Rappahannock Station, Pleasanton telegraphed Hooker authentic and highly important information of the concentration of large bodies of infantry and cavalry about Culpeper. Operator Emerick wrote in his diary, "Plenty of cipher work; wounded being brought here rapidly, many of them with saber wounds, which prove the nature of the engagement." Mr. Doren's building party began, on the eleventh, to restore the line from Falmouth to Bealeton, and Hooker commenced moving forces to checkmate Lee's undeveloped purposes. A loop was built from Warrenton Junction to Pleasanton's headquarters near by, on the twelfth, and Joe S. Kirby, an operator, was detailed from a cavalry regiment to assist there. By the fourteenth, it was clearly ascertained that Lee was indeed moving north with a view, it was believed, of entering Maryland and perhaps Pennsylvania.

On the thirteenth, the evacuation of Falmouth and vicinity began; the great army by many routes was in motion. Operator L. A. Rose was the last to close office at the front and took the last train for Acquia Creek and Alexandria. Operators S. H. Edwards and E. A. Hall, with General Sigel at Stafford

Court House, where they had delighted in Hope Landing baked shad, stuffed with beans, were ordered to General Meade's head-quarters, to reach which they had to ride through the Eleventh Corps. While they were passing, the soldiers started the rhyme:

> Let the telegraphers through;
> They'll get us into a fight,
> And out of it, too.

The army was now fairly *en route*, no one knew just whither. A. H. Caldwell and J. B. Pierce were with Hooker; C. C. McConnell and F. N. Benson were with one of the corps; E. A. Hall with Howard, S. H. Edwards with Hancock, and A. H. Bliss, R. H. Ryan, T. H. Fonda, H. W. Cowan, J. D. Tinney, J. H. Glazier, R. F. Weitbrecht, L. A. Rose and F. T. Bickford either accompanied corps head-quarters or reached their respective stations *via* Washington and Poolesville in apt time. David E. Rand, who had served with his cavalry regiment since October, 1861, and participated in the battles of Front Royal, Cedar Mountain, second Bull Run and Chantilly, and was captured near Philmont, Virginia, and sent to Libby, being exchanged, was detailed in June and ordered to relieve operator Bliss at General Abercrombie's head-quarters at Centreville. While there, the northward movement of the army caught him and he, by direction, joined General John Newton's division of the Second Army Corps, followed it to Poolesville and Frederick City, from which latter point he was sent to Baltimore and thence to Fort Stevens, one of the defenses of the capital. G. W. Baldwin and J. W. Sampson were at this time operating at General Schenck's head-quarters, in Baltimore; C. W. Moore and S. K. Rupley, at Harpers Ferry, Va.; R. R. McCaine, at Winchester, Va.; Chas. C. Ways, at Martinsburg, Va.; D. C. Aughinbaugh, at Hagerstown, Md, besides whom there were sixty-six other operators variously stationed throughout the department. Emerick with Pleasanton started on the seventeenth, indirectly for Aldie, making a ride of about sixty miles. On the twentieth, being hopelessly distant from any telegraph, he returned to the War Department, where he reported late operations of the cavalry against the enemy, and was complimented by Major Eckert for the performance of recent services which were very trying—so much so that Emerick was prostrated for about three weeks.

The enemy very suddenly appeared about Winchester, in the valley, where Milroy was in command of about seven thousand five hundred men. While strong indications pointed to a general movement of the enemy down the Shenandoah, it was quite uncertain, until Milroy (who, it seems, was not well advised by his superiors in command) felt the pressure of the Confederates as early as the twelfth, but from disbelief remained until hemmed in. On the night of the thirteenth, the telegraph was cut, and nothing more was known of Milroy until a part of his force having, after a severe fight, cut their way out of the rebel environment, reached the Potomac on the fifteenth. The loss of communication with Winchester and Martinsburg—for that line also went down at seven, P. M., of the fourteenth, while the operator was reporting the progress of an attack upon that place—was severely felt by General Hooker and the authorities at Baltimore and Washington. All plans were entangled by an "if." "If," says the President to Hooker, by wire, "the head of Lee's army is at Martinsburg, and the tail of it on the plank road between Fredericksburg and Chancellorsville, the animal must be very slim somewhere. Could you not break him?" "To proceed," replied Hooker, "to Winchester, and have him make his appearance elsewhere, would subject me to ridicule." However, by the evening of the fifteenth, positive evidence was at hand that the enemy was moving toward the Potomac.

The Martinsburg garrison having reached Harpers Ferry, the force there was swelled to about eleven thousand men. Except the railroad line, the only wire intact from the Ferry ran direct to the war office. The troops evacuated the town and took position across the river, on Maryland Heights. The pontoon was taken up and the planking on the railroad bridge removed. Operator Moore and two mounted orderlies alone held the post. Major Eckert had ordered Moore to remain as long as the line was "O. K." This he did until about eleven, P. M., of the sixteenth, when General Tyler, commanding on the Heights, sent over an orderly with the countersign and a note, asking "what the devil," he, Moore, "wanted in Harpers Ferry after all the troops had left?" and directing him to report at head-quarters immediately. On hearing this, Eckert, at one, A. M., allowed Moore to go. His little party forded the river safely, and an hour later the enemy were in the town. Some hours before this, however, General Jenkins' Confederate cavalry had dashed into Chambersburg, Pa.

The Government was now thoroughly aroused. The prospects were gloomy enough. Harrisburg, Philadelphia, Baltimore and Washington were especially threatened. Throughout the North, crowds hung about the bulletin boards and telegraph offices, eagerly devouring the latest news. The President telegraphed the Governor of Maryland for ten thousand troops, Pennsylvania for fifty thousand, New York for twenty thousand, Ohio for thirty thousand, and West Virginia for ten thousand. These calls were again telegraphed to every city and town in those States, by their respective Governors. The telegraph was busy everywhere. Business was regulated by the ebb and flow of the Union fortunes. Newspapers in those days were very largely composed of telegrams and telegraphic correspondence, but at this time when every hour was pregnant with great events, only government messages had the right of wire. In the expressive language of operators, the lines were kept "red-hot" with military despatches. Death messages were rushed through, not to say smuggled, whenever possible, but ordinary business and reports for the press were delayed hours after the contending armies were silent in sleep. Offers to bribe operators to prefer one press report over another were not unknown. The newspaper rivalry became very great. This was natural, as every hour brought news more dire than the preceding, and millions of men and women were seeking to allay their fears by telegraphic reports.

A portion of Hooker's army crossed the Potomac on the 26th of June. About the twenty-fourth, the Second Corps, encamped in the vicinity of Thoroughfare Gap, was ordered forward. Its commander, General Hancock, having been in telegraphic communication with an officer (probably Hooker), at Gainsville, and having said good-bye at its close, the Gainsville operator, to complete the circuit with Washington, connected his ground wire with the main line, as Thoroughfare Gap office was then to be abandoned; but after riding a short distance,

Hancock recalled an important matter which he had forgotten, and asked Edwards, his operator, if he could raise Gainsville office again. Edwards replied that if the ground connection was imperfect, he might succeed, but the chances were against it, and, moreover, he did not like the looks of the "Johnnies" in the woods beyond. Hancock directed him to go and do what he could, adding, "My men will cover you." It was arranged that, if Edwards succeeded, he was to raise his hat as a signal, when a force was to be sent to get the line in a proper place for the General. But Edwards was unsuccessful, and as he was climbing a hill on his return, the Confederates opened fire on him, which was promptly and vigorously returned by Union troops. "I never felt so grateful in my life," writes Edwards, "as I did then to General Hancock, for shielding me. The tears came. I could not stop them, and when the General said 'You 've got lots of grit, for a civilian,' I felt more thankful than before."

The position of the seven army corps, respectively, and of the cavalry forces, was frequently despatched to the War Department telegraph office, where Mr. Lincoln, Mr. Stanton and General Halleck spent much of their time. Cipher operator, A. B. Chandler, in the War Department, observed, during this movement, that Mr. Lincoln was especially anxious to keep himself informed of the position of the Fifth Corps, then commanded by General Meade. Hooker's telegrams were unsatisfactory to the Government during this march of his army. He complained of lack of men and means, and on the 27th of June, sent Halleck a telegram, saying that his original instructions required him to cover Harpers Ferry and Washington ; that he had then imposed upon him, in addition, an enemy in his front, of more than his number ; begged to be understood that he could not comply with such conditions with the means at his disposal, and requested to be relieved from the position he then occupied. Whether he intended to ask to be relieved from command, or from the treble task he said was imposed upon him, was not perfectly clear, but assuming the former, the Government relieved him by placing General Meade in his stead. Meade immediately telegraphed from Frederick, in brave and soldierly, yet modest language his acceptance. While Chandler was translating this cipher message, the President was looking over his shoulder, watching with great interest the words as Chandler deciphered them, and seemed to feel great relief when he knew that Meade was in command. Relieving a commander of a large army while an important battle was impending, has been spoken of by able officers as extremely dangerous in practice, if not historically unprecedented; but McClellan was superseded by Burnside on what was thought to be the eve of battle, as the former expected to move on the enemy at Culpeper near by. However, in these days of railroads and telegraphs, precedents have to give way to many innovations.

The wires converging towards Meade's head-quarters in Maryland, in June, were, from Washington *via* Poolesville to Point of Rocks, Maryland, Point of Rocks to Frederick City, and thence three miles to the toll-gate. Other lines followed the route of the Baltimore & Ohio Railroad to Frederick City and more yet on the main line of the Northern Central road to Hanover Junction and Harrisburg. From the Junction the line originally connected Gettysburg *via* Hanover, but the rebels destroyed that for many miles. From Poolesville a line was constructed to Leesburg, Virginia, and from Point of Rocks to Harpers Ferry. On the 28th of June, General J. E. B. Stuart, commanding the bulk of the Confederate cavalry, crossed the Potomac and sweeping north to the right of Meade and by way of Westminster, succeeded, after dark, in severing all of the telegraphs by which Meade could speedily communicate with General French, then commanding at Harpers Ferry, Schenck, at Baltimore, Couch, who commanded a new department with head-quarters at Harrisburg, and the authorities in Washington. For several days, Meade's telegrams were forwarded to the nearest office by couriers, one of whom was killed June 30, near Glen Rock, while bearing a long cipher despatch, detailing Meade's plans and operations. Arrangements were also made to run an express train every three hours to Westminster with despatches for Frederick City, But Meade and the authorities in Pennsylvania, impatient at these necessary delays, sought to re-establish the wires. In a general order of June 30, Meade directed that "a telegraph corps * work east from Hanover, repairing, and all commanders * work repairing the line in their vicinity between Gettysburg and Hanover." About the same time, L. D. McCandless, of the Telegraph Corps, left Frederick for Hanover Junction, to work west with his builders, but the redoubtable Colonel Thomas A. Scott, assisted considerably by General Haupt, both eminent railroad officials, succeeded in restoring the line to Hanover in time for Meade's use at the close of the battle, if not earlier.

On the first day of July, Meade's head-quarters being about Taneytown, he received from Generals Couch and Schenck his first information that Lee's army was not marching for the Susquehanna, but was likely to concentrate at Gettysburg. One of these despatches was received at Frederick City over the newly repaired line to Baltimore at one o'clock on the morning of the first, and given to operators Ten Eyck H. Fonda and L. A. Rose, with instructions to carry it to General Meade as rapidly as possible. These young men performed their task so faithfully that on their return they were complimented by the Secretary of War. Over roads unknown to them, from one o'clock A.M., until five fifteen, they dashed at the rate of eight miles an hour until they found the General, to whom they delivered the message. Major Eckert, in an annual report, says:

> At the beginning of the fiscal year, the Army of the Potomac, under command of Major General Meade, was in the vicinity of Gettysburg, Pennsylvania. Telegraph wire, instruments, material, etc., together with a full force of builders and operators, were on hand with the army, but the Commanding General did not think it expedient to have telegraph lines established to the army. Communication was kept up, however, by means of a line of couriers from Frederick City, Maryland, and from Hanover, Pennsylvania, to both of which points we telegraphed directly from the War Department.

While Meade was at Taneytown, thirteen miles distant, contemplating the line of Pipe Creek, where he proposed to receive the enemy, the battle of Gettysburg, about ten A. M., of the first, unknown and unanticipated by either commanding general, was ushered in. Precious time, involving precious lives, and the demoralization of two corps of the Federal army, was passing, while Meade was being informed by courier of the thrilling events transpiring beyond his power of direction, and incalculably momentous to his army, the Government, the Union, the world. When the battle began, Meade's army was located about as follows: the Third Corps at Emmitsburg, the Sixth at Manchester, the Second at Taneytown, the Twelfth at Two Taverns, the Fifth at Hanover, the First and Eleventh at Gettysburg, with a portion of the cavalry. Too much credit can not be awarded Meade and his corps commanders for their diligence in concentrating at Gettysburg, when that new line of defense was determined upon. For three days, these great armies, nearly equally matched, for they had respectively something over seventy thousand men, strove to destroy each other. Whichever forced the fighting was doomed already. In the providence of God, the Confederates took the offensive on the first and continued it throughout. The night of the first, Buford's cavalry, which had made itself immortal, the gallant First Corps and the unfortunate Eleventh, rallied on Cemetery Hill, back of Gettysburg, whither they had been driven in confusion by the advance of the enemy.

The meagre tidings of this day's encounter that reached the tempest-tossed North were eagerly devoured by a news-famishing people, whose appetite it but whetted, even as the salt draught from the ocean increases the thirst of the ship-wrecked mariner. The second day came, and ere its sun set behind the distant hills, the victims of another day's dreadful conflict lay motionless or helpless on the sod where they had valorously dared, and yet the result was in the balance. All night long the nearest telegraph offices are busied with orders, private messages and correspondence. Operators of the various companies in the land are alert to catch each passing report. "Any news?" ask the distant points along the Pacific coast, while yet the conflict rages. "Any news?" asks the sympathetic Canadian, while yet the sun is high. "Give us news; we must have news," say New York, Pittsburgh and Washington. "Tell us of the hour; is our city safe?" demand those of Harrisburg and Philadelphia, which is re-echoed from Baltimore. Henceforth, until the end comes, business must suspend. Mothers, whose sons are there; wives, whose husbands are there; sisters, whose brothers are there; maidens, whose hearts are there, and the patriot, whose hopes center there; all, all join in beseeching God to be there also; to grant a glorious victory and save those nearest and dearest. "No news is bad news" the saying goes, and as the second day's battle did not fairly begin until four P. M., many were the downcast looks of those who believed the War Department was withholding evil tidings. At eleven P. M., General Meade wrote the following telegram for General Halleck, but for some reason it did not reach Washington until 5:15 P. M. of the third:

> The enemy attacked me about four, P. M., this day, and after one of the severest contests of the war, was repulsed at all points. We have suffered considerably in killed and wounded. Among the former is Brigadier General Paul Zook, and among the wounded, Generals Sickles, Barlow, Graham, and Warren slightly. We have taken a large number of prisoners. I shall remain in my present position to-morrow, but am not prepared to say, until better advised of the condition of the army, whether my operations will be of an offensive or defensive character.

Brave Paul Zook! He was an old-time telegrapher, having learned the art at a time when telegraph magnets weighed one hundred pounds instead of about one pound, as now made, and the comprehension of electrical matters compared with the present was about an inverse ratio to the weight of metal used. Zook was in that part of the conflict which Swinton, in his "The Twelve Decisive Battles of the War," speaks of as "where hot battle boiled and bubbled as though it were some great hell-cauldron."

The third came, bringing new hopes and fears, responsibilities and dangers—bringing on afresh the great battle which was ushered in by the use, some say, of more cannon than had ever before been employed in any single battle. With the repulse of the brave eighteen thousand, massed for the charge to break the Union lines, the star of the Confederacy began to wane. Happy North! Independence day brought news of a great victory. It was, indeed, a glorious Fourth. Bonfires, the ringing of bells, fireworks, all betokened it. In one short day the news traveled across the continent and made glad all the people in the settled districts north of the Confederacy. On that day, also, General Pemberton surrendered his army and Vicksburg to General Grant. As the news from Vicksburg was received at

Memphis, the tidings from Gettysburg went down from Cairo. Passing onward by despatch boat to Cairo, the news of Grant's remarkable achievements flashed to the East by the way of Cincinnati, to the North from Chicago, and to the West from St. Louis, and officially to all points from Washington, turning the people from joy to wild delight, from gladness to grateful thanksgiving. The armies of the Potomac and Tennessee had broken the clouds that hung like a pall over the Union, and the telegraph in the hands of several thousand operators had spread the great news so quietly and unostentatiously that the people thought not of its wonderful workings, nor of the patient endurance of its operatives, nor, in their excess of zeal to spread the great news, did the commercial telegraphers themselves then, nor have they since, thought of any other recompense or reward, than the consciousness of having unburdened many a heavy heart and relieved many a doubt-sick soul.

At the head of the Government, however, was an appreciative soul, over-burdened with solicitude. Speaking of him and passing events, Mr. Chandler, before quoted, said:

> I shall never forget the painful anxiety of those few days when the fate of the nation seemed to hang in the balance; nor the restless solicitude of Mr. Lincoln, as he paced up and down the room, reading despatches, soliloquizing and often stopping to trace the map which hung on the wall; nor the relief we all felt when the fact was established that victory, though gained at such fearful cost, was, indeed, on the side of the Union.

Of the losses in the three days fighting, General Meade reported 2,834 killed, 13,709 wounded, and 6,643 missing—23,186 in all, on the Union side; and a careful estimate of the Confederates, places them at 5,500 killed, 27,500 wounded, and several thousand prisoners, aggregating according to yet another estimate a loss of 36,000 men—a total loss to the two armies of nearly as many men as were engaged in reforming governments of Europe at the battle of Waterloo.

Immediately after the battle, operators Edwards and E. A. Hall were sent to Hanover, Pa., seventeen miles north of Gettysburg, to open an office on the line belonging to the railroad company, which was run into General Schenck's head-quarters office, at Baltimore, from whence it connected with the War Department. The operators arrived about five, A.M., of the fourth, and, finding the doors of the station fastened by great padlocks, they broke open the window with stones and clambered in. While in the act of connecting their instrument, the Superintendent of the road, Mr. Eichelbergher, came running to the depot with fifteen or twenty men, and demanded of the operators, in a loud voice, "By what authority do you break into this depot?" Edwards presented a blank U. S. M. T. envelope, as being the best evidence he had of his authority, but that proved enough. As the Confederate States troops had torn down the line running thence to Gettysburg, strenuous efforts were made to rebuild it, which was first done temporarily—the repairers using for the purpose scraps of stove-pipe and odd pieces of wire. These were fastened to the necks of old bottles that were placed in the crotches of trees, where they held and insulated the line. Edwards started from Hanover with the builders, but on reaching a small stream near Gettysburgh, where a bridge had been burned, it was found that the wire had shrunk so that the two ends could not be brought together. In hunting about, Edwards discovered in the stream some old soda bottles and an iron rail. The latter was propped up so as to rest on the bottles, and the ends of the line connected with those of the rail. Thus was communication first established between Gettysburg and the North, after being cut by Stuart. Several operators remained at Gettysburg some time, receiving and transmitting a great number of messages to and from the wounded in hospitals there.

On the morning of the fifth, as it was evident that Lee was retreating, the Sixth Corps began the pursuit, followed the next day by the army to Middletown, where it lost a day awaiting supplies, and then crossing South Mountain and moving *via* Boonsboro, it met the enemy in strong position on the twelfth, when a council of war was held at Meade's head-quarters, where it was voted not to immediately attack Lee's forces which had been unable to cross the Potomac, owing to recent rains and the destruction of pontoons.

As Meade's army moved from Frederick toward Hagerstown and Williamsport, the telegraph was promptly extended to the head-quarters of the General and the Corps Commanders.

These lines were worked continually until after the rebel army had evacuated Williamsport, when the main force of Meade's army moved to Pleasant Valley, Md., the head-quarters of the General being established at Knoxville. Lines were then built to the army, connecting with the Harpers Ferry wire. General French, having evacuated Maryland Heights, was at Frederick City during the battle of Gettysburg, which being decided, General Kenley's command repossessed Harpers Ferry, driving out a few rebels. Operator Moore crossed in the fourth boat, and re-established communication with Washington.

While the battle of Gettysburg was in progress, the President was deeply impressed with the incalculable importance of destroying Lee's army before it could recross the Potomac, and felt the utmost confidence in the ability of Meade's army to accomplish it. On the 8th of July, General Halleck telegraphed Meade:

> There is reliable information that the enemy is crossing at Williamsport. The opportunity to attack his divided forces should not be lost. The President is urgent and anxious that your army should move against him by forced marches.

Meade replied, giving the position and condition of affairs from his standpoint, adding assurances that he would do his utmost to push forward his army, and stating that his information as to the crossing of the enemy did not agree with Halleck's. On the twelfth, Meade telegraphed the position of his own army and that of the enemy as near as he could ascertain, and said, "It is my intention to attack them to-morrow unless something intervenes to prevent it." The President was becoming exceedingly anxious lest the enemy escape without further punishment, and upon the receipt of the above message in the Department telegraph office, where he spent so many of his anxious hours, the President called cipher operator Chandler to a map, on which he traced the various positions occupied by portions of each of the armies; expressed the greatest astonishment that an attack had not been made before; said it seemed to him the rebels were *being driven across* instead of being *prevented* from crossing, and as he walked across the floor said bitterly, "They will be ready to fight a magnificent battle when the enemy are all over the river and there is nobody left to fight."

His fears proved well grounded. Meade called a council of war, as stated, and the next day telegraphed that "five out of six" corps commanders "were unqualifiedly opposed" to attacking the enemy to-day, and consequently Meade did not feel authorized to attack until after he had made a more careful examination of the enemy's position, strength and defensive works. Halleck replied at 9:20 P. M., of the same day (thirteenth):

> You are strong enough to defeat the enemy before he can effect a crossing. Act upon your own judgment; make your generals execute your orders; call no council of war. It is proverbial that councils of war never fight. Reinforcements are pushed on as rapidly as possible. Don't let the enemy escape.

Next morning it was discovered that Lee had indeed successfully crossed the river with almost his entire army. Halleck promptly telegraphed the President's disappointment as follows:

> I need hardly say to you that the escape of Lee's army without another battle has created great dissatisfaction in the mind of the President, and it will require an active and energetic pursuit on your part to remove the impression that it has not been sufficiently active heretofore.

In consequence of the foregoing expression, Meade immediately asked to be relieved, but his request was not granted. On the fifteenth, the President, in reply to a message from the Hon. Simon Cameron, of Pennsylvania, said:

> I would give much to be relieved of the impression that Meade, Couch, Smith and all, since the battle of Gettysburg, have striven only to get the enemy over the river without another fight. Please tell me if you know who was the one corps commander who was for fighting in the council of war on Sunday night.

Meanwhile, as soon as Lee's movement into Pennsylvania had become certain, General Dix, in command at Fortress Monroe, organized a movement for the capture of Richmond, which was undertaken, but proved abortive.

In his report, ending June 30, 1863, Major Eckert said:

> It has been my duty to superintend the construction, equipment, operation and management of the Military Telegraph in my department, and to such purpose I have devoted my undivided energy

and attention. It is a matter of congratulation to me, and I trust to the Government, that from the uniform interest manifested by the telegraph employés in their several vocations, and the willingness on their part to undergo privation and hardship that the welfare of the Government and its cause might be served, but little delay in the construction of new military lines, when required, has occurred, and but slight interruption has occasionally existed to the daily operation of the several lines in the department. The general service which the telegraph performs is especially important to the Government and imperative in its character. The successful operation of army movements often depending upon the reliability and promptness of the telegraph and its operators. The amount of Government business transmitted over the military wires has been enormously large, and in its general purport, of the most vital natnre. It is a matter of gratification to me and a merited compliment to my subordinates, to be assured that this vast amount of important labor has been performed to the satisfaction of the honorable Secretary of War and his department.

CHAPTER II.

THE TELEGRAPH IN NORTH CAROLINA, SOUTH CAROLINA, FLORIDA, AND ALONG THE COAST.— THE CORPS' SUCCESSES AND SACRIFICES.

Thus far we have at best merely alluded to operations against the Confederate seaboard defenses. It is time to make amends, that continuity may not be greatly wanting.

As heretofore shown, Forts Hatteras and Clark, Confederate defenses of Hatteras Inlet to Pamlico Sound, were taken in the summer of 1861. Mention has also been made of the capture in February, 1862, of the forts on Roanoke Island, which were intended to protect Pamlico and Croatan sounds from Union vessels. General Burnside reported "that a combined attack upon this island was commenced on the morning of the seventh by the naval and military forces of this expedition, which has resulted in the capture of six forts, forty guns, over two thousand prisoners and upward of three thousand small arms."

Operators Jules F. Guthridge, George W. Norton and John B. Stough accompanied this expedition, taking telegraphic material for the erection of military lines. Only five miles, however, were constructed, and that was upon Roanoke Island, where the operators remained until Burnside embarked for Fortress Monroe, whence he went to Fredericksburg, to Pope's relief.

We have also shown the capture and occupation of Newberne, Morehead City, Beaufort, Washington, Plymouth and Elizabeth City, N. C., and Norfolk and Suffolk, Va.

October 29, 1861, a grand naval and military expedition, consisting of seventy-seven vessels of all classes, under Commodore DuPont, and twenty thousand men, under General Thomas W. Sherman, sailed from Fort Monroe, to operate against defenses constructed along the South Carolina, Georgia and Florida coasts. Weathering a terrible gale, most of the vessels were able to reach their destination off Port Royal, S. C., situate between Savannah, Ga., and Charleston, S. C., and about fifty miles from the latter. The principal out-lying islands, which serve as defenses against the sea between Savannah and Charleston, are Cockspur, on which was Fort Pulaski, forty-seven guns, at the mouth of the Savannah River, and commanding its main entrance, and northwardly, in this order, lie Tybee, Hilton Head, St. Helena, Edisto, Kiawah, Folly and Morris. Port Royal Inlet, between Hilton Head and Phillips and St. Helena, opens the interior by many water routes, one of which leads to Savannah, twenty-two miles distant. Forts Walker, twenty-three guns, on Hilton Head, and Beauregard, twenty guns, on Phillips, and their out-works, constituted the defenses of the Port Royal Channel.

November the seventh, Commodore DuPont's fleet of war vessels circled between Forts Walker and Beauregard, subjecting them and Confederate ships of war to a furious cannonade, driving off the latter and capturing the forts. The northernmost point of Tybee Island is sixteen hundred yards from Fort Pulaski. DuPont and Gilmore next turned their attention to Pulaski, operating from Tybee. While Gilmore was arranging his land siege guns, DuPont and Sherman possessed themselves of the Florida coast as far as St. Augustine. April 11, 1863, Pulaski was forced to surrender after sustaining a galling fire from siege guns and war vessels for two days. May 20, the Bay of Stono River was entered by Union gunboats, and Confederate works destroyed. This gave the Federals Kiawah and Folly Islands. North Edisto, to the south of Kiawah and north of St. Helena, had not been seriously defended, and was possessed in February. The lower end of Morris Island was occupied July 10, 1863, after a splendid attack by the combined land and naval forces upon the Confederate works commanding Lighthouse Inlet, between Morris and Folly Islands. At the upper end of Morris, near Cummings' Point, which is sixteen hundred yards from Fort Sumter, in Charleston Harbor, the Confederates had strong earth-works, called Batteries Gregg and Wagner. These islands and other minor ones were famous for the production of "Sea Island Cotton," and under Federal encouragement were populated and cultivated by runaway slaves and stay-at-home negroes; for the owners of island plantations left them entirely on the approach of the "Yankees." Thus the blockade was rendered quite effectual; safe anchorage obtained for Union vessels in distress; and permanent lodgments effected on the soil of these States.

It will surprise most readers to learn that these outer islands were connected by a land and submarine telegraph, the southernmost end of which was on Tybee Island, and the northern at Battery Gregg on Morris. It was not until after two sanguinary and heroic assaults upon Wagner had failed, and after months of costly, laborious and dangerous service in preparation for a third one, that early in September, 1863, the works were occupied, unopposed; the enemy having evacuated the night previous.

Lemuel F. Sheldon and his brother George D. entered the United States telegraph service about the same time (early in 1862), the former at Fortress Monroe, and the latter at Wheeling, W. Va., afterwards at Martinsburg and General Schenck's headquarters, from which he drifted also to the Fortress, where he became chief operator upon his brother's appointment as assistant to Major Eckert, without rank. Lemuel thereupon proceeded to his new field, reporting to Commodore DuPont and General Sherman in the Department of the South, which and North Carolina were in Eckert's field.

In January, 1863, James R. Gilmore, whose term of enlistment in a volunteer regiment had expired, re-entered the telegraph service at Hagerstown, on special duty, but was soon sent to Fort Monroe. It will be remembered that Gilmore succeeded David Strouse, the first Superintendent in the East, and was himself superseded by Captain Stager. After about three months employment at the Fortress, Gilmore concluded to accompany Sheldon to his new field of labor—the scenes of his brother, the General's splendid military operations.

Sheldon and Gilmore laid a cable across Calibogue Sound (eight miles), from the south end of Hilton Head Island to Tybee, near the Martello Tower; from Tybee across Lazaretto Creek to Cockspur Island, establishing an office in Fort Pulaski. Another cable was laid from the north-west point of Hilton Head Island across Broad River to the south end of Parry's Island, in the Port Royal entrance; also across Beaufort Creek to Beaufort Island, and across Port Royal Creek to St. Helena. These being connected by land lines, made a circuit from Fort Pulaski to Beaufort City, and thence to the naval depot on St. Helena Island, nearly opposite Hilton Head and across Port Royal Bay. Thus Sherman, on Hilton Head, was in communication with his troops and the naval commander. This was the first Federal military telegraph in South Carolina. Cables and land lines completed the circuit from Fort Pulaski to General Gilmore's new quarters at Cummings' Point, where he was, conjointly with the navy, operating for the reduction of Fort Sumter, in Charleston Harbor. This line was sixty-three and one-half miles long. A. G. Safford operated at Gilmore's head-quarters until February, 1864. Frank H. Lamb was chief operator. J. M. Nye was sent from the War Department office to Fort Pulaski, where he found H. S. Martin. Milton E. Roberts, a detailed soldier who could read the Bain alphabet by sound, but not the Morse, although he could make its characters, worked at Fort Wagner, nearly opposite the enemy's Fort Moultrie, also in the harbor. Roberts was greatly aided by Lamb, who could write Bain, though he could not read it. E. H. McGintey was at Hilton Head, and William Forster at Beaufort. These were, in time, reinforced by J. B. Barry, J. R. Dunlap, D. W. Smith, Ransom Phelps, T. and W. T. McGinty, A. Chamberlain, M. G. Kirkman, M. S. Andrews, A. R. Chamberlain, Paul C. Havens, Alex. K. McMurray, Thos. Roche, F. W. Kingsbury, J. F. McIlvaine, D. N. Bryant, J. J. B. Frey, J. L. Cherry, and L. W. Wortsman, a reformed Confederate, captured near Jacksonville, Fla. But it must not be supposed that all these operators were present at any one time—fourteen being about the average number.

Major Eckert, in an annual report, correctly stated that

> Invaluable service was rendered the army and navy by use of these lines, as, the important posts and stations being situated upon different islands, couriers could not be employed for purposes of communication; consequently, but for the telegraph, great expense would have resulted in the use of steamers for this service.

Nothing, but the importance of the work performed, could have induced the operators to remain on those hot, sandy and largely barren islands, where even the water was unhealthful.

For instance, when Smith went to General Terry's head-quarters on Folly Island, he pursued the usual practice of sinking barrels for water which was very poor at best, but after a few weeks it became intolerable, and while cleaning the well the soldiers discovered a dead man's elbows projecting into it at the bottom edge of the barrel. Then it was recollected that the Charleston authorities had been accustomed to quarantine against small-pox and yellow fever on this and Morris Island. which accounted for the many remains interred there.

The line running across Morris Island, especially between Forts Wagner and Gregg, opposite Sumter, was many times broken down by the enemy's shells. On one occasion, while two repairers were up telegraph poles, a shell struck the intermediate one, shivering it and jerking the repairers from their respective positions to the sand below. McGintey, operating at Battery Gregg, was in an extremely unpleasant place, owing to the bombardment from Sumter, and consequently had to spend much of his time in the bomb-proof. Smith's duties frequently required him to inspect the line along Folly and Morris islands. His horse, an old campaigner, named Gothic on account of his remarkable build, being a strange combination of bone arches and curves, had fixed notions of military proprieties, some of which caused his rider considerable uneasiness. Thus, while under fire from the enemy's guns, no amount of urging could induce Gothic to move faster than a parade canter. If Smith continued to press the beast after that, he would go through a series of exercises on the spot, which showed agility enough, but frightfully misapplied, for there is no mark the cannoneers like to try so well as a man on horseback. Smith's only consolation lay in the pride he took per-force, that others would think him remarkably cool. He ended his service there as too many others did, in sickness caused from malarial exposure. Phelps, his comrade, who served at Beaufort, and the Sanitary Commission people, nursed him from that death which the doctors had said was inevitable.

About the middle of September, 1863, William Forster, a native of New Brunswick, N. S., operating at Beaufort, volunteered to tap the enemy's main line leading from Charleston to Savannah, which of course would prove a very dangerous undertaking. The Commanding General had considered the matter some time, and as Forster was anxious to undertake the risk, he was sent, although Captain Sheldon refused to order him to go. Forster succeeded in tapping the wire a little west of Pocotaligo. On this errand he was led by runaway negroes, who were familiar with the country, and it is probable that he reached his destination *via* the Combahee River. His pocket instrument was inserted in the circuit by means of fine copper wire covered with silk, which was hidden by moss that grows so beautifully in that country. For two days he listened undisturbed to the passing messages. One of these was an order for a night attack, or else a distinct mention of such an assault, to be made upon General Terry's forces on Folly and Morris islands, for which the troops at Savannah were to concentrate with those at Charleston, and overcome the southern flank of the Union troops on those islands by crossing Folly Creek and suddenly attacking in overwhelming force. General Terry had noticed that the enemy on Johns and James islands, opposite his left, was unusually active, though he could not divine the cause; but when a faithful negro courier brought the telegram which disclosed the plan of operations, all was clear to the General. It is well here to repeat that the Confederates did not resort to their cipher system except rarely. This was especially so in 1863 and might well have been so throughout the war, for their method was not ingenious.

The Union commanders under Terry and Gilmore were at once advised of the intentions of the enemy, who seem to have noticed counter-preparations, for they desisted from the attack. Perhaps the unfortunate capture of Forster, on the third day of his anomalous position, had something to do with the abandonment of the plan. Forster was imprudent enough to walk out to the track, which was very straight for a long distance, and while there he was seen by the engineer of an approaching engine. Probably his hurried retreat excited suspicion. Be that as it may, the engine was stopped at the spot, the wire discovered and a vigorous search made by a company of soldiers. Bloodhounds scoured the woods and neighborhood. Poor Forster was found sunk to his armpits in a swamp. His negro attendants succeeded in escaping, but Forster never saw home again. He died in Columbia, where he was taken, a few months after.

To make but one special chapter on affairs in the Department of the South, reference will here be made to those in 1864. On the 6th of February, General Seymour, with about five thousand men, embarked for Florida in twenty-eight vessels, occupying Jacksonville the next day. Forty miles advance along the Tallahassee road was made by Seymour's cavalry with excellent results. But the telegraph, which the Confederates used to post themselves as to the progress of the Unionists, was now turned to account in their turn by the latter. Colonel Henry, of the cavalry, having found the enemy too strong in his front, telegraphed General Seymour from near Lake City, and that officer thereupon hurried troops from Sanderson. Operator Lamb started with Seymour to supervise telegraphic affairs, but a kick by a horse at Baldwin Junction caused his return, leaving Milton E. Roberts chief at the front, and J. M. Nye at Jacksonville. On the advance from Jacksonville to Baldwin, a telegraph operator was captured at his station, and under the influence of a revolver at his ear, compelled to telegraph in his own style (for most operators have peculiarities of manipulation), a train order which in due time brought a complete train into the Union lines, where it was of course captured. Contrary to Gilmore's directions, Seymour pushed on and was whipped in the battle of Olustee near Lake City, losing about two hundred and fifty killed and one thousand, two hundred and fifty wounded, while half those numbers would cover the enemy's loss. It is said that Gilmore, by signal from the mouth of St. John's River to Jacksonville and thence by telegraph, warned Seymour of his danger. Gilmore also sent a staff officer to Jacksonville with orders, but it was too late, and Seymour, on their receipt, replied, "We have met the enemy and are retreating; a devilish hard rub." The telegraph was of great service on this trip and probably hundreds of Union messages passed over this No. 8 English wire, which was taken down on the retreat and used along the coast.

Work for the reduction of the defenses of Charleston continued. Gilmore's shells, after traveling four miles, fell into the city, doing no little damage, and as to the great fort itself, it crumbled under a steady fire, but the stars and bars continued to wave as proudly as when first planted upon its solid ramparts. Sailors attempted to scale its walls, and suffered great loss for the few engaged.

The Confederates also relied on the telegraph. Beauregard had here caused, perhaps, the only strictly military telegraph in the Confederacy, to be erected, from Fort Sumter to Battery Bee, *via* James Island to the south, Charleston to the west, and the battery on Sullivan's Island on the north. An original and interesting character, named G. W. McCann, worked at Sumter. J. W. Kates, the first, and W. R. Cathcart, the last Superintendent, had an office in the city, where William Bryan was operator, and on Sullivan's Island was another, named —— Peacock. Thus were the troops and forts united by lines of telegraph.

Among the many projects for blowing up Sumter was one which occupied the attention of Captain Sheldon and operator A. G. Safford. While General Foster was in command (July, 1864), he asked Sheldon if he could explode powder by electricity through submerged wires, and if so, to arrange to blow up Sumter. Sheldon possessed some field cable, turned over by the Signal Corps, which he tested, and reported that it would stand submersion successfully for twelve hours. He then procured a large number of tin canteens, and had them made air-tight, to use as floats to his cable. Ten miles of cable on one reel, and five of manilla rope on another, were arranged to pay out equally. The rope was attached to a scow containing ten barrels of powder. The two ends of the cable were placed in a barrel of the powder, fixed on a piece of coke with platinas half an inch apart. The other ends of the cable went to a battery on deck of a monitor. The project was to float the scow with the flood tide against Sumter, and when in position, a signal from Morris Island was to indicate that the proper time had come, and Fort Sumter was to disappear forever. The whole arrangement was satisfactory after practical tests, and at eleven A. M., on the day agreed upon, the monitor stood inside the Charleston Bar, but the storm then brewing grew harder and fiercer until the sea ran over the deck. At seven, P. M., the storm still prevailing, the Admiral notified the Captain of a postponement. That was the last the navy had to do with the affair. An explosion from the shore subsequently took place, but beyond eliciting a lively response from thirty or forty guns, produced no effect.

After Mr. Gilmore had assisted Sheldon in laying the Port Royal Inlet cable, the latter, whose territory also included North Carolina, sent him (June, 1863) to Newberne, in that State, to erect and personally superintend lines in that district, then commanded by General Palmer. The offices to be opened on this line were Morehead City, Newport Barracks, Newberne and Bachelor's Creek. This line, forty-two miles long, was built in a substantial manner, fully equipped, and in operation in one week from the date at which Gilmore arrived in Newberne with his six builders. John Lock was, at that time, Gilmore's only operator, but Charles H. Lithgow, Herman Waterhouse, Douglass Kent, Robt. B. Vanderhoof, D. C. McGaughey and B. F. Gilmore arrived in good time. Of this line, Major Eckert reported:

> In the District of North Carolina, forty-two miles of line have been operated during the fiscal year, under the direction of James R. Gilmore. * * * Great credit is due Captain Gilmore, [before this report was made, he was commissioned Captain—AUTHOR], for his management whilst in North Carolina. Offices were kept open in Newberne, Morehead City, Beaufort, Newport Barracks and Bachelor's Creek. By this means, General Palmer, commanding the district, was kept in telegraphic communication day and night with four of the most important military posts in his district, and incalculable service has been furnished the Government. Besides this the navy has been assisted very greatly in its operations by having telegraphic communication between the blockading fleet off the coast, and the squadron in the Sounds.

General Foster succeeded to the command of the Department of North Carolina, but Burnside having withdrawn so many troops, Foster could not act aggressively to any considerable extent. After Foster took twelve thousand troops to the South Carolina coast, it was still more certain that advances were hardly to be counted on from that quarter.

On the 1st of February, General Pickett, commanding the Confederates in North Carolina, appeared at Bachelor's Creek, where D. C. McGaughey was operator. Pickett was intent on

3

retaking Newberne. He might have made better progress, had he captured McGaughey, for he was an invaluable man ; always ready, day or night, for any service, whether in the office or out on the line. He accompanied every scouting expedition and was several times selected by the Commanding General, to carry despatches which it was deemed unsafe to entrust to ordinary couriers. He rode like a Comanche, and was happiest in the performance of extra-hazardous service. In short, he was an emergency man. While Pickett was approaching Newberne, McGaughey was stationed about ten miles out with a cavalry, escort, to give warning of the enemy's approach in that direction. He was furnished a pocket instrument, and provided himself at a farmer's house with a double-barreled shot-gun. At midnight, his escort deserted him, retiring to Newberne. But though alone, McGaughey would not quit his post until three, A.M., and then only because the enemy had cut the line between him and that place. He made his way on foot through the rebel forces, and reached Newberne safely at daylight. Pickett was afraid to assault the defenses with vigor, and nothing further was attempted until April, when Plymouth was captured after a splendid defense by General Wessell's force of twenty-four hundred men, who were killed, wounded or captured. There was no telegraph there. In his report of affairs about Newberne, General Palmer wrote as follows :

HEAD-QUARTERS ARMY AND DISTRICT OF NORTH CAROLINA,
Newberne, N. C., February 17, 1864.

MAJOR GENERAL B. F. BUTLER,
Commanding Department of Virginia
and North Carolina, Fortress Monroe :

GENERAL : * * * * * *

Extract.—At no time, I suspect, during the present war, has the utility of the Military Telegraph and the Signal Corps been more fully demonstrated, than during the late attack. The attack had scarcely commenced at the out-posts, when the telegraph had not only informed me of all that was going on in front, but the whole line of posts to Morehead was put upon its guard, and during the day when the enemy were immediately around the town, the Signal

Corps kept us advised of the smallest movement of the enemy at any point of the line. I can not think too highly of these two corps.

* * * * * * *

J. N. PALMER, *Brig. Gen.*,
Comdg. A. and D. of N. C.

In forwarding a copy of the above to Washington, Mr. Gilmore added: "On two like occasions since then, our line has done work the value of which to the Government, General Palmer says, 'can not be estimated in money.'"

Exposure on a raid against the Weldon Railroad compelled Gilmore to go north to recruit. No sooner had he returned with his brother, another operator, than the yellow fever broke out in the district, and three of the operators were stricken, poor McGaughey first. Herman Frank Waterhouse was convalescing in the hospital at Newport Barracks, where he had suffered from bilious fever, when, hearing of McGaughey's sufferings, he volunteered to nurse him. Thus Waterhouse took the dread disease. McGaughey, Waterhouse, and Douglas Kent another operator, fell victims of the scourge, and on the pay-rolls, which alone indicate that these men were in the service of their country, is written opposite their respective names the simple word, "Discharged." An eternal discharge, indeed; but who would suspect from that word the honorable service rendered. *Yet that is the only thank ever vouchsafed Union operators, living, or dead, for their services, sufferings and sacrifices.*

In response to repeated requests for men to take the places of these "discharged" ones, Mr. Gilmore was ordered by Major Eckert to close the lines and protect himself and men as best he could, as he did not consider that the exigencies of the service required additional sacrifices.

JAMES R. GILMORE was on the third day of November, 1864, commissioned captain and assistant quarter-master, and the next month relieved Captain Sheldon at the latter's request. We have heretofore had many personal glimpses at Mr. Gilmore's career, which was one of marked devotion to the cause, whether recruiting or serving as a private soldier, operator or officer. Wherever he went, his presence was manifested by an unflagging zeal and determined purpose. Whether conducting the first telegraph at Fortress Monroe, building a line from Washington to Banks' lost army, ferreting out treasonable schemes, aiding on the hot sands of the Southern coast, conducting telegraphic affairs in North and South Carolina, or, as will appear later, restoring the lines in several Atlantic States, he was equally indefatigable and successful

JAMES R. GILMORE.

When relieved of the management of the telegraphs in the Department of the Potomac by Captain Stager, late in 1861, he assisted in organizing Company A, One Hundred and Twenty-sixth Pennsylvania volunteers, with which he served in the battles of the second Bull Run, Antietam and Fredericksburg, but since the war he has been in the National Engineer Department, constructing fortifications in New York harbor and on the South Atlantic coast, besides improving rivers and harbors in South Carolina, Georgia and Florida.

CHAPTER III.

THE TELEGRAPH IN THE DEPARTMENT OF THE GULF.—PORT HUDSON, RED RIVER, AND OTHER CAMPAIGNS.

Very little occurred in this department of noteworthy consequence, in a telegraphic point of view, until after the coming of Charles S. Bulkley, who was commissioned captain, assistant quarter-master and superintendent of military telegraphs in January, 1863. Indeed, no military or naval warfare took place here until about the last of April, 1862. Admiral Farragut, with the largest naval force that up to that time had ever sailed under the flag of the Union, after expending in his five days' bombardment of Forts Jackson and St. Phillips, river defenses of New Orleans, five thousand, five hundred and thirty-two shells, on the night of the 24th of April passed the forts with most of his fleet, under a terribly galling fire, and after silencing shore batteries and capturing and destroying numerous armed vessels, appeared before the city on the twenty-fifth. General Butler, with less than fifteen thousand men in the department, occupied the place about May 1st.

There was a private *ante bellum* telegraph line connecting at least the Head of the Passes of the Delta with the forts and New Orleans. This line belonged to the Good Intent Tow-boat Company of New Orleans, and was the medium by which, during the bombardment, the citizens were kept posted as to its progress, but about two hundred of Butler's men, avoiding the forts by indirection, fell upon this line shortly after the fleet had mainly run the gantlet, and cutting the wire left the city in a gloom of suspense.

The troops in one of the forts mutinied, and the surrender of both speedily followed. Farragut soon moved up the river and bombarded Vicksburg. He even ran its batteries and joined the Federals above. An unsuccessful expedition ventured up the Yazoo in July. In August, the Unionists (nine regiments), under General Williams, who was killed at the moment of victory, were attacked by General Breckenridge's forces (thirteen regiments) at Baton Rouge. Federal and Confederate gunboats played a small part in this affair, and the rebel ram "Arkansas" was disabled and destroyed.

No adequate arrangements had been made for telegraphic advantages when General Butler entered New Orleans. Fred J. Grace, an expert telegrapher, accompanied Farragut, even in running the forts, and Lieutenant I. Elliot Smith, who served as superintendent of Butler's telegraphs, was doubtless an operator, and it may have been these operators for whom Butler sent to the fleet almost as soon as he entered the city. We are not aware of any report that shows what was done by telegraphers under Butler. E. Von Eye, who was detailed as a messenger in the city office, September 2, 1862, has, however, informed us that the operators were detailed from the ranks and had to learn the art, as none could be had who were proficient; most of them did very well. They were taken from Vermont, New Hampshire and Connecticut regiments. The lines extended north-west *via* Carrolton and Camp Parapet to Pass Manchac bridge; north, a little east, *via* Fort Macomb to Fort Pike; the Balize line *via* Algiers, Quarantine, the forts, Head of the Passes, South-west Pass, and later to Point La Hoche and Pass á l'Outre; west, to La Fourche crossing, Boutte and Bayou des Allemands. These lines, except that down the river, were extended from time to time, as we shall see. Captain Bulkley brought a number of operators with him, and yet others soon followed. Among these were Edward Conway from the Potomac, W. A. Sheldon, late captain of a cavalry company, S. B. Fairchild, formerly manager of New Haven, Connecticut, office, B. B. Glass from Department of the Missouri, S. L. Griffin, J. J. Wilkie, —— Pitfield, —— Foster, David Elphick and Henry Stouder.

December 16, General N. P. Banks succeeded Butler in the department, which then included Texas. Banks' forces (Nineteenth Army Corps), about thirty thousand, were expected to aid in opening the Mississippi and Red rivers and to restore Texas. Baton Rouge was reoccupied late in December by ten thousand Federals, and the telegraph erected thereto *via* Bonnet, Carre and Donaldsonville. Banks' first expedition of moment to meet the enemy was a joint land (under Weitzel) and naval (under Commodore Buchanan) move up the Teche Bayou. January 14, 1863, quite an affair occurred at Carney's bridge near Pattersonville, in which Buchanan lost his life while engaged with two rebel gunboats, which, during the engagement, were sunk near other obstructions. The infantry drove off over eleven hundred Confederates, Weitzel's force being forty-five hundred; but the bayou was impassable and Superintendent Bulkley was commissioned to open the channel He subsequently reported his operations, as follows, to Colonel Stager:

I have the honor to report in regard to the operations in Bayou Teche, that I was ordered by Major General Banks, to proceed up the above named bayou, with the necessary apparatus and with the aid of a colored regiment in the engineer service, to remove the obstructions, consisting of two sunken vessels filled with brick, secured by piles driven around them, and the iron-clad gunboats Cotton and Hart, making in all four barriers completely closing the bayou. The water being shallow and the vessels imbedded in the muddy bottom, it was impossible to operate with powder inside their hulls, owing to the slight resistance the shallow water would give us above the charge, nor could the charges be successfully placed beneath them, with the means at my disposal. The only course left was to place them as low as possible alongside the hulls. The first charge of eighty pounds was exploded near the bow of one of the vessels filled with brick, which moved her bodily twenty feet, tearing down the piles and discharging parts of her brick cargo. With twenty-five-pound charges, she was then broken up and hauled to the banks. Vessel number two was removed in the same manner, but with less powder, not being so thoroughly filled with brick. The gunboat Cotton was found loaded with her heavy machinery. Rather than risk the chances of dropping this in the bayou, we removed her stern only. In this case, one charge of eighty pounds was used inside her hull with great effect, considering the shallow water—only six feet in depth. Small charges of twenty-five pounds alongside the fragments completed the removal.

The gunboat Hart also had her machinery on board, and three large boilers, which were under water, securely bolted to her hull and connected with large boiler-iron pipes. We succeeded in placing a charge of fifty pounds under them, near the farther end from shore. By this explosion they were torn from their fastenings, and landed near the bank of the bayou, besides shattering the hull.

Our next charge of four hundred pounds was placed alongside, directly amidship, in water nine feet in depth. This removed her center from side to side completely, and her ends were readily hauled near the bank. This cleared the bayou and rendered it navigable for our steamboat transportation. In removing these, we expended seven hundred and fifty pounds of powder and used three cups of Grove battery to ignite the charges. Our conducting wires were two thousand feet in length—the electric current passing from this over a small platina wire fixed in a cartridge in the case containing the charge. This conducting wire is part of a lot captured in New Orleans, of Confederate manufacture, rather imperfect, and intended for exploding torpedoes in the Mississippi River In compliance with the order of Major General Banks, the United States Military Telegraph Department has furnished the necessary apparatus, material and superintendence for this work. The colored regiment, Colonel Robinson commanding, rendered the most willing and efficient aid.

In February, the Union ram Queen of the West and gunboat DeSoto were captured up Red River. The Confederate ram Webb, the Queen of the West and other war vessels then attacked the Indianola, one of the best river boats, capturing it. Thus the Confederates obtained control of the Mississippi between Vicksburg and New Orleans, besides a number of navigable tributaries. Farragut having again passed Vicksburg, co-operated with Banks to regain control. Banks advanced against Port Hudson with twelve thousand men, while it was defended by sixteen thousand. Farragut ran its batteries about the middle of March with some vessels, losing one and others being disabled, wherupon Banks returned to Baton Rouge.

The telegraphers had followed him with the line, and of course took it down on his retirement. Banks again moved westward from New Orleans. This was early in April. His forces being divided, a part fought near Fort Bisland, after which Confederate General Richard Taylor retired to Opelousas. The Queen of the West was also destroyed in Grand Lake. Taylor, driven from Opelousas and Alexandria, retired with a remnant of his troops to Grand Ecore and Shreveport. Lines were extended along the railroad from Brashier City up the river immediately in the rear of Banks, and offices opened at intermediate points.

Having scattered the enemy's forces in this section, Banks now hoped to reduce Port Hudson and join Grant in besieging Vicksburg, or at least to hold the garrison at Port Hudson until Grant could capture Vicksburg and assist him. Accordingly, he marched across the country, and, joined by Augur's division from below, invested Port Hudson, May 25, making a very unsuccessful but heroic assault two days later. The telegraph was quickly opened at Banks' head-quarters, where Ed. Conway was manager. Siege lines were built from Springfield Landing along the fronts of Generals T. W. Sherman's, Augur's, Paine's, Grover's and Weitzel's troops, and offices opened at the officer's head-quarters. This line was fourteen miles long.

During the siege of this stronghold, the Union wires were lengthened, and offices advanced towards the enemy's works, whenever the Federal forces moved forward. It was pending one of these changes that operator B. B. Glass was sent from General Augur's to the extreme left, to open a new office. This was probably during the fierce assault of June 14. The supply of field wire failed about a quarter of a mile from the point to which he was sent, and as his orders were to follow the wire, he reached the end of it, when, expecting it to be immediately extended, he took his direction from the troops, but soon emerged from the woods to find himself almost on the line of battle, a short distance from the enemy's works. Glass, who was mounted, had to pass an open space of about three hundred yards, to reach shelter from the storm of bullets, shot and shell. He tried the Indian dodge of hugging the safe side of his horse. As that was a new trick under trying circumstances, he nearly tumbled, but hanging by the pommel with one hand, and the mane with the other, he managed to keep one limb free. The spur on the boot of that one was dextrously plied. The soldiers along the route found rare sport in the operator's equestrián tactics and many were their remarks, which all seemed to appreciate but poor Glass, at whom they were directed. In this dash the horse received two slight flesh wounds. Soon after, Glass's office was opened, for lack of wire, uncomfortably near the enemy and just behind an Indiana battery. Consequently, the operator oftentimes lay very flat in a three-foot ditch, while his instrument was on a cracker box within hearing. After awhile, Glass arranged a crude *bomb-proof* (?) out of rails and *débris*, which, fortunately, was not struck. At this office he staid night and day, dodging shells by day and fighting gallinippers by night. We believe that S. L. Griffin and W. A. Tinker, operators, also worked on this siege line.

About 3:30 A. M., July 6, Glass received despatches for Banks from General Grant and Commodore Porter, announcing the surrender of Vicksburg; but Conway at Banks' had fallen asleep from exhaustion, and the messages were sent by orderly. Two days after, the originals were forwarded to Confederate General Gardner, as proof of the facts stated, and on the ninth, he surrendered six thousand, four hundred and eight men, fifty-one pieces of artillery, two steamers and much other valuable property.

Once more the Mississippi River, under Federal control, ran "unvexed to the sea." Glass, who was among the first to enter Port Hudson, secured as a trophy General Gardner's head-quarters flag, made with cross-bars on a red field, and blue stars on the bars, which he retains to this day.

During the siege of Port Hudson, the line to New Orleans was cut one afternoon, north of Bonnet Carre, and the officer commanding the guard there, whose duty it was to patrol the line for about twenty miles every day, telegraphed General Emory, in New Orleans, that two hundred guerrillas were reported just below Donaldsonville, and had torn down the wire. The message ended with a request for permission to withdraw the guard into the works at Bonnet Carre, and being authorized so to do, he soon reported his command safely within the works. W. A. Sheldon, manager of the city office, and Department chief operator, had had much experience as a cavalry officer, and doubting the report, obtained an order from Emory to take command of the patrol, about one hundred strong, and, if possible, restore communication with Banks. The officer in command pleaded indisposition, but thought if he could ride up on a *gun-boat*, he could take command in case of a fight, if landed in time. This novel *horse-marine* idea the Captain was left to develop while Sheldon and the troopers hurried off. About fifteen miles distant, a large party was seen coming toward them. Sheldon took them for guerrillas, but they were soon proven to be passengers, who the evening before, had left New Orleans on the same steamer that landed Sheldon. This steamboat had been fired into a few miles below Donaldsonville by a battery on the opposite side of the river, striking the steampipe, whereupon the steamer was run to the east side and the passengers landed. A gunboat soon after drove off the battery and saved the steamer. Sheldon discovered a quarter of a mile of wire missing, near Donaldsonville, which an old negro reported to have been cut by an overseer, who threw it into the river. Of course the line was restored, and the overseer imprisoned.

It was a cruel necessity that compelled Banks to abandon Alexandria and the Red River country to besiege Port Hudson, and many are the sad stories of misery and worse inflicted on Unionists, particularly negroes, who were left to the enemy. General Taylor, thereupon, being reinforced, swept away or captured all the outposts west of Algiers, Louisiana. In this advance, many Federal prisoners were taken and millions of dollars worth of property captured or destroyed. While Colonel Stickney was gallantly defending La Fourche crossing, repulsing two attacks late in June, 1863, a shell exploded on the roof of the telegraph office, where the operator was telegraphing to New Orleans an account of the progress of the fight. Stickney retired to Algiers. Taylor could not cross the river and therefore moved north, his advance being repulsed at Donaldsonville, with a loss of three hundred and twenty-four killed, wounded and missing, which was off-set, a few days later, by a defeat of Federals six miles inland. This occurred shortly after the fall of Port Hudson, and Taylor fled westward.

During the siege of Port Hudson and Taylor's operations, the operators throughout the department became greatly exhausted from long vigilance. In only one office was there more than one operator, and that was New Orleans, the focal point in the department, where Sheldon was zealously assisted by Henry Stouder, at that time a very ordinary operator. While General Taylor was operating against Fort Butler and Brashier City, Sheldon, about midnight, granted permission to the operators at La Fourche, Baton Rouge and Donaldsonville to retire until six

A. M., as they had been on continuous duty for two days and nights. Sheldon, himself, also went to sleep. An hour later he was awakened by heavy knocking of two young aides of Emory's, one of whom in a gruff voice ordered him to get up and send a message at once. Sheldon protested that the operator at the other end was asleep and the message could not be sent. Angry words followed and a guard was placed at Sheldon's door and he held a prisoner. But when General Emory heard of the affair, he roundly berated the officers for their conduct. This instance recalls the fact that in the preceding May, while Banks and Captain Bulkley were away from New Orleans, General T. W. Sherman, in command, placed one of his officers in charge of the telegraphs, whereupon the two city operators rebelled, and being threatened with imprisonment, resigned, and refused to reconsider their resignations on Bulkley's return to remedy matters.

Immediately after the capture of Port Hudson, Banks sent a steamer up to Vicksburg after additional telegraphers, and returned with William Foley, G. W. Baxter, E. H. Johnson and H. W. Nichols. Baxter and Foley stopped at Port Hudson just in time to serve operator Griffin a good turn. It was during the summer of 1863, General Herron sent an orderly late at night with a telegram for New Orleans. Tinker, the night operator, sent word that the line was down and the message could not be sent until it was repaired. The next morning, General Andrews, post commandant, sent for Griffin, the manager, who appeared in the absence of that officer, and having returned, was escorted back by a file of soldiers, because he declined to close his office and go again. Griffin was put in the guardhouse, but soon released, as Baxter and Foley refused to work unless he was set at liberty.

After Port Hudson, General Franklin operated unsuccessfully, in conjunction with the navy, along the coast of Texas, and as a diversion, General C. C. Washburn was sent up the New Orleans, Opelousas & Great Western Railroad as far as Opelousas; but on retiring was beaten by Taylor and Green, losing seven hundred and sixteen killed, wounded and missing. In October and November, Banks conducted operations which gave the Unionists control of all the coast from the Rio Grande to the Brazos, even penetrating far inland.

W. A. Sheldon having been attacked by fever shortly after the capture of Port Hudson, Sidney B. Fairchild became department chief operator and Ed Conway, manager of lines. These were extended to Morganzia, where A. W. O'Neil operated. On General Banks' attempting his Red River Campaign in March, the telegraph was advanced to the mouth of the Red River. In doing this, Bulkley's men were constantly annoyed by small parties of cavalry, some of whom were captured by Bulkley's command, but he reports that they killed two of his men in one of their attacks. This line to New Orleans was interrupted so much by the enemy, but mainly by bushwhackers, that between Baton Rouge and Port Hudson it was changed to the west bank. General Banks then ordered that whenever the wire was molested, every house or habitable domicile within, it is said (but we doubt it), twenty miles of the injury, should be destroyed, and in one instance houses were burned, pursuant to the order. After that the line was unmolested.

March 13, Banks' army (seventeen thousand) about Franklin started for Alexandria, on the Red River, where it met A. J. Smith's troops (ten thousand) which had gone up the river convoyed by Admiral Porter's fleet, and about the same time General Steele left Little Rock with fifteen thousand troops. These forces, co-operating, were expected to destroy the Confederate armies under Dick Taylor and Kirby Smith, and to converge about Shreveport. Edward Conway accompanied Banks' army, in charge of the field telegraph, by which it was expected to maintain immediate communication with the different parts of the army during battle. Some disagreement occurring between Bulkley and Conway, the latter resigned, and Fairchild was placed in charge. The other operators were W. H. Munro, Wm. Foley, Horace W. Nichols. The telegraph train consisted of five wagons, three of which were laden with wire on reels, and were drawn each by four-mule teams. At New Iberia, Conway rejoined and took charge as before. Banks severed his telegraphic connection with New Orleans, at a point seven miles north-west of New Iberia, March 15, when the following, the last message vouchsafed an anxious world before plunging deeply into the wilds, morasses and dangers of North-west Louisiana, was sent

To Captain Bulkley, New Orleans:

Head-quarters in the saddle for the last two days. Hind-quarters sore. (Signed) S. B. F.

At one time the telegraph party lay asleep in tents, having had no notice that the army would move early, and was consequently actually left so far behind that bushwhackers came up and fired upon them, one ball striking a tent pole; but fortunately Banks had asked for his cipher operator, Conway, when it was discovered that the party was not up with the army. The Second New York Cavalry was sent to their rescue, arriving just in time to drive off the guerrillas. At Vermillion, Foley and Fairchild were sent spinning down a hill by a careless driver, the reels following after them and imparting a kind of real life, which, to be enjoyed, "needs but to be *seen*."

At Alexandria, the operators boarded the steamer Laurel Hill, for Grand Ecore, where teams and all arrived on the 5th of April, and the next day, Fairchild, Munro and Foley, having obtained permission, started to visit Natchitoches, four miles distant and beyond the Union lines. At the edge of the town, after passing through a grove, Munro, who was familiar with the people, discovered that his party was followed by guerrillas. These fired upon the operators, whereupon they returned rapidly to camp, and, owing to their report, Natchitoches was occupied.

April 8, the advance, at Sabine Cross Roads, was violently assaulted and badly beaten. At Pleasant Grove, three miles in the rear, on the same day, Emory's division barely saved the army and, consequently, the vessels in the river. The next day, Banks was again attacked at Pleasant Hill, fifteen miles further back, where the Confederate charges were repulsed with great loss. In these conflicts the enemy numbered about twenty-two thousand. Banks was unable to concentrate, at any one time, over fifteen thousand men, and that number only at Pleasant Hill. His loss in these battles was three thousand, nine hundred and sixty nine, of whom about half were taken prisoners.

General Steele's advance had not reached supporting distance. At Grand Ecore, Banks heard from him by a cipher message brought across the country. Fairchild labored long upon it, as it was erroneously made up, but finally succeeded in translating it.

Banks' expedition, owing to numerous causes, but especially the defeat on the eighth, and the falling water in the river, had become a positive failure, and the great problem now was to reach the Mississippi with troops and flotilla. Conway left Grand Ecore, May 20, with Admiral Porter, on the tin-clad "Cricket," taking batteries and other electrical apparatus for use in blowing up gunboats in case they failed to pass over the falls near Alexandria ; but he lost all his apparatus with the boat itself, which was literally riddled with shells and balls, above Alexandria, whereby half of her crew were killed and wounded. It is written "that there was scarcely any of her left." After many conflicts great and small, between the Union soldiers, the navy, and the Confederates, the former reached Simmsport, May 16, and Porter's fleet of fifty vessels, less a few gunboats and transports lost, arrived at the mouth of the Red River, after a harassing retreat, which has few parallels. Banks' campaign losses exceeded five thousand, and the enemy's was probably about the same.

The operators shared the hardships, privations and dangers of the undertaking. Captain Bulkley officially commended Conway for "efficiency and creditable conduct throughout and during the campaign." When Banks had reached Alexandria, he found reinforcements, from the Texan coast, which had been mainly abandoned in April.

During that month, Bulkley built a military line from Brazos to Brownsville, which was the first Federal military line ever erected in Texas.

On the 6th of May, at midnight, Colonel Stager was ordered by Secretary Stanton, to meet and proceed with General E. R. S. Canby (who was about to relieve Banks) to Cairo, Ill., to arrange "for prompt transmission and receipt of intelligence between that point and the forces on Red River." Stager promptly executed this order, and received soon after Stanton's commendation, as follow : "Accept my thanks for your prompt and energetic action."

Canby became Department Commander on the 20th of May, 1864, and Bulkley, at his own suggestion, was relieved about

August 1, to take a prominent part in the erection of an international telegraph line, *via* Behring's Straits. Stager endorsed Bulkley's resignation, as follows:

Captain Bulkley has served for nearly two years as Assistant Superintendent of the United States Military Telegraph with great zeal and ability, and the successful management of military telegraphs in the Department of the Gulf has secured him highest testimonials of Generals commanding in that Department.

During the first six months of 1863, two hundred and forty-four miles of telegraph, connecting Bonnet Carre, Franklin and New Iberia, were constructed by the Military Telegraph Corps, and two hundred and four miles of telegraph were erected within this Department, from July 1, 1863, to June 30, 1864, at which latter time the total force—operators, linemen, messengers and teamsters—was but ninety-six, to work over thirty offices. The number of military telegrams sent during the year was one hundred and forty thousand.

Although Steele's movement from Little Rock was co-operative, as stated, it, too, was unsuccessful. Probably Theodore Holt accompanied Steele, as cipherer, but there was no telegraph train. Captain Clowry, Superintendent of Telegraphs in Missouri and Arkansas, reported that,

On March 14, 1864, General Steele started south with his army to co-operate with General Banks on Red River, leaving but a very small force on the line of the Arkansas River. The telegraph line to Fort Smith did not work after the army withdrew south. The guerrillas became so numerous that it was impossible to keep it working west of Lewisburg. Three of my men, Alexander Kane, Jacob Richard and Thomas Jones, were killed by guerrillas while repairing the line near Clarksville. Their bodies were horribly mutilated before life was extinct. Their escort had camped and were surprised, but they all got away, leaving my men behind with a wagon and five horses. The escort was from the First Arkansas cavalry. I continued to send repairers out, but the line was cut as fast as fixed up, so I discontinued all efforts to keep the line up west of Lewisburg, in March, 1864.

April 20, we received news of Steele's occupation of Camden, he having made a demonstration on Washington. The rebels evacuated Camden and he, by forced marches, got in their rear and occupied it. Camden was thoroughly fortified and navigation opened on the Ouachita, on which river it is located; but no boats came to General Steele's succor. After the defeat of General Banks, the rebels fell on Steele in great numbers, capturing his supply trains and forcing him to evacuate Camden, which he did successfully, marching toward Little Rock, closely pursued by the enemy, who overtook him on the south side of the Saline River with fifteen thousand men, Kirby Smith in command. Steele repulsed them severely, capturing three pieces of artillery, two of which were taken by the Second Kansas colored troops. General Solomon's division did most of the fighting, Brigadier General Rice's brigade bearing the brunt of the battle. Rice was wounded and has since died from its effects. On April 30, General Steele's army reached Little Rock.

During May and June, I had great difficulty in keeping the lines up to DuVall's Bluff, Pine Bluff and Lewisburg, but managed to do so by keeping men and teams out constantly with escorts.

From the temporary abandonment of the Fort Smith line until late in the fall of 1864, the situation of the troops at the fort under General Thayer, and of the telegraph party, consisting of W. H. Woodring and J. L. Sears, operators, and James Lane and J. K. Bear, repairers, was quite critical at times, as the enemy concentrated in considerable force in that vicinity and attacked outlying posts. In August, Thayer was so apprehensive that he caused Woodring to put in cipher a message for Steele, advising the latter of the situation. This report was given to two soldiers, dressed as citizens. Avoiding the highways as much as possible by day, they started on their perilous errand. *En route*, they fell in with a citizen, and, from motives of policy, became quite as violent in their expressions of Secession sentiments as he. They remarked upon the condition of the telegraph to this stranger, who had become quite communicative, and he boastingly mentioned the part he had taken in killing the three repairers, Kane, Richard and Jones, who, as we have seen, were most brutally murdered. Not content with their death, the captors cruelly tore out Kane's eyes while yet alive, and otherwise foully dealt with him and the other two. These facts the trusty soldiers well knew, and no sooner had the guerrilla revealed his complicity than they bound him and hung him to a pole with a piece of broken wire, and there left him with a card pinned to his coat, showing why he was hung.

VanBuren office, distant four miles from Fort Smith, was operated during these times by Joseph Hansen, a brave ex-soldier, who had fought in several engagements, to be captured at last by a pair of beautiful eyes that grew in that vicinity. It was his last engagement; since which time he has been out on his parole of honor.

CHAPTER IV.

THE TELEGRAPH IN THE DEPARTMENTS OF THE CUMBERLAND AND OHIO.—RAIDS IN KENTUCKY.—OHIO INVADED.—MIDDLE AND EASTERN TENNESSEE OCCUPIED BY FEDERALS.—CHICKAMAUGA.—MISSIONARY RIDGE.—DEFENSE OF KNOXVILLE.

In the Department of the Cumberland, after the battle of Stone River and before another campaign was initiated, as usually happens when large opposing armies long confront one another, raiding expeditions were freely indulged in; thus, on the part of the Federals, a foraging command, under Colonel John Coburn, numbering nearly three thousand men, moved out of Franklin, Tenn., March 4, 1863, fighting most of its way to near Spring Hill, where it was opposed by Generals VanDorn and Wheeler. After a loss of about two hundred killed or wounded, Coburn and three regiments were captured. On the eighteenth, Colonel A. S. Hall advanced from Murfreesboro, to intercept General John H. Morgan's command, resulting in a fight of several hours at Milton, when Morgan retreated, leaving four hundred men on the field to fall into Federal hands. Less than two weeks after, General D. S. Stanley drove Morgan to McMinnville, inflicting considerable loss.

In April, Colonel A. D. Streight moved up the Tennessee to Eastport, to operate in conjunction with General Dodge from Corinth, upon the enemy's communications, stores and foundries in Northern Alabama and Georgia, but he seems, strangely enough, to have omitted taking an operator with him. Having separated from Dodge, after a running fight of several days, he surrendered to General Forrest his command of seventeen hundred men, then in the neighborhood of Rome, Ga. Expeditions under Generals Reynolds and Stanley, and Colonel Watkins and others, resulted advantageously to the Unionists.

On the part of the Confederates, late in January, Wheeler, Forrest and Wharton, passing through Rosecrans' army, *via*

Triune, and escaping the Federal cavalry and infantry with a loss of three hundred and fifty men, attempted, on the 3d of February, the capture of Fort Donelson, defended by the Eighty-third Illinois—Colonel A. C. Harding—which beat off the enemy's repeated assaults, with, it is said, the enormous loss to the assailants of eight hundred killed or wounded and one hundred captured; being a total of two hundred more than were defending the fort. The Union loss was eighty-three killed, wounded and missing. Peter Fowler or Wm. A. Thayer operated there at that time. Operator J. A. Fuller chanced to be on a steamer, about this time, which Wheeler captured at the shoals near by. Until Chaplain Gaddis, also on board, prevailed on Wheeler to let the boat go, Fuller, who took to his bed when the boat was over-hauled, was to all intents and purposes sick unto death—*in extremis*—but as soon as he heard of Gaddis' success, he was as healthy as any. The ruse was a good one, for it saved him from being paroled.

He went immediately to Brentwood, Tenn., nine miles from Nashville, where, on the 25th of March, he was awakened early by a negro who told him that Forrest was coming, and the Yankees were fixing for a fight. Fuller immediately telegraphed the facts to Nashville, and while the skirmishers were deploying, threw his instrument under the house and started on a run for the city, knowing full well that the force at Brentwood could not defend the place. Colonel Trueblood, commanding there, soon surrendered, and Fuller, from his hiding-place in the woods, saw captors and prisoners pass on the pike. After that, he was about to cross an open field near by, when happily he discovered a body of the enemy preparing to charge across the same field to the woods beyond, where a large number of negroes were cutting wood for the Government. Consequently he changed his course to the Granny White pike, and thus reached Nashville at the head of the frightened contrabands.

Rosecrans, with the advantage of high water in the Cumberland, besides his railroad communications, was nearly six months preparing for another campaign against Bragg, notwithstanding he was repeatedly urged, if not commanded, to advance, lest pending his inaction, General Johnston, Bragg's superior officer, should detach by rail a sufficient force to compel Grant to loosen his hold on Vicksburg. Bruner and Mullarkey having left Rosecrans' head-quarters, operators John C. Holdridge, late of Western Tennessee, Jesse H. Bunnell, recently from the Department of the Potomac, and Robert M. Talbot, who had served under Generals McDowell, Burnside, Wright, and others, took their places. The telegraph had long been busy with orders for supplies. VanDuzer collected a large quantity of telegraph stores, including about one hundred miles of wire, for the grand movement which was ordered on the 23d of June, 1863. The army advanced, feigning a direct attack, but in fact striking at Bragg's flanks, particularly his right, causing him to retire on the twenty-ninth, and cross the Tennessee River soon after.

Thus Middle Tennessee and so much of Northern Alabama as lies north of the river, was again in the virtual possession of the Unionists, who now bent their energies to effect the restoration of the railroads and telegraphs as indispensable to further progress. VanDuzer reported his operations as follows:

> The line kept pace with the advance of the army from Murfreesboro, and my party entered Tullahoma and opened an office at General Rosecrans' head-quarters three hours after the establishment of such head-quarters, having opened offices at Fosterville and Wartrace. On the fourth, the line was extended to Elk River and an office opened at the camp of the Pioneer Brigade; on the fifth, to Decherd, and on the seventh to the head-quarters of the Twentieth Army Corps, at Winchester. No further progress was attempted southward on this line during the month, but a branch line was rebuilt to McMinnville from Tullahoma, connecting the left wing of the army and the head-quarters of the Twenty-first Army Corps with general head-quarters, which, towards the end of the month, were moved to Winchester, and a line was rebuilt along the turnpike from Franklin to Columbia, Tenn. Early in August, I constructed a new line from Cowan to Tracy City, Tenn., and extended the line south along the Chattanooga Railroad to Bridgeport, Ala., opening offices at Anderson and Stevenson; at department head-quarters, near Stevenson; at head-quarters Fourteenth Army Corps, near Bolivar, Ala., and at Bridgeport, Ala. At the same time, the line was extended south from Columbia to Pulaski, Tenn., and west from Stevenson, Ala., to Brownboro, or more properly to head-quarters Second Cavalry Division, on Flint River, with offices at Scottsboro and Larkensville, Ala.

Thus the lines of telegraph abandoned by Buell one year previous, were mostly restored. Many of the operators who served in this section under Buell were replaced in their former positions, or otherwise located at railroad stations or other points. Among those thus employed were J. T. Joyce, A. M. Nichols, R. B. Hoover, Geo. C. Peirce, John P. Lathrop, T. E. Rawlings, Geo. Railton, John Lonergan, E. W. Atwater, W. W. Burhans, Hugh Craig, Robert Wagner, J. C. Gregg, W. H. Hartman, W. R. Plum, F. S. VanValkenburg, W. H. Miller, Wm. Patterson, E. J. Wilson, J. T. Rabbeth, I. C. Showerman, Ed. Schermerhorn, M. C. Baldwin, Con Dwyer, S. Ford Perdue, H. R. Mapes, Frank B. Tyler and H. W. Plum, the last five of whom were serving in Nashville. On the line leading to Louisville were J. N. Brooks, Sam Barth, Martin Barth, J. A. Cassell, A. C. and J. Jones, W. S. Pierson, C. H. Griffith, James Forker, E. W. Atwater, John D. Richardson, and others. Geo. W. Bell was at the Glasgow end of a new line, built from Cave City, in June, to watch such raiders as Morgan.

From the outbreak of the rebellion, the Federal Government had striven to possess and hold East Tennessee. Never was such possession so important to the loyalists there residing, as since the sweeping Confederate conscription acts, which dragged the Union citizen into the armies, or drove him into the mountain fastnesses, where he was hunted by armed men. General Burnside recently from the Army of the Potomac, with a view to the repossession of the district of East Tennessee, vacated by Federal General Morgan, was, in March, 1863, given the command of the Department of the Ohio, with head-quarters at Cincinnati, O. Charles Jacques managed his telegraph office there, assisted by two others. The military line, from Burnside's office by way of Lexington, connecting Nicholasville, Ky., was extended so as to place Burnside in immediate communication with his main forces and out-posts, rendering surprise in force at Lexington and Frankfort impossible.

As Chattanooga was the objective of the Army of the Cumberland, so Knoxville, one hundred and twelve miles north-east of Chattanooga, was that of the Army of the Ohio, which was organizing to advance simultaneously and in co-operation with

Rosecrans' command. While the authorities were fearing that Rosecrans would not prevent another invasion of Kentucky in force, Confederate General Pegram, late in March, 1863, entered that State through the Cumberland Mountains, ostentatiously proclaiming that Breckenridge was following his (Pegram's) cavalry with infantry, for the redemption of the State. General Carter was in immediate command of the Federals, whose advance guard was at Crab Orchard, where John Lonergan was operating. Twenty-two soldiers of the First Kentucky and Seventh Ohio Cavalry and First and Second Tennessee Mounted Infantry were posted there, to keep up communication with General Carter's command in the direction of Somerset, General Welch's stationed at Liberty, Colonel Gilbert's at Mt. Vernon, and Colonels Alexander and Bird's at Lancaster. This was the general location when Pegram entered Crab Orchard about daylight, *en route*, as he said, to rescue Kentucky from her oppressors. Lonergan, in his night clothes, narrowly escaped capture by hiding under a chicken coop in the rear of Carson's Hotel. A negro woman rolled a log of firewood in front of the coop, completely hiding the man in white from the woman in black, and Pegram in particular. Pegram pressed forward upon Danville, where operator Benner was exposed to his musketry fire, and from whence Carter was driven across the Kentucky River, by which time it was discovered that the raiders were unsupported, and were in fact driving a larger force than they possessed. So the tide changed.

Burnside, in turn, in June, sent a cavalry force into East Tennessee, which greatly injured the railroad, and captured many prisoners and arms.

About the same time that Rosecrans advanced from Murfreesboro upon Tullahoma, General J. H. Morgan, with a view, at least in part, of so engaging the attention of Burnside as to prevent his threatening Bragg's rear by an advance into East Tennessee, essayed from Sparta the most daring movement yet undertaken. Crossing the Cumberland near Burksville, July 1, with a mounted force of over two thousand men and four guns, he was defeated at a Green River bridge, which he attempted to capture. Undaunted by this failure, Morgan pushed on in the direction of Lebanon, Ky. His operator, Ellsworth, failing to deceive the telegraphers on the Stanford line, took a small command over to the Lebanon branch railroad, near St. Mary's, and by cutting off Lebanon, succeeded in deceiving operator E. W. Atwater, at the Junction. Circumstances conspired against Atwater, for the night previous, Bennett, the night operator at Lebanon had invited him to spend the Fourth in that place, and as the latter had just received a bran-new and showy uniform, which he was anxious to exhibit, he resolved to take the 8:30, A. M., train for Lebanon. About ten, P. M., the line ceased to work. It seems that Morgan had taken the precaution, before appearing in front of Lebanon, to cut this wire. About six, A. M., of the fourth, Ellsworth connected his instrument with the main line, which he ran to the ground by his side, when the following conversation took place. "Z" was the call of Lebanon Junction; "B," that of Lebanon.

"Z"—"Good morning, Bennett; where have you been all night? 18." (Meaning, what is the matter.)

"B"—"Been on duty all night; line O. K., S. E. of here; noticed it open between us; seems to be O. K. now."

"Z"—"Any news about guerrillas?"

"B"—"There was a report of some in the country yesterday, but everything quiet here."

"Z"—"I will be there about noon."

"B"—"That's right; will have something set up for you."

At 8:30, the Lebanon train arrived at the Junction from Louisville, with about thirty passengers, of whom five were women. The conductor asked Atwater if he had heard of guerrillas on the branch, adding that a message *via* Danville and Lexington stated that guerrillas had burned a bridge near Lebanon. But the operator satisfied the conductor that all was well, and boarding the train Atwater and all started. At New Hope, they learned that a culvert near St. Marys, five miles from Lebanon, was burned and the track torn up. Thirteen soldiers boarded the train here as a guard. It proceeded, and notwithstanding the engineer was cautioned, the engine and several cars were thrown from the track in a cut just beyond St. Marys. Immediately, about thirty of Morgan's men fired into the train from the right, killing a Union soldier and wounding an Irish passenger in the shoulder. The latter rushed into the baggage car in wildest excitement, exclaiming, "My God! My God! We are all kilt, shure. Ivery mither's son uv us will be murthered till we're dead. Howly Mither, have mercy on us." The guards leaped from the car windows on the left side and fought under cover of the train until the enemy decamped; but they soon halted, as if to return to the train.

At this stage of affairs, Atwater was sent for aid and had proceeded but a short distance when he was approached by three mounted men, who fired at him until he waved a copy of the Louisville *Journal* in token of surrender, when the following conversation occurred:

Ellsworth.—"Where did you get on the train and where are you going?"

Atwater.—"At the Junction; going to spend the Fourth with a friend in Lebanon."

E.—"What is your name?"

A.—"Atwater."

E.—"You are an operator at 'Z' and I was talking with you over the line this morning, having cut the wire and connected this instrument. I found out what I wished and I am the one who invited you to Lebanon. I am Ellsworth, John Morgan's operator."

The reader is left to imagine Atwater's confusion when Ellsworth informed him how completely he had been deceived, but mopping the perspiration from his face, Atwater replied, "I congratulate myself, my dear friend, that I have fallen into such good hands. If I must be a prisoner, let my captor be an operator; and now let us have those cool drinks." "All right," said Ellsworth, as he drew a revolver upon Atwater, "give me your revolver and holster and follow me." It was of no use for Atwater to protest that he was only an operator. His new uniform, Ellsworth insisted, made him a fit subject for Libby prison. But a company of Union cavalry that left Lebanon on a scout before the attack, had heard the affair at St. Marys and were now seen in the distance, hurrying to participate; consequently, Ellsworth and his comrades rapidly rode off.

Learning from three prisoners captured that the enemy was a detachment from Morgan's command and that he was moving on Lebanon, Atwater, by tapping one end of the line on a ground connection, telegraphed to General Boyle at Louisville an account of the brush at St. Marys and the news of Morgan's near presence.

When Morgan appeared, on the fifth, the Federals at Lebanon were measurably well prepared to receive him. Hanson's brave four hundred fought seven hours before surrendering. George Purdon and —— Bennett made good their escape, saving their instruments. It was an exciting day for the people at Lebanon. Plunder, pillage, murder and insults continued until the setting of the sun. At dusk, Morgan's command began moving upon Bardstown. Bennett, who had secured a refuge for the night in a muddy trench just out of Lebanon, connected his instrument and telegraphed direct to General Boyle what had occurred and was transpiring, the Louisville line being connected at "Z" for that purpose.

After capturing Bardstown, where T. H. Smith operated, Morgan struck the Louisville & Nashville Railroad, capturing James Forker, operator at Bardstown Junction. Ellsworth interviewed Forker at his office in the following energetic style: "Hello, sonny," said Ellsworth, handling a cocked revolver, "move one inch except as I direct, and you are a dead man." At this time a passenger train left "Z" for Louisville and was duly reported. Morgan captured the stockade at Salt River bridge, six miles north of "Z," and stopped the train with his cannon. The doors of the coaches being guarded, the passengers were regularly allowed to surrender their money, watches, jewelry and baggage, after which the conductor was permitted to return with his train to "Z." Operator Lenhart, in the railroad superintendent's office in Louisville, asked Forker, "Has the train passed north yet?" "Tell him yes," said Ellsworth, which Forker was obliged to do. He was then mounted on a mule and sent to camp. The following evening he appeared at "Z" without hat, coat or boots. Next day the line was restored.

Morgan moved on Brandenburg, where he captured two steamboats, by which he ferried over the Ohio his forces, now much increased, and after burning one of the steamers, he moved northwardly to Salem, Indiana. By this time the civil and military authorities were doing their utmost to effect his capture. From Salem, Morgan went north-east *via* Vernon to Harrison, on the Ohio State line, and around Cincinnati, moving quite near there, to Miami; thence easterly through Batavia, Sardinia, Jasper and Piketon to the Ohio River at Pomeroy; thence eight miles up the river to Buffington's Ford, where eight hundred of his men were captured. Over three hundred swam the river at Belleville and escaped; over a thousand surrendered there; but Morgan and perhaps two hundred others, pushing on north-eastwardly, struck the river again above Marietta; thence they moved to New Lisbon, where they surrendered, July 26.

Soon after crossing the Ohio, Morgan captured a passenger train, and among those robbed was Patrick Mullarkey, a Federal operator, on his way to Grant's department.

The Western Union and railroad telegraph operators in Southern Indiana and Ohio certainly deserve a good share of the credit which was accorded to the troops under Hobson, Shackleford and Judah, and to the gunboats that patrolled the Ohio, for the capture of Morgan and his command, for by no other means could the enemy have been so harassed on all sides, as they were, by concerted action of the regular troops and militia; nor could the gunboats have been kept so well advised of Morgan's movements Of course Morgan destroyed the telegraph wherever met, but cutting the lines as he advanced did not greatily interfere with telegraphing to towns ahead of him by circuitous routes, of which there were many.

In June, a line was built by Lehr from Lexington to Mount Sterling, Kentucky. Samuel B. Roberts operated there until 1864.

On the 16th of August, General Burnside began his great march upon Knoxville, having a force in hand of about twenty thousand. The route of his main army was from Crab Orchard through Mount Vernon, London, Williamsburg, Chitwood, Montgomery, Kingston, to Knoxville. Charles W. Jacques, who came with the General from the East, and C. H. Johns, accompanied Burnside as cipher and telegraph operators, while Lehr, chief operator in the district, advanced by the old route, opening an office every night at Burnside's head-quarters, until Barboursville was reached, when he accompanied the General to the main army at Williamsburg. But the builders under W. L. Tidd advanced the line with De Courcy's forces directly upon Cumberland Gap. Lehr struck the rebel East Tennessee line at Loudon, capturing the operator there and finding everything in good working order, also many useful Confederate telegrams. Jacques took temporary charge of the office, and Lehr, going with the Federal cavalry, captured the operator and instruments at Campbell Station, when he pressed rapidly on with the Fifth Indiana cavalry, reaching Knoxville unannounced late in the afternoon of the 1st of September. Hastening to the city telegraph office, he entered the same, just as the two operators there retreated through a back door and escaped, without having done the slightest injury to any telegraphic apparatus or material. Lehr found in the office a large number of important messages, and just outside a good main battery. Communication was opened immediately with Burnside at Loudon. Johns, who also came with the cavalry, was left to operate the Knoxville office, and Lehr and the troopers advanced up the valley of the Tennessee, past Strawberry Plains and Newmarket to Morristown, where the operator was captured in his office, and the superintendent of this line in the town. Pressing on again, through Bull's Gap to Greenville and Jonesboro, the cavalry drove General Sam Jones into Virginia. The operators at these places, like the one at Morristown, were captured and pressed into the Federal telegraph service; George E. Jones being one and perhaps Sam Reese another. At each of these offices, a Union lieutenant was left in attendance and two soldiers on guard, the former provided with a cipher key, under the cover of which all messages were transmitted. And in this way one hundred and twenty-six miles of telegraph were opened by Lehr *within five days after striking the Loudon bridge;* a feat not equalled even by Crittenton's remarkable operations in 1862, under Mitchell, in Northern Alabama. General Burnside, who, on every occasion, manifested the highest regard for the telegraph service, was greatly pleased at this result.

From Morristown Burnside turned his mounted force toward Cumberland Gap, which he now sought to reduce. On this route was found a poorly constructed line leading to the Gap, where General Frazier was in command of about two thousand Confederates and fourteen guns. Lehr, with fifty cavalrymen, repaired this wire in Burnside's wake. That General sent couriers around the Gap to De Courcy, and learned of his arrival and that he had artillery in position to begin shelling. Frazier, soon after, finding himself effectually cut off from all aid, surrendered to Burnside on the ninth, at noon. The telegraph was connected across the mountain on the same day, making nearly two hundred miles of Federal telegraph built or recovered in East Tennessee in nine days, and placing Burnside not only in communication with his own exposed posts, but with the authorities in Washington and with Rosecrans, who was now again marching against Bragg, as we shall soon discover. A few days later, Burnside returned to Knoxville and had the pleasure of a telegraphic confab with President Lincoln, the General-in Chief Halleck, and Secretary Stanton, who were in the War office at Washington. During this interesting talk, Burnside was congratulated upon his great success. In these telegraphic operations, Burnside's chief of staff, General Hartsuff, an old-time operator, rendered valuable aid.

This was the condition of affairs when, September 10, William L. Gross, late of Fuller's department, was appointed by Colonel Stager, "Assistant Superintendent of the United States Military Telegraphs for that portion of the Department of the Ohio embracing Central and East Kentucky and East Tennessee." This new position had been tendered to E. B. Gorton, of Cincinnati, but was declined by him on the ground that he was satisfied where he was. Lehr, chief operator, resigned, and Captain Bruch's brother, Adam, was appointed in his stead, with head-quarters in Knoxville, Gross establishing his at Danville. Gross received a commission as captain and assistant quartermaster, October 27, 1863. We have already noted his service within Grant's department, beginning as manager of Cairo office, August 12, 1862. Discovering his disposition to work and executive qualifications, VanDuzer at first, then Bruch and afterward Fuller added to his labors until it became necessary to remove him entirely from the key, of which he was but an indifferent manipulator. Fuller made him chief clerk, in which position he exercised much discretion regarding the management of telegraphic affairs in that department. How successfully he performed his part may be gathered from his appointment as captain, and how harmoniously, from the presentation of a gold watch by his co-laborers in Memphis, as related in the next chapter. Captain Gross's appointment was solely the result of meritorious service in the telegraph department — mainly clerical and executive in character. Doubtless his study of the law, but especially his contact with others, while teaching school and practising law, gave him a degree of confidence better calculated to inspire others with respect for his abilities than those enjoyed who had not been equally schooled by the friction of mind with mind.

WILLIAM L. GROSS.

WILLIAM L. GROSS was born in Herkimer County, New York, February 21, 1839. His father, a minister, moved to Illinois in 1844, and in 1848 settled on a farm in Knox Co. After a public schooling and an academic course, William, at seventeen, prosecuted his law studies while teaching. Afterward, he learned to telegraph, and took charge of the railroad telegraph office at Dwight, Illinois, but not with a purpose of continuing in that business. In 1862, at Springfield, Illinois, he formed a partnership with his brother, Eugene, who subsequently acquired a splendid reputation at the Bar, but soon after, the war fever overcame William's civic ambitions. Having obtained official permission to raise an infantry company, he labored night and day in Dwight and vicinity, making private and public appeals for recruits, with marked success. But a strife arose as to commissions, and Gross, not to compromise himself in an unseemly struggle for rank, accepted the position of manager of Cairo military office, then proffered him by his friend, John C. VanDuzer. He is now a leading lawyer at the capital of Illinois.

Nothing unnoticed of special interest occurred in this district up to the middle of November, if we except a raid of seventy guerrillas into Danville, resulting in scaring a good many, routing Benner out of bed unceremoniously and disarranging things in his office. But the guerrillas were careful not to destroy any thing. It was the poorest conducted raid on record. Two hours after the guerrillas left, Benner had one thousand cavalry there from Camp Nelson. Perhaps it should be stated that Burnside became greatly angered at Superintendent Gross, because the line was down so much, but he insisted that it resulted from army teamsters breaking down the poles, and Captain Bruch represented that "this was no new thing. They (teamsters) have actually given us more trouble than the rebels. They break down poles and cut up the wire and use it for mending broken wagons; make lines to hang clothes on; tie up fodder and make pot-hooks, besides numerous other things. I *know* this, for I have seen it myself, and made complaints to generals in command, but have never seen any attempt made to stop it."

The following shows the lines, stations and operators in Gross's department at this time ·

	MILES.		MILES.
Lexington to Mount Sterling, *via* Winchester	23	Lexington, *via* Nicholasville and Camp Nelson, to Danville	35
Lebanon to Columbia	40	Kingston to Loudon (about)	15
Lebanon, *via* Danville, Stanford, Crab Orchard, Mount Vernon, London, Barboursville, Cumberland Gap, and Tazewell, to Knoxville	216	Clinton to Knoxville	15
		Loudon, *via* Lenoirs, Knoxville, Morristown, and Greenville, to Jonesboro	126
		Stanford to Somerset	25
			495

These lines were operated by Adam Bruch, C. W. Jacques, C. H. Johns, Joe Anderson, J. R. Brown, Geo. Cole, John T. Wilber, Geo. J. Goalding, Geo. Purdon, C. H. Pond, A. W. Howell, F. C. Cook, Sam B. Roberts, W. H. Wilson, Thos. M. Sampson, T. G. Harvey, Jos. H. Gilkerson, W. H. Sisson, Sam Carey, J. S. Lyle, J. M. Spencer, Chas. A. Keefer, J. E. Nagle, G. D. Wilkinson, J. A. Torrence, S. W. Scott, E. T. Chapman and Thos. H. Smith. Sam Reese and Geo. E. Jones, Tennessee operators, who had convinced Lehr of their loyalty, were also employed on these lines.

As Burnside's movements were designed to be in concert with Rosecrans, it has seemed best to disclose their operations in the same chapter. We will, therefore, again turn attention to Tennessee and Northern Alabama. Having repaired his railroads and awaited the ripening of corn, Rosecrans advanced from various points, on the 16th of August, crossing the Tennessee at Stevenson, Bridgeport, Shell Mound and Capertons, and after shelling Chattanooga from Waldron's Ridge, he crossed a force at Harrison. Those crossing below at once, and the others subsequently, converged on Trenton, or vicinity, from whence McCook's corps pushed on through the mountains to Alpine, near Rome, Ga. Bragg evacuated Chattanooga to confront Rosecrans, and determine the issue in battle. The Union General, greatly deceived by Bragg's movements, imperiled his own army by stringing it out until the extreme ends were forty miles apart. Gen. Bragg, instead of retreating, was massing to crush Rosecrans in detail, which it is highly probable he might have done, had his orders been better executed. While Bragg was concentrating his whole force about La Fayette, nearly opposite Rosecrans' center, the latter, who had made Chattanooga his head-quarters, learned with amazement that the enemy was actually preparing for a general assault. This was about the 11th of September, and from that time to the 18th, the Union army may fairly be said to have been at the mercy of the Confederates, who were greatly strengthened by veteran forces from Virginia, Georgia and Mississippi, besides Buckner's corps from East Tennessee.

Instead of Burnside co-operating with Rosecrans, he had done him a positive injury by driving Buckner from East Tennessee and furnishing Rosecrans no equivalent.

Concerning telegraphic operations conducted under Captain VanDuzer's superintendency since the advance, he officially reported:

Early in September, the army crossed the Tennessee River, and I pushed the construction of a line along the railroad to Wauhatchie, at which point Brigadier General Wood, with a division of the Twenty-first Army Corps, threatened Chattanooga by Lookout Pass. Opening an office at the Running Water bridge, and opening a branch line through Murphy's Hollow to Deer Head Cove, with an office at Department head-quarters, at or near Trenton, Ga. On the ninth, this branch line was taken up, and the enemy having withdrawn to La Fayette, we pushed into Chattanooga and opened an office in the rooms just vacated by the rebel telegraphers.

Before quoting further, we will notice some incidental matters. By the Captain's orders, George C. Peirce had opened an office at Shell Mound, where Reynolds' division was crossing the river, and hearing from some of Wilder's cavalrymen that Bragg had evacuated Chattanooga, Rosecrans, September 5, ordered Reynolds to move to Running Water, keeping up telegraphic communication. Peirce was accordingly directed to take four men and necessary supplies, and proceed along the railroad, repairing the lines. He started from Shell Mound about two o'clock in the afternoon of that day; on a hand-car, with the men. They found little to do as the line was in good order. The army was marching along the highway, parallel with the railroad, but somewhat obscured from view by the trees and hills. Peirce proceeded much faster than the troops, as it subsequently appeared. Meeting citizens along the track, he gathered information which seemed to confirm previous reports of the *non*-existence of any rebel force that side of Chattanooga at least, and in fancied security the party moved along, and in due time reached a point within a short distance of Running Water bridge. Leaving the men to unload the car, Peirce proceeded along the track and stood upon one of the piers of the bridge, the superstructure of which had been burned by the rebels. He had hardly surveyed the splendid scenery about him, when he was startled by hearing a voice exclaim: "What you all doing up there" Turning to his right, not thirty feet distant, he saw a rebel cavalryman, with his carbine leveled at him. A glance showed that resistance would be of no use, as four more Confederates also put in an appearance. He replied: "I was just looking around." "Come down from there," was the next thing he heard. He complied promptly, and was placed upon one of their horses and told to hang on to its rider for life. The squad having been out on a scout had arrived at the bridge at the same time Peirce did. During the ride he exchanged clothing with them and surrendered about everything of value he had, except twelve dollars in currency. By the time he reached the enemy's picket line, to all appearances he was as much of a rebel as any of them. They went direct to Chattanooga, which was *not* evacuated. Peirce was taken before General Bragg, who questioned him for more than an hour, trying to ascertain the strength of the Federal forces and location of their supplies. The operator professed such profound ignorance upon those subjects that about midnight he was sent to the guard-house; supposing he would go directly through the Confederacy, *via* Richmond, and in two or three weeks be back again. In this he was mistaken. From Chattanooga he was sent to Atlanta; from there direct to Richmond, where he was confined on Belle Isle until the following March, the seventh, when he was exchanged.

When the Murphrys Hollow line to Trenton was abandoned, on the ninth, Lonergan and Forbes were left without transportation at the signal station on Sand Mountain, where they had been forwarding the reports received from six connecting signal stations, which were variously located with the army. Forbes having manufactured a telegraphic order from General Thomas on Colonel Iredell at Bridgeport, for an ambulance for Lonergan, whose sickness was as genuine as the order which was duly honored, the two proceeded in the direction of Trenton, visiting houses by the way for something to eat, and realizing as they progressed that an army had preceded them. Stopping at a house late in the evening, a paroled Vicksburg Confederate officer startled them with the information, the truth of which the author doubts, that a large body of rebels was just ahead, and, at the suggestion of the officer, they turned up a hog-path, when their trials began, increasing as they advanced along the mountain and into the entanglements of the by-way. Darkness set in and hunger increased, but they pressed on as best they could, putting their shoulder to the wheels when the mule was stalled. About one A. M., they were halted by a vidette guard of the Fifteenth Pennsylvania cavalry under Lieutenant Kelley, who had nothing for them to eat. At daylight, Forbes shot and roasted a pig, after which the operators rode into Chattanooga with Wood's division.

VanDuzer further reported:

> On the 17th of September, Department head-quarters being near Crawfish Springs, I received an order from General Rosecrans to connect him by telegraph with Chattanooga, which order I obeyed, completing the line after the opening of the action on the morning of the nineteenth. Offices were opened at General Granger's head-quarters at Rossville; at a point in the rear of General Thomas' head-quarters and at Department Head-quarters, which latter office was moved at about noon to the house owned by Mrs. Glenn, to which point General Rosecrans had removed his head-quarters, where it was maintained during the day and until six o'clock A. M., of the twentieth. From six until nine, A. M., of the twentieth, the office nearest the front was the one in the Dry Valley, in General Thomas' rear, and Department Head-quarters office was only reopened for a few minutes before the driving in of the right of the line forced us back upon the road to Rossville, and no success attended my efforts to reopen farther to the front than the Dry Valley office before mentioned. This line to Rossville and the office at Rossville was kept open until the withdrawal of our line on Monday night (twenty-first and twenty-second) to the intrenchments of Chattanooga. During these two days (nineteenth and twentieth), my men, operators, builders and repairers, were often under fire, nearly all of them for the first time, and I take pleasure in declaring their uniform coolness and good behavior. I am especially indebted to Messrs. William Patterson, John C. Holdridge J. H. Bunnell, J. A. Fuller, W. W. Forbes and W. H. Miller for active, intelligent and fearless co-operation and support.

When the operators were driven from widow Glenn's, where they had rendered invaluable service, they opened an office on a stump at the entrance to Dry Valley. After the Union right was driven beyond the stump office, Captain VanDuzer, who, besides his telegraph duties, served Rosecrans as aide, directed Holdridge to open an office on the Rossville road. He had just succeeded in cutting the wire for that purpose when a brigade of infantry formed in line of battle immediately in front, and opened fire on the advancing enemy; bullets were flying as numerous as hailstones in a cold storm. Bunnell called out for a ground wire and Holdridge cried, "Give me a ground wire; who's got a bayonet for a ground? I'll go find one," and off he started. A few moments later, the brigade gave way and John was swept on to Rossville, whence he probably accompanied Rosecrans to Chattanooga. Two days after, Bunnell met him at the office door in Chattanooga and asked, "Have you found that bayonet yet?"

Forbes and Fuller reported to General Granger at Rossville and opened an office in a tent. Granger directed the boys to mess with his orderlies, which they refused to do, as they thought the General should treat them as well as he would an officer. Shortly after the battle began, Patterson, chief of the field construction party, wanted a trophy, and induced Fuller to go with him to the battle-field to pick up a revolver. Leaving Forbes to attend office, they started on their foolish errand, but unexpectedly soon found themselves among Union soldiers in battle array and near a battery just going into action. About the same time, other Federals were seen coming in full retreat. A general confusion ensued, and the operators suddenly concluded that they did not want any trophies. This was the beginning of the rout of the right wing of the Union army on the twentieth, and as Fuller was well mounted, he reached his office ahead of the crowd and telegraphed Chattanooga his startling news. As division after division passed his office, reporting that the rebels were just behind, the operators were advised to fall in line for Chattanooga, but they would not go without proper orders. Night came at last, and theirs was the only tent standing. Telegrams to and from Rosecrans, Garfield, Steedman, Granger and others—general officers—were being sent or received nearly all that night.

The sanguinary conflict around widow Glenn's house and near the Chickamauga, beginning on the morning of the ninteenth and ending with the next day, wherein about fifty-six thousand Federals fought an equal number all day on the ninteenth—and about two-thirds of the national troops contended, at least from noon of the twentieth, with Bragg's army, reinforced the night before by Longstreet's corps from Virginia—was maintained on the part of the Confederates with gallantry unsurpassed; but the Fourteenth Corps (Thomas') and the few brigades from Crittenden's and McCook's corps, that withstood the assaults of the victorious hosts that charged again and again, ever moving to the Federal right until Thomas' command was almost entirely enveloped and the Rossville road nearly gained, just as Granger comes to the rescue, barely in time to save the army from disintegration ; such heroism, such gallantry ! oh, how grand ! Had McCook and Crittenden stopped at Rossville ; had Rosecrans telegraphed his orders therefrom to Chattanooga, instead of going in person to attend to pontoons, and had these three officers reformed the shattered and retreating (four) divisions behind Gordon Granger's and fallen upon the enemy, as Longstreet had long feared would be done, how different might the issue have been !

The Federal loss in the two days' battle was about 2,000 killed and 9,412 wounded, or 16,336 killed, wounded and missing, besides fifteen thousand small arms and thirty-six guns, while the Confederates suffered the frightful loss of two-fifths of their army of whom only about two thousand were prisoners.

At seven, A. M., September 22, Rosecrans' army was in position at Chattanooga. The night before, Captain VanDuzer, at the head of his corps of operators and linemen, worked unceasingly in running trip-wires just above the ground and outside of the Federal defenses around Chattanooga, which were intended to throw the enemy charging the Unionists, and thereby confuse their ranks and render the charges abortive.

On the twenty-fourth, the Confederates took possession of the south bank of the Tennessee River and of Lookout Mountain, from which shells were fired into the Union camps. They were also in possession of the railroad bed, east of Bridgeport, and all communication by water, by rail and telegraph, with that point, was cut off. Rosecrans' beaten army lay besieged, with mountains and ridges in front ; river, ridges and mountains in rear ; three thousand wounded, suffering or dying for want of medicines and nourishment ; the army itself on short allowance, with scarce ammunition enough for a day's battle, and only one practicable roadway open for supplies, and that sixty miles long, precipitous, circuitous, wild and forbidding, by way of Jasper to Bridgeport, to which latter place thousands of horses were sent, leaving ten thousand other animals in and about Chattanooga, that died of starvation.

From the twenty-fourth to October 1, VanDuzer was preparing to construct a telegraph in part over this cruel road—the route adopted being from Bridgeport, *via* Battle Creek, Jasper, Pryor's and Haley Trace, over Waldron's ridge to the river road at the Suck, and thence to Chattanooga, in all forty-two miles. Operator N. S. Townsend and a company of the Sixth Ohio (sharpshooters) deserve great credit for their laborious and hazardous work on this line, the completion of which, wrote Van Duzer officially, was delayed until the 17th of October, on account of the passage of wagon trains, laden with subsistence, and the loss of his outfit and dispersion of his builders by Wheeler's raid through the Sequatchie Valley, and by the presence of rebel sharpshooters on the south bank of the river at various points where the line ran necessarily close to the north bank. For nearly four miles, between the foot of Williams' Island and the Suck, the whole distance was under fire of rebel sharpshooters, armed with Mississippi rifles, themselves completely covered by rocks and trees, and their average distance from the road at the foot of the cliff along which the line ran, less than four hundred yards. This part of the line VanDuzer was compelled to build and repair in the night, and so watchful was the enemy that the stroke of axe or blow of hatchet would often bring a volley from a dozen rifles, well directed and dangerous.

This condition of things continued until the seizure of the south bank, at Brown's Ferry, by the forces under command of Brigadier General W. F. Smith, and the occupation of the Lookout Valley by the Eleventh and Twelfth Army Corps. In building and maintaining this line, the Captain reported himself as much indebted to John C. Gregg, operator, who had lately joined him from the Army of the Potomac, and who volunteered to go over and repair the line when every builder and repairer, enlisted man and civilian employé, alike refused to go, deterred by the danger to be incurred. During the month of November, the line from Bridgeport to the Running Water bridge was rebuilt, and a new line constructed over Raccoon Mountain to Kelly's Ferry, where an office was opened; thence to General

Hooker's head-quarters in the Lookout Valley, and thence, *via* Brown's Ferry, to Chattanooga, abandoning the wire on the north shore of the river. During all of November, a party was kept at work putting a second wire upon the poles between Nashville and Bridgeport, and immediately after the actions on Lookout Mountain and Mission Ridge, soon to be noticed, both the lines were put up along the railroad to Chattanooga, and that over the mountain, *via* Kelly's Ferry, was abandoned.

About the 1st of October, the Eleventh and Twelfth Army Corps, from the Army of the Potomac, arrived, and were located mainly along the railroad from Bridgeport to Nashville.

Pursuant to instructions, General Grant arrived at Cairo, October 16, and reported by telegraph to Secretary Stanton, who immediately replied, "Meet me at Louisville." The Secretary at the same time directed Colonel Stager to accompany him, and gave the Colonel the following *carte blanche* order to arrange for the meeting:

WAR DEPARTMENT, WASHINGTON,
ORDERED: October 16, 1863.

That all Railroad Companies be and are hereby authorized and directed to furnish a special train to Colonel Stager, Superintendent of Military Telegraphs, and give him the utmost dispatch to the points designated by him. (Signed) EDWIN M. STANTON,
Secretary of War.

So efficiently was this order executed, that Stanton and Stager met Grant and his staff, including operator Beckwith, at Indianapolis, Ind., from whence they traveled to Louisville, and after about two hours additional consultation in the Gault House, Beckwith was given (nineteenth) two despatches for transmission, one of which telegraphed, by operator John J. Flanagan, relieved Rosecrans of his command, and the other appointed Geo. H. Thomas in his stead. At the same time, the Departments of the Tennessee, Cumberland and Ohio, over one thousand, three hundred miles in extent, were consolidated into the Military Division of the Mississippi, under the command of General Grant. The next message, dated at 11:30, P. M., of the same day, was no less important, and was, in part:

Hold Chattanooga at all hazards; I will be there as soon as possible.

To which brave Thomas answered, before Grant retired that night:

> I will hold the town till we starve.

It was because Bragg fully expected to starve the army that he did not attack it. Grant went to Chattanooga, says General Badeau,

> Telegraphing to Halleck and to Sherman, to Porter and to Thomas and to Burnside on the way, attending to the supplies and directing the movements of his three armies; ordering vegetables to Chattanooga, securing gunboats to protect Sherman, and directing working parties to re-make Thomas' roads.

We will see in the succeeding chapter that Sherman, with his own (Fifteenth) corps, and a part of Hurlburt's, proceeded from Vicksburg *via* Memphis and Corinth, to the relief of Rosecrans as soon as possible after the disaster on the Chickamauga. The line from Stevenson to Huntsville was abandoned when the enemy repossessed Lookout Mountain; but now that Sherman was advancing and Bragg threatening Knoxville, if not intending an advance on Nashville, it became a matter of vast consequence that communication be reopened along the Huntsville road, as Sherman was expected to emerge at Decatur. General Crook, commanding the Second Cavalry division, was sent to scout that line as far west as Maysville, Alabama, and John Lonergan, operator, was sent to Crook on an engine with a Captain Craig and citizen Sheets, a scout, and a guard of eighteen men. At Woodville, the engine was dropped into a culvert by the explosion of a torpedo, breaking Sheets' ankle. The guerrillas under Frank Gurly immediately fired upon the party, which formed behind the engine and repelled the enemy. Darkness set in with a heavy storm. The Fourth Ohio cavalry at Paint Rock bridge was the nearest Union force, and it was decided that Captain Craig, conductor Hammell and Lonergan should undertake the dangerous task of apprising Major Patton, commanding at the bridge, of the situation, while the soldiers defended the engine and tender.

For two miles they followed the track, by the flashes of lightning, but total darkness was preferred, as a volley was expected at every flash, for they were as near Rhoddy's troops as their own. In passing near Paint Rock bridge, through a long stone cut, in which a strong stream of water ran, making, with the rain, so much noise that a voice was hardly audible, Lonergan led the party, moving in single file, when they were ordered to halt a number of times before hearing the command. Finally Lonergan, as ordered, advanced alone, carrying his instrument, and when just in front of the videttes, the lightning revealed four men covering him with their guns. They proved to be of Patton's command. The sergeant stated that several times he had his men level their carbines in the direction of the three, intending to fire, and as often told them to recover arms. The danger was all the greater as, only a few hours previous, the troops had repelled an attack by rebel scouts in that same cut. Relief was immediately dispatched and at midnight Paint Rock bridge was in communication with Chattanooga, when the following despatch, dated October 24, from Grant for Sherman, who was restoring the railroad, was received:

> Drop everything east of Bear Creek and move with your entire force toward Stevenson until you receive further orders. The enemy are evidently moving a large force toward Cleveland and may break through our lines and move on Nashville, in which event, your troops are the only forces at command that could beat them there.

This message was given for delivery to that prince of scouts, Corporal Pike, whose escape down the river to Tuscumbia was narrow as his venture was hazardous and successful. When Sherman received this telegram, he ceased building the railroad and crossed the river at Eastport. At Waterloo, Wayne H. Parsons, Sherman's operator, who had on foot overtaken the General, was furnished a horse. At Florence, he was ordered to push on with despatches to Deckard, Tennessee, seventy-five miles. Escorted by Lieutenant Audenried, with one hundred troopers, he reached Deckard after a dangerous and tiresome ride of eighteen hours and a loss of two men killed by bushwhackers.

Parsons, without a word of explanation, and to the greatest astonishment and concern of little Jimmy Lowe, who operated there with a paper instrument, and was consequently unable to tell what devilish tricks that mud-bedaubed, rain-soaked and weather-beaten stranger was up to, transmitted Sherman's messages to Grant and received the following in reply:

> Come on to Stevenson and Bridgeport with your four divisions; I want your command to aid in a movement to force the enemy back from their present position and to make Burnside secure in his.

Leaving Jimmy Lowe to attend his little sutler shop, by which he turned an honest penny, and to ask Chattanooga all sorts of questions about the stranger, who for the fun of the thing, vanished as unceremoniously as he had entered, Parsons returned to Sherman at Fayetteville.

On the twenty-third, most of Sherman's troops crossed the Tennessee north of Chattanooga.

Late in October, Hooker, concentrating most of his troops, possessed Lookout Valley in force, after a severe engagement with Longstreet. Thus the river, the railroad and direct wagon roads were entirely within Federal hands up to about nine miles of Chattanooga, and no longer was there any thought of starvation. November 7, Grant, who had become greatly exercised over Burnside's situation, owing to Longstreet's departure to crush the Federals in East Tennessee, directed Thomas to attack, but that officer was wholly unable to do so as he had no horses to move his artillery. Again Bragg detached troops to aid Longstreet, but Grant could not stir until Sherman came up, when Thomas obtained animals; but Sherman's movements were unavoidably delayed and Grant was restive under his restraints, fearing that Burnside would be starved into a capitulation or retreat, or that Bragg would escape without punishment until he could reunite his army. A general attack was ordered for the twenty-first, but Sherman did not come; the twenty-third, he was only partially ready, owing to broken bridges, but the battle was begun, Sherman on the left, Thomas on the center and Hooker on the right. Orchard Knob, in the center, was carried on the twenty-third, Lookout Mountain, by Hooker, on the twenty-fourth, and then the whole line of rebel works along Missionary Ridge, the next day.

The South was paralyzed by the blow, generally regarded as

impossible. The Confederates were pursued beyond Ringgold. It is stated that Grant's forces in action numbered sixty thousand to Bragg's forty thousand. The Federals lost seven hundred and fifty-seven killed and four thousand, five hundred and twenty-nine wounded. Doubtless the enemy suffered near the same in casualties, but six thousand, one hundred and forty-two prisoners and forty-two guns were captured from them. In short, their rout was as complete as it was terrible. Parsons accompanied Sherman and was under fire nearly all one day, counting at one time, the smoke of nine shells that burst over their heads simultaneously. Beckwith was with Grant from first to last.

The glad tidings of the great victory spread faster than the winds, more rapid than sound, speedier than light. The lightnings heralded the triumph. Jesse Bunnell, M. C. Baldwin, W. H. Miller and William Patterson, at the Chattanooga end ; H. W. Plum, F. S. VanValkenburg, Cass G. Sholes, Ellis J. Wilson, and others, at Nashville, worked untiringly, night and day, to forward the news, official and reportorial, as well as the great number of private messages announcing all manner of fortunes and misfortunes incident to a battle. Even Dana's description of the battle above the clouds sped over that metallic thread, to an eager people.

Let us now turn again to East Tennessee. Longstreet, with twenty thousand men and eighty guns, arrived at Sweetwater, November 13, and, moving northward, obeyed Bragg's instructions to keep open telegraphic communication with him, as the army advanced. Reaching Loudon on the fourteenth, about five thousand men crossed the Tennessee near Le Noir, Longstreet's main command remaining on the south side, but connected by a pontoon bridge. Such was the enemy's position while Burnside and his generals were holding a council of war, at which it was decided to make a night attack upon the isolated five thousand. General Jeff. C. Davis, at ten, P. M., was about leaving Burnside's quarters, to begin the assault—in fact, he had reached the door on his way, when a messenger rushed up and handed Jacques, the operator, a cipher despatch, just received. Davis awaited its translation. It proved to be an order to Burnside from Grant, to draw Longstreet across the river and as far north toward Knoxville as possible. This changed the whole aspect of affairs, and Burnside who had but five thousand men in hand was obliged to make forced marches to reach Knoxville with his trains, ahead of Longstreet, who took the shorter road.

From the time Burnside first learned from Grant of Longstreet's movement, which was about five days before Longstreet put in an appearance, up to eleven o'clock, A.M., of the nineteenth, except perhaps a few hours on the twelfth, Grant and Burnside were in telegraphic communication. Burnside, by the use of the telegraph, concentrated his troops at Knoxville to the number of twelve thousand—Longstreet in the meantime having been reinforced. The rebels tapped the Gap line on the seventeenth, but allowed the Federals to work to Knoxville until the forenoon of the nineteenth, doubtless in hopes of obtaining valuable information. The above statement is predicated upon the fact that cipher telegrams, to and from Grant during those days, were published in Southern newspapers, with a request for their translation. A number of the operators at posts contiguous to Knoxville entered that city with the garrisons. Captain Bruch reported that "during the siege, lines were built to all the forts and operated by them day and night. Great praise is due to Mr. Adam Bruch, Chief Operator, and his assistant operators, for their untiring energy and bravery in constructing and operating these lines; every man working almost constantly day and night, and under the fire of the enemy during the siege."

Early on the morning of the first day of the partial investment, Jacques volunteered to serve as aide on the staff of General Sanders, and rode out to the front with him and others—a gay and happy party—little thinking that before returning they must pass through a fight involving a Federal loss of one hundred killed and wounded. But it so happened, General Sanders himself being one of the killed. Jacques and another of the party were wounded, the former slightly. Another lost a horse, killed under him.

As soon as the enemy began to close in on Burnside, all the telegraph wire within reach was taken down and fastened to stakes about ten inches above ground, in front of the trenches and forts, to trip the enemy in case of a charge. This worked admirably, for when Longstreet's troops, brigade after brigade, were hurled against Fort Sanders, about daybreak on the 29th of November, many fell over these wires ; the front ranks were thrown into confusion near the fort, from which grape, canister and bullets were reaping a harvest of casualties, numbering near one thousand before the assaults were abandoned. Many of the assailants thus tripped, supposed they were shot, and others lost their arms. Charles Jacques was operating at Fort Sanders at this time. Burnside's office in the city was telegraphically connected with this fort and the quarters of General Shackleford and others. After the enemy had twice planted their flag on the parapet of Sanders, and each time been driven off with dreadful slaughter, of which Burnside was constantly informed, Jacques telegraphed that the enemy were coming again. Just then a shell cut the wire, and as it was certain death for any one to venture out to repair it, Burnside, who of course heard nothing further from the fort, supposed it was captured, and rode out to where his reserves were posted, expecting every moment to see the enemy coming over the hill.

The following operators were stationed in Knoxville during the siege, viz., Adam Bruch, C. W. Jacques, C. H. Johns, George Cole, and perhaps others. Joe Anderson was with Colonel Bird's command, cut off at Kingston, where several affairs took place with the enemy, under General Wheeler.

After Grant had driven Bragg from Missionary Ridge, he received despatches from the Federal advance telegraph office at Tazewell, where J. A. Torrence operated, notifying him that Burnside could not hold out longer than until about the 1st of December ; hence it became a matter of gravest consequence to apprise that officer of the fact that Gordon Granger and Sherman had been sent to his relief, and would probably arrive by the 2d of December. Accordingly, Secretary Stanton, who was keenly impressed with the importance of informing Burnside, telegraphed for Colonel Stager to come to the key. Stager had retired, but it was his custom to have an instrument in circuit by his bedside, for any emergency, and repeated calling awakened him. Stanton in Washington asked Stager, who was in his bedchamber at Cleveland, Ohio, to forward the message to Burnside from four separate points by the most trusty means. The Colonel called up Captain Bruch, in Louisville, Ky., and the latter the operators at four of the offices nearest to Burnside. Thus it happened that in the dead of night, four telegraph repairers, each with a copy of the cipher message notifying Burnside of the approach of Union troops, were started from their respective stations on their important and perilous journeys. Stanton had promised to do every thing in his power for either that might fall into the enemy's hands, but it is believed all reached Burnside safely. Thus was the news of succor carried to Burnside, whose army was saved. East Tennessee was redeemed, and the telegraph rebuilt by Gross and VanDuzer, from Knoxville to Chattanooga and the strategic localities of East Tennessee, and Chattanooga thoroughly reclaimed to the Union, to be harassed indeed thereafter, but never again to be surrendered to a hostile army.

CHAPTER V.

TELEGRAPHIC AFFAIRS IN WEST TENNESSEE IN THE FALL OF 1863.—NICHOLSON'S DEATH.—GENERAL SHERMAN'S ESCAPE. — FORREST'S RAID. — MEMPHIS. — PRESENTATIONS.

The history of the army telegraph service is fringed with many sad pictures. We are now to relate the misfortunes of John Nicholson. John was the eldest of the seven children of his very poor but worthy parents, who resided in Woodstock, Canada West. He and George B. Cowlam, whom we have noticed in connection with the "Merrimac" fight, were taught to telegraph by James Izard at Woodstock, in 1859. H. P. Dwight very kindly gave Nicholson a small office; but as he was the main-stay of the family and a mortgage on his father's place was maturing, with no prospect of its payment from the salary young Nicholson could earn in Canada, he entered the United States Military Telegraph service, May 7, 1863, in the ardent hope of saving enough money to cancel the mortgage, but later, he even aspired to the erection of a comfortable house for his parents, brothers and sisters. He was sent to Glendale, Mississippi, but his quarters were so different from what he had been accustomed to that his health was soon impaired. On June 25, he wrote:

> I have got an old shed that the rain just pours through; might as well go out of doors on a rainy day. Every night that it rains the water pours down on top of my bed, which is not very comfortable. * * * The doctor says that unless I get a better place to stay in, I will be in the hospital before long.

The trouble was, that the Colonel commanding there said it was the post quarter-master's duty to repair the office, and he shirked it, saying it was the Colonel's. When Captain Fuller heard that Nicholson was sick, he placed him in the officers' hospital in Memphis, where he received the best of care; but August 25 he died. His parents were stricken with grief, and the poor mother was greatly cast down because she was unable to defray the expense attendant upon a removal of the body that she might once more look upon her son's features. These facts coming to the knowledge of his comrades on the Memphis & Charleston line, they each subscribed five dollars, which, with Messrs. Fuller and Gross's subscription, paid for embalming and expressing the remains to Woodstock, leaving one hundred and fifty dollars, earned by deceased, to go entirely to his father. Such acts of kindness are ennobling; but a more sympathetic and generous body of men than those of the Telegraph Corps never lived. The subscribers were C. H. Spellman, Ellis Stone, L. B. Spellman, Joseph Blish, Jr., J. J. Egan, Fred W. Snell, R. S. Fowler, W. A. Thayer, C. McReynolds, J. W. Atwell, W. B. Somerville, Ed Kinney, R. J. Hughes, E. F. Butler and repairers James Rail and John Pursel.

Rosecrans having suffered a terrible defeat at Chickamauga, Tennessee, General Halleck sent Slocum's (Eleventh) and Howard's (Twelfth) corps from the East, under Hooker, and late in September, 1863, Sherman was ordered to move by water to Memphis and thence east, repairing the Memphis & Charleston road east of Corinth, as far as Athens, where he was to report to Rosecrans. Sherman reached Colliersville, twenty-six miles from Memphis, by train, (a small force being also on board,) about noon of October 11; some of his troops having already reached Corinth, and yet others were in the rear on foot. Among the latter was Corse's division, about eighteen miles back. Sherman's train had gone perhaps a half mile farther, when picket firing began and the train stopped. A member of General Chalmer's staff demanded the surrender of Colliersville and its forces. General Sherman, in his "Memoirs," says:

I saw Anthony, Dayton and the rebel bearer of the flag in conversation, and the latter turn his horse to ride back, when I ordered Colonel McCoy to run to the station and get a message over the wires as quickly as possible to Memphis and Germantown, to hurry forward Corse's division. I then ordered the train to back to the depot and drew back the battalion of regulars to the small earth redoubt near it. The depot building was of brick and had been punctured with loop-holes. To its east, about two hundred yards, was a small square earthwork or fort, into which were put a part of the regulars, along with the company of the Sixty-sixth Indiana already there. The rest of the men were distributed into the railroad cut and in some shallow rifle-trenches near the depot. We had hardly made these preparations when the enemy was seen forming in a long line on the ridge to the south, about four hundred yards off, and soon after two parties of cavalry passed the railroad on both sides of us, cutting the wires and tearing up some rails. Soon they opened on us with artillery. * * The fighting continued all round us for three or four hours, when we observed signs of drawing off, which I attributed to the rightful cause—the rapid approach of Corse's division, which * * arrived * * on the double-quick.

Edwin F Butler was the operator at Colliersville, and bravely did he maintain the reputation of the Telegraph Corps. He telegraphed Corse at Germantown and Hurlbut at Memphis, as directed by General Sherman, remaining in his office as long as possible. Much of the fighting occurred about the depot and train, which were captured. The line was repaired that evening, when Butler sent Captain Fuller, at Memphis, the following telegram:

Please send a relay and key. When the line was cut and they were making a charge on us, I took the instruments and all but the key was taken and a shell came through the office. I pulled the key loose and broke it; the relay was stepped on and broken in the fort. I got a shot through my left arm while shooting from the fort, but think I will not be laid up.

During the preceding February, while Grand Junction was threatened, W. H. Parsons, operating there, went to the fort and obtained a Springfield musket and forty rounds of ammunition, with which he took position by the soldiers, prepared to do what Butler did at Colliersville; but the enemy did not attack.

On the 16th of October, Generals Matthias, at Burnsville, and Osterhaus, at Iuka, both of Sherman's command, were very much in want of operators. Sherman telegraphed for two to accompany his advance, but especially asked for Wayne H. Parsons, who alone was sent. C. O. Whitney left Cam. Culbertson and Samuel D. Cochrane in charge of Iuka office, about this time, to accompany a Federal raid from Corinth.

Captain Fuller, some time in October, received the following despatch:

> The Secretary of War directs the immediate repair of line between Memphis and Cairo. Can you secure military escort and protection for working parties and repairers? Lose no time in opening line.
>
> (Signed) A. STAGER,
> *Colonel, Supt. U. S. Mil. Telegraphs.*

Generals Sherman, Hurlbut and Dodge agreed, however, that the line could not be maintained with the force in West Tennessee, and it was not undertaken.

Rebel General Forrest, who worked such harm to the corps north of Jackson, in December, 1862, reached Okolona, Miss., from Bragg's army besieging Chattanooga, in the middle of November, and organized another raid into West Tennessee, for the particular purpose of enforcing the conscription laws, and generally to do all possible injury to the Memphis and Charleston Railroad and telegraph, and to carry off as many animals as possible. General S. D. Lee was chief of cavalry in Mississippi, having under him Slemmons, McCulloch, Chalmers, and other brigadiers. Lee aided Forrest materially, although few if any of his troops accompanied the latter beyond the Memphis & Charleston road. At four, P. M., December 2, the telegraph was cut near Saulsbury, and after a short encounter the Federals were driven off and most of the town burned. The next day, another column, at 12: 30, P. M., cut the line east of La Fayette, and three hours later drove off the handful of troops there, and with them William A. Thayer and presumably his bride whom he had recently received from the North and married at Memphis. About the same time that LaFayette and Thayer's instruments were captured, the pickets at Colliersville were driven in, but the line was not cut. Moscow was attacked the same day; George E. Spellman operating there telegraphed to Memphis, at six, P.M., that the enemy had been driven off. Lee then retired south, and Forrest penetrated as far north as Jackson, ridding the country of many guerrillas by making them join his forces. On the return LaFayette was again captured, but this time Thayer saved his relay and register. A few months later, Captain Fuller wrote: "Forrest and Chalmers constantly harass our lines, and generally succeed in cutting out and carrying off a quarter to a mile every week or ten days." Indeed, owing to Hurlbut's forwarding troops South and East, the matter of evacuating the posts from Memphis to Corinth was, early in November, seriously contemplated. These positions were, however, retained. Operators Egan, Ingle, Hughes, Stone and Olmstead were at Corinth, at this time, when not out with the cavalry on raiding expeditions.

As yet we have scarcely mentioned the telegraph affairs in Memphis. Very different were appearances there than in Cairo. The city, having a population of about thirty thousand, presented many evidences of thrift and culture. Its buildings were strikingly dissimilar, but not a few were imposing and architecturally pleasing. The finer structures were in the main named probably after their first owners: thus there were the "Ayers Building," "Jefferson Block," "Clay Building," "Gayoso Block" and "Irving Block," and the like. The telegraph office was in the "Telegraph Building," of which the entire third story and four front rooms of the second were seized for the United States Military Telegraph service. A restaurant was underneath. The building was located fronting the "Court Square," a beautiful little park, in the center of which, protected by an iron railing, was a bust of Jackson. The pedestal once bespoke his devotion to the Union in his own words: "The Federal Union, it must be preserved;" but some miscreant had nearly effaced the word "Federal." The park, surrounded by an iron fence, was tastefully designed. It was intersected by graveled walks, between which were handsome evergreens and numerous and various other ornamental trees. Besides, there were about thirty grand old oaks, much older than the city, and in them were numerous gray squirrels which little children delighted to feed. Truly, the Memphis telegraphers could say, "Our lines have fallen in pleasant places." Alvah S. Hawkins succeeded D. T. Bacon as manager, and so continued to act until he went to the office of General J. D. Webster, the efficient and talented Superintendent of military railroads in that department. The trains were run largely on telegraphic orders from Webster's office. A large number of the department operators served in the city office temporarily.

Captain Fuller made Memphis his head-quarters from about the first of February, 1863, having with him, as his chief clerk, W. L. Gross, from the Cairo office. Gross had been of invaluable aid to VanDuzer, who recommended his appointment as manager in the department. Captains Fuller and Bruch also spoke highly of him, and consequently he was made manager of the telegraphs under Bruch in General Burnside's department, with head-quarters in Danville, Ky. Before he left Memphis, many of the telegraphers united in the purchase of a gold hunting-case watch, which was presented to him at the dinner table. Pack, the colored cook who catered for the operators, had prepared a splendid repast. That finished, some Presbyterian ague cure that Captain Fuller was keeping for that or some other malarial occasion, was produced for "the stomach's sake." Gross supplemented the medicine by a basket of Longworth's Elixir of the Tower of Babel, as being best fitted to make lasting impressions. There sat at the table Major Benner and Lieutenant Dempster, of General Hurlbut's staff; operators S. D. Cochrane, who taught Gross to telegraph; R. S. Fowler, James Pitton, A. S. Hawkins, Lew. Spellman, W. D. Hawkins, William B. Somerville, Edward Fuller, Gross' successor; W. G. Fuller, Ed. Kenney, manager of the city office, and the wordy editor, Bingham, of the Memphis *Bulletin*. Touching speeches and glasses were followed by the delivery of the watch, on which was engraved "Presented to W. L. Gross as a token of esteem, from his friends in the United States Military Telegraph Corps, Department of Tennessee, Memphis, September, 1863. The strongest proof the author has discovered of there having followed a confusion of tongues, lies in the nature of the case and the incoherent report printed the next morning.

On the 27th of October, Mr. Fuller was appointed by the President, as captain and assistant quarter-master; whereupon another presentation was arranged, an account of which, taken from the *Bulletin* of the nineteenth, is as follows:

Last evening, the private rooms of the Military Telegraph Corps were the scene of a most genial incident. A little group of the friends of Captain W. G. Fuller assembled to the number of a score or more, and sat down to an elegant supper. This *recherche* repast did infinite honor to the *cuisine* of the Captain, and would be difficult to excel. Game, oysters and all the delicacies of the season were served in a most tempting manner. After the claret had been removed and the champagne began to circulate, Mr. W. B. Somerville, a member of the Telegraph Corps, presented a beautiful gold watch to his chief, in a neat speech on behalf of his comrades and associates. His remarks consisted of a few sentences only, but they bore the distinguishing stamp of cordiality and good feeling, and in this is the secret of eloquence. * * Captain Fuller responded in a neat manner, winning a round of applause. A toast was drunk to General Hurlbut, the chairman of the occasion, to which he responded in a spirited, happy style, materially increasing the good humor of the evening.

The watch was a magnificent Jules Jurgensen chronometer, and bore the following inscription: "Presented to Captain W. G. Fuller, A. Q. M., U. S. A., and Superintendent of Military Telegraph, as a token of esteem, by the Telegraph Corps of the Department of the Tennessee."

Even Palmer did not escape the contagion, as shown, a little later, by another printed item:

Sol Palmer, the excellent foreman of the Military Telegraph construction party, was yesterday presented with a heavy silver hunter watch, purchased at a cost of $150, bearing the following inscription: "Presented to Solomon Palmer, Foreman of the U. S. Military Telegraph builders and repairers, as a token of esteem. Memphis, Tenn., Feb. 9, 1864." Captain W. G. Fuller added to the token a heavy silver guard chain. Sol is a worthy man, and we are glad he can carry these mementos of appreciation.

Besides the operators already mentioned in this department during the year 1863, there were Augustus Tyrrill, John R. Frank, H. J. Fish, E. B. McNairn, D. N. Stewart, C. W. McReynolds, Charles Wallace, William Gibson and K. B. Howe.

CHAPTER VI.

THE SIGNAL CORPS AND ITS EFFORTS TO CONTROL THE ELECTRIC FIELD TELEGRAPHS. — STANTON PRAISES THE U. S. M. T.

In a former chapter enough has been explained to show the purposes and, generally, the methods adopted in armies for signaling information or orders, without elaborating or particularly restating here. In a word, army signaling, as practiced at the outbreak of the rebellion, consisted of the "motion of a single flag, attached to a staff, held and worked by the hands of one man, in the day; or, by the similar motions of a lighted torch fastened to the staff, instead of a flag, at night; whereby, a single man is converted into a semaphore, useful for any distances at which the signs made are visible, either with the naked eye or with telescopes."

This branch of the service, when the war began, was rather experimental than otherwise, and like the army itself, had been neglected. Consequently, when McDowell advanced, in July, 1861, to fight the battle of Manassas, he had no field telegraph, visual or electric, and the only sign given or received was the signal gun, which announced that the attack should begin.

In June, the chief signal officer, Major Albert J. Myer, began organizing a small party at Fortress Monroe, for the signal service, and instructing them in the mysteries of the code. Other camps of instruction were subsequently established, but soon all were concentrated in the one at Georgetown, D. C. By November, 1861, a large party of detailed officers was instructed and attached to the Army of the Potomac, and, we are told, they at one time signaled from Georgetown, D. C., to Maryland Heights, *via* intermediate stations. At that time, such officers were much in demand, and as "other armies were formed or expeditions were prepared, skilled officers and men, sent from the parent camp, formed with these armies."

The signal officers so sent created a nucleus in the several departments, which rapidly grew, so that the average number of officers on duty with the armies, for the year following October, 1862, was reported at one hundred and ninety-eight; but, perhaps, more impartially speaking, since there were two detailed soldiers with every officer, who did as much work, in quite as much danger, but without equal pay or notice, the average number of signal men was five hundred and ninety-four. That these enlisted assistants should be thus considered, follows from the official report for the year preceding October 30, 1863, which shows a loss during the year from wounds, capture and death by disease of sixteen officers and thirty-nine enlisted men, one of the latter having died from wounds. The succeeding year's report shows a loss of fourteen officers and sixty-two enlisted men, of whom one officer and five privates were killed. The corps increased with the armies; even recruiting stations were opened in various States. The expense of the service for the first three years of the war will best illustrate its growth. For the year ending June, 1862, that was $192,523.40; to the end of June, 1863, $545,612.64; and to the end of June, 1864, $848,584.00; a total in three years of $1,586,720.54; or, doubtless, over $2,000,000 before the war ceased.

That for this enormous expenditure much meritorious service was obtained, can not be gainsaid. Many, doubtless most, of its *attachés* were daring, skillful and useful. The author has a personal and pleasing recollection of some of them. Within the legitimate sphere of its service, this corps was probably as efficient as, considering its experience, it could be, and, judging from the laudatory reports of its chief, much more so than any other signal corps in any army in Europe. There does not appear to be anything improbable or unreasonable in this, because in every war improvements are made, and in ours they were manifest in every department.

Had its chief been content to leave telegraphing by wire to electricians, who, by the way, were in the field in and about Washington, Western Virginia and St. Louis, while the chief signal officer was journeying from the land of the Navajos, the telegraphic and signal corps would have vied with one another as mutual aids to a common purpose, quite as cheerfully as did the different arms of the army proper.

In no former war had the electric telegraph advanced with the army to any considerable extent. Telegraphy is an art for speedy communication ; that was precisely the main purpose of the signal system, but it was so expensive as to be afforded, with few exceptions, only for national purposes. The really important messages, transmitted by the use of the flag, probably cost the Government many thousand dollars each. But it is not difficult to single out some such that well repaid their cost ; and while eminent soldiers have spoken slightingly of the service as a whole, there are not wanting the certificates of numerous others to its efficiency. The good that it did will be preserved, for its chief, in his reports, which are most excellent, in many respects, seems to have striven to record only its benefits, and to direct attention to those officers whose usefulness gave him the items. One of those reports represents truly that,

> As early as August 6, 1861, the Signal Officer of the Army proposed, with the permission of the Secretary of War, to organize a Telegraphic or Signal Train, to accompany the army on the march ; the wagons of this train to carry all articles needed for temporary telegraphic uses ; that is, apparatus and supplies for the use of both electric and aerial telegraphs, rockets and composition night-signals, carefully prepared, packed, numbered and arranged for instant use. Four flying field telegraphs were to be carried in the train, which was to be accompanied by, and be in charge of suitable officers and men, to each of whom his duties should be assigned, and of whom a proper proportion should be electric telegraphists, fully instructed in the use of the telescope and aerial signals ; and who, employed for the war, should be sworn to the faithful discharge of their duties.

Major (now* General) Albert J. Myer meant by this that he wanted a "movable telegraphic train," equipped with instruments invented by Mr. G. W. Beardslee, of New York. "They work," said he, "without batteries, and can be used by any one who can read and write, after a day's practice." These instruments and Myer's insistence caused infinite trouble. The scheme was to monopolize army telegraphing by the use of frictional and dial telegraphy. In Major Myer's enthusiasm, even commercial telegraphing was threatened. It is not surprising that

* This chapter was written while General Myer was alive.

his efforts were at first thwarted, nor that the Secretary of War gave non-committal replies to requests to organize such trains. Finally, one was obtained and taken to the Peninsula. It was in the neighborhood of the Chickahominy for about two months, to little purpose. Then "improved" machines were purchased. Congress legislated early in March, 1863, in behalf of the organization of the corps, and the War Department was more liberal in fostering the development of the system of field telegraphs, than had been requested, and Beardslee's experiments thus quickened resulted in the constructing also of "a sounder, by which the signals transmitted are addressed to the ear;" but it is believed these were only experimented with by the signal officers.

During the year ending June, 1863, there were thirty field frictional electric telegraph trains, which were distributed as follows:

In the Army of the Potomac,	5
In the Department of the Cumberland,	5
In the Department of the Gulf,	3
In the Department of North Carolina and Virginia,	3
In the Department of the South,	2
In the Department of the Tennessee,	6
In the Department of the Ohio,	2
At Signal Camp of Instruction, Georgetown, D. C.,	3
At U. S. Military Academy, West Point, N. Y.,	1
Total,	30

Concerning the working of the Beardslee instrument, the reports show at best but very indifferent success, even after Secretary Stanton had liberally supplied the corps with funds. It was not claimed as possible to work over twenty miles, and seldom attempted over five. If it operated at all five miles, the signalists were happy. Before proceeding with the disastrous history of this machine, it is best to explain it by reproducing the cut and the chief explanatory directions:

Section showing the wire Connections.

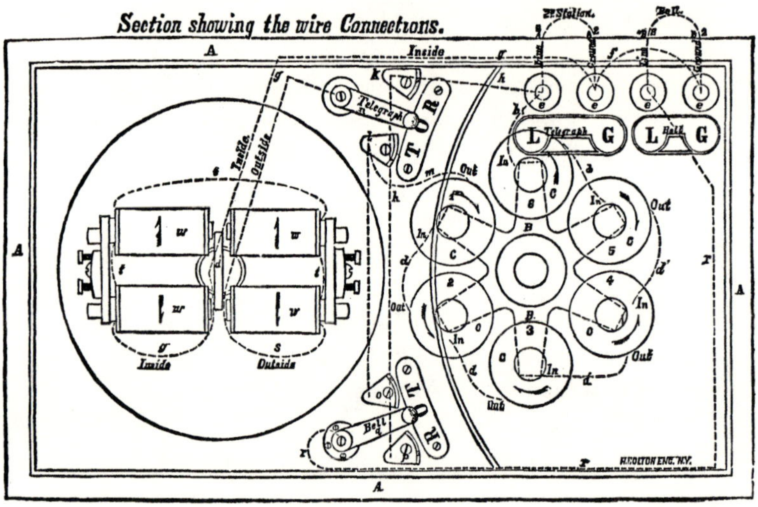

DIRECTIONS.

A, A, A, in the drawing, represents the box or case of the instrument ; B, the radiating magnet, which, with its radial points, is revolved beneath the spools, c, c, c, there being six poles, alternately, three north and three south poles. d, d, d, are the terminal wires of the magneto-spools, showing the manner in which they are connected together. The outside wire, m, of No. 1 spool, is connected with the lug l, of the telegraph switch, extending to the lug o, of the bell switch. The inside wire of the same spool is connected with the outside wire of the next spool, No. 2, and so on through the series of spools to No. 6, which has its inside wire, h, connected with the L hitcher, e, above the word telegraph, and extends to the lug, k, of the telegraph switch, and then to the lug, p, of the bell switch. w, w, w, represent the electro-magnet spools, whose terminal wires, t, t, t, are connected, throughout the series outside with an inside wire. The terminal outside wire, s, is connected with the telegraph switch, n, and the inside terminal, g, is connected with the G hitcher, e, over the word telegraph.

A^2 $A^2$are the wires connecting with the hitchers, e, e, and extending to the second station or other instrument.

f, is a wire connecting G hitcher, e, over theword "Telegraph," with G hitcher, e, over the word "Bell."

r, is a wire which connects the bell switch with L hitcher, e, over the word "Bell."

d is a vibrating magnet between the cores of the electro-magnet spools.

The arrows indicate in what direction the wires of the several spools are wound.

B^2 B^2 are the wires connecting the bell spool with the hitchers, e, e (L and G), over the plate marked "Bell."

The bell and telegraph switches have each a plate with the letters T, for "transmit," O, for "open circuit," and R, to "receive" message.

The switches of all the instruments in a given circuit, when not in operation, have their telegraph switches placed over O, and the bell switch over R.

When a message is to be sent from either of the stations, the operator places his bell switch over T, makes one turn of the handle from the adjustment point of the dial to the same place and stops, and repeats this movement till the number of the station to which a message is to be sent is reached ; he then places his bell switch over R, when the operator at the station called will place his switch over T, and repeat the call.

When this is done, the operators of all the instruments in the circuit will place their telegraph-switches over R, their bell-switches over O. The operator who made the call will then move his telegraph-switch over T, and proceed to transmit his message as before described.

The telegraph-machines are all numbered consecutively from one upward, and before leaving the factory they are all adjusted in the following manner, which must at all times be borne in mind, whenever the machines are taken apart and put together again.

In all machines marked with even numbers, the indicating handle being placed at the adjustment-stop at the left of the dial, the north poles of the radial magnet must be placed under the spools marked 1, 3, and 5, and in the machines marked with odd numbers, the indicating handle being at the same division of the dial, the north poles of the radial magnet should be under the spools marked 2, 4, and 6. In putting the machines together, this adjustment is readily effected by simply lifting the intermediate cog-wheel (which is between the driving-wheel and the one on the shaft of the radial magnets) and placing the indicating handle as stated, and turning the radial magnets until the poles are in the position stated, and then replacing the intermediate cog-wheel.

When a message is being transmitted from a station, the current is generated in the magneto-spools over the radiating magnet by the passage of the several radial poles of the magnet in close proximity to the ends of the cores of the several spools.

When the poles of the magnet are approaching and leaving the cores, magnetism is temporarily induced and discharged from the soft iron cores of the spools. This peculiar action in the soft iron induces electric impulses in the surrounding insulated coils of wire, which impulses are transmitted through the wires connecting the electro-magnetic spools of the instruments, and the lines connecting the stations.

The effect of these impulses of the electric current passing around the soft iron cores of the electro-magnets, induces magnetism in them, which causes the vibrating magnet to be attracted or repelled alternately in opposite directions, as the current is alternately reversed.

When the north poles of the magnet are passing spools Nos. 1, 3, and 5, and the south poles Nos. 2, 4, and 6, the current is in one and the same direction, with the power and strength of all the impulses of the six spools united.

When the reverse is the case, that is, when the south poles are passing where the north poles have just left, and the north poles where the south poles were, the current is induced in the opposite direction.

This change of direction causes the vibrating magnet to be alternately repelled and attracted in opposite directions, giving a pendulous motion to the magnet and pallets, which causes the needle to be moved around the dial, to indicate the letter intended, each impulse moving it the distance of one division of the dial.

When the instruments are in operation, the telegraph-switch of the receiving instrument being placed over R, the currents induced in the transmitting instrument are conducted along the wire h, through the hitcher L, e, and line wire A^2 to the receiving instrument, through the lug k, of that instrument, and the telegraph-switch n, along the wire s, through the electro-magnets, w, w, w, along the wire g, g, to the hitcher G, e, over "Telegraph," along the wire or ground A^2, to the same hitcher in the transmitting instrument, then along the wire g, g, through the electro-magnets w, w, w, and the wire s, and the switch n, the lug l, and the wire m, through the magneto-spools back to the terminal wire h, of spool 6.

Thus it will be observed that the electric current passes from the generating instrument only through the electro-magnetic spools of the respective instruments, and does not follow through the bell-spools as they are switched open, nor through the magneto-spools of the receiving instruments as they are switched out of the circuit.

When the bells are operated, it will be observed that when the switches are placed as before described for that purpose, all the electro-magnetic spools are switched out, as well as the magneto-spools of the receiving and transmitting instruments.

If from any cause the spools of the electro-magnet require adjusting, this can be done by the set-screws and the holding-screws, but in no case should the ends of the cores be placed so near the range of vibration of the vibrating magnet as to permit it to come into contact with them, but so as to be just clear when the pallets are seated down in the escapement ratchet-wheel of the needle-shaft. Care should be taken to give the like adjustment to all the cores of the electro-magnets.

The instruments may all be moved from their cases by removing first the holding-hinge, then the velvet cover, the beads around the dial-top, then the six screws in the bottom of the box; care being taken, when replaced, that all is made tight and firm, and that no dirt or other matter is allowed to remain in case or instrument, and particularly no particles of iron are left in, or allowed to get in, as these will be attracted to the vibrating-magnet, and get between that and the cores of the spools, and prevent its vibrations, and thereby prevent the instrument from working.

It will be observed that the radial magnet-shaft rests in a step, which may be raised or lowered by a set-screw, which is then secured in place by a jamb-nut. This magnet should always be so adjusted as to revolve as near as possible to the ends of the cores of the magneto-spools without touching; and to prevent their being lifted into contact by the magnetic attraction, a set-screw is provided on top to hold it down, which will be required to be adjusted for that purpose, whenever it may be necessary to do so.

It is also important that whenever more than two instruments are placed in the same circuit, if it be desired that the messages shall pass any given instrument or instruments, it or they can be closed out of the circuit by uniting the two hitchers, e, e, over the plate marked "Telegraph," by means of a wire.

In addition to the above, there were over seven pages of other explanatory print, showing, among other things, that each wagon contained one "machine," which, in use, should be on a level; a tool box, five reels of insulated wire, lances for poles, with one sharpened end and the other notched to receive the line; crowbars for ground connection, and if that was ineffectual or inconvenient, it was recommended that spikes be driven in trees, "and attaching the conducting wire to the spikes; or, instead, * * the bark may be cut off all around the tree and a copper wire wound around it."

In transmitting a message, which is done by moving the handle in succession to, and making a short pause at, the letters required to spell successively the words of a message, the operator, at the end of each word, must move the handle to, and make a short pause at, one of the stops, either to the right or left of the dial. To transmit a message, the switch must be at T. As soon as transmitted, the switch must be shifted to R, to be ready to receive a message, if one should be sent from the other station.

To transmit figures, the operator should first move the handle to the stop marked, "Stop for figures," at the top of the dial, then proceed the same as in sending letters, there being a circular row of figures, as well as letters, on the dial.

To facilitate transmission, the following abbreviations, and many others of like character, were used:

Aq.	All quiet.	Brig.	Brigade.
A.	After.	Cav.	Cavalry.
Art.	Artillery.	Com.	Communicate—tion
Adv.	Advance.	Cov.	Covered.
Ans.	Answer.	Co.	Company.
A. R. L.	Advance to reinforce our line.	Concl.	Conceal.
		C.	Can.
B.	Before.	Cld.	Could.
B. O.	The batteries are ours.	Cps. D. A.	Corps D. Army.
Bat.	Battery.	Chg.	Change.
Bag.	Baggage.	Comd.	Command.
Btn.	Between.	C. F.	Cease firing.
Bn.	Been.	C. R.	Cover the retreat.

C. F. B.	Cease firing on batteries.
Dir.	Direction.
Div.	Division.
Eny.	Enemy.
E. F.	Enemy are in front of us.
E· R.	Enemy are retreating.
E. A. L	Enemy are advancing on our left.
E. A. R.	Enemy are advancing on our right.
F. B.	Fall back.
Far.	Farther.
Ft.	Front.
Fwd.	Forward.
Flk.	Flank.
F. O.	Fire over us.
F. L. L.	Fire a little to the left of last shot.
F. L. R.	Fire a little to the right of last shot.
Gen.	General.
G. A.	Good aim; your range is correct; keep it.
M. T. L.	Move troops to the left.
R. N. H.	Reserve needed here; move forward to this point.
Tps.	Troops.
Val.	Valley.

Each train consisted of

5 miles vulcanized rubber insulated wire,
150 lances, 15 feet long,
50 lances, 18 feet long (6 x 12 feet), and

Two wagons, each containing

1 Beardslee's Patent Magneto-electric Field-telegraph machine (portfolio inside),
5 reels, with 4 small and one center reel stand,
25 notices and snaps,
4 bearers for carrying reels, with 3 straps,
2 galvanized crowbars with binding-screws for ground connection,
1 iron crowbar,
1 grease-pot and 1 pail,
2 lanterns,
2 poles with hook and fork, 18 feet long,
1 ax.

Tool box containing

3 spikes for ground connections in trees, etc.,
100 small iron spikes to hang wire on trees, etc.,
2 pounds No. 16 copper wire,
34-inch $\frac{3}{4}$ rubber sleeve to run the wire through the hand,
1 set climbing irons.

For making connections, joints and repairs:

10 insulated binding-screws, 10 feet rubber for joints, 1 flask rubber cement, 1 pound gutta percha in sheet, 12 sheets sandpaper, 1 spool No. 30, 1 spool No. 23 copper wire for making joints, 1 ball

twine, 1 pair cutting pliers, 1 screw wrench, (large), 1 do. (small), 1 screw-driver, 1 file, 1 machinist hammer, 1 hatchet, 1 pruning-knife, 1 auger, 1 saw, 1 oiler.

For soldering:

1 spirit-lamp with cap and shield, for melting solder, 1 bottle soldering acid, 1 soldering-iron, 2 bars solder.

Duplicate articles:

1 small gear, 1 linchpin and washer, 5 iron binding-screws, 10 iron washers (galvanized), 1 large gear, 2 iron rings for reel stands, 1 handle for reels.

Such was the complicated system worked after "one day's practice," to generate St. Vitus electricity; revolutionize Morse telegraphy, and drive Voltaic telegraphers from the field. That it was an expensive failure need be proved only to those unfamiliar with the science of electric telegraphy, but even electricians may be interested in some of the evidences of its *non*-success.

We are informed by the reports of the chief signal officer, that "at the first battle of Fredericksburg, field trains were, first in the history of the war, used on the battle field, under the fire of the enemy's batteries," but that is not unlike his report of seventy-six pages on the Peninsular campaign, in which he utterly ignores the military telegraph, which was there worked on the field of battle repeatedly. Of the Fredericksburg operations it was reported that "the trains in use contributed something to the success of these movements." "At Chancellorsville, several lines were extended. The shorter worked successfully. * * Success cannot always be commanded;" also that "these lines materially aided" the concentration of the army before Gettysburg, and were successfully employed at the siege of Charleston, S. C. Finally it is reported that the expense of "this service is trivial, compared with the results it is capable of accomplishing," and that with proper care it may be made "one of the cheapest, as well as effective branches of the service." Concerning its use at the battle of Fredericksburg, Samuel H. Edwards, late chief operator Atlantic & Pacific Telegraph Company, in New York, wrote·

During the engagement quite a rivalry arose between Myer's Field Telegraph and our (U. S. M. T.) system. DeWitte (W. K.) was at the pontoon crossing, I, in the woods an eighth of a mile back, with General Sedgwick. Myer's men were stationed, one by DeWitte at the pontoon, and one beside me at head-quarters. I asked him to give them fellows (Myer's men) a taste of what we could do, and he did. We beat them so badly that that one day's fight settled the Myer's system as far as General Sedgwick was concerned, and from that time, I never saw it in service during a battle.

Another operator, lately Superintendent of the New York Fire Alarm Telegraph, writing of affairs at the battle of Chancellorsville, says:

At that time winter quarters near Falmouth were in connection with head-quarters in the field, on the south side of the Rappahannock, near United States Ford. The Signal Corps people having put up the wire, and it being turned over to Morse men between these points, the Signal Corps people then extended the line further toward the front, to near Chancellor's house, where General Hooker remained during the battle. All business from Falmouth, where part of the staff remained, and from Sedgwick (Sixth Corps), who crossed at Fredericksburg lower down, came to James Murray and myself. We handed it to the Signal people for re-transmission. For the two days, business was very heavy, and A. Harper Caldwell and Jacques at Falmouth, sent it to us so rapidly that we handed the Signal folks as much every five minutes as they could transmit in an hour. We were obliged to send most of the messages to Hooker by orderlies; half a company being detailed for that purpose. The consequence was, the Commanding General and his subordinate commanders became totally disgusted at the inadequacy of the Beardslee telegraph, and it was immediately after turned over to our people.

Captain Buckley, Superintendent of the United States Military Telegraph in the Gulf Department, an accomplished and able officer, being, in the fall of 1863, asked to give his opinion, did it as follows:

I have the honor to acknowledge the receipt of your communication with the enclosed "Beardslee's Military Telegraph" pamphlet. He commences by asserting that a portable telegraph has long been a *desideratum* which became more apparent during the early stages of the Rebellion, and that the then known system of telegraphing was tried with indifferent success. Admitting this, we start from the same point, with the same facilities for establishing portable lines, but with the possession of a most perfect portable instrument in our favor. Galvanic batteries adapted to actual service in the field are not only possible, but accomplished facts. They are light, simple in construction, and can be transported on pack animals. In the Department of the Gulf, we have "soldier operators," and the "click of our instruments" has been heard and read by civilians "in the roar of cannon and the din of war," full as near the open batteries of Port Hudson as the most enthusiastic admirer of Beardslee's machine ever steadily manipulated his instrument. If the "enemy can connect with the wires of the Union lines" on poles, he can certainly connect more readily with those on the ground, or if "a telegraph station is taken by the enemy," the result would be the same whatever the instrument used. The statement that electric currents from our batteries "require very large conducting wire," can only deceive those unacquainted with telegraphing. Our smallest relay magnet wire, that is no larger than a horse hair, can be worked by our system ten times the distance that he can work one of any size.

It is a proved and established fact that magneto electric currents will not give the certain steady and continuous flow over long lines, that is produced by the galvanic current generated in a battery. Material for galvanic batteries is not necessarily "carried in glass or earthen vessels," but can be transported in "metallic vessels," or even in gunny bags. The most destructible material used in the construction of our field batteries, is sheet copper, and the "considerable supplies" would amount to one pound of material per day for a line of one hundred miles in length. If "officers or even soldiers from the ranks become expert operators with these instruments after a few hours practice;" to be without good operators must be inexcusable folly, hence the frequent errors in the transmission of their messages must result from imperfection of the instruments.

In the last movement of our army up the Bayou Teche, I was ordered to make New Iberia the end of our line. From this place Beardslee's instruments were used with the field wire eighteen miles in length, towards Vermillionville, which they worked in two circuits of nine miles each. The errors in the messages transmitted were so frequent and annoying that we were obliged to send mounted orderlies with all given them. At Port Hudson these instruments were so inefficient and their performance so unsatisfactory, that Major General Banks gave me orders to take and work their wire of three and one-half miles in length, which we connected with our own camp lines and worked to the end of the siege. Major Myer, signal officer, in his communication, states that there is "occasional liability of going to the next letter to the one intended."

The train in the Department of the Gulf, for which they made requisition over the Opelousas Railroad for transportation, consisted of twelve wagons, thirty-eight mules, thirteen horses, fifty-two enlisted men and ten teamsters, with eighteen miles of wire, of which the parties in charge informed me. Our force, which has not been exceeded, consists of fifty-two white men, including operators at twenty-three stations, thirteen negroes, four horses, ten mules, two army wagons and one ambulance, with five hundred and ten miles of line now working. Our lines have been generally kept up with the advance, giving the other system but little opportunity to display the superiority claimed.

In the field before Port Hudson, the Commanding General had an operator at his side who was connected with important points throughout the whole extent of our lines. We are using the same field wire as the Beardslee system, with as much ability to establish lines rapidly. Our perfect instruments are less than two pounds in weight, and their reliability is proved by the success of the Morse telegraph during the past fifteen years. Our whole working apparatus is light, simple in construction, and certain in its operations.

The evidences of the unreliability of the Beardslee machine were fast accumulating, and coming, usually, from sources that could not be charged with partiality, they had great weight. Here was an emergency, and the Chief Signal Officer aimed to meet it by employing and commissioning telegraphists. Among those thus engaged were D. Wilmot Smith and John Thomas. As explanatory of Major Myer's purposes, Mr. Smith writes:

About the last of October, 1863, I received an invitation from the Signal Service to come to Washington and be examined for a commission in that branch of the service. On my arrival there, I learned that it was proposed to try and have the Military Telegraph combined with the Signal Service, and under its control and direction. It was proposed, if I passed examination, that a number of operators throughout the country should be carefully selected, examined, commissioned, and a Military Telegraph organized as part of the Signal Service. I had been telegraphing since 1850 ; for several years superintendent of lines ; had built lines, and was extensively acquainted with operators throughout the country. I was to receive the first commission, and thereupon enter upon the organization of the department and invite operators of known ability to join.

Mr. Thomas' evidence is to the same purport:

I called upon Colonel Myer and had a very pleasant interview. I passed examination successfully. Colonel Myer requested me to remain at the Signal camp on Georgetown Heights, for a time, and make a thorough examination of the telegraphic apparatus. * * I examined the wires, instruments, etc., of the Signal Corps, and speedily settled it in my own mind that, to do effective work, some skilled telegraph operators and Morse instruments were needed.

Of course the facts concerning the inefficiency of the Signal Corps' electric apparatus and of the efforts to make amends by the employment of Morse operators, soon came to the knowledge of Colonel Stager, who saw in the latter move all the evil consequences, if persisted in, that he expressed in the following recommendatory communication to the Secretary of War:

HEAD-QUARTERS U. S. MILITARY TELEGRAPH,
Washington, D. C., October 27, 1863.

HON. EDWIN M. STANTON,
Secretary of War.

SIR: In consequence of the embarrassment already experienced and the complications likely to arise from the organizing of Field Telegraphs by the Signal Corps, I beg leave to submit for your consideration the propriety of placing the Field Telegraphs under the immediate direction of the Military Telegraph Department, and thus avoid the existence of two organizations in the same grade of service. This change can be made without material loss of time, and will not in my opinion impair the operations of the Field Telegraph, but will be made more efficient in the hands of expert telegraphers. It will also relieve a large number of inexperienced men now employed for this work by the Signal Corps. The Military Telegraph Department can successfully operate the field and military electric telegraph in conjunction, using telegraph experts for both as occasion may require, and without detriment to any other branch of the public service, and at a considerable reduction of expense in the working force and machinery required for the two organizations.

The Field Telegraph of the Signal Corps is at present operated by men unacquainted with the practical working of the electric telegraph, and it has failed to meet the requirements of the military authorities. On several occasions where the emergency called for its use, the Commanding General directed the operators of the military telegraph to take possession of the field telegraph wire and operate the same, which was done with entire success, and this, too, after the Signal Corps had repeatedly failed to transmit with accuracy or promptness the important despatches entrusted to it. To overcome these failures and to make the Field Telegraph a feature of the Signal Corps, it has become necessary for it to adopt a system which requires the practical experience of military telegraphers, and the Signal Corps is now making efforts to secure the best electricians in the service by offers of rank and increased pay, which it is enabled to do through its military organization, an advantage not possessed by the Military Telegraph Department.

Two distinct organizations, employing the same machinery, the same class of experts and operating in the same Military Department, can not be conducted without clashing and confusion, nor without doubling the expense necessary for the performance of the telegraphic service. I, therefore, respectfully recommend that the management of all field and military electric telegraphs be confined to the organization known as the United States Military Telegraph Department, or, that that Department be abolished, and the whole business placed under the control of the Signal Corps.

Very respectfully, your obedient servant, ANSON STAGER,

Col. and Supt. U. S. M. T.

From this excellent statement of the case, it would seem that but one intelligent conclusion could be reached, and Mr. Stanton, having arrived at that, with his usual promptness, did not hesitate, by an original and two amendatory orders, to promulgate his judgment, which forever placed electrical operations in the management of electricians who had dearly won his confidence.

WAR DEPARTMENT, ADJUTANT GENERAL'S OFFICE,
Washington City, November 10, 1863.

SPECIAL ORDER 499, *Extract* 20.

Colonel A. J. Myer, Signal Officer United States Army, is hereby relieved from duty in charge of the Bureau of the Signal Corps, and will repair without delay to Memphis, Tenn., reporting his arrival there by letter to the Adjutant General of the Army. He will turn over his office to the next officer in rank of the Signal Corps. All magneto-electric field signal trains and apparatus will be turned over to Colonel A. Stager, Superintendent United States Military Telegraphs, to whom all officers on duty with the same will at once report by letter at this place.

November 13, Special Order 502, Extract 30.—Paragraph 20, Special Orders, No. 499, of November 10, from this office, is amended in its last sentence, as follows: Major Nicodemus, Signal Corps, will turn over all property, trains and apparatus, belonging to the Magneto-electric Field Telegraph Signal Trains, to Colonel A. Stager, Superintendent Military Telegraphs.

By order of the Secretary of War. E. D. TOWNSEND,
Assistant Adjutant General.

Special order No. five hundred and four, extract twenty-nine, further amends four hundred and ninety-nine, by requiring "all officers and enlisted men now on duty with the Signal trains," to report for "temporary duty, to Colonel A. Stager," and on "being relieved from duty," they "will report to the commanding officer of the Signal corps." Thereupon, in the Department of the Gulf, ninety officers and men, fifteen wagons, forty horses, forty mules, twenty-one saddles, nineteen bridles, twenty-eight revolvers, and "*about twenty miles of field wire*," and large quantities of harness, tools, camp equipage, etc., were turned over to Captain Bulkley of the United States Military Telegraph Corps.

Having now the whole field, and the grave responsibilities which expectation and field requirements imposed, Colonel Stager, February 23, 1864, issued a circular order to his assistants, from which the following is extracted:

The object in changing the supervision of the Field Telegraph from the Signal Corps to that of the United States Military Telegraph, was to make the Field Telegraph a more efficient and useful branch of the military service, by placing it in the hands of expert telegraphers. Such officers as are now in charge of field telegraph trains, must see that they are organized in the best possible manner for spring operations. There should be no delay in this respect. It is designed to use the Caton or pocket, or field instrument ; to use number fourteen wire for the field, and to have the wire and batteries carried by pack mules, the wire to be run off from reel on the mule, and the batteries to be worked, as they are packed, from the backs of the mules. Small screw insulators are to be used in putting up field lines, instead of lances, as far as may be possible. These insulators can be taken down very readily. Officers must organize their trains, ascertain how many miles of wire, number of screw insulators, pack saddles, reels, cups, battery, etc., they will require for field service, and make proper requisition. Do not let a want of interest or energy in this matter reflect discredit upon the Military Telegraph.

The United States Military Telegraph Department never used the Beardslee machine. What finally became of those (about forty) owned by the Government, may safely be left to conjecture. It was hinted about Washington that Colonel Myer was seeking recourse to Congress, but if so it is believed it was not pressed. There was a time (about December, 1863) when Secretary Stanton seriously contemplated giving Major Eckert, of the Telegraph Department, the command of the Signal Corps, with rank of Colonel, but it blew over. Such is the history of the electric telegraph operations of the Signal Corps, of its surrender of its adopted calling, and of the assumption by the Telegraph Department of the responsibilities incident to such relinquishment, and the manner, in the main, in which that work was henceforth to be carried forward.

It seems most fitting to close this chapter with quotations from the Report of the Secretary of War, made to the President of the United States, December 5, 1863, and by him submitted to the first session of the Thirty-Eighth Federal Congress touching the services of these two branches of the military service.

The Signal Corps was organized under an Act of the last Congress. The average number of *officers** on duty is reported as one

* Ita ics by author.

hundred and ninety-eight. The full capacity of this branch of the service has not yet been developed, and different opinions as to its value seem to be entertained by commanding officers. In combined land and naval operations it has been considered useful, and commanding officers of Western armies have commended it with favor. The continuance of the corps is recommended by this Department, with proper restrictions upon the number of officers, and limiting them to their proper scope of duties.

The Military Telegraph, under the general direction of Colonel Stager and Major Eckert, * *has been of inestimable value to the service, and no corps has surpassed, few have equaled, the telegraph operators in diligence and devotion to their duties.* From the Superintendent's report, it appears that the military telegraph lines required by the Government have been constructed over an extensive and scattered territory, embracing the District of Columbia, parts of the States of Pennsylvania, Ohio, Indiana, Illinois, Maryland, Delaware, Virginia, North Carolina, South Carolina, Louisiana, Mississippi, Alabama, Arkansas, Tennessee, Kentucky, Missouri, Kansas, and the Indian Territory. Under the immediate direction of Major Eckert, assistant superintendent of the Department of the Potomac, three hundred miles have been constructed during the year. Under the direction of Captain Smith, assistant superintendent of the Department of the Missouri, five hundred and forty-eight miles have been constructed during the year, one mile of which was submarine. Under the direction of Captain Bruch, assistant superintendent of the Department of the Ohio, Cumberland and Mississippi, five hundred and ten miles have been constructed during the year. Under Captain David, assistant superintendent of the Department of West Virginia, ninety-seven miles have been constructed during the same period. Under the direction of Captain Bulkley, assistant superintendent of the Department of the Gulf, three hundred miles have been built, one mile of which was submarine. On the first day of July, 1862, there were three thousand, five hundred and seventy-one miles of land and submarine lines in working order. During the fiscal year, one thousand, seven hundred and fifty-five miles of land and submarine lines were constructed, making a total number of miles of land and submarine military telegraph lines in operation during the year, five thousand, three hundred and twenty-six, being length of line sufficient to girdle more than one-fifth of the circumference of the globe. By a

*Italics by author.

close estimate, it appears that at least one million, two hundred thousand telegrams have been sent and received over the military lines in operation during the fiscal year ending June 30, 1863, being at the rate of about three thousand three hundred per diem. These messages have varied in length from ten to one thousand words and upwards, and generally were of an urgent or important character.

CHAPTER VII.

THE *STATUS* OF FEDERAL AND CONFEDERATE OPERATORS AND THEIR SALARIES.

The position of the National military telegraphers was vexatiously anomalous. They were neither citizen nor military, and yet they were both. This paradoxical statement which explains nothing, may seem unwarranted, but the truth remains that such a contradiction of facts existed. In the Northern armies as in the Southern, the operators were chiefly *non*-commissioned. The exceptions, which were rare, were where officers of not exceeding the rank of captain, were detailed at their request. A very large number of operators were detached from the ranks. Probably every general in the army made such details for temporary service, and until a permanent release could come from the War Department. Some operators like Charles G. Eddy, were detailed against their wish, and compelled for some time to serve at soldier's pay. Eddy thus worked one year for thirteen dollars per month, and perhaps forty cents per day extra, by the side of others, who not having enlisted, were paid seventy dollars and upward. Yet others, like W. H. Woodring and J. B. Morgan, serving in Missouri, were deprived of all bounty because they had not served the required term as "soldiers," notwithstanding they were detailed because they were more useful as telegraphers. Some soldiers were discharged absolutely from the army to enable them to serve as operators. Among these, Milton H. Bassett, Charles I. Fancher, F. H. Bassett and S. S. Ulmer, in the Department of the Gulf, discharged August 17, 1863, contracted in writing, on receipt of their discharges, to serve the Government to the end of the term of their enlistment, at the rate of forty dollars per month, but even that was better than James C. Denny and some other soldier operators received, for they obtained a private's pay. Over one hundred and fifty operators and a large number of repairmen were permanently detailed by the War Department, from the army, to enter the telegraph service.

With few exceptions, soldier-operators were glad to escape the duties of a private, corporal or sergeant for the responsibilities and labors of the army telegrapher, which, while being better remunerated as a rule, were far less monotonous. These soldier-operators, having been disciplined to obedience, and experienced the *ennui*, not to say worse, of camp life, to which they might at any time be returned if disobedient, were usually more reliable than those who could resign at pleasure and enter the service elsewhere or engage with any of the private companies in the free States. The privilege of resigning was always accorded civilian operators as well as officers of the army, but any attempt to quit in a body would not be tolerated on the part of either.

The urgent necessity for operators was felt very early in the war, and, as many entered the armies as soldiers or officers, and new offices were opened, the want of expert telegraphers became very great; even experienced line-men were becoming scarce. Consequently, as early as August, 1862, Colonel Stager successfully represented the evil consequences to result if these were to be subject to conscription. The Secretary of War, thereupon, issued the following order, which was telegraphed by the Associated Press all over the country:

WAR DEPARTMENT, Washington, August 5, 1862

ORDERED:

That, the use of the telegraph lines being required for military purposes, all persons actually employed in constructing and operating telegraph lines, at the date of the order calling for three hundred thousand men, be exempt from military duty so long as they remain in such service.

By order of the President.

(Signed) EDWIN M. STANTON,
Secretary of War.

Although it is believed this was never modified, yet in subsequent drafts it sometimes happened that the marshals included the names of operators. Doubtless, these could have secured immunity therefrom, had they in apt time claimed exemption under the order of the President; but this, some at least appear not to have been advised of. Thus operators *while actually engaged in the Federal service, in exposed parts of the theater of the war*, were actually enrolled for draft, and some drew the prize (?). Telegraphists in the corps at Little Rock, Arkansas, Vicksburg, Mississippi, and perhaps other advance points, were thus enrolled, and some in the Department of Missouri were actually forced to proceed to conscript camps for muster in the army. These proceedings reflected greater disgrace upon the service than anything else during the war, and it is to be regretted that telegraphers did not at that time insist upon the exemption of their drafted co-laborers, irrespective of the foregoing order and on the sole ground of the military service they were in when conscripted. It was the opportune moment for an understanding. A volunteer soldier would have resented enrollment for draft in unmistakable terms. An operator who had been detailed from the ranks, because of the greater need of his services as telegrapher, would have felt equally indignant; why should not a *volunteer* operator, for whom and whose, no bounty or pension laws were ever passed? Conscripts were not looked upon with much favor in the army. To draft one was to say to him, "If you will not serve your country in this hour of its peril, we will make you." A Federal operator ought to have been allowed to reply, "I am serving my country." Duncan T. Bacon, who was drafted in New York, took the proper course when he insisted on an exemption from enrollment on the ground that, as an operator, he had already served in the army. This is the only instance of the kind that has come to my knowledge, and it is pleasing to note that the officers of the Government credited him with his service the same as if he had been mustered in as a soldier.

Captain Fuller was officially advised that Federal operators from Ohio were entitled to vote with the soldiers, because they were "in the actual military service of the United States."

There is always a certain amount of looseness in matters of rule and routine in armies, during actual war, and hence, *all* operators were not required to take the telegrapher's oath, but nearly all did take it, and it was Colonel Stager's direction that all should. No harm ever came from any omission, however.

These oaths were mainly formulated by the department superintendents, one of which is as follows:

I do solemnly swear (or affirm) that I will bear true allegiance to the United States of America, and that I will support and sustain the Constitution and laws made in pursuance thereof, as the supreme law of the land, anything in any State Constitution or laws to the contrary notwithstanding. That I will not take up arms against the United States, or give aid and comfort to the enemies thereof, or to any authority or pretended authority that is or may hereafter be engaged in armed hostility thereto; and that I disclaim all fellowship with the so-called Confederate States and Confederate armies.

I do further swear that I will not reveal to any person or persons the contents of any despatch, report or other communication, either directly or indirectly, that may come to my knowledge through my connection with the telegraph, in any manner whatever. That I will not reveal or divulge to any person or persons any cipher that may be given me for United States military purposes; and that I will faithfully keep and observe this my solemn oath of secrecy and allegiance to the Government of the United States of America.

Subscribed and sworn to by..............................
Before me, this........day of.............186

More complaint has been uttered by telegraphers on account of their anomalous position in the service, and its consequent evils, than all other causes united. Performing tasks of extreme responsibility and delicacy—tasks that required extraordinary care and diligence and such skill and confidence as only comes from long practice at the key—the Federal military operator was but a citizen, and in most departments so dressed, although surrounded by all the trappings of war and located at the headquarters of its "pomp and circumstance." Intellectually, at his age, he had few superiors. His bravery was so splendid that the rare exceptions were overwhelmed by the multitude of heroic acts he performed. It is unnecessary to compare the importance of his work with the grand service of brave officers, for they concede it. *No other service was so confided in.* The secret of a contemplated movement of a large army is a great trust. Mr. Shanks, a newspaper reporter, whose business it was to "worm out" interesting items by all the recognized arts of the profession, wrote that the operator was "generally considered proof against fevers of excitement."

A commission, it is true, was not needed to give authority, nor was it extremely essential as promoting discipline, but it was warranted by the service, and might at least have been given as a fitting reward of merit. When Mr. Stager recommended an organization of the Telegraph Department, the statesmen, North and South, believed the war would cease within six months; wherefore the problem was, how to serve the country under the laws as they stood, rather than to organize, by virtue of special legislation to be had, a corps theretofore unknown. We have seen that, in the plan recommended, it was not contemplated that any body, not even the chief, should be commissioned, but that, upon his appointment, it appearing necessary, to protect the Quarter-master General, Mr. Stager was made Captain and Assistant Quarter-master at the suggestion of General Meigs. The other original appointments that followed long after were to secure Colonel Stager, not to reward anybody, though the appointees deserved it; nor to enhance discipline in the service, but as a protection to the Chief in advancing large sums of Government money, and to facilitate business with local quarter-masters. Operators never found fault with these appointments, but rather rejoiced in them. The *gravamen* of their complaint lay in this, that they were as patriotic and serviceable as army men, and yet oft-times treated as mere civilians, although doing purely military duty, and in the conduct of their work, officers of little rank not infrequently by disparaging remark or conduct induced generally by jealousy, made them quite sensitive and dissatisfied. To remedy some of these petty ills, it was proposed by operators in the West, to obtain authority to wear a uniform, without distinctive marks representing rank, a thing probably quite as strange in an army as was the indefinite position they occupied. General Thomas was requested to authorize the wearing of the uniform agreed upon, which he cheerfully did as follows:

HEAD-QUARTERS DEPARTMENT OF THE CUMBERLAND,
Chattanooga, Tenn., March 26, 1864.

GENERAL ORDER NO. 51.

The Telegraph Operators in this Department are hereby authorized to wear an undress uniform, described as follows: Blouse, dark blue; trousers, dark blue cloth, with silver cord one-eighth of an inch in diameter along the outer seam; vest, buff, white, or blue; forage cap, like that worn by commissioned officers, but without any distinctive mark or ornament; buttons, like those worn by officers of the general staff.

By command of Major General Thomas.

W. D. WHIPPLE, *Asst. Adjt. Gen.*

July 5, 1864, General McPherson, commanding the Department of the Tennessee, issued a like order, except that it authorized wearing a small silver cord around the cap band.

Thus dressed, the operators were regarded as officers by all not personally acquainted, and the result was highly gratifying for a time. One evil result was that about the time they obtained their new uniform, which was very pretty, they took to homesickness. A short visit, or slight delay, restored contentment. It was extremely amusing at times to note the embarrassment of officers and soldiers in their vain efforts to recognize operators by their proper rank. To err complimentarily, if mistake they must, they often gave operators much higher rank than they ever aspired to. This occurred the more often because they rode with the General's staff, and, being in the *best* company in the army, even a colonelcy was sometimes vouchsafed these knights of the key and professors of mysterious ciphers, who were not even "high privates." While this state of things was novel, it was entertaining; but in time too many knew that the operators were in borrowed plumage, and, although loaned by high authority, it began to seem like "putting on airs." Field operators, however, found it necessary to continue wearing the uniform, but many post operators discarded theirs, to avoid embarrassing questions, such as "What is your rank?" What regiment do you belong to?" and the like.

For these and other reasons, the question of commissions broke out afresh and with renewed energy. In the Department of the Cumberland, Captain VanDuzer, who always writes vigorously, wrote to his superior, Captain Bruch, as follows:

Our men don't belong to anybody when in the field. They have no recognized *status*, and every officer, of whatever grade, who happens to command a post where an office is maintained, has to be taught that the operator is not a servant—not subject to his (the officer's) orders in regard to his duties. Another trouble, and that no small one in practice, though it may seem so in a complaint, is the manner in which operators are too often treated, with regard to quarters and messing arrangements. I have had an operator (a gentleman, too), whom I had sent to the head-quarters of a Major General, in the field, assigned to the quarters and mess of the orderlies, very cavalierly informed that the "other messes were full." One of my best cipher men, attached to a general, in eminent command, was forbidden by a staff quarter-master to ride in the car with the staff, and ordered to ride with the servants. I am glad to say that the order was disobeyed by my man, and revoked by the chief of staff. I laugh at these annoyances when I hear of them, and, by making light of them, keep the men from taking them up seriously, but they have a very depressing effect upon the service, I assure you. I am not asked to suggest a remedy, but as I am volunteering the complaints, I may as well proffer one, and that is, to give us a thorough military organization. Make operators officers, cadets, second and first lieutenants.*

Captain Bruch referred this to Colonel Stager, endorsing it as follows:

I would earnestly request that the matter be laid before the Secretary of War, and that some system be adopted which will enable us to keep good operators in the military telegraph service, and which will secure us from the many abuses which operators are at present forced to submit to from superior officers in the army, who are not in the telegraph service.

Prior to this, Captain Bruch had, in an annual report, recommended the consolidation of the Telegraph Department with the Signal Corps; but which organization should control, and how operators were to be affected, was left to be inferred. Captain Fuller, about the same time (September, 1863), in his report, said:

In my opinion, the efficiency of the Telegraph Corps would be improved by giving the operators a *status* in the army, affixing a severe penalty for desertion of duty or malfeasance. Their duties are very confining, giving them but little opportunity to provide for their personal wants, and more definite instructions to commanding officers to provide suitable quarters for offices located at outpost stations, should be given. In several instances, my opera-

* Assistant Superintendents were but captains then.

tors have been denied messing privileges with the officers and been compelled to beg the privilege of messing with the private soldiers. These instances have been the exceptions, however. Officers, generally, have treated us with courtesy and respect.

Some time anterior to these and other like communications from other officials, the operators in the Department of the Potomac united in the following petition to Colonel Stager:

WASHINGTON, D. C., Dec. 3, 1862

COLONEL ANSON STAGER,

General Superintendent U. S. Military Telegraphs, Washington, D. C.:

DEAR SIR: The undersigned operators, employed in the United States Military Telegraph Department, respectfully submit to your consideration the following: The Honorable Secretary of War, in his late report to the President, says, in reference to the Telegraph Corps, "*In diligence, fidelity and important aid, they have been unsurpassed by any branch of the service.*" You are, no doubt, aware that the operators have performed the more important part of these duties under your supervision, overcoming great difficulties and doing service under disadvantages never before experienced in American telegraphy; yet, as you know, we are but poorly remunerated for these important services, especially when compared with other branches of the public service. For instance, clerks in the Quarter-master, Commissary, or any other department, ever so unimportant, have received but seldom less than seventy-five dollars per month, while the majority have over one hundred. Telegraph operators have but sixty dollars, and that under the disadvantage of often changing quarters and paying high prices for everything. While we are ready, as heretofore, to use all means in our power to perform our duties, we are also compelled to call your attention to the steady advancing prices of all necessaries of life, and that with the compensation now allowed us it will soon be impossible to respectably maintain our position. While most of the employés of the Government in work-shops, navy yards, arsenals and also employés in private establishments, have successfully demanded an increase of wages for their services, on this account we have forborne to do so until now, when we think the present session of Congress presents an opportunity for you to recommend a *telegraphic organization* similar to the Engineer Corps or other branches of public service, and be remunerated according to the value of our services. We appeal to you as our "head," and trust to your magnanimity to devise means to comply with our request. We beg you to transmit a copy of this to the Secretary of War.

(Signed) Very respectfully yours,

Jos. Schnell, Jr.,
Geo. W. Nail,
R. Cunningham,
J. H. Nichols,
Albert W. Orton,
G. J. Lawrence,
T. H. Fonda,
W. H. Eckman,
E. Rosewater,
John Allen,
Henry H. Matlock,
John S. Kerbey,
Richard Graham,
Edward Conway,
Geo. Garland,
F. Lamb,
W. K. De Witte,
J. L. Cherry,
Wm. C. Hall,
C. W. Jacques,
Frank Drummond,
A. H. Bliss,
F. T. Bickford,
John Winthrup,
C. W. Moore,
R. F. Weitbrecht,
H. C. Buell,
Edward F. Chapman,
Henry W. Cowan,
C. H. Lounsberry,
H. P. Royce,
Charles I. Brown,
Geo. E. Cromwell,
M. V. B. Buell,
O. H. Dorrence,
Thomas N. Loucks,
R. H. Ryan,
W. F. Holloway.
M. Y. Holly,
P. H. Nunan,
Thos. Morrison,
Silas C. Burns,
W. N. Embree,
R. O'Brien,
Thos. E. Moreland,
J. H. Glazier,
J. W. Parsons,
J. H. Bunnell,
A. H. Kanode,
J. W. Sampson,
J. O'Brien,
C. L. Snyder,
J. H. Emerick,
Chas. R. Roberts,
W. G. Baldwin,
D. McCarty,

It is true that the subject of salaries embraces the major part of this petition, but that was because it seemed to be the only point requiring argument. Owing to representations like these, and corroborating observations made by Colonel Stager and Major Eckert, those gentlemen, accompanied by P. H. Watson, Assistant Secretary of War, had an audience with Mr. Stanton on the subject of a more perfect organization, whereby operators would be commissioned and remunerated according to their respective deserts. The Secretary listened to reasons adduced and then said that he was well satisfied with the present efficiency of the service; that he did not want the operators commissioned, because that would only enable every superior officer to command them; a result he was most anxious to avoid. Of course, the Secretary being unconvinced, his views prevailed, but had he and President Lincoln lived, there is no doubt ample honors would have been accorded the operator at the close of the war.

Several officers endeavored to obtain commissions and staff appointments for their operators. Among these was General Burnside, who tried hard to have his operators, Jacques and Johns, commissioned and assigned to his staff.

The salaries of operators were raised from time to time, so that at the close of the war they were paid from seventy-five to one hundred and fifty dollars per month and were well provided with horses, wagons, tents, fuel, rations, servants, etc.

It is believed that but one strike was attempted among Union military operators, and that was at a time in the Department of the Ohio when even the highest salaries would have been a poor recompense for dwelling in the poorly protected and benighted sections of Kentucky and East Tennessee, where most of the complaining operators were located. The fact that they *could* resign and leave such lonesome positions favored their unrest. The further fact that they were regarded as mere civilians was also disquieting. So they presented this petition:

DEPARTMENT OF THE OHIO,
January 17, 1864.

W. L. GROSS,

Captain and Assistant Superintendent Military Telegraphs, Danville, Ky.:

WHEREAS, we, the telegraph operators in the military service in the Department of the Ohio, having a more important and laborious duty to perform than most any other branch of the military service, and our present salary does not sufficiently compensate us for the many inconveniences, too numerous to mention, under which we labor, therefore, *Resolved*, first, that we, the undersigned, telegraph operators in the military service of the United States, at work in the Department of the Ohio, hereby solemnly affirm that we will not and can not possibly work for less than one hundred dollars per month. *Resolved*, second, that we ask to be immediately relieved unless our request is complied with. *Resolved*, third, that we sincerely hope it will meet with the approval of our Superintendent.

(Signed) J. F. WILBER, operator, Barboursville, Ky.,
J. R. BROWN, " Barboursville, Ky.,
J. G. GOALDING, " Cumberland Gap,
GEORGE PURDON, " Tazewell, Tenn.,
W. T. STRUBBE, " Tazewell, Tenn.,
JOE ANDERSON, " Strawberry Plains, Tenn.,
GEORGE COLE, " Loudon, Tenn.,
G. D. WILKINSON, " Knoxville, Tenn.,
G. JONES, " Mossy Creek, Tenn.,
W. H. SISSON, " East Tennessee,
SAM ROBERTS " East Tennessee,
T. M. SAMPSON, " Crab Orchard, Ky.,
T. G. HARVEY, " Stanford, Ky.,
J. M. SPENCER, " Lexington, Ky.,
J. W. CURTISS, " Mt. Sterling, Ky.,
A. W. HOWELL, " Somerset, Ky.,
JOHN NAGLE, " Mt. Vernon, Ky.,
JOE H. GILKERSON, " Columbia, Ky.,
F. BENNER, " Danville, Ky.

This petition was referred to Captain Bruch without comment, and likewise by him to Colonel Stager, who returned it with this endorsement, dated January 30, 1864:

> Communications bearing the letter and spirit of the one returned herewith, will not be entertained and can be looked upon in no other light than as a conspiracy by the parties, against the Government. The resignation of these individuals will be accepted singly, whenever you are able to supply their places. If operators or employés attempt to quit the service in combined bodies, they shall be put under military arrest and held as conspirators. Operators will be paid salaries according to the grade of their competency, as per authority given you in my letter of January 28, 1864.

In that letter, Colonel Stager stated that if there were any competent operators in the service who were paid fifty, sixty or seventy dollars per month, and whose services were worth more to the Government in such capacity, their pay might be increased according to the grade of their competency. "Please ascertain who originated the 'strike.'" The story of this attempted strike is almost ended without its beginning—so short is it. The idea of being treated as conspirators brought the boys to their senses, and right or wrong in their demands, they saw that the course adopted was the worst conceivable, in a land held by armies which depended so largely on the telegraph. Whether Colonel Stager's letter *ante*-dating but two days his endorsement, was in anticipation of the petition I know not, but whether fortuitous or studied, it worked well, and many of the operators did find it "possible" to live for less than one hundred dollars per month, although none who have tried it there would do it again, except for the "cause."

The recital of the foregoing affair is a reminder of a much more interesting and elaborate experiment across the "bloody chasm." Southern telegraphers had for some time prior to the organization of the "Southern Telegraphic Association," been smarting under grievous wrongs which they, true to the style of the "Declaration of Independence," recounted in the Preamble adopted by them in Augusta, Ga., October 26, 1863. Chief among the evils therein complained of, were these: * * "Our rights have not been respected by the various telegraph companies, and they have recently used the conscript law of the Confederate States as a means to intimidate us to succumb to demands we consider unfair and tyrannical." We are "required to enter on duty at seven or eight A. M., and are never off until ten P. M., and often kept on till two to five A. M., without extra compensation, * * also * * to work on the Sabbath from the hours of eight to ten A. M., and from seven to nine P. M." Our "salaries are entirely inadequate to defray the expenses of living in the simplest manner;" we claim "as an inalienable right, the privilege of resigning our positions on one line, to accept one on another; and denounce and will resist any attempt to prevent members of this association from changing their employers, either by a refusal to accept their resignations, conscripting or ordering them to their commands *in transit*, as unjust and illegal, as shown in the late decision of his Honor, Judge MaGrath, in South Carolina." This association was regularly organized in all respects. Its chief officers were, a President, a Senior Vice President, one Junior Vice President from each State, a Secretary and Treasurer. The organizers were Charles A. Gaston, of Mobile, J. Sisson Clarke, of Charleston, Charles F. Barnes, of Augusta, E. J. Saville, cf Mobile, W. H. Clarke, of Savannah, William Sandford, of Mobile, J. W. Kates, of the military lines about Charleston, W. A. McMurchy and perhaps some others.

This society, in the terms of its organic laws, probably took a bolder stand against capital than any other similar telegraphic association in this country. It undertook to obtain lists "of the operators in their respective States; their salaries; where stationed;" to interdict its members from working "in any office requiring more than one operator, unless the other operators in such office are members of this Association;" to prohibit students from acquiring the art unless they obtained leave from the President and paid the association one hundred dollars, more or less, at that officer's discretion, and agreed to join the association when qualified; "to sustain as far as may be right, its members in all difficulties" with employers; to arrogate to itself obedience as against "any order, emanating from whatever source, which shall conflict with its constitution or by-laws," and in case such obedience, in spite of conflicting orders from employers, works injury to the telegrapher, every operator on the line whereon he served was to be directed to "immediately cease sending or receiving" commercial business. It also arranged the following schedule of working hours, viz: from October 1, to April 1, 8:30 A. M., to 8 P. M.; April 1, to October 1, 8 A. M., to 7:30 P. M.; Sundays, 8:30 A. M., to 9:30 A. M., and 6 to 7 P. M.; same for holidays, and no private business to be done on the Sabbath. When the Government exigencies required longer hours, the companies should pay at the rate of fifty cents per hour therefor, and finally it sought to "maintain a rate of compensation." As the Southern Telegraph Company, of which Doctor W. S. Morris was President, was said to be the chief instrument of oppression, so the membership of this association, which was very large, was obtained principally from the employés of that company, and by virtue-thereof, such of its members were extremely gratulant and confident.

At this time operators' salaries in the Confederacy, ranged about five hundred dollars per month. One would think they ought to have been contented, but they lived so extravagantly that their board cost four hundred dollars per month; their elegant cow-hide boots, two hundred dollars per pair, and their tasteful and *real* American home spun suits, seven hundred dollars. Other essentials in proportion. It was a rare era in that land when everything but money was a luxury.

President Charles A. Gaston, confidential operator with General Lee, issued an excellent address to "all managers of telegraph offices and telegraph operators in the Confederate States," exhorting co-operation. In that address, we note the following, which deserves to be preserved, and is therefore here inserted: "There is no agency known to society that so fully challenges the confidence of man. It is entrusted with every one's secrets, and employed in the furtherance of every one's concerns, no matter whether public or private, trivial or momentous." Doctor Morris did not long hesitate before ordering all employés to withdraw from the Association, or be dismissed. As a dismissal would leave the operators liable to do military duty, and that as conscripts and among strange regiments, a dilemma was at once presented, but notwithstanding, the Acting President (Gaston being with the army) retaliated by forbidding members from receiving or transmitting any telegrams, except such as were on government, railroad, or express business. Thus an issue was fairly made. A dead lock ensued, lasting a week or ten days. Operators were dismissed, and for a time working operators were as scarce in the large offices as during the great strike on the Western Union lines, in 1870, when business was refused by the managers. The name of every discharged operator was sent to the conscript officer in his vicinity, and, it is said, in some cases a guard entered offices and escorted the operators to camp, where they earned seventeen dollars per month, musty bacon and sour corn meal, as conscripts. But this state of things could not last. The Confederacy had not operators enough to make soldiers of them in large numbers, and it was soon evident that Doctor Morris could obtain their detail, and thus *compel* them to serve at soldiers' pay. It was this, more than being forced into the army, that put a *quietus* on the strike. Most of the operators returned to duty. It was the first and last strike in the Confederacy, and it was also the end of the Southern Telegraphic Association.

Two other organizations took root later in the war, and died much more innocently. These were the "South-western Telegraphic Association," and the "Confederate States Telegraphic Club." The former, started by E. J. Saville, Geo. W. Trabue, John B. Morris and E. H. Hogshead, was born in Meridian, Miss., May 1, 1864. Its purposes were merely mutually beneficial in a general way, with no thought of antagonizing any thing or body, for the operators and officers of the South-western (C. S.) constituted a model happy family, in which mutual confidence, esteem and courtesy prevailed. The following were its members: E. L. Marchant, operator at Jackson, Miss.; J. H. Henderson, at Grenada; G. B. McMurchy, at Meridian; John B. Morris, at Corinth; Geo. W. Trabue, at Mobile, Ala.; E. H. Hogshead, at Meridian; E. J. Saville, at Mobile, Ala.; E. Howard, at Meridian; F. C. Whitthorne, with Forrest's Cavalry; Geo. M. Dugan; M. Flippen, at Brandon; Charles St. John, at Brandon; A. M. Record, at Morton; W. H. Young, at Buckatunna; C. I. DePew, at Citronelle; E. Garaudy, at Enterprise; James M. Osborn, military operator; George W.

Beard, at Canton; Thomas F. Marshall, at Grenada; E. A. Fulda, at Jackson; John A. Galbreath, Jr., at Jackson; Wm. A. L'Hommedieu, at Jackson; A. G. McN. Russell, at Jackson; N. C. David, at Oxford; B. W. Collier, at Oxford; C. W. Montgomery, military service; L. T. Lindsey, military service; M. W. Barr, military service.

At the first annual meeting a quorum was wanting, notwithstanding the efforts of its president, the lamented and talented John B. Morris, ably seconded by E. H. Hogshead, the secretary.

The "Club" was of Charleston (S. C.) origin, and was formed January 9, 1865. The constitution and by-laws of the "Southwestern," somewhat modified, were adopted, and T. M. Bryan elected President, E. N. Woodhouse, Vice President, John D. Harbers, Secretary, Samuel S. Turner, Treasurer. The three Directors were W. D. S. Anderson, J. D. Westerveldt and R H. Pegues. This also died while teething.

A number of Confederate operators were commissioned and held staff appointments. When the civil telegraph force proved insufficient, these officers assisted. They usually accompanied the cavalry leaders, with a view in part to tapping Union wires; bnt others went with infantry columns, doing telegraphing incidentally only, When General Hood advanced against Thomas, at Nashville, it was reported by a captured Confederate operator, that *Major* M. William Barr was chief of the telegraphic operations conducted for Hood's army; that William Allen accompanied the cavalry under General Wheeler; J. M. Powers, that under General Roddy; and Whitthorne, that under General Forrest. There were other army operators, some of whom were commissioned. These made telegraphing but an adjunct to their other duties, and were not carried on the rolls as operators. Southern telegraphers were subordinated to a greater extent than Union operators, to the direction of army officers, for the reason that there was no cohesion in the telegraph branch of the Confederate service—no general head—at best, only department chiefs, and they, it is believed, never reported to the War or Postal Departments, in Richmond, but only to the officer commanding the District. Some of them at least were obliged to take an oath of fealty, but how generally this was practiced, we have not inquired. To a considerable extent, they were excepted from the rigors of conscription. Even in those dark days of the Confederacy, when its armies were fast crumbling by desertion, disease and the bullet, we find that the telegraphists were exempted from actual service; but this immunity seems to have resulted from a wise discretion, rather than by virtue of any positive enactment, in which respect it accorded with the conduct of the national authorities. General Withers, commanding the Reserves, with head-quarters in Montgomery, Ala., ordered all, or nearly all, of the operators in Alabama, to join the forces in camp, and, in excuse for so doing, wrote that "the Secretary of War decides that all able-bodied men shall be put into the army." Lieutenant General Richard Taylor, foreseeing that such measures would isolate him from his Department, telegraphed as follows:

SELMA, ALA., November 10, 1864.

To HON. SECRETARY OF WAR, Richmond, Va.:

Please authorize Major General Withers, commanding conscripts and reserves in Alabama, by telegraph, to use that discretion which General Kemper has been permitted to use, in Virginia, relative to sending to camp all telegraphers and steamboat men. Unless this is done immediately, he will literally execute his orders, the effect of which will be to stop all telegraphs and steamboats in this Department, and work great inconvenience and perhaps disaster to the service. Please reply by telegraph. Respectfully,

R. TAYLOR, *Lieut. Gen.*

In Virginia, North Carolina, South Carolina and Georgia, the Secretary of War permitted the generals of reserves to exercise a proper discretion in the matter of taking telegraph operators from their offices. In some cases they were detailed and returned to duty in their offices. In Virginia, no details were issued, but in lieu thereof was the following, a copy of which each Southern operator received:

HEAD-QUARTERS RESERVE FORCES OF VIRGINIA,
Richmond, October 5, 1864.

Telegraph operators will not be interfered with under the orders for organizing the local forces of the State. By order of Major General Kemper. R. H. CATLETT, *A. A. G.*

General Taylor's request was granted so far, at least, as operators were concerned.

CHAPTER VIII.

THE TELEGRAPH IN EASTERN VIRGINIA FROM JULY, 1863, TO JULY, 1864.—SAVED FROM EXECUTION BY TELEGRAPHIC REPRIEVES.—MOVEMENTS OF THE ARMIES OF THE POTOMAC AND JAMES.—WILDERNESS.—SPOTTSYLVANIA.—COLD HARBOR.—PETERSBURG.

General Meade's army was not destined to remain long about the banks of the Potomac. July 19 and 20, 1863, it crossed near Berlin, Maryland, and immediately moved to Gainsville, Virginia. But the telegraph was extended, on the eighteenth, from Berlin to Lovettsville, where the General had his head-quarters. This office was worked, until the twentieth, by operators Caldwell, Emerick and Rose, Pierce having left on the sixteenth, for a short time. After a five days ride over a beautiful country, during which they enjoyed the splendid blackberries that abound there, and the excellent repasts that the superintendent of construction, D. Doren, foraged, the head-quarters operators having reached Warrenton were, thanks to the energy of Eckert's linemen, enabled once more to communicate direct with the War Department office. Within a few days, the army reached the line of the Orange & Alexandria Railroad, and two wires were kept in constant working order from Washington to the front, one being used exclusively for military and the other for railroad business. Besides these, other lines were built from General Meade's to all the corps and some of the division head-quarters. They proved to be of very great value to the army, keeping all its parts in quick communication with each other and with the National capital. In August, L. A. Rose (and C. Douglass from the eighteenth) operated at the head-quarters of the Second Corps at Morrisville; H. W. Cowan and W. C. Hall, at Bealeton; J. H. Glazier, J. D. Flynn and E. A. Hall, at Manassas Junction; D. E. Rand, at Burkes Station; R. H. Brigham, at Centreville; J. D. Tinney and A. H. Bliss, at Rappahannock

Station; R. F. Weitbrecht and W. H. Shreffler, at Kelley's Ford; R. R. McCain and J. A. Murray, at Hartwood; R. H. Ryan, at Warrenton; S. H. Edwards and R. Graham, at Warrenton Junction; and W. D. Wood and F. N. Benson, at Fairfax Station. Messrs. Bates, Tinker, Chandler, McCarty, Orten, Kanode, Morrison, Burns, Safford, Barron, Nail, Nichols, Dwight, Loucks, Moreland, Kettels, Ney and C. L. Snyder were variously stationed in Washington, and Dorrence, Maiden, Power, Flack and Garland in Alexandria offices.

It was about this time that sentence of death was to be executed upon a soldier charged with desertion. His case had been submitted to the tender-hearted President, who granted a pardon, a few hours only prior to the time when the soldier was to be executed. But before the President's message could be sent, telegraphic communication was interrupted. O. H. Dorrence, chief of the railroad operators at Alexandria, took the despatch by special train to the break in the line west of Burkes Station and telegraphed it just in time to stay the execution of the soldier.

Similar occurrences are by no means unknown. Mr. Shaffner, in his "Telegraph Manual," relates with thrilling particularity how he succeeded, while in St Louis, Mo., in obtaining from President Millard Fillmore, in 1851, a telegraphic respite in behalf of a young Indian, who, to save his aged father from an ignoble death for a murder the parent was innocent of, had confessed himself guilty of the dreadful crime, of which he also was guiltless, and being convicted, mainly on that confession, was on his way to the scaffold when the order suspending the sentence, which reached St. Louis by three telegraphic routes, was placed in the hands of the marshal, at Jefferson City, Mo.

In the early part of the war the Confederate military authorities at Knoxville, Tenn., convicted an East Tennesseean of bridge-burning, for which he was sentenced to be hung at two o'clock in the afternoon of a certain day. On the morning of the fatal day, a daughter of the condemned man telegraphed from Knoxville, to President Davis, at Richmond, beseeching him in pitiful terms to have mercy on her father, and save him from the terrible doom fast approaching. D. E. Norris forwarded the message to Lynchburg, where it was relayed to Richmond, by operator S——, at half past nine, A. M. About noon, "Rd" (Richmond) called "Bg" (Lynchburg), saying: "Here it is! Rush it through." S—— hurried with it to the Western instrument, called "Kd" (Knoxville) twice, and was answered, but before the reprieve could be told, the operator at Dublin Depot put on his ground wire, thereby practically severing the line, only one end of which ran to his instrument; that unfortunately was the western end, and the purpose doubtless was to enable the Dublin operator to dispose of his telegraph business westward without interruption from Eastern operators. Perhaps he had to receive important train orders, after which he probably forgot that his ground wire was on. At any rate, S—— remained at his instrument until fifteen minutes before the time for the execution of the condemned, growing more and more nervous as each moment passed, and when the last flickering ray of hope was about to expire, the ground wire was removed, when S—— quickly called Knoxville. He answered. Davis' message, staying the execution, was rapidly transmitted, but Norris said he feared it was too late, as the officers of the law had left the jail with the prisoner sometime before. However, the message was received in time, and the father was ultimately restored to his affectionate child.

Another instance: During General Hunter's raid through the valley of Viginia, two private soldiers—substitutes—exhausted by long marches and scant food, left behind, fell into the enemy's hands. It is related that, after being captured they were heavily ironed and closely guarded for some days; that the Confederates, to prevail upon them to enter their armies, threatened them with death unless they joined the Confederate service—a dread *ultimatum*, which induced acceptance in the secret hope of escaping. They accompanied the command of the notorious Harry Gilmore, in a raid upon the Baltimore & Ohio Railroad, and were captured by the Federals and taken to Harpers Ferry in irons. There they were recognized, accused of bounty-jumping, desertion and bearing arms against the Government. Appearances were all against them, and though they told a consistent story as above related, it was disbelieved, and they were sentenced to be shot. The day preceding the time fixed for the execution, a Catholic Chaplain heard the prisoners' story and the corroborative statements of a rebel prisoner, whose evidence was not heard by the court martial, whereupon Father F——, the priest, telegraphed the President his understanding of the case, and asked mercy for the prisoners. D. J. Ludwig, operator at the Ferry, anxiously awaited for eighteen hours a reply from the President, but at daylight all the wires to Washington and Baltimore were down. At noon they were yet as still as death. No current could be "tasted." The prisoners had but one hour yet to live. Sadly and anxiously Ludwig sat adjusting his instrument for the slightest current. He saw the procession pass by, in which were the condemned, sitting upon their coffins. He heard the band play the "Dead March," but though his hope waned, his vigilance did not relax. At ten o'clock that forenoon, W. D. Gentry, operator in Baltimore City office, received a telegraphic reprieve for the condemned men, who were eighty miles away. Feeling the trying responsibility thrust upon him, Gentry tried every means then known to telegraphy to raise Harpers Ferry office, but was unsuccessful. At a quarter past twelve he heard the armature of the magnet on that line splutter as if somebody was at work with two ends of the wire; an instant more the lever of the sounder fell with a heavy click, and before Gentry could make "H. F." twice, Ludwig answered and received the respite and dispatched his mounted orderly therewith before stopping to give O K to Gentry. The place fixed for the execution was some two miles away, on the heights beyond the little village of Bolivar, and the road for nearly a mile extended up the side of a mountain. The orderly lost no time, but whipped his horse into his fastest gait, and maintained it to the end. As he approached the party having the dread work in hand, and discovered the soldiers who were to fire, were getting into position, the orderly, yet several hundred yards away, redoubled his exertions to urge his horse, which, when within a few yards of the prisoners, gave out and fell to the ground, *dead*. His com ing was however, heard by the soldiers, who stood awaiting the signal to fire. They rested their guns, and the doomed men were saved.

The army having left the neighborhood of the Potomac, operator Douglass, at Poolesville, Md., was in constant danger of being carried off by guerrillas from the Virginia side. But

Douglass, accustomed to take such chances—for it was here that he, Weitbrecht and the repairer held councils of war nearly every night during the Gettysburg campaign, and Cherry had had numerous adventures—had become less nervous than formerly. About midnight, however, Aug. 16, the enemy came and found Douglass in bed. They took the telegraph instrument and mounted the operator behind one of his captors. The guerrillas broke open store doors and Douglass assisted in stowing goods into the bags the rebels had brought. They soon became so deeply interested in their work, that the operator, keenly intent on escaping, was enabled quietly to pass out and enter an adjoining corn-field, where he remained until the raiders left town, after which he returned to his office; connected the wires and went to bed again.

After several Federal cavalry incursions, one of which resulted in the re-capture of two gun-boats in the Rappahannock, a part of the army about the middle of September, advanced to the Rapidan, and the rest to the neighborhood of Culpeper Court House. W. C. Hall and Rose, with General Warren, opened an office at that place in the brick building just vacated by the rebel operators, and Meade's telegraphists, reinforced by James Caldwell, were located in Wallach's house, half a mile from the village. Other offices were opened at the front.

Meade was about to cross the Rapidan with his whole force, when he was directed, notwithstanding his protests, to detach the Eleventh and Twelfth Corps, from his army in aid of the troops at Chattanooga, whither Lee had previously sent Longstreet's corps, but being reinforced in October, Meade would have advanced, had not Lee already begun moving on his right with the intention of getting position between the capital and the Union army. This move was foiled by Meade's rapid retreat, in consequence of which, Lee again fell back, destroying the railroad and telegraphs. During this retrograde movement, the telegraph performed an important part, keeping Meade constantly informed of the position of his forces and transmitting intelligence of the enemy as the same was obtained. An animated telegraphic discussion between Generals Halleck and Meade followed the retirement of Lee; Meade proposing to possess the heights of Fredericksburg with his army. So sharp was this correspondence, that a change of commanders by Meade's desire, was seriously proposed, but it was averted and he continued in command of the army of the Potomac to the end of the war.

October 19, when the Union forces again started toward Culpeper, the telegraph was actively employed. Doren was indefatigable in his efforts to construct the lines. On the twenty-first, head-quarters' office was open at Warrenton, and on the twenty-third, the Second Corps at Turkey Run was connected. At 10:30 P. M., the twenty-fourth, Emerick placed French (Third Corps) in the circuit at Catletts. Shreffler and Glazier relieved Emerick the twenty-sixth, and head-quarters being moved to Auburn, he kept the Sixth Corps at Warrenton in communication until the twenty-eighth, when Ryan and Edwards relieved him. While Edwards and escort were *en route* to this place, they were attacked by bushwhackers and some of the party injured. On the thirty-first, an office was opened at Bealeton for Generals Buford and French; the day before, for the Third Corps at Licking Run; the 8th of November, at Rappahannock Station and Kelley's Ford; the tenth, at Brandy's Station and on the twelfth, once more at Culpeper Court House.

November 26, everything being in readiness, Meade's army crossed the Rapidan at Germania, Culpeper and Ely's Fords and Jacob's Mill, and on the twenty-eighth, the Union army lay in front of the enemy's position. Meade was in telegraphic communication with Washington and, by field telegraph, with both wings of his army. It was arranged on the twenty-ninth, that Warren was to attack the enemy's right, but not doing so, he was, on the thirtieth, ordered by telegraph to do it. Warren however, replied that the enemy's position was too strong to be assaulted, and protested against the useless loss of life in attempting to carry it, whereupon Meade, who had given the order upon previous representations, examined the position in person and annulled the command. Such had been the disposition of the forces, that further efforts to dislodge the enemy at that time, were deemed unwise, and hence the army retraced its steps across the Rapidan. The losses were probably one thousand men on each side.

Some of the operators at the front, notably W. C. Hall and

Rose, by this time found it necessary, on account of the gray-backs which infested their blankets and clothing, to go to Washington to renovate and make purchases; but the Spanish itch, which broke out in some of the camps, and particularly at Bealeton, was infinitely more disagreeable. Operators Edwards and E. A. Hall, at this place, found it a distressing subject of frequent jest. The disease appeared on the fingers and every part of the person that was clothed. The body was covered with running sores. At first, physicians said it was caused by salt meats, and ordered an *anti*-scorbutic diet. Later, the doctors declared it was a blood disorder, and directed free applications of sulphur and lard, used as a salve; but, to use the language of another, the "insects fattened on the pomade," thus necessitating a trip to Washington.

There was a long *interim* between Gettysburg and the Wilderness, which was now impending. The War Department had manifested marked vigilance. Throughout the fall and winter of 1863–4, the energies of all departments were strained to place the military power upon an irresistible footing. On the second day of March, Ulysses S. Grant was confirmed as Lieutenant General and Commander, under the President, of all of the armies in the field. He determined to make his head-quarters with General Meade's army, and to inaugurate, about the 1st of May, co-operative movements of all the armies against a common enemy. Consequently, the telegraph was of infinite importance to the Commander, who, from his tent in Virginia, was to move his men upon the great continental chess-board of war understandingly. But once during the ensuing struggle did he go to any other of his commands. Badeau, in his "Military History of U. S. Grant," says (p. 41, vol. 2):

> From this spot (Culpeper), Grant issued his orders to all the soldiers of the Republic. Hither came to him reports from Banks, on the Red River, and from Sherman, at Chattanooga; from Butler, at Fort Monroe, and from Sigel, in West Virginia. His staff officers were sent to all these generals, to carry instructions that could not so well be given in writing; and the telegraph, that revolutionizer of modern war, brought him despatches daily of the doings of his armies a thousand miles away.

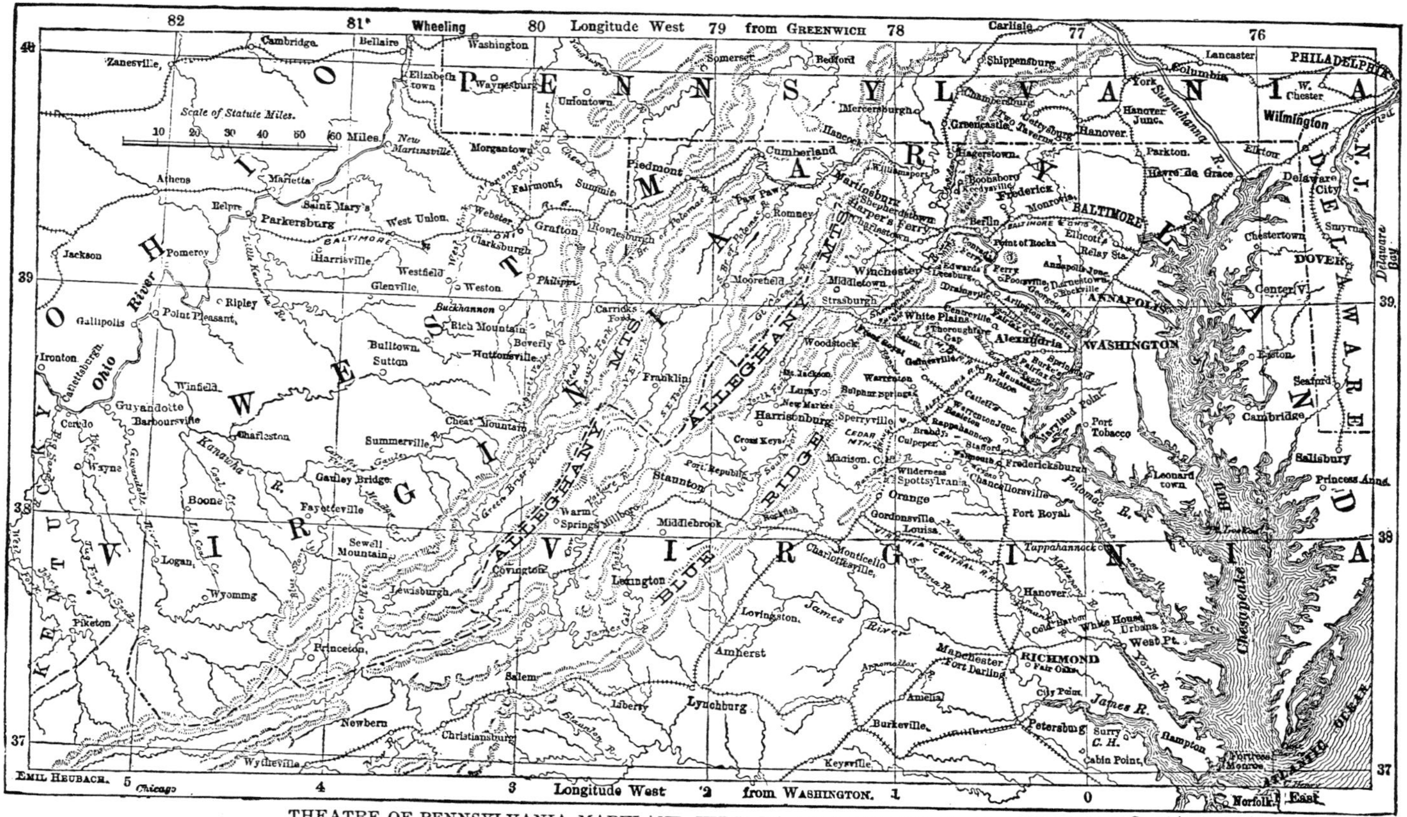

THEATRE OF PENNSYLVANIA, MARYLAND, VIRGINIA AND WEST VIRGINIA CAMPAIGNS. Copyrighted, 1882, by W. R. Plum.

February 27, Custer's cavalry, fifteen hundred strong, moved to the west of Lee's army, *via* Madison C. H., to near Charlottsville, with the view chiefly of distracting the enemy's attention from Kilpatrick's cavalry movements, looking to the liberation of Federal prisoners in Libby. Forces from Fort Monroe were started to co-operate on the Peninsula ; but the only success of the raid consisted in the capture of about five hundred of the enemy and destruction of railroad bridges, military stores, etc. Kilpatrick's cavalry, less one hundred and fifty men lost, including Colonel Ulric Dahlgren killed, returned by water, *via* Fort Monroe. Operator James A. Murray accompanied Kilpatrick, mainly for the purpose of stealing information from the enemy's wires. He had an interesting and eventful experience.

In March, 1864, by direction of Secretary Stanton, a telegraph line was being built from Washington, D. C., *via* Port Tobacco, to the new depot for rebel prisoners of war, at Point Lookout, Md., and offices were to be opened at those places and St. Mary's, the head-quarters of the Potomac flotilla, and at Leonardstown. The use of this line proved of immense advantage to the Government in the coming campaign.

We have seen that the Beardslee Field Telegraph System, under the control of the Signal Service, was supplanted by the Morse Telegraph, under the general management of Colonel Stager. Major Eckert, in contemplation of the spring campaign of the Army of the Potomac, constructed an apparatus for speedily paying out fine telegraph wire, which consisted of reels fitted to pack saddles borne by mules ; each reel carrying one mile of wire. The construction party, organized, was under the immediate supervision of D. Doren, and all the field operators were subject to the orders of A. Harper Caldwell, chief operator at Meade's head-quarters. The electric current for charging the wires was supplied by a portable battery, consisting of sixteen sections, of six cells each. The cells were made of copper, about four inches in diameter, by nine in depth, and contained blue vitriol in solution. In the liquid was placed a leather cup, one-half of the diameter of the copper, containing a zinc plate and water. Each cell was insulated by a casing of thin sheet rubber, and fitted on the top was a bone-rubber cap, thus making the cells water-tight. Each section was enclosed in a strong box, and the whole securely packed in a common army wagon, which also carried a quantity of blue vitriol, the only article except water necessary to replenish the battery and keep it in working order. The operator's table, instruments, tools, battery, etc., were placed in one wagon, thus making a complete telegraphic outfit in one conveyance, sufficient for working any requisite number of lines at a moment's notice.

On the night of May 3, the Army of the Potomac, numbering 97,273, crossed the Rapidan at Ely's, Germania and Culpeper fords, and advanced to the vicinity of Chancellorsville, where the battle of the Wilderness occurred. Meade's army was reinforced by Burnside's corps, exceeding twenty-two thousand men, who took part in the battle.

There were at Meade's head-quarters, for the campaign now opening, operators A. H. and J. Caldwell and Emerick; at Warren's, Edwards, George Henderson and E. A. Hall; at Hancock's, Rose and Shreffler, besides whom was operator C. H. Beckwith, whose duties consisted chiefly of cipher work for General Grant.

Co-operating with the Army of the Potomac, were General Butler's forces, at and near Gloucester Point, Va., about thirty thousand men; General Sigel's, in the Valley of Virginia, of about ten thousand; and Cook and Averill, in West Virginia, about eight thousand, while Sherman, in Northern Georgia, was co-operating to the extent that he was enabled to keep the forces in his front from reinforcing Grant's immediate adversary, Lee. Writing of the position of affairs just after crossing the Rapidan, Badeau says:

> Meade's head-quarters were near those of the General in Chief, and at night the two commanders discussed the plans of the morrow. While they were thus engaged, telegrams arrived, announcing that Sherman and Butler and Crook had all advanced. It had never happened before, in the history of war, that one man directed so completely four distinct armies, separated by thousands of miles and numbering more than a quarter of a million soldiers, ordering the operations of each for the same day, and receiving at night reports from each that his orders had been obeyed.

It will be recollected that a telegraph line was constructed to

Fortress Monroe, through Maryland, and by cable across the Chesapeake, and thence to Yorktown, for McClellan's Peninsular operations. This line, continued in use to Fort Magruder, was extended from Yorktown to General Butler's, at Gloucester Point, where, to deceive the enemy as to his real objective, his forces were mainly stationed, awaiting the order of the Lieutenant General. With General Butler were chief operator Richard O'Brien, and J. Hervey Nichols and J. W. Collings, assistants. The latter was detailed from the ranks, and sent in June to the army of the Potomac. Other operators were stationed in this vicinity, as follows: At Fort Monroe, chief, George D. Sheldon, John A. Sheridan, C. L. Snyder and Silas C. Burns; at Newport News, F. B. Knight; at head-quarters near Portsmouth, Wilbur F. Halloway; at Norfolk, John E. O'Brien; at Fort Magruder, Robert J. Murphy; at Yorktown, H. N. Snow.

As soon as General Butler received his telegraphic orders to move, he embarked his forces, consisting mainly of the Eighteenth Corps, General W. F. Smith, and the Tenth, General Gilmore; the latter from South Carolina, and landed at Wilson's Wharf, Fort Powhatan and City Point, immediately possessing Bermuda Hundred in force. On the seventh, the Petersburg & Richmond Railway was struck, and the enemy met in some force. Two days later, a larger force broke the railway and advanced to within three miles of Petersburg.

As the Confederates were wholly unprepared, without the Confederate telegraph system connecting Richmond, Lee's army, Lynchburg, Savannah and the Carolinas, it would have been utterly and unqualifiedly impossible to defend Petersburg; but messages couched in hot words spread the feverish anxieties of the defenders, the momentous importance of the issue and the cries for help. Assistance came as rapidly as possible from all quarters, especially Beauregard's command from the South, by reason of which and incorrect reports that the Army of the Potomac had won a great victory, Butler turned his forces northward to within ten miles of the Confederate capital, where he was met by the enemy in strong position and force.

As soon as the transports started from Gloucester Point, a telegraph building party began reconstructing the Fort Monroe line used by McClellan in 1862, from Williamsburg to Jamestown Island in the James River, from whence messages were soon forwarded by despatch boats, which were kept running between the Island and Bermuda Hundred. Jamestown Island office was worked by W. N Embree and W. S. Logue. It was, at that season of the year, a very disagreeable place. Most venomous reptiles in great numbers were there, but what proved infinitely more annoying were the mosquitoes and other winged tormentors, that compelled the operators, while on duty at night, to wear gloves and cover their heads and faces, much as one would to defend against a swarm of bees. Field lines were also constructed from Bermuda Hundred to the head-quarters of General Butler on Proctor's Creek, and to both wings of his army, which proved invaluable to that officer in the conduct of his operations. These lines were under the direction of Richard O'Brien, who, says Major Eckert, was "indefatigable in his exertions to render them of service to the Government." Operating in this district at this time were John D. Tinney, at City Point; O'Brien and Nichols, at Butler's head-quarters; H. N. Snow and W. H. Wilson, detailed from Ohio regiments, at Bermuda Hundred; O. B. Vincent, at General Gilmore's, and at Smith's on the right, near Fort Darling, C. A. Homan, whose office consisted of a cracker box under a tree.

As already seen, Grant's army had crossed the Rapidan. As the morning broke, Lee's signal officers on Clark's Mountain espied the advancing columns and promptly notified the General, whose signal fires upon the heights attracted the watchful sentinels along the whole line. Lee's army then knew that Union legions were pushing forward, and it bravely concentrated to resist, even unto death. The 5th, 6th and 7th of May were sanguinary days in the Wilderness, where Lee's inferior forces were more easily maneuvered than Meade's bulky army. It was a "give and take" battle all along the line for five miles. Grant's only information came by staff officers and couriers. He could see but little. No attempt to construct the field telegraph appears to have been made during this engagement, but in all subsequent ones it was greatly relied upon. The Union loss in the three days battle was 2,261 killed, 8,785 wounded and 2,902 missing, while those on the other side must also have been very great. As no official report was made, they are computed to be from about seven thousand to twenty-five thousand men, according to the ruling bias or purpose of the historian.

During the afternoon of the seventh, telegrams were received, stating that Sherman expected to fight Johnston on that day, and that Butler's force reached City Point on the fifth. These and numberless other cipher despatches were received by mounted scouts from long distances, as small bodies of the enemy had gotten behind Grant's army. These were sometimes successful in capturing despatch-bearers. On the eighth, Emerick, after riding all night, was obliged to assist Beckwith in translating and putting up ciphers, as, owing to the uncertain ties of the route to the nearest office, all telegrams that would afford any important clue to Grant's operations or purposes were put in cipher.

General Grant, failing of Lee's destruction in the drawn battle of the Wilderness, moved the army past Lee's right to the neighborhood of Spottsylvania Court House, where he was again confronted by his wily adversary. Warren, who took the lead, held the Union center, Hancock, the right, Sedgwick, the left, and on the next day (tenth) Burnside took position on Sedgwick's left. The engagement here began on the morning of the eighth and continued with some intermission through the ninth, tenth, eleventh and twelfth, the losses on both sides far exceeding those of the Wilderness. On the eleventh, late in the afternoon, it was determined to move Hancock's corps to the rear of the Fifth and Sixth, and connecting with Burnside, with a view on Hancock's part, particularly, of a daybreak attack on the twelfth. In bringing about this sudden change, the field telegraph was, if not indispensable, at least of great service, as the "details," wrote an operator at Meade's head-quarters in his diary on the eleventh, "could not have been arranged with sufficient promptness without it." About nine o'clock, P. M., the Second Corps, an army in itself, moved in a darkness that could almost be felt, over muddy roads, three or four miles to the left. Operator Rose accompanied Hancock along the line of the previous day's terrible battle and almost within hearing of the enemy. Not a loud word was spoken as the long column slowly and silently moved to the point from whence, at four in the morning, it was to charge and surprise the enemy behind the works he had struggled so stoutly to hold. At two A. M., of the twelfth, Rose, accompanied by an officer of General Mott's staff, rode somewhat to the rear to find the telegraph wire and connect his instrument. By the light of a candle the line was discovered and the office opened on another cracker box. Hither also came Colonel Morgan, chief of Hancock's staff, who had a talk with General Burnside (who was at his office on the left), giving the position of the Second Corps and learning that of the Ninth. General Meade was also the recipient, at this time, of several despatches from Morgan. A heavy mist obscured the Union operations, when at 4:30, A. M., the most successful charge since crossing the Rapidan was made upon the enemy, resulting in the capture of about four thousand prisoners, over thirty colors, several thousand stand of small arms, horses, caissons, etc. Major General Edward Johnson and Brigadier General Stuart were among the prisoners. The assaulting troops pressed on for half a mile, when they met another line of works that resisted further advance. And now in their turn came the Confederates, charge upon charge. Slowly but surely, by reason of their reinforced masses, they drove Hancock's men back to the first line they had captured, beyond which they would not budge and could not be driven. Burnside's corps had already become heavily engaged and was fighting valiantly. While the enemy, constantly increasing in numbers, were pressing Hancock hard, he had recourse to the wires. To General Meade he telegraphed that "if the Sixth Corps" (on his right) "did not attack at once he could not hold the works he had captured." Meade promptly telegraphed Warren on the Beverly plantation, and in ten minutes after this despatch was sent, that corps did attack and Hancock held his own.

Edwards and Ed Hall, operating for Warren, had very uncomfortable quarters, owing to the fact that the enemy seemed to take special delight in shelling the large farm house in which their office was situated. During two successive days that it was fired at, a number of shells went clear through the building. It often occurred that the line was thrown down by the enemy's shells. Then it became the duty of the operators to cause it to be promptly repaired, a work which they often undertook themselves. At one time before Spottsylvania, E. A. and W. C. Hall, Henderson and Edwards were together, when they tossed up a penny to see who should repair a break thus made near their office, but the line was broken so often that these operators took turns in splicing it, always remarking before starting out, "If I stop a shell, send my things home."

With the twelfth, the action at Spottsylvania ceased and the horrible details of it were given to the telegraph, from which history may be compiled. Not the least important of the despatches, is one of the eleventh, from Grant to Halleck. The following excerpt therefrom, promptly given the public, reanimated the nation, already in gloom over what it feared might prove useless sacrifices, viz.:

> We have now ended the sixth day of very heavy fighting. The result to this time is much in our favor. But our losses have been heavy as well as those of the enemy. We have lost to this time eleven general officers, killed, wounded and missing, and probably twenty thousand men. I think the loss of the enemy must be greater, we having taken over four thousand prisoners, whilst he has taken from us but few, except a few stragglers. I am now sending back to Belle Plain all my wagons for a fresh supply of provisions and ammunition, and *propose to fight it out on this line if it takes all summer*.

After the battle of Spottsylvania, Fredericksburg was taken possession of by Union troops, and constituted a temporary base of supplies. We have already stated that a new line was built from Washington *via* Port Tobacco, Md., to Maryland Point. This was extended across the Potomac, by a submarine cable two miles in length and the wire continued thence *via* Belle Plain to Fredericksburg, from which place a line of couriers was established with the army. The office at Maryland Point was opened May 16, and those at Belle Plain and Fredericksburg within a day or two thereafter.

On the night of May 20, the Army of the Potomac moved from Spottsylvania Court House to Bowling Green and Milford Station, on the Richmond & Fredericksburg Railroad, and a temporary base was established at Port Royal, on the Rappahannock below Fredericksburg. Another line was constructed from Belle Plain to Port Royal, the office at the latter place being opened May 24. After a short halt, the army moved forward from Bowling Green and vicinity, to and across the North Anna where several spirited engagements took place. The field telegraph was now so thoroughly organized, that though every corps of the army moved daily, yet the head-quarters of each was in communication with General Meade every night. The field line consisted of seven small, soft iron wires coated with rubber and strung on lances or trees. On the twenty-seventh, Harper Caldwell, Doren and Emerick remained with their battery wagon some time after Meade had left, and when they started, moving parallel to the North Anna, where it was impossible to retreat, the enemy opened a brisk fire on them. The mules becoming unmanageable, ran against a gate post, badly injuring the wagon, but the driver bravely remained with his team, bringing all inside the picket lines and out of danger long after the operators and Doren, who were well mounted, had reached that haven.

On the twenty-eighth, Meade's army withdrew from its position south of the North Anna and marched to Hanover Town, on the Pamunkey, where it crossed and took position about Cold Harbor. The offices at Belle Plain, Fredericksburg, Port Royal and Maryland Point were closed about May 30, the base of supplies having been changed to White House. Consequently the line from Fortress Monroe to Yorktown was extended along the north bank of the York from Gloucester Point to West Point, submarine cables being used to cross the York and Mattapony Rivers. From West Point a line was built on the north bank of the Pamunkey to White House. The office at West Point was opened June 2, and at White House on the third. Between the latter place and Cold Harbor, a line of couriers was established. William McIntosh and party of builders, accompanied by operator Homan, constructed about twelve miles of line from Gloucester Point. In the construction of the line from Gloucester to West Point, McIntosh's or some other building party had several serious skirmishes with guerrillas, in which the latter were driven off, but of the builders, two were killed and several wounded. The following operated at these several offices: at Port Tobacco, C. Weber; across from Maryland Point D. E. Rand; at Belle Plain, F. T. Bickford; at White House, Clinton S. Barrett and later D. E.

Rand and Bickford; at White Plains, W. K. Applebaugh, assisted doubtless by some others. Cowen, J. E. Bliss and Rand worked the White House office during its busiest days.

Applebaugh, detailed from the Sixteenth Pennsylvania cavalry, in which he was Sergeant-Major, while at White Plains, started with a small party in a tug-boat for the mouth of the Mattapony to repair a break in the line. The guerrillas fired at the party from the river banks on the downward trip, and on reaching Little York, one of the boat hands abandoned the tug, consequently Applebaugh by request of the captain, took charge of the wheel on the return trip and piloted the little vessel up the river sixteen miles, notwithstanding the firing from the banks was renewed. The bales of hay surrounding the pilot house, proved a good protection to the novice at the wheel. When the office at the White House was being removed later, owing to a new base being established, General Fitz Hugh Lee hurried operations greatly by shelling its locality until driven off by Sheridan's force that had come after him. At Bethesda church, the enemy succeeded in driving back a portion of General Warren's corps, thus leaving exposed for some distance, the field telegraph line. This, the rebels gathered, coiled up and rammed into one of their guns and shot back to the Union camp, whither it came whirling through the air with a hideous noise, until caught in the bushes and trees. The operators joked over the affair a good deal, and used to ask one another, why the Confederates didn't send an operator along with it.

Every day's advance brought on a battle, not a general engagement to be sure, but a serious conflict. June 2, Meade's head-quarters were moved to a point near Cold Harbor. Doren and Emerick remained with the light battery wagon at Via's house (where a field office had been opened the day before for Meade) until new head-quarters were established. The enemy's cavalry were close by this house when the office was closed. The operators and the superintendent of construction escaped, but Sergeant Corrigan, late of the Signal Corps, but then under Doren, with a wagon and guard, went back to take down the wire. Before reaching Via's they were fired upon and driven off. The sergeant however, was shot through the knee and wounded so badly that he died the next day, greatly regretted by all the telegraphers with the army, especially at Meade's, where he was highly esteemed. At another time hereabouts, Doren, whose bravery was as marked as his energy, instead of sending a subordinate as one less courageous would have done, started at night with a lantern to find a break in the field wire and repair it. Meade's quarters had become too uncomfortable, in consequence of the excellent range the artillerists had obtained, and it was while head-quarters were being removed rearward that he advanced with his light, which the Confederate gunners soon began to fire at. But undaunted by their concentrated fire, he pressed on and mended the line. Just then his spirited horse was hit by a minnie ball. Before Doren could control the terrified beast, he had carried his master *beyond the Federal picket lines* and nearly into those of the enemy. Doren dismounted, and stealing his way back, succeeded in reaching camp with the dawn of day.

The battle of Cold Harbor occurred on the 3d of June, and was even more hurtful to the Union army than Spottsylvania and not nearly so damaging to the enemy. In Greeley's "American Conflict" (Vol. 2, p. 582) appears the statement that "twenty minutes after the first shot was fired, fully ten thousand of our men were stretched writhing on the sod, or still and calm in death; while the enemy's loss was probably little more than one thousand." Doubtless this is somewhat overwrought as to time and the enemy's losses, but it indicates forcibly the fearful character of the contest on the Union side, and the remarkable courage of men, armed with a just cause, who, regardless of their love of life and health, of the necessities of living to those dependent on them in the far-off Northern hamlet, town or city; regardless of all ties that bind man to earth, rushed, for their country's sake and for posterity, through marsh and mire, over hill and dale, upon an enemy as daring as they, behind works erected by soldiers as skillful as they, into the vortex of death and wounds, into a seething cauldron of missiles.

In a diary of an operator on the field, I find: "Very heavy fighting indeed. The telegraph played an important part; offices being opened at head-quarters in the rear center and at each corps head-quarters. Operators at Ninth Corps head-quarters on the right were under heavy fire, and one, if not two of them, retreated in decidedly bad order." (The only instance during the campaign of an operator being frightened from his post.) "Operator Sam Edwards was ordered to go there, and did so fearlessly, although exposed to great danger." Edwards and E. A. Hall especially distinguished themselves for bravery on many occasions during this campaign. Rose and W. C. Hall's office, like most of the others, was within easy reach of the enemy's guns, and sustained quite a heavy shelling during the action at Cold Harbor. One shell passed directly over their telegraph tent and cut off two legs of a mule standing near by. On four other occasions their office was under fire, and during one of them Captain McCune, provost marshal at head-quarters, lost a leg. The Eighteenth Corps having reinforced Grant's army, George Henderson and C. K. Hambright, operators, kept its commander, Smith, in telegraphic communication with Meade's quarters. The enemy's shells killed a soldier close by the operators' office, near Cold Harbor. C. J. Ryan arrived shortly after and relieved Hambright at a time when head-quarters was subjected to a galling fire.

Grant's next move was, by marching cautiously and keeping constantly in readiness to repel attack, to cross the Chickahominy on the enemy's right, and then the James, effecting a new base of supplies at City Point, with the view to the speedy capture of Petersburg and operations against Richmond from the south side. The passage of the James was effected magnificently, telegraph offices being opened at the crossings. Pending an engagement near the Avery plantation, Edwards, whose office was a cartridge box on the ground, under an apple tree, read to General Warren, Meade's telegraphic instructions, although at the time the operator was subjected to a heavy musketry fire, most of the enemy's shells passing over. One of the minnie balls struck close to Edwards' feet. Warren drew his sword and with it dug up the bullet, saying as he found it, "A close one; I'll save it," and put it in his pocket. Edwards preserved as mementos a piece of shell which, at this point, cut away a part of his horse's tail; also a musket ball that passed between himself and Ed Hall, and within two feet of either

Meade's army advanced against Petersburg, which was attacked on the sixteenth, and on the eighteenth, assaulted by the combined armies of Meade and Butler. It was the old story of the same defenders behind formidable works, resulting as had nearly all previous assaults by either army against an entrenched enemy. The Union loss was fully six thousand men. Telegraphers at each corps and army head-quarters did constant and most important service during this battle. General Grant was at City Point, a few miles away, but in telegraphic communication with Meade, who made his head-quarters, during the fight, at the Ninth Corps head-quarters, near the right of the line. Emerick was at Meade's office, receiving occasional telegraphic instructions from Grant to Meade, and transmitting Meade's to the corps commanders. Thus the telegraph did staff duty more satisfactorily than horse and rider ever could.

Pending Grant's operations against Lee, Butler's army, which as we have seen, advanced toward Richmond, met, on the 16th of May, a severe repulse, and until June 10, remained substantially inactive at Bermuda Hundred, *i. e.*, all except Smith's corps, which, during the last two days of the month, joined Meade's army. On the 10th of June, General Gilmore's corps and Kautz's cavalry made an unsuccessful effort to capture Petersburg. This was renewed four days later by Smith's returned corps, but that too was unsuccessful, owing to causes which produced a war of words. Wilson's and Kautz's cavalry destroyed about fifty miles of Lee's railroad communication, but was terribly harassed on their return, losing fifteen hundred men.

During all these operations, Grant's eye was on his other commanders also. In General Sherman's "Memoirs" is printed:

> The value of the magnetic telegraph in war can not be exaggerated, as was illustrated by the perfect concert of action between the armies in Virginia and in Georgia, in all 1864. Hardly a day intervened when General Grant did not know the exact state of facts with me, more than fifteen hundred miles off, as the wires ran.

In his annual report for the year ending June 30, 1864, Major Eckert said:

> A telegraph line was built on the south side of the James River, from City Point to Swan's Point, there connecting with a submarine

cable to Jamestown Island; but owing to interruptions by guerrillas, this line was not worked successfully until June 24, at which time a sufficient force was stationed along the line to protect it from guerrilla raids. From City Point, lines were built to General Meade's head-quarters, two and one-half miles south-east of Petersburg, and to General Butler's head-quarters, at Point of Rocks, on the Appomattox, crossing at that place with submarine cables. Lines were also constructed to all corps head-quarters and to our advance works. During the operations at Spottsylvania, on the North Anna, at Cold Harbor, in the march from Cold Harbor to City Point, and in the battles in front of Petersburg in June, the field telegraph lines were worked with great success, and invaluable aid was thus rendered to the Government. General Grant and General Meade were kept in almost constant communication with each other and with the different corps of the army. In the above-mentioned operations, *one hundred and fifty miles of field telegraph were constructed and worked*, and when the army moved, were taken down, thus making it necessary for the construction party to travel a distance of three hundred miles. To D. Doren, Superintendent of Construction, A. H. Caldwell, Chief Operator, and the men under them, is due much of the success attending these lines. They have worked many times in the face of the enemy, exposed to fire, without shelter; have been kept up day and night whenever required, and have had innumerable difficulties which can never be known to but few. They deserve the highest commendation.

In a letter written by Major Eckert, June 15, 1864, is found the following additional statement:

My field telegraph continues to work like a charm. Instead of letting down, it has improved every day since we left Brandy Station, and is complimented by all. Doren has *built and taken down* an average of twenty-four miles *daily*. Most of his work has been after night, and under very great disadvantages. *All corps* head-quarters and many brigades have been in constant connection with Generals Grant's and Meade's head-quarters, during *every engagement*. Also, every reconnoissance that has been made in force has had telegraphic connection with head-quarters. Last, but not least, connection has been kept up while on the march. This was accomplished by making a halt at stated times (intervals of thirty minutes to one hour), reporting any change with the advance that might occur, or any change in orders from head-quarters to the advance or rear. We have lost thus far one man killed and one wounded (builders), ten horses and four mules killed. In Longstreet's attack upon Hancock after leaving Spottsylvania, we lost our battery wagon. A solid shot struck the running gear, demolishing that portion of it completely. The sections of battery were soon transferred to a common army wagon, and put in working order in less than twenty minutes. Fortunately, not a single cup was broken. During this dash, we lost seven miles of covered wire in one line, and one mile in another.

Considering the novel use and vital importance of the telegraph on this campaign, is it not marvelously strange that a staff officer, at general head-quarters, participating in this campaign, could write a large volume thereon, in which every noteworthy disadvantage that General Grant labored under is brought out in strong light, and yet make no mention whatever of the striking military help which the field telegraph afforded?

CHAPTER IX.

THE TELEGRAPH IN THE SHENANDOAH VALLEY, WEST VIRGINIA AND THE DISTRICT.—SIGEL DEFEATED.—CLOYD'S MOUNTAIN.—PENNSYLVANIA, MARYLAND AND THE DISTRICT INVADED.

In the Valley of the Shenandoah and West Virginia, the Union forces were, as stated, to advance contemporaneously with Meade's movement over the Rapidan. General Sigel, in the Valley, moved as far as New Market, where, on the 15th of May, 1864, he was defeated by General Breckenridge, and retired to near Strasburg, wiser by a loss of seven hundred men, six guns, etc.

To defeat Sigel, the forces opposing Crook and Averill in West Virginia, were greatly diminished. Nevertheless, Averill, with two thousand cavalry, was roughly handled by General John Morgan's force, at Wythesville. The remainder of Crook's command, comprising three brigades of infantry and about three hundred cavalry, left Fayetteville, West Virginia, during the first week of May, and proceeded to Lexington, destroying there the camp and garrison equipage of two rebel regiments, which had been left in charge of a guard. On the evening of the eighth, the expedition arrived at Shannon's Cross Roads. A Confederate telegraph line near by was tapped by the cipher operator, W. K. Smith, accompanying the expedition, but owing to the precaution of the Confederate operators, but little information was obtained.

One of Smith's chief objects in tapping the wire was to ascertain who commanded the opposing forces, as that might indicate the extent of the reinforcements the enemy may have received. When Smith "cut in" his instrument, he found no circuit, but wisely waited half an hour before trying a ground connection. Failing at the end of that time, he tried his ground wire, when the operator at Dublin Depot said, "Sign," but as Smith did not know the office calls, that was just what he could not do without establishing his identity ; consequently, he instantly took off his "ground," and waited another half hour, hoping the line would become "OK ;" but it did not. Smith then again attached his ground wire, when the Dublin operator, who somehow had learned the name of the telegrapher with Crook, said: "Oh, Smith, we know you are there, you d— Yankee." Thereupon Smith began a conversation, during which the rebel said: "You will be captured, sure, but I will speak to *General Jenkins* and have him parole you. Yes, I will do more than that, I will. I will take you around and introduce you to my sweet-heart. Do you want Richmond *Press* news?" But Smith had found out who was in command and could not stay longer, so he and escort hurried to General Crook.

On the morning of the ninth, the Union troops reached Cloyd's Mountain, five miles from Dublin Depot. General Crook at once made an attack upon the rebels, who were posted behind entrenchments, and after two hours resistance, they fled in utter confusion, leaving their dead and wounded. During the engagement, the cipher operator acted as aide-de-camp to the General, and received a complimentary notice from that officer, in his official report of the affair. General Crook pushed on for Dublin Depot, and when within three miles of the town, met a force of Morgan's men, numbering about one thousand, who had come from Saltville to reinforce the Confederates at Cloyd's Mountain but were not in time to effect a junction. After a skirmish the rebels retreated, and the Federals, about three, P.M., of the ninth, entered Dublin, a station on the Virginia and Tennessee Railroad. The depot buildings, Gevernment property and telegraph office here were destroyed. Operator Smith hurried to the telegraph office, but his rebel friend had run away, having first partially atoned for this discourtesy by engaging the station agent to be hospitable to him. It is a pleasing fact that most of the opposing telegraphers would not, or could not, resist those sentiments of fraternity which have ever characterized the business.

Crook pushed on to burn the long railroad bridge ten miles from Dublin Depot which, after an engagement of two hours, was accomplished. The object of the expedition having been attained, the command started upon its return *via* Salt Pond and Peters Mountains, through Union, across Green Briar River to Meadow Bluffs, and reached there May 23, having marched about three hundred miles.

Captain David, the superintendent in this department of West Virginia, resigned the service January 20, 1864, on account of ill health, and was succeeded by Captain Lynch, who was commissioned October 27, 1863.

T. B. A. DAVID, 1862.

THOMAS B. A. DAVID, for eight years preceding the war, had been manager of the telegraph office at Wheeling, Va. His character was established, and in selecting him, Col. Stager well knew that his orders would receive scrupulous attention. David was another graduate of the Pittsburgh office, which is *alma mater* to perhaps more men who began as messengers and have distinguished themselves since, than any other office in the country. His was uphill work from a very early beginning, for owing to the demise of his father, he was, at ten years of age, thrown largely upon his own resources. On his father's side he was French, but his mother was of Puritan extraction. Grandfather David was a prominent Frenchman, and owned a large property in San Domingo. An insurrection on the island destroyed his property and a new *regime* in France exiled himself in America, where David's father was born. The latter, who was highly educated, in the schools of Paris, at the time of his death, was in the service of the Federal Government in the ordnance department, for which position he was indebted to General La Fayette, a friend of the family. David's mother was a daughter of the gentleman who conceived and executed the plan of making the famous "Springfield" musket of interchangeable parts, so that a fractured portion was easily duplicated. The subject of this sketch was, from early boyhood, the *man* of his mother's family, consisting of six children. He was born in Pittsburgh, Pa., in 1836, where he began his labors in the telegraph service at the age of thirteen years, as messenger. Three years later he was employed there as operator, and the next year he became operator and manager of the "Lake Erie Line" office in Pittsburgh. A year later he was appointed manager of the Wheeling, Va., office, a position he held when the war began. Ill health has been a great drawback to Mr. David. Doubtless had his growing days been less severely tasked by the necessities of his position, he would have developed a more robust habit, and remained in the military service to the close of the war.

STAFFORD G. LYNCH, David's successor, was born in Waterloo, Seneca Co., N. Y., September 9, 1832, and after his common and academic schooling, like so many other officers of the corps, learned the art of printing. For associate compositors at the office of the *Cayuga Tocsin*, in Auburn, N. Y., where he began, he had Ed F. Barnes and C. M. Stebbins, who since became well known telegraphers. In 1846, Stafford became a messenger in the Auburn telegraph office, then among the first opened. There he learned to operate. Having been a typo and messenger before becoming a telegraphist, his success as an electrician was, according to many examples, only a question of time. Those who succeeded without these prerequisites, especially those who had neither, must have had genius thrust upon them. The Little Falls, N. Y., office was Lynch's first, but in the course of a year or so, he went to Cleveland, Ohio, thence to Detroit, and subsequently to Philadelphia offices. While he was at the latter city under James D. Reid, superintendent of the Western lines, now a telegraph patriarch, revered by surviving "old timers," whom he aided by advice and deed, but two instruments were used. These registers indented in Morse characters all the Western business for Philadelphia and the cities East and South.

In a year or two Lynch returned to Cleveland, where he took charge of the O'Reilly line office, which in 1853 he resigned to become chief clerk and cashier in the post office there, which position he retained until the commencement of the war.

While contemplating how he could best serve his country, Mr. Stager, who had just been placed in charge of military telegraphs, invited his services. Accordingly he went to Western Virginia, where the Unionists had recently obtained possession of the western divisions of the Baltimore & Ohio Railroad. His first office was at New Creek, the base of the mountain range on the south branch of the Potomac. While here he distinctly heard the guns of that historic battle field, the first Bull Run, seventy-five miles distant. When General Kelly advanced to Romney, T. B. A. David built a line there where Lynch operated until the approach of Stonewall Jackson's troops made the Federal position untenable. Then the operator pressed a Secesh horse and buggy into his service, and in a storm, at midnight, started for Green Spring Run with the command under General Lander. Lander and Lynch went to Paw Paw, where the former, a brave officer, a kind hearted man and a devoted husband, died, it is believed by some, for want of proper treatment and care. His wife was in Washington at the time, forced to content herself with his affectionate messages and to console him solely by the loving telegrams she sent.

STAFFORD G. LYNCH.

From Paw Paw, Lynch was ordered to the telegraph office in the War Department, where from the spring of 1862, he remained until the fall of 1863. During a part of 1863, he served there as Censor of the Press in addition to his other duties. Those were days that tried men's souls; tested their physical endurance and taxed their patriotism. After working all day with scarce an intermission for meals, the operators were very often required to continue until two A. M., and even later, but they were incited thereto largely by the remarkable zeal displayed by the great war secretary, Stanton, who also kept long hours; he who proved himself an invincible barrier to the cormorants who sought to rob the Government through fat contracts; he who devoted himself more arduously than any other cabinet officer to destroy the Rebellion. Mr. Lincoln also encouraged the telegraphers by his daily visits, his greetings and intense solicitude, for as his eyes ran over the telegrams, his countenance bespoke the pleasure or sorrow they gave him. The President's kindness led him sometimes to trust important secrets to others, who betrayed his confidence, and hence it sometimes happened that a military necessity was believed to exist for withholding from his observation, despatches concerning important army movements. Whether such necessity was real or imagined, the fact remains that telegrams were kept from his eyes on account of such actual or assumed essentiality.

While McClellan was at the gates of Richmond, and its fall regarded as certain, Post-master General Blair wrote Mr. Lynch virtually promising him the Federal post office in that city, but the Union forces did not enter it. The following fall, Lynch was offered the chief clerkship of the New Orleans, La., office, at a salary of sixteen hundred dollars per annum, but declined it, and a year later he was made captain in the telegraph service and assigned to duty at Cleveland, Ohio, as purchasing quartermaster for the corps, where he handled for Colonel Stager all the money which was transferred in his name to his assistants, and when Captain David resigned, his duties also devolved upon Captain Lynch, but the care and operation of the telegraphs in West Virginia was left almost exclusively to the discretion of Charles O. Rowe, who had the reputation then as he has now, of discharging the trusts confided to him in an able, thorough and commendable manner. Perhaps no greater trust devolved upon any assistant officer of the corps than that which was so honorably and ably executed by Captain Lynch.

For the year ending June 30, 1864, the Union forces within the Department of West Virginia were expected to maintain a series of posts along and adjacent to the Baltimore & Ohio Railroad and the Kanawha River, as far up as Gauley, Virginia. Captain David transferred to his successor a line from Hamden to Fayette, one hundred and fifty-one miles; one from Galipolis, Ohio, to South Point, Ohio, forty-five miles; one from Clarksburg, Virginia, to Beverly, Virginia, fifty-eight miles, and another from Clarksburg to Bulltown, Virginia, forty-nine miles; total, three hundred and three miles. Captain Lynch built a line from Gauley to Lewisburg, sixty-two miles, and from Green Spring to Springfield, seven miles, making a total in operation in May, 1864, when the general advance was begun, of three hundred and seventy-two miles of line. These wires were all operated by a force averaging forty persons, including repairers. At least one hundred and twenty thousand telegrams were transmitted over them during the year preceding June 30, 1864. Among the operators in this department during this time were Charles O. Rowe, W. F. Allen, D. Colestock, N. De Bree, G. K. Smith, C. D. Tull, J. Jolls, R. A. Furr, W. H. H. Lancaster, C. C. Starling, D. W. Warner, M. D. Graham, G. H. Curtiss, M. T. Seymour, John Morgan, C. J. Thomas, J. W. Vermillion, A. T. Brenaman, Joseph S. Keith, R. C. Humphrey, Curley Berry, J. C. Florence, W. H. Wilson, L. B. Dennis, S. M. Shurr, D. H. Johnson, F. B. G. Miller, C. C. Phillips, J. P. Klunk, Robert G. Worth, A. T. McKelvey, A. K. V. Hull, John W. Gregg, J. S. Williams, S. G. Lynch and Alfred Winder.

Many incidents occurred during the year 1863 and the forepart of 1864, which indicated that the spirit and temper of the operators along the Baltimore & Ohio Railroad, west of Cumberland, and also south of the railroad, among the barren mountains, in what they then called a God-forsaken country, was the same as that which marked those in the other departments where loyalty and fidelity were the watchwords.

Alfred Winder, on the Clarksburg and Gauley line, whose office was in Scott's house, about midway between Sutton and

Summerville, had a guard of twenty soldiers. Early on the morning of January 3, 1864, half an hour after a supply team and eight men had left to return to Sutton, he started towards that place to repair the line. Shortly after, five men of Jenkins' cavalry stepped into the roadway from adjoining woods, and with cocked guns persuaded the operator to halt and surrender to Captain Duffield. He was ordered to proceed up the mountain adjoining and keep his eyes to the front until directed to stop. The guard, taking the bridle rein of Winder's horse in his left hand and a cocked pistol in his right, walked behind the captive for a few hundred yards, when they sat down and waited. Soon one or two shots were heard, and four of the eight Union guards, whom Winder had passed before his misfortune, together with the driver and team, were captured without resistance, as the soldiers' guns were in the wagon ahead of them. The remaining four, luckily, were in a ravine near by, and escaped unseen. For eight days, Winder and his fellows were compelled to march over the mountains, through snow-storms, in Indian file, with a led horse and man in advance to break the way through the deep snow, until the man was exhausted, when his place was taken by another. Their first stop, but for a short time only, was at the house of a Union man named Rose, who escaped on their coming. His patriotic daughter, a pleasing girl of sixteen, ran to Scott's house to inform the guard of the capture, but the lieutenant in command took the wrong road. Captain Duffield proved a good protector of his captives, as several bushwhacking residents vainly urged upon him their policy epitomized in the adage, "Dead men tell no tales." Until beyond danger of recapture, each Unionist was allowed, at night, to ride behind a captor. The prisoners were taken before General Jenkins at Staunton, who lectured those that stated they were from Southern States, adding, as he struck his fist upon the table, "You have no State pride; if Virginia seceded from the Confederacy to-day, I would go with her." He dismissed all but the telegrapher, whom he questioned closely as to the forces at Summerville, Sutton, Birch River and Gauley Bridge. In answer, Winder purposely greatly understated the garrisons at these posts. After six days visit to Castle Thunder, they were taken to Libby Prison, where the soldiers were promptly paroled and forwarded to the Federal lines by flag-of-truce boat; but Winder was marched to the citizens' room, to remain until called for. This room, 70 x 60 feet, contained a large number of men of various callings. There, black-eyed peas or boiled rice, bread and mule or horse meat constituted the general diet. After a month's incarceration, Ross, the chief clerk, asked, "Where is that operator? I've a partner for him." It was none other than R. F. Weitbrecht, captured with General Stoughton at Fairfax Court House. Weitbrecht, unaccustomed to horseback riding, had been indulging it for three days and nights, and when he entered, he felt like exclaiming, "I'm glad there are no chairs in this room." Six weeks later, when an Englishman and the operators were the only remaining prisoners in the room, clerk Ross started the latter to their respective departments.

General Hunter succeeded to Sigel's command, and with a larger force, moved up the valley of the Shenandoah, to encounter the enemy under W. E. Jones, Breckenridge having gone to Lee. On the 1st of June, General Crook started with his command for Staunton, Va., *via* Lewisburg and Hot and Warm Springs, through Panther's Gap, into Augusta Valley. At Panther's Gap a soldier in rebel uniform was captured, claiming to belong to Imboden's command, but upon being brought to headquarters, was found to be the bearer of cipher messages prepared by R. R. McCaine, operator, accompanying General Hunter, which, when translated by W K Smith, Crook's operator, commanded that general to join Hunter at Charlottsville. The order was however, afterwards changed and the two commands united at Staunton. From there the expedition moved *via* Lexington and Buckhannon, crossing the Blue Ridge at foot of Peaks of Otter, *via* Liberty, for Lynchburg. On arriving in the vicinity of Lynchburg, Hunter successfully encountered the rebel force. At night the enemy was reinforced by troops under General Early. After resisting the Confederate attack on the second day, the Union forces fell back under cover of the night and started for the Shenandoah Valley, but finding that the rebels were pursuing actively, and that Hunter probably could not get out in the direction taken; the column turned toward the Kanawha Valley and marched out *via* Salem, Saint's Springs and

Lewisburg, to Gauley Bridge. At the latter place, orders were received to remove the troops to Charleston, where they took transports for Parkersburg, and from thence by railroad to Martinsburg, Va. From Martinsburg the troops were ordered to Harpers Ferry, and into Maryland after General Early, who had come down the Shenandoah Valley and crossed into that State. Union General Wallace was defeated near Frederick City, Md., losing 1,959 to the enemy's—as they say—600, and rebel Johnson's cavalry pressed on toward Baltimore.

By this time the usual current of horrible rumors had set in. Baltimore and Washington and the cities of Southern Pennsylvania were frightened at Early's twenty thousand men, duly magnified into a great army. It was now the time for the North to become again despondent and decry the management of the armies. In the campaign from the Rapidan to Petersburg, including the assault on the 16th, 17th and 18th of June, the Army of the Potomac proper had lost, at the most favorable calculation, forty-five thousand men. True, at the end of the month, including Butler's troops, the army, owing to large reinforcements, numbered 107,419 present for duty, but Lee's additions had been considerable also, and hence the situation *was* disheartening. Gallant Sheridan had penetrated the enemy's lines several times; had whipped Stuart's troopers, that would not stay whipped, and wounded that splendid officer unto death; had torn up railroads and telegraphs; cut off supplies; made captures and even entered some of the defences of Richmond; but the great losses made the nation bleed, and the people in their Northern homes could not comprehend the claim of advantages commensurate with the sacrifice, while they could understand that Lee held the Army of the Potomac at bay, and still had spare men enough to endanger the Union capital and Northern cities.

General Grant forwarded troops to Washington, to resist Early, but held on determinedly where he was until a better hold could be gained. The Sixth and Nineteenth Corps, the latter a reinforcement from the Gulf intended for City Point, were sent to Washington. Early appeared before that city; severed the lines of telegraph leading from the capital, except that to Point Lookout, Md., and for three days all telegrams for the armies with Grant and for the North were sent *via* the Point, and thence by dispatch boat to Fort Monroe. Field lines were rapidly built to the forts around Washington, and offices opened at Forts Lincoln, Totten, Stevens, Reno, Corcoran, and at Chain Bridge and Arlington, which were kept open and were of great service until the enemy retreated. Early dared not attack, but moved off across the Potomac, and through Snicker's Gap toward Winchester, Va. The Sixth Corps returned from its pursuit, leaving Crook to move on Winchester; but Early turned on him at Kernstown, July 24, and drove him back to Bunker Hill, where the operator, Smith, was busy all night long, putting the General's telegrams into cipher for transmission, from the nearest office, to Washington and other points. Smith took part in several engagements hereabouts. Pennsylvania was again menaced and Chambersburg burned. The people all over the free and border States were intensely excited at the position of affairs, and crowded the telegraph offices. Never since telegraphy was known, were operators in such demand. In this trying military emergency, Sheridan was sent to command in the Valley. The clouds began lifting. It was the darkness that precedes the dawn.

CHAPTER X.

THE TELEGRAPH IN KENTUCKY AND EAST TENNESSEE IN 1864.—PERILOUS OPPOSITION TO ENLISTMENT OF NEGROES IN KENTUCKY.—MORGAN'S LAST RAID.—STONEMAN'S SUCCESSFUL RAID OWING LARGELY TO THE OPERATOR WITH HIM.

Affairs in this Department continued comparatively quiet from the time of Bragg's retreat, to the close of 1864. There were no great battles, nor were there large armies on Kentucky's soil during this period. The Rebellion had been effectually pushed farther back. Partisan forays and organized raids there were indeed, but Kentuckians were settling to the general conviction that their State was forever committed to the Union by the masses of her people, besides being welded thereto by the hot purpose of the North. Nevertheless it was a season of domestic turmoil; the foment was natural and usual in such cases; it was not probationary, for that implies conviction too general; it was transitionary. Old truths were crumbling before the logic of arms, if not the philosophy of higher morals than had been generally accepted in slave territory. Lincoln's proclamation of emancipation was being acquiesced in, when the next step, that of arming the negroes, seemed quite as harsh as the former, to this pro-slavery people.

On the first day of January, 1863, the President's proclamation, manumitting the slaves within districts in armed rebellion, went into effect. This was followed by Acts of Congress, directing the enlistment of colored men. Thomas E. Bramlette, elected Governor in August, 1863, by a Union majority of over fifty thousand, was stoutly opposed to the enrollment of slaves in Kentucky for the army, and disposed to resist the enforcement of the laws requiring such enlistment. The Provost Marshals in the State, nevertheless, began enrolling the colored males between the ages of twenty and forty-five years. At a public meeting in Lexington, March 10, 1864, the Governor was on the platform with Colonel Frank Wolford, when the latter denounced the President as a tyrant and traitor, and counseled Kentuckians to resist the enrollment of their negroes. Captain Gross, of the Telegraph Corps, promptly telegraphed to General Schofield at Knoxville, the purport of Wolford's remarks, whereupon that officer was arrested and ordered to report at Knoxville, and about the twenty-fifth, he was dishonorably dismissed from the service by the President. Bramlette's appearance on the speakers' stand indicated sympathy with Wolford's remarks, and the uneasiness felt on that account, was painfully increased early the next month, when Lieutenant Governor Jacobs—subsequently sent into the Confederate lines—made an address quite as violent. The next development was a telegram from the Governor to the Provost Marshal at Danville, as follows: "I have telegraphed the President demanding that he stop the enrollment of negroes in this State. I await his answer. If he does not stop it, I will." A copy of this message was also forwarded to Schofield. That was Saturday, March the twelfth. On Monday, owing doubtless to the refusal of the President to suspend the enrollment, Bramlette telegraphed Doctor Breckenridge to "come to Frankfort prepared to go to Washington on an important mission, which might save the State much evil." At that time the Reverend Doctor Robert J. Breckenridge, an uncle of General John C. Breckenridge (C. S.), was one of the most influential men of the State in religion and politics. The Governor next telegraphed General Burbridge, who commanded the District of Kentucky, that he had sent for Breckenridge. He also wrote in strong terms of his purpose to prevent the enrollment, and the extraordinary means he proposed to adopt if it became necessary. Bramlette's letter determined Burbridge to see the Governor; Doctor T. S. Bell, a prominent loyalist, of Louisville, accompanied him. Tuesday morning Breckenridge and, at his request, J. M. Kelly, of the Telegraph Corps, went to Frankfort, where they found that the Governor had prepared and put in type a proclamation of an inflammatory character, calculated to array the State against the enlistment of the negroes. Burbridge and Bell were piloted by Kelly to Breckenridge's room, when the latter remarked, "I am glad you have come, for we are on the eve of open rebellion. I have declined to go to Washington to try to induce the President to resist an act of Congress." Kelly was sent for the Governor, who came with Colonel Hodges, editor of the *Commonwealth.* The Governor read his proclamation from the proof sheets he was correcting, and added the remark that he had received a bushel of letters from different counties, saying that they would prevent the enrollment by force. After some discussion, during which the Governor continued to correct his proofs, Doctor Breckenridge asked Burbridge what his orders were, and how many troops he had. The General replied that he was instructed to carry out the act of Congress; that he had twenty-five thousand men, and was directed to call upon Governors Morton of Indiana and Brough of Ohio, for twenty-five thousand more each if needed, and in reply to further inquiry, threatened dire consequences to the Governor if he issued that proclamation. Bramlette who heard all, soon turned to the gentlemen, saying, "I am in the hands of my friends and will do as you advise." He handed the proclamation to Burbridge to revise. This was done by eliminating the most objectionable features and allowing the rest to stand as a compromise. This work was completed about four o'clock in the morning, and immediately telegraphed to Louisville and Cincinnati for publication. It cannot be doubted that this result was a happy one for Kentucky; brought about mainly through the instrumentality of Burbridge and Breckenridge, who, perhaps more than all others, should be credited with having avoided an armed conflict between the State and Federal authorities, which would have greatly strengthened the Confederates, notwithstanding a purpose to that end may not be fairly imputed to such men as Bramlette, Jacobs and Wolford, whose daring in the Union army made them conspicuous. After the proclamation, the Governor, Senator Dixon and Colonel Hodges visited Mr. Lincoln without effecting the result they desired.

After General Sherman raised the siege of Knoxville, in East Tennessee, in December, 1863, Longstreet retreated toward Rutledge and after some heavy skirmishing, posted his command in North-east Tennessee, with head-quarters in Rogersville, where he spent the winter. In the spring he removed into Virginia, rejoining General Lee, and Breckenridge's command, consisting in part of Morgan's, Vaughn's and Duke's forces, occupied that section. During January, 1864, field telegraph offices were opened toward Longstreet at Mossy Creek and Powell's River Bridge, but before the end of the month these and Tazewell, Strawberry Plains and Kingston offices were closed, owing to the withdrawal of the Federal forces therefrom. 'Captain Gross' operations in the Department of the Ohio for the first half of 1864, are shown by him in his most excellent annual report as follows:

During the months of January and February, a new line was extended from a point six miles south of Camp Burnside, Kentucky, to Clinton, Tennessee, a distance by the line of one hundred miles. The severity of the weather, the nature of the country, rough, mountainous, heavily timbered and almost uninhabited, the difficulty of procuring the means of transportation for the necessary material, and subsistence for the men and animals engaged in the work, on account of the extraordinary demands for the same, to subsist the regular military forces in the field, renders this a very difficult and laborious work. I was only able to work the line, a few days, beyond a point between Chitwoods and Jacksboro, distant from Camp Burnside about sixty miles. The first failure to work was owing to depredations committed upon the line by the Twelfth Kentucky Regiment, while *en route* from Knoxville to Camp Burnside, by shooting off the insulators and cutting down the poles for fire-wood. Added to this, heavy storms of wind, sleet and rain passing over this section, trees in many instances were blown down upon the line, crushing it to the ground, and passing trains of army wagons, becoming entangled in the wires, rendered its destruction complete. Evil-disposed citizens in the vicinity of Pine Knob Tavern, and elsewhere along the line, damaged it severely by cutting down the poles and cutting the wire, some portions of which were entirely removed. Strenuous efforts were made to repair these damages. Parties for that purpose were organized and sent out to aid the stationary repairers in getting the line again in working condition; but no sooner were repairs made in one part of the line than difficulty was found to exist in other portions. One serious cause of annoyance was the total disregard paid to the line by the large parties of laborers sent out to work upon the military road, along which the wire was extended. They repeatedly felled trees across it, and in many instances cut down the poles and trees to which it was attached for support, even where no necessity for so doing existed. Prompt reports of these depredations were made to the military authorities at Camp Burnside, and protection asked as soon as it could become known. The country through which the line passes is almost a wilderness, uninhabited save by a few wretched mountaineers, and was entirely destitute of either forage or subsistence. The greatest difficulty was experienced in procuring enough forage to keep the animals alive while the work of putting up the line was going on. I lost five out of twenty-two horses from starvation, and my men were, at times, two and three days without anything to eat. A full report of these facts was made to General Schofield, commanding the Department, and the abandonment of the line from Camp Burnside to Clinton ordered by him. Subsequently, portions of this abandoned line was reclaimed and used again, but the greater part of it was hopelessly ruined for any purpose whatever.

On the 1st of February, the line along the Virginia & East Tennessee Railroad, from Knoxville to Charleston, seventy miles, was transferred by me to Captain John C. VanDuzer, Assistant Quarter-master and Assistant Superintendent, for the reason that it was found to be much more easily supplied and managed by railroad from Chattanooga than by wagons over the mountainous country between Danville, Kentucky, and Knoxville, Tennessee, and also for the reason that at that time I was unable to keep telegraphic communication open between Cumberland Gap and Knoxville by the old route, viz., *via* Tazewell, Bean Station and Morristown, because parts of it were in the possession of the enemy.

Burksville, on the Cumberland River, thirty miles south-west of Columbia, Kentucky, had been for some time held as a military post, and communication with it by telegraph was very desirable. In obedience to an order from General Burbridge, commanding the District of Kentucky, through Captain Samuel Bruch, in March, I extended the line from Columbia to that point, thirty miles. The line from Columbia to Jamestown, eighteen miles, not having, for a number of months, been in actual use, although kept in repair, was taken down and the material used in the construction of the line to Burksville. In April, by running a second wire from Danville to Stanford, eleven miles, I was enabled to connect the two circuits of Camp Burnside and Mount Sterling and throw them into one; thereby lessening the amount of "repeating" done at Danville, and making communication between Camp Burnside and Camp Nelson complete, which was very desirable in view of the great business between those two important points. In that month, the line was extended from Cumberland Gap to Knoxville, sixty-three miles, by Captain VanDuzer and myself; his party starting from Knoxville, mine from Cumberland Gap, and a junction formed at Fincastle, thirty miles from Cumberland Gap, thus giving Knoxville the advantage of a double northern connection; one, through my own department, *via* Cumberland Gap and Danville, to Louisville or Cincinnati; the other, *via* Chattanooga, Nashville and Louisville. Between Fincastle and Knoxville, Captain VanDuzer was enabled to connect at Clinton with the line put up by me in February between Clinton and Jacksboro, about nineteen miles, thus making the line actually put up between Knoxville and Fincastle, fourteen miles. I desire to acknowledge the valuable and timely assistance rendered me by Captain VanDuzer in the construction of this line. The distance, the nature of the intervening country and the extremely slow means of transportation by army wagons for the necessary material, made all my operations laborious and seemingly dilatory.

During the month of May, no demand was made upon me for new lines, and my efforts were directed toward repairing such of my lines as had been longest built. Many of the poles, none too good when first put up, had partially decayed and were easily broken off by wind and storms, and the insulators (Wade), from constant strain and exposure, were fast becoming worthless for purposes of insulation, especially in wet weather. These defects were removed by replacing the rotten poles and damaged insulators with new ones.

Early in the month of June, the enemy under the notorious, ubiquitous John H. Morgan, entered the State and made their first demonstration upon Mount Sterling, the terminus of my line in that part of the State. After stubborn resistance, the town was finally occupied by the enemy, and the country to the westward, through which my line ran, to a point within eleven miles of the city of Lexington. While held by the enemy, serious damage was done my line. The operator at Mount Sterling, Mr. J. W. Curtiss, remained at his post until the evacuation of the place by the Union forces, and then only saved himself by flight, carrying with him his instrument, despatches, etc. For six days he lay concealed, and finally stole through the enemy's lines and reached Lexington in safety, without the loss of anything valuable connected with his office. His conduct, under such trying circumstances, was exemplary and received my approbation. The enemy were, in a few days, defeated; first, in the streets of Lexington, and afterwards, at Cynthiana; their organization broken up and the entire force scattered and put to flight.

No sooner had the enemy disappeared, than prompt repairs were given the line, and the office at Mount Sterling re-opened. To aid in the repulse of the enemy in this important move, Colonel Weatherford's command, the Thirteenth Kentucky Cavalry, was called in from Burksville. A portion of the command was held at Green River Bridge, about forty miles from Burksville, and the operators on duty at Burksville and Columbia were ordered in to that point, thus for the time closing those two offices, and abandoning the line south of Green River Bridge. But no sooner were the enemy's forces scattered, than the Union troops again occupied Burksville, the line repaired and the abandoned offices re-opened as before. It was found that but little damage had been done this line, thus temporarily abandoned, and the work of putting it again in order was but light. After the defeat of the enemy at Cynthiana, small parties of his forces were traversing the State in all directions, stealing, marauding and working destruction wherever they went. Wherever they came in contact with my lines, only slight damage was done, and the interruptions thereby occasioned existed but a few hours. Such was the case between Camp Nelson and Nicholasville, between Stanford and Crab Orchard, and between Danville and Lebanon.

The following tabular statements are respectfully submitted·

LINES CONSTRUCTED, JANUARY 1, TO JUNE 30, 1864.

	MILES.
January—From six miles beyond Camp Burnside, Ky., to Jacksboro, Tenn.,	65
February—From Jacksboro, Tenn., to Clinton, Tenn.,	36
March—From Columbia, Ky., to Burksville, Ky.,	30
April—From Danville, Ky., to Stanford, Ky.,	11
April—From Cumberland Gap, Ky., to Fincastle, Tenn.,	30
Total,	172

LINES ABANDONED, JANUARY 1, TO JUNE 30, 1864.

	MILES.
January—From Tazewell, Tenn., to Powell's River Bridge, Tenn.,	7
January—From Knoxville, Tenn., to Strawberry Plains, Tenn.,	22
January—From Loudon, Tenn., to Kingston, Tenn.,	25
February—From Camp Burnside, Ky., to Clinton, Tenn.,	100
February—From Powell's River Bridge, Tenn., to Cumberland Gap, Ky.,	5
February—From Knoxville, Tenn., to Charleston, Tenn.,	70
March—From Columbia, Ky., to Jamestown, Ky.,	18
June—From Mount Sterling, Ky., to Lexington, Ky.,	35
June—From Burksville, Ky., to Green River Bridge, Ky.,	40
Total,	322

ABANDONED LINES REPAIRED, JANUARY 1, TO JUNE 30, 1864.

	MILES.
June—From Mount Sterling, Ky., to Lexington, Ky., . .	35
June—From Burksville, Ky., to Green River Bridge, Ky., - -	40
Total, - - - - - - - - - -	75

LINES IN OPERATION JUNE 30, 1864.

	MILES.
From Mount Sterling, Ky., to Camp Burnside, Ky., - - -	127
From Lebanon, Ky., to Knoxville, Tenn., *via* Danville and Cumberland Gap, Ky., - - - - - . - . - -	216
From Lebanon, Ky., to Burksville, Ky., . - -	67
Total, - - - . - - - - -	410

* * * * * *

The estimated number of despatches per month sent and received over the lines under my direction is 17,300, and, for the six months embraced in this report, 103,800. The number can not be ascertained to a positive certainty, for the reason that at times of imminent danger operators had been instructed to destroy their despatches to prevent the possibility of their falling into the hands of the enemy, and thereby valuable information given to the enemy. But the number herein stated is certainly within the actual number.

* * * * * *

One glaring abuse of the privileges of the Military Telegraph to which we are at all times more or less subject, and which ought to receive official condemnation, is the practice, too common with the officers of the army, of conducting their entire correspondence by telegraph, when the same might be much more satisfactorily transmitted through the regular mails. This practice is, as I conceive, based upon a misapprehension of the uses and purposes of the Military Telegraph. The beauty and utility of the telegraph as a means of communication is its rapidity ; but, as any one can see, there must be a limit to the capacity of any telegraph line, and to have it lumbered up with long, unimportant despatches at a time when the utmost celerity is demanded for really important despatches, is a perversion of its use which ought not to be sanctioned. The prevalent idea that he who sends most despatches is most efficient, is as untrue as it is absurd.

The very efficient service which the Military Telegraph has rendered the military authorities, within this Department, has been from time to time acknowledged in an unofficial manner ; but justice demands that these acknowledgements should receive official sanction. Spread out, as these lines are, over the very heart of this State and extending to its borders, we have been enabled to give the military authorities prompt and timely notice of the approach or appearance of the enemy. In times of actual invasion or imminent danger, no persons in the employ of the Government, whether they be officers, soldiers or citizens, or with whatever branch of the service they may be connected, have been more faithful, attentive or sleepless than the Military Telegraph operator, who, after working hard all day in the discharge of his ordinary duties, often has sat by his instrument half, two-thirds, or even all night long, faithfully watching, ready to send or receive the important despatch that should give timely notice of the enemy's movements, or foil his most daring schemes.

The general condition and working of the lines under my charge has been greatly improved since the 1st of January. I doubt if lines can be found anywhere, which are not extended along railroads, upon which fewer interruptions occur from natural causes, or where less time is lost in re-opening communication when an interruption does occur, or which works clearer in all kinds of weather, than the military lines within this Department.

The raid of Morgan, who escaped from the Ohio Penitentiary, began from East Tennessee, June 1, with twenty-five hundred men. He moved *via* Paintville, Hazel Green, Owingsville, Flemingsburg and Maysville, capturing Mount Sterling, Paris, Cynthiana and Williamstown, doing much damage, and, with a detached force of three hundred under Colonel Giltner, captured General Hobson with sixteen hundred men. Concerning this, historian Greeley wrote: "It is to be hoped that they paroled their prisoners not to serve again during the war, unless on their side." General Burbridge, however, drove Morgan back, the latter losing nearly all of his command. Morgan was surprised in Greenville, East Tennessee, September 3, 1864, and killed while attempting to escape.

No other operation of any considerable forces occurred within this Department (Ohio) during the war, except the defeat of General Gillem, November 13, 1864, near Bull's Gap, whereby Breckenridge captured five hundred Federals, and the expedition under Generals Stoneman and Burbridge, organized by direction of General Thomas, to penetrate South-western Virginia. Stoneman with four thousand, two hundred men, started on the 9th of December, and moving as far east as Wythesville, Va., destroyed the railroad, salt and lead works, seven railroad trains, a vast quantity of ammunition and much other property, and captured many prisoners and ten pieces of artillery, driving Breckenridge's remaining troops into North Carolina. Operator E. T. Chapman, who had served at Fortress Monroe and elsewhere in the East, accompanied this expedition as operator and cipherer. His report of his operations is as follows:

LEXINGTON, January 2, 1865.

CAPTAIN W. L. GROSS, A. Q. M. and Asst. Supt. Mil. Telegraph:

CAPTAIN:—I have the honor to submit the following report of Major General Stoneman's raid into Western Virginia, so far as relates to the operations on rebel telegraph lines. We entered Bristol on the morning of the 14th of December, at two o'clock, capturing without resistance, the property of the Virginia & Tennessee Railroad, all Government property and the telegraph office on a line of which, Mr. Morris, of Richmond was President. Our arrival was so unexpected that the telegraph operator, Mr. Fred Montgomery, formerly of Lynchburg, and the book-keeper were taken prisoners while engaged in playing cards in their office. By order of the General commanding, I took charge of all books, papers and other property belonging to the telegraph company. A main battery (Grove's) was located here and was in tolerable good order, their only deficiency being glass tumblers, in lieu of which, they used earthen jars. Brevet Major General Burbridge had, previous to our arrival, at Bristol, instructed me to make any rebel operator whom we might capture, do all the telegraphic manipulating, and in case he said anything over the line not dictated by me, to shoot him instantly. At eight A. M., the hour for opening their offices for business, Abingdon called Bristol and asked if any "Yanks" were, or had been there. From the fact that a party of Confederate soldiers had escaped capture and gone to Abingdon, I deemed it best to send the following despatch: "A small Yankee scout appeared in town this morning about two o'clock, remained but a few moments and retreated in great haste toward Kingsport, Tenn." A similar inquiry was made by Lynchburg, to whom the same answer was given, when the business of the day was commenced, Abingdon and Lynchburg sending a great number of messages to rebel Government officials (supposed to be at Bristol) and private citizens.

Among the despatches on file in the office when I entered, were some from General Duke to Breckenridge, Breckenridge to Duke, and Vaughn and Duke to Breckenridge, which, together with the information obtained over the line during our occupation of the office gave us full and accurate information of the enemy's strength, position and intended movements. I may here state that while working the line, I received an Associated Press report from Richmond and also General Breckenridge's orders from the Confederate War Department.

From what I saw of the telegraph line, it was well constructed, and according to Mr. Montgomery, the rebel operator, the best working line in the Confederacy. The wire was about No. ten, and insulated with glass "egg" insulators. The main battery was about thirty cups.

At three (3) P. M., all Government railroad and telegraph property was fired, when we proceeded toward Abingdon, arriving there about midnight. I proceeded at once to the telegraph office, but the operator had left. We however, captured the instrument and one very important despatch from Duke to Breckenridge. As nothing could now be done in the way of telegraphing, the rebels being apprised of our approach, we proceeded to destroy the line. Owing to the limited facilities for transportation we were obliged to destroy all the telegraph instruments. After leaving Abingdon, a portion of the command proceeded to Wythesville, where they captured and destroyed an instrument and battery. After our fight with Breckenridge, near Marion, we proceeded to Saltville, at which place there was no telegraph office.

From Mr. Montgomery I found that commercial operators in the Confederacy were receiving two hundred dollars (Confederate money) per month; army operators ninety dollars and rations. In conclusion, I have the satisfaction of saying that the General commanding acknowledged to me that the *success of the expedition was attributable in a great degree to the information derived from the rebel telegraph line.* Very respectfully, your obedient servant,

E. T. CHAPMAN, *Operator.*

Captain Gross officially reported that "there can be little doubt that the entire success of the expedition resulted from the invaluable information as to the position and numbers of the enemy, which Mr. Chapman obtained while holding the telegraph office at Bristol. For several hours he held the office while despatches were passing to and from General Breckenridge and his subordinates, which Chapman carefully copied and laid before General Stoneman."

Nothing else occurred to telegraphers within this Department during the year, requiring special mention, excepting the operators' strike, mentioned in a chapter on the *status* of military operators; the presentation by the telegraphers in his department, to Captain Gross, of a beautiful sword, sash, belt, revolver, flask, pair of spurs and gauntlets, costing two hundred and seventy-five dollars; and a dash by guerrillas into Danville, Ky., driving Gross and Chapman into the woods, where they eluded the enemy. Perhaps other operators were driven from their posts; such incursions were at least threatened along the whole line every week in the year.

CHAPTER XI.

THE FORT DONELSON LINE.—BECKWITH DISMISSED FOR OBEYING GENERAL GRANT'S ORDERS: IS REINSTATED.—INTERFERENCE WITH CIPHERS AND CIPHER OPERATORS BY STAFF AND OTHER OFFICERS FORBIDDEN.—ATLANTA CAMPAIGN TELEGRAPHS.—VARIOUS BATTLES.—A TERRIBLE ACCIDENT.—A STRANGE BUT PROVIDENTIAL PREMONITION.—CONFEDERATE RAID INTO EAST AND MIDDLE TENNESSEE.—ATLANTA TAKEN.

Although the last chapter concerning affairs in this Department included all noteworthy *army* operations occurring during the last six months of 1863, and this one will record those of the succeeding year, yet it is deemed proper to commence retrospectively, so far as concerns the military telegraph, in that district through which the line ran that connected Clarksville, Fort Donelson and Smithland, with Cairo, Ill., and Nashville, Tenn. Our last particular mention of this line was in connection with the news of the battle of Stone River. When Rosecrans' Department was enlarged, in the summer of 1863, to include territory east of the Tennessee, as far north as the Ohio, Captain Fuller's jurisdiction over the Fort Donelson wire ceased, and VanDuzer's obtained. He at once recalled W. R. Plum from Columbus, Ky., and placed him in charge of the line south-east of Paducah, with directions to make it more reliable than it had proved—a command easily given, but difficult to execute, as the line for fully one hundred miles ran through a densely wooded country, sparsely inhabited and frequented by guerrillas, not to mention other troubles, such as the destruction of miles of line at once by sleet. H. J. Fish, and later, H. W. Nichols operated at Paducah, and E. G. Fish at Smithland; A C. Van Gilder, at Fungo; J. R. Thompson, at Fort Henry; John Q. Mason and Richard H. Elliott, at Fort Donelson, and J. J. Wilson at Clarksville, on this line. July 17, Fort Henry was abandoned by the

Federals, and poor Thompson, who was dying of consumption, resigned. Mason, one of the best operators in the Department and who had, as a member of the Chicago Mercantile Battery, participated in the Tallahatchie march and the battles of Chickasaw Bluffs and Arkansas Post, being a detailed operator, was sent to Fort Donelson, because the army regulations would not allow him enough to pay expenses of living in Nashville, where VanDuzer preferred to employ him. As it was necessary to have a test station in the neighborhood of Fort Henry, little Charley Bush opened an office at the house of one Phillips, the only Union resident in that section.

Three times Plum rode over the route of the line, accompanied only by a division repairer. The second time, after he and repairer Joice, a citizen of this section, had inspected forty miles in one day, stopping in the evening at a house three miles from Phillips' and connecting his instrument, he was urged to continue on to Phillips' by Bush, who offered to get up a dance in his honor, but as it was too dark to see the line, the invitation was fortunately declined. Very early the next morning, Jack Hinston's guerrillas captured Bush and Phillips, destroyed the telegraph and decamped with their prisoners. Repairer Johnson, located there, started east by daylight to repair the line, which had been cut the night before, and thus escaped certain death, as Hinston would surely have killed any native captured in the federal service. Shortly after Hinston left, Plum and Joice arrived at Phillips', and, repairing the line, telegraphed Colonel Brott, at Fort Donelson for cavalry to pursue Hinston, but Brott's few mounted men, he replied, were already *en route* for Phillips'. All day they were awaited, but came not, and it was decided by the three, Johnson having returned, to remain over night. Their horses were, consequently, hid in the woods. About dark, the tramp of horses crossing a rivulet near by was heard. Joice went to ascertain the cause, and reported guerrillas. The three could not defend the house, such was its construction, and Johnson was unarmed, so they fled to a fence corner near the woods, within hailing distance of the house where old Mrs. Phillips remained. There was a pale moonlight and nearly all night the tramp of horses was heard, sometimes close, at others distant. One of the three watched while the others slept. It was Plum's turn about one A. M. After an hour's listening and frequently punching Johnson to moderate his snoring, which was extremely provoking, he saw what appeared to be a man a little in front, motionless, but apparently staring right at the trio. With an understanding that neither was to fire until sure of discovery, Joice, who on being awakened, declared they were discovered, crept to the left and Plum to the right, each with cocked revolver, intent on taking the enemy on the flanks; cautiously each neared the enemy; their hearts crowding Adam's apple and beating the advance. A brave charge was saved, by the discovery that the enemy was a wooden man.

Plum in the retreat from the house, at Joice's request, had discarded his white straw hat. Mosquitoes attacked by regiments, brigades, divisions, corps and armies, Johnson sounding the war-whoop; it was dreadful. The dull tramp of horses and an occasional voice or shrill whistle, was enough, without the steady attack of these insects. The next morning Mrs. Phillips said the guerrillas had disturbed her, but of course offered no violence. The operator and repairers then proceeded to the Fort. Some days after, Bush returned paroled, having walked fifty miles barefoot and without coat. Hinston stole his coat, boots and eighty dollars in money. Joice, Johnson and another repairer were subsequently, but at different times, killed while in the service on this line. James Maloney, another repairer was captured by Hinston near Cumberland City, not far from Clarksville.

On one occasion, the chief operator and thirty men left the Fort at eight P. M., to make a night attack on the guerrillas, but returned the next morning without having found them. At another, with a like force, he rode fifty miles one day and forty the next to come up with them, but they crossed the Tennessee River at Chaudits Landing half an hour too soon. Scarcely had the troopers returned to Chaudits, when a force under Captain Bush, arrived from Smithland, and taking the Fort Donelson party for the guerrillas, Bush said "Now we have them, prepare for action," and was about to charge, when the operator dismounted, laid aside his arms and waving his white handkerchief ran quickly to them, where with difficulty he succeeded in convincing Bush that the Donelson party were Unionists. Thus another of those unfortunate mistakes whereby friends destroy one another was barely avoided. At another time, operator Van Gilder and repairer Smith were driven from Fungo into concealment in the woods.

Notwithstanding these and similar vexations, Captain Van Duzer officially reported in November, 1863, as follows:

> On the fifteenth of the same month (June) the charge of the line from Clarksville to Paducah, was given to me, and during June I succeeded by the aid of Mr. W. R. Plum, chief operator, and by putting in a set of repeaters at Fort Donelson, and increasing the battery, in getting that line in condition that it could be depended on as a means of communication with the North and West; a condition in which it yet remains.

After Plum was ordered to Chattanooga, in May, 1864, John P. Lathrop took charge of that line and the one from Clarksville to Hopkinsville, built by Burnside's order, in June, 1863, in which charge, Lathrop was greatly assisted by that excellent foreman of repairers, William Moak, but more especially by the following order, printed, and circulated along the line:

HEAD-QUARTERS DISTRICT OF TENNESSEE,
NASHVILLE, TENN., July 7th, 1864.

Special Orders No. 158.

EXTRACT:

* * * The telegraph line between Nashville and Smithland being continually molested by guerrillas, and other evil disposed persons, the disloyal citizens living within five miles of the line, in either direction, will hereafter be held responsible for its preservation; and whenever the wire is broken or otherwise injured, such disloyal citizens may be assessed to pay damages and required to make all necessary repairs, or assist in making them when notified. The citizens can and must prevent the continual breaking of the line.

By command of Major General Rosecrans.

B. H. POLK,
Major and Ass't Adj't Genl.

Up to early in May, 1864, no movement of special moment occurred in the Department not hereinbefore noted, except the advance of the Army of the Cumberland, so as fairly to locate the enemy in front of Dalton, Ga. On the 18th of March, General Grant, as Lieutenant General, was invested with the command of all of the armies of the United States, and going East a month later, he left to Sherman the command of the Military Division of the Mississippi, which extended westward to and included the Department of Arkansas, under Steele, whose headquarters were at Little Rock. Sherman was in telegraphic communication thereto *via* Fort Donelson, Cairo and St. Louis, and also *via* Louisville and St. Louis, thence through the Boston Mountains to Fort Smith, and along the river to the State Capital.

Before General Grant left Nashville, where he located his head-quarters shortly after the battle of Missionary Ridge, he had another difference with the Chief of the Military Telegraph. Late in December, 1863, Grant went to Knoxville to arrange with the Department Commander to drive Longstreet from East Tennessee and omitted to take his cipher operator, Beckwith, with him. Being overtaken by cipher telegrams, which he could not translate, he naturally felt much annoyed, and on his return to Nashville, where Beckwith was with the only key of that number in the Division, at the suggestion, it is said, of Colonel Comstock, of Grant's staff, he ordered Beckwith to impart the secret to that officer, but the operator protesting that he would like first to communicate with Colonel Stager, the General replied, "Give Comstock a copy first, then if you like, you can tell Stager what you have done;" whereupon Beckwith handed Comstock the key to *study out* and at once advised his chief, which resulted in an interesting correspondence, but before giving which, we will exhibit a recent order which was unknown to General Grant when he directed Beckwith as above. General Grant always acted a reasonably independent part, but it is believed he never purposed disobedience.

War Department,
Washington City, January 1st, 1864.

Ordered:

That the cipher books issued by the Superintendent of Military Telegraphs be entrusted only to the care of *telegraph experts*, selected for the duty by the Superintendent of Telegraphs, and approved and *appointed* by the Secretary of War for duty at the respective head-quarters of the Military Departments, and to accompany the armies in the field. The ciphers furnished for this purpose are not to be imparted to any one, but will be kept by the operator to whom they are entrusted, in strict confidence, and he will be held responsible for their privacy and proper use. They will neither be copied nor used by any other person, without special permission from the Secretary of War. Generals commanding will report to the War Department any default of duty by the cipher operator, but will not allow any staff or other officer to interfere with the operators in the discharge of their duties.

By order of the Secretary of War. E. D. TOWNSEND,
Official: T. S. BOWERS, *A A. G.* *A. A. G.*

UNITED STATES MILITARY TELEGRAPH, WAR DEPARTMENT,
Washington, D. C., January 21, 1864.

MAJOR GENERAL H. W. HALLECK,
General-in-Chief, Washington:

I beg leave to offer the following in explanation of my message to Captain Bruch, referred to in General Grant's communication of last evening. The information furnished me led me to believe that the request of the staff officer for copy of the cipher was without General Grant's authority, and as a new cipher had been arranged expressly for Mr. Beckwith's use at General Grant's head-quarters, and the order of the Secretary of War, recently issued, that the operators for this duty should be held responsible for strict privacy in its use, I indited the message referred to, not thinking it would come in conflict with General Grant's orders or wishes—the General having recently expressed his entire satisfaction with Mr. Beckwith's services I am extremely mortified at the result, as my only desire was to furnish the most reliable means of communication to General Grant, with the War Department. The new cipher was arranged with the view of being used by telegraph experts, and, it is believed, can not be used with any success by other than telegraphers. A great number of errors have been made by staff officers working ciphers, owing to their lack of experience in telegraphic characters, and it is believed that greater accuracy can be secured by placing ciphers in the hands of experts selected for this duty. The new cipher differs in many respects from those formerly used, and the one arranged for General Grant should not be known by any other party; hence my anxiety to keep it in Beckwith's hands. I sincerely regret that General Grant is led to believe that it is willful interference on my part. I am, General, very respectfully, your obedient servant,

ANSON STAGER,
Col. and Supt. Mil. Tel.

HEAD-QUARTERS OF THE ARMY.
Washington, January 22, 1864.

MAJOR GENERAL GRANT, Chattanooga:

I enclose herewith a copy of a note from Colonel Stager, in regard to his instructions to Mr. Beckwith respecting the new cipher. Your telegrams in regard to Lieutenant Colonel Comstock's orders to Mr. Beckwith have been submitted to the Secretary of War. It was known that the contents of telegrams communicated by means of existing ciphers, had been made public without authority. As these ciphers had been communicated to a number of persons, the Department was unable to discover the delinquent individual. To obviate this difficulty, a new and very complicated cipher was prepared for communications between you and the War Department, which, by the direction of the Secretary of War, was to be communicated to only two individuals—one at your head-quarters, and one in the War Department. It was to be communicated to no one else—not even to me or any member of my staff. Mr. Beckwith, who was sent to your head-quarters, was directed by the Secretary of War to communicate the cipher to *no one.* In obeying Colonel Comstock's orders, he disobeyed the Secretary, and has been dismissed. He should have gone to prison, if Colonel Comstock had seen fit to put him there. Instead of forcing the cipher from him in violation of the orders of the War Department, Colonel Comstock should have reported the facts of the case here, for the information of the Secretary of War, who takes the personal supervision and direction of the Military Telegraphs. On account of this cipher having been communicated to Colonel Comstock, the Secretary has directed another to be prepared in its place, which is to be communicated *to no one, no matter what his rank,* without his special authority. The Secretary does not perceive the necessity for communicating a special cipher, intended only for telegrams to the War Department, to members of your staff, any more than to members of my staff, or to the staff officers of other Generals commanding geographical departments. All your communications with others were conducted through the ordinary cipher. It was intended that Mr. Beckwith should accompany you wherever you required him — transportation being furnished for that purpose. If by any casualty he should be separated from you, communications would be kept up by the ordinary cipher until the vacancy could be supplied. It is to be regretted that Colonel Comstock interfered with the orders of the War Department in this case. As stated in former instructions, if any telegraphic employé should not give satisfaction, he should be reported, and if there be a pressing necessity, he may be suspended ; but as the Corps of Telegraphic Operators receive their instructions directly from the Secretary of War, their instructions should not be interfered with, except under very extraordinary circumstances, which should be immediately reported. Very respectfully, etc., H. W. HALLECK,

General-in-Chief.

P. S. Colonel Stager is the confidential agent of the Secretary of War, and directs all telegraphic matters under his orders.

H. W. H.

HEAD-QUARTERS MILITARY DIVISION OF THE MISSISSIPPI,
Nashville, Tenn., February 4, 1864.

MAJOR GENERAL H. W. HALLECK,
General-in-Chief, Washington, D. C.:

Your letter of the twenty-second, enclosing a copy of Colonel Stager's, of the twenty-first, to you, is received. I have also circular, or order, dated January 1, 1864, post-marked Washington, January 23, and received on the twenty-ninth. I will state that Beckwith is one of the best of men. He is competent and industrious. In the matter for which he has been dismissed, he only obeyed my orders, and could not have done otherwise than he did and remained. Beckwith has always been employed at head-quarters, as an operator, and I have never thought of taking him with me, except when head-quarters were moved. On the occasion of my going to Knoxville, I received Washington despatches which I could not read until my return to this place. To remedy this for the future, I directed Colonel Comstock to acquaint himself with this cipher. Beckwith desired to telegraph Colonel Stager on the subject before complying with my directions. Not knowing of any order defining who and who alone could be entrusted with the Washington cipher, I then ordered Beckwith to give it to Colonel Comstock, and to inform Colonel Stager of the fact that he had done so I had no thought in this matter of violating any order, or even wish, of the Secretary of War. I could see no reason why I was not as capable of selecting a proper person to entrust with this secret as Colonel Stager; in fact, thought nothing further of the matter than that Colonel Stager had his operators under such discipline that they were afraid to obey orders from any one but himself, without knowing first his pleasure. Beckwith has been dismissed for obeying my orders. A better man can not be selected for the position. I respectfully ask that Beckwith be restored. When Colonel Stager's directions were received here, the cipher had already been communicated. The order was signed by himself and not by direction of the Secretary of War. It is not necessary for me to state that I am no stickler for form, but will obey any order or wish from any of my superiors, no matter how conveyed, if I know or only think it comes from them. In this instance, I supposed Colonel Stager was acting for himself, and without the knowledge of any one else.

I am, General, very respectfully, your obedient servant,

U. S. GRANT, *Maj. Gen.*

Thus amicably ended the matter, Beckwith being duly reinstated.

During the five months that intervened between the battles of Missionary Ridge and Lookout Mountain and the advance of Sherman's forces into Georgia, all that energy, ingenuity, patriotism, art and wealth could do to strengthen and prepare the three armies concentrating at Chattanooga, for movement against the graneries, the workshops and communications and the Confederate army of Tennessee, was done and well done quickly. With Grant at the head of the armies to second intelligently Sherman's demands, he was enabled to assemble at Chattanooga, well equipped and provided in all respects, the Army of the Cumberland, under Thomas, numbering sixty thousand, seven hundred and seventy-three; the Army of the Tennessee, recently along the line of Huntsville Railroad, under McPherson, numbering twenty-four thousand, four hundred and sixty-five, and the Army of the Ohio, under Schofield, who succeeded Foster, the successor of Burnside, numbering thirteen thousand, five hundred and fifty-nine. These armies had two hundred and fifty-four guns, and were about to be reinforced by additional cavalry when they advanced. The railroad *via* Huntsville and Decatur to Nashville, commonly spoken of at Chattanooga as "Around the Horn," was re-opened, as was also that

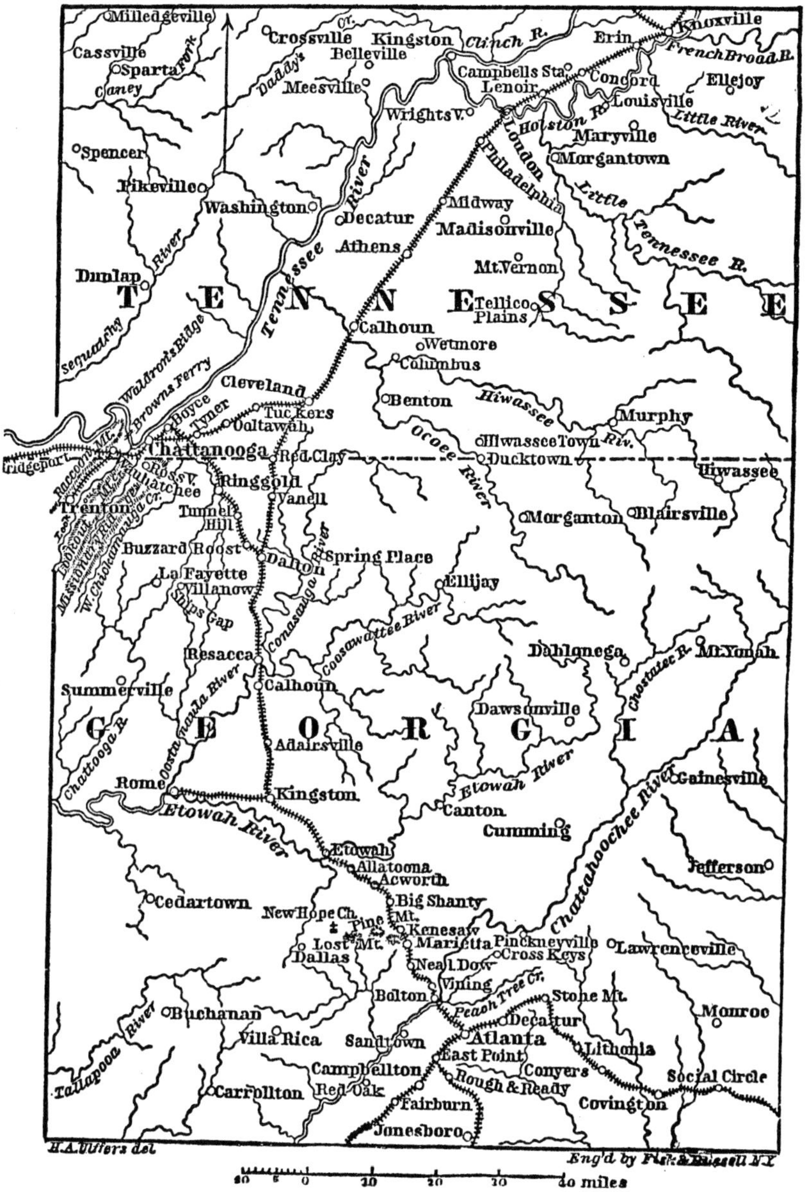

THEATRE ABOUT KNOXVILLE, CHATTANOOGA AND ATLANTA.

from Chattanooga to Knoxville. In January, 1864, the line between Pulaski, Tenn., and Flint River Bridge *via* Athens, Decatur and Huntsville, Ala., was rebuilt and a new line constructed between Chattanooga and Ringgold, Ga., the Union advanced position. In February, Captain Gross turned over to VanDuzer the property appertaining to, and the men employed upon, the line on the East Tennessee & Georgia Railroad. He extended it up the valley as Longstreet withdrew, and opened offices at various points. In March he re-built the line from Knoxville to Jacksboro and constructed a new one from Jacksboro to a point between there and Cumberland Gap, where he met Captain Gross' builders, thus completing a new wire *via* Cumberland Gap to Louisville direct, and giving three wires from Chattanooga to Louisville. In April, VanDuzer added a third wire to the line between Chattanooga and Stevenson which, connecting with the wire *via* Decatur, gave three lines from Chattannooga to Nashville. He also established a wire from Huntsville to Whitesburg, Ala.

Thus prepared, on the 28th of April, General Sherman, leaving a part of his staff at head-quarters in Nashville, where C. W. Jacques operated, started for Chattanooga, accompanied by F. S. VanValkenburg, there to await the signal from General Grant, who was in person at the head of the Army of the Potomac, which was about to move upon those sanguinary fields where it suffered such frightful losses. On his way to his army, Sherman received this order by telegraph:

CULPEPER, VA., April 28th, 1864.

MAJOR GENERAL SHERMAN:

Get your forces up so as to move by the 5th of May.

U. S. GRANT, *Lieut. Gen.*

From Chickamauga to Mission Ridge, was an anxious *hiatus;* a season, as we have seen, of doubts, fears, starvings and solicitous misgivings; one of energetic preparation however, fraught, as shown, with high honors and glorious victory. From Mission Ridge to the advance of all the armies; in May, 1864, was another *interim* of ceaseless preparation. Men and war material were furnished almost as certainly as requested. Life and treasure were not too dear to throw into the balance against disunion, and when the time arrived for Sherman to be ready to move, he was ready and moved.

The Federals in apt time began their difficult march against the forty-five thousand troops under J. E. Johnston, acting on the defensive. Captain VanDuzer on his part was well prepared, having experienced builders under Solomon Palmer and N. S. Townsend. With Sherman, went operator F. S. VanValkenburg; with Thomas, Jesse H. Bunnell, Cass G. Sholes, John Wilbur and Joseph Anderson; with McPherson, Wm. B. Somerville and later, Jacques; with Schofield, G. W. Wilson. C. D. Whitney was with the cavalry, and Wm. Patterson and John Lonergan were general utility men, for whom there was constant demand. VanDuzer himself, accompanied the army, which mainly advanced direct against the enemy in front of Dalton, but McPherson's command passed through Snake Creek Gap to destroy the railroad in Johnston's rear at Resaca. It is believed he might have succeeded in this, and thus have forced a general battle on an equal field, but Johnston headed off the movement and withdrawing, after inflicting heavy loss on the main body of the Unionists, reached Resaca May 13, where to the sixteenth, conflicts ensued resulting in great sacrifice and in Johnston's retreating further. The railroad was quickly repaired, for by it Sherman's army was supplied throughout the campaign. The telegraph accompanied the advance of the forces from Ringgold, and offices were opened as near the front and Department headquarters as possible, viz.: at Tunnel Hill, during the reconnoissance of Rocky Face Ridge and Buzzard Roost Gap, and pending the flank movement by Snake Creek Gap; at Dalton, within two hours of the evacuation of that place by the rebels; at Tilton and at a point on the railroad near the Union left, during the three days' fighting in front of Resaca. Jacques coming to the front with Lewis B. Spellman, opened the Dalton office.

About this time occurred a most terrible railroad accident. The water tank where engines took their supply, was about a mile north of Dalton and up a heavy grade, on account of which, the practice prevailed of detaching the engine at the Dalton station and taking it alone to the tank for water. A large train filled with wounded soldiers from the Resaca battle field arrived at Dalton, and the engineer started his engine up the grade for water, although he knew that the engine of another train lying at Dalton had preceded him. When near the tank he discovered the forward one returning, and to avoid a collision, reversed his engine, opened the throttle, and both he and the fireman jumped off. The two engines met without injury, when the south engine, under full steam, backed down the hill, acquiring a fearful momentum in the mile it ran before plunging into the train it had just left, crushing car after car, and mangling unto death many wounded and helpless soldiers.

A few days later, Jacques relieved Somerville, whose health was failing, and Charles Spellman assisted his brother. On the seventeenth, the army again advanced, a part arriving near Kingston the next day, and on the nineteenth, another portion at Cassville, after engaging the enemy spiritedly near Adairsville.

A good story is told at the expense of Cass Sholes in connection with the Adairsville fight. Cass never having witnessed a battle, requested VanValkenburg to send an orderly for him when one was in prospect. Accordingly Sholes was sent for and Van proceeded to baptize him with fire. After changing from one bad position to a worse one, the two were sitting on horse under a tree near the road, when a battery of artillery was planted abreast of them. In their rear, also mounted, were two general officers and their staffs. So many horsemen in an open field, of course drew the fire of the enemy's guns. A canister shot exploded in the tree directly overhead of the operators, scattering all around and wounding several staff officers and horses. It is possible that Sholes left first, but more probable that both operators took time by the forelock. As soon as distance lent enchantment, Sholes as pale as death, turned to Van and inquired if he looked white. Van laughed inordinately and replied "Yes," to which Sholes rejoined, "I am teetotally cow-kicked if I can't taste lead." Indeed, he imagined he felt a bullet in his stomach. To this day, he stoutly asserts that he did taste lead.

At Cassville, Johnston was reinforced by French's Division of Polk's Corps, but being dissuaded from giving battle, retired across the Etowah to Allatoona Pass, a position too strong for Sherman to attack. Accordingly, he moved to the right, with twenty days' rations, meeting the enemy at Dallas, and gradually extending his army to the left, struck the railroad again, June 5, after a severe but not general engagement. The enemy was now posted on Kenesaw, Lost and Pine Mountains. The telegraph had been opened at or near Sherman's head-quarters every night during the march from the Oostanaula to the Etowah, and at Kingston it was halted by direction, and communication had by courier with the army during the movement by Burnt Hickory and Dallas, to turn the Allatoona Range—a delay of about twelve days. Early in June, a branch line was opened from Kingston to Rome, and when the enemy retired to Kenesaw, Palmer advanced the line, first, to Ackworth where the army was resting, and then to Big Shanty where the Federals confronted the enemy and prepared for a direct assault. Field lines were quickly run along the entire front of the troops, from General McPherson's head-quarters on the left, to Schofield's on the right, with intermediate offices at Sherman's and Thomas' and such other points as Sherman wished. This line was from day to day extended and retracted, reaching out a loop to the signal station on Pine Mountain when that was taken, and another to Sherman's field quarters close to Kenesaw; stretching away to the right with Schofield as he tried in vain to turn the rebel left; pushing up after the assaulting column on the twenty-seventh, thus enabling, says VanDuzer, "General Sherman at all times to communicate with his subordinate commanders." It was worked until the enemy abandoned his position on Kenesaw Mountain, when the whole Union force was pushed forward in pursuit to Marietta, Ga.

Reports of Union losses show 9,299 killed, wounded and missing, during the month of May, and that the Confederates during the same period lost, killed and wounded, 5,393, to which should be added, according to an estimate by General Sherman, 3,245, prisoners captured by him. These figures demonstrate the character of the advance the first hundred miles into Georgia, nearly every mile of which was the scene of strife, and every day added to the sacrifice. At Ackworth, Sherman's losses were made up to him by reinforcements. Johnston's army then numbered 64,456. June 10, Sherman advanced to Big Shanty, at the foot of the Confederate strongholds, of which Kenesaw was the principal. Here General Sherman was extremely anxious for the completion of his field telegraph, and asked Townsend when it would be at his quarters, to which he answered, "In an hour and one-half," to make sure not to disappoint the Commander; but in thirty minutes Sherman heard the instrument near his tent, and Townsend thought he was under a heavy fire when Sherman asked: "What did you mean, sir, by telling me 'an hour and one-half?'" After that, the shortest, instead of the longest, time was given to such inquiries.

General Sherman, in his "Memoirs," says:

> "The 27th of June was fixed as the day for the attempt" (to break the rebel center), "and in order to oversee the whole and be in close communication with all parts of the army, I had a place cleared on the top of a hill to the rear of Thomas' center, and had the telegraph wires laid to it. * * * About nine, A. M., of the day appointed, the troops moved to the assault, and all along our lines, for ten miles, a furious fire of artillery and musketry was kept up. At all points the enemy met us with determined courage and in great force.

The assault, lasting two hours and a half, was a decided failure, the Federals losing twenty-five hundred to the enemy's eight hundred.

By this time, it had become necessary to send the wounded back as far as Nashville. When great calamities are averted on the brink of their consummation, the saved, regardless of former conviction concerning an overruling Providence, instinctively bless God for the rescue. Instances of premonitory impressions, resulting beneficially, however rare and unaccountable, are well authenticated. As a long train, laden with wounded from Chattanooga, moved northward, Robert B. Hoover, day operator at Normandy, Tenn., was impressed more and more deeply, with the coming of darkness and the train, that some great catastrophe, what, he knew not, was impending. Every train that passed north carried wounded men, and regular hospital cars were fitted up not unlike sleeping coaches, except that a common car was so arranged with hammocks and bunks as to accommodate the largest possible number. On the night in question, Colonel Ketchum, of New York, Commandant at Normandy, called Hoover's attention to the fact that the north-bound midnight train had for several nights past failed to stop as ordered; wherefore, Hoover directed the night operator to stop the train with a red light, and about nine o'clock retired to his cot in the baggage-room of the depot. A hundred nights before he had slept peacefully and soundly, without even knowing of the passage of trains; but this night a feeling of nervousness and suspense made sleep impossible. The measured tramp of the guard along the outer platform, the challenge of relieving patrols and the plaintive voice of the whip-poor-will attracted his attention. Having fullest confidence in James A. Howard, the night operator, Hoover was, nevertheless, so keenly persuaded that the train *must be stopped*, as to rise from his cot and await the coming of this midnight express. Finally, an engine whistled for a water station half way down the seven-mile grade, at the foot of which lay Normandy. About five hundred yards south of the office, the road crossed Duck River and when the locomotives were not to stop at the station, they customarily whistled off brakes while crossing the bridge, and naturally enough Hoover was listening intent for the signal. A hospital train came thundering down the grade. It struck the bridge, and two sharp blasts ordered off the brakes and startled Hoover, who exclaimed to his fellow operator: "Quick, Howard! the red light!" A few moments later, the train was at a dead stand by the depot. "What's wanted?" asks the conductor, as he approaches. The operators explain that the time-card required No. 7 to stop regularly, which it had neglected to do, and consequently they used the lantern. "Well, that's all right, boys," he replied, "but I am not running No. 7. Its engine gave out at Tullahoma, and my special train, full of wounded, was ordered ahead and to carry flags for No. 7. They will be along soon. There is no harm done, so good-night. Keep a sharp lookout for old Wheeler. All aboard!" As the ten coaches passed the station—coaches full of helplessness and misery—the operators watched until the rear red light was just turning the curve near the north end of the side track, when they heard first a shrill call for brakes, then a few dull thuds followed by an ominous silence, which transfixed Hoover at the station door, for he felt that whatever it was he had dreaded so many hours, had come to pass. In a few minutes the conductor returned with his lantern, pale, but self-possessed, and looking at the boys with an expression not to be forgotten, in a husky voice, said: "Come and see what your red light saved us from." Arriving there in silence, it was discovered that in the darkness the rebels had broken the switch lock, turned the rails from the main track and fastened a bar of railroad iron across the track with a telegraph wire. The road at that point turned sharply to the left around a high rock point, while to the right was a steep embankment running down to the deep, black water of Duck River. As the conductor pointed down to the water he said: "Boys, if you had not flagged us to-night, we would have struck this obstruction at forty miles an hour, and the whole train with its three hundred sufferers would have gone down into that river." Such of the wounded as were able to walk that night crowded around the telegraphers, and, with tears of thankfulness, expressed their gratitude; but Hoover insisted that 't was not he, but an overruling Providence, that saved them.

After the Kenesaw disaster and after the June losses of seven thousand five hundred and thirty killed, wounded and missing Federals and about six thousand Confederates, the army, July 2, began a movement to the right, pursuant to a plan arranged by telegraph, for flanking Johnston out of his stronghold. General McPherson, who was to take a leading part, was in Jacques' office, in electric communication with Sherman three hours before starting, and it is probable that Thomas and Schofield also took part in the telegraphic interview. Johnston, upon discovering the movement, fell back to the Chattahoochee.

It was during these movements that W. R. Plum, long detained in Chattanooga, was ordered to the front, and stationed first on Signal Hill, overlooking the river and the three bridges across it, which were fired by the rebels July 9, the night of his arrival and after most of the enemy had crossed to the southern side. About this time VanDuzer made his annual report, wherein he wrote:

In the erection, maintenance and working of these lines, Messrs. G. W. Wilson, C. G. Eddy, F. S. VanValkenburg, Cass G. Sholes, C. W. Jacques, Joe Anderson and C. D. Whitney, operators, were unwearied assistants, and to their hearty co-operation I owe much of the success that has attended my efforts to satisfy the demands of the armies for telegraphic communication. What the measure of that success was, I leave for others to say.

J. A. Fuller opened the office at Marietta, and soon after joined the field operators, working at the river ferries, Thomas' and Stanley's head-quarters, and Chattahoochee Bridge. Van-Valkenburg left the vicinity of Marietta for the North, after which C. G. Eddy assisted Sholes at Sherman's. On the sixteenth, Grant, having telegraphed the failure of the enemy's designs in Maryland and the consequent possibility of heavily reinforcing Johnston, Sherman moved at once to cross the river to the left of the railroad, *via* Roswell and the ferries, particularly Paice's and Power's. The railroad was immediately repaired to Vining's Station. Telegraph lines were promptly built to the right from Paice's Ferry to the Sweetwater, and to the left from Paice's to Power's Ferry, and from Marietta to Roswell—in all about forty-five miles. These lines were all constructed to facilitate the crossing of the Chattahoochee. While that movement was progressing, the few operators at the front were on duty all day and night throughout the seventeenth and eighteenth. On the twentieth, General Hood, who superseded Johnston on the seventeenth, attacked Sherman in force, but was driven to the defenses of Atlanta. On the same day the linemen crossed the river with the wire, building along the railroad to the third mile post from the city, being but three-quarters of a mile from the rebel defenses. From this point, a field line to the right and left connected the head-quarters of Generals Sherman and Thomas with those of such officers as they desired and with the points of observation established by the Signal Corps. This line extended to the left as far as the position south of the Augusta railroad which was attacked by the enemy on the twenty-second, and to the right as far as that wing was extended before the abandonment of the attack in front and the movement on Jonesboro. This line was at one time twenty miles long, and was continued in use twenty-eight days—offices having been opened and worked at twenty-three different points, more than one-half of which were exposed to the fire of the enemy.

On the twenty-second, Hood sallied forth from the city, making a fierce attack on the Federal left and rear, and on the twenty-eighth savagely assaulted the right, meeting severe repulse both days. McPherson was killed during the first attack. No officer in the army appreciated the value of the telegraph and the importance of the operator more than brave McPherson, who was pleased on all occasions to recognize its utility and to treat his operators with consideration.

General Sherman shows the Confederate losses on this campaign, during July, to have been ten thousand eight hundred and forty-one, of whom but two thousand were prisoners. The Union losses in killed, wounded and missing were nine thousand seven hundred and nineteen.

During the month, General Rousseau's cavalry destroyed many miles of railroad about Opelika, Ala., and General Garrard's did great damage to the Augusta road.

Every night the rebels threw shells over and beyond Thomas' office, but sometimes they failed to go far enough, and one killed an orderly at head-quarters, where Bunnell, Anderson and Plum then operated. Bunnell, on the sixteenth of August, resigned on account of ill health, and the service lost one of its ablest and bravest operators. Richmond Smith, fresh from the North, succeeded Bunnell as manager, to the dissatisfaction of some who had been long in the service. Thomas' was always the main or general office, and consequently was the most important.

About the middle of August, Wheeler's cavalry cut the line near Calhoun, and again about Dalton, which place was attacked, but the cavalry were held at bay until Steedman, from Chattanooga, came down and punished the rebels badly, after which they moved through East Tennessee, *via* Strawberry Plains, crossing the Clinch a few miles east of Jacksboro, and passing between Clinton and Jacksboro, crossed the Sequatchie Valley and moved on Murfreesboro, *via* McMinnville, intending to work great destruction to the Chattanooga road, while General Roddy with another force was besieging Athens with a view to operate, after its capture, against the Decatur road. It will be remembered that the telegraph route from Danville, Ky., to Knoxville, *via* the Gap, had by this time been abandoned, and a new line erected *via* Burksville, Chitwoods, Jacksboro and Clinton. Wheeler's scouts cut this line south of Jacksboro before he crossed in force. James Jones was operator there, and James Palmer repairer. Palmer, approaching Clinton to repair the break, learned that the rebels had appeared on the south side of the river in force ; that one soldier was drowned while attempting to swim the river to get the ferry boat at Clinton, where the negro ferryman had left it. The repairer also met a woman who had recently returned from a visit at her father's, on the south side of the river. She reported having talked at her father's house with a number of Wheeler's men, who stated that they were *en route* for Kentucky, and would cross the mountains at Jacksboro, or two miles below, at Wheeler's Gap. These reports created great excitement in Jacksboro. The few troops and male citizens there at once looked about for places of safety in the woods, leaving Jones and Palmer about the only men in the town. They kept alternate watch all night, intending to telegraph North should the enemy appear, and thus afford ample notice to the commanders in Kentucky. Superintendent Gross telegraphed Jones : "If you should have to leave there, do not go farther than to secure your safety, and return as soon as possible. Try and report where they cross the mountains ;" to which Jones replied : "You may rely upon it, the rebels shall not cross the mountains without my reporting them." The rebels not appearing, Palmer was sent to learn their whereabouts. He ascertained, and reported that Wheeler's force crossed at Winter's Gap, and also the general officers, regiments and guns, and their destination. Jones telegraphed these facts, closing : "They will strike the railroad at or near Murfreesboro, Tenn., in three days from the time of crossing ;" and so it proved.

Wheeler struck the railroad about Smyrna, a little north of Murfreesboro, doing a little damage, but Rousseau commanding at Nashville, and Steedman at Chattanooga, being timely warned, drove him off quickly in the direction of Lawrenceburg and Florence, where he crossed. Captain Gross acknowledged the importance of Jones' report at the time, saying, "The information which you were able to give of the numbers and direction of the invading force under the rebel General Wheeler, was promptly laid before the proper military authorities, and was the earliest information received of that force of the enemy."

No military commander however, ever thought of acknowledging this service. The telegraph operators at Smyrna, John T. Duell and his assistant were captured. Wheeler took them well on toward Florence, but being hard pressed by Rousseau, turned them loose to foot it back.

Finding the defenses of Atlanta too strong to assault, and the cavalry expeditions under McCook, Stoneman and Kilpatrick respectively, against the communicating railroad unsuccessful, or nearly so, Sherman essayed another flank movement to the right, leaving the Twentieth Army Corps at and near the Chattahoochee Bridge. During this movement, Eddy and Sholes accompanied Sherman, Whitney the cavalry, Jacques, General Howard, commanding the Army of the Tennessee, recently under McPherson, and Plum, General Thomas. Although these operators carried their instruments, their going was mainly as cipherers. No lines were built, but many miles were destroyed. The other telegraphers returned to the river. On the way to Jonesboro, cipherer Sholes while stooping at a spring to drink, lost his criptograph key; the only one used at Sherman's headquarters. It was not until some hours after the rear guard passed this point, that Sholes discovered his great loss. While wondering what the General would do to him and becoming more frightened as he pondered upon his misfortune, the brave boy determined to take his chances of capture by riding back to the spring, where he concluded the key must have dropped from his coat pocket. Mile after mile he rode, unattended and without meeting a human being. But no success attended his return, so he hastened back and stammered out to Sherman a report of his loss. Great was his relief when that officer cheeringly replied that the cipher was too old anyhow, and a new key ought to be made. That night, however, the key was returned to the operator by a member of Howard's staff, who found it at the spring. August 31, Howard's forces struck the Macon Railroad near Jonesboro, having already destroyed many miles of the West Point road, and on the same day, Stanley and Schofield to the left, also struck the Macon road. Quite a battle ensued September 1, mainly between a part of Thomas' troops and the enemy, in which ten guns and one thousand Confederates were captured. During the subsidence of this battle,

Thomas and his staff and operator, having dismounted, were sitting side by side, when a bullet passed behind each of them in a line about an inch from their necks. During the day, but before the battle, a soldier exhibited to Thomas the rim of a canteen, which, while on the soldier's person, had been pierced squarely in the center by a small cannon ball, without injuring the wearer.

The Confederates were pursued to Lovejoy, where, hearing of the evacuation of Atlanta by the corps left to hold it, Sherman declared the campaign ended and turned back to that city on the eighth. He estimates the Federal losses since leaving Chattanooga at thirty-one thousand, six hundred and eighty-seven killed, wounded and missing, and the Confederate at thirty-four thousand, nine hundred and seventy-nine.

A few hours after the enemy left Atlanta, the telegraph party at the bridge, went in with the troops and opened communication with the North, which at this time was in the throes of a Presidential canvass. The Democratic party had just resolved against "four years of failure to restore the Union by the *experiment* of war," and before the delegates had returned to their homes, the telegraph spread the news throughout the North of the capture of Atlanta and Mobile Bay. Never was rejoicing more heartfelt. Secretary of State Seward, in a public speech declared that "Sherman and Farragut have knocked the bottom out of the Chicago nominations."

As Sherman's army encamped to the east and west of Atlanta, the telegraph was connected with its parts; thus a line was built by William Patterson from Roswell, through Cross Keys and Decatur to Atlanta, and thence *via* East Point to Campbellton, with offices at head-quarters of the army corps and cavalry commanders. This work was completed almost as soon as the army encamped. Then, Superintendent VanDuzer, who had been constantly at the front for nearly five months, returned to his quarters in Nashville. Besides the operators heretofore named, employed south of Chattanooga, there were in the month of September, the following: Thomas Williams, James Bryant, John E. Clark, Alfred Winder, William H. Hartman, —— Waterhouse, N. S. Townsend, and probably some others.

Within the Department of the Cumberland there were in June, three hundred and eighty-three telegraph employés, of whom one hundred and forty-two were operators, one hundred and forty-four laborers, sixty-three repairers, and the rest made up of foremen, clerks, teamsters, wagon-masters, wagon-makers, blacksmiths, messengers and saddlers. And there were one thousand, five hundred miles of telegraph in constant use. It was with special reference to the telegraphers' service in this campaign, that Sherman in his Memoirs, wrote:

> The value of the magnetic telegraph in war cannot be exaggerated, as was illustrated by the perfect concert of action between the armies in Virginia and Georgia during 1864. Hardly a day intervened when General Grant did not know the exact state of facts with me, more than one thousand, five hundred miles away, as the wires ran. So on the field, a thin, insulated wire may be run on improvised stakes or from tree to tree, for six or more miles in a couple of hours, and I have seen operators so skillful that by cutting the wire they would receive a message with their tongues from a distant station. As a matter of course, the ordinary commercial wires along the railways form the usual telegraph lines for an army, and these are easily repaired and extended as the army advances, but each army and wing should have a small party of skilled men to put up the field wire and take it down when done. This is better far than the signal flags and torches. Our commercial telegraph lines will always supply for war enough skillful operators.

CHAPTER XII.

THE TELEGRAPH IN THE DEPARTMENT OF THE TENNESSEE (1864) AND GULF TO SEPTEMBER.—SHERMAN'S ADVANCE FROM VICKSBURG.—A PERILOUS RIDE ON A FLAT CAR.—W. S. SMITH'S RAIDING FAILURE.—FORREST'S SUCCESSES AND DEFEAT IN WEST TENNESSEE AND KENTUCKY.—FORT PILLOW MASSACRE.—CAPTURE AND REMARKABLE ESCAPE OF AN OPERATOR.—STURGIS' AND A. J. SMITH'S RAIDS.—AN OPERATOR RIDDLED WITH BUCK-SHOT.—FORREST'S DASH INTO MEMPHIS; CAPTURES AN OPERATOR; HIS EXPERIENCE.—VICKSBURG MATTERS.—GRIERSON'S RAID.—CAPTURE OF FORTS IN MOBILE BAY.

Compared with 1863, the new year (1864) opened in the Department of the Tennessee with few telegraph offices. These, operated under Captain Fuller, were located as follows: At Paducah, Ky., Edwin Peel, operator; Metropolis, Ill., Charles Wallace; Cairo, Ill., George Gallup, W. S. Lewis and Richard H. Tracy; Columbus, Ky., Jacob V. Hill and William Gibson; Union City, Tenn., Edgar B. McNairn; head-quarters, Memphis, Tenn., Lewis B. Spellman, William H. Merkley; Memphis, W. B. Somerville; R. R. depot, C. W. Pearson; Germantown, F. W. Snell; Colliersville, Edward F. Butler; LaFayette, R. B. Griffin and Wm. A. Thayer; Moscow, Chas. H. Spellman; La Grange, Peter Fowler; Grand Junction, Ellis Stone; Camp Davis, Miss., Robt. J. Hughes; Salisbury, Tenn., James Pitton; Pocahontas, Tenn., Jno. Q. Olmstead; Middleton, Tenn., Jno. O. Ingle; Chewalla, Tenn., Joseph Blish, Jr.; Corinth, Miss., Cambridge Culbertson and John J. Egan; Vicksburg, Miss., Robert S. Fowler, Samuel D. Cochran; R. R. depot, Edw. D. Butler; Black River Bridge, Stephen L. Robinson; and at Hebron, Miss., John R. Frank.

The Federal hold on West Tennessee, except along the river and the Memphis & Charleston Railroad, had for months been nominal only. The great body of this section was given over to guerrillas, and, as if to invite organized expeditions against the river towns, it was concluded to vacate the line of the railroad also. Accordingly, Corinth was evacuated January 26, 1864; Grand Junction, the twenty-eighth; LaGrange, the twenty-ninth. In short, all east of Germantown was given up, and that also, March 26. Offices were opened at Fort Pickering and at Colonel Hoges', out on the Pigeon Roost road, a few miles from Memphis, to prevent surprises.

This was the work of General Sherman, who reached Memphis the 10th of January, bent on concentrating there a cavalry force, seven thousand strong, under General W. Sooy Smith, to overwhelm Forrest, while he, Sherman, with McPherson's Vicksburg forces and two divisions of Hurlbut's, moved east from Vicksburg against Lieutenant General Polk. In a great measure both expeditions failed. John J. Egan, operator, went with the army from Vicksburg, but found little to do. He reported the capture of Frank C. and Mervin Jones, Confederate States operators, who took the oath of allegiance and announced their purpose of going North. Sherman held them as prisoners of war. Operator Compton was also captured and brought to Vicksburg. Nat. Walton escaped from Clinton with his instruments, two hours before the Union troops arrived.

The expedition, consisting of about twenty-three thousand men, left Vicksburg, February 3, and was feebly retarded by Polk's force, exceeding ten thousand men. Meridian, Miss., was reached, and great destruction was wrought wherever the army went, especially to the railroads. Finally, the troops returned leisurely, without being met by the cavalry force from Memphis, which was expected at Meridian.

When Sherman passed Jackson, the anxiety at Meridian was indicated by hasty and general preparations for departure. Even the telegraph material, not in actual use, was packed. On the night before the evacuation, Major Fleming, General Superintendent of the Mobile & Ohio Railroad, arrived with an engine, one coach and box cars. Fleming and the telegraphers expected to leave at ten, A. M., the next day, but Sherman had pressed his cavalry forward during the night, and by daylight Confederate General Adams had passed through the city, in retreat. Emmet

Howard, operator there, snatched his instruments from their places, connected the through lines, and hurriedly boarded the train. Then the engineer, probably fearing ambush, refused to run the locomotive, but the yard-master consented after an anxious delay. The train moved southward with great rapidity, passing the three-mile post just as Federal cavalry rode up, after which the trip was uneventful to Enterprise, and thence south of the Alabama line, where an office was opened in a box car.

When Sherman retired, Superintendents David Flanery, from the South, and James Pressly, from the North, restored the line to Meridian, using for that purpose every piece of wire, however short, that could be found. Not long after this, a most disastrous freshet washed portions of the Mobile & Ohio Railroad in scores of places, and many bridges south of Meridian were carried away. For miles nothing could be seen above the water along the road but the iron rails and narrow embankment supporting them, and in some places the rails were held together only by fish-bar connections. Of course, the telegraph was much damaged, many poles being carried away. To ascertain the extent of the injury, Superintendent Pressly and operator Howard left Meridian for Mobile, with orders for the use of engines and hand-cars wherever found. At Enterprise, an engine was obtained, and the driver prevailed upon to quit a saloon and run the locomotive to the first break. Pressly, Howard and a Mr. David Cummings boarded a flat car attached to the tender. The moment the engineer mounted the engine and opened the throttle, his condition was discovered; but it was too late, and and the little train was soon moving at a speed of fifty miles an hour. What made it worse was the fact that the engine was pushing, not drawing the empty flat, which afforded nothing except a few loose standards or uprights along its sides to hold on by. It is believed that this situation was better calculated to an inspection of one's sins than the telegraph. At one time, Pressly proposed jumping. Cummings tried to reach the tender, but the oscillations of the car made this difficult and dangerous for a novice. However, he finally succeeded, and, throwing the engineer aside, the train was brought to a stand. It stopped within fifty feet of the Chicahasaha River, the bridge over which had been washed away. Had Cummings been a few seconds later in reaching the engine, none of the party would have lived to relate the story of that dreadful ride.

Wm. Sooy Smith's expedition, which was to co-operate with Sherman's march on Meridian, left Memphis one day after it was due at Meridian. Operators Lew. B. Spellman and Fred W. Snell accompanied Generals Smith and Grierson, respectively. The command returned February 26, having met Forrest, who with an inferior force, near West Point, defeated and sent it back to Memphis in hot haste, with a loss, according to a Confederate writer, of six hundred killed and wounded, and three hundred prisoners. Forrest's loss is stated at fifty killed and one hundred and fifty wounded.

It was now Forrest's turn to move, and he did so with his usual success. He prepared a new expedition into West Tennessee, consisting of two columns, one moving by way of Pocahontas and the other entering near Corinth. March 23, he sent a detachment under Colonel Duckworth against Union City, Tenn., which was guarded by about four hundred and seventy-five troops under Colonel Hawkins. Approaching the place early the twenty-fourth, the Confederates were received by Hawkins, who made a vigorous defense for some hours, but Duckworth pretended the presence of Forrest in overwhelming force, and reinforcements not arriving, Hawkins surrendered. Some hours before the enemy appeared, scouts reported their near presence, when Hawkins roused Edgar B. McNairn, a young operator from Canada, to telegraph General Brayman, commanding the Columbus, Ky., district, for aid. General Brayman ordered operator Pearson to proceed with him from Columbus to Union City. Three regiments of infantry and one battery loaded on flat cars, were started by rail to succor Hawkins. Before this, the line was cut (three A. M.) near Union City; proof of the presence of the foe, consequently the train moved very slowly, fearing ambush. Too tardy as we have seen. About six miles from Union City an old negro reported the fall of the place, and Brayman hurriedly returned. McNairn, who had taken position behind the defenses with the soldiers, was of course among Duckworth's prisoners. We will follow the military operations to the conclusion of this raid, and then see what befell McNairn. Forrest's main force moved directly for Paducah, where (twenty-fifth) Colonel Hick's troops, aided by gun-boats, repelled several assaults upon Fort Anderson. The Confederate loss was about three hundred killed and wounded, and the Federal, sixty. Ed Peel was operator at this place at this time. From Paducah, Forrest moved southward, leaving Buford at Trenton, instructed to overrun the territory with his division, conscripting and foraging. Hendricks, the telegraph repairer at Blandville, was among the conscripts. The telegraph north of Columbus was cut and unused for over two weeks. April 10, the rebels in small force dashed on the pickets at the latter place, killing one and capturing another. On the thirteenth, Brayman was summoned by Buford to surrender, but instead, he opened his big guns of Columbus on the enemy, who decamped. On the fourteenth, Buford appeared again at Paducah, but retired about the expiration of the time allowed for the removal of the women and children. Meanwhile a small body from Memphis was whipped by a brigade at Bolivar, *en route* to join Forrest. Grierson in Memphis having organized a large force, still other cavalry was sent to Forrest. His command doubtless now exceeded five thousand men.

While Buford was approaching Paducah, Forrest was making a reputation at Fort Pillow, which place he captured on the 12th of April. Out of a garrison of five hundred and seventy-five, about forty-five per cent. were *killed*. The excuse for the remarkable disparity between the loss of the assailants not under cover, viz.: fourteen killed and eighty-six wounded, and the defenders behind works, few of whom were wounded and about two hundred and fifty killed, is that the *negroes* and *renegade* Tennesseean defenders fought *heroically*. It is extremely difficult to fitly appreciate such heroism if we consider the paucity of results. It savors too much of *heroically* selling one's life as *cheaply* as possible. Suicides and self conceited martyrs are sometimes found to have preferred being dead cowards to living heroes, but instances where so *many* of either class, valiantly elect to fall by an enemy's hand are unknown, unless Forrest's apologists are to be believed. Soon after he moved into Mississippi.

R. B. Griffin and C. W. Pearson were operating at Columbus during these operations, and Jacob Volney Hill, the manager, was dying in the small-pox hospital. He entered the corps early in the war and rendered invaluable service, incurring great risk and suffering severe exposures. His energy, daring and ability attracted the personal attention of Colonel Stager, who complimented him for his achievements, and ordered Captain Fuller to increase his pay. Hill's was a hard fate.

A force of about twelve thousand left Memphis April 30, to operate under General Sturgis against Forrest's troopers, before they could return to Mississippi. They quickly left Bolivar on Sturgis' arrival, and effected their escape. Operator Fred W. Snell accompanied the Federals until, owing to his being retarded while mending the wire, the troops left him and his little party of repairers behind, liable at any moment to be captured or shot by the prowling banditti that infested the line of the Memphis & Charleston Road in that district. Consequently one of the men stood watch while the others worked. To all requests for a guard and orderlies, the General deigned no reply. He even took up his pontoons over Wolf River, thus compelling the operator and builders to construct a raft on which to cross the building material. As the party was sent out by General Washburne's order, to keep Sturgis in communication, the latter's utter disregard of the operator and constructionists seems strange. A raft of logs was made two miles from the railroad, the material carried to it from the hand car, was at the end of a hard day's work, landed on the other side the evening of May 3, when the line was extended to Moscow. Sturgis sent back important despatches for Washburne, but just as Snell began sending them the wire was severed, he knew not where, but believing it was at Colliersville, he secured five negro soldiers, and with Joe Wooten, a brave repairer, started for that town, seven miles from Wolf River. They cautiously approached the place, found the wire on the ground, connected the instrument and sent the despatches. They remained there, lying flat on the ground, while telegrams were being sent and received, over two hours. Happily it was quite dark, for a squad of the enemy's cavalry scarce twenty rods away were discussing the quality of a jug of Memphis whisky. When Washburne's last message was received the party returned by the hand car that brought them, as rapidly as possible. As the horrors of Fort Pillow were fresh in their minds, the venture was the more commendable. Among the despatches to Sturgis was one ordering a guard for the telegraph party, which was sent.

E. B. McNairn, captive operator at Union City, began his great southward tramp at midday, reaching Trenton at two P.M. the following day. *En route* to Jackson some of the prisoners began to lag; McNairn among them. Colonel Wilson, commanding the guard, made a pass at him with a sword, which, owing to the rashness of McNairn, who thereupon seized a gun, came near resulting in the death of both; for had McNairn fired, Wilson's men would have made short work with the operator At Jackson, it was arranged by some of the prisoners to club the guard and escape; but as the Federals were encircled by a fire, the plan was abandoned. From here to Pocahontas, under Colonel Wilson, and thence to Ripley and Kelly's Mills, under the considerate Lieutenant Colonel White, these prisoners were marched leisurely or hurriedly by long or moderate marches, according as the air was infused with rumors of approaching Federals. From Kelly's Mills to Aberdeen, quite a number of prisoners escaped. After a tramp of about two hundred and fifty miles, West Point, Miss., was reached, from whence by rail the prisoners proceeded to Mobile, Ala., two hundred and thirty-two miles. Here they boarded a steamer, and ascending the Alabama River twenty miles, were debarked to march to Montgomery, and thence to Camp Sumter, in Anderson, Ga., where they were appalled, on arrival at daylight, by the sight of the great prison pen. Into this cruel prison — this bastile where were immured at one time full thirty thousand Federals, from which, during its occupancy, twelve thousand nine hundred and twenty Union dead were taken—McNairn and most of his comrades, who had survived their long marches, were lodged. But McNairn's stay here was short, and in the county jail, at Macon, sixty miles north, he soon met certain officers of his party, who had not been sent to Anderson. Altogether, there were one hundred and twenty-three officers confined in the jail on McNairn's reception.

Thoughts of escape were soon dominant. Colonel Hawkins succeeded in making an instrument from wire, by which he could unlock the door, but he seems not to have profited by it. Captain Moore had saved a pocket map of the United States, which was frequently consulted in debating a proper route of escape, on which point there were many opinions. McNairn secreted a case knife which was converted into a saw, and therewith the plank floor, after six days patient toiling, was cut. Six persons, confined in one apartment, were engaged in the effort to escape, which, as planned, involved the digging of a tunnel under the three foot thick wall of the prison, and twenty feet beyond. The parties engaged in this work were Captains Murray and Galloway, Lieutenants Ayers, Oats, Allander, and operator McNairn. Each man in his turn labored in digging down below the foundation. After succeeding in this, divesting themselves, one at a time, of most of their clothing, they in turn entered the hole and tunneled with the case-knife saw as long as the impure air would allow, and, having clasped his arms around the loose earth, a comrade would drag him out. The dirt was then thrown under the floor. By the 14th of May, every thing was in readiness for the great venture. It was arranged that one party, consisting of Galloway, Oats and Allander, should pass first. These were to go north and enter Sherman's lines about Dalton, Ga. The other three, starting a little later, *i. e.*, two, A. M., were to try and make General Asboth's quarters at Pensacola, Fla. Oats and McNairn finished the tunnel up to the surface, making a small opening through which stars could be seen. Oats led the way, but as he reached the opening, was alarmed by a sentinel within the yard (the tunnel opened into the public square), not over three feet from him. The sentinel was poking the earth and stones with his bayonet, doubtless believing the noise he had heard was made by rats. This caused the lieutenant to return and report that they were discovered, and that all was lost. For an hour all listened intently. The suspense grew each each moment and became oppressive. The first party refused to proceed, but the second determined to take their chances.

Ayers led the way out. McNairn followed in a few moments, and then came Murray. The night was clear and beautiful. The lights were burning in the streets. Across the open square was a guard house, from which came snatches of song and bursts of drunken laughter. A solitary man was seen walking a deserted street. The trio, having rubbed their boots well with onions to prevent being tracked by the hounds, took the Southwestern Georgia Railroad, and walked about four miles to a copse, where they lay all day — the trains passing them on the right, and negroes working in the fields near by to the left. Even the noise of hounds was heard, and when night came they were wearied with excitement. At dark, providing themselves with stout cudgels, they started on their long, perilous and adventurous march for Pensacola, flanking each station when possible and avoiding as much as practicable their greatest enemy, mankind. Night after night they walked, oftentimes losing their way when off the railroad, sometimes nearly drowned in swamps and quicksands ; again, nearly famished, or crazed with thirst, they, nevertheless, under the impetus of hope and the excitement of their undertaking trudged on, sleeping by day on the bare ground, and finding too frequently that they had located their resting place unfortunately near some farm house or village, in which event anxiety prevented sleep. On a Sunday morning they were awakened by singing in a contiguous meeting-house. The rations which these men had stinted themselves to save for this trip while in jail, gave out before reaching Fort Valley, which place they tried to leave on the left, but took a road which led them right into the town. The next day Murray was in great agony from thirst, which was finally slaked from a mud puddle where McNairn and Murray drank themselves sick. Following the Columbus road, they crossed the Flint River and turned south-westerly toward Eufaula, sometimes barely living on the fish they caught, or the fowls they foraged ; again, surfeiting themselves at the table of some generous hostess or in a negro's cabin, or while lying in the swamps where Unionists sent them food. They were enabled eventually, after passing through Lumpkin at midnight, to reach and cross the Chattahoochee River at Florence, near Eufaula. At one farm house, where were three women and one man, the venturers, in the *rôle* of furloughed Confederates, were enjoying hospitalities. During the conversation which ensued, the three claimed to have met the son of the old gentleman in the army of Virginia. This son had been captured at Gettysburg, and well treated while a prisoner near New York. On his return he wore a blouse quite similar to McNairn's. This fact caused the son's mother, while the party were at table, to lay her hand on McNairn's arm and, looking him squarely in the face, say "I believe you are a Yankee." His blushes and hesitation carried conviction, and it was well, because out of gratitude for her son's treatment, the three were well provided for continuing their journey. Passing Clayton, they struck the Pea River, intending by boat or raft to follow it to the gulf, but dams and waterfalls prevented. While lying in a swamp by the river, they separated and hid from a party approaching, whereby Murray became lost until near nightfall. Crossing the river, they continued their journey to Slip-up, where, as furloughed rebels, they dined with a Captain Harne and another Confederate officer. Pressing on over Winslow's bridge, past Geneva, they reached Hewitt's Bluff, Fla., where, while resting at midday in fancied security at a Union man's house, they luckily discovered the approach of troopers, who rode up to the house just as the Federals, leaving all behind, rushed out at the rear and ran rapidly toward a canebrake, called "tight eye," which was gained only after the pursuers on horseback had wildly fired several shots at the fugitives. The horses could not enter the swamp, and their riders drew off to get their dogs. At night a guide took them to an island, where were a great many Unionists avoiding conscription. On this island the three obtained a boat in which, after several nights hard work, they emerged at the mouth of the Choctawhatchee, where the owner of a fishing smack received them and carried them to the gunboat Bloomer, Captain Christie, near East Pass. The next day a tender conveyed them through the Santa Rosa Sound to the frigate "Potomac," lying off Pensacola, where they were landed and taken to General Asboth's head-quarters. Murray was too sick to start North immediately; Ayers went *via* the Atlantic coast and McNairn reached Memphis *via* the Mississippi, on the 2d of July, one hundred and one days from his capture, after traveling near three thousand miles, two thousand through the heart of the Confederacy, about seven hundred of which was on foot, and when he reached Memphis he was reduced from one hundred and fifty pounds to ninety-nine and one-half.

It may here be noted that in May, 1862, Pensacola was evacuated by the Confederates, and the telegraph wire connecting that city with Northern ones was reeled up as far inland as Camp Lomax, thirty miles from Pensacola. At this point of observation, J. G. Thornton, operator and officer, communicated with the interior. Major T. J. Myers commanded the battalion of three companies of the Fifteenth Confederate cavalry, known in all the region thereabouts, which they scouted, as the "bloody Fifteenth," and it is not improbable that McNairn's visitants at Hewitt's Bluffs, were of this command. Federals from Pensacola frequently pursued the "bloody Fifteenth," and at one time drove them back to Pollard, where General Tyler, accompanied by Confederate States operator Frank U. McMullin, from Montgomery, with one thousand, five hundred reinforcements, checked the advance. In the retreat on Pollard, Thornton narrowly escaped capture or worse, in the conflicts which ensued, and during which, as an officer on Confederate General Clanton's staff, he was much exposed. C. H. Edwards was then operating at Pollard. The line was abandoned after this raid.

General Sturgis having signally failed to punish Forrest, was nevertheless allowed upon his return to Memphis, to organize another powerful expedition, which left that city early in June. Operator Snell accompanied Sturgis, but before starting, General Washburne gave an order directing the commanding general or any other officer to afford required assistance when solicited by the operator. The telegraph was repaired twenty-seven miles out, where the troops debarked from cars and began their south-easterly march to strike the Confederate cavalry and prevent its operating against Sherman's communications. Near Salem, Snell and three hundred cavalrymen were ordered to operate along the Mobile & Ohio Road. Passing through Ripley, they crossed the Hatchie River, captured some scouts and moved rapidly toward the railroad at Danville and Rienzi, where the depots were burned, as were also several trestles and one other bridge. On the fifth, Snell and fifty cavalrymen took another tack, which was enlivened by a running fight of nine miles, driving the enemy across the Tishimingo Creek with a small loss. Joining the larger party two days later, they narrowly escaped capture by the timely arrival of a reinforcement of five hundred. Snell was unable to effect anything on the Confederate wires.

Forrest warmly received Sturgis at Guntown, where ample

preparations for whipping the Federals had been made. On the tenth, the Union cavalry encountered the enemy's pickets, which were driven back, and the Confederates discovered in position. After four hours' fighting, the Federal cavalry was driven back, and the infantry began. All the forces Sturgis had in hand were soon engaged in a desperate struggle. Cavalry, army wagons and infantry became inextricably mixed, and about five P. M., the Unionists began giving ground. The negroes remembering Fort Pillow, fought with desperation, even clubbing their guns. That night, Sturgis' command fell back to Ripley, about twenty-five miles, a defeated and demoralized mass—nine thousand in all. The Confederates pressed their advantage nearly to Colliersville, by which time the Union loss was three thousand, five hundred men, many guns and wagons.

The monotony of being whipped or out generaled every time, grew burdensome to the Nationals in this section, and in part to retrieve success, yet another expedition was organized, but its great purpose was co-operative. General E. R. S. Canby at New Orleans, directing operations south of Cairo, was aiming at Mobile. Washburne's forces (including A. J. Smith's), Steele's in Arkansas, and the lower river troops, including Slocum's at Vicksburg, were to co-operate. The plan in this regard, which I deduce from messages in my possession, written by Canby himself at New Orleans, and put in cipher, was for "A. J. Smith's troops and any other that Washburne could reach, to pursue, and if possible destroy Forrest's command," and to that end wrote Canby, June 27, "I have placed under his (Washburne's) control, all the militia from the North-western States that were ordered to report to me, and several regiments of old troops from Missouri. This will enable him to give Smith an effective force of twelve or fifteen thousand men, and leave a reserve of five thousand for other operations. I will start the expedition against Mobile four days later. A cavalry expedition will start at the same time for (from) Vicksburg, for the purpose of distracting the attention of the enemy from Smith's operations and those in this neighborhood. A large cavalry force will leave the river above Baton Rouge, for the purpose of cutting the Mobile & Ohio Railroad." Steele at Little Rock, was directed "to make a demonstration in the direction of Price's forces," to "prevent their coming down upon the river while these operations are pending." Slocum was ordered "to send all your (his) cavalry force that can be spared, for the purpose of making a demonstration in the direction of Jackson or beyond, for the purpose of distracting the attention of the enemy as much as possible from the other operations."

Smith was to start on the 2d of July and Canby the sixth. Steele was too weak to press on, and called for help to save his communications, and Canby sent a brigade and two batteries from Morganzia to the mouth of the White River, to be commanded by General Gordon. Price moved and entered Missouri in September. Five days before Canby was to begin the great movement, on which he had set his heart, he received an order for a great part of his troops to be sent to the Army of the Potomac, to which without a murmur he promptly responded as follows:

HEAD-QUARTERS MILITARY DIVISION OF THE WEST, Mississippi.
NEW ORLEANS, July 1, 1864.

MAJOR GENERAL HALLECK, Chief of Staff of the Army, Washington:

Your despatch of the twenty-fourth has just been received. I will send twenty thousand men. All sea-going steamers now here will be taken up, and the movement commenced at once.

(Signed) E. R. S. CANBY, *Maj. Gen.*

And the next day six thousand were on board. Even Texas, except twelve hundred at Brazos Santiago, was ordered to be abandoned.

But Canby determined to demonstrate all along the river line and Smith started early in July. With him was Stephen L. Robinson, operator and cipherer. Skirmishing was incessant, and the heat intense. It is said that from three to four hundred of Smith's soldiers were sunstruck during the first two days of the advance from Salisbury. Tupelo was reached, the enemy falling back, but to advance again for a speedy conflict in order that the calls for aid now coming from Mobile, which place Canby had thoroughly aroused, might be met. Generals S. D. Lee and Forrest were badly beaten on the fourteenth, but Smith's forces were out of ammunition, and what was worse, of food also. Indeed the soldiers were nearly famished while returning.

On nearing Ripley, Smith determined to send Grierson on to Memphis with the cavalry and rest the infantry on the line of the Memphis Railroad, and accordingly Grierson soon advanced some miles beyond.

The forces were thus divided, when Smith concluded to let Robinson go on also, and writing despatches for Washburne, which were handed one of the two orderlies who accompanied the operator; the three rode ahead to overtake Grierson. To shorten the gap, they took a by way, which was skirted by woods. While riding single file over this road, the coarse voice of a man entirely hidden in the brush, called "halt!" As if by preconcert, the three riders spurred their horses at once. The two soldiers being ahead, escaped unhurt, but Robinson was behind and being dressed as an officer, except shoulder straps, he was at once singled out as *the* one of the three, and not stopping, one guerrilla, who had a musket, shot Robinson's horse in the right shoulder, while the other with a two barreled gun, loaded heavily with buck-shot and slugs, shot Robinson; one bullet lodging in the joint of the right knee; two others penetrated almost through the leg near the thigh, stopping just under the skin; two entered the hip, where they are to this day; one cut through the shoulder and another passed through the arm above the elbow and striking a rib, ran around to the spine, where it was afterwards found. Robinson's blouse and pants were literally riddled. His horse fell against a sapling as soon as struck, throwing the operator to the ground, and immediately after hobbled off, following the orderlies, who rode in haste to overtake Grierson. Of course Robinson was entirely at the mercy of the two bushwhackers, and especially was he impressed with this fact on reaching for his revolver in his right boot leg, for he then discovered that his arm was paralyzed. As his telegraph instrument was in his saddle-bags, it was saved, but his cipher key, number twelve, and all his valuables, except a chainless watch in his pants fob, were taken from him, as was also his coat and vest. The captors then ordered him to get up and follow them, but Robinson replied after feigning an attempt to comply, "You see I cannot walk." "Well," one said, "we will soon come after you," and off they started, but no sooner were they out of sight than Robinson tried to run away. He made some progress, tumbling every few rods, like a drunken man, and finally reached a small creek, which he tried to cross. He walked out on a log and fell off into the water. It seemed as if even the elements were against him, but in truth the ducking was a blessing in disguise, and a few minutes later, the orderlies and some of Grierson's men were riding up at a rapid pace. All of the remaining shot or slugs were cut out, except those in the thigh, and Robinson was one of the few of the wounded of that expedition, who recovered. Gangrene was more fatal than lead. If it be asked why operators were sent on these expeditions, we reply, first, they were the only cipherers in the army, and it was customary to send back reports and receive orders in cipher; and second, it was the operator's duty to attach his instrument to the Confederate lines and thus obtain evidence of the enemy's movements whenever possible. Snell and Robinson seem not to have been very successful in tapping the enemy's wires, while accompanying the several raids just related, but they were *liable* to be of greatest usefulness.

Yet another movement from Memphis was undertaken; this time as before, under A. J. Smith's immediate command. His forces (twelve thousand, largely infantry), advanced in August, south of Grand Junction on Oxford, along the railroad. Chalmers, and later Forrest also, opposed this movement, but feebly and finding Smith too strong, Forrest determined on a dash at Memphis itself. The telegraph was erected as far south *via* Grand Junction as Smith's main force went, but Forrest doubtless had foresight enough to cut the line before flanking the Federals. At any rate, starting from a point south of Oxford at five P. M., August 18, with over two thousand men and riding as rapidly as possible, he astonished the natives at four A. M., of Sunday, the twenty-first, by appearing in force in the streets of Memphis. Generals Hurlbut, Washburne and Buckland were supposed to be sleeping at the Gayoso House, and thither the advance of Forrest's forces rode as hurriedly as possible, intent on capturing these and other officers. The "others" were there, but the Generals having enjoyed their evening out, had unsuspectingly lodged elsewhere. July 13, Superintendent Fuller placed R. S. Fowler in charge of the Memphis district, which included all territory in telegraphic communication with that city, and specially enjoined on him to have "a good operator on hand *at all times* at Washburne's office," and it so happened that C. W. Pearson was operator at general head-quarters, on the night and morning of Forrest's approach. R. S. Fowler, J. R. Dixon, C. H Spellman and J. O. Byrnes, also head-quarters operators, escaped. Perhaps they were out late that night also. Pearson getting somewhat drowsy, was startled by firing on Union Street, and looking out, saw quite a force of cavalry, which he mistook for Union troops, and was about returning to his seat, when the Judge Advocate ran through the hall and up stairs, reporting the presence of rebels. Pearson thinking the building would be fired, hastened to the back stoop and jumped off, never thinking that it was eighteen feet high. No sooner had he landed, than he was called to a halt by a Confederate, but avoiding him, he ran through the lower hall and right into a nest of the enemy, who took him under their protecting care and hurried off with him out of the city. There were five thousand soldiers in Memphis, and Forrest was no match for them, so after about an hour's stay he hurried off with three hundred prisoners, losing about two hundred of his own men by the dash. About five miles out, the General lay in a fence corner on a poncho, when Pearson approached to ask to be paroled. The General plied him with questions about Smith's plans, forces, etc., and receiving evasive replies, Forrest became vexed and made Pearson take his place with the other prisoners and march to Tupelo, seventy-five miles, in a two days' rain, where, after a sound sleep in the mud, he boarded the freight cars. By this time his clothes, except an undershirt and pair of pants, had been appropriated by his "guides," and now one of them enforced an exchange of his pantaloons for a pair big enough for Barnum's giant. One of the fellow prisoners spoke out, "let him keep his pants, that is all he has left of the blue," but the rebel replied, "can't do it, his belly is blue, that will answer" Pearson's excursion included Grenada, Canton, Jackson, DuVall's Bluff, Ala., Selma and then to that great resort, Cahaba, where soon after came Pettit and Ludwig. Thinking his chances of exchange as operator very poor, Pearson reported himself at Cahaba as a member of Company M, Second Illinois cavalry, and as such was finally exchanged.

Connected with the prison was a yard where the cooking was done. The prisoners could pass out to this yard, but were forbidden to stop within a circle which they were obliged to go through and which was marked off by sticks. This circle was the *dead line.* Pearson was, one morning, passing through it, when some one behind called to him, whereat he thoughtlessly stopped and looked behind; but no sooner had he done so than another cried out, "For God's sake, Pearson, get out of that!" He jumped as if to escape a bullet. It is sad to relate that the guard did indeed fire, and that the ball struck a prisoner—a mere lad, who was passing near by at the time. Pearson got leave to watch over the wounded boy and keep off the flies; but it was a short duty, for in twenty minutes the young soldier, after saying "Boys, if you ever capture a rebel soldier, remember me," died on the rude stretcher he was being carried on to the hospital.

Another time the operator, for some real or imagined disobedience, was made to mark time on the burning sand, the sun beating hot upon him, so that his feet were covered with blisters. Pearson, lighter by thirty pounds was not recognized at Washburne's head-quarters when he came "marching home again," with the seventy-six exchanged head-quarters' troops.

General A. J. Smith's forces were supplied at Abbeville by rail. W. D. Hawkins operated in the depot in Memphis; Snell and McNairn at LaGrange; G. E. Spellman at Colliersville; James K. Gulliher at Moscow, and Edward F. Butler at Holly Springs.

Chalmers now commanded the Confederates, and being steadily pushed southward, and fearing defeat north of the Yocona, crossed most of his forces, but left his adjutant general, Captain L. T. Lindsey and a small scouting command, to watch the Federals, who were nearing Oxford. Lindsey, being an operator, kept Chalmers constantly advised of affairs in front. But Smith was now needed to save Missouri from Price's advancing army, and all of this section was again vacated, and September 1, the line from LaGrange to Germantown was abandoned, and, later, to White's Station. Washburne, taught a lesson by Forrest's dash, determined to have telegraph "offices on the main roads leading into Memphis, just outside the city."

During Smith's advance on Oxford, the guerrillas in the

north-west corner of Tennessee were very annoying. Captain Samuel Bruch, having relieved Fuller of the Cairo District, wrote General Paine, at Paducah, who then (August) commanded the District of West Kentucky (with whom, by the General's special request, was John A. Cassell, telegrapher), giving the names of fourteen "guerrillas who infest the country between Blandville and Columbus, and who are in the habit of cutting the telegraph." * * "I have no doubt," he said, "but a little wholesome shooting will be very beneficial to the telegraph as well as the service generally." Among the fourteen were "John Price; lives ten miles from Columbus; swears will kill Dicus, telegraph repairer, if ever catches him." "George Griffie, now prisoner at Paducah. He captured Government horse from telegraph repairer at Blandville, and told Hendricks, the repairer, would shoot him if ever caught him putting up line again." Bruch also gave the names of five guerrillas who "came to Dicus' house a few days since, five miles from Columbus, hunting for him." Paine's severe orders, which followed, produced the desired effect.

Vicksburg is now to be paid a hurried visit. We last noticed this place in the fall of 1863. John C. Sullivan was District manager then. His operative force was soon reduced to E. D. Butler, John R. Frank, S. L. Robinson, K. B. Howe and James K. Gulliher. The office in the city in the "Wirt Adams Bank" building was pleasant and roomy until a quarter-master seized it by authority, sending the telegraphers as angry as hornets into the top story. Outpost offices were established, and monotony like constantly falling drops of water made its deep impress on the operators. Sullivan was unfortunately relieved in September, 1863, by Robert S. Fowler, of Chicago, Ill., who had never seen any military service. In consequence, "Yankee," the peer of the best, resigned and left the department, as he ought to have done. Fowler was much liked after the first feelings of jealousy wore off, and was the next year transferred to the management of the Memphis District, where he was relieved in September by E. F. Butler. Samuel D. Cochrane succeeded Fowler at Vicksburg, and George Gallup was sent to take Cochrane's place. Gallup hung around for two weeks, prospecting, and then took a boat for New Orleans, and Cochrane, against his wish, retained the management. E. D. Butler was made superintendent of the Big Black River Railroad. Various, but except as already related, mainly unimportant, were the expeditions out from Vicksburg in 1864, one of which under General Dana brought in two telegraph instruments, taken at Woodville, Miss. This expeditionary force was one of three which started late in the year, to prevent Confederates from being sent East to operate against Thomas or Sherman. Dana's troops met the enemy in force on the Big Black, and defeating them, began destroying railroad property. General Davidson led another from Baton Rouge to Tangipahoa, Franklintown and West Pascagoula, creating great alarm in Mobile; but it was Grierson's raid from Memphis, with three thousand and five hundred cavalry, that wrought most damage.

Grierson started December 21, taking with him Stephen L. Robinson, operator, now fully recovered from his wounds. He struck the Mobile & Ohio road at Tupelo, intending to so destroy it as to render it utterly useless to Hood's army, which was then flying from the scene of its overthrow (Nashville) by Thomas' army. The enemy were in Corinth, and the idea was to break the road as far north as possible. Grierson sent Robinson with about two hundred troopers from Ripley, to strike as near Corinth as they could reach, and, moving south along the road, destroy it as badly as possible, while overtaking the main body. Riding hard, Robinson's party, under a major, dashed into Rienzi, twelve miles south of Corinth, at four, A. M., capturing a picket post. Robinson connected his instrument with the line, but the hour was too early, hence no information of consequence obtained. The torch was then applied to bridges, etc., when the party moved toward Tupelo, but was soon overtaken by Confederates who harassed them greatly.

While feeding their horses on a roadway which ran through a swamp, the troops were not a little excited over a remarkably perfect mirage, by which Grierson's whole camp, fifteen miles away, was distinctly seen *in the sky*. Every move in that distant camp was clearly defined by this strange *fata Morgana*.

Two days and nights Robinson was in the saddle before overtaking Grierson. At Okolona, important telegrams announcing the coming of reinforcements, were intercepted. The Federals promptly attacked the enemy at Egypt, near by, and while whipping the force in front, other troops turned against and repulsed another which had just arrived.

Robinson frequently connected his instrument with the main line as Grierson moved on toward Meridian. One night, taking two orderlies, he went beyond the troops, and attached his pocket instrument to the wire, and listened for two hours to the passing telegrams. In this way the operator was of very great service, as he gained much valuable information.

During this trip, Grierson, Captain Woodard, Robinson, and perhaps six others, chanced to precede the command several miles, to find themselves suddenly confronted by a company of rebel scouts who would surely have captured the little party, had not the General audaciously commanded a charge. The Confederates, like sensible men, supposed the Federals had more coming, and hurried off, as they ought, leaving Grierson and his party to lionize themselves.

Striking across the country south-westerly, the Mississippi Central Road was badly damaged, and finally after several minor conflicts, the force reached Vicksburg with five hundred prisoners, eight hundred beeves and one thousand negroes.

An immense drove of hogs was captured *en route*, which the soldiers tried to drive into Vicksburg, but the swine were obstinate; they preferred all ways but the right one and once in a while it seemed as if the devils had again entered these "unclean" beasts; now they are in the very midst of the force, tumbling horse and the rider, who usually rose with blasphemy escaping from his lips, while the laughter of the more fortunate drowned the wickedness, so that no record was ever made against the *fallen* ones. Anon, the quadrupeds were in the bush where horses could not go, and then his hogship began rooting with provoking *nonchalance*. It was ultimately decided to corral the animals in a lot and shoot and burn them. Accordingly they were killed, the fire applied, and that great mass cremated.

In the Gulf Department where General E. R. S. Canby was located, and concerning which, we have in this chapter had glimpses, but little (unrelated heretofore) of special moment occurred until the summer of 1864. Canby was energetic, but crippled as we have seen, when about to advance. However, he expressed himself to General Washburne as determined to make the most of the forces left him, and as a consequence, extended his actual jurisdiction in all directions, thereby requiring of Superintendent Bulkley the building of many miles of telegraph.

About the 1st of August, 1864, Captain Bulkley at his own suggestion, was relieved and furloughed indefinitely without pay or emolument, to enable him to enter upon the *quasi* public work of building the American portion of an international line called "The Collins Overland Telegraph," which, east of the Amoor River, was to be constructed under the auspices of the Western Union Telegraph Company, and from the mouth of that river to St. Petersburg, Russia, seven thousand miles, by the Russian Government. It was estimated that this would give an European telegraphic connection sixteen thousand miles long. At this time no cable had been successfully laid across the Atlantic Ocean. Bulkley took with him from the Gulf Department his chief assistant, Frank N. Wicker, Scott R. Chappell, head-quarters cipherer, Ed Conway and J. W. Pitfield.

At the time of Captain Fuller's formal assignment (August 1) to this Department, Bulkley had constructed six hundred and ninety-five miles of telegraph, as follows:

	MILES.
From New Orleans to the mouth of the Red River, . .	205
New Orleans to Pass Manchac, ...	39
New Orleans to Lake Port, ...	7
New Orleans to Fort Pike, ...	32
New Orleans to South-west Pass, ...	113
New Orleans to Brashear City, ...	80
New Orleans to New Iberia, ...	133
Fort St. Philip to Fort Jackson, ...	1
Head of Passes to Pass a l'Outré, ...	20
La Fourche to Fort Butler, ...	40
Baton Rouge to Port Hudson, ...	25

The lines from Brashear City to New Iberia, forty-one miles, and Morganzia to Red River Landing, twenty-five miles, were temporarily abandoned.

The offices in this Department for the month of June, for example, were operated as follows: New Orleans, by H. W.

Nichols, Henry Stouder, S. B. Fairchild, Scott R. Chappell, Daniel Elphick, G. W. Baxter, B. H. Upham, John Hefferin, I. A. Sherman and Wm. Anderson; Port Hudson, by E. H. Johnson and Miltón H. Bassette; Morganzia, by A. W. O'Neal and Wm. A. Tinker; Baton Rouge, by Charles W. Moore; Thibedeaux, by Alvin A. Howard; Brashear City, by S. S. Ulmer; Algiers, by F. H. Bassette; Fort Pike, by Charles I. Fancher and F. A. Sanborn; Boutte, La., by Henry Bowerman; Camp Parapet, by Patrick Moloney; South-west Pass, by W. H. Munro; Quarantine, by Edward VonEye; Bonne Carre, by L. J. Hebard; Pass a l'Outré, by Robert C. Loomis; Fort Jackson, by D. W. Hunter; Cubitts, by Richard Cubitt, Jr.; Fort Macomb, by John H. Meehan; Head of Passes, by H. S. Valentine; Pass Manchac, by J. W. Hynes; Napoleonville, by Chas. O. Conner and Geo. S. Hanna; Lake Port, by C. Lindauer; Terre Bonne, by Edward E. Bates; Tigerville, by C. S. Johnson; Point Celeste, by J. Deslondes; at General Canby's head-quarters, S. P. Kimber was chief.

Many of these offices were located near cypress swamps, where it was found impossible for operators to stay and live any considerable length of time. D. W. Hunter, operator from West Chester, Pa., became a victim of malaria and died at Fort Jackson, October 12, 1864, leaving a wife and three children in Mifflin County, Pa.

Recurring once more to military affairs, it seems proper by way of epitomizing the situation in the midsummer of 1864, to exhibit a report from Canby, telegraphed from Cairo, as follows:

HEAD-QUARTERS MILITARY DIVISION OF WEST MISSISSIPPI.
NEW ORLEANS, LA., August 4, 1864.

MAJOR GENERAL H.W. HALLECK, Chief of Staff, Washington, D. C.:

On the nineteenth ultimo, General E. K. Smith and Major General Magruder reached Alexandria from Texas, and on the twentieth, General Smith left for Shreveport. On the same day the divisions of Magins, Walker Wharton and Polignac commenced moving in the direction of the Atchafalaya. The advance reached Semmesport on the twenty-first, and from three to five hundred mounted men were thrown across that river. The forces at Morganzia were reinforced by four thousand troops from this city, and twenty-three hundred from Vicksburg, and on the twenty-eighth, a force under General Ullmann encountered and, after some sharp skirmishing, dispersed the rebel force that had crossed. The casualties on either side were not large, but Colonel Chrysler, Second New York cavalry, a valuable officer, was severely wounded. Captain McNally, an active partisan on the rebel side, was killed. On the twenty-ninth (29th) the rebel force at Semmesport had disappeared, and the whole force appeared to be moving in the direction of Alexandria. On the first instant, Hardeman's brigade had crossed the Red River and was moving in the direction of Vidalia. Walker's division was moving in the same direction. A scout sent out from Natchez on the third of last month, has just returned and reports a general movement of the rebel troops on Red River to the eastward. The troops on the lower Teche are also moving in that direction. The scout reports that General Smith had been ordered to cross the Mississippi, and reinforce the troops opposed to Sherman. Mobile papers of a late date, state that this order has been given and that Smith's force was expected to cross soon. The recent movements indicate this intention, or to make an attempt upon General Steele's lines. General Gordon, who had been sent temporarily to White River, where General Steele's communications were threatened by Shelby, has now been instructed to remain, if the movements of the enemy should indicate the line of the Arkansas as the real point of attack. Price was still at Camden on the thirty-first; Fagan at Monticello; Marmaduke on the Arkansas River, near Red Fork, and Shelby east of White River. General Steele reports sharp skirmishing on the line of his communications, with, in the aggregate, about equal losses. None of these forces have been materially reduced, although small parties have been undoubtedly sent into Missouri for the purpose of collecting recruits. General Buford reports sharp fighting at Simm's Ferry, on Big Creek, with severe loss on both sides. His detailed report has not yet been received. The difficulties in Missouri can only be controlled by the inhabitants who should be constrained to array themselves for the government or be regarded and treated as enemies. I have ordered (2) two regiments of cavalry from Memphis to report to General Rosecrans. I do not think it expedient to send more, even if they could be spared, unless there should be an organized invasion of the States and we can prevent that, as long as we can hold the line of the Arkansas. Admiral Farragut will make an attack upon the forts at the entrance of Mobile Bay in the course of this week, and I have sent Major General Granger with two thousand men to occupy and hold Dauphin Island in order to secure the Admiral's communication with the sea, if he should succeed in passing the forts. The troops from Texas are now arriving. I have notified the collector of the port of the temporary abandonment of Brownsville, but I presume that the same notification should be given by such authority in Washington, as will prevent clearances from foreign ports to Brownsville.

E. R. S. CANBY, *Major General.*

All eyes were now turned toward Mobile. With Granger, went operator B. H. Upham and later followed John R. Dixon, recently with Gen. Reynolds at the mouth of White River, and Dorsey Berry. Admiral Farragut having completed preparations, moved his fleet consisting of four gunboats and fourteen wooden ships of war, against Forts Morgan, Gaines and Powell August 5, 1864. The Confederate ram Tennessee and other war vessels were captured; Fort Powell abandoned; Fort Gaines surrendered on the seventh and Morgan the twenty-third; thus 104 guns and 1,464 prisoners were taken and Mobile effectually blockaded. The Federal monitor Tecumseh was blown up by a rebel torpedo and lost with nearly all her crew.

Gen. Canby, September 5, ordered a telegraph cable laid from Morgan to Gaines, which was accomplished on the eleventh. The distance is twenty-three thousand five hundred and thirty feet. Operators Israel A. Sherman and M. H. Bassette were sent to Morgan and George R. Penn and C. S. Johnson to Gaines and later, Orrin K. Newton to Bon-Secour Bay to the north-east of Morgan. These operators had but little to do for the most part, except dig oysters during the ebb tides. The Confederates had worked a line from Gaines to Powell and Powell to Cedar Point on the main land and thence to Mobile. Only about fifty yards of the water route was cabled, the rest being on poles sunk in the shallows. A large quantity of very good cable was captured at Fort Powell, which in later operations was made useful.

The end of the rebellion in the valleys of the mighty rivers running through the south western territory was evidently approaching, but a brave hope and a great pride were procrastinating it, honor at least, it was cruelly urged, required additional sacrifices.

When a people continue a fratricidal war on mistaken notions of the dictates of chivalry, it is to be regretted that the sacrifices are not confined to one side.

CHAPTER XIII.

THE TELEGRAPH IN ARKANSAS AND MISSOURI DURING THE LAST HALF OF 1864.—PRICE'S INVASION OF MISSOURI.—VARIOUS BATTLES, AND INCIDENTS.—MAJOR SMITH RESIGNS IN FAVOR OF CAPTAIN CLOWRY.—A SUMMARY OF THE SERVICE.

Notwithstanding the fact that there were sections of the seceding States where the Confederates more than held their own, on the whole, it was evident by the fall of 1864, if not prior thereto, that the rebellion was becoming exhausted. Perhaps a cessation of the struggle might reasonably have been first expected where it last occurred, *i. e.*, in the sparsely settled trans-Mississippi States where poverty was the rule. It is to this division, once more ablaze from the Gulf to St. Louis, that we now invite attention.

August 1, 1864, the only lines in Arkansas in operation, were from Little Rock to Lewisburg, fifty miles; Little Rock to Pine Bluff, forty-five miles; Little Rock to DuVall's Bluff, forty-seven miles; Fort Smith to Van Buren, four miles, and sometimes to Fayetteville. Although in the preceding fall every county but fourteen was represented at Little Rock in constitutional convention, whereby the State was reorganized, and hence recognized by Congress, yet by the defeat of Banks in the Red River country and his retirement, the great Confederate wave which began on the Texan coast now swept northward, retaking two-thirds of Arkansas, separating the Federal strongholds by destroying communication, conscripting and punishing Unionists who had avowed their sentiments, raiding over the State almost at will, and even threatening St. Louis, Mo., itself. The whole trans-Mississippi country was influenced decidedly by Banks' withdrawal. Confederate cavalryman Shelby harassed the Unionists in various sections of Arkansas, though at times he was himself overmatched by pursuing troopers. The rebels in considerable force even ventured to the neighborhood of Helena, Ark., where they were defeated after a severe conflict, by five hundred and fifty Unionists. Late in August Shelby captured a Federal regiment between DuVall's Bluff and Little Rock. Price was now organizing the last great expedition that ever invaded Missouri. It was composed of his own, Fagan's and Shelby's forces, released by Banks' retirement, and was to be reinforced in Missouri by twenty-three thousand Knights of the Golden Circle and Sons of Liberty. General Rosecrans assumed command of the Missouri Department January 28, 1864, but his troops were reduced by calls elsewhere to but a few thousand home guards, some hundred-day men whose time had expired, and a few volunteers; perhaps twelve thousand all told, variously located. Price moved northward *via* Batesville and advanced on Pilot Knob *en route* for St. Louis. This was late in September, 1864.

Let us now take a glance at the telegraphic facilities then afforded Rosecrans, for he fought this coming campaign from his telegraphic offices in St. Louis and Jefferson City.

It is proper to premise that up to October 1, 1864, all telegraphic operations in Missouri and Kansas were under Major G. H. Smith's immediate direction, after that under the care of Captain Clowry, his successor. March 9 Rosecrans ordered the Sedalia line extended from Warrensburg to Kansas City, following the route of the Pacific Railroad. Under this order, the telegraph was erected *via* Pleasant Hill and Weston to Fort Leavenworth, Kansas, one hundred and eight miles, thus placing St. Louis on a direct circuit with Fort Leavenworth; and under an order of the 28th of March a line from Allen *via* Glasgow to Booneville where it connected with Syracuse and Tipton, was built. In June, by order of General Fisk commanding in North Missouri, St. Joseph was connected with Weston (forty-five miles), thus giving the former two telegraphic routes to St. Louis.

Late in April the department commander directed the rebuilding of a line *via* Pilot Knob, Cape Girardeau and Sikeston to New Madrid. This was completed in May, seventy-four miles being reconstructed. Another line was built connecting Cape Girardeau with Bloomfield (forty-five miles) and the Pilot

Knob wire extended southward (twenty-eight miles) to Patterson. Altogether there were in operation including the wire (one hundred and twenty miles long) from Fort Leavenworth to Fort Scott, Kansas, one thousand three hundred and twenty miles of military telegraph under Major Smith's direction, which included also General S. R. Curtis' Department of Kansas; that officer having assumed command with headquarters at Fort Leavenworth in January, 1864. Very soon after taking command Curtis wrote Smith as follows:

> I am obliged to you for the early efforts you have manifested to supply my command. I must have telegraphic communication with Fort Blunt (old Fort Gibson) and Fort Smith. The Arkansas River must be lined with posts, and department commanders must have connection with them everywhere. I hope that you will be prepared to add from two to three hundred miles of line to the Fort Scott line as I expect to get the privilege of so extending it. I have ordered out a full company in place of a small detail to push the work to Fort Scott, and your superintendent (Byington) is actively, and I think successfully, carrying on the work.

But the proposed extension was not approved by the Secretary of War.

July 13, 1864, Major Smith wrote Colonel Stager, "On the first of September next I shall have been two years in the service and propose at that time to resign my commission." He also recommended Captain Clowry, then in Arkansas, for the management of the Corps in Missouri, Kansas and Arkansas. Much as the Major's retirement from a field, where he had served so usefully, was regretted, the fact that Col. Stager coincided with his views as to his successor in case of his quitting, made it certain that the members would not abate any of their zeal for the service. Upon Major Smith's taking leave of those who had been pleased to serve under him so long and well, many were the heartfelt expressions he received.

George H. Smith was born in Norwich, New York, June 23, 1833. Expecting to pursue a collegiate course, he prepared therefor in an academy located at home, but an untoward turn of fortune's wheel reduced his father's means. Instead of obtaining a collegiate education, George, at seventeen, entered a drug store as clerk thereof, bringing with him, however, his Greek and Latin grammars, which he studied at intervals. In 1851, he began business in a modest way for himself, as proprietor of a paper and book store. Next year, the "Utica and Oxford Telegraph" line, one of the many little companies then springing up to die young, brought Norwich into the telegraphic world, and Mr. Smith was selected as the one best adapted to attend to the operating. Accordingly, he was sent to Utica to acquire the art, but after learning the alphabet and getting some idea of the manipulations, he was obliged to return to his own business, where, however, an instrument was placed in circuit. It was not long before he taught himself to operate, and to read by ear. Practice made him proficient. In 1857, he constructed and superintended a line from Terre Haute, Ind., to St. Louis, Mo. This line ran to the consolidated office in St. Louis, of which Mr. Smith became manager. The other wires were those of the "Western Union," the "Illinois and Mississippi," and "Missouri and Western" companies. Mr. Smith had had about four years service at the telegraphic head-quarters of Missouri, when the war began. He knew, perhaps better than any other telegrapher, who, in that service, were trustworthy, for some were bitterly opposed to the Union. He knew many of those men who plotted treason in the State, and, having had years of practical instruction in every branch of telegraphy, and being by reason and birth, of pronounced Northern sentiment, and in the prime of life, it was not unnatural that General Halleck, as shown in a

GEORGE H. SMITH.

preceding chapter, preferred his appointment to that of any other, however well qualified, loyal and energetic such other might be. Smith's position in the consolidated office brought him in contact with Union officers in Missouri, who had entrusted him with various telegraphic commissions. Foreseeing the necessity of organized effort to maintain telegraphic communication with the gathering armies in Missouri, Mr. Smith, who early won the confidence of General Fremont, submitted to that officer his plan for departmental telegraphic organization, which was substantially approved by Fremont, and partially carried out, as shown elsewhere, when orders came from Washington to disband for want of authority of law, and because another plan had been adopted, which placed all the war telegraphs of the Union, under one management. Smith's appointment as captain and assistant quarter-master followed properly and naturally from the position he held when it was found necessary to commission the superintendents. His territory, overrun as it was by guerrillas, was at least as difficult as any to work, but he devoted his energies with patriotic zeal to the labor, and as a result, the telegraph in the Department of Missouri proved vitally important to the Union. His victories were bloodless. They even saved lives of Union soldiers, and materially contributed to the restoration of the trans-Mississippi country.

October 1, Captain Clowry relieved Major Smith and appointed Charles A. Hammann assistant for Missouri, Theodore Holt for Arkansas, and re-appointing Dwight Byington for Kansas.

While telegraphic extensions were progressing, raids, bushwhacking and other minor skirmishing took place in various parts of the State, but as a rule it had become much quieted, except that in a *semi*-peaceful way many of its people were organizing a fresh outbreak, when Price should appear in force to encourage and support them. Indeed, aid was confidently expected from members of treasonable orders as far east as Ohio.

General Washburne at Memphis telegraphed Rosecrans, *via* New Madrid, that Price was moving toward Missouri. That was before the movement was fully developed, but none too soon, as A. J. Smith's veteran division, on which hung the fate of St. Louis, was on transports, ascending the river to join Sherman in Georgia. Telegrams at Cairo diverted Smith's course to St.

Louis. On came Price, with a force ranging from ten thousand to twenty thousand men, largely cavalry. Scouts reported at the various southerly telegraph stations the advance on Patterson. September 28, Price assaulted Pilot Knob, where John H. Byrne was telegraphing. General Thomas Ewing, commanding a force of twelve hundred men, repulsed the assaults, suffering a loss during the day of two hundred men, and inflicting on his adversaries near five times greater injury. But Ewing's only safety lay in retreat on Rolla, and accordingly he spiked his guns and moved therefor, reaching it in safety after an exhaustive march and being harassed by Shelby. At Mineral Point, sixty-one miles from St. Louis and twenty-five north of the Knob, a good part of General Smith's command was concentrated to meet a portion of Price's troops expected there. Smith called in his out-posts, planted his guns and awaited attack. A train laden with soldiers and refugees, including the Irondale operator, was delayed in consequence of injury done the road near the Point. The attack on the train which followed was repulsed, the track repaired, and the train saved. By this time the woods were filled with Confederates, and picket firing began. Miss Louisa Volker, operating at the Point, having been at her instrument continuously for two days and nights, was relieved by the Irondale operator.

At noon of the twenty-eighth, General Smith was telegraphed to fall back, and by three, P. M., the last train started. Every male citizen, fearing conscription, left also. Miss Volker and sister remained to protect their father's home from destruction. After hiding all evidences of her employment, and placing a pistol in her pocket, with a fixed purpose of defending herself and sister against violence, she overlooked the little village from her window, and discovered Confederate cavalrymen, ragged and dirty, with "lean and hungry" looks, suddenly possess the place and begin their ravenous search for food, not to mention their hunt for plunder. This rabble was composed of men, barefooted, but spurred; others clothed in gaudy-colored curtain damask; all manner of hats and caps; some in Federal uniform, and strapped to their saddles was all kinds of plunder—calico, domestic, shoes, boots, tin pans, bed quilts, etc. Volker's house was soon filled by men who stole blankets and clothing, and helped themselves to the edibles at the same time. Miss Volker now discovered the depot, tank and engine-house in flames. Mineral Point and Coles bridges were also destroyed. By five o'clock, the enemy had all passed north, and the silence that prevailed in that deserted village was more trying than the presence of the dreaded enemy. The dead bodies of three negroes, killed by the cavalry, lay where they fell until the hogs had nearly devoured them. Night approached, and darkness and imagination multiplied terrors in Volker's house, at least. The two young ladies, armed with pistol and their father's shot-gun, stood in the center of a room, still as death, listening intently. Morning brought report that St. Louis was captured. Not long after, an unfounded rumor that Indians had deluged Potosi in blood, stampeded the women and children from the Point.

When Price moved north-westerly, R. C. Volker rebuilt this line and re-opened the Pilot Knob office. M. E. Washburn took the Patterson office. Price having advanced *via* Potosi to Franklin on the Pacific Railroad, thirty-seven miles from St. Louis, burned the depot, Government warehouses, machine shops and other property. It was unfortunate for the Government that C. S. Barrett, operator at this station, was East on leave of absence, but it was fortunate that the line from Allen (near Moberly), on the North Missouri Railroad to Sedalia, gave Rosecrans the benefit of constant and direct communication with his forces about Jefferson City, in Price's front. A. J. Smith followed in Price's wake, accompanied probably by J. Blish, Jr., or B. McMurtrie, operators. Generals Sanborn (Ewing L. Armstrong, operator) and McNeil came up from Rolla and joined Generals Fisk and Brown at the capital, ahead of Price, who being afraid to attack, moved toward Kansas *via* Booneville and Lexington, followed now by Generals Pleasanton, Sanborn, Brown, Fisk, Smith and Mower; the latter coming up from Arkansas. When the Booneville line was cut, the direct one to Jefferson City was repaired, and the indirect one *via* the North Missouri Railroad, St. Joseph, Western and Independence was still open. This was of the greatest consequence, as Generals Curtis, commanding the Kansas Department, and Blunt must co-operate with Rosecrans or keep entirely aloof. Shelby captured over a regiment across the river at Glasgow. Price was now recruiting and foraging successfully. Lexington was entered and Blunt driven west of Independence. Rosecrans at St. Louis, telegraphed Pleasanton and Smith to follow Price to Lexington. They pursued and that was all, as Price burned bridges and thus easily avoided them, striking Curtis and Blunt a hard blow as he advanced. Rosecrans ere this reached Jefferson City, *i.e.*, he took the *field* by telegraph, as before. Price reached Independence and received a good pounding from Pleasanton, but a severer one soon after on the Big Blue, where Curtis was detaining him for Pleasanton. Price's retreat was now a flight. *En route* to Arkansas, he lost by capture ten guns, Generals Marmaduke and Cabell and many other prisoners. It was his last raid. Grant was dissatisfied with the conduct of the pursuit by Rosecrans, and historians who always *know*, say Rosecrans erred in abandoning the westward route to follow to Lexington.

At Jefferson City, Rosecrans issued an order as follows:

JEFFERSON CITY, MO., October 18, 1864.

General Field Order No. 2.

For obvious reasons no person will be permitted to remain in telegraph offices except the employés of the office, and no person will be allowed within hearing of the instrument, except on business, which must not be made an excuse for lounging in the office beyond the necessary time. Operators will be held strictly responsible for the execution of this order.

By command of Major General Rosecrans.

J. F. BENNETT, *A. A. G.*

The Rolla, Springfield line was repaired October 12. Colonel Albert Sigel commanding at Rolla, entered the telegraph office and demanded the news of W. S. Forsey, the operator. Forsey replied that he knew none, but would report any that came. While the operator was sending a telegram to Sanborn, Sigel wished Forsey to telegraph Mrs. Sigel that he was safe and well. Forsey insisted on sending military messages first, and got into trouble. A sergeant and two soldiers marched the operator to the guard-house. Forsey's assistant telegraphed Captain Clowry and he Rosecrans, with the following result:

To Colonel Sigel:

Your conduct in arresting Mr. Forsey is in direct violation of order No. 2. Release him at once, and report the facts to these head-quarters. W. S. Rosecrans, *Genl. Comdg.*

and so the the matter dropped.

Farther down the line R. N. Howell, operator at Cassville, and Harden Case at Fayetteville, were having their experiences. These men went to their respective stations in the spring of 1864. Lieutenant Colonel Cameron of the Second Arkansas cavalry, commanded at Cassville, and Colonel M. LaRue Harrison of the First Arkansas cavalry, at Fayetteville. Howell had seven repairers, whose duty it was to keep up the line on both sides, especially toward Fayetteville. They succeeded poorly for awhile. In the summer, notice was posted on poles that any repairer caught mending the line would be shot on the spot. Nevertheless they were undaunted until one forenoon, they barely escaped with loss of tools, because they were better mounted than the guerrillas, who chased them into Cassville. The line was cut as fast as repaired, and abandoned south during the summer. The Cassville office was in a prison cell in the Court House, which General Price had previously used to stable his horses. When Price entered South-eastern Missouri, Coffee, with about seven hundred men moved direct for Cassville, but passed it, moving to the east. Colonel Harrison and wife and other officers and their ladies, accompanied by several companies of Harrison's cavalry, were out on a pleasure and foraging expedition, and reached Cassville when Coffee's movements were reported. Owing to Coffee's presence in the vicinity, the expedition halted at Cassville. General Sanborn at Springfield, at four P. M., telegraphed an order directing the troops at Cassville to move at once, to Springfield, and as this meant evacuation, Howell prepared to go too. But Harrison insisted on his remaining. Major Melton in command (Cameron was at Springfield) protested against it, but Harrison put Howell in his office and a guard at the door. At eleven P. M., the next day, Howell discovered to his surprise, that all of Harrison's men were mounted and in a few minutes they were off, leaving the operator *the only man in Cassville*, which was fifty-five miles from Springfield.

After cogitating over the situation some time, Howell telegraphed Sanborn the state of affairs, ending his message "for God's sake, send some relief to get me away, as there is not a horse, mule or cow here to ride on." Falling asleep at last, Howell was rudely but happily awakened by Major Melton himself, who had come with his force to the operator's rescue.

About this time Gregg, operator at Springfield, found himself a bran-new wife, and it so happened that they, Howell and six

THE ORIGINAL SEVEN SLEEPERS.

others stopped over night at a farm house, where there was but one spare room; in that were two beds. Gregg and wife took one of these and Howell and six soldiers lay cross-wise of the other, leaving their feet well out "for the chickens to roost on." These were the original seven sleepers, Mark Twain to the contrary notwithstanding.

Colonel Harrison reached Fayetteville safely, but was besieged from October 28, by Colonel Brooks, with two thousand Confederates, until Fagan's division of Price's army joined Brooks, only to be easily, speedily and handsomely beaten off by Curtis, who came to Harrison's relief. Harden Case, better known as Hard Case, operator, was at Fayetteville during the two weeks' siege.

Thus ended the last great raid into Missouri. A. J. Smith's forces hurried to Thomas' aid at Nashville. Price, beside losing heavily by desertion, lost one thousand, nine hundred and fifty-eight prisoners, most of his wagons and other material; moreover, his remaining army was now only a disintegrating crowd.

The Fort Scott line was extremely useful during Price's movements, but it so happened that just before they began, Anthony R. Walsh and Joseph K. Porter, operators on this circuit, at Paola, were imprisoned on charge of disloyalty. They were tried at Fort Leavenworth, and, to use the language of Byington's report, written shortly after, "were proven loyal beyond a doubt, and the principal witness against them was impeached." This trial cost each of them nearly three months' salary.

As signally illustrative of the anomalous position the military telegrapher occupied, it will be interesting to note that Charles A. Hammann, who was assistant superintendent in Missouri, as we have seen; Edgar H. Brown, and James D. McGuire (prominent military operators for over three years), Neil Stewart and Clinton S. Barrett, while actually continuing in the Federal Military Telegraph service, as operators, were drafted late in 1864. It was an outrage on the entire corps. Search warrants, so to speak, had been sent to army commanders, to ascertain if any operators were in the ranks — generally resulting, in 1864, in reports like this one: "I am directed by Brigadier General Solomon, commanding Third Division, Army of Arkansas, to inform you that, after a careful investigation, no telegraph operators were discovered in this Division." No, they had been furloughed, or discharged to enter the telegraph service. Even Major Eckert urged Clowry so strongly to send operators East, that he complied by forwarding W. H. Woodring, Charles H. Mixer, Joseph L. Sears and John W. Gregg, at the very time (Price's raid) when they were in greatest need in Missouri. It was while Price was before the State capital, that Hammann, Brown and McGuire were drafted and mustered into Company H, Eighteenth Missouri Volunteer infantry, to *be detached at once*, by order of General Rosecrans, that they might serve as operators. The ignominy of being drafted, however, attached to them. In Little Rock, the Assistant Superintendent, Theodore Holt, mustered all his employés "for the draft, and locked the roll *in the safe*, where," he wrote, " it will stay until the draft is over," which was quite right.

Late in November, efforts were made to re-establish communication with Fort Smith, but hardly had it been accomplished when General Canby ordered the destruction of the fort and the evacuation of towns on the Arkansas, above Little Rock.

General Steele left Little Rock, December 17, to be succeeded by General Reynolds. General Carr was in temporary command. Carr accompanied Steele to Pine Bluff, and the next morning manifested his new authority by arresting operator James A. Shrigley. Superintendent Holt reported the affair at the time, as follows: " Carr went over to the Bluff yesterday with the General; came into Shrigley's office about six, A. M., this morning, to ascertain the whereabouts of the train, and (as Shrigley says) became angry because he could not raise ' B ' at that hour, and placed Shrigley under arrest and came away. At 7:30 (office hour) ' B ' was on hand, and Shrigley did the work *under guard.* He was released by General Andrews as soon as Carr left on the train. I have instructed Shrigley to make a sworn statement of the affair, which I will forward to you, and which, if possible I hope you will carry through, at least far enough to show Carr that he don't own the Telegraph Corps." Assuming the above to be a true statement of the affair, it falls short of the idiocy and cruelty of a major general in Knoxville, Tenn., who told Captain Gross, notwithstanding the line was down, that if he did not telegraph a certain message to a distant station within an hour, he would be shot.

An idea of the magnitude of the Military Telegraph service in Missouri, Arkansas and Kansas for twelve months, ending July 1, 1864, may be gathered from the fact that there were telegraphed two hundred and thirty-two thousand messages—over twenty thousand per month—an average of six hundred and sixty-six *per diem.* Four hundred and twenty-three thousand six hundred and ninety were sent the succeeding year. Seven hundred and forty-nine miles of new line were constructed up to July 1, 1864, making a total in operation at that time of thirteen hundred and twenty miles, which does not include abandoned lines and those in Arkansas, turned over to Captain Clowry, aggregating two hundred and twenty-seven miles more. The entire cost of this work, for the year ending June 30, 1864, was $137,153.44. During the fall of 1864, Captain Clowry operated seventeen hundred and two miles of wire. A quarter of a million of messages were sent the last six months of 1864. Such figures leave no room for a supposition that the war in those States was not planned with reference to and conducted largely by means of the telegraph.

CHAPTER XIV.

THE TELEGRAPH IN TENNESSEE, ALABAMA AND GEORGIA.—FORREST CAPTURES TROOPS AND OPERATORS AT ATHENS.—HIS RAID IN TENNESSEE.—OPERATOR'S EXPERIENCE ABOUT ATHENS.—EXPERIENCES IN CAHABA PRISON.—JOHNSONVILLE BURNED.—HOOD'S ARMY ON SHERMAN'S RAILROAD.—HOW ALLATOONA WAS SAVED.—MANAGER OF ATLANTA OFFICE ARRESTED.—SHERMAN STARTS FOR THE SEA.—FRANKLIN.—OPERATORS CAPTURED; THEIR ESCAPE.—HOOD'S WATERLOO.

The free and border States were struggling over election issues in a heated and bitter Presidential contest, when we took leave of General Sherman on his triumphal entry into the city of Atlanta, Ga., in September, 1864. If it was a prize, it was a troublesome one, for no sooner had Sherman entered the city, without seriously crippling the Confederate army, than he discovered that it would be ruinous to remain; that to retire, even in pursuit of an army on his communications, would count for lost that which had been fairly won, and to advance, he must sever his railroad connections in order to go far enough to do service; in short, he must proceed to Mobile, Savannah, Charleston, or Augusta and the rear of Richmond. It is not our province to discover here why he preferred Savannah.

Before any move was deemed feasible, he regarded it necessary to collect thirty days rations, at Atlanta, for his whole marching army. In no preliminary service was the railroad and telegraph more important than pending Sherman's preparations. The enemy felt this to be so, and tried to destroy these advantages, as we shall see. A second line of telegraph was built from Chattanooga to Atlanta (one hundred and thirty-eight miles), to aid the railroad officials, who by the use of the wire and long switches at stations, nearly doubled the capacity of the track. Colonel W. W. Wright, the Military Superintendent of this road, earned the thanks of the nation for his comprehensive, energetic and skillful management of this railway ; but nothing is clearer than the fact that without the telegraph, the road, even if protected could not have supplied daily more than the army ate.

As preparation for war often ensures peace, so the fact that the presence of an enemy at any given station in the rear of Atlanta was at once communicated to all points, made raids by infantry detachments impracticable, and even cavalry incursions produced but a *minimum* of damage; results by no means commensurate with the preparations of the enemy.

Hood, who had left Lovejoy for Palmetto, Ga., some time before, crossed the Chattahoochee late in September, intent upon striking Sherman's communications and drawing the Union army into Tennessee or Kentucky.

Co-operating with this movement, General Forrest with eight thousand cavalry — the Confederates were always strong in the saddle—started for middle Tennessee, expecting to work great destruction. The Federal cavalry north of Chattanooga was far inferior in number and considerably scattered. Stockades could not withstand artillery. Posts were garrisoned by few troops. In short Sherman's Tennessee roads were seriously threatened; by them, all of his forces south of Nashville, and even in East Tennessee, were supplied almost from hand to mouth. Sherman understood Forrest had gone to Mobile, where in truth some of his forces had been sent, and was therefore surprised when on the 24th of September the telegraph announced to him that, the garrison at Athens, Ala., about six hundred strong, and later on the same day, two relieving regiments, had surrendered to Forrest. Generals R. S. Granger at Decatur, Steedman at Chattanooga and Rousseau at Nashville, District commanders, were also apprised. General Thomas was sent from Atlanta to attend to Forrest, and Morgan's Federal division followed the same day. Telegrams were hurried to Memphis to send infantry and cavalry to Eastport, Miss. Rousseau came down to Pulaski and headed off Forrest, who turned toward Dechard on the Chattanooga road, but before reaching it in force, he ascertained that Steedman was awaiting his coming. Thus baffled, Forrest sent about half of his force under Buford to Huntsville, Ala., and again with the remainder struck the Decatur road, this time at

Spring Hill, October 1. At this point he entered the telegraph office. His operator claims to have here received several official telegrams concerning the positions of pursuing forces, and to have ascertained from one of them that Steedman had turned down the road to intercept Buford's troops. Forrest tried by misleading messages to deceive Rousseau, who had gone back to Nashville to take the other road, but soon found that officer after him with infantry in wagons. Chattanooga and Nashville were in telegraphic connection *via* Cumberland Gap when the more direct wires were cut, and thus Forrest's and Buford's respective forces were watched. They were driven, the one by Rousseau and the other by Steedman, across the Tennessee, and the roads and telegraph promptly repaired, but Washburne's troops from Memphis, in recrossing at Eastport, were severely handled by a portion of Forrest's.

As illustrative of the telegrapher's experiences along the more exposed portions of the Decatur line in the summer and fall of 1864, we will briefly mention those of the operators at Athens. James E. Pettit who, as a soldier, had participated in the battle of Falling Waters, Va., and as a quarter-master's clerk had been present at the siege of Vicksburg, took charge of the Athens office in June, 1864. Murdock Platt Burr, chief operator of this line, was located at Athens, and John F. Ludwig, after a short service at Brownsboro and Larkinsville, became assistant operator. Sam Wireman, a cavalryman, was messenger. One day in August Burr and Sam, returning by horse from Decatur, were about five miles from Athens when they espied five unmounted bushwhackers just emerging from the woods. Burr, whose life had been threatened by prowling banditti, on discovering them, quietly said to Sam: "We must not be taken prisoners; hold your pistol on the off side and do as I do." As they approached the bushmen cried, "Halt!" and one leveled his shot gun at them, whereupon Burr pleasantly remarked, "Come, my friend, lower that gun; it looks ugly and might go off." This request, so happily made, was as graciously acceded to. By this time Burr and Wireman had come up. Then Burr, uttering a terrific yell, plunged his spurs deep into the flanks of his horse and at the same time began discharging his revolver at the guerrillas. Wireman was quick to do likewise. One of the five was thrown to the ground by one of the horses, another is believed to have been wounded, and the remaining three did not recover from their surprise in time to use their shot guns effectively. It so happened that Wireman's horse threw a clod of dirt against his back as the enemy fired, which caused him to turn pale and cry out, "Burr, I'm shot!" Much inquiry as to the condition of his wound was subsequently made of Wireman, to his great embarrassment.

On another occasion Burr and Ludwig started at four P. M. to repair the wire north near Sulphur Trestle, where Colonel Hall commanded. Having mended the line and supped with Hall, they started after dark in a rain-storm to return. The roadway was, for the most part, skirted by timber, through which there were many wood-roads. After making a circuit, when they thought they were going south, and being halted by Hall's pickets, they were again started on the right road and again took a by-way, this time to discover their mistake at the end of the road deep in the woods. Retracing their steps until they thought they had struck the main road, they, in fact, diverged through another by-way, finally to pull up in front of a fence along a cleared field. It was now in the small hours, and the storm subsiding, they turned their horses into the field, and using their saddles for pillows wooed Morpheus. The mosquitoes prevented sleep. Near morning, however, Burr, when almost unconscious, was suddenly startled, and, rising, cocked his revolver as he called out, "Halt! who goes there?" After listening intently for some minutes he was answered with a grunt, and they knew it was a hog. After daylight they entered the main road near the trestle and reached home a sorry looking couple.

Once when General Roddy raided Athens, Pettit hid his instrument under the hotel in which he had his office, and then ran to safer quarters. At another time General Wheeler passed by the place with two thousand cavalry, not daring to attack. Ludwig was sent to Sulphur Trestle just before Wheeler got on the track, but as that officer cut the line on both sides of the stockade at the bridge, Ludwig could do no good. Wheeler passed on without attacking.

When Forrest captured Athens as above shown, he found

Pettit and Ludwig at the Fort. The former not believing that a large force was at hand, left his instrument and cipher key at his office, but Ludwig carried one set to the Fort, where he destroyed it to prevent capture. These operators had arranged to perform a soldier's part when an assault appeared imminent, and were deeply chagrined at being surrendered, as the troops were strongly posted. They were taken to Meridian, Miss., placed in a stockade and after a short stay, removed to Selma, Ala., and from thence to Cahaba, Ala., where they remained until March 14, 1865. During their stay at Cahaba, both were sick with fever a long time—Pettit for two months. The prison, called Castle Morgan, consisted of an old cotton shed with half of the roof blown off. It was located on the banks of the Alabama River, and there were twenty-eight hundred prisoners confined there at one time. The unexampled rains of that fall season caused the river to flood the prison. The shallowest place was over three feet deep, and in some places it was more than twelve feet. The prisoners floated in cord wood and built platforms on which they slept, there being bunk accommodations for only eight hundred. The water remained in the prison over eight days, and when it subsided, there were carried out three wheelbarrow loads of dead rats. The filth that collected within the prison at this time was poisonous and nauseating. Of course the noxious gases were destructive of life, especially to the sick. In the ward where Pettit and eleven others lay sick, *the eleven died.* Three boys, the eldest not eighteen, being wounded, cared for themselves, rejecting assistance; two of them died from gangrene. A soldier afflicted with the scurvy pealed off a perfect scab cast of the calf of his leg; another died of consumption, having lain in one position nine days and nights; the bare bones protruding through the skin at each shoulder blade and hip. Frozen feet were the rule not the exception. Sometimes during the coldest nights of the winter, all of the prisoners were on their feet to keep from freezing. Over a dozen captives were shot by the guards before March, 1865. Beef bones were cut into fanciful shapes and sold to the guards. Occasionally a few barrows of white bread were brought in and sold at three dollars per loaf, Confederate money, and sweet potatoes for a dollar apiece. When we next meet these operators, we will find one at Confederate head-quarters in direct and regularly telegraphic communication with Union head-quarters; the only instance of the kind during the war, where the connection was permanent.

After Forrest had refitted his force, he moved down the Tennessee River on the west, to the vicinity of Johnsonville, Tenn. capturing there one gun-boat, one steamer and some transports, which he destroyed, causing the Federals to burn other gunboats and transports, besides about $1,000,000 worth of other Government property. It is supposed the operators there decamped with the colored troops.

Of course Hood was not inactive all this time. October 2, Stewart's corps, the van of the Confederate army, struck the railroad near Big Shanty and destroyed it for ten miles north. That very day General Sherman, who had notice of Hood's approach, was in the main city telegraph office a large portion of the time, and anticipating the destruction of the telegraph he *conversed* quite freely with officers in Washington, and the author's recollection is that these were President Lincoln, General Halleck and Secretary Stanton. While such telegraphic talks were unusual over such great distances, they were not new. As early as the time when Buell's headquarters were in Louisville and Halleck's in St. Louis, General McClellan, their chief at Washington, was thus in direct communication at once with Buell and Halleck at their quarters and Commodore Foote at Cairo, Ills. Each of these officers heard all that was said by all the others. The distance by the line was over thirteen hundred miles, and the conversation is said to have been maintained from eleven o'clock in the morning to six in the evening with the promptness of a personal interview; calling forth enthusiastic praise of the telegraph from all of these officers.

Of course many messages were sent to Sherman's post commandants on the Chattanooga road, and to Thomas, who had been sent after Forrest in Tennessee. At about five P. M. the line was cut, and Marietta became the most northerly station on the circuit. The next day Sherman left with his army, except the Twentieth Corps, commanded by General Slocum, and started in pursuit of Hood.

Allatoona, held by a small Federal brigade, was a fortified depot of supplies, and French's division was directed against it. From the Signal Hill at Vining's Station, on the fourth, Sherman ordered General Corse at Rome to reinforce the command at Allatoona. This important message was signaled to Allatoona over the approaching enemy, telegraphed to Kingston, where W. H. Drake was operator and train-despatcher, and, the line west being cut, sent by him by locomotive to Corse at Rome. This is perhaps the only instance during the war in which optical signaling, electricity, and steam united in the conveyance of an important order. There is no doubt but that had either miscarried, the stronghold at Allatoona, its defenders and supplies would have been captured. As it was, Corse was enabled to reach that place on the fourth and repel an attack the next day in which French lost over one thousand and the Federals seven hundred and seven, out of a force of nineteen hundred and forty-four. It is said that G. M. Farnham, operator at Allatoona, retired to a very disagreeable place, where he remained until the fighting was over. Perhaps there were no pleasant places thereabouts at that time.

Hood, moving to the west of Rome, appeared before Resaca on the eleventh, but deeming it imprudent to attack, moved north, destroying the road and telegraph for twenty-three miles and capturing Dalton and Tilton, but was driven off the road about the fifteenth, going into camp near Gadsden. Sherman pitched his tents at Gaylesville. The telegraph was completely restored on the twentieth and the railroad eight days later.

On the 1st of October W. R. Plum relieved Richmond Smith as manager of Atlanta office. On the eighth an operator near Allatoona informed Atlanta that signal officers had said, "We have reports that Grant is fighting three miles from Richmond. So far as we can learn all was progressing favorably." This report, like many other rumors that prevailed in those days, eagerly listened to but always discredited, was given to a newsdealer who bulletined it. General Slocum, who had taken command in the city about five days before, and who was personally unknown to the operators, felt outraged because he received the news from his *orderlies* who had read the bulletin, and to appease his mighty wrath, caused two armed soldiers to march the manager to his quarters, whence, after an exhibition of official temper, the telegrapher was incarcerated among a lot of deserters, thieves, spies and the like about two hours, and until liberated at the personal solicitation of General Brannan, Colonel Parkhurst, Major Hoffman and Captain Willard, of General Thomas' staff, they having remained in Atlanta. Cass Sholes, chief operator, was refused an interview with the prisoner, and thereupon Sholes, John O. Ingle and C. W. Jacques, in the city office, and Joe Anderson and John Egan, in the railroad office, notified Slocum that until Plum was released they would close their offices. They also telegraphed General Sherman, signaling part of the way, but it was all over before a reply could come. It was an interesting day to the fraternity in Atlanta.

After driving Hood to Gadsden, Sherman, intent on his contemplated march to the sea, was preparing to move sixty thousand infantry, five thousand five hundred cavalry and sixty-five guns, leaving Hood with nearly fifty thousand, including Forrest's cavalry, to march into Middle Tennessee *via* Guntersville and Florence, Ala. On the 7th of November, the last train left Chattanooga for Atlanta.

The next day F. M. Speed was appointed chief operator of the Atlanta line, and Plum started to rejoin General Thomas, leaving in Atlanta nine operators, who were to accompany the army wheresoever it might go. About noon of the twelfth, one of the Chattanooga operators, probably Thomas Sampson, began sending the last despatch to be sent over the line. It was from Thomas to Sherman, and was being received by C. G. Eddy at Cartersville, when the wire was severed. Sherman's army was shut off from the North. His troops began the destruction of the railroads and telegraph south to Atlanta, from whence they were to march and fight, they knew not where, nor against what forces, and recked as little. Full of confidence, they were quite ready for any venture, for such it was really considered by many in the secret. On it hung untold hopes and fears of the Federals.

We will now attend to the affairs transpiring under General Thomas, and in another chapter chronicle the March to the Sea. After losing about one thousand men in efforts to capture Decatur, Hood moved west to Florence, and crossing the Tennessee moved against the Nashville & Decatur road by various routes leading to Columbia. This was about the 19th of November, and it soon became evident that he would not follow Sherman as that officer had predicted, but that Thomas, with the odds and ends of Sherman's army rallied around Schofield's and Stanley's troops, about twenty-five thousand available for the field, must beat back that army which Sherman with his one hundred thousand had failed to destroy. Grant, with evident uneasiness, had consented to Sherman's move. Neither Grant, Sherman nor Thomas foresaw the delays that were yet to embarrass the latter in concentrating and equipping a suitable force to repel the invasion. As the enemy rapidly approached by many roads he was met and his advance hindered to enable reinforcements to come to Schofield, who was retiring from Pulaski and removing his army supplies and material largely by rail. The last week in November was one of supreme interest. The anxiety of the Administration, of Grant and Halleck, and of the people, was intense. From the first, Thomas had been watching and parrying Hood by telegraph. The despatches grew in frequency, in interest and earnestness, as the Confederates swept boldly on. Pulaski was abandoned. The entire line therefrom to Decatur and from Decatur to Stevenson was given up; Columbia was evacuated; the enemy was already with two corps of infantry well in the advance of many of the retiring Federal forces, and liable at any moment to strike them. All night the troops and wagon trains marched within hearing of the Confederates. That night march saved the trains and doubtless the army. Minor but severe conflicts were frequent.

Thomas, at the capital, anxiously watched the movements on Nashville. The next day, November 30, Schofield had passed his trains over the Harpeth, at Franklin, when, at about four, P. M., Hood emerged from concealing woods and ravines, and fell with mighty power upon Schofield's command. It was then, or never! If the army, scarce half assembled, without suitable defenses, could not be thoroughly beaten, behind the works of the capital, reinforced by A. J. Smith's twelve thousand and others, it surely could not. Both sides fought for Nashville at Franklin. In the onset the Southerners were amazingly successful. Two brigades were swept back, disorganizing others, and batteries were taken. Then, under Stanley and Opdycke, the latter's brigade rushed into the breach with fixed bayonets. The strife was too harrowing for detail, when not required by the scope of the work ; hence we spare it. The Federal line was regained, eight guns retaken ; but the assaults were renewed with a valor unexcelled anywhere. It was courage *plus* desperation. Some said Hood had given his troops whisky and gunpowder. Doubtless, that was a slander. His were brave troops. They needed no false courage. The battle lasted, with occasional intermission, until midnight, and when the next day's sun arose, a sad scene was visible. Seventeen hundred and fifty Confederates lay dead on that field. Of these, five were generals. Thirty-eight hundred were wounded, of whom six were generals. One hundred and eighty-nine Federals were killed, and ten hundred and thirty-three wounded.

Nashville was saved to the Union. In that city there was no sleep. Before, its sixteen thousand citizens and ten thousand Government employés had been on the *qui vive* as its fate hung on the issue of battle. That was when Stone River was fought. Twice, earlier (Donelson, 1861, and the siege, 1862), it was doubtful which party would possess the city, but now a great struggle nearer than Donelson, nearer than Stone River, quite as fierce, but less protracted, was progressing. Steedman, with five thousand of Sherman's soldiers and a brigade of colored troops, had not arrived from Chattanooga. A. J. Smith's advance troops were debarking at Nashville while the battle was progressing. Donaldson's armed civilians were not fully prepared. Indeed, Thomas had but five thousand ready in the city. If the people were oppressed with the portents of the hour, how was it with Thomas on whom the grave responsibilities rested? How, with Military Governor Andrew Johnson, whose war record at least was above reproach? These men and Medical Director George E. Cooper were in the St. Cloud Hotel, counting the chances, when E. C. Boyle, then manager of the Nashville office, delivered Schofield's telegram to Thomas, giving the particulars of the battle. Johnson, more excitable than the others, evinced his joy by grasping Boyle's hands and shaking them cordially, saying at the same time, "The news is glorious."

The next day the enemy began investing Nashville, and telegrams increased in number and urgency. Nashville office was worked under a steady pressure. D. O. Dyer, George Everett, B. B. Glass, Ellis Wilson, John Thode, E. M. and G. H. Shape, and other experts, worked early and late. Almost as soon as Schofield's forces were posted about Nashville a field telegraph line, around the defenses, was constructed, connecting Thomas' head-quarters with those of his principal generals. which line was in constant use until the 16th of December.

Forrest now set about destroying the Nashville and Chattanooga road, beginning with the block-houses near Nashville, and including those at LaVergne and Smyrna. The latter was taken during the night of the fifth while the operators were asleep. It was an important part of the duty of Forrest's men to conscript every citizen they could find. The Smyrna operators, T. E. Rawlings and John T. Duell, were lodging at Doctor Kennedy's when three cavalrymen went there in search of a brother of the doctor's, who escaped. But they found the operators in bed and compelled them to mount, each behind a cavalryman, and ride twelve or fifteen miles before sunrise to General Bates' quarters, where they were severely catechised concerning Federal movements, after which the boys were turned over to the provost captain as prisoners of war. The first food they received was boiled beef, dealt out to each prisoner in pound pieces, without salt or bread. Rawlings threw his in the ditch—to regret it subsequently, as he received nothing more that day, except a bit of corn-pone at sundown. On the sixth, seventh, and eighth they witnessed portions of the fighting between Forrest's and Rousseau's forces at Murfreesboro, during which Forrest lost heavily. The guard with the prisoners hurried off in the direction of Smyrna, where, unable to find head-quarters, they camped by themselves regardless of the other troops, and building a log fire lay on the ground to sleep. That night a cold, drizzling rain fell and froze, covering trees and ground for many miles with a heavy sleet on which it was almost impossible to travel. Towards morning, the fire having deadened, Rawlings awoke nearly benumbed with cold, and looking about discovered that some of the prisoners had escaped and only one of the guard was awake, and he engaged parching corn. Nudging Duell, Rawlings whispered "now is our time," but Duell did not believe it and Rawlings made his escape alone and went directly to Doctor Kennedy's, where he took quarters in the cellar where also was the doctor's brother John Kennedy, hiding from the conscript officers. And there the two remained about a week, when they were liberated by the Federals as we shall see. One day General Bates made his head-quarters in the doctor's house, and that evening and the next morning the good mother of the Kennedys took no victuals below, but by noon Bates left and a well filled basket was at once taken to the cellar. Duell was taken to Pulaski and confined upstairs over an old store-room, from which he escaped by means of a trap door in the rear room to the first floor and, personating a waiter carrying a pail (it being dark), he passed the guard standing at the door without difficulty, and then went to the residence of some one whom he either knew or knew of, and was hidden in the garret until relieved by the Union troops.

General Lyon crossing the Cumberland with a cavalry force, captured a train of cars on the Nashville Road near Clarksville, driving operators A. S. Hawkins and Martin E. Griswold in on the engine that was detached just in time to escape. Lyon pushed on through Hopkinsville, Elizabethtown, Burksville and into Alabama, having been badly used up by McCook, La Grange and Steedman.

At Nashville all was activity. Every night was robbed to lengthen the day and many a day was indeed of twenty-four hours duration. Thomas himself set an example of utilizing the small hours. His despatches bear dates including all times of the night. But there came a series of telegrams from City Point and Washington urging the General to attack. It was in vain that that officer replied that even if Hood was whipped, the cavalry being dismounted, he, Thomas, could not reap the fruits of victory; it was in vain that he urged the impossibility of the enemy's crossing the Cumberland patrolled by gun-boats and very high; in vain, that he replied that the roads were such that neither man nor beast could travel; in vain, that he offered to serve under any other commander; "Papa Thomas," the hero of Mill Springs, the commander of the center which was *not* whipped at Stone River, the Rock of Chickamauga and the grand soldier with whom Grant and Sherman had been pleased to counsel, who had by telegraph conducted a most masterly retreat from Pulaski in the face of overwhelming numbers, Thomas the "true and prudent; distinguished in council, and on many battle fields celebrated by his courage," was to be judged and condemned at Washington and City Point. He was judged and he was condemned. In view of the astonishing patience of the Administration, when other generals full handed, delayed action for months, it will ever seem strange that Thomas could not have two weeks after one battle before beginning another.

December 9, the order appointing Schofield, Thomas' successor, was issued by direction of General Grant, but the authorities at Washington not being really willing to execute it, Grant was again asked if it should go forward, and it was withheld. On the thirteenth, General Logan was ordered to Nashville and General Grant started for there to take personal command. Beckwith as usual accompanied him. Arriving at Washington, on the fifteenth, Grant received Captain VanDuzer's telegram detailing Thomas' success of that day, and returning to his hotel, said to Beckwith, "Well, I guess we won't go to Nashville, Thomas has given Hood a good thrashing."

And so indeed it proved, for on that day, the six days' sleet disappearing, Thomas attacked. All night long, Thomas' despatch, date of nine P. M., of the fifteenth to Halleck, modest almost beyond precedent, was being repeated from one line to another, from the Atlantic to the Pacific, and at daylight, the entire North was rejoiced, for that Thomas had said, "Attacked enemy's left this morning; drove it from the river, below the city, very nearly to Franklin Pike, distance about eight miles." That night Thomas' chief operator was ordered by the General to accompany him the next day, which he did. On the sixteenth, about daylight, the troops again advanced against Hood, and before four P. M., his army was in full retreat, pressed on rear and flanks by infantry and cavalry. The Confederates having lost during the two days' battle, fifty-three pieces of artillery, thousands of small arms, twenty-five battle flags and four thousand, four hundred and sixty-two prisoners, besides the killed and wounded. Their army was now a routed mass, hurrying by night and by day to the Tennessee, which, not more than one-half of his army re-crossed and that half in no condition for effective resistance.

VanDuzer and F. S. VanValkenburg were untiring in their zeal to open telegraphic communication with Thomas, at each bivouac. On the sixteenth, after following the enemy to Franklin, Thomas went into camp in the mud and rain at Brentwood, and there, as at every other camp as far as Pulaski, except one, the telegraph was O K at head-quarters. On the night of December 20, Thomas went into camp at Rutherford's Creek. After a ride from early morn in a drenching rain, with half a dozen orderlies and a lantern, the operator started for the railroad, to find the line and transmit important telegrams for Washington. One of the mules in the darkness fractured a leg, thus compelling the party to advance on foot. The rain fell in torrents on the ice-covered ground. The line was found, but no current, and after marching northward hoping to find the break believed to be near, as the party reached the creek their light went out and there was not a dry match to be had. It was therefore impossible to climb over the ruins of the bridge, and the party, cold and wet, felt their way back to camp to report the first and last telegraph failure of the campaign. On this trying advance, besides VanValkenburg, operators C. D. Whitney, C. D. Montanye, F. M. Speed, John T. Joyce, John Ingle, John Egan, C. C. Starling, J. W. Wickard and G. D. Wilkinson rendered efficient service. All of the lines abandoned on Hood's approach, were re-opened during the month of December, except that to Johnsonville, which was not repaired until January. Superintendent VanHorn, of the Confederate telegraph service, soon after wrote that the Decatur line was "left up," implying that no attempt was made to destroy it—a strong indication of the utter demoralization of the army, or of the hopelessness of the cause.

General Thomas returned to Nashville early in January, and soon started the final expeditions of the Department. All now began to think of peace. The North hailed Thomas' triumph with unrestrained enthusiasm. The glad tidings were overwhelming. Strong men wept for joy. Sherman's success was accomplished in Hood's defeat. It was like two great victories from one battle. Sherman was justified, but Thomas—he had gained the greatest success thus far achieved on the open field between two such forces. His own losses in the entire campaign were about ten thousand, many of whom were slightly wounded. It was the last grand sacrifice of the war in the West.

The telegraphers during this time lost Frank B. Tyler and Robert Reaser from sickness. The surviving operators subscribed about one thousand dollars for those left dependent, and Captain VanDuzer installed Mrs. Reaser, an operatrix, in her consort's place, and thus she cared for his two children at Waverly, Tenn. The Government erects head-stones over the graves of soldiers, and pensions those dependent on them. Not so for the operator. Why?

CHAPTER XV.

SHERMAN'S MARCH TO THE SEA.—ATLANTA IN RUINS.—A GREAT PICNIC.—TAPPING THE ENEMY'S WIRES.—A VIEW WITHIN THE CONFEDERATE LINES.—STORMING FORT McALLISTER.—INTERESTING INCIDENTS.—SAVANNAH TAKEN.—TELEGRAPHIC CONNECTIONS.

We left General Sherman at Cartersville, Ga., November 12, 1864, standing by his operator, C. G. Eddy, who at the time was receiving General G. H. Thomas' final message. Before its completion, the line north was severed by Union troops, who had begun to destroy the railroad and telegraph. Thenceforth, for many anxious days, an army exceeding sixty thousand in number was to the Northern people, as it were, in utter darkness. It was worse than temporarily lost, because besides a natural suspense, anxiety was aggravated by reports of the destruction of that great body which represented so many households. This expeditionary movement was to be of infinite value in calling peoples' attention to the beneficent advantages of the telegraph, which, like health, is most appreciated when lost.

Just before Sherman left Kingston, Ga., he telegraphed Captain Poe, Chief of Engineers, to prepare for destruction all round-houses, depots, store-houses and buildings that might be used as store-houses. Fire, powder and twenty-one foot railroad iron battering rams, swung from ten foot wooden horses and worked by a swinging chain fastened to the center of the bar, caused fearful havoc. After the Chattanooga Railroad round-house had been thus in two hours battered down and that on the Macon road fired, Poe turned his attention to the large passenger depot. Then came the stone ware-house and brick depot; the former was mined, then operator Joe Anderson's office and buildings around it. Then the Trout house, the main telegraph office building, and nearly all Whitehall street besides not a few residences. Such fires always extend beyond one's wishes, so this exceeded Sherman's. A prominent Confederate telegrapher who arrived at Atlanta *via* Jonesboro shortly after the army left, at the time thus described appearances:

The road between Jonesboro and Atlanta bears the marks of many a fierce struggle. As you approach Atlanta from the south, at every mile you strike first the Confederate breastworks and rifle pits, then the similar defenses of the Yankees: the torn shrubbery, the scarred oaks, the hastily constructed grave with its humble pine head-board, giving the name and regiment of the men of both armies, scattered here and there in twos and threes along the road, all tell the tale, if the burned houses, the desolate chimney stacks, and deserted dwellings, once the scene of peaceful domestic happiness, failed to bring the horrors of war before one's mind. But Atlanta caps the climax. Huge fortifications of red dirt wind snakelike around its whole extent; dead mules and horses and fragments of slaughtered beeves lie around; the absence of all life in the town itself, with its smouldering walls, tall naked chimney shafts, with only an occasional glimpse, as I rode along, of a solitary individual slowly picking his way over the masses of brick and *debris* in the streets, leaves an impression on the mind not easily described. It is appalling: Atlanta now stands like a grinning skull on the wayside; a fragmentary memento of a former life and greatness.

Colonel VanDuzer had obeyed instructions in directing nine operators to report for service on this march. These were furnished, each, with a cipher key and pocket telegraph instrument. They were, Richmond Smith, chief operator, generally at the headquarters of the army; C. G. Eddy at Sherman's; Joseph Anderson with Jeff C. Davis who commanded the Fourteenth Corps; John Lonergan with Osterhaus who commanded the Seventeenth Corps, and later with General Howard's (right wing) where C. G. Sholes was located. D. T. Berry went with Slocum, commander of the left wing; P. J. Murray probably with General Blair (Seventeenth Corps); —— Waterhouse with General Williams (Twentieth Corps) or the chief of artillery, and William H. Drake with Kilpatrick, who commanded the cavalry, about six thousand strong. The operators were well prepared for a long march. VanDuzer had even taken the precaution to send Edward P. Whitford with funds to pay them off. Whitford, a bright lad for one so young, occupied the responsible position in Nashville of cipherer and assistant clerk at VanDuzer's headquarters, where most of Governor Johnson's and General Webster's telegrams were ciphered.

Scarcely had the operators left the city, when they entered upon the enjoyments of a sort of authorized pillage. It is proverbial that there is no fruit like stolen fruit, so applying the saying to all the substantials, and many of the delicacies of the season, the reader will appreciate that the telegraphers and army also, had entered upon a "promised land." Turkeys, geese, chickens, pigs, bacon, sheep, cattle, sweet potatoes, peanuts, corn meal, honey, molasses, wines, milk, eggs, fruits and scores of other articles, not to be thought of until discovered.

The march resulted in two things; first, a vast foraging expedition through the heart of the Confederacy, and second, a change of base. From Atlanta to Savannah, no enemy was met which Sherman's cavalry could not probably have routed unaided. Chief Smith and two signal officers, Cole and Jones, were (November 19) hunting for horses in the advance, and just as they were about to enter Newborn, Ga., they were fired upon by guerrillas, but the latter knew that a great army was behind and stopped not to see the whites of any eyes, consequently no bones were broken. Lieutenant Jones at least was not deterred, for that day he brought in two horses and five mules. Captain Cole could hardly bring in himself, having found six bottles of blackberry wine somewhere. All this time, *i. e.*, for several days back, the chief operator had been barely able to ride at all, owing to his horse being lame. Those offered him seemed more undesirable. The planters were careful to run off or hide their best horses, but a "warrior bold" on a limping horse does not attract so much attention as Joe Anderson did several days later when advancing from Sandersville. Joe's saddle was too good to keep, and somebody *foraged* it. For some miles he rode bare back, cutting of course an extremely undignified figure; but ere night he "borrowed" a lady's side-saddle and rapidly grew conspicuous. General Davis about this time tied up a couple of soldiers for stealing silk dresses, which he compelled them to wear during a day's march behind a wagon with the word "stolen" posted in large letters on their backs. Had the General given Joe one of these to go with his saddle, the soldiers of the Fourteenth Corps would have restrained the jocularity, with which they pelted the smooth faced operator. As it was he said nothing, but unutterably looked *Ride si sapis.*

November 22, General Sherman entered Milledgeville, Georgia's capital. Richmond Smith hurried to the telegraph office, but operator Berry had preceded him by one day. Even Berry failed, not however because the Confederate States' operator had run off leaving only a lightning arrester and some valueless messages from Governor Brown, for W. H. Drake had preceded him and tapped the line between Macon and Milledgeville. Kilpatrick's cavalry had cut the wire on the Milledgeville side, else Drake might have listened successfully; as it was, his attempt to deceive the Macon operators was abortive also. It will be remembered that Drake was the operator who spelled out "Lebanon Junction" (Ky.) for General John Morgan's operator, Ellsworth, consequently he was exceedingly anxious to turn the joke somehow on the Confederates. Soon after putting his instrument "in circuit," Drake heard some one calling "Md." It was Macon office and it had important messages to send. Drake correctly surmised that "Md." meant Milledgeville, and answered "ii, ii, Md." That meant, here am I, go ahead; but Macon was suspicious and asked as a sort of countersign, "what is your name?" That was a hard question, for Drake's name wasn't a pass-word on that wire and he did not know that of the Milledgeville operator. So after a little safe bravado the dialogue closed and Drake re-entered the saddle; then Kilpatrick's men began destroying the line on the Macon side. Arriving at Sandersville, Ga., Eddy tried his hand on the wires running to Augusta. He discovered in the office what every "plug" operator keeps, but never refers to, and what a good operator seldom has, viz.: a list of the "calls" of the offices on the line. It was found that "A" meant Augusta, "9" Millen and "F" Tenille. The wire west had been cut. Eddy connected the main line with his ground wire and circuit "closed." Immediately "A" began calling "F." Eddy said "i i, i i, F." "A" thereupon repeats the natural question, "Who is it?" but as Eddy had heard that Barney Hughes was the fugitive operator, he had a ready answer. Still Augusta was suspicious, but was at last convinced that it was Hughes indeed. "A" says, "What news from our (rebel) forces, and where have you been all day?" Eddy replied, "The Yankees drove me out this morning, and I came back after they had retired. Have heard nothing late from the forces at Sandersville." Things were working finely and doubtless news and telegrams would have soon come to Eddy, but at this moment General Jeff. C. Davis' troops struck the wire between "A" and "F," and thus ended that effort.

That day (twenty-sixth) the Augusta *Constitutionalist* in an editorial inquired, Where is Sherman? and called on the people to rally and make the Oconee Sherman's river of death. At this very time his army was east of the Oconee. Surely Sherman's great command was lost to the Southern and Northern people alike. Throughout the entire country there was now extreme solicitude about these Federal forces. The North feared the army would be overwhelmed and the South was in distress lest it should not be. Neither side had any telegraphs — Sherman had left his own and was fast destroying those in his wake —only, the wake came *after* the demolition. Of course new lines or advanced portions of others were constantly met intact, but there are so many outriders to such an army that their presence is but little indicative of the whereabouts or route of the main columns.

December 1 the Union forces were in the vicinity of Herndon, Ga. Portions of the army had had a brush near Macon, and Kilpatrick's cavalry another near Augusta, but that was about all worthy of note, here.

These things indicate what a picnic it was to the main army; they also show what a sweep that army was making. It started from Atlanta by destroying, it is said, two millions of dollars' worth of property. Every day's march added to that great sum. Georgia was feeling the evils of war. A mock legislature had been held in her very capitol by Federal officers, who after a heated discussion passed a resolution restoring the State to the Union.

About this time General Howard proposed sending a scouting party of fourteen far out from the army and Sholes gladly acceded to his request to accompany it. Howard's idea was for the party to ride by night and hide by day until well out, when

Sholes was to tap the enemy's line. After riding until very late the first night the party unexpectedly dashed into a stream which crossed the highway, when a stentorian voice across cried "Halt, who goes there?" "Federals," responded one of the party, and instantly the picket fired. The Federals retreated in hot haste followed some distance by the enemy, numbering, as was learned the next day, two regiments.

It was near Millen that Smith and Eddy, after going into camp walked to the telegraph office and testing, found current from Savannah, but neither of the operators had their instruments. Waterhouse was also in camp nearer. When his instrument was connected the "circuit" was gone. Probably the soldiers had torn down the line a little ahead to make pot hooks. Waterhouse however had better success on the Augusta line. He worked with that office until two A. M., but the Confederate messages were frauds, intended for the "Yanks"; an uncommon but by no means untried ruse to deceive an enemy. One of these was a telegram nominally at least from General Longstreet in Augusta to General Wheeler at some point south, saying "I will leave here in the morning with ten thousand men." Other messages partook of the same character. These were new tricks in war. The Rebellion first developed them. Every war opens up some new notions to deceive. Deception is legitimate. It is a part of the *art* of war. The very essence of military science consists in getting what, in anything else, would be an undue advantage. Even Christian soldiers, like Howard, must concede that the ninth commandment has no application in war so far as *their* side is concerned; hence a good telegraphic lie is much sought after. Ellsworth manipulated them successfully for the wily General John Morgan, and Whitthorne served General Forrest in like manner. But the story, currently believed, that on the occasion of the capture, in August, 1862, of Manassas Junction, Va., by General Stuart's cavalry, his operator, in Union General Pope's name telegraphed, to General Rucker in Washington, D. C., for tents and other supplies, which—as the report goes—were at once forwarded by rail and used by the Confederates, is wholly unfounded. That other yarn that comes across the Atlantic, that Morgan, when he invaded Ohio, procured horses from Cincinnati on a like requisition, is also an interesting invention. Stoneman and Gillem—Union generals—raiding from East Tennessee; Grierson and others in Mississippi; in short, most of the very prominent cavalry leaders on either side have profited by the deceptions of their telegraph operators.

Operator Lonergan tapped the enemy's line in the rear of General McLaw's head-quarters, near Station seven (Scarboro). Major Whittle, assistant provost marshal general, Captains Taggart and B——d and Lieutenant Sampson, signal officer, probably all staff officers, were present. Lonergan intercepted messages from Hardee, the opposing commanding general, to General McLaws, giving the movements of Bragg, on Sherman's rear from Augusta, also those of the cavalry; the number of rations forwarded and other information. All of which was highly important eavesdropping; so much so that General Howard directed Lonergan to make an official report thereof for General Sherman. When Lonergan and the others were about to quit the line, Captain B——d dictated a despatch in General Howard's name, which Lonergan sent to Hardee in Savannah, inviting the latter to remain there and dine with him (Howard) on Christmas. General Howard was greatly provoked at this un authorized use of his name, because socially he had great respect for Hardee; B——d was allowed to go North *via* Savannah very soon after reaching there, but there seems even more reason in General Sherman's dissatisfaction, which was, that it showed too clearly his whereabouts. Of course by this time the fact that Savannah was the objective, was quite clear and so much of the message as requested Hardee to await Howard's coming, may not have been hurtful. Lonergan got off with a good scolding, but the Savannah operators were even more abusive.

Having followed Sherman's army about two hundred and twenty-five miles, and to within seventy of Savannah, let us withdraw the Confederate curtain, and, looking at the *débris* which marked the heavy tread, see its effects also on the Confederate mind. By the time General Sherman left Atlanta, the Confederate telegraphers had well nigh exhausted all sources of material supply. Even in the city of Petersburg, Va., horses could not be procured for line repairers except by *borrowing* for each trip of the quarter-master. Superintendent Merriwether, at Mobile, sent out to Pascagoula to take all insulators and brackets off that wire to Hall's Mills, in order that he might overhaul and repair the line from Tensas River to Montgomery. Nearly all the telegraph wire in the Confederacy was already on poles. Much of that was nearly valueless, owing to rust. Copying paper, very like that used for printing newspapers, was three dollars per pound. Zinc, in which was a preponderance of type metal, was two dollars and fifty cents per pound. Sulphuric acid was twenty dollars per pound. Peanut oil was used for light in lamps that cost fifteen dollars each. It was while the supplies in the Confederacy were about exhausted, that Sherman's army began a march which involved the tearing down of five hundred miles of telegraphs in Georgia alone. What could more effectually work that end than the means used, of heating the centers of the railroad iron, and twisting the bars around telegraph poles. At this time J. B. Tree was General Superintendent of the seaboard lines, which included nearly all in Sherman's path. L. R. Hoopes was manager of Columbus, Ga., office; Wm. M. Nettles, of Montgomery; D. E. Norris, at Macon, assisted by operators A. H. Moffatt and A. W. York; J. A. Brenner, assistant superintendent at Augusta, Ga.; and C. G. Merriwether, assistant superintendent at Mobile, where William Sandford was manager of the city office.

Two days after Sherman's army began its march, G. T. Beauregard, General-in-chief in the West, telegraphed as follows:

IUKA, November 17, 6:30 P. M., 1864.

Forwarded from Selma to

LIEUTENANT GENERAL TAYLOR, Macon, Ga.:

Adopt Fabian system. Don't run risk of losing your force and guns to hold any one place or position, but harass the enemy at all points. *Hannibal held heart of Italy sixteen years and then was defeated.* Be cool and confident, and all will yet be right. I will join you soon as possible. (Signed) G. T. BEAUREGARD, *General.*

This, of course, meant the abandonment of Macon, should Sherman appear in dead earnest to want it. Consequently, Merriwether was sent to that city to pack up all surplus materials ready for instant shipment to Montgomery. By the twenty-second, all communication *via* Millen with Augusta and Savannah was cut off. On the twenty-third, Tree telegraphed Brenner, of Augusta: "Keep your offices open night and day. If you have to fall back, take it coolly and gather up the operators, instruments and material as you retire. If the enemy diverge from the Central or Georgia Road, establish an office at the end of the break and send your business through by couriers. We will do the same at the Macon end." This message went *via* Columbus, Ga., Tallahassee, Fla., and Savannah—the only route then open. Next day, that line was cut south of Macon. Then operator Conner started out with a relay to try and flank Sherman's army on the direct route to Savannah. He must have found it a serious undertaking. The great cry now, from Macon to Mobile and thence to Corinth, was "Where is Sherman?"

Beauregard, not thinking that Fabius Maximus harassed Hannibal without any telegraphs, was himself paralyzed for want of one, although Sherman, like Hannibal, also without any, was whipping everything he could reach. Leave Sherman's bummers sixteen years raiding in any country, and it would not raise sorrel!

C. T. Campbell, chief operator of Montgomery office, and —— Logan, an assistant, were hurried off to Atlanta to collect wire from the Chattanooga Road, to reconstruct the line to West Point. One wire from Montgomery to Opelika was taken down for the same purpose. John K. Mingle was sent up the Blue Mountain line toward Hood's command, to see what could be done in that direction. Merriwether went to Jonesboro, Ga., to hurry the work on to Atlanta by the Macon Road. The old Demopolis line was taken down, and December 3, the Tallahassee wire was being debated. Beauregard was demanding a line to Augusta, *via* West Point and Atlanta. Tree was pushing everybody and everything, but he was manacled. No railroads—little wire—insulators wanting—wagons hard to get, and soldiers would not work. Merriwether started east from Macon early, rebuilt the line for nine miles, when he was driven back, and idled five days for the hostile cavalry to leave. December 4, he reached Oconee, but no one yet knew where Sherman was, and the great inquiry among telegraphers west of the armed simoon was, "Have the

Yanks destroyed the line to Millen?" Next day, it was learned that both wires on the Central Road, beginning three miles east of Macon to Millen, had been destroyed. Out of the *débris* and collected wire, Merriwether was trying to patch a line to the latter place. In time he succeeded, but three hundred insulators were wanting. What a line for a quadruplex instrument!

The Federals now moved out from Pensacola to Pollard, the junction of the Mobile & Great Northern Railroad with the Alabama & Florida Road, and at eleven, A. M., of December 15, Mobile was cut off. The next day, report comes *via* Macon and Albany Railroad, that the Yankees have cut the Gulf Road at or about the Ogeechee River, and destroyed the bridge. Here is a Confederacy cut into three pieces. Augusta is at one point of the triangle, Savannah at another, and Richmond the third. Within these lines is one Confederacy, but sorely perplexed, and Grant is pounding at the head while Sherman cuts the tail. In the great middle Confederation, Thomas is driving Hood pell-mell out of Tennessee, and Captain VanDuzer is restoring Union telegraphs far south. To the west, Mobile by the Gulf lies in great danger, and Corinth is to be speedily abandoned. Then, there is another long stretch of country, from Meridian to a point far west of the Mississippi and far south of Missouri, where a third federation barely lives. All of these acknowledge but one head, it is true; but that is the best and most generally abused head in all the South. If the Fates did not then foretell, events must at least have cast a warning shadow. Such had now become the feeling in Georgia, that residents along the route of the Federal line from Atlanta to Rossville, a distance of twenty-six miles, had appropriated all the wire and refused to surrender it; but a guard of Confederate soldiers was effective. At Ackworth were many coils of abandoned wire, but the *Confederates* were afraid of bushwhackers and dared not go after it. It was January 1, before Brenner's men from Augusta completed the line to Millen. Then, for the first time in two weeks, was Richmond in communication with Mobile by any route. For a long time the Government freight route was from Augusta, *via* the Georgia Railroad to Greensboro, thence by wagon to Eatonton, and thence by rail, *via* Milledgeville and Gordon, one hundred and forty-five miles. The repair of the Central Road east of Gordon was abandoned.

From Scarborough to near Pooler, close by Savannah, it was only a goodly march, except over to the left where Kilpatrick again attacked Wheeler successfully, though it was quite an affair. Richmond Smith, while riding just behind the last of a body of cavalry, discovered by the roadside a horse *minus* one foot, which had been blown off by a torpedo buried in the road bed. It seems that four torpedoes had been placed in a line crosswise of the road. The two center ones had exploded. A colonel on General Blair's staff and the adjutant of the First Alabama cavalry rode up and began searching for others by pushing loose dirt from the top. The colonel uncovered one, but the adjutant not finding any so close to the surface as was that just found, supposed there was none and began treading about carelessly, when suddenly the last one exploded. Operator Smith was within eight feet of the adjutant at the time and the colonel even nearer, but both escaped unhurt, though somewhat demoralized. No wonder, for the adjutant lay bleeding and in intense pain. A piece of the exploded shell had carried one foot entirely away. It entered the heel and came out at the knee. General Sherman soon came up and set rebel prisoners at work digging for shells.

Matters began to look less like a picnic. The enemy were being crowded to where they had to shoot. This was December 8. The last time that day the enemy opened their guns on the Unionists, Smith was riding with General Blair and staff. The Confederates getting the range, a solid shot fired at the crowd, passed within two feet of General Blair's head and striking a shoulder of the Quarter-master of the first Division of Blair's corps, it came out under the opposite arm. He was dead at once.

The army pressed forward, mainly bending to the right, toward Ossabaw Sound. Slocum's corps pushed up against the city. Howard went to King's Bridge, about fourteen miles south-west of Savannah with a division and his engineers. Near there a little fight took place. While it was going on, Lonergan tapped the "Gulf" line at Miller's Station, in full view of the rebel line of battle. Howard was present with his operator, but they did not stay long. A battalion from Augusta nearly captured the General and his party The timely arrival of the Seventh Illinois mounted infantry, alone prevented it. The wire toward Savannah had been destroyed by Sherman's troops, but it was O K the other way.

Richmond Smith's diary will sufficiently explain the further operations resulting in the fall of Savannah.

DECEMBER 13.

General Sherman has been *very anxious to establish a base of operations.* From the day we first established our lines in front of Savannah, our right has been gradually extended and pushed in toward Ossabaw Sound, until it rested upon a point three miles distant from Fort McAllister. Last evening, General Sherman disappeared from head-quarters, and it was soon ascertained that he had gone with General Hazen and his (the Second) Division of the Fifteenth Corps, as far as King's bridge, the only point where the Ogeechee could be crossed, and went into camp that night. He made his appearance, however, early the next morning, on his way to our extreme right. At that place there is a large rice mill on the plantation of Doctor Cheeve, from which Fort McAllister could be plainly seen. Hearing about what was going on, I hurried to the mills, twenty-one miles distant from head-quarters. I reached there just in time. With the use of a field glass, I could see the fort, rebel flag, and even the men, very distinctly. Affairs were evidently drawing to a focus rapidly, as the General was beginning to get somewhat excited. Our skirmishers were getting so close to the fort that they could do little or no service at firing on the enemy. All at once the guns of the fort burst forth with a fearful fire of grape and canister. Our boys shot out of the woods towards the fort over the *chevaux-de-frise*, and quickly mounted the parapets and poured volley after volley into the rebels. This was the moment the General looked for. We could no longer discern our boys, but the old flags, two of which were held upon the parapets by our boys, could be plainly seen, and we knew as long as they stood up so proudly, all was well. Volley after volley continued to be showered unmercifully upon the rebels, still they did not surrender ; nor did they until every gun was taken from them. They were evidently among the best men the South could boast of. Their manner, habits and language spoke much in their favor. The first evidence that the fight had ended was our flag waving to and fro on the parapets a sign evidently agreed on by the generals. General Sherman was almost beside himself. The importance of the victory at this point can only be estimated when I say that some of our boys were so short of rations that they offered a dollar apiece for hard crackers, and in many cases they could not be obtained at any price.

This was the key to the river, and in fact the only obstruction to our communication with the fleet. We could not have formed a base at any other point, short of four or five days ; hence the value of the victory at this point and at this time could not be overestimated. The Signal Corps here did good service. Just as the capture of the fort had been effected, a vessel hove in sight, but so far off that it could barely be seen ; but as they gradually neared us, they perceived our signal officers calling them, which they answered by asking who was at that station. We replied: "General Sherman." A message was then addressed to General Sherman, asking what they could do for him, signed J. G. Foster and Commodore Dahlgren. The General replied that General Hazen had just assaulted and carried Fort McAllister, and wished them to go to their assistance. This, however, they could not do, as the river was supposed to be full of torpedoes, as well as having obstructions in the shape of a row of piles across it. The fort was manned by about two hundred men, having seventeen guns in position, nine of which were heavy siege guns, and eight field pieces.

Our loss, including those killed and wounded by torpedoes, I learn, does not exceed ninety ; that of the rebels, twenty killed and the usual proportion wounded.

Operator Lonergan, who was also at Cheeve's Mill, says that General Sherman was at that time loud in his praise of the Signal Corps and electric telegraphs, but thought they should be united.

The day after the assault, Captain John Hays of Kilpatrick's staff, got permission to take ten men and cross the Ogeechee. On Hays' invitation operator Drake went also. They rode south-east to Sunbury, at the head of St. Catherine's Sound. While there they caught a negro, who told them that a company of Confederate foragers was at a church ten miles distant. Hays and Griffith, the latter Kilpatrick's big dare-devil scout, were much elated at the prospect of a scrimmage. All started and found and captured twice their force and five wagons loaded with hams and bacon. The rebels were kept in the church until morning, when Hays and Griffith purposely within hearing of the prisoners, began an argument as to their disposition. Griffith, talking only for effect on the Confederates, insisted upon killing them, urging that there was not enough to eat in camp —which was true. Hays was for more merciful action. The decision was reserved, while Drake, of course unknown to the prisoners, drafted a parole broad enough to take in the whole squad at once. It read something like this: "We each and all, and each for the other do hereby sincerely promise and solemnly swear that we will not take up arms against the United States, nor give aid and comfort to any of its enemies, until regularly exchanged. So help us, God."—"N. B. The penalty for a violation of this, is death." Hays and Griffith who had withdrawn, now approached the church with Drake. The poor fellows inside looked all that they feared. It was a sorry joke, but Hays told them the decision to parole. It was joyful news indeed. Drake stepped inside and laid the writing on the table for signatures. The first man that came forward, stopped short before Drake and ejaculated, "The devil and Tom Walker!" Then Drake looked the man squarely in the face and exclaimed, "The devil and Tom Walker!" Each recognized the other. The rebel was the lieutenant in the First Georgia cavalry in General Wheeler's command, whom Drake, as we have seen, had captured in 1862 near Boston, close by Lebanon Junction, in Kentucky, while General Bragg was in that State. They shook hands and then took something.

Sherman's operations were now within the Department of the South, in which Captain James R. Gilmore was chief of the Federal telegraphs. Wherefore Smith, Drake, Murray and Waterhouse by Sherman's consent, December 19, left for the North. Drake had obtained from the soldier who stole it, rebel Colonel McAllister's copy of "Hogarth," and was taking it home with him. Out on Ossabaw Sound Murray got seasick and spoiled "Hogarth." It was quite difficult for Drake to leave Kilpatrick's, where he was placed on the pay rolls at one hundred and fifty dollars per month as copyist, notwithstanding Captain VanDuzer was paying him fully another hundred as operator, but there was in the North a nice pair of blue eyes, which he afterwards married. His heart was not set on money —and in the end it was all for the best, for those eyes tenderly watched his welfare for fifteen years, and then day and night as he slowly but peacefully sunk into a consumptive's grave.

On the night of the twentieth, General Hardee, going by the only open road, that leading into South Carolina, evacuated Savannah, which was occupied in the morning by the Federals.

On the tenth, Captain Gilmore reported to General Sherman for orders, and in a day or two had line material in abundance at King's Bridge. His work was speedily begun and when finished, his line connected Savannah with Hazen's head-quarters near Fort McAllister; with Kilpatrick's on the Ogeechee road; Fort Pulaski on Cockspur Island at the mouth of the Savannah River, which since June, 1863 was in telegraphic communication with Beaufort, on Port Royal Island. A line was also, by order of General Foster, commanding the Department of the South, built from Devereaux Neck to the head-quarters of Brigadier General Hatch, commanding provost brigade, to facilitate operations against the enemy at Pocotaligo, S. C.; and on its occupation, that place was put in communication by telegraph with Port Royal Ferry. "The building and operation of these lines," Gilmore reported, "in this Department, extending as they do, through swamp and river, was attended with almost insurmountable difficulties and the work was greatly retarded by the troops, who frequently destroyed the line for miles, using the poles for firewood, doubtless in the belief that these were rebel wires. Captains Gilmore, of the Telegraph Corps, and Churchill, United States Navy, superintended the removal of wrecks and other obstructions in the Savannah River, thus enabling vessels to approach the city.

Savannah was gained; the "shell" broken. Thomas' success in Tennessee made Sherman's doubly glorious. That great soldier, in his "Memoirs," says: "I wrote on a slip of paper, to be left at the telegraph office at Fortress Monroe for transmission, the following:"

SAVANNAH, GA., December 22, 1864

TO HIS EXCELLENCY, PRESIDENT LINCOLN,
Washington, D. C.:

I beg to present you, as a Christmas gift, the city of Savannah, with one hundred and fifty heavy guns and plenty of ammunition ; also about twenty-five thousand bales of cotton.

W. T. SHERMAN, *Major General*

"This message actually reached the President on Christmas eve, was extensively published in newspapers, and made many a household happy on that festive day."

CHAPTER XVI.

THE TELEGRAPH IN THE SIEGE OF PETERSBURG.—THE MINE EXPLOSION.—OPERATORS COURT-MARTIALED; NOT GUILTY.—DANGERS OF THE SERVICE.—ADVANCES TOWARD RICHMOND CHECKED.—NARROW ESCAPE AT REAMS STATION.—A MILITARY RAILROAD.—GRANT'S WIRES TAPPED; IMPORTANT RESULTS.—HATCHER'S RUN FIGHT.—WEST VIRGINIA AFFAIRS.—NUMEROUS TELEGRAPHIC INCIDENTS.—THE REBELS OUT-WITTED.—SHENANDOAH OPERATIONS UNDER SHERIDAN.—A GREAT VICTORY.—OPERATOR McCRICKETT AND OTHERS KILLED.—OPERATIONS IN NORTH CAROLINA.—FORT FISHER.—SHERMAN'S ADVANCE FROM SAVANNAH.—INCIDENTS OF THE MARCH.—SHERMAN AT GOLDSBORO.

Petersburg, a city of about twenty thousand citizens, is situated on the Appomattox, at the head of sloop navigation. It lies twenty-two miles south of Richmond, and there, in 1864, centered various important railroads, wagon ways and telegraphs. When Meade's army began its *semi*-siege—for the city was never completely invested—the Confederates held the Weldon, Lynchburg, and Richmond railroads. The forces under Grant, viz., Butler's and Meade's hereabouts, on the 1st of July, 1864, held a semicircular line from the Weldon Railroad, about Reams Station, to Deep Bottom.

Lieutenant Colonel Pleasants having completed a tunnel and lateral branches running under the rebel works in front of Cemetery Hill, mounting sixteen guns, for the purpose of exploding large quantities of powder and thereby making a breach through which it was hoped Burnside might push his corps of fifteen thousand men, and thus enter the city and divide Lee's forces, it was arranged, mainly by telegraph, that the mine should be exploded and the assault made at 3: 30, A. M., on the 30th of July. Anticipatory of this assault, movements of other forces to the north of the James, countermarched by night to be returned again in open day, had succeeded splendidly in inducing Lee to move all of his troops, except three divisions, also north of that river, which being effected, most of the Union troops were again moved by night to the support of the assaulting corps. Thus far the excellent plans had been admirably executed. Burnside advanced his head-quarters temporarily to Battery 14, and the field telegraph was continued thereto from his regular quarters. Operators accompanied Burnside to this new scene of danger, and thus every corps commander, of either army and Grant, Meade and Butler were brought within speaking distance of each other. General Meade made his head-quarters, as announced in the order of battle on the day previous, at the permanent quarters of the Ninth (Burnside's) Corps, and was accompanied by General Grant. General Meade subsequently testified that he adopted this position in consequence of its being a central one, and in telegraphic communication with all parts of the line where officers were stationed, with whom it was necessary to communicate.

Owing to the use of fuse which proved defective, instead of electricity, which should have been employed, the eight thousand pounds of powder placed in the mine could not be exploded until 4:46 A. M., when great masses of material, earth, timbers, cannon, soldiers and caissons were hurled near two hundred feet high as in one body, shaped like an inverted cone, through seams of which rushed the seething smoke and gases, whose pressure held or seemed to hold this confused mass, poised on high for an instant, when it fell with a great noise. Instantly one hundred and sixty Union cannon and mortars opened fire upon the enemy. The three South Carolina regiments stationed about the mined works were either buried in the ruins or suddenly awakened, to be frightened into uselessness. By five o'clock the first line about the crater was occupied by the assaulting party, which it is generally believed might then have taken Cemetery Hill and insured complete success, but instead of advancing as they ought, the troops huddled together in the crater in inextricable confusion ending about ten A. M., in an order of withdrawal and a loss, killed, wounded and missing of four thousand, four hundred men. During the progress of this assault, Meade had frequent occasion to telegraph to the various head-quarters' offices, and if he was not so fully advised of operations on Burnside's front as he should have been, it was not for lack of telegraphic facilities, nor has anyone so intimated. Between the hours of five and ten o'clock, "I received and transmitted," said General Meade, "over one hundred despatches and orders, averaging one every three minutes." "The vast number of despatches and the frequency with which they were sent and received," is another form of expression adopted by General Meade subsequently, to explain why his memory might not recall the exact order of incidents occurring during the assault.

Meade was disappointed at what he regarded as Burnside's misconduct, and after Grant returned to City Point it was the subject of telegraphic correspondence between the Lieutenant General and Meade. Burnside obtained somehow information of the character of this correspondence, greatly to Meade's displeasure. Meade accused Burnside's operators, Jerry Flynn and Charles K. Hambright, with the disclosure, and caused their arrest and confinement with other prisoners, in a guard house called the "bull pen," near his head-quarters, but they were allowed under guard to eat with Meade's operators. They were tried by a court martial convened at head-quarters of the Army of the Potomac, and defended by Major VanBuren of Burnside's staff. Burnside himself appearing as a witness. The judgment of the court was, "not guilty," but Meade, nevertheless ordered Flynn and Hambright sent beyond the lines of the army, which was done. Number one operators were in great demand, and it is said that upon one occasion subsequently when Meade applied to Major Eckert for more operators, he was informed that Flynn and Hambright were his only available ones, wherefore they were returned to service with that army.

In the month of July, the Second (Hancock's), Ninth (Burnside's) and Fifth (Warren's) Army Corps of the Army of the Potomac and the Eighteenth, of the Army of the James, were moved so often in the vigorous attempts of Grant to make headway against the enemy, that they can scarce be said to have had any particular camping ground, but wherever they were, they were in telegraphic communication with each other. At Meade's head-quarters, were operators Caldwell, Emerick and Pierce. At Warren's in the Avery House, Edwards and E. A. Hall until the latter part of the month, when the former, from severe exposures more dead than alive, went home it was generally believed, to die, but thanks to his constitution and nursing he recovered. His friend Dealy, operating at Fort Monroe, met him shivering on the hurricane deck of the steamer John Brooks and administering warm comfort internally and externally, started him again for home where he recovered in the fall, when he was offered by Major Eckert, the position of chief operator in the Army of the Potomac. But owing to sickness in his mother's family, he was unable to accept, and thus the Telegraph Corps lost one of its ablest and most heroic servitors. At Hancock's in the "deserted house," were operators Rose and Shreffler, except during the thirtieth, when Burnside's mine was exploded, at which time they were at Battery 14, where the noise of the conflict was so great that the instrument could be heard only by placing the ear close to it. At Burnside's were Charles K. Hambright and Jeremiah Flynn; at the Eighteenth (Smith's, afterwards Ord's), David E. Rand and J. E. Bliss; but the latter was relieved late in the month by Charles Douglass. The Tenth Corps stationed along the James about Jones' Neck and Bermuda Hundred, was in line of several offices, one at Jones' Neck, where W. S. Logue and O. B. Vincent operated; another at Bermuda Hundred, W. H. Wilson, operator; another at General Brooks' Head-quarters, Bermuda Hundred Front, W. K. Applebaugh; and another at Point of Rocks, a pontoon station across the Appomattox, where Maynard A. Huyck was stationed part of the time. At Butler's head-quarters, Bermuda Hundred, were Richard O'Brien, J. H. Nichols and Maynard Huyck; at General Grant's, City Point, C. H. Beckwith, cipher operator, J. D. Tinney, manager and F. T. Bickford and George Henderson, assistants.

Telegrams for the North from these points were relayed at City Point, which was in direct communication with Washington *via* Fort Powhatan, Jamestown Island, Yorktown, Fort Monroe, Cherrystone and Wilmington. L. M. Painter operated at Fort Powhatan until relieved in August by T. Q. Waterhouse. W. N. Embree at Jamestown Island until relieved in August by T.

N. Loucks. At Fort Monroe, were George D. Sheldon, C. L. Snyder, W. J. Dealy, C. A. Homan and S. C. Burns. This line, from Swan's Point, opposite Jamestown Island *via* Surrey Court House, was frequently interrupted by guerrillas, and as a sufficient force could not be spared permanently for its protection, a cable was laid in July from the Island to Fort Powhatan, twenty-two miles, which, with but few interruptions occasioned by anchors, worked successfully to the end of the war. The total number of miles of cable between Washington and City Point was fifty-two. A repair boat was arranged to raise and under-run the cable whenever it proved defective. The land line between Swan's Point and Fort Powhatan was maintained as well as practicable to mislead the enemy, who were content with cutting that.

Some incidents connected with the siege of Petersburg will best illustrate the uncertain tenure of life which the operators hereabouts held. Rand and Douglass were stationed a mile and a half from Petersburg and within plain view of the town clock, from which they regularly obtained the time of day. Head-quarters here were in plain view of the enemy, and received a large share of his attention. A member of the Seventy-ninth New York, the Highlanders, whose service had nearly expired, was acting as orderly or messenger at this office, and had a presentiment that he should never see home again. While sitting with his back to the enemy engaged in checkers on a hard-tack box marked out for the purpose, and within ten feet of the telegraph tent, he was struck in the middle of the back by a bullet which caused him such intense pain that three men were required to hold him. He died several days after. An operator's horse was tied to a picket rope about twenty feet from the tent. A shell struck the ground near by, richochetted and killed that and another horse. The operators inside declared that they could hear the breaking of the bones. A soldier guarding commissary stores, was sitting on a vinegar barrel, when it was pierced by a solid shot passing between his legs. This was usually referred to as the time the rebels drew vinegar from the Eighteenth Corps head-quarters. Shortly after Ord took command, the quarters were moved further back.

At Hancock's office, affairs were not much better. Late in

June, as the Second Corps was straightening and advancing its line, the enemy broke Gibbon's division line, and made some captures, which were duly reciprocated. During the fight, head-quarters camp received a number of shells, causing a general scattering and a removal of the camp. The telegraph operators, however, remained until all had gone, when they were ordered to leave also. An operator at Meade's head-quarters, speaking of the dangers incident to the telegraphers with Hancock, probably in the "deserted house," says "They were located in a house near the front, which was shelled day after day, some of the shells actually passing through the house."

In front of Burnside's corps, was a thirty-two pounder, which sent its missiles into the city every five minutes, with such regularity that it became known as the "Petersburg Express." The enemy, greatly annoyed, used to return the compliment, frequently overshooting, much to the disturbance of Flynn and Hambright.

The Confederates amused themselves often at night, attempting to shell Meade's head-quarters, but never got the exact range. Mortar shells, however, frequently fell very near, and quite disturbed the operators' sleep. A stray "Whitworth" shot dropped into their neighbors' tent one morning, while the telegraphers were eating breakfast.

August 9, from some unknown cause an ordnance boat at the City Point wharf exploded, destroying it and another boat, overturning buildings and scattering shells, shot, bullets, timbers, human limbs, etc., about the bluffs. Grant's head-quarters camp received a shower of all conceivable missiles, besides pieces of a pilot house and smoke stacks. Over sixty persons were killed, and one hundred and thirty wounded. Operator Beckwith's bed-clothing was set on fire by a shell, while he sat on the opposite side of his tent uninjured.

A little later, operator Laverty took his horse down a ravine near Fifth Corps head-quarters, where H. W. Cowan and another operated, to give it some water. While the animal was drinking out of the barrel set in a spring, a shell fell a few yards in front and immediately exploded, killing the horse and knocking the operator off. A piece of the shell made a large wound across his forehead, knocking one eye nearly out of its socket.

On the 13th of August, new movements were initiated with the triple purpose of preventing Lee from sending reinforcements to Early in the Shenandoah Valley, and threatening Richmond on the south and Petersburg around the rebel right. Butler laid a pontoon across the James at Big Bottom, and W. H. Wilson opened an office on the north bank at General Foster's headquarters. The Second Corps debarked there, and for several days it and the Tenth Corps engaged the enemy along Bailey's Creek near by, Wilson at the time being at Hancock's headquarters, which were in telegraphic communication with Grant and Butler. These forces being unable to break the enemy's lines, were ordered to recross the river during the night of the twentieth. General Hancock instructed his operator Wilson to open an office at the pontoon bridge; so mounting his horse he made post haste and reopened Deep Bottom office at eight o'clock P. M., when it was dark and raining hard. His office consisted of an empty cracker box standing on one side. His instrument and paper were placed inside, and to use either he had to lie flat on a blanket spread upon the ground and reach inside the box to copy or operate. His ground wire was inserted in a hole made with his pocket knife. It was thus, that General Grant was kept advised for several hours of the movements of the forces north of the James.

On the left, meanwhile Warren's corps, succeeded in striking the Weldon railroad, which, owing to the destruction on the South Side and Danville roads by Wilson and Kautz not yet wholly restored, was considered by Lee as nearly vital to his position; hence he made many and desperate efforts to regain the route. The last one resulted in the battle of Reams Station, to the left of Warren, on the 25th of August, when Hancock's corps was severely handled, losing altogether twenty-one hundred and fifty-three men. His telegraph office on this occasion, worked by Rose and Shreffler, was in Mr. Spier's house near the railroad. The fighting was continued until after dark; Hancock's men falling back, in the main, steadily and fighting and resisting every foot. There was disorder in parts of the line, owing to the fact that many of the troops were new recruits. The telegraphers being to the right of the fight, were protected merely by a picket line, which was withdrawn rapidly without the operators' knowledge. Their messenger was in search of Hancock with important despatches. In this condition of affairs a member of Meade's staff with escort and drawn revolver rode up and demanded who was there. He said, on ascertaining, that the operators were cut off and must not stop a minute, but take to the woods for the rear. At that moment the orderly returned and reported that Hancock's corps was in retreat, but he thought he could pilot the operators out in safety. A severe thunder-storm deluged the ground as the three, passing near the rebels, succeeded in escaping through cornfields and woods, and after wandering until 1:30 A. M., they reached Jones' house and found the General and staff—except Captain Brownson, mortally wounded while rallying the men, and Lieutenant Colonel Walker, captured while trying to find General Miles. During all of these operations Hancock, Warren and Ord, Meade and Grant, and probably several division commanders, had frequent occasion to use the telegraph, the local lines of which must have been fully forty miles in extent.

Grant's forces had already extended so far to the left that a military railroad, deemed necessary, was built, connecting with the City Point railroad about six miles from that place and running around to Meade's Station near the center front, and thence to Warren's, Hancock's and Patrick's Stations, striking the Weldon railroad a little north of Reams. This line from City Point was superintended by George Huntington, an ex-operator from New York State, A. G. Safford operated for a time at City Point depot, R. C. Laverty at Meade's, R. D. Williams at Patrick's, John F. Sabine at Hancock's and V. H. Wells at Warren's Station.

It was at this time that the most successful wire tapping of the war (in point of time) was accomplished by Charles A. Gaston, one of General Lee's operators. Lee had been exhausting his resources to ascertain the purposes of Grant, when the idea of tapping the Fort Monroe line was suggested. An independent company of men known as Reide's Scouts was directed to assist Gaston, who was chosen for the hazardous undertaking of connecting his instrument to the line in the vicinity of Doctor Richards', near Surray Court House. Gaston was conducted to the locality I believe by a few men under General Roger A.

Pryor, and left with the scouts, who for the most part were, so far as appeared, peaceable citizens engaged in chopping wood *i. e.* when they were in danger of discovery, but a careful watch was kept that they might not be discovered at all. Gaston, before leaving Richmond caused an insulator to be prepared so as to hold the two ends of the main line in place on the pole so as not to attract attention. To these he connected his fine silk covered wire and running that under the bark to the bottom it was extended a considerable distance along the ground into the adjoining woods and well covered with dry leaves. Thus Gaston opened his office on the military line connecting City Point with the War Department. His office, none the best, was as good as the woods afforded. Two men were on guard along the edge of the woods all the time. If seen they made the best of it, being careful not to appear unwilling to meet the discoverers, thus very little attention was paid them. It was arranged that when there was special need of caution or preparation for the Federals the guards along the roadway where the line ran should make a shrill whistle, which was repeated by another watch nearer and so on to the main camp. Thus for *six whole weeks* the Confederate operator was on Grant's wire. While that indicates what a brave operator may accomplish under perilous circumstances, the fact that but one message received by Gaston during all of that time proved of any value to the Confederates, reflects infinite credit upon the Corps for its telegraph cipher system. Gaston copied a great many cryptographic despatches which were sent to Richmond but *were never translated.*

The one message however that the enemy did profit by proved of incalculable value. That one not being marked "cipher" was sent plain about the 12th of September. It was from the Quartermaster in Washington and requested a guard, to meet twenty-four hundred and eighty-six cattle at Coggins Point, where they would be landed, and convoy them to City Point. At that time Lee's commissariat was struggling with the question of food. Meat was especially needed and difficult to obtain. This message pointed out a new way. Accordingly Wade Hampton on the fifteenth with a large body of cavalry left the vicinity of Reams Station, and making quite a detour arrived at Coggins Point the next day in time to receive the cattle and convoy them *to the Confederate* army, which they supplied for about forty days. He also captured three hundred prisoners, two hundred mules, thirty-two wagons and forty telegraph builders.

When a new party of builders reached the vicinity of Surray, Gaston knew his wires would be discovered and accordingly he destroyed as much of the telegraph as possible and returned to Lee.

Again the word came for a forward march. October 1, Generals Warren and Parke, the latter having succeeded Burnside in command of the Ninth Corps, with four divisions and cavalry pushed westward to the Squirrel Level Road, adding so much to Meade's permanent lines, but losing twenty-five hundred men.

The Tenth Corps under Birney and the Eighteenth, had already advanced north of the James by the Newmarket and Varina roads respectively; Kautz' cavalry preceding also on the right. Fort Harrison on the Varina road was captured and held in spite of gallant assaults to retake it. David E. Rand opened an office at General Birney's head-quarters and after the capture of Fort Harrison was relieved by W. K. Applebaugh whose office was in a small log house about six miles from Richmond, where he maintained communication with City Point until the evacuation of the Confederate capital. Rand opened an office at Jones' Landing near Dutch Gap, assisted by C. D. Hammond.

General Grant on the 27th of October, causing Butler to demonstrate on his front, sent Hancock's corps to the extreme left to the neighborhood of the Boydtown crossing of Hatcher's Run. Rose and S. K. Rupley accompanied Hancock, starting on the twenty-sixth and again at three A. M. of the twenty-seventh, and by eleven o'clock attempted to tap the rebel line on the Boydtown road, but found no circuit. The field telegraph for some reason was not built with this advance, and Rose and Rupley consequently, and owing to the proximity of the enemy, found it extremely difficult to locate themselves. A shell burst over their heads dispersing the cavalry advance, to the woods. The operators took to the rear and left, when a shell passed just over them from the *rear;* they were between two fires. Next, they joined Hancock. Grant and Meade with their staffs soon came up. So many horsemen in the open field drew the enemy's fire, and in the general scatter the operators went to an improvised hospital where they attempted to rest. But about three o'clock the enemy charged and turned Hancock's right. The stragglers in large numbers and an ambulance train came rushing by; the operators hurried to the open field to discover the cause and saw the enemy's successful advance, which however Hancock soon turned, to their great discomfiture. As that was not an insurable risk, the operators galloped to the extreme left, but there an attack upon the cavalry was made by Hampton which caused much confusion. Again the telegraphers started for a safe place, several shells hurrying them on. They were turned back from their new direction by a party who told them it was foolish to go that way, so they took the only road open to the rear which was found filled with pack-mules, stragglers, non-combatants, ambulances and wounded men. It grew very dark and rained hard; the operators got lost in the woods, lay down and slept in the pelting rain and dreamt of home, sweet home. At 1:30 A. M. they joined the retiring corps and in due time were again telegraphing at the "deserted house." The Union losses in this affair were killed, wounded and missing twelve hundred and eighty-four, but their captures exceeded that number. It was the last important advance Grant made during the year against the enemy in Meade's or Butler's front.

Many new operators joined these armies between July and January. Their names are William H. Woodring, Joseph L. Sears, Charles H. Mixer and John W. Gregg loaned to Major Eckert by Captain Clowry of the Missouri Department for six weeks, and Charles D. Hammond, George Warren, George F. Durant, Edward W. Mason, John F. Sabine, R. A. Shutt, James N. Allen, F. W. Davis, James Caldwell, Ed. Schermerhorn, John D. Thurston, James A. Glazier, S. C. Holbrook, F. C. Long, Edward W. McKenna, and M. S. Andrews. James A. Murray, D. C. Aughenbaugh, Thomas Morrison, T. N. Loucks, W. C. Barron, T. Q. Waterhouse, P. H. Nunan and John A. Sheridan were brought here from other parts of the department. These men were busily employed, as the practice continued throughout the war, of having all important offices open night and day.

It is now in order to relate the operations in West Virginia and the Valley of Virginia. The Confederates before this had seemingly resolved that there was nothing in the mountainous region of West Virginia justifying them in putting forth strong efforts to re-capture it, but were content to watch Federal operations with guerrilla bands and an occasional raid in force. Consequently the Union side was held by small detachments variously stationed along the Baltimore & Ohio Railroad, along the Kanawha to Fayetteville and out from Clarksburg to Beverly and Bulltown. On these routes the telegraph, under the general management of Captain Lynch, and the immediate charge of Charles O. Rowe, chief operator, was maintained by a brave and lonesome corps of operators and repairers.

On the 26th of September, 1864, the military telegraph office at Weston, on the Bulltown line, was taken possession of by Confederates under Colonel Witcher. R. D. E. Rowe, the operator there escaped by jumping out of a second story window and running to the woods while the enemy were firing on him. Witcher had with him a telegraph operator who, pretending to be the regular Union telegraphist at that post, transmitted in the name of the captured commandant there, a message to the commander at Clarksburg, stating that three thousand rebels under General Basil Duke were advancing on Weston by the Parkersburg pike and asking how many troops were at Clarksburg, and how many could be sent to Weston. Charles J. Thomas, M. B. Graham, Daniel Colestock and George K. Smith were operating at Clarksburg at the time, and the one of them who chanced to answer the call became satisfied from the peculiarity of the telegraphic manipulations at Weston, that the telegram was written by the enemy and transmitted by a stranger, consequently he informed Colonel Wilkinson, commanding at Clarksburg, of his conviction, but wisely O Ked the message when sent, without indicating to the rebel operator his suspicions. Wilkinson promptly replied, stating that two thousand troops had just arrived by rail and that more were expected during the night. It subsequently became known that Colonel Witcher's command nine hundred strong, after plundering Weston, retreated south-easterly, striking the Beverly line at Buckhannon, although his original intention was to capture Clarksburg and destroy the large amount of Government property at that depot, which he could have very easily accomplished, as there were but two companies of troops at Clarksburg at that time. This is a clear case of being beaten at their own game, but it is only one of the many bloodless victories won by the telegraph and heralded by nobody. If the place had been saved after great sacrifices in battle, some one would have been promoted.

On the morning of October 29, an infantry force three hundred strong, under Captain Hill, was repulsed at Beverly. After they retreated Philip Reger, the military telegraph repairer, stationed there, having a thorough knowledge of the country, piloted the pursuing Union forces over a shorter route—a by path—so that they were enabled to charge the advance of the retreating enemy who were driven across a creek with a loss of ninety-three prisoners, besides whom, the Federals re-captured forty of their own men taken at Beverly. Shortly before, Witcher captured Buckhannon (September 27, 1864), when C. D. Tull, operator there, escaped, this same repairer started for that place to mend the line. After riding a few miles he came within sight of a body of Witcher's men and immediately turning back was pursued by the cavalry. Reger dismounted and took to the woods. While running through them, he felt something strike his boot-leg, and looking down found he was dragging a large rattle-snake whose fangs had become fastened in the leg of his boot. Of course he dispatched the reptile whose fourteen rattles he brought to Clarksburg.

The next January, J. Jolls operating at Beverly was secreted by friends while the enemy held possession of the town.

About 11 A. M., November 28, rebels in the uniform of Federal soldiers, under General Rosser, surprised the Union force at New Creek, Va., and took possession of the place. Rosser had with him a considerable force of cavalry. The military telegraph office was seized before the operator, W H. H. Lancaster, could escape. After being robbed of his valuables and clothing, he was compelled to march barefoot to Harrisonburg, with nothing to eat until the third day of his captivity, when he was given three-quarters of a pound of fresh beef, which was all he had until the evening of the fifth day. He was confined in Castle Thunder, Richmond, where he kept from freezing by using half of the blanket of a kind comrade in misfortune. He was released in about four months.

The telegraph line from Weston to Bulltown ran through a very rough country, thickly wooded, and hence the wire was mainly strung from tree to tree. The circuit was frequently interrupted by bushwhackers and others. William K. Smith, formerly a member of the Seventh Michigan Volunteers, who had seen much service under McClellan, was wounded at Antietam, discharged in consequence, re-enlisted in an Ohio regiment, was detailed in the fall of 1864, and sent to Bulltown to operate. While there he discovered that the wire was frequently tampered with at a certain point in the woods on the Weston road, which determined him to investigate the cause. Arriving one night near the location where the line followed a rough and rugged hillside, at the base of which meandered a small stream, he led his horse to a dense thicket and tied it. He then walked to the roadside and by the aid of climbers took position in a high tree from which he could view the line on either side for some distance. He had not long to wait when three armed men rode up almost beneath his retreat. It was there arranged, that one of the party should ride some distance ahead and another as far behind to watch, while the third cut the wire. The third man rode up the steep bank underneath the wire which sagged at that point. The two guards having taken their positions, the other, standing on the back of his horse, attempted to sever the line with some kind of instrument, when Smith, whose revolver had been cocked all the while, took deliberate aim at the bushwhacker and fired. The bullet missed the man and shivered a glass insulator which had become detached from a tree and hung by a wire. The sudden report of the revolver and breaking glass startled the horse into a run, throwing the rider who rolled down the bank toward the road, but getting upon his feet as quickly as possible, he ran off rapidly. Nothing was seen afterward of any of the three men, thereabouts, and the line was seldom thereafter molested in that neighborhood.

In the Shenandoah Valley there was earnest war. Sheridan

was by nature thoroughly aggressive. His force, the Sixth and Nineteenth Corps and Crook's cavalry, equalled his enemy's under Early and Anderson—about twenty-two thousand men. The War Department had for some time telegraphed official bulletins for the information of the people, who had been deceived so greatly theretofore by unscrupulous brokers anxious to affect market values, and incautious correspondents, who were too apt to transmit mere rumors (which were always floating) as ascertained facts. The bulletin, of September 10, contained Sheridan's despatch to Grant, which was put in cipher at headquarters by operator McCaine, in Rachel Wright's house, Winchester, September 19, at eight, P. M. The most important part, translated, read as follows:

> I have the honor to report that I attacked the forces of General Early, on the Berryville Pike, at the crossing of Oquoquan Creek, and after a most stubborn and sanguinary engagement, which lasted from early in the morning until five o'clock, completely defeated him, capturing about two thousand prisoners, five pieces of artillery, nine army flags and most of their wounded.

Three days later, Early was again routed at Fisher's Hill.

October 9, the Confederate cavalry was most effectually beaten near Strasburg. Notwithstanding these defeats, Early, heavily reinforced, took advantage of Sheridan's absence and routed his army; but on his way to join the command, Sheridan heard of the disaster, which, by his rapid ride, celebrated in verse, he turned into a decided Union victory at Cedar Creek before night. In short, from September 1, to January, 1865, Sheridan's forces captured one hundred and one pieces of artillery, thirteen thousand prisoners and disabled nine thousand Confederates, besides destroying millions of dollars worth of property in the Valley, which could have been turned directly or indirectly to the advantage of the Southerners. Thus — save an encounter in the following spring, with a small force under Early —ended the war in the Valley.

The telegraph lines frequently connected with Sheridan's head-quarters, and he was almost always within a short ride of the wire when he had no office with him. Grant and the country were thus kept advised of operations in this vicinity.

While Sheridan was operating so near Front Royal early in October, the authorities at Washington deemed it prudent to reopen the Manassas Gap Railroad and rebuild the telegraph to Front Royal, having in view the supplying of his army by that route, and his retreat on Washington in case of defeat. So long had this road from Manassas Junction been unused, that it was overgrown by tall weeds and grass, completely hiding the rails. A large construction party, accompanied by Richard Power, a telegrapher of much experience and great usefulness, then and until after the war closed a despatcher of trains at Alexandria, started from Alexandria early one morning, escorted by eighty soldiers under Colonel Gallupe, to restore the road and wire. Advancing slowly, McIntosh of the Telegraph Corps constructed the line while the railroaders were rebuilding the bridges and repairing the track. The party worked their way through Thoroughfare Gap, reaching the south end of it at Broad Run late one evening, and halted. Power then opened an office on a barrel head, and reported progress to M. J. McCrickett, Superintendent at Alexandria of that and the Orange & Alexandria Railroad. McCrickett was a graduate of the telegraph, and richly deserved his new and responsible position. Next morning work was resumed, and, on nearly reaching White Plains, a couple of cannon balls flew through the air, screeching "Halt," and as one barely missed the bell of the locomotive, it was considered highly expedient to obey the warning voice. Colonel Gallupe and Power were on the engine at the time, and the officer remarked that the range was excellent for the first shot, but he thought he could dislodge the enemy; looking straight up the track, he discovered fully one hundred and fifty more on horse, whereupon, by direction of Gallupe, the train and construction party returned to Broad Run. Reinforcements were sent, to the number of two hundred infantry, to aid Gallupe, who meanwhile was not attacked—the enemy doubtless having counted the one hundred and fifty constructionists as part of the armed force. The enemy awaited and severely repulsed an attack, capturing sixty or seventy Union soldiers, but the position about the cars was held. Next morning the Confederates were gone. That afternoon, White Plains was reached and the guard largely strengthened. McCrickett and operator W. H. Glasscott came with the reinforcements, stopping *en route* a few moments at Manassas Junction to give operator H. P. Bull some messages. Power returned to Alexandria. During the next day McCrickett, accompanied by Glasscott and Colonel Fuller of the Sixteenth Michigan Volunteers, went beyond White Plains, taking with them two engines, as one was unable to overcome the obstructions that had grown over the track. Edwin J. Bolt and G. W. Fuller, conductors, two engineers, and firemen Charles Brooks and Richard Cowig were the only other persons on the train. Proceeding slowly, about one and a half miles out, they were fired at from the woods, which caused the engineers to put on steam and the train to move rapidly to a steep embankment just ahead, where Kinchloe's guerrillas had drawn the spikes and fastened telegraph wire to the rails which were not displaced until the engines were near, when the scoundrels in the woods pulled on their end of the wire, displacing the rails and throwing the engines down the embankment. Immediately the bushwhackers opened a galling fire, and the wonder is that any escaped. McCrickett, Bolt, Fuller, Brooks and Cowig were killed; the former was horribly mangled and recognized only by his clothes. His remains were taken to Detroit for burial. McCrickett was about twenty-three years of age when murdered; of an unusually bright intellect; a splendid railroad officer and greatly admired by all with whom he came in contact. He left Detroit to join the United States Military Telegraph Corps, but being an accomplished railroader, the Government railroad officials first secured his services. The remainder of the party escaped, and doubtless made good time to White Plains. E. L. Wentz, chief engineer and general superintendent of military railroads in Virginia, in a special order dated October 11, 1864, justly said:

> The unfortunate victims of their latest atrocity were faithful and valuable public servants—men who, knowing the danger incident to their duty, fearlessly encountered it, and lost their lives in executing the trust reposed in them. Though not equipped with the implements of war, and sent upon the battle field where fame is won, they have none the less sacrificed their lives upon the sacred altar of our country, and deserve a place in our memories among the honored dead who have fallen in its defence.

Beside Power, McCrickett and Glasscott at Alexandria, O. H. Dorrence, G. M. Huntington, A. H. Holden and John Thomas and other graduates of this same office in Alexandria, deserve honorable mention for their splendid telegraphic service on the military railroads of Virginia.

The Government repaired the railroad from Harpers Ferry to Winchester; O. H. Dorrence was made superintendent of the branch from Harpers Ferry and Thomas King, another operator, was appointed train despatcher. These two men thus became largely responsible for feeding, clothing and equipment of Sheridan's army in the Valley.

Wilmington, N. C., now the remaining chief Confederate *entrépôt*, was next to fall and with it the series of coast defenses to give way, at most points, to a cordon of successful vessels of war and the legions of troops that under the stars and stripes were henceforth to maintain the defenses along the Atlantic and Gulf shores.

Admiral Porter and General Butler attempted the capture of Fort Fisher, on the north bank of Cape Fear River, that furnishes navigation to Wilmington. Butler's explosion of an immense quantity of powder in a vessel floated near one of the faces to the Fort, proved a complete failure, and his subsequent reconnoitering late in December, 1864, equally fruitless, but General Terry and Porter co-operating more cordially, succeeded in storming the works successfully. It was a gallant fight on both sides. The Union loss being one hundred and ten killed and five hundred and thirty-six wounded, to which should be added two hundred more killed and one hundred wounded the next day by an explosion of a magazine. Over two thousand prisoners, one hundred and sixty-nine cannon and much other property were captured.

Next month General Schofield arrived from the Department of the Cumberland with the Twenty-third Army Corps and took command; his force amounting to about twenty thousand men; with these he was able to take Wilmington and Kingston.

Richard O. Brien, chief operator with General Butler, and a party of telegraphers were sent with Schofield to take charge of the telegraphs in this new field. To assist in the operations of the force advancing against Wilmington *via* Fort Fisher, a line was built from the fort up the peninsula towards the city, and immediately after its capture was extended into the city and an office opened February 23. The line from Morehead City *via* Newberne to Bachelor's Creek, built by J. R. Gilmore in the summer of 1863, was still intact. As Terry moved from Wilmington and Schofield from Newberne, the lines from those places were restored as these generals advanced on Goldsboro. General Sherman was already *en route* from Savannah, Ga., on his great march through the Carolinas, and Schofield was to co-operate from Newberne in the direction of Goldsboro, N. C. Let us now note the progress of Sherman's grand army of sixty thousand men newly equipped at Savannah, with everything that was needed to make it most potent in war. The end was approaching. The converging forces of Sherman, Schofield, Grant, Sheridan and even Thomas were threatening positive ruin to the Confederate cause.

February 1, 1865, Sherman's army was on its way from Savannah, Ga., for Goldsboro, N. C., *via* Midway, Columbia, S. C., and Fayetteville, N. C. Charles G. Eddy accompanied the General as cipher operator. Cass G. Sholes and John Lonergan were with O. O. Howard, who commanded the right wing of the army, and D. T. Berry was with General Slocum, commander of the left wing. Joseph Anderson went with Kilpatrick, who commanded the cavalry. All of these operators took with them pocket instruments and cipher keys, Eddy alone having a special key designed for communication with the War Department, Grant or Schofield. These operators also frequently performed duty as aides-de-camp to their respective Generals; their duties seeming to invite such commissions. Indeed it was quite usual for operators to serve in that capacity when occasion presented itself. At Rivers Bridge, Orangeburg, Cherew, Fayetteville and Bentonville, Lonergan was thus exposed to the enemy's fire. During the march he prepared in cipher a number of communications for the Union navy officers along the coast, giving information of the progress of Sherman's army; the resistance met and expectations for the future.

Negroes were hired with a promised reward of one hundred dollars in gold in case they successfully carried the messages, but it usually took more intelligent couriers to succeed, and hence Union soldiers were sent—disguised of course—when the matter was urgent. At midway, Lonergan tapped the line leading from Charleston, S. C., to Augusta, Ga., and Eddy chanced to be on the same mission four miles west at the same time. But Sherman's enterprising bummers, as his foragers had come to be called, were usually too far ahead to give the operators with the infantry much of a chance, for every time the bummers struck a line or a railroad they broke it, without awaiting orders.

Shortly after Sherman's army reached the Santee River, J. K. Witherspoon, about forty-five years old, operator at Camden, was aroused at two A. M. by a party from Beauregard with despatches for Hardee at Charleston, but as there was no night operator at Kingsville and no direct circuit to Charleston, Witherspoon bade them wait till daylight when he thought the Kingsville telegrapher would be on hand. And so he was, but the only telegraph route open from Kingsville that morning was *via* Florence to the north-east. After sending the messages, Witherspoon asked John F. Miller, the operator at Kingsville, where the Yankees were. Miller did not know, but thought near Gadsden depot, five miles above. In spite of the excellent advice he received, Miller insisted that Sherman could not catch him. When Witherspoon returned from breakfast, Miller had been captured and compelled to fire his own office and then march along.

The day before the soldiers entered Camden, some young men, claiming to belong to Wheeler's command opposing Sherman, came to the telegraph office, where they and the operator indulged in much buncomb regarding their splendid daring and insisted loudly that they only awaited the coming of the Union troops to enable them to demonstrate their bravery and skill. But when reports came that the army was scarce twenty miles away, the operator began to arrange for escaping. His wife however manifested the greatest zeal to assist. That night Witherspoon slept in the woods. Next day the militia captured a few bummers who laughed their captors in the face and said

Sherman's men would soon liberate them. The militia officer wanted Witherspoon to telegraph Sumter for reinforcements, which he tried to do, but soon returned to put his instrument in its hiding place. A number of muskets just then announced the dreaded foe and Witherspoon mounted his mule, which he whipped for miles and until certain woods were reached. As he passed his door, Mrs. Witherspoon stood with saddle in hand for him, but he had no time to put it on his animal, and so he went barebacked, developing his magnificent courage. The young men who had only waited an opportunity to exhibit their prowess were Yankee scouts whom the good citizens of Sumter feasted as Wheeler's troopers.

Lonergan while out early one morning with a cavalry escort, expected to outride the bummers and find the wire intact, but on nearing the railroad Lonergan's party was attacked by some of Hampton's cavalrymen and would have had a hard time of it, but for the enterprising bummers who had followed close behind.

One of the first communications received by Sherman from Schofield was brought by a scout to Lonergan at Fayetteville. He could not translate it and as it was addressed to Eddy it was turned over to him. The anxious moments awaiting its gradual translation may well be imagined, for what fortunes or misfortunes might have occurred during the interim from Savannah to Fayetteville all longed to discover.

Joe Anderson with Kilpatrick was having a livelier time, but had no very serious mishaps. On the 9th of February Kilpatrick ordered Anderson to take an escort of fifty men and tap the enemy's line at Windsor, S. C., but he found no circuit, and meeting with Griffin's scouts all pushed on four miles where four of the enemy were found, two of whom were killed, one wounded and one captured unhurt. The Unionists fired two grist mills and returned to head-quarters without other successes. On the twenty-second Anderson again tried near Winsboro, to connect with a line wire, but failed, and for want of better employment accompanied a flag of truce five and a half miles beyond the Union advance. Soon after, he with some officers of the staff went to a handsome house, where, representing themselves as belonging to General Butler's Confederate cavalry, they were handsomely feasted. At daylight of the 10th of March Hampton's cavalry dashed through the Union lines and completely surrounded Kilpatrick's quarters, capturing several of the staff. Kilpatrick himself barely escaped. Only he, Major Estes and Captain Brink were up. They hurried to the swamp and returned shortly after, driving away the enemy who meantime had captured Lieutenant Potter and Captain Brookfield while trying to get out of the house. Anderson and the others were up-stairs dressing as rapidly as possible and it was while they were so doing that the battle of Kilpatrick's headquarters, North Carolina, took place, resulting in killing eighty-six of the enemy, who carried off their wounded; quite a number of Confederate prisoners were taken.

On the 16th of March a more pretentious affair occurred near Averysboro, where Slocum's wing lost seventy-seven killed and four hundred and seventy-seven wounded, but Hardee's troops which had evacuated Charleston to oppose Sherman, were driven further on.

Johnston, commanding the concentrated forces from Savannah, Charleston, Augusta and other places, vigorously attacked Sherman's troops about Bentonville on the nineteenth, the attack resulting in a repulse. The Union losses were fifteen hundred and eighty-one killed, wounded and missing to Johnston's at least twenty-three hundred and forty-three. This inconsiderable struggle was the last of any note, between these contending forces, which however continued their respective tactics. Sherman's army moved to Goldsboro where it joined Schofield's from Newberne and Terry's from Wilmington. His force was thus raised to near ninety thousand men and one hundred guns.

Sherman was now in telegraphic communication with Newberne and Wilmington. Terry McGintey was operator at Fort Fisher; F. S. Clarke and R. Clapp at Wilmington, J. L. Sponagle, C. L. DeForrest and Carl Gifford at Kingston; L. M. Shurr, E. F. Wortman, H. B. Chamberlain, C. I. Thomas and R. B. VanDerhoef at Newberne; H. K. Clarke at Bachelors Creek; John E. O'Brien and J. L. Sears at Goldsboro; and James Powers, G. C. Felton and William Grasby at Moorhead City. Anderson with Kilpatrick at Mount Olive Station, seventeen miles from Goldsboro and the nearest camp to the enemy about Smithfield, opened an office on the 24th of March. This office was destined to prove of great usefulness pending negotiations during April.

Having completed our tour to the various armies most nearly co-operating with those of the Potomac and James, we will soon be able to observe the death throes of the Rebellion, which had already passed beyond the purview of possible success. Grant's policy was to act offensively everywhere except about Petersburg, as he did not want to drive Lee upon Sherman or Sheridan. Accordingly Stoneman's cavalry of Thomas' command was sent east from East Tennessee toward Lynchburg; Wilson south from Eastport, Miss., to Selma and elsewhere; Canby on the Gulf was pushing vigorously against the defenses of Mobile, and Sheridan was sent February 27 on his last raid up the valley, during which he captured the enemy sixteen hundred strong confronting his force of ten thousand, destroyed the James River Canal and Virginia Central Railroad, and moved to White House on the Pamunkey, ready to take part in the last act of the great tragedy, which we reserve for another chapter.

CHAPTER XVII.

THE TELEGRAPH IN THE DEPARTMENTS OF THE GREAT SOUTHWEST.—STONEMAN'S RAID.—REBEL WIRES TAPPED.—CONFEDERATES DISHEARTENED. — GUERRILLAS. — WILSON'S RAID.—SELMA, MONTGOMERY AND MACON CAPTURED.—SHERMAN ORDERS WILSON BY CONFEDERATE WIRES.—TELEGRAPH CASHIER ROBBED AND NEARLY MURDERED IN MISSOURI.—*RESUME* OF AFFAIRS IN MISSOURI.—CAPTAIN BRUCH'S DEATH.— MONUMENT.—CABLE LINE FROM NEW ORLEANS TO MOBILE BAY.— OPERATIONS AGAINST MOBILE.—ITS CAPTURE.—THE BAY CLEARED OF TORPEDOES BY THE CORPS.—THE RAM, WEBB, NEARLY ESCAPES TO THE SEA.—A TELEGRAPH FROM FEDERAL TO CONFEDERATE HEAD-QUARTERS.—TERRIBLE ACCIDENT.—A VENTURESOME RIDE ; NARROW ESCAPE.—DEATH OF AN OPERATOR. —SOME SERIOUS JOKES.—THE WAR CLOSED IN THE SOUTHWEST.

The year 1865 was the antithesis of A. D. 1861. 1861 was productive of disorganization ; 1865, of re-organization. The spring of the former ushered in civil war ; that of the latter announced its close. The former will ever be memorable for the evil it gave date to ; but no less the latter, for that it brought peace to a suffering people. On the twelfth day of April, 1861, the disunionists opened fire on the national defenses of Charleston, the first coast city in the South to be seized *vi et armis.* On the twelfth day of April, 1865, the Unionists repossessed Mobile, Ala., the last Confederate coast city east of the Mississippi. In April, 1861, the war spirit throughout the South was at fever heat ; in April, 1865, it was chilled through and through. In the beginning, the southerner was audaciously confident without victory, except at Sumter ; in the end, he was amazingly subdued, without calamitous defeat, except at Nashville. It was not one grand climax of defeat, but a moral conviction that such a catastrophe was impending, that produced a surcease of strife, for nearly two hundred thousand Confederates were in arms when the year began, comparatively few of whom were killed, captured or wounded. They confronted Grant about Petersburg, Sherman in the Carolinas, Sheridan in the Valley of Virginia; Stoneman, Gillem and Wilson, on their raids; Gilmore, about Charleston; Canby, about Mobile; Steele, in Arkansas, and numerous smaller commands. But everywhere the inferiority of the Confederates became more and more apparent as raiders brushed away opposition and armies pressed their sieges, or the foe in the field, until, at last in utter despair of the wicked cause, the end came, and foemen "shook hands over the bloody chasm."

It is our purpose in this chapter to pass from one department to another of the great South-western and Gulf States, and give as it were a panoramic view of the final struggles which culminated in a restoration of that territory to Federal authority. Beginning with East Tennessee, we shall discover the sweep of Union armies far into the Confederacy. We shall see expeditions start from Knoxville and Nashville, Tenn., and Eastport, Miss., New Orleans, La., and other places. We shall note their universal triumph and mark how Union success, so long uncertain, was now acknowledged by the vanquished as certainly as by the "boys in blue" and "those they left behind them."

Department of Ohio.

March 13, 1865, General Thomas proceeded to Knoxville with his staff and operator to organize a raiding expedition. There the General held interviews with Generals Stoneman, Tillson and Gillem, Governor Brownlow and others, on the fifteenth and sixteenth, returning the next day; and on the twenty-second General Stoneman began his great raid in aid of Grant's and Sherman's operations east. With three brigades of cavalry he reached Jonesboro on the twenty-sixth, which place Confederate General Jackson evacuated hastily. Not to detail the bold operations of the Federal force or detachments thereof, suffice it to say, that it penetrated the Confederate egg-shell a great distance, doing incalculable damage to railroads, telegraphs and depots of supply; occupying Boone, Wilkesboro, Mt. Airy,

Germantown, Danbury, Greensboro, Salisbury, N. C., Hillsville, Wythesville, Max Meadows, Jacksonville, Salem, Christianburg, Martinsville, Taylorsville, and Danville, Va. At Salisbury, N. C., Generals Gardner and J. C. Pemberton were utterly defeated, losing thirteen hundred prisoners and immense quantities of arms and supplies. Detachments having broken the railroad nearly to Lynchburg, Stoneman returned to East Tennessee, reaching LeNoir April 15; and in a few days Gillem took the division and effectually drove off all opposition far into North and South Carolina; almost reaching the route of Sherman's cavalry under Kilpatrick. These expeditions resulted in capturing forty-six guns, of which twenty-one were subsequently abandoned, and over six thousand prisoners.

John J. Wickham, who had been operating at the southernmost point on the Atlanta road (Dalton), since Sherman's march for the sea was summoned into East Tennessee shortly before Stoneman started and assigned to duty as cipher and field operator with that officer. His services with Stoneman and later with Gillem, were exceedingly useful. At the capture of Christianburg, Va., he captured the operator O'Neil, who was located at that point, by riding with two orderlies at midnight a mile ahead of the Federal column, into the town and up to the telegraph office. Quite a number of valuable telegrams were found there, and O'Neil was forced to give Wickham the office calls, which enabled him to deceive Lynchburg and other offices as to Federal operations. Near Thomasville, he narrowly escaped capture while tapping the telegraph. At this time he was not less than six miles from the command. Subsequently, while serving at General Thomas' head-quarters, he received this official acknowledgment of his services on these expeditions:

NASHVILLE, TENN., April 13, 1866.

MAJOR GENERAL THOMAS, Commanding Mil. Div. Miss., Nashville:

General:—I beg leave to recommend for an appointment as Second Lieutenant United States Army, J. J. Wickham, telegraph operator. Mr. Wickham accompanied my division on the raid into Virginia, North and South Carolina and Georgia, in the spring of 1865. Mr. Wickham attracted my attention by his alacrity, intelligence and gallantry. Not only did he perform the duties of his own department in a most satisfactory manner, but was at all times ready to perform any service in his power, and often acted as A. D. C. He is young and educated.

Very respectfully, your obedient servant,

ALVAN C. GILLEM, *Maj. Genl. U. S. Vols.*

Telegraphers were otherwise of service to the military in this section. In March, in anticipation of Stoneman's movements, the line between Strawberry Plains and Bull's Gap was again (the fourth time) rebuilt, and preparations were made to add a second wire between Chattanooga and Knoxville to facilitate the moving of trains on the United States Military Railroads, and generally to expedite military operations. During April, this second line was constructed, and owing to Federal successes the telegraph along the East Tennessee and Virginia Road was rebuilt as far as Carter's Station, one hundred and ten miles from Knoxville. This line was operated in April, 1865, at Strawberry Plains, Bull's Gap, Jonesboro, Greenville and Carter's Station. Other telegraphic service in this section was of a routine character.

DEPARTMENT OF THE CUMBERLAND.

The great Union victory in front of Nashville in December, 1864, effectually settled the question as to which was the stronger power in this section, and that conclusion became more and more evident each day after Hood's overthrow, owing to frequent desertions. The total number of prisoners captured, and deserters who took the oath of allegiance from September 7, 1864, to January 20, 1865, was fifteen thousand, three hundred and ninety-six, besides there were captured seventy-two guns. As might have been expected, there remained however, "last ditchers" enough to make quite serious opposition to Federal detachments striking at long distances from their bases. The following telegram from a superintendent of the seaboard telegraph line to its president in Richmond, Va., dated at Milledgeville, Ga., January 10, 1865, clearly represents the Southern situation:

The feeling in Macon, Columbus, Atlanta and Gordon indicates that the ordinary Georgian thinks there is no further use in prolonging the contest in which they are sure to be beaten at last. The whole State is filled with deserters. Alabama is not a whit better, and the Legislature of Mississippi refuses to allow the militia force to go into an adjoining county of its own State to repel an attack. There is no doubt the men of those States, already in the army will fight to the last, but the cry of the "stay-at-homes" is that it is useless to fight longer and they are no doubt encouraged in this idea by the Governor of Georgia, who it appears, is forever hunting up a pretext to fight Jeff Davis. It is reported to me by one of the superintendents of the Southern Express, upon the authority of a Judge in Tennessee, that Bragg had one hundred and three thousand men on his muster roll at the battle of Missionary Ridge, of which number sixty thousand were deserters, thus leaving him forty-three thousand men to bear the brunt of the fight. John Butler told me that Wheeler's muster roll calls for twenty-three thousand men, and that his pay-master told him he could only find seven thousand men in active service. Here now is a force of seventy-six thousand men at large, hiding in the swamps and fastnesses of this and contiguous country. From the feeling of the country people I am convinced no information would be given as to their hiding places.

Notwithstanding these facts, the ubiquitous guerrilla was abroad by night as in time when desertion was unpopular. February 22, of this year, about midnight a party of the latter, heavily armed, rode into Wartrace, Tenn., where Will F. Ware and another were operating, and captured Ware, an express agent, and Townsend, line repairer, whom they relieved of considerable money, watches and clothes; but Ware fortunately eluded them soon after, and called out the squad of soldiers stationed there, who rescued the others and killed the captain of the gang, whose home was near La Vergne, Tenn.

March 10, Normandy was raided in like manner, and operator Shaffer was captured and taken about five miles. Of course, he was robbed of everything but his shirt and pants. The instruments Shaffer found hidden in a hollow stump near his office. Operator R. B. Lines barely escaped being shot while riding on an engine on the Johnsonville road. Two of his comrades were not so fortunate. But in truth even guerrilla raids grew less frequent as Federal successes multiplied.

General Thomas left Pulaski (whither he went in pursuit of Hood) January 2, 1865, for Eastport, Miss. Stopping at Nashville until the ninth, to fit out the expeditionary fleet which accompanied him, he reached Paducah on the tenth, where he counseled with Admiral Lee, with whom was cipher operator William Whitney. John O. Ingle and William Martin and a party of builders from Nashville, were landed at Johnsonville—the operators to work there, and the others to reconstruct the telegraph toward Nashville. On the fourteenth, Clifton was reached by the General. Schofield's Twenty-third Corps, in camp there, was ordered to Annapolis, Md., and J. J. Cassell, Schofield's operator sent to General Paine at Paducah. A. J. Smith's corps was found at Eastport, and Thomas, who went into camp there, having dispatched Smith's troops and Knipe's cavalry to aid Canby in capturing Mobile, and arranged for General J. H. Wilson's cavalry expedition, returned to Nashville, as we have seen, to start Stoneman's. John J. Egan, Smith's operator, accompanied him to the Gulf.

Although Wilson's movements were co-operative with Canby's, who was operating against Mobile, yet as these forces did not unite, we will treat the expedition as entirely distinct, for it was co-operative only because it was diversionary. Wilson's three divisions under McCook, Long and Upton, started from near Eastport, Miss., March 23. Confederate General Forrest's cavalry was mainly at West Point, Miss., Roddy's at Montevallo, Ala., and Hood's infantry at Tupelo, Miss., under Dick Taylor's command. Moving by nearly parallel routes in three columns, not far distant, to Elyton, whence Croxton's brigade was sent to Tuscaloosa, the main column pressed on through Montevello toward Selma, defeating Roddy and then encountering Forrest. Croxton was engaged with Jackson. Upton and Long, in a pitched battle defeated Forrest on the Selma road, driving him that day twenty-four miles. The situation was anything but pleasing to the Confederates. Henry F. Lines United States Telegraph and Cipher operator, accompanied Wilson, and doubtless tried his hand at tapping Confederate lines. March 30, J. B. Tree, of the American Company's lines (South), telegraphed to its President, as follows:

AUGUSTA, GA., March 30, 1865.

To DR. W. S. MORRIS, President, Richmond, Va.:

Nettles reports the Blue Mountain line was tapped by a Yankee operator. He obtained no information. Found him out too soon.

J. B. TREE, *Gen. Supt.*

And later another:

AUGUSTA, GA., April 1, 1865.

To DR. W. S. MORRIS, Richmond, Va.:

Nettles reports we have been defeated above Selma, and there are serious apprehensions of the safety of that place. The enemy are now advancing on Selma, following the line of the Alabama & Tennessee Railroad. The struggle for the possession of the place will, it is thought, take place to-day.

J. B. TREE, *Gen. Supt.*

Later, telegrams announced that "Nettles reports the situation brighter. Superintendent Merriwether reports heavy fighting to-day at Spanish Fort and Blakely." "The Montgomery office is informed by General Roddy's operator, who came on the line on the east side of the Alabama river, that the enemy flanked us on our left at Selma at dark this evening, capturing the place. General Taylor has gone in the direction of Demopolis." William M. Nettles referred to above, was managing operator at Selma. It is reported that he acquired during the war, the sobriquet of "Peach-tree Nettles," owing to his remarkable haste to perform some service, while the Yankees were in close proximity. Instead of quietly untying his horse which was hitched to a standing peach tree, he carried it off with him.

The day after defeating Forrest, General Long's division appeared before Selma. Long and Upton lost no time awaiting reinforcements, but brilliantly assaulted the works, and captured the town, losing forty killed and two hundred and sixty wounded. Operator Lines "put up" the following cipher for Israel A. Sherman, cipher operator with General Canby, who was then working against Spanish and Blakely Forts, defenses of Mobile:

This cipher, which is here shown, was carefully sewed in the lapel of a citizen's coat, worn by a soldier who was sent to Canby with it. Translated, it reads as follows:

HEAD-QUARTERS CAVALRY CORPS.
SELMA, April 4, 1865,
10 A. M.

For MAJOR GENERAL E. R. S. CANBY, Mobile.

General: — My corps took this place by assault late on the evening of the second. We have captured twenty field guns, two thousand prisoners, besides over two thousand in hospital, and large quantity of military stores of all kinds. The arsenals and foundries with their machinery are in my possession intact. I shall burn them to-day, with everything else useful to the enemy. I have already destroyed the iron works north of here, eight or ten in all, and very extensive. Forrest, Dick Taylor, Adams, Armstrong and Roddy succeeded in getting out in the darkness and confusion, following the assault, by wading the swamp east of the city. The place is strongly fortified with two continuous stockades on the glacis, extending from river above to river below the city. They were defended by

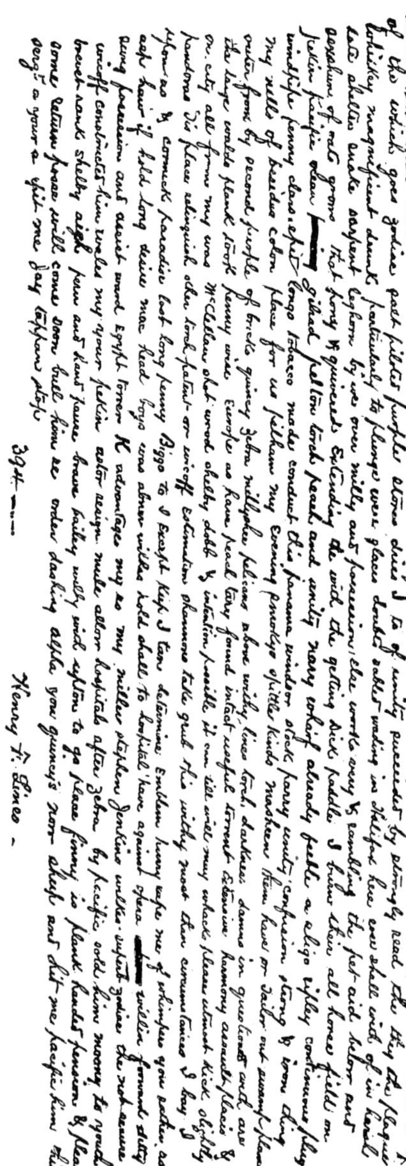

FAC SIMILE CIPHER.

four brigades of cavalry and all the first and second class militia of this section, from six to nine thousand men. The conduct of my troops, particularly that of Long's division, which made the attack, was magnificent. General Long was wounded slightly in the head; Colonel Dobbs, Fourth Ohio, killed; Colonels Miller, McCormick, Biggs, wounded. It's my desire and intention to hold the place as long as possible. I shall not relinquish my hold upon it except to secure other advantages. If I can keep Forrest west of the Cahawba till I have constructed bridge over the Alabama, I will move against him or Montgomery, as circumstances may determine operations westward rather than forward. Montgomery will, in my estimation, assist you most. Please hurry forward your gunboats and transports with the utmost rapidity. Take possession of this place, and allow me to finish up Forrest's cavalry. Upton is now after him with Second Division. This will be handed you by Sergeant Bailey, Fourth Michigan Cavalry, a brave and dashing soldier. Please order him a handsome remuneration, and return him to me as soon as possible.

(Signed.) J. H. WILSON,
Brevet Maj. Gen'l.

The Selma victory rendered the success of Wilson's further operations certain, and moving eastward, he captured Montgomery on the twelfth. At this time John K. Mingle was manager of the telegraph office there, and William H. Benton, Alf. Saville, Jerry McLaughlin, Jerry Reid, Barney Hughes and J. G. Thornton were operators. These men did not stay to make Wilson's acquaintance. Thornton and Benton retreating kept well to the rear of the Confederate troops, and every little while telegraphed to Columbus, Ga., the progress of the enemy. There a useless resistance was made when the Unionists approached. Twelve hundred prisoners and fifty-two guns were captured; also operator Thornton, who was robbed and paroled and probably L. R. Hoopes, chief operator at that place and others, as the town was taken at night. Macon was captured April 20.

The fact of an armistice between the Eastern armies was here communicated to Wilson, and his operator, Lines, put in cipher, a message to be sent over Confederate wires to General Sherman at Raleigh, N. C., inquiring if it applied to Wilson's forces, and was advised in like manner that it did. Croxton's force and other detachments subsequently arrived at Macon after doing great damage. The success of Wilson's expedition had been remarkable. Immense quantities of cotton, stores and war material and hundreds of miles of railroads and telegraphs, and many steamboats had been destroyed. Stoneman, Gillem, Sherman and Wilson had severed and bankrupted the Confederacy. However, the armistice was disapproved and Wilson's command was ordered forward.

Kansas, Arkansas and Missouri.

Let us now turn our attention Westward. Guerrillas there were, and plentiful in Arkansas and Missouri, but they did not interrupt telegraphic communication, now fully restored between Little Rock and St. Louis, so much as formerly. From October 1, 1864 to June 30, Captain Clowry succeeded in keeping his lines "almost constantly in operation day and night to the full satisfaction of all officers from the division to post commanders." As the country became more settled and peaceable, he gradually reduced expenses. His one thousand, seven hundred and two miles of line in the States of Missouri, Kansas and Arkansas were operated, constructed and repaired at an average cost of $15,000 per month. Considering that these lines were single, *i. e.*, one set of poles for every circuit, thus necessitating a full force of men on every line, and the country overrun with desperate characters, Clowry's success and expenditures compare favorably with others.

The following lines were in operation in this Department under Clowry; that to Fort Smith and Little Rock having been restored shortly after Price's defeat in the fall of 1864:

	MILES.
St. Louis to Fort Smith, Ark.,	403
St. Louis to New Madrid, Mo.,	207
St. Louis to Macon, Mo.,	168
St. Louis City Lines,	10
Pilot Knob, Mo., to Patterson, Mo.,	28
Bloomfield, Mo., to Cape Girardeau,	45
Jefferson City, Mo., to Syracuse,	43
Allen *via* Booneville and Weston, Mo., to St. Joseph, Mo.,	296
Fort Leavenworth, Kansas, to Fort Scott, Kansas,	120
Little Rock, Ark., to mouth White River, Ark.,	152
Little Rock, Ark., to Pine Bluff,	45
Little Rock, Ark., to Fort Smith,	185
	1702

Four hundred and twenty-three thousand, six hundred and ninety telegrams were sent over the lines during the year ending June 30, 1865.

The following sworn statement explains itself, and illustrates the dangers on the highways, notwithstanding the complete dispersion of organized troops:

STATE OF MISSOURI, } ss.
Barry County. }

Robert Willis, being duly sworn, deposes and says, that he is employed as traveling cashier of the United States Military Telegraph, Department of the Missouri, and while acting in that capacity, to wit: on the evening of July 25, 1865, whilst traveling from Fayetteville, Ark., to Cassville, Mo., in company with Reuben Carter, telegraph repairer, and another person whose name is to this affiant unknown, but who was acting as escort to this affiant, they overtook three mounted men who are unknown to this affiant; that said three men during all of that day continued in company with said escort until about five o'clock in the afternoon, when the said escort upon their own suggestion, returned to Fayetteville; that shortly thereafter, this affiant being in advance of the three men aforesaid, heard a pistol fired, and upon looking back, became convinced that he was the person fired at, and being unarmed, he then spurred his horse in the hopes that he might elude his pursuers, who continued firing and shortly came up with this affiant, and when quite close fired again, the ball this time passing through the body of this affiant, and that shortly thereafter, the horse of this affiant fell with him and thereupon the said men came up with this affiant, and the one who had fired upon this affiant then robbed this affiant of over eleven hundred dollars, which this affiant carried in a bag suspended from his neck on the inside of his shirt. This affiant further states that the said telegraph repairer well knew that this affiant so carried his money in said bag, and from the conversation carried on between said men, this affiant is convinced that the said repairer informed the said men that this affiant had money and where he carried it. This affiant further says, that the person who had wounded him wished to kill him and said he would do so, but that his companions would not agree to this proposition, and thereupon the said men having so robbed this affiant, left him upon the road. This affiant says that after being so shot and robbed he rode on to the first house upon the road, where he was met by two men from Cassville, who had been sent to meet him. Further this affiant saith not. ROBERT WILLIS.

Attest, S. V. HOWELL, Operator U. S. Mil. Tel.

Sworn to and subscribed before me, this twenty-sixth day of July, 1865. A. S. STRATTON,

Captain Second Ohio Cavalry, Comdg. Post.

Willis recovered. Colonel M. LaR. Harrison scoured the country with the First Arkansas cavalry and caught the robbers, two of whom were Iowa deserters. Of course they were severely dealt with.

As late as the middle of May it continued necessary to patrol the lines even as far north as Lexington, Warrensburg and Independence. But without a pretext the guerrilla was rapidly disappearing.

The military operators in these three States had accomplished wonders. To name a few would be to wrong the many. If sometimes a telegrapher, living at a distance from scenes of activity, from laudable pleasures, from society, from friends, from Unionists, and immured in the backwoods of Missouri, Kansas or Arkansas, where he was fed on corn dodgers and pork, where the staple drink was horrible coffee or worse whisky,—if he, sometimes, to drive off *ennui*, indulged in that which was worse, it is not, perhaps, more surprising than that he voluntarily exiled himself, for the good he might do while living the life of a recluse. Let him who harshly condemns these solitarians for sometimes trying the "inordinate cup," whose "invisible spirit cheers the sad, * * makes weariness forget his toil and fear her danger," first try to serve his country in times of passion and crime, far removed from sympathetic souls, with no rank to maintain, no courts-martial to restrain and no prospective promotion to attain. The sinless man who was never tempted, may have angel wings as a deer does antlers, and be entitled to as much credit. But, resisting in the main, all untoward influences, the operator in these parts zealously, uncomplainingly and unremittingly strove to perform his duties, and was successful. Near half a million of telegrams transmitted in twelve months showed in part the active work, but nothing but the wan face, "sicklied o'er with the pale cast of thought," evidenced the dry rot that was slowly undermining the vigor of youth, in the solitudes of this Department.

To Theodore Holt, at Little Rock, Dwight Byington, at Leavenworth, and Charles A. Hammann, at St. Louis, Mo., Assistant Superintendents, belong a large share of the credit due the employés under Captain Clowry, that model officer who began his career as Superintendent in these States, under Colonel Stager, by forbidding speculation in employés' vouchers, whereby the Shylock business received a merited quietus.

Department of the Tennessee.

Affairs within this Department remained substantially unchanged from the time of Grierson's raid to that of Wilson's. Corinth continued in possession of the Confederates even later. Captain Bruch's personal supervision over the Memphis district was arranged for in the spring of 1865, and Captain Fuller, of New Orleans, was relieved thereof. Consequently, Bruch went to Memphis late in March to examine into matters there, intending to return, when they were well arranged, to Louisville. It so happened that General Thomas, who was commander of the Military Division of the Mississippi, went to Memphis that month, arriving at General C. C. Washburne's on the twenty-sixth, and leaving the next day for Nashville. It was on this trip, and at Washburne's, that the author last saw Captain Bruch, who intended returning with us to Nashville, but was detained by business, and three days later he died.

No sooner had word passed from place to place, that Captain Bruch was dead, than with one accord the Corps in the Division of the Mississippi, which embraced his territory, began to take action looking to the erection of a suitable monument. Upon being advised of the desires of the employés, his widow, in view especially of the strong feeling of her late husband against gifts that are virtuallv forced from others by reason of the subordinate positions they occupy, making the donors timid about refusing what perhaps they could not afford to spare, consented to the erection of such a monument, *provided* that no assessment be levied therefor, but every contribution be voluntarily given. In Captain Gross' department alone, four hundred and fifty-five dollars was thus raised. In the Departments of the Cumberland and

Tennessee, a much larger sum was collected. In the beautiful cemetery at Canton, Ohio, overhung by graceful trees and surrounded by well trimmed evergreens, at the grave of Captain Bruch stands the imposing monument, which the telegraphers purchased and placed *in memoriam*, where, let us hope, it will tend to indicate the worth of the deceased and the appreciation of his friends many centuries after all of us shall have rendezvoused on like camping grounds. Engraven on the shaft is the following: "Erected by the officers and employés of the U. S. Mil. Telegraph Corps, Military Division of the Mississippi;" on the pedestal is engraven, "Capt. S. Bruch, Asst. Supt. U. S. Mil. Telegraph Mil. Div. Miss. Born in Stark Co., Ohio, Dec. 30, 1831. Died in Memphis, Tenn., March 31, 1865;" and on the base, "Bruch." Captain Van-Duzer succeeded to the superintendency of the Division of the Mississippi.

BRUCH'S MONUMENT.

SAMUEL BRUCH was born near Canton, Ohio, December 30, 1831. His parents were of German descent, and natives of Pennsylvania. When Samuel was thirteen years old, his father died. His mother was left with six young children and, owing to the long illness of his father, with little property, except the unpretentious home in which they lived. Until Samuel was sixteen years old, he worked at any labor he could get to do, and every penny earned was given to his mother. He did not go to school after his father's death. At the age of sixteen, with his mother's consent, he decided to learn the tailor's trade, and went to Canton in search of a situation in one of the shops. He visited every one in the town, but being a stranger was refused the work he sought. "No boy wanted." Much discouraged, he returned that night to his mother's house tired and foot-sore, having walked that day fully thirty miles. Soon after this, a lady who was a stranger to him but who had observed his faithfulness to duty, and knew of his unsuccessful attempt to get work in Canton, sent for him and gave him a letter to a gentleman of influence in C., commending him as "a boy who was trying to help himself and his widowed mother." By the kindness of this gentleman, he soon found work as apprentice to a tailor. Beside his board and lodging, he received from his employer twenty dollars per year, with which to clothe himself. He worked all day and much of the time until nine o'clock at night. Any extra money was made by working "over hours," either in the early morning or late at night.

It was while he was learning his trade that the first telegraph line, erected along the old stage route from Pittsburgh, Pa., west, reached Canton. The office was across the hall in the same building as the tailor shop. Long after, Mr. Bruch said that "from the time the office was established, and he could hear the tick, tick, ticking of the instrument, he felt that he *must* know *what* it was saying." Out of his poverty, by overwork he paid the the operator fourteen dollars for teaching him what he could of the art of telegraphy. Of course, the time for practice had to be found outside of the time belonging to his employer. Bruch still expected to be a tailor, and was much surprised two or three months before his apprenticeship expired, at being offered the position of operator in the Canton office. He arranged to buy the remainder of his time from his employer, and entered the Canton office at a salary of twenty dollars per month, a great sum in those days for lads to earn.

He remained there seven months, and then accepted an offer to go to Waynesboro, in Southern Middle Tennessee, where there was a junction of three or four lines, belonging to different telegraph companies. By working before and after office hours in Canton, he "made up" the new clothes he was to take with him. He also provided himself with such books as he thought would be of use to him in the pursuit of the knowledge which he was not permitted to obtain in school. He left Canton on May

1, 1852, for Waynesboro, Tennessee, a journey of nearly one thousand miles, most of the way by steamboat, with twenty-six dollars, for traveling expenses and any other that might be incurred before receiving pay for work in the W. office. He suffered much from loneliness and homesickness at first, and felt keenly the difference in the way of the country and people, but determined to stay and overcome his feelings. His salary there was seven hundred dollars per year. He remained in Waynesboro about three years, when the lines were consolidated and became known as South-western Telegraph Company's lines.

SAMUEL BRUCH.

He was then offered a situation at Nashville, Tenn., where he served two years, at the end of which the company offered him a situation in Memphis, Tenn., where, as in Nashville, he named his own salary. During the seven years spent in these offices, he made use of leisure time in improving his mind by reading standard works, also by hard study of text books, such as grammar, philosophy and chemistry. During all his life he was a student, and accumulated a fund of knowledge superior to many who pass all their youth in school.

By the time of his marriage (1859), Bruch had the feeling, which sooner or later comes to all operators, *i. e.*, that some other business was more desirable, and hence he intended the following summer to quit telegraphy and locate in Galion, Ohio, where he had already bought property and entered into partnership. Soon after marriage he was offered the place of chief operator of South-western telegraph lines, at Louisville, Ky., and decided to remain there at least one year. During that time he was offered the position of Superintendent by the telegraph company, but declined it, not wishing to be absent so much of the time from his home, and looking forward to leaving telegraphy very soon. He often said he did not wish to continue in a business where there was so little opportunity for growth or promotion, and from the *first*, determined, by economizing and careful investment of surplus means, to make himself *independent* of the companies employing him—serving them as faithfully as he could, yet keeping himself in a position to decline to go to any place or office that might be distasteful to him. This very disposition or characteristic made him valuable to the company. A subservient spirit never rises any higher than a fault-finding one. In the summer of 1860, he again expected to go to Ohio, but his partner persuaded him to remain in Louisville one year longer. During that year the war of the Rebellion began. Determined to leave Kentucky should that State secede from the Union, he remained in the office at Louisville, acting as censor for the Government over all telegrams for the South, during the summer of 1861. The night that Confederate General Buckner invaded Kentucky with an armed force, General Robert Anderson, commanding at Louisville, sent for Mr. Bruch to come to the depot of the Nashville Railroad and take charge of the telegraph lines. Bruch did what he could to repair damages to the wires, and as the troops advanced south, communication between them and the city was established, and kept up largely by his aid. Anson Stager, manager Military Telegraph in McClellan's great department, in a subsequent interview with Doctor Norvin Green (President South-western Telegraph Company), said to him: "The United States Government is not disposed to interfere with the management of your lines, unless it becomes necessary to do so. Have you any one in your employ whom you can recommend as manager for the Government?" Doctor Green answered: "We have one man in our employ whose *executive* ability I can not recommend, for he has not been tried in that capacity, but he has two qualifications necessary in one filling the place, he is *loyal* and *honest*, that man is our chief operator, Sam Bruch!" When Stager accepted Bruch as his assistant in managing for the United States, the South-western Company made him its Superintendent also. For nearly one year Bruch acted nominally as a sort of clerk for Stager, though really managing the details in this district, as much as he afterwards did. In August, 1862, Bruch was commissioned by United States as captain and assistant quarter-master.

There are so many misleading roads on which operators have ever been found, that in reviewing thus concisely Bruch's career up to the date of his commission, we involuntarily ask, may not this brief biographical sketch serve as a guide to point out better ways.

Department of the Gulf.

While General Thomas was preparing Wilson's expedition, General Canby was organizing a larger force to operate directly against Mobile. It had long been evident to the Confederates that Mobile was next in order for attack, it being the only coast city of magnitude (except Charleston, S. C., which surrendered in February, 1865), yet under Disunion rule. They knew that preparations on a grand scale were making to reduce this place, already effectually blockaded. Our last view showed Federal possession of the outlying sea forts, with telegraphic connection by cable. In anticipation of land operations, Captain Fuller was directed late in 1864, to connect New Orleans with Forts Morgan and Gaines, but the enemy held the land routes. It was coast or nothing, and after a careful examination, a possible but doubtful way was selected, as follows: from New Orleans to Fort Macomb, thence along Chef Menteur Pass and the swamps to Alligator Point; thence by cable across Lake Borgne, nine miles, to Point Aux Marchettes and across bayous Catherine, Biloxi, Little Cable, Dixon, False Bay, Nine Mile Bayou, Boudreaux, Three Mile Bayou, Lagoon Jones, Grand Pass, Blind Bay, Johnson's, Dead Man's Bayou, and the pass between Isle Du Pied and Cat Island, four and a half miles; thence about four miles of land line on Cat Island and cable connecting Ship Island, six and one-half miles; using twenty-three miles of cable, besides the land lines wherever possible. This was further extended from Ship Island across to Horn Island, twelve miles and land lines over Horn Island, thence to Petit Bois Island, four miles by cable, and land line over this island. The cable supply here gave out and the Dauphin Island line connecting with Fort Gaines could not be laid. The laying of these cables was rendered extremely hazardous by the severe south-easterly gales called in that section "three day storms." Fuller and his colaborers, including operator John R. Dixon, several times narrowly escaped drowning in the surf. The line worked to Ship Island from March 5, to the eleventh, when the cable gave out and from the seventeenth to April 14, but connection was never completed to Fort Gaines, consequently despatch-boats were used. Captain Fuller was indefatigable in his efforts and felt conscious that everything was done that promised success, to accomplish the work of keeping New Orleans and the army advancing up Mobile Bay in telegraphic communication, but the frail nature of the armor wires covering the Red Sea cable that was used, and the action of the currents and quicksands in the island passes, could not be overcome, nor could the constant and severe south-easterly gales which prevailed to a very unusual extent during this season. To these causes alone was attributed the slow progress of the work and final failure of the scheme. William T. House, operator, was located at Ship Island, Alvan A. Howard at Cat Island and Charles O. Connor on Horn Island.

Canby rendezvoused his forces in January at Kennersville, about ten miles above New Orleans. General Steele was in command there, where William T. House was operator. Algernon W. O'Neal was at this time and some months later chief operator with the "Pack Mule Train." This organization was about to carve a name for itself in the swamps, sands, muck and clay along the telegraph routes in Alabama, Mississippi, Louisiana and Texas. With it in February and March, were operators G. W. Baxter, Z. P. Hotchkiss and John R. Frank. Hotchkiss, who is *now* an author and poet, *then* made this train the subject of his muse, in part to the following descriptive effect:

> Spirit of ancient Morse, awaken !
> Too soon you'll slip the mortal coil.
> Half thy honor now is taken
> By those who former plans would spoil,
>
> Now, through the camp O'Neal and staff
> Display the newly conceived rules

For field work, and a telegraph
  Carried upon the backs of mules.

What! they, amid the cannon's rattle,
  The clang of strife, the flash of fire,
Steal from 'neath the cloud of battle
  And chain the lightnings to a wire?

Yes! When foe with foe contending,
  Man against man and steel 'gainst steel,
Then, from loaded mules, outsending
  An iron wire from off the "reel,"

To join this with other stations,
  By dreaded chains of Yankee wire,
O'er which t' dart communications,
  Swiftly as Heaven's electric fire.

See, the work is now completed;
  Hark! 'T was the trembling magnet spoke:
"Gen'ral! Here! the foe 's retreated
  From 'neath our fire through yonder smoke.

"Quick, charge your caissoned batteries!
  Stand ready, boys, to fire the mine;
Pause not for thankful flatteries,
  But charge—charge along the line!"

Dots, dashes, dots—the word 's spoken
  From corps to corps, from post to post:
"Hip, hip, hurrah! their line is broken;
  We have beaten the rebel host."

Along ours, the red fire pouring
  On center, rear—on front and flanks—
Though they'd withstand the big guns roaring,
  Yankee lightning breaks their ranks.

Everything being ready by the middle of March, Canby, with twenty-five thousand men or more, aided by Admiral Thatcher's fleet of gunboats, began operations. A. J. Smith, with whom was J. J. Egan, operator, landed his corps at Fort Gaines, and Knipe's cavalry marched overland, accompanied by operator M. H. Bassett. Israel A. Sherman was with Canby, and with General Steele who moved up from Pensacola, was operator B. H. Upham. Operators John R. Dixon, Louis H. Korty and George R. Penn were general utility men. Charles W. Pearson and Stephen L. Robinson were located at the temporary army quarters at Navy Cove, ten miles from Fort Morgan, on Bon Secour Bay, which place was connected by a land line five miles long, and cable of equal length. Here also came operators C. B. Smith and Benjamin Dorsey, *alias* Dorsey Berry, awaiting the advance of the army; and here, too, came the pack mule train and its *attachês*, including O'Neal, Hotchkiss, Frank and Baxter.

Steele, moving from Pensacola, met and defeated General Clanton, capturing and dispersing his force of eight hundred. Canby advanced along the eastern shore of Mobile Bay, against Spanish Fort, at the same time Steele invested Fort Blakely. These proved to be the defenses of Mobile, and if taken, the city must surrender. The pack mule people built a line as the army advanced from Navy Cove thirty-five miles to Fish River, which was soon after abandoned.

On the Confederate side, the military telegraphs about Mobile were under the superintendence of C. G. Merriwether and William Sandford, the latter then and now manager of the city office. The field lines radiated from the city connecting with, among other stations, Spanish Fort, where James Stinson operated until relieved shortly before the siege by J. L. Adam, a member of Colonel Patton's Regiment. Fort Blakely was also in circuit, as were Batteries McIntosh and Montrose, at which latter place was operator Philip Delgado. Delgado and Adam used to crack their jokes over the wire in French, much to the annoyance of the other operators, who could only make out the "ha ha's" and their own names. Besides repairing and operating, Delgado managed a little pine-knot still of his own, at Montrose, where he made that which smarts like pepper on a raw throat. Adam fared better, as Crane came to his aid just before the arrival of the Federals, and Post Surgeon Clanton, a brother of the General's, joined the telegraph mess, bringing thereto as an appetizer, the best hospital whisky. Crane was an industrious fellow. He induced his assistant to do nearly all the telegraphing, while he busied himself making shoes for the soldiers at about one hundred dollars per pair. In the Mobile city office were operators Louis F. Zantzinger, John Bohanna, Thomas Eastland, Ed. Leloup and one or two others. Zantzinger is one of the oldest living telegraphers. He operated in

Washington, D. C., as early as 1845. He saw much active military service under Beauregard and Bragg from April 19, 1862, to August 24, 1862, as "Military Superintendent of Telegraphs," but on account of his situation, prospects and family ties, his service at Corinth, Tupelo and Chattanooga was rather to earn a livelihood than to forward the cause. Bohanna was a bright, intelligent and cheerful fellow. He served in the Columbus, Miss., office in 1861, going to Mobile in the fall of 1863, and after the war became agent there of the New York Associated Press. Thomas Eastland was another genial companion, but physically unfit for the exacting hours of a siege. He died of consumption since the war, at his home in Huntsville, Ala., mourned by all who knew him.

Montrose was an out-post station, and Delgado hastily retired *beyond* the Fort. On came Canby and the Pack Mule Train. March 29, 1865, O'Neal wrote from in front of Spanish Fort, that he was working a siege line of nearly seven miles, with four offices on it, and perhaps would have to open more offices at the other head-quarters in a day or so: That the Corps beat the Signal folks in getting the wire working just one day, and that General Canby seemed to think he had no use for an office on their wire; consequently the telegraphers operated the only line having an office at Canby's head-quarters. April 1, the Signal Corps put up a line, but it did not work right, and the operators did all of the business.

April 7, O'Neal reported:

> I am now working fifteen miles of siege line surrounding Spanish and Huger Forts and Fort Blakely and numerous water batteries, rifle pits, etc. The offices are at our base of supplies, Stark's Wharf, head-quarters Thirteenth Army Corps, head-quarters Sixteenth Army Corps, head-quarters of the army and head-quarters General Steele's command. Our lines are worked night and day and the operators are all giving out. Upham at head-quarters Thirteenth Army Corps can not read by sound and he is the only man I can get. As it is, business is delayed by sending so slow to him, for the line is kept red hot sending messages all the time and in a delay of fifteen minutes, business accumulates at all the offices. Can you send two or three sound operators? They are needed badly and I am afraid the boys will give out now before they ar-

rive. The way we work it now Colonel Christensen has a man detailed at each office to set up nights and when the instrument works, he wakes the operator. Nearly every night two or three offices are kept busy all night so that the others are awakened almost every minute. The General and Colonel Christensen (chief of staff) say this telegraph is everything to them. The Colonel says he don't know how they could get along now without it.

About sundown April 8, Canby intent on an assault upon Spanish Fort advanced a few troops to intrench and sharpshoot, but the brave soldiers found it necessary to press on without stopping and being promptly supported, captured the key to the fort which determined its commander to evacuate that night.

The following telegrams indicate the general character of messages sent over the field wires and explain passing events:

FLAG SHIP STOCKDALE, 8:45 P. M., April 8, 1865.

MAJOR GENERAL CANBY, Commanding Army, Department Gulf:

Your despatch of eight P. M. just received. I congratulate you on your success this P. M. A splendid shelling you gave them. Could not be surpassed. I am now hard at work and have been all last night and all day and shall be all this night clearing the ground for the monitors to advance, and I will try to put them within shelling distance to-morrow. Torpedoes very numerous, one hundred pounds powder in many of them. I shall persevere. I have a one hundred pound Parrott *en route* from Pensacola, all fitted.

(Signed) H. K. THATCHER, *A. R. Admiral.*

HEAD-QUARTERS THIRTEEN ARMY CORPS, April 9, 1865.

MAJOR GENERAL A. J. SMITH, Commanding Sixteenth A. C.:

Your mortar firing last night was excellent, and relieved my line considerably. Two mortar shots from your line at about one o'clock this morning were especially good, but fired a trifle more to the right would have done even better.

(Signed.) G. GRANGER, *Maj. Genl.*

FLAG SHIP STOCKDALE, April 9, 1865.

MAJOR GENERAL CANBY, Commanding Army:

I have just received your despatch, and have to thank you for the information relative to the number of prisoners and guns captured in the Forts. I trust we may be as fortunate at Blakely, which must, I think, compel them to yield the city without further fighting. So soon as matters are settled on this side, I propose to cross the bay with most of the naval force. Do you think favorably of a demonstration on the western shore and a demand for surrender of the town? I am General, etc.,

(Signed.) H. K. THATCHER, *A. R. Adml.*

HEAD-QUARTERS THIRTEENTH ARMY CORPS, April 9, 1:35 A. M.

MAJOR GENERAL A. J. SMITH, Commanding Sixteenth A. C.

You were late! We had possession of Fort Alexis and Spanish Fort an hour before we heard from you. *Our men are asleep inside now.* (Signed.) G. GRANGER, *Maj. Genl.*

C. S. operators, Crane and Adam, just before this had trouble to escape Federal shells, one of which upset their office. When they moved to the plateau fronting the river, they were still within range of the Union guns. When Spanish Fort was evacuated, the troops silently crossed on a treadway which was built from shore to shore, and hidden by weeds. The soldiers' boots were removed to prevent noise. The operators, Crane and Adams, were ordered to remain until about midnight, when they were to take a skiff for Battery Tracy. As they and certain others reached the boat, the Nationals having discovered the evacuation were approaching with shouts of victory. Only the darkness afforded any protection. Wading out to the skiff one man was shot dead, perhaps by a chance fire, may be by design. A moment's halt enabled the party to lay the poor unfortunate on the shore, and then with muffled oars the operators pulled out into the darkness and out of range, two miles away, to Tracy.

April 9, the word went over the wire, "charge along the line in front of Blakely," and about 5:30 P. M. came the tug of war. The carnage was awful. One thousand Unionists and five hundred Confederates were killed or wounded in ninety minutes, but Blakely was captured, and with it three thousand prisoners.

Mobile itself was next, and with it and its additional defenses, one hundred and fifty guns and other valuable arms and property. It was evacuated on the eleventh, and surrendered the next day. General Maury, commanding, retreated up the

Alabama River with nine thousand men. Thus fell the last coast city east of the Mississippi. It cost the Federals twenty-five hundred in killed and wounded.

Besides Zantzinger, Eastland, Bohanna and Leloup, operators at Mobile, George S. Hanna, Charles M. Sabin, Henry E. Nichols and J. M. Worden took the oath of allegiance, and were employed by Captain Fuller.

Admiral Thatcher having applied to Captain Fuller for aid in "blowing up submerged obstructions in Mobile Bay," including torpedoes, that accomplished electrician David Elphick was sent from New Orleans with proper appliances and a steam tug for that purpose. In the bay, seconded by other operators, Elphick, by use of the "plan," located the infernal machines and sinking cans of powder alongside, connected by insulated wire with the battery on the tug Blossom, exploded one torpedo after another until the bay was free and clear.

Superintendent Tree's instructions to Merriwether not to leave a single instrument in Mobile in case it became necessary to evacuate, had been strictly obeyed. The office was completely gutted.

Negotiations for the surrender of the escaping Confederates were carried on largely by telegraph. For this purpose, by mutual consent, the wires were repaired, and Federal operator Robinson actually took charge of the Citronelle office before the surrender of the troops there. The following are two of the messages sent over southern wires at this time:

HEAD-QUARTERS, MOBILE, May 3, 1865.

COLONEL RITTER, Whistler, Ala.:

General Canby and party of about six officers leave the city at eight to morrow morning, to meet General Taylor (Confederate) at Citronelle.

(Signed.) C. T. CHRISTENSEN, *Lt. Col. and A. A. G.*

MERIDIAN, MISS., May 2, 1865.

By Telegraph and Flag of Truce.

To COMMODORE I. S. PALMER, U. S. N.,

Commanding Naval Forces, Mobile:

Sir:—I have the honor to request a personal interview with you at such time and place as you may request for the purpose of arranging terms for surrender of the forces of the Confederate States Navy now commanded by me. Learning that Major General Canby and Lieutenant General Taylor will probably meet under a flag of truce very soon, I respectfully suggest that if it suits your convenience our proposed meeting shall occur at same time and place which Major General Canby may appoint for his meeting with Lieutenant General Taylor. I have the honor, etc.

(Signed) E. FERRAND, *Flag Officer*,
Comdg. C. S. Naval Forces.

The surrender of Taylor's and Ferrand's commands took place on the 4th day of May at Citronelle, Ala.

It was now evident to the most obtuse rebel that the end was at hand and that restoration would soon be in order. Turning from this locality for the present, to preserve the chronological order of events in this department, let us observe, first, the dying gasp of the Confederate navy.

The C. S. ram W. H. Webb, ran the blockade at the mouth of the Red River at night, laden with merchandise, and hurrying down the Mississippi passed New Orleans April 24, in *broad daylight.* The way it happened that telegraphic news had not preceded it was this. Every little while between Baton Rouge and Donaldsonville and the latter place and Bonnett Carre, and Bonnett Carre and New Orleans the boat was stopped and a gig carrying a long rope was rowed ashore; one end of the rope was fastened to the telegraph and the other to the stern of the ram which then starting on tore down great lengths of line, breaking insulators and poles and finally the line itself. Bonnett Carre was reached and passed. L. J. Hebard operator there discovered at sunrise that the line north was down though he thought it but an accident, for want of reason for suspecting something else. He promptly sent a repairer to mend it. While the repairer was away, a soldier reported to Captain Kimberly commanding the Post, that the "Webb" had passed. Kimberly was about telegraphing the fact to New Orleans when Colonel Everett rode up, walking his horse for fear of creating an excitement, and being informed that the line was "O K" to New Orleans, talked about arresting the man who brought the news, which the Colonel seemed not to believe, and rode away. Ten minutes later the Colonel sent to know if the wire was all right yet. The operator replied that he had just had New Orleans, but soon after the *line was cut.* This was indeed a calamity. The report *was* true and there must be a sacrifice. The punishment of an officer involved a trial; not so of a telegrapher. Everett reported that but for the absence of the operator, the "passing of the Webb would have been communicated before wires were cut." Hebard was discharged. He had been absent a few minutes only and was near the office, returning when sent for. He was in the office long enough to have telegraphed a number of messages before the line went down. If we possess all the facts in the case and we have the statements of three witnesses, the department superintendent made a great mistake in allowing Everett's request. The ram reached New Orleans about noon, flying the United States flag at half mast. Her true character however was discovered and Admiral Thatcher sent the tender, Holly Hock, carrying two guns after her. The fleet had not steam up. On sped the Webb; New Orleans grew excited. The telegraph was again sought. Forts Jackson and Philips must be warned or there will be another privateer on the high seas. William Mathews operated at Fort Jackson, no one at Philips, and Mathews who had been employed since the fall of Mobile and was probably in the service of some Confederate States telegraph company prior thereto "*could not be raised.*" Quarantine office five miles north was resorted to. ii, ii, ii, said Operator VonEye. "Take a message from General Banks to the commanders at Jackson and Philips and deliver in person," said S. B. Fairchild the chief operator in New Orleans. The message was sent. Von Eye, crossing the river, took to horse and rode as did Paul Revere, from "Charlestown shore" just about ninety years before. Forts Jackson and Philips were duly warned and officers with rockets sent up to Quarantine to announce the passing of the Webb. She never came. There was a "Richmond" in the (naval) field. This gunboat passed Quarantine for New Orleans that morning. Near English Bend the watch called out "Steamer ahead with stars and bars," for the Webb had donned her own clothes. There was no escape and the Webb was run ashore and blown up. Her crew taking to the swamps. Mathews served fifteen days in all and it seems was justly discharged. General Banks complimented VonEye.

About Vicksburg the exchange of prisoners was regularly proceeding under the cartel. The Federal camp was four miles out from the city and the Confederate at Big Black River bridge. Operators Pettit and Ludwig, who were captured at Athens, Ala., in September, 1864, and imprisoned in Cahaba were sent to the parole camp for exchange, in March, 1865, but instead of their being released, the wire was repaired from Vicksburg to the bridge, and Pettit stationed at the rebel head-quarters of exchange and Ludwig in the Federal camp. This was March 21, 1865, and while the war was in full blast. It is an interesting and doubtless isolated fact, that thus by mutual consent, Confederate and Union head-quarters were linked together, and *both* offices worked by Union operators. Exchanges were greatly facilitated by this means, but it was a great hardship to both of the operators, who had suffered greatly from chronic diarrhœa. Indeed Pettit worked most of the time in a hammock. Exposure here produced a relapse, which, while it nearly destroyed, saved his life. He and Ludwig were to go North with two thousand, one hundred other exchanged prisoners on the steamer Sultana; this relapse prevented Pettit's going, and General Dana refused leave to Ludwig, as he was greatly needed to assist Samuel Cochrane, the telegraph manager of the Vicksburg district, who was slowly dying, when he should have gone North to recuperate. The Sultana blew up at two A. M., April 28, near Memphis and took fire. About *fourteen hundred* home-sick soldiers were thereby killed or drowned. May 9, Pettit and Ludwig were exchanged. I find the following official endorsement concerning these operators:

HEAD-QUARTERS DEPARTMENT OF MISSISSIPPI,
VICKSBURG, June 2, 1865.

* * The men named within, Pettit and Ludwig, were of great service here at a time when it was impossible to have done without them. * * Ludwig was detained here when Pettit left, by order of Major General Dana, there being no telegraph operators to be had, and rendered very important services in building the line to Jackson. * * (Signed) G. K. WARREN,
Major General.

Dana ordered Ludwig to construct a line from the bridge to Clinton, which, aided by negro soldiers and Irishmen, he did, over the hills, bridgeless streams, the silent battle field of Champion Hills and through heavy timber, in a scorching sun. The obstacles were so great and Ludwig so weak that Dana, growing impatient, sent one hundred more soldiers, and later also Louis H. Korty, operator, who had just arrived from Mobile. The soldiers reached Clinton just as the line was being extended into that place. This was in the evening of May 19. The circuit was thus completed *via* Meridian to Mobile. Let us now follow Korty, who on that day reached Black River Bridge. He left the river in company with Captain Turner, of a New York regiment. Before starting, they had discussed the propriety of only two persons traveling by night over a country, infested by guerrillas and other bad characters, who had left their commands with their arms, without waiting to be paroled. Many of their homes were in the trans-Mississippi Department, and there E. Kirby Smith commanded, and was determined to fight to the last extremity; thither also, others, moved by the same unconquerable will, were tending, and hence it *was* a dangerous ride that Korty and Turner proposed, but being badgered by others who thought the two dared not attempt it, they mounted their horses and were lost in the darkness of that silent night, to brood over imaginary evils as they slowly picked their way. Turner had fought over this ground, and was wounded at the battle of Champion Hills, which locality, about midnight, as the moon escaped the clouds, the two reached safely. Turner's adventures hereabouts, which he related *en route*, the apprehensions of evil, the wrecks of caissons, pieces of shells; the hour, the loneliness, the silence; but above all, the whitened bones of slain animals, the rude burial boards announcing the names of the fallen, and other evidences of mortal conflict upon which the moon's pale light gave a sepulchral hue—these things, at a time when "graveyards yawn," strung the nerves of the riders to the highest tension.

After contemplating the scene a little, they pressed on through thick woods, and on emerging, were suddenly completely surrounded by eight or ten men coming out of a thicket with drawn revolvers, calling out, "Get down, you Yankee

--- - - ——--." One caught Korty's bridle, and another leaped for Turner's, but the latter, who was remarkably brave, opened on his assailant and others with his revolver. Thus began the second battle of Champion Hills, with greater odds against the Unionists than Grant had against Pemberton. Before Korty could draw his weapon, his horse, a spirited, unmanageable animal, reared up and, dashing headlong, broke from the grasp of the bushman and was out of range with its rider. But Turner was fighting *alone;* however he was fighting gallantly, emptying two revolvers and shooting at least two of the assailants, and then leaving unharmed amid a shower of bullets. Korty, in the meantime, had checked his horse and was returning to Turner's aid, when the latter rode up. Their first idea was to press on to Clinton, but then they heard the neighing of a horse, and concluding that the men were mounted (perhaps familiar with the country) and would pursue, they determined to take an elevated position among the hills, where, tying their horses, they threw up a barricade, confidently expecting to be besieged. If so, they determined to re-baptise this historic ground. But the guerrillas, discovering the preparations, or for some other reason, let them alone. Waiting, watching, listening, until sunrise, made a long night of it, but at last, daylight came, and the two entered Clinton in safety. At breakfast, a professor, who before the war conducted a female seminary there, related the circumstances of his finding the bodies of two men who had been murdered two nights before, near where Korty and Turner were attacked.

Samuel D. Cochrane manager of Vicksburg District caught cold from exposure in building a line to Black River Bridge early in March, from which he never entirely recovered and July 1 died in the harness at his post and, wrote James K. Gullihur who succeeded him, "received a charity burial" July 2. Thus another offering was made, and he, one of the best and most devoted. *Dulce et decorum est pro patria mori*, is a pretty expression, but one which no man ever uttered from experience. It is a staple ingredient in Fourth of July oratory and monumental design. Next to volunteering one's wife's relations, is sacrificing others less deserving thereof, on their country's altars. I never heard that "sweet and glorious" expression on a battle field and my manuscripts do not show that it was used in Cahaba, Libby or Andersonville. It may have been the consolation of the sixteen thousand Federals who lie buried about Vicksburg. It may have been uttered there by operators Cochrane and Booth, but the *Government not having provided headstones for any of its dead operators*, no such sentiment is preserved to their credit.

On the 9th day of May the line was finished from Mobile to East Pascagoula where Benjamin Dorsey opened an office in General Twiggs' house. Two days later Israel A. Sherman, operator at Spring Hill, on that line (John R. Dixon having succeeded him at General Canby's) pretended to be General Wade Hampton's operator *en route* with Hampton's cavalry for Texas, pursued by Grierson. C. B. Smith the manager of Mobile District military telegraphs, hastily reported Sherman's representations, which ought not to have deceived anybody and Sherman was ordered to report *in arrest* at Mobile. One would naturally suppose that his written report aided by Captain Fuller's recommendation for leniency and Smith's request to withdraw the charge and re-employ Sherman, would have prevailed, but Canby ordered him "discharged" with "permission to go North." The following is Sherman's report:

MOBILE, ALA., May 11, 1865.

LIEUTENANT COLONEL C. T. CHRISTENSEN:

Colonel:—At about one o'clock this P. M. in my office at Spring Hill, the line being at rest, I commenced a conversation with the operator at Mobile, and without any malicious intent whatever, I represented myself to be a rebel operator with Wade Hampton. After talking in this manner for not more than ten minutes I disavowed the truth of the statement and explained the conversation, as the operator in Mobile will testify. I, by this, prevented, as I supposed, it from going any farther than intended—simply a joke upon the operator. Indeed the conversation was of such a nature, that it was believed for an instant, is surprising to me. The Pascagoula operator discredited it from the beginning, making laughable remarks. If it had been really malicious any person conversant with telegraphy can explain to you, that I could have very easily avoided disclosing my name, and thus have escaped detection. Knowing that hostilities in this district had ceased, that Wade

Hampton had surrendered; that no cavalry from his command could under the most favorable circumstances have reached this section of the country: that a rebel operator would not have held any such conversation as mine and that my writing in telegraphic characters has a peculiarity which is easily recognized, I am very much astonished and I need not say pained that it was believed and that it has resulted thus, especially after my prompt disavowal, and as it was done in perfect innocence, unconscious of any serious result; I respectfully, but most earnestly ask that all proceedings be dropped and that I be restored to duty or allowed to resign and go North.

Early in the war, Jesse H. Bunnell, a great wit, very young, and so small that he had to sit on a stool to reach the instrument (in Pittsburgh office, I believe), perpetrated a joke on Wheeling newspapers, to the effect that a great naval disaster had occurred to the Unionists off the Rip Raps. Of course Jesse was dismissed as he ought to have been, but offers of employment came from all quarters. It was one of those cases where the employer *had* to discharge, even if he immediately re-employed the malefactor elsewhere at advanced wages.

In the Confederacy a cruel joke was inflicted by an operator near Selma, who telegraphed West that Chattanooga had been captured. This was when the Disunion army was on Mission Ridge. Confederate scrip and spirits went up with Southern hats and huzzas, but they fell deeper than before. This imposition provoked the following message:

JACKSON, MISS., January 23, 1864.

TO MERIDIAN OFFICE:

Your attention is called to the fact that Mr. ——— operator at ——— who sent us that bogus despatch about Chattanooga, is debarred from all courtesies of the profession on the lines west of Meridian. Messages from or to him must be paid, and all messages signed "operator" from ——— must not be sent free until you ascertain that such messages are not from ———. I shall require a strict observance of this, and you ought to be as sensitive on this subject as myself and require no reminders.

(Signed) D. FLANERY, *Supt.*

But a Federal brigadier general (Merrell), whose head-quarters were in Macon, Mo., tried his hand at this game early in the war. Frank S. VanValkenburg was Merrell's operator and the General, not having much else to do, tried to learn the art of telegraphy. Having learned the alphabet, he sometimes practiced on the (N. M.) line, pretending to be a rebel operator. The only harm done was to interfere with the regular business. R. C. Clowry was chief operator in St. Louis, and J. B. Clarke, train despatcher at Macon (R. R.) office, and Charles York, assistant. Clarke reported the cause of the trouble to Clowry and it was agreed to repay the General in kind. The next time he came in, Clarke went to the battery room and there telegraphed the General some startling reports that worried him, but it was York who succeeded best. He went to the Hannibal & St. Joseph office and pretending to be Callao operator, told Merrell that there was a large force of rebels marching from south-west on Macon, and would cut the railroad west and south. The General looked at the paper, put on his spec's and looked again. While making sure of what he had read, another report came apparently from Clarence, twelve miles the other side, to the effect that a large force was moving from the east, destroying the railroad, to capture Macon. It was now time for action. The General sent for his staff; ordered reinforcements of infantry and artillery from Brookfield, and prepared to fight. In due time all came but the rebs. York used to laugh in his sleeve over this, but never dared *know* anything about it.

TRANS-MISSISSIPPI.

This department was all that remained of the Confederacy. Not a hostile force was left on the cis-Mississippi side, for Grant and Sherman had ended the war in the East. General E. K. Smith pleaded with the Disunion governors of Louisiana, Arkansas, Missouri and Texas, to aid him. They were powerless, if not hopeless. Then he addressed his army, to "protract the struggle." An affair, known as the battle of Palmetto Ranche, did occur in Texas, on the 11th of May, but it is only notable in that it was the last combat of the War of the Rebellion. Canby went up Red River to Shreveport. Sheridan came to New Orleans and organized an expedition against Smith, but the forces of the latter disbanded or surrendered, and the war *was over*.

In all of the seceding States, governmental reconstruction

and general pacification were matters of supreme moment The most potent agency in this work of peace, the telegraph, lay flat on the ground for hundreds of miles. Before the rivers could be confidently navigated, before the railroads could be repaired, the telegraphs, by which confidence alone could speedily be restored and reconstruction fairly begun, must be re-established. The work of re-opening telegraphic communication throughout the South, was a great one, and in it the Confederate States and United States telegraphers vied with one another. Their services will receive deserved attention hereafter.

CHAPTER XVIII.

THE TELEGRAPH IN VIRGINIA DURING THE LAST CAMPAIGN.—GRANT'S ANXIETY.—FIVE FORKS.—GENERAL ASSAULT.—EXPOSURE OF OPERATORS.—LINCOLN AND DAVIS HEAR THE NEWS OF VICTORY AND DEFEAT.—PETERSBURG AND RICHMOND EVACUATED.—LINCOLN VISITS PETERSBURG.—HOW THE NEWS OF THE CAPTURE OF RICHMOND WAS SENT AND RECEIVED—LINCOLN IN RICHMOND.—AN OPERATOR RECEIVES APPLAUSE INTENDED FOR GRANT.—THE PURSUIT OF LEE.—SAILORS CREEK.—FARMVILLE.—APPOMATTOX.—HOW THE NEWS OF THE SURRENDER OF LEE'S ARMY WAS SENT AND RECEIVED.—THE NEWS OF LINCOLN'S ASSASSINATION.—CAPTURE OF THE ASSASSIN.—LINCOLN'S STORIES.—JOHNSTON'S SURRENDER.—RECAPITULATION OF THE SERVICE OF THE TELEGRAPH.

At the opening of the new year, there were in Major Eckert's department alone one hundred and eleven operators and about the same number of persons employed as builders, repairers, clerks, messengers, orderlies, foremen, teamsters, cooks and battery-men. These operated and maintained about eight hundred miles of telegraphs, centering at the capital of the nation. From all fronts in Virginia and intermediate points was daily gathered the news of interest, which, from this focal point, was spread from ocean to ocean. In the War Department were C. A. Tinker, A. B. Chandler, F. Stewart and G. W. Baldwin, exclusively employed in putting up or deciphering telegrams of the greatest moment and secrecy, from the commanders in the various departments and many minor officers. There were also in that office, telegraphers, some of whom have since been eminently successful as they were notably faithful in those stirring head-quarters of civil war. George C. Maynard, T. A. Laird, George A. Low, J. A. L. McKenna, J. N. Bentley, J. H. Dwight, William E. Kettles and W. J. Bodell were there.

On the line to Winchester, *via* Harpers Ferry, were D. J. Ludwig, H. P. Bull, J. Minnehann, W. W. Shock, J. J. Sabin, George J. Lawrence, J. D. Flynn, R. R. McCaine, J. H. McCaine and S. C. Holbrook.

On the eastern shore of Maryland line to Cherrystone, were M. V. B. Buell, chief operator; W. T. VanGesel, J. W. Winthrup, C. R. Roberts, George T. Kay, L. J. Blades, C. H. Lounsberry, H. R. Browne, J. W. Boyd, W. A. Dunn, H. P. Darlington and George E. Cromwell.

Other operators, variously stationed north of Fort Monroe, were John Horner, at New York City; E. W. Culgen, at Pittsburgh; J. F. Stevens, at Cincinnati; C. C. Northrup, H. H. Atwater, D. C. Aughinbaugh, H. H. Bishop, George B. Davis, John W. Carver, A. F. Childs, Richard Powers, W. L. Ives, A. G. Hancock, W. S. Maynard; and on the Fort Monroe and City Point wire, George D. Sheldon, chief; John D. Tinney, A. G. Safford, Thomas Morrison, R. J. Murphy, C. A. Mixer, M. S. Andrews, H. D. Frank, J. E. O'Brien, V. H. Wells, T. Q. Waterhouse, and C. A. Homan.

With the armies in the field were W. K. Applebaugh, W. B. Wood and Ed. Schermerhorn, Twenty-fourth Corps; W. H. Wilson and E. W. Mason, Twenty-fifth Corps; P. H. Nunan, and James H. Glazier, Fifth Corps; D. E. Rand, and S. K. Rupley, Second Corps; A. C. Snyder and H. D. Franks, Ninth Corps; J. D. Thurston, W. N. Embree and J. J. Carroll, Sixth Corps; in February, at head-quarters Army of the Potomac, A. H. Caldwell, James Caldwell and Richard Graham; at head-quarters Army of the James, commanded by General Ord, J. H. Emerick, J. H. Nichols, F. W. Davis, M. A. Huyck, Charles H. Mixer; at General Sheridan's was R. R. McCaine.

At various stations about Petersburg and Bermuda Hundred, were J. A. Munson, J. J. Mulligan, L. M. Painter, J. H. Sheridan, R. D. Williams, C. D. Hammond, T. N. Loucks, R. C. Laverty, F. C. Long and James M. Allen.

Lee assaulted Fort Steadman in front of Petersburg, March 25, meeting at first with splendid success, even capturing the Fort, but the Union soldiers re-formed, captured over two thousand of the enemy, re-took their positions and gained others, suffering a total loss, however, of nine hundred men.

Badeau, whose opportunities for obtaining the figures were extraordinary, places Grant's available force in March, 1865, at one hundred and eleven thousand men to Lee's seventy thousand. Sheridan was brought from the White House and his available troopers, increased by others, now numbered twelve thousand. Leaving twenty odd thousand men in the trenches, Grant projected Sheridan to Dinwiddie Court House, with instructions to strike out north-westerly against the South Side Railroad, now the great hope of Lee, and without which he could not feed his army in Petersburg. At the same time the Fifth Corps, Warren's, was sent to the Union left to within about four miles of Dinwiddie Court House, when the Second Corps joined it on the right. Other corps also took part in the general movement of swinging to the left, thus greatly attenuating the opposing lines.

At this time Kautz' cavalry was to the right of Weitzel in the suburbs of Richmond, and from there to near Dinwiddie Court House, were opposing forces ready to receive, if not to give battle. It was doubtless the longest stretch of battle array ever beheld on the American continent—probably thirty miles in extent. All night long Grant watched the operations by telegraph. Every few minutes he penned an order to Meade or Ord, or a telegram to Lincoln or others. All night long Meade was busy ordering by telegraph, dispositions for the morrow. Emerick, chief operator in O'Brien's place (the Army of the James), had recently ordered fifteen miles more of field wire and Gulick came with it and took charge of a building party, which was to push on to Richmond as soon as that city fell. The line already extended from Bermuda Hundred to Varina Landing, head-quarters Army of the James, Twenty-fifth and Twenty-fourth Army Corps and Kautz' head-quarters, the latter about six miles from Richmond, where John A. Torrence was operator.

Sheridan and Warren met some resistance the first day. Next day the roads being almost impassable and the rain continuing, much advance was impracticable. Lee by the morning of the thirty-first, had massed eighteen thousand men against Warren's corps, and while they were being effectively hurled against the devoted Fifth, Sheridan pushed on to Five Forks; then they fell upon Sheridan, cutting his force in two and driv-

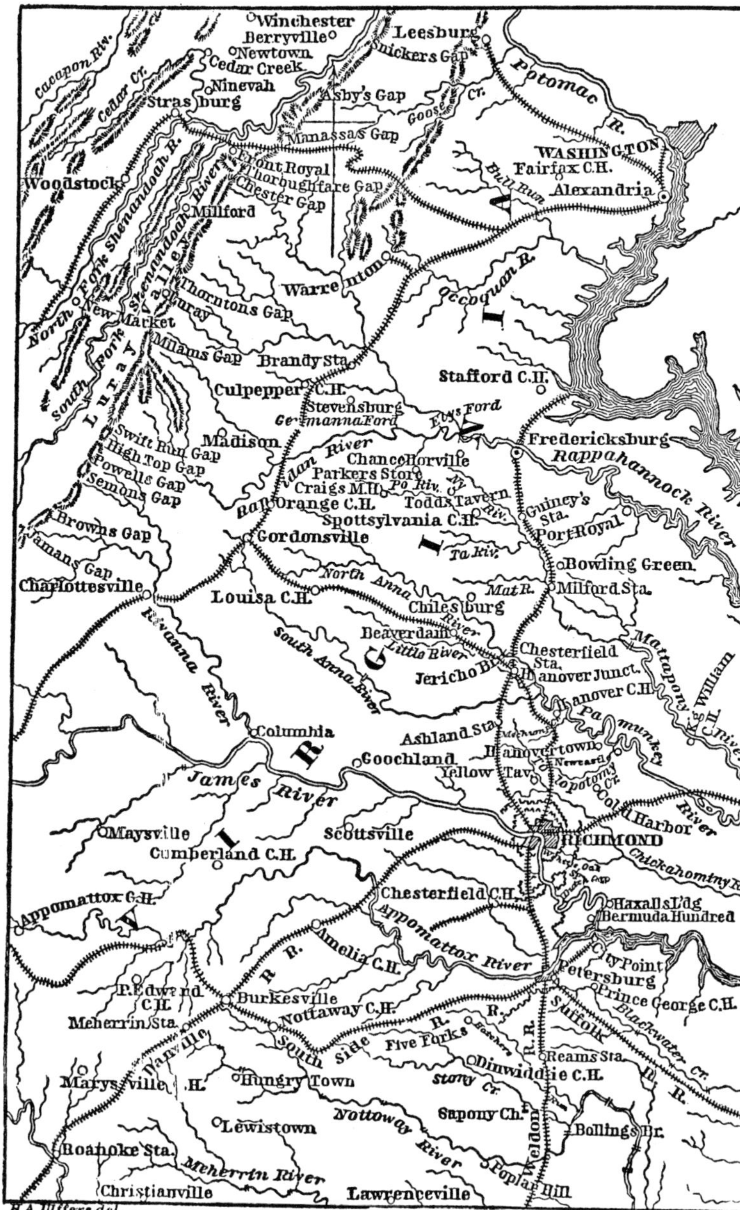

H.A. Ulffers del.

Eng'd by Fisk & Russell, N.Y.

10 5 0 10 20 30 0 miles

THEATRE ABOUT RICHMOND AND OTHER POINTS IN VIRGINIA.

ing him back to Dinwiddie. But Sheridan, though driven, resumed the offensive with considerable success, and re-united his columns. That night Grant's head-quarters were at Dabney's saw mill, southwest of Petersburg. It is said, that that (thirty-first) was the most anxious night that Grant ever experienced—that night, which brought Sheridan back to Dinwiddie, Warren to a standstill after a partial rout; that night, so dark and with roads so muddy, when the results and the elements were discouraging; when Warren was reported cruelly derelict, and officers of great reputation were recommending a return to the trenches of Petersburg; the third night since starting out on the series of hazardous movements pregnant with overwhelming joy or deepest grief—what would it not have been to the great chieftain, had he been dependent alone upon staff officers or miry roads, for reports all along the line. Sleep came not to Dabney's Mills that night, where Ed. Schermerhorn and R. C. Laverty were busy transmitting or receiving orders and important messages. Probably John D. Tinney was there, also engaged in cipher work, for Beckwith had reported to President Lincoln at Grant's, and accompanied him to City Point, but before leaving, the President visited head-quarter's operators and made many inquiries about their work, manifesting here, as he had ever done in Washington, a sincere interest in their welfare. On leaving, he was careful to shake hands with all the telegraphers at Grant's. Sheridan's operator, McCaine, alone was beyond the telegraphic circuit which continually re-echoed from Warren's to Kautz', the "all is well," so dear to a commander whose energies and abilities are so severely taxed, as were Grant's during those dark hours that preceded the dawn of his and the Nation's final triumph. Meade was no less active. He despatched suggestive messages to Grant, mandatory ones to Warren, Humphreys and others, and inquiring telegrams to all points. The efforts of the telegraphers during that night so to aid the officers that perfect order and assurance might come with the opening of the day, may never be half told. They are locked in the vault of death, to which no man alive knows the combination. There lie the two Caldwells, Tinney, Mason, Sheridan and McCaine; others are scattered, and can not be reached.

New offices were being opened and reliefs required, wherefore, by General Ord's order, Emerick reduced his force north of the James, sending Snow, McKenna and Nichols to the active front.

April 1, General Sheridan, now commanding his own troops and Warren's corps about the Forks, succeeded in defeating Lee's right, capturing nearly six thousand prisoners, and connecting with the main army by moving eastwardly. It was a great victory, and was proudly hailed by the entire North, for it opened a certain way to an open attack upon the enemy with assured success, or else he would be harassed in a general retreat, after giving up the Confederate capital. Lee's separation from Johnston was now secure. Grant spent no more over-solicitous nights. He telegraphed Wright, Parke and Ord to assault at daybreak. With dawn, the great guns opened, all along the line; the various corps moved forward, some very successful, others partially so; but when the day was done, Lee had failed to restore his lines, and was being hemmed in.

Some of the telegraph operators were at times greatly exposed this day, but all escaped unhurt. At the Rainey House, at the intersection of the Boydtown plank road, Humphrey's head-quarters office, where Rand and Rupley operated, may be said to have been on the field of battle. A battery in action was in front of the house firing over the soldiers' heads. This house was not over five hundred feet from the battle array. Humphreys sent the telegraph line guard to the rear, so as not to attract too much attention. Of course all sorts of missiles fell hereabouts. It not infrequently happens that the greatest danger is just behind the battle line, as in battle overshots are more frequent than undershots, and either may outnumber those that are rightly aimed.

Every office has a ground connection for use in case the wire breaks on either side. That at the Rainey House was a sword, which was left in its place when Rand moved to the new headquarters at ten o'clock that night. Rupley had gone forward with Humphreys early in the day.

All day Lincoln waited at City Point, the news from Sheridan, Meade, Ord and Weitzel. Grant frequently telegraphed, and every additional message brought news of fresh victories.

At about eleven o'clock A. M., Davis, the President of the Confederacy, received his first information of the approaching crisis. It recommended that arrangements be made for evacuating Richmond that night, with a view of concentrating near the Danville road. The enemy now, had but one wire route intact out of Petersburg, and that led to Richmond, but doubtless there were fortification lines, as Colonel Taylor, of Lee's staff, has written that the Federals having broken through a portion of the lines, "General Lee communicated to the authorities at Richmond his intention of evacuating his lines that night. * * During the whole day he was engaged in issuing orders and sending despatches by couriers and by telegraph, in preparation for this event. Early in the forenoon, while the telegraph operator was working his instrument at head-quarters under the supervision of the staff officer, charged with the duty of transmitting these orders, a shell came crashing through the house and the operator declared himself unable longer to work his instrument. He was ordered to detach it, and as the staff officer and the operator emerged from the house, they with difficulty escaped capture at the hands of the Federal infantry, which just then advanced up and drove away the battery of artillery, which had been placed in position around the house to assist in delaying the advance of the enemy." Lee's head-quarters, thus captured were two miles from Petersburg in Mr. Turnbull's house.

Twelve thousand prisoners were captured from Lee that day and more soldiers cut off. It was two P. M. before the disaster was publicly known in Richmond. Davis and his cabinet, except one officer, secretly escaped by train at eight. Before the public in Richmond knew that ere another night the Union troops would be in possession of that city, it is probable that Grant's first assuring messages to Lincoln had reached the great marts of trade throughout the Northern States.

Orders went forth that night for all of Lee's troops to concentrate at Amelia Court House about fifty miles west of Petersburg.

It was an anxious night throughout the National lines. Would Lee escape or compel an assault more terrible and desperate than any the army had ever yet made? For unless he fled the assault would follow. Between three and four A. M of the third, it became apparent that Lee had withdrawn, and a little after four o'clock the National troops took possession of Petersburg unopposed. As soon as word could be gotten to the nearest office the fact was telegraphed generally. Grant sent the news to Lincoln. It was too good to keep and at breakfast time in the cities, the newsboys were doubtless crying their extras. Many of the operators were able to visit the city. Schermerhorn and Laverty at Grant's could not leave, but sent an orderly in after tobacco. He soon returned with enough for all hands, for more than a month. Doran, chief of construction, now, was the busiest man in the telegraph service. It was his duty to connect telegraphically the various corps advancing to intercept and overcome Lee. Fortunately most of the poles on the South Side Road were standing and upon these he hastily strung his wire. At three P. M. of the third Grant's head-quarters were ten miles out from Petersburg and at midnight the first message to Petersburg from Grant was sent, addressed to President Lincoln who had come down on the third and visited that city. The line along the South Side railway was restored as rapidly as the troops advanced, excepting always Sheridan, who was a long way out at the start and whose troops were mostly mounted.

On the north side of the James, Weitzel was equally alert and active. At 8: 30 A. M. he entered Richmond which was in the hands of the mob and largely in flames. The roughs there were brutal and fiendish and the better classes so frightened that the advent of the hated Yankees was hailed as a protection, which indeed it proved to be. Add to riot, burglary, to burglary, arson, to arson, murder, to murder that which is more shameful; multiply these until a city of a hundred thousand people is terror stricken and you have Richmond during the night that neither the stars and bars nor the stars and stripes represented anything there, and do you wonder that law and order in any command was hailed in quiet thanksgiving? W. K. Applebaugh accompanied Weitzel and was the first Union operator to enter that city. Passing up the main street all were compelled by the heat from the burning buildings to protect their faces with the capes of their coats.

Palmer Gulick and E. A. Eckert and their party hurriedly

constructed their line along the Osborne pike to Richmond; City Point office connected Fortress Monroe and Wilmington, Del., and the War Department, Washington, and once again, but by a new and untried route, Richmond was brought into the great telegraphic union as well as into the political union of the States. Emerick could hardly keep his operators within bounds when they heard of the fall of Richmond, they were so wild with joy; so were all—the army pressing after Lee,—the troops about Petersburg, City Point, Bermuda Hundred, and on to Richmond. A reckless, devil-may-care-now, spirit of comfortably exhilarating, of delightfully intoxicating joy, prevailed and overcame all desires for work, even in furtherance of duty, for the country *was safe now*, even if slight neglects were permitted. Nevertheless the operators remained at their posts and did all that was needful. Emerick hurried up Gulick and before night, the line to the city of Richmond was complete, but only for a short time. In that hour the country was electrified by the intelligence that the Confederate capital, now re-possessed by the Federals, was telegraphically connected with the National head-quarters. But before the circuit was completed over the Osborne road, a mounted courier, who had been to General Butler's head-quarters with Weitzel's telegram and there discovered that the operator had moved on toward Richmond, rode in hot haste to the office at Twenty-fourth Army Corps head-quarters. For four long years the Northern people had been waiting, watching, and praying, while brave ones were fighting, for the reduction of Richmond. This orderly carried the glorious news of its fall. No wonder he spurred his horse to its utmost speed.

> "Each muscle and vein
> Of his charger knots with the nervous strain,
> As with head stretched forward and streaming mane,
> It bends to the pace, its nostrils red,
> And flecks of foam on its breast and head,
> Galloping free, with the ringing sound
> Of the iron hoofs on the solid ground."

W. B. Wood, operator, was about to close his office and move on, when he espied the horseman, and thinking from his furious riding that possibly he bore important messages, Wood awaited his approach. Operators Dealey and Homan were on duty at Fortress Monroe, which office being quickly informed of the fall of Richmond, connected the War Department office direct, so that a copy of the telegram might be taken there while it was being received at the Fortress for Norfolk. Little Willie E. Kettles, a precocious boy and an expert operator in Di (War Department) office — he who had worked the line to Harpers Ferry and Winchester while an engagement was progressing at those points, and Lincoln, Stanton and Halleck were sitting by, eagerly listening to each word as it was read from the instrument, or cautiously suggesting plans of operations that the reports called forth—this boy answered, ii, ii, Di. "Sharpen for Richmond," said Dealey. Then Wood, who was five miles therefrom, called, and Di answered. For about four years Richmond had not been heard in Washington. Wood's office was almost within the limits of the former city, and Di was wild with joy. The following message then came:

RICHMOND, VA., April 3, 1865.

To BRIG. GEN. GEO. F. SHEPLEY,
Military Governor, Norfolk, Va.,
We entered Richmond at eight o'clock this morning.
(Signed) G. WEITZEL, *Brig. Gen. Comdg.*

Bodell was standing by Kettles and no sooner heard the purport of the telegram than he rushed into Eckert's room and told him. Probably no other message was ever received at the National capital, that caused such rejoicing. The surrender of Lee and Johnston created perhaps profounder joy, but not such boisterous pleasure. "I took the message," said Kettles, "the Lord only knows how, and upsetting inkstand and instrument, I hurried to Major Eckert's room, where he or another snatched it and ran to Stanton's office." Charles A. Tinker came into the operating room at the same time, and espying a friend, the only person near, on the broad ground in front, called to him and reported the glorious news, which he was fairly aching to tell somebody. While the other operators were calling up distant offices to spread the glad tidings, Secretary Stanton rushed in and snatching Kettles in his strong arms, actually held him at arm's length out of the window, to the view of the fast gathering crowd which Tinker's friend below, by his uproarious conduct, was drawing.

NEWS OF FALL OF RICHMOND.

Stanton addressed the crowd a few remarks complimentary to Kettles, and repeated the news of the downfall of the rebel Government, as the capture of Richmond was interpreted. The crowd called on Kettles for a speech, but he was too much disconcerted, and retired amid the loud huzzas of a jubilant multitude. Before Stanton, who was in his shirt-sleeves, retired, and while the tears of joy and gratitude were falling (a strange thing for that man), he said to the operators: "*Boys, I consider the telegraph my right arm, and if ever I can do anything for you, don't hesitate to ask it.*" Below, the exultant crowd was still gathering, and hats flew high in air at the inspiring intelligence. That night the Nation's heart palpitated in pride and rapturous transports. All felt positive at last that a happy ending was in the near future.

Next day the President visited Richmond, and Beckwith, who was a little behind in Grant's carriage which was brought up for the President's use, received loud huzzas from the excited populace as the elegant equipage came rolling up to the house where the President was—the house that Davis had so recently left. "Young man," said Lincoln, with a twinkle in his eye as the operator came up to report, "I am afraid you have been stealing somebody's thunder." While out walking later, Beckwith heard for the first time of the presence of Grant, who, in fact was miles away. It was Grant's thunder that Beckwith had had thrust upon him. This reminds that earlier in the war, while operators Shreffler and E. H. McGintey were off duty in Washington a couple of days, the latter donned the military clothes of his brother-in-law, a colonel and member of Congress, when Shreffler took pains to introduce Colonel McGintey to all his acquaintances.

Schermerhorn and Laverty retired about 1:30, A. M., of the fourth, but on waking about five, were minus their tent in which they had gone to bed. Torrence had come over the James, too, and while they slept soundly, he had removed the tent to his own office, which he established further on. That night Grant's head-quarters were twenty-seven miles from Petersburg, and Doren's men built a loop from the main line on the South Side railroad to head-quarters. The promptness with which the telegraph reached Grant's camp pleased him and he so expressed himself.

Messages to Lincoln, Stanton, and probably Sherman, *via* Fort Monroe, were soon on their way. The next night Laverty opened office at Burksville Junction—his partner coming up shortly after, in an ambulance. Here they remained until the afternoon of the seventh, when Torrence came up and they went on toward Farmville.

In the afternoon of the fourth, Sheridan was in front of Lee, who was at Amelia C. H. At the Jettersville telegraph office, five miles from Amelia C. H., an unsent telegram was found, signed by Lee, ordering two hundred thousand rations from Danville. Sheridan promptly sent it, probably by his operator McCaine, to Burksville, whence it was telegraphed as prepared by Lee, who must have been surprised at Sheridan's appearance in his front so quickly; but Sheridan did not get the rations. On the sixth, Lee's army was nearly surrounded. Sheridan manœuvered the troops in the battle of Sailors Creek, capturing seven thousand of the enemy. Other forces on other quarters had taken two thousand more. This was glorious news for Lincoln and the people at the North. That night, however, Lee eluded Ord in his front, and reached Farmville when Ord struck him in rear and flank. Humphrey's Second Corps checked the enemy's progress during the day, but Lee's forces gained two inconsequential victories with portions of his troops.

That night, Grant, at Farmville, called on Lee to surrender. Lee asked terms the next morning, and Grant said he insisted only upon the prisoners being disqualified for taking up arms against the government, until exchanged. This vigorous march was cruel to man and beast, but it was skillful war. The operators, less used to exercise than most others with the army, were severely taxed. They stole snatches of sleep as best they could, and pushed on. The movements were so often changed, and in such winding ways and counter directions, that a field telegraphic system was utterly impracticable, but the main line to Burksville Junction, and thence to Farmville and on toward Appomattox, was pushed vigorously. It often served for field orders. Schermerhorn was on duty at Grant's on the night of the eighth; Lee had not surrendered, nor had Grant relaxed his hold. Sheridan's cavalry again encountered the advance, this time at Appomattox. Infantry came up, and again the enemy was cut off. After a sharp little fight, a truce was called, which ended in the surrender of Lee's army, now numbering twentv-seven thousand, five hundred and sixteen.

Grant at 4: 30 P. M. of the ninth wrote a telegram to Stanton saying;

> General Lee surrendered the army of Northern Virginia this afternoon on terms proposed by myself. The accompanying additional correspondence will show the conditions fully

These allowed each officer to retain his side arms, and all were permitted to take their horses home with them. Laverty and Schermerhorn were called to General Grant's presence and the foregoing message and others were given them with directions to find the wire. They soon lost the way but quickly righting their error they reached Appomattox station to which point the wire was already repaired, and selected for their office a box car laden with bags of sand intended for defensive purposes. As soon as a ground wire was laid the operators called Petersburg office. The operator on duty there was prompt to reply. After telling the glorious news, Petersburg connected the War Department direct. Congratulatory replies from the President and Secretary of War were received at near five o'clock. So quickly was this good work accomplished, that Grant was both surprised and pleased. Within two hours after the surrender, the line was extended to Grant's at Appomattox Court House. The intoxicating news spread with lightning rapidity all over the land. Bells were rung, cannon fired, rockets sent up, bonfires built and meetings held. From that box-car at the station in the midst of the armies, had been forwarded the few words that made tears of joy gather in the eyes of millions, who now saw the clouds break entirely away, leaving a clear sky for the great future. No one doubted the messenger. No one said it was a stock-broker's ruse and no one suspected the telegraph of infidelity. There was nothing to mar the victory except the loss of the ten thousand killed, wounded and missing since the 29th of March when the movement was first begun. It was the last great sacrifice of the war; that dreadful war in which the Federals lost by disease and wounds nearly three hundred thousand men and their erring brethren buried, perhaps as many more; besides whom four hundred thousand more were wounded, but not fatally — one million American citizens — what misery ! ! !

Measures were at once taken to repair the telegraph lines leading west and south. An office was opened at Lynchburg April 16, at Danville April 21. Caldwell, chief operator, and Doren, superintendent of construction and those under them deserve, says Eckert, "great credit for the skill and energy displayed in establishing and maintaining communication with the advance of the army."

On the night of the fifteenth, as Laverty about midnight entered operator Sheridan's office at Burksville, after a walk of eighteen miles to repair a break in the line, he heard the instrument spell out those dreadful words, "President Lincoln was shot at Ford's theater." Meade's operator on duty near by, had just copied them for that officer. Wilson, operator with General Wright of the Sixth Corps was startled by the same despatch; he too copied it, his hand trembling so that he could hardly write. This was handed his partner Embree, but not a word was lisped until Meade officially announced it to his officers. At General Gibbon's request Nichols and Snow with sixteen cavalrymen restored the line from Appomattox to Lynchburg. There they found seven Confederate States operators and there on their arrival they heard the sad news. C. D. Hammond in Richmond received a message for General Ord announcing the assassination of the President. It was kept from the people all that day. The Secretary of War ordered the arrest and confinement in Libby prison, of Judge Campbell and all paroled Confederate States officers, but at Ord's solicitation the order was countermanded. On the seventeenth the news went over the parole line from near Vicksburg, the Union headquarters where J. F. Ludwig operated, to the Confederates on the Big Black, where J. E. Pettit was stationed. John Q. Mason at Cairo had received and forwarded it. F. H. Campbell at Chattanooga, first heard of the dreadful tragedy and sent it on to the East, West and South. The author in Nashville, Tenn., heard the direful news on opening his office at headquarters and stepping into an adjoining room where General

Thomas sat reading his mail, informed that grand soldier. "Is that so, Mr. Plum?" is all he could say. At Webster, W. Va., D. W. Smith half asleep heard, "The President died at 7:22 A. M.," which startled him into a wretched wakefulness. Grant and party were at Bloodgood's restaurant in Philadelphia. It was near eleven P. M. the fourteenth. Somehow great excitement preceded and foreshadowed the evil news. An operator edged his way to the General with the message announcing the fate of Lincoln. The crowd groaned, but in suffused tears and with choking hearts could hardly more than say, "Isn't it awful?" Grant's face was like marble, perhaps his lips were a little more firmly compressed, but no other change was observable as he read. Not a sound escaped his lips, but his wife sitting by, roused him from his silent, thoughtful self, long enough to reply, "The President has been assassinated." During the rest of the journey to Burlington, N. J., whither he was bound, but few words were spoken. The following telegram however indicated Stanton's further apprehensions:

WAR DEPARTMENT, WASHINGTON, April 14, 1865.

S. H. BECKWITH:

The Secretary wishes you to see that a pilot engine is sent in front of your train both in going to and coming from Burlington. Do not inform the General of this as he may consider it unnecessary. (Signed) THOS. T. ECKERT,

Assistant Secretary.

Just as General Sherman was about to enter his railroad car at Raleigh for near Durham Station, where he was to meet General Johnston to discuss terms of Johnston's capitulation, the operator at Raleigh whose office was up-stairs in the depot-building ran down and reported that he was that moment receiving a cipher message from Morehead City which the General ought to see. Being soon translated and brought to Sherman it revealed the fact that Lincoln had been assassinated. "Dreading the effect of such a message at that critical instant of time," writes Sherman in his "Memoirs," "I asked the operator if any one besides himself had seen it; he answered no. I then bade him not to reveal the contents by word or look till I came back, which I proposed to do that same afternoon." And so on throughout the Union the news spread, and the nation so recently raised in ecstasies of delight at the prospects of peace was bowed down as one parent viewing the corpse of a beloved son. The nation had by slow degrees come to look upon Lincoln as its strength and support, and now that his broad and charitable mind was most needed, it was suddenly disenthroned forever. Fully five thousand offices then in telegraphic connection responded before noon "too bad, too bad!" Thousands upon thousands of extras advised the general public, and business ceased as by a command from the Almighty. Men, women and children, ready now more than ever to do homage to Lincoln, joined in the universal requiem of sobs. Perhaps in the Providence of God, the time had come when it was necessary to temper bitterness with sorrow. The arm that pointed death at the nation's head struck down in an instant more animosities than were unedged by any one act since the wounds on Calvary. His life was likened unto that of a few of the greatest and best, but the influence of his death could be compared to that of no human being alive, dying or dead.

"Within fifteen minutes after the murder," says Detective Baker, "the wires were severed entirely round the city, excepting only a secret wire for Government uses, which leads to Old Point. I am told that, by this wire the Government reached the fortifications around Washington, first telegraphing all the way to Old Point, and then back to the outlying forts." Be this as it may, it is certain that but little of the dreadful news went North that night. Intense excitement prevailed at the Capital, and the Department wires were frequently called into requisition. The following telegrams, received by H. H. Atwater, operator at the Navy Yard, will indicate some of the activity that prevailed:

WAR DEPARTMENT, April 14, 1865.

TO COMMANDER PARKER:

An attempt has been made this P. M. to assassinate the President and Secretary of State. The parties may escape, or attempt to escape, down the Potomac. J. H. TAYLOR,

Chief of Staff.

NAVY YARD, WASHINGTON, April 14, 1865.

S. NICKERSON, a Vol. Lieut., St. Inegoes:

Send the fastest vessel you have, with the following message, to Commander Parker.

T. H. EASTMAN,
Comdr. Potomac Flotilla.

TO COMMANDER PARKER:

An attempt has, this evening, been made to assassinate the President and Secretary Seward. The President was shot through the head and Secretary Seward had his throat cut, in his own house. Both are in a very dangerous condition. No further particulars. There is great excitement here.

T. H. EASTMAN,
Lieut. Comdr. U. S. Pot. Flotilla.

The operators in and around the city were on duty all night, and the next night special orders were given to keep offices open.

WAR DEPARTMENT, April 15, 1865.

TO COMMODORE MONTGOMERY, Navy Yard:

If the military authorities arrest the murderer of the President and take him to the Yard, put him on a monitor and anchor her in the stream, with strong guard on vessel, wharf and in yard. Call on Commandant Marine Corps for guard. Have vessel immediately prepared ready to receive him any hour, day or night, with necessary instructions. He will be heavily ironed and so guarded as to prevent escape or injury to himself.

GIDEON WELLES, *Sec. Navy.*

Operator Beckwith was hurriedly sent to Port Tobacco, where he tapped the Point Lookout line, April 23, and, it is said, informed the War Department of the whereabouts of Booth, the assassin. For his services in aiding in the taking of Booth, Beckwith received five hundred dollars of the reward offered by the Government.

The best friend the military operator ever had in high office, was stricken down. Wherever he went, he not only was, but manifested himself, friendly to the telegrapher. The operators in the War Department were all on friendly terms with the President, who was not too big a man to know them, and yet a greater, this country may not have produced. To these, but especially the cipherers, whom he met most frequently, he appeared greatly inclined to sadness and melancholy, and it seemed at times that he indulged in levity on purpose to divert his mind. He had, however, naturally an exceedingly keen sense of the ridiculous and many very serious events reminded him of nonsensical and absurd sayings and anecdotes, to which he was pretty sure to give expression, whoever was present, and he apparently told his stories to the cipher operators, with as keen a relish as to more distinguished auditors. But Mr. Stanton did not appear to relish such trivial talk, evidently regarding the terribly serious business on hand as needing more dignified and too deep attention to permit it.

Among the stories related by the President to the cipherers in the War Department, are the following—the first of which I have seen substantially published elsewhere:

Hearing some newsboys crying, "Here's yer Philadelphy Inquiry," he minced their peculiarly funny pronunciation and tone, and then said to the operators, "Did I ever tell you the joke the Chicago newsboys had on me?" They replied that he had not, so he said: "A short time before my nomination, I was at Chicago attending a law suit. A photographer of that city asked me to sit for a picture, and I did so. This coarse, rough hair of mine was in a particularly bad tonsure at the time, and the picture presented me in all its fright. After my nomination, this being about the only picture of me there was, copies were struck to show those who had never seen me how I looked. The newsboys carried them around to sell, and had for their cry, "Here's yer Old Abe, 'll look better when he gets his hair combed." As related by himself, this struck the operators very humorously, and he laughed over it as heartily as if it was a good joke on somebody else, and had just occurred.

It so happened that for several days, Major Eckert had been out whenever the President came into the office. Coming in one day and finding the Major counting money at his desk, Mr. Lincoln remarked that he believed the Major never came to the office any more, except when there was money to count. Speaking of the coincidence of Eckert's always being out when the President happened to come in, reminded the Major of a story which he related to the President as follows: A certain tailor in Mansfield, O., was very stylish in dress and airy in manner.

As he was passing a shop-keeper's door one day the merchant puffed himself up and gave a long blow expressive of the inflation of the gay little tailor, who indignantly turned around to him and said, "I'll learn you not to blow when I'm passing." The shop-keeper instantly replied "And I'll teach you not to pass when I'm blowing." Mr. Lincoln said that was very good —very like a story which he had heard as follows: A man driving through the country in an open buggy was caught at night in a pouring shower of rain He was hurrying forward towards shelter as fast as possible. As he was passing a farm house, a man laboring under the effects of bad whisky thrust his head out of the window and shouted lustily, "Hulloo! Hulloo!! Hulloo!!!" The traveler stopped and asked him what he wanted. "Nothing of *you*," was the reply. "Well what in the devil do you shout hulloo! for when people are passing?" angrily asked the traveler. "Well what in the devil are you passing for when people are shouting hulloo," coolly answered the wretched inebriate.

Major Eckert asked Mr. Lincoln if the story of his interview with the complainant against General Grant was true. The story was: a growler called on the President and complained bitterly of General Grant's drunkenness. The President inquired very solicitously, if the man could tell him where the General got his liquor. The man really was very sorry but couldn't say where he did get it. The President replied that he would like very much to find out so he could get a quantity of it and send a barrel to all his Major Generals. Mr. Lincoln said he had heard the story before and it would be very good if he had said it, but he did not, and he supposed it was charged to him to give it currency. He then said the original of this story was in King George's time. Bitter complaints were made to the King against his General Wolfe in which it was charged that he was mad. "Well," said the King, "I wish he would bite some of my other Generals then."

Mr. Lincoln also told a pun he had heard, which was gotten off by somebody during the riots in New York, in which the "Paddies figured most conspicuously." Here it is: "It is said that General Kilpatrick is going to New York to quell the riot, but that his name has nothing to do with it."

On another occasion he came to the office, accompanied by a short, thick-set man, whose appearance presented quite a striking contrast to Mr. Lincoln's. "Good evening, Major Eckert," said he. "This man that I have got here with me, is Mr. Judd, of Chicago, the man that I have been having over at Prussia. I don't know that it's any compliment to him to say he is a friend of mine." "Perhaps it's a failing," Mr. Judd remarked, bowing and shaking hands with the Major, and then adding: "We hear it reported at the hotel and on the streets, that Grant has been fighting all day at Chattanooga. How is it, Major?" "I'll tell you how that is," said the President, "because I keep no secrets from you, Mr. Judd. They have made an advance down there and had a little fray, captured about two hundred of the rebels, and gained a point or two. No general engagement as yet reported, and even what has been done is not known to those who make up street reports." He then showed Mr. Judd some despatches, and the conversation turned upon the general condition of affairs—the rebel style of sending in flags of truce in the midst of an engagement, to turn attention, or, as recently done at Chattanooga, to intimidate the Federals by warning them to remove all non-combatants from the town, indicating that a fierce attack upon it was about to be made. Mr. Judd remarked to the President, that from his stand-point, five thousand miles away, with all the circumstances known there, he had often thought the Federal Government not severe enough, not retaliatory enough. "Well," replied Mr. Lincoln, "it is well to be no more severe than we have power to enforce. I have often thought of the man who had made all of his arrangements to go over and whip Billy Grimes, and when about starting was quite taken aback by being asked, 'Suppose Billy Grimes whips you?' 'I declare,' said the man, 'I never thought of that.'"

On New Year's morning, 1864, he came into the office, followed closely by Mr. Chandler, one of the telegraphers. Both noticed a woman crying in the hall, near the telegraph door. Mr. Lincoln asked Eckert, "What is that woman just outside crying about?" but Eckert did not know. "I wish you would go and see," said the kind-hearted President. Eckert learned, and reported that she had come, expecting to visit her husband in the army of the Potomac, but finding she could not go, was nearly out of funds and had her child with her, she felt very bad. The President thereupon, in his frank and awkward way, said: "Come, now, let's send her down; what do you say?" Major Eckert replied that the strict orders of the Department were, that no woman could go to the army, and suggested a leave of absence for two or three days, so that the man could come to Washington. "Well, come, let's do that. You write the message," said Mr. Lincoln. But the Major suggested that it would be better to make it official, and have Colonel Hardee, A. A. G., prepare it, to which the President replied: "Yes, yes; let Colonel Hardee do it, and let the man come up." After reading the messages on hand, he went away, having at least made one heart glad on this New Year's day, just one year from the day on which he sent forth the great proclamation of emancipation.

In the midst of the great conflict, the Government had an Indian war also on hand, but vexatious as were the great problems, Mr. Lincoln did not overlook many smaller matters. A prominent army officer reported by telegraph from the Northwest with much detail, the operations resulting in the capture of many Indians who had been guilty of barbarities. The report was very long and expensive. It gave every Indian's full name. Some of these, operator H. P. Jones at Philadelphia, who sent it to operator Rosewater at War Department, declares were a line and a half long. Mr. Lincoln saw this despatch and soon the following was sent to the officer by direction of the President: "Hereafter forward such communications by mail."

In North Carolina affairs were still on a war footing. On the 11th of April, news having come on the sixth of the successes of the Army of the Potomac, Sherman's army was again well under way, now for Raleigh. On the thirteenth, operator Anderson with Kilpatrick's cavalry entered Raleigh, where he found three Confederate States telegraphers and a good battery, but pushing right on to Morrisville he tapped the line leading to Greensboro and succeeded in obtaining valuable information by taking off a telegram to General Johnston stating that Federal General Stoneman who started from East Tennessee with cavalry was fighting at Salisbury, N. C. On the fourteenth Johnston sent word intimating a desire to surrender. Durham's Station was reached by the cavalry next day and now began negotiations in earnest looking to Johnston's capitulation. Anderson opened an office here and maintained telegraphic communication throughout the parley with Sherman, who was much of the time at Raleigh. At two of the meetings of Johnston and Sherman between the armies Anderson was a fortunate spectator. The parley continued until April 26 when Johnston agreed to surrender on the same terms made by Grant. The line from Raleigh to Durham was very useful to Sherman who thus transmitted many despatches for Johnston and received others from him. On the twenth-ninth J. C. McCutcheon, a military operator captured at Burgow, N. C., came into the Federal lines looking decidedly the worse for his experience. The thirty-six thousand eight hundred and seventeen prisoners of war who surrendered to Sherman were permitted to go home, and now the problem was to restore civil authority and general concord throughout the Southern States. When the surrender was made, the telegraph was at the very head-quarters of the officers in Goldsboro where the paroles were being prepared. Major Eck ert justly reported that "Mr. O'Brien is deserving of special notice for his energy and perseverance in establishing prompt communication by telegraph in this Department and the men under him, for their vigilance and faithful attention to the interests of the service."

The telegraph had proved of inestimable advantage in the war and was now destined to reap the last victory; the bloodless one which should restore the civil supremacy and re-unite the shattered sections of this Nation under the old constitution, reinvigorated by important amendments involving largely the results of the war. What the Military Telegraph did in this peaceful direction will be told next. It was the crowning glory of all its operations and in the more legitimate sphere of its marvelous workings. Its first telegram, "What hath God wrought?" like "Let us have peace" had filled the people with gratitude.

Ours was the first telegraphic service for army operations carried on, on a large scale. The success achieved by the telegraphers with the Union army more than any other one thing, caused every nation in Europe to embody it in one form or another, as they deemed best, as a part of their regular army service. During the four years' war over fifteen thousand miles of telegraphs were built for army purposes, which, if one continuous line, would reach from China *via* the United States and great oceans, beyond Palestine. Many miles of these lines were built under the fire of an enemy and very many more when there was constant danger of being shot from the poles or along the wayside by the numerous bushwhackers, who took special delight in breaking the telegraph, and then skulked from view of mortal like a coyote that preys by night, but hides himself in his hidden hole by day. Multiply a bushwhacker by six or more and they become dangerous as guerrillas, and it is in this form that they caused greatest danger to the builder or repairman, for they often outnumbered the unguarded builders and besides being better armed, succeeded in surprising the laborers. The service performed by these men, but comparatively few of whom I have been able to mention individually, will never be written. There is no practicable means of obtaining the facts. But the great truth remains, that in danger and out, they labored with a zeal deserving a reward which an unthanking, but not thankless nation has never vouchsafed them. Not a few lost their lives, some in battle, more were wounded and yet a greater number from malaria and other evils incident to constant out of door life, lie unmarked on the soil where hostile armies trod. They were poor and without other ambition than to advance the service they were engaged in. It was their way of fighting for their country. There was no bounty in this way of theirs; there were few promotions; there was constant necessity, and ever attendant danger. Many soldiers were detailed from the ranks to serve thus when civilians could not be obtained in sufficient numbers, thus fixing a measure of the worth of the lineman who may trot his children on his lap as the soldier does, but whose growing boy can not say "my father was in the army too," because the rolls that he was paid on were quarter-master's instead of adjutant general's.

These fifteen thousand miles of line and their operation from May 1, 1861 to June 30, 1865, cost the Government $2,655,500, but single messages have not been wanting that amply repaid this entire sum. The probable aggregate number of telegrams sent over these lines during the war, is six million, five hundred thousand, by which dividing the entire cost, we have an expense to the Government of about forty cents per message. Thus the message to Corse at Rome sent part way by telegraph and which saved Allatoona, cost about forty cents; the messages sent during the battle of Mechanicsville, which perhaps saved the Army of the Potomac, cost about forty cents each; those from the Wilderness to Appomattox calling for troops to this point and that, cost respectively about forty cents; the message that turned A. J. Smith's troops to the protection of St. Louis, cost forty cents; the telegram announcing the fall of Richmond, cost forty cents; but why reiterate? Every chapter is in itself conclusive that under the wise administration of Colonel Stager the Government was getting more for its cost *than from any other war service.*

Side by side with the "wonder working wire," often ran that other revolutionizer of war, the steam engine. Like a team well yoked, they always worked well together. There always subsisted a fraternity of feeling between the employés, due in part to the intimate relations they sustained and in part to the anomalous position of the *attachés* respectively; for the railroad man too was beyond the purview of military honors. That splendid fellow who, on the Atlanta Road, at General Sherman's request, ran his engine into the enemy's lines to draw the fire of his cannon and thus disclose their position, returning it is true, in safety after sustaining a terrible fire from many guns; and that other gallant engineer who, on the Orange & Alexandria Railroad, stopped his engine because guerrillas were firing at him, and began throwing chunks of wood at them; these and many another, whose conduct I might mention, were among the citizen defenders of the Union, *for whom the Government reserved no honors.* General McCallum and other railroad officials spent their best energies and most wearisome hours in transporting, feeding, arming and clothing the army. The long and numerous trains necessary were, under the direction of the telegraph, passing at nearly every switch, and thus only, were single tracks on long routes, able to maintain great armies hundreds of miles from their base. In this way Sherman supplied his host in Atlanta from Louisville.

Well done, thou good and faithful servants.

CHAPTER XIX.

RECONSTRUCTION.—THE CORPS DISBANDED.—ITS OFFICERS PROMOTED.—LINES TURNED OVER TO PRIVATE COMPANIES.—RECAPITULATORY STATEMENTS.—OFFICIAL PRESENTATION TO TEN MEMBERS OF THE CORPS.—DEATH OF TWO MEMBERS.—A FEW APPRECIATORY LETTERS AND OTHER EVIDENCES OF THE VALUE OF THE CORPS—CANADIAN OPERATORS.—FINALE.

Triumphs of peace are ofttimes more beneficent than those which are war-stained in human misery. Suffering is an essential in war, which not infrequently has produced no public good to either belligerent, but the victories of science and art, untinged by the glamour of heroic strife "on the red sands of the battle field," while they may not occasion ostentatious triumphs, result only in absolute good. The day of the conqueror for conquest merely, has passed; that of the philosopher, scientist and artist is supplanting it, and all cognate evils. Men may for diversion delight in Homeric tales to the end of time, but Ulysses will never supplant Haydn; Hector will not overmatch Angelo; Cæsar will not o'ertop Luther; Napoleon will not outshine Fulton; even Grant's history may be laid aside for that of Samuel Finley Breese Morse; and no man aims so high as to hope to shadow the great Antetype.

If Grant's will and combination during the war justly made him illustrious, the genius of Morse which made his strategy possible, also crowned the Union victories with a speedy national reconstruction. No sooner had the clash of arms ceased at Smithfield, N. C., than with merciful haste over strips of line then repaired, Sherman's message to General Wilson, who was killing and burning in the heart of Georgia, called a halt. The white flag was quickly raised all along the line, even west of the Mississippi, where there were yet men, who, as a *dernier ressort* dreamed of a trans-Mississippi republic; they too, soon accepted it as denoting peace.

The Federal Government, alive to the importance of the telegraph in the last crisis; the crisis of dawning peace, edicted that all lines within the States lately at war with the Government should be restored and operated, or at least controlled, by the United States Military Telegraph Corps. Hence, no sooner was the end at hand than Gilmore, about Charleston, S. C.; O'Brien, about Raleigh; Caldwell, about Richmond; Lynch, in West Virginia; VanDuzer, in the Division of the Mississippi; Clowry, in the Department of Missouri; Fuller, of the Gulf, and Gross, in Kentucky, were directed to restore communication, by the main routes, throughout the entire territory whose people were so lately in rebellion. It was a great work, that of rebuilding in many places and repairing generally, the telegraphs of the Confederacy which, for lack of funds, of help, but most of all, of material, had become almost useless. Yet it was a labor of love in which operators and other employés on either side joined with an alacrity that reflected infinite credit to those engaged. It was, moreover, destined to serve a peaceful purpose; one, of all others, its inventor most delighted in. The final victory after all was not to be one of arms, but of applied science; one of art. An art which would make marauding dangerous and guerrilla warfare impossible, for now that the Confederate armies were disbanded, the authorities feared that small bands of desperate men would make a hell of that splendid country, so well adapted to become a paradise. And so it would have been had not the tell-tale telegraph revealed their hidings, until these land pirates without a flag, became solicitous only for personal safety.

When the war ceased, the trans-Mississippi wires were badly broken south of Little Rock and west and south of Memphis. East of the Mississippi River, Canby's forces had worked great harm to the lines, while Wilson, in Georgia, Stoneman and Gillem, in Western Virginia, North and South Carolina, and Sherman, in Georgia and the Carolinas, had also wrought great destruction. To repair these injuries and open the entire country as speedily as possible, Colonel Stager now ordered all hands to the work. Accordingly, in the Gulf Department, including

Texas, (Captain Fuller, and after him, his successor, Captain Gross, who came early in July, 1865,) we note the important fact that over thirty-four hundred miles of telegraphs were put in order, including about five hundred miles of new lines. The Shreveport and Alexandria telegraph line extending from Shreveport, La., to Mansfield, La., forty miles, recently constructed by D. L. Holden for the Confederates, was not working. The Arkansas State Telegraph line, running from Camden, Ark., *via* Magnolia, Shreveport, Marshall, Jefferson, Henderson and Rusk, to Crockett, Texas, nearly three hundred miles, was found working from Shreveport to Crockett, one hundred and seventy-five miles, but between Shreveport and Camden, it was broken in many places, and so poor had the company become that all its net earnings for two years were forty thousand dollars, due it by the collapsed Confederacy. Hence, L. C. Baker, its Superintendent, who had followed it beyond its original limits farther into the Confederate wilderness, as the successive Union victories drove the enemy away, was without the necessary funds to restore the wire. It was no longer the Memphis, Little Rock and Fort Smith line, but one which C. P. Bertrand, its President, and H. A. Montgomery, its General Superintendent, had been forced to remove and relocate. Baker, who followed it to protect the company's interests, was made Military Superintendent. The Southwestern telegraph line continued the connection from Crockett to Houston, one hundred and sixty miles, under the direction of D. P. Shepard. Only about one hundred and twenty miles of this wire was in working order in fair weather. This line was likewise the result of Federal successes, for originally it ran from Vicksburg, Miss., to Camden, Ark. The Star State Telegraph, another line, connected Houston, Texas, with Galveston and Liberty, one hundred miles, under the management of D. L. Davis, after C. C. Clute went to Mexico, just before the final collapse. It was built of different sized wire, and would not bear the necessary strain between poles. Louis B. Spellman was sent to General Granger's Military District, head-quarters at Galveston, and S. L. Robinson to General Herron's, at Shreveport, as assistant managers, to oversee the restoration and workings of these several wires. In their efforts they were greatly aided by the respective superintendents and operators found on the lines, all of whom took the oath of allegiance, and were glad indeed to work for "greenback" money, as heretofore their salary in Confederate scrip would not pay for their board. About July 1, these lines were in workable condition. Besides these and lines connecting New Orleans, new ones were built by order of General Sheridan, to connect him with the distant parts of his command, viz., from Hempstead, Texas, to San Antonio, two hundred and five miles; from Brazos Santiago, Texas, to Brownsville, thirty-five miles; and from Vicksburg, Miss., to Shreveport, La., two hundred and ten miles. Among the operators found on the old lines and employed by the Government, were Wm. Davis and E. Brooks, at Galveston; W. A. Wherry and F. V. Daniel, at Crockett; G. M. Baker, at Shreveport; Philip H. Fall, Lee S. Daniel and M. Harris, at Houston, and Frank Thompson at Anderson. Federal operator A. W. O'Neal, at a later date, came from Alabama, and served in the Rio Grande District very acceptably as assistant manager, and John R. Dixon, at Austin, and Louis H. Korty, at Shreveport, and Edward P. Whitford, at San Antonio, acted as cipherers.

In the States of Arkansas, Kansas and Missouri, no new telegraph territory was opened within Clowry's district, and hence his seventeen hundred and two miles of line answered every purpose, connecting with St. Louis, beside many intermediate points, the distant ones of St. Joseph, Kansas City, Fort Scott, Fort Smith and Little Rock. Thus the territory west of the great river was more perfectly than ever before brought within the jurisdiction of the telegraph.

The Cairo and Memphis wires were put in order once more, and Captain VanDuzer, at Nashville, started numerous parties in three general directions, to meet Fuller's men from the south-west, Eckert's from the east, and Gilmore's from the south-east. VanDuzer's three thousand miles and over of telegraphs were soon added to Fuller's and Clowry's. Gilmore's territory was extended westward to Montgomery, Ala., and northward to Atlanta, Ga., and included South Carolina and Florida.

We have noted in a previous chapter the completion of a line from Mobile, *via* Clinton, to Vicksburg, and from Decatur, Ala., and Corinth, Miss., to Memphis, Tenn. At Montgomery, Ala.,

J. J. Egan was doing his best early in May, to restore that line to Mobile. Confederate States operators Hughes, Reed, Thornton and Mingle were there, also anxious to take the oath and assist. M. H. Bassett was sent to Selma to work from there. Superintendent C. G. Merriwether co-operated with Captain Fuller's assistant, A. W. O'Neal, effectively about Montgomery, as Manager Sandford did with Smith at Mobile, and Superintendent J. H. Presley and Ed. Burke about Aberdeen. Montgomery was, in May, in communication with Macon in the East, points on the Memphis & Charleston Railroad on the north, Vicksburg on the west, and Mobile to the south, and scores of intermediate places. John F. Ludwig, R. B. Lines and Frank Benner at various times, at Jackson, Miss., C. W. Pearson, at Columbus, C. W. Moore, at Mobile, C. A. Keefer, at New Orleans, W. H. Munro, at Baton Rouge, T. J. Rodgers, at Grenada, acted as cipherers.

It was at Montgomery that General A. J. Smith made his head-quarters, and from there he distributed troops to many towns. Among the Southern operators employed hereabouts were, at Montgomery, Barney Hughes, James G. Thornton, John K. Mingle, Jeremiah L. Reed, Samuel E. Holt, Milliage Hoffman, Alfred F. Brannon ; at Uniontown, J. R. Adam ; Demopolis, R. D. Hoot ; Selma, W. M. Nettles ; Talledega, C. E. Lacey ; Greenville, James D. Porter ; Evergreen, W. P. Jones ; Pollard, F. N. McMullen ; Columbus, W. T. Aude ; Jackson, E. L. Marchant, A. G. M. Russell and John Galbraith ; Meridian, E. Howard, G. Heiss, G. W. McMurchy and W. H. Cody ; Shubuta, W. H. Patton ; Mobile, J. Bohanna, Taylor Adams, R. B. Weeden, W. Henley, H. B. Taylor, Dr. John C. Thomas ; Blakley, J. M. Worden; Osyka, J. T. Alleyn and George B. Hennesy; Lauderdale, Ed. C. Burke; Gainsville Junc., Ed. H. Hogshead; Columbus, W. E. Flippen; Brandon, Nat. Flippen; Water Valley, N. C. David; Oxford, B. W. Collier ; Holly Springs, B. H. Lucas; Panola, W. M. Marshall ; Senatobia, J. F. Richardson ; Clinton, Miss., Nat. Walton ; Hazelhurst, W. H. Lanbright ; Brook Haven, C. B. Harvey ; Summit, Chas. St. John ; Amite, H. C. Davis ; Clinton, La., J. E. Mansker ; Woodville, W. H. Chisholm ; Columbus, Miss., S. E. Smith ; New Orleans, P. Molony and J. L. Adam ; Enterprise, S. C. Thielgard and W. O. Kelly ;

Citronelle, C. I. Depew; Buckatuna, W. H. Young; Gainsville, W. H. Bush; Macon, Miss., J. R. Coburn; Okolona, J. R. Coburn and —— Emanuel; West Point, Miss., F. W. Mitchell; Lake, A. Grimes; Morten, A. M. Record; Canton, G. W. Beard; Goodman, L. C. Presley; Durant, E. J. Marshall; Vaiden, J. T. Poindexter; Grenada, T. F. Marshall and James H. Henderson.

In the Augusta district, J. A. Brenner superintendent, was especially active and his corps of operators found more line building to attend to than ever before. Among these were Southern telegraphers, W. H. Turner, chief operator in Augusta, T. L. Brown, A. Tweedy, W. J. Evans, V. F. Campbell, J. K. Freeman and E. O. Martin of the same city. The latter took charge of builders on the Savannah route—M. H. Callum Columbia, S. C., and S. P. Evans, of Florence, S. C. offices. J. R. Ligon rebuilt the line towards Charleston. E. M. Burch operated at Sumter, S. C. C. T. Mason acted as assistant superintendent, rebuilding lines on Wilmington and Manchester road between Kingsville and Wilmington and Florence and Charleston. S. C. Dodge was at a way office between Augusta and Columbia. E. R. Dodge and P. Sledge at Milton. G. B. Huyler at No. 10 Central Railroad, a test station between Macon and Augusta. S. D. Mundle at Marion, S. C.; J. R. W. Johnston at Aiken, S. C.; A. Fulda, G. H. King, W. R. Johnson at Winsboro, S. C.; D. B. Northrup at Opelika; J. D. Potts at Wilmington; E. K. Wilson at West Point; G. Winter at LaGrange; J. H. Wade at Newman; J. B. Aveithe at Winnsboro, Ga.; P. H. Ward at Marion, S. C.; C. Y. Brennecke at Chester, S. C.; J. T. Quarles at Pocotaligo; and T. M. Bryan at Darlington. In North Carolina F. W. Kennan, W. E. Dulin, S. W. Whitaker, Lee Angel, John W. Brown, N. B. Topping, N. A. Lee, Arthur Daniel, S. A. Howard, J. C. Duncan and J. E. Sheppard, also Southern operators, were, under Richard O'Brien, rendering such aid as lay in their power. I am well aware that, chiefly for lack of the facts, many able Southern telegraphers are not named above; that not a few of them were men of rare worth; such, for example, as John VanHorn, J. B. Tree, George Trabue, D. Flanery, John B. Morris and Thomas Johnson in the West, J. W. Kates, Joseph McGovern, Charles

A. Gaston, E. T. Smith, W. R. Cathcart and William Barr who during the war were in the East.

O'Brien and Caldwell, Eckert's assistants, were not behind hand with their two thousand miles of wire. H. B. Plant and R. B. Bullock of the Southern Express Company that owned about four hundred miles of line and Dr. Norvin Green of the "Southwestern" rendered every assistance in their power. Lynch's lines were intact. Gilmore brought his twenty-five hundred miles of telegraph promptly into the great network and the month of May that terminated the war, did not end until from Houston, Tex., to Washington; from St. Augustine, Fla., to St. Louis, Mo.; from Louisville, Ky., to Mobile, Ala., the territory was enwebbed by a network of wires operated by an able corps of telegraphers, thus making the restoration of peace seem more real, than it were otherwise possible.

Thus once more the whole country was brought again to the National Capital and to every newspaper office in the land. Relatives and friends, unwilling to await the reconstruction of the railroads and re-establishment of the mails, had recourse to the wire. It was the first time in four years; and in that period what untoward events might not have occurred. The work of the civilian authorities was now paramount. The autonomy of the States instead of being curtailed was to be enlarged by the autonomy of the individual. Military governments arose and in apt time were displaced by those of the States, rehabilitated to their *ante bellum* privileges. These things and their incidents occasioned a vast amount of telegraphing throughout the territory whose people were so lately in revolt. A free interchange of opinions could only be possible by the use of cipher keys, wherefore the cipherers were useful to the end.

With the policy of the Government in reconstructing the States, the telegraph had nothing more to do than with the plans for the conduct of the war. Its mission was an agency merely. But it was that kind of an agent that brought the power of the Government speedily to all disturbed points with happiest results.

By direction of the Secretary of War, Union operators were placed in the principal offices and exclusively handled the cipher keys. J. H. Emerick in Richmond, J. H. Nichols in Petersburg, George C. Maynard and Frank Stewart in Washington, C. A. Homan in Wilmington, N. C., George Henderson in Newberne, W. J. Dealey at Fort Monroe and W. F. Holloway in Norfolk were among those who served as cipherers in the East. The unimportant offices were filled by Southern operators who took the oath. In cases where the Government did not require the exclusive use of lines, their legitimate and usual business was not interfered with, but all Government despatches took precedence. Managers of private lines were authorized to collect tolls and required to pay most of the operators and operating expenses.

The Federal control of all the telegraphs in the South continued until December 1, 1865. The Secretary of War having directed that all Southern telegraph lines belonging to commercial telegraph companies be turned over to the respective companies owning the same, those companies assumed control of their properties December 1, but the Assistant Superintendents under the direction of Colonel Stager, continued on duty at their respective posts, in charge of all military telegraph property in their respective departments and continued to exercise a supervision of the business, prohibiting the transmission of anything improper, so far as the Government was concerned and held themselves ready to resume entire control of the lines at any time they might be directed so to do. Consequently on the 30th of November, all operators not employed upon strictly military lines and not retained at chief cities as cipherers were discharged. One by one the cipherers left the service or were discharged until finally but one office remained; that was presided over by B. R. Bates (1869) at the stairway landing in a hall of the War Department, where Colonel Scott had in 1861 opened the first military telegraph office.

The following official showing of the condition of affairs when the Military Telegraph was substantially wound up, is taken from General Stager's last annual report for the fiscal year ending June 30, 1866:

My report for the fiscal year 1865 was forwarded to your Department September 15, 1865, accompanied by the reports of the several officers then on duty in this service, and acting under my immediate orders. During the past year, I have continued to be stationed at Cleveland, O.: as chief of United States Military Telegraphs; issuing from time to time, to the officers serving under me such directions and orders for the guidance and maintenance of the Military Telegraph, as the operations of the same within the several Departments, demanded or the exigencies of the military situation required. The decisive and glorious result of the brilliant campaign of General U. S. Grant before the city of Richmond and the crushing blow the rebellion received by the gallant and remarkable march of Lieutenant General W. T. Sherman, from Atlanta to the seaside, enabled the Government to immediately commence a retrenchment of expenditures incident to the military operations in nearly all branches of the public service. The Military Telegraph however was an exception, to a great extent, to the immediate application of this principle, so desirable to all, and important to the public welfare. That the rebellion had received its death-blow, was undeniable, but the important and momentous question of the restoration of the rebellious States, was yet to be considered and decided upon. All commercial lines of telegraph, within the Southern States, that had been usurped by and were in the possession of, the rebel authorities, were taken possession of by the United States in the latter part of the fiscal year 1865. So long as these States continued under United States military control, and until their civil governments had been restored, to a certain extent the lines of telegraph throughout such States remained, by order of the Secretary of War under the direction and control of the officers of the United States Military Telegraph.

On the 1st day of December, 1865, by direction of the Secretary of War, the commercial telegraph lines above referred to, were restored, with wise restrictions, to the telegraph companies claiming and establishing ownership thereto. It was thought expedient however, to retain in certain localities expert telegraphers as cipherers, through whom the Government would at all times have a reliable, rapid and confidential medium of communication with its officers and authorized agents. By order of the Quarter-master General, under date of February 27, 1866, all United States Military Telegraph lines and appurtenances within the States south of the Ohio River, were turned over to the several commercial telegraph companies owning the "telegraph patent right" within such territory. This was in consideration of the relinquishment by the telegraph companies of "all claims against the United States for the use of their patent," for the use of their lines preceding their final restoration, and for all losses sustained by said companies by the exclusion of commercial business from those portions of their lines, which have been in possession of the United States. Under the provisions of the order of the Quarter-master General above alluded to, such United States Military Telegraph lines as were in operation north of the Ohio River, were sold from time to time to the commercial telegraph companies owning lines in the States in which the Military Telegraph lines were constructed, and the proceeds arising from said sales properly accounted for, by the officers under whose direction the property was disposed of. In further obedience to the order of the Quarter-master General previously referred to, as soon as the Military Telegraph lines under charge of the several officers, had been disposed of, the officers turned over to the Quarter-master's Department, all other public property then remaining in their hands, and reported to the Quarter-master General, the complete execution of his order. The Quarter-master General had no further duties to assign to the officers, thus reporting. An order issued by the Secretary of War, under date of Washington, D. C., May 28, 1866, honorably mustered out of the service the following officers, taking effect upon the date following each name, respectively:

Brevet Lieut. Col. R. C. Clowry, Capt., A. Q. M., May 31, 1866.
Brevet Lieut. Col. S. G. Lynch, Capt., A. Q. M., May 31, 1866.
Brevet Lieut. Col. W. L. Gross, Capt., A. Q. M., June 5, 1866.
Brevet Lieut. Col. J. C. VanDuzer, Capt., A. Q. M., June 5, 1866.
Brevet Lieut. Col. J. R. Gilmore, Capt., A. Q. M., June 5, 1866.
Brevet Capt. L. Whitney,* Capt., A. Q. M., May 31, 1866.

At the close of the fiscal year 1866, the only officers remaining in the United States service connected with Military Telegraphs, were the chief of this Corps on duty at these head-quarters, and Brevet Brigadier General Thos. T. Eckert, on duty at the War Department, Washington, D. C. I have the gratification and honor to state, that all of the officers connected with United States Military Telegraph during this fiscal year, received from the President of the United States, for meritorious service, promotion of two or more grades of brevet rank, as follows:

Colonel A. Stager, A. D. C., A. Q. M. to Brevet Brigadier General.
Major T. T. Eckert, A. D. C., A. Q. M. to Brevet Brigadier General.

*Note.—Whitney's head-quarters were in Washington, where he rendered Major Eckert very valuable aid.

Captain S. G. Lynch, A. Q. M. to Brevet Lieutenant Colonel.
Captain W. L. Gross, A. Q. M to Brevet Lieutenant Colonel.
Captain J. C. VanDuzer, A. Q. M. to Brevet Lieutenant Colonel.
Captain R. C. Clowry, A. Q. M. to Brevet Lieutenant Colonel.
Captain J. R. Gilmore, A. Q. M. to Brevet Lieutenant Colonel.

The following tables show the amount of money for which I have been accountable during the fiscal year 1866:

On Hand July 1, 1865.	
Received from officers,	$ 12,561.10
Received from U. S. Treasury,	563,900.00
Received from other sources,	3,737.00
Total,	$580,198.10
Expended During the Year.	
Transferred to officers,	$567,637.00
On hand June 30, 1866,	12,561.10
Total,	$580,198.10

I have the honor to submit the following *résumé*, showing the number of miles of United States Military Telegraph lines constructed yearly during the continuance of the rebellion—Total number of miles constructed—number of miles in operation each year—the amount of funds applied yearly to military telegraphs, and the total amount of funds furnished by the United States for the operation and maintenance of United States Military Telegraphs from May 1, 1861 to June 30, 1866:

	LAND LINES.	SUBMARINE.
No. miles constructed up to June 30 1862,	3,508	63
No. of miles constructed during fiscal year 1863,	1,755	
No. of miles constructed during fiscal year 1864,	3,692	15¾
No. of miles constructed during fiscal year 1865,	3,246	68¾
No. of miles constructed during fiscal year 1866,	2,010	30½
Total No. of both classes,		14,389
Addition of temporary field lines constructed during war,		1,000
Total No. of miles U. S. Military Telegraphs, field, land and submarine, lines constructed,		15,389

No of miles of United States Military Telegraph lines in operation during

	LAND.	SUBMARINE.
The fiscal year 1862,	3,658	63
The fiscal year 1863,	5,798	28
The fiscal year 1864,	6,911	55½
The fiscal year 1865,	8,501	121½
The fiscal year 1866,	8,253	81

In addition to the eight thousand, six hundred twenty-two and one-half miles of Military Telegraph lines in operation during the year 1865, there were about five thousand miles of commercial telegraph lines within the Southern States under the control and supervision of the officers of the United States Military Telegraph.

Amount of money received from the U. S. Treasury up to June 30, 1862,	$ 271,500.00
For the fiscal year ending June 30, 1863,	418,000.00
For the fiscal year ending June 30, 1864,	606,000.00
For the fiscal year ending June 30, 1865,	1,360,000.00
For the fiscal year ending June 30, 1866,	563,900.00
Total received and expended or transferred by me as chief of U. S. Military Telegraphs,	$3,219,400.00

The only official recognition ever vouchsafed members of the Corps, other than by mere words of appreciation, occurred shortly after the war, and was confined to a few of the most prominent in the East, as the following letter of presentation, written by General Eckert, who had been appointed Assistant Secretary of War, will show:

OFFICE UNITED STATES MILITARY TELEGRAPH.
War Department, July 31, 1866.

D. H. BATES, Assistant Manager Department of the Potomac; CHARLES A. TINKER, Chief Operator War Department; ALBERT B. CHANDLER, Cipher and Disbursing Clerk, War Department; A. H. CALDWELL, Chief Operator Army of the Potomac; DENNIS DOREN, Superintendent of Construction, Department of the Potomac; FRANK STEWART, Cipher Clerk, War Department; GEORGE W. BALDWIN, War Department; RICHARD O'BRIEN, Chief Operator Department North Carolina; GEORGE D. SHELDON, Chief Operator Fortress Monroe, Va.; M. V. B. BUELL, Chief Operator Army of the James—

GENTLEMEN: I have been instructed by the Secretary of War, to present to each of you one of the silver watches which were purchased and used to establish uniform time in the Army of the Potomac, marked "U. S. Military Telegraph," as an acknowledgement of the meritorious and valuable services you have rendered to the Government during the war, while under my direction as an employé of the United States Military Telegraph. It gives me great pleasure to comply with these instructions, and I will take this occasion to thank you for myself for your faithful performance of the important trusts which have been confided to you in the various capacities in which you have served, and especially as cipher operators. Yours, very truly,

THOS. T. ECKERT, *Asst. Sec. of War,*
and Superintendent U. S. Military Telegraph.

It is to be regretted that in this closing chapter there should be occasion to refer to further misfortunes. John A. Sheridan, a noble-hearted young man, died at Burksville Junction, Va., in the summer of 1865, from congestive chills and gross neglect by the post surgeon, who, although often sent for did not call on the sick man once during the day he was attacked. Toward evening, Sheridan became unconscious, and at 7:30, P. M., died. His brother operators caused his remains to be embalmed and sent home to his widowed mother, in charge of operator L. M. Painter, who was with him when he died.

A. Harper Caldwell entered this service in August, 1861, in charge of General McClellan's head-quarters office, and continued from that time, to the end of the war, chief operator of the Army of the Potomac, accompanying it on every march, and serving during every siege. On the 1st of September, 1866, he left the Department of the Potomac for General Sheridan's head-quarters in New Orleans, where he remained as cipher operator until his death, November 25, 1866. At a meeting of the operators in the District of Columbia, immediately held, proper resolutions were passed and many telegrams of sympathy and condolence from operators already scattered, but who felt that they must give expression to their sorrow, were read. His remains were sent to Zanesville, Ohio, for interment.

In an appendix we have noted the names of many of the military operators who have since gone to that "undiscovered country from whose bourne no traveler returns." Some of them died, fighting an enemy more terrible than any human foe—that dread disease, yellow fever, that almost depopulated Memphis and other river cities, and spread terror throughout the South, and caused deepest solicitude to pervade the United States. In such times, when the carrying of mails is prohibited and only medicines and sanitary goods are conveyed, the telegraph becomes the only messenger from the stricken parts to an anxious public. In such emergencies, ex-Union operators and Southern telegraphers have stood side by side at their respective posts, receiving and transmitting the dread news of fresh burials and the urgent demands of a dying people. If it were within the province of this work to detail the intrepid conduct, the heroic self-denial, the extraordinary disregard of self during those great public calamities, to the end that humanity might be subserved, it would afford the author great pleasure to chronicle those splendid efforts that cost so dearly—that laid John D. Tinney and Edmund W. Barnes, Alvah S. Hawkins and Hugh Irvine, and scores of others, many of them volunteers, in martyr graves.

The Corps lost by death at the hands of the enemy — not counting soldiers temporarily employed to assist or protect—at least eight members; many more were wounded; probably twenty-five operators died in the army, and very many more from diseases there contracted, while yet others are to this day suffering from the exposures which they underwent; seventy-one others were captured, many of whom were imprisoned in Southern hostelries of death. About one in twelve of the operators engaged in the service were killed, wounded or captured or died in the service from its exposures. If these losses, the risks and the service rendered, do not entitle the Corps to military honors and Governmental recognition, it is difficult to understand wherein the Signal *corps d' armee* and staff officers obtained such preference. On the Peninsula especially, and other fields that might be named where the Signal Corps could not operate effectively, the field operator and his orderlies performed better staff service than it would have been possible for rider and horse alone to do. I have seen fit to address five or six of the Union Generals for some word evidencing their estimate of the War Telegraph Service and received the following replies:

HEAD-QUARTERS MILITARY DIVISION OF THE MISSOURI,
CHICAGO, September, 24, 1879.

W. R. PLUM, ESQ., Chicago, Ills.:

MY DEAR SIR:—Your letter of the thirteenth instant is at hand, and I have carefully considered its contents. I fully agree with you that the duties assigned to tne Military Telegraph operators during the war, were, as a general thing, well performed, and the men were almost universally trustworthy. In my own experience, I found them invariably active, brave and honorable; and I am glad to know that some one who is cognizant of their work, and who can speak from personal experience, proposes to write their military history. I have no doubt, but that individual acts of heroism can be found, that will show them to the public in a new and surprising light, and place them, and their work, on a higher plane than they hitherto have held in the estimation of the loyal citizens of the country. I am, dear sir, truly yours,

P. H. SHERIDAN, *Lieutenant General.*

SAINT PAUL, February 6, 1879.

MY DEAR SIR:—Upon my return from Washington to-day I find your letter of October 1 unanswered. I have not by me the memorandum to enable me to give the dates of the marches and battles of the Price raid campaign. My official report on file at Washington will give all and if you can get a copy of that or get access to it you can supply all dates. It seems to me that I can not write anything that will be interesting to the outside world upon the "Military Telegraph." All officers commanding in chief in the field of course testify alike to its inestimable value. The location and movements of the hostile army were determined by this means almost as by magic when the wire was cut; and when it was not cut the knowledge of all officers concerning the enemy wherever such officers might be, became at once through this means the knowledge of the commander in the field. On many instances Armstrong cut the wires at points remote from all telegraphic offices and got information and orders from points hundreds of miles away, and from Department Head-quarters, the movements of Rebel and Union troops in all portions of the Department, and to those of us who had their field operators, it seemed a mystery how campaigns on a large scale, had ever been conducted without their aid. There can be no doubt that in the late war, thousands of lives and millions of treasure were saved by the field telegraph operators. So far as my observation extended they were as a class brave, energetic and faithful young men who deserve well of their country. Truly yours, JOHN B. SANBORN,

Late Bvt. Major Genl. Vol., Comd. in the Field.

HARTFORD, CONN., Feb. 20, 1879.

MY DEAR SIR:—I have received your letter of the twelfth inst. The duties of the telegraph operators who accompanied armies in the field during the late war, were so well performed, and the men themselves were so modest and unobtrusive, that their merits have not received sufficient notice from the generals with whom they served. I know no class of men in the army who were more faithful and energetic. The fact that they were not military men who depended upon proper notice from their commanding generals for their reputation and promotion, is probably the reason for this neglect. In my own case, I testify that I always found them alert, intelligent and courageous, and it gives me pleasure to certify to this fact, even at this late day. I do not now recollect any incidents that would be of interest in your forth-coming work. But I recall that it was always a surprise to me, and a very pleasant one, too, to learn, often within an hour or two after getting into a camp after a long day's march, that telegraphic communication was open with head-quarters, several days' march in rear, thus saving the lives and time of couriers and horses, and giving a comfortable feeling of security, which would otherwise have been wanting, as well as enabling the Commanding General to receive and impart important information. The telegraph operators of that time remind me of a first-class military small arm. The man who uses it is not particularly struck with its merits while he has it, but deprive him of it or give him an inferior one, and he immediately appreciates the merits of the one he has lost. Very truly yours,

W. B. FRANKLIN.

UNITED STATES SENATE CHAMBER,
Washington, March 3, 1879.

MY DEAR SIR:—Your kind letter of the twenty-fourth inst. is at hand. I have no hesitation in saying, that the "Telegraph Corps" was of infinite service during the late war, and I am free to say that I never knew a body of men who possessed more integrity, industry and efficiency than the operators with whom I was thrown. I wish you great success in your work. Am glad to hear such good accounts from Jaques. He is a nice fellow.

Yours truly, A. E. BURNSIDE.

STATE OF NEW JERSEY, EXECUTIVE DEPARTMENT.
Trenton, March 8, 1879.

MY DEAR SIR:—Your letter of the twenty-fourth ulto has reached me. I am very glad to learn that a competent person has undertaken the task of doing justice to the services of the Military Telegraph operators during the war, and I share with you your feeling of surprise and regret that the work has not already been done. I do not think that any one appreciates more highly than I do, the value of those services, and the loyal and invaluable devotion so constantly displayed by the men of whom Caldwell was so excellent an example. If I can at any time be of service to you in carrying out your plan, please let me know.

In great haste. Very truly yours,

GEO. B. McCLELLAN.

Shortly before their deaths General Joseph E. Hooker remarked that the Corps "had never received proper recognition at the hands of the Government," and he was glad its history was being written, and General George H. Thomas wrote incidentally: "I feel deeply indebted to the telegraph operators of my late department, for the efficient service they rendered me in the campaigns we passed through."

Secretary Stanton at one time actually spread a splendid lunch in the War Department for his operators, and presided at the repast. General Halleck, Major Eckert and other prominent officers also sat down. In an official report Stanton said:

The Military Telegraph, under the general direction of Colonel Stager and Major Eckert, has been of inestimable value to the service, and no corps has surpassed it.

In our recent memorable struggle for the preservation of free Government, spoke General N. P. Banks at the Morse Memorial reception, the telegraph performed an important and patriotic part. * * It was the constant telegraphic communication from the Government and army, that re-assured the people, bound the loyal States together and stimulated civil and military authorities to greater exertion, by the almost limitless contributions of men and money to the cause of the Union. * * It was in truth an electric nerve that united them, consolidated their power, inspired them with courage and hope and *finally led them to victory.* It was the trusted agent in all great movements, civil or military; the foundation of every triumph, partial or general. Thirty million people were united by new and unaccustomed ties; electric chains of celestial fire, flashes of etherial and supernal light, touched, thrilled, instructed, fortified, illumined every soul. It was the chord of the national heart; overburdened by sorrows, it might have been broken, but with this support it was always ready for greater sacrifices, and beat with higher and nobler aspirations for the cause of universal liberty. The humblest citizen of the republic will remember it as the instrument that, morning, noon and night, renewed the determination and courage of the people, and every soldier will gladly attest to the great aid it gave the defenders of the nation from the Satanic assaults on Sumter to the surrender of the Confederate armies to General Grant, which ended the war.

Comte De Paris, whose history of the war so far as finished, is undoubtedly one of the best ever published, although a *foreigner*, is the only historian who attempts fitly to notice the telegraph service. He says:*

A single example will show the importance of the Military Telegraph. Without counting the lines already in existence, of which possession was taken, the employés of the Government constructed five thousand, two hundred kilometers during a single year of the war, and they forwarded nearly one million, eight hundred thousand despatches; and sufferings and dangers were not spared those men whose merit was the greater, in that it was less conspicuous. More than one among them, shivering with fever in an unhealthy station, lay down with his ear against the instrument to write with a trembling hand under dictation, some important despatches whose secret he would confide to no one. Many paid with their lives for their boldness in setting up their instruments under the very fire of the enemy, and one fact almost incredible, bears testimony to the dangers to which they were thus exposed. During the siege of Charleston, the wire which connected the besieging batteries ran so close to the rifle pits of the Confederate skirmishers, that it was frequently cut by their balls.

It would be at least unkind not to notice before closing, the fact that many of the Union operators came from Canada and other northern provinces, and in all cases performed their duties with a fidelity that perhaps should have been expected only of those who jeopard their lives for their own country. The names of such now recalled, are: Edward Conway, Frank Benner,

*History of the Civil War in America, Vol. I, 280.

Frank Drummond, M. J. McCrickett, W. H. Glasscott, A. W. Howell, R. M. Howell, William B. Somerville, John Nicholson, J. M. Nye, William Forster, J. Blair, John H. Bracken, Hugh Craig, A. B. Ellison, Geo. N. Norton, Charles A. Keefer, W. H. Merkley, Edwin Peel and George Railton.

War ended, and reconstruction, despite its errors and consequent crimes, became an accomplished fact. The Union once more complete, strange as it may seem, is the cherished condition of the whole people. The telegraph may not now run to the war office; what matters it? Its twelve thousand offices and twenty-five thousand operators in North America continue to disseminate information and further the interests of mankind. The armed legions so lately massed are scattered forever; the right has prevailed; and the forces now surviving are mainly pursuing peaceful vocations. Soon they even will be mustered out of that service. But the telegraph will be left, and under the guidance of others than those of whom we have written, it will continue conquering peace and territory until every continent and every hamlet thereon will have its "wonder working wire," that great missionary of the Most High, which even now counts almost a million miles of line and probably twenty-five thousand stations. So long as men read and admire its operations during the late war, they will rejoice in the skill and fidelity, vigilance and energy, aid and devotion, exercised and displayed by the officers and men of the Military Telegraph Corps. Its achievements, silently performed, were as silently accepted. Its conquests were unheralded; its victories found voice in no general order from department or army head-quarters, and rhetoric and metaphor lent none of their charms or graces to sound abroad its bloodless conquests. Its slender swaying iron thread stretching over plain, winding through the silent forest, creeping over the rugged mountain, leaping across the oozing swamp and bayou, stealing beneath the turbid river, along the busy railroad, noiseless, yet always speaking, silent, yet eloquent, mysterious, yet ever obedient, bore alike the stern military order, the first word of victory to an anxious, waiting nation, or the sad, sad message of defeat or death. Denied recognition commensurate with its work, the Corps without organization, except such as was possible with volunteer civilians, quietly, but none the less surely, wrought out the mighty problem—the annihilation of time and space in war—and modified the whole character of the conflict and all subsequent wars the civilized world over. While the war progressed the pulse of the people was by it daily taken. If at times of frequent disaster the beats ran high, the hand that recorded the throbs was as steady as when fear and heart-ache gave way to loud acclaim. It produced a concentrated power that caused the world to marvel; it made possible co-operative blows with such fatal power as to seem like divine command. Half a million of men were in the palm of the hand of a master commander. It was an Ægis, which enabled him from City Point to protect the entire border, thousands of miles in extent, for it guarded the armies in the field and proved a faithful sentinel all along the line. It re-assured the private soldier and civilian and all the military and civic functionaries of the Union States. It protected great cities and immense storehouses from the fire-brand of secret emissaries. It prevented many seizures upon the high seas. It even thwarted the purposes of some to seize vessels and make privateers of them, and it intercepted others upon the ocean. It captured traitors in the North at critical moments, and proved the dreaded enemy of secret societies, organized with hostile designs. It shielded secrets upon which hung the successes of the armies, for it insulated them from prying eyes and spying ears. It reduced the cost of the war by many millions. It shortened its duration by many months, perhaps years, and it saved thousands of human lives. Finally, it spread, as upon angels' wings, the tidings of returning peace, and was the confidential messenger in re-uniting a suffering people. It had sounded the tocsin call to arms, and now re-echoed the benediction upon the dismissed armies assembled in final review.

APPENDIX.

Annual Reports of Anson Stager to the Quarter-master General.

Washington, D. C., February, 1862.

General M. C. Meigs, Q. M. G.:

General: The cable will be shipped from New York on Thursday, of this week, to Fort Monroe, and if no unforseen delay or accident occurs, it will be laid in ten days after it reaches the Fortress. The only cable that could be obtained on so brief a notice, was the light one to be manufactured for this purpose. It may prove too light for the service, and I would urge upon the Government the purchase of a large cable that will resist anchors. The light cable may work for a few months.

In the latter part of December, a written proposal was made to the Secretary of War, by N. A. Zabriskie, for building a line to Fort Monroe, to cost thirty-five thousand dollars. The Secretary believed the work should be done by parties already employed by Government, and requested me to make an estimate of the cost. I stated that the work could be done in a substantial manner at about twenty thousand dollars, using a cable costing three hundred dollars per mile. It was decided to construct the line with the least possible delay. The work was commenced about the 15th of January. The land line, from Wilmington, Del., to Cape Charles, will be completed and in operation the last of this or first of next week.

Including the line to Lewes, Del., which is not included in the foregoing reports, I found about twelve hundred miles of line had been constructed by Government, and one hundred and eight offices opened and maintained, employing one hundred and sixty-three operators and fifty-one builders and messengers.

The bills for material, construction and service had been paid in most instances by the telegraph companies, and the labor and responsibility of settling these claims devolved upon me. I carefully investigated all charges, and believe those that have been settled are just and proper. In no case have I allowed profit, or commission over the original bills of purchase, nor has it been asked. The prices in the original bills have not exceeded the most favorable rates paid by the leading telegraph companies. The telegraph companies, in many instances, furnished material and facilities at a considerable sacrifice to their own lines.

The lines in each Military Department radiate from the center, or head-quarters of the general commanding, to each camp, or outpost. The effect of this is to render it the most reliable for ordinary communication, and at the same time valuable for field purposes. The interruption from any cause of either of the radiating lines does not affect the others, and the contents of communications to one division or camp, can not be known to the other, except by the direction of the officer at the central office. These wires will be extended as fast as the army advances. The commanding general, if near either of the radiating lines, can be connected at a moment's notice with either or all of the divisions of his army, by the aid of the central office. Some of the advantages of this plan are that the lines are easily protected and maintained, being located on routes occupied by by our troops. No extra force or organization is required as would be in an exclusive "Field Telegraph Corps." The danger of interruption by crossing fields, ravines, creeks, etc., is avoided.

The military lines and offices are open for business day and night. During the night the operators are required to report themselves every half hour, so that an interruption of the line or neglect of duty can be known. At each important post or camp an efficient operator is employed, whose duties are in connection with the operating department to put in cipher and translate all important despatches. Each department has been supplied with an amount of wire, insulators, instruments, field cordage and submarine cables sufficient for an emergency, and the extension of lines to follow the march of each division of the army. A considerable force of builders are in readiness with full equipments, who with the aid and protection of troops can push forward the lines ten or fifteen miles per day.

The safeguards placed about the army lines has, it is believed, prevented the disclosure of any important information while passing over the wires. With frequent changes of the cipher now in use and the difficulty that experts would have in obtaining accurate copies of the same, it is believed that despatches of the most private nature can be safely entrusted to the lines as now directed.

A large number of general and special orders issued by commanding officers are transmitted by telegraph to each camp and division. By this means a considerable clerical labor is saved, to say nothing of the saving of courier service and the delay by the ordinary means of communication.

Great care is exercised in regard to the loyalty of all persons connected with this Department. The oath of allegiance is required of every employé. The offices are guarded from intrusion of persons not immediately connected with its operations.

Up to the present time the Government has not taken possession of any telegraph line operated or controlled by corporations or individuals, although the owners understand that their lines are subject to the order of the Government for special or extraordinary service at any and all hours, and that preference shall be given to all Government business. Upon the lines where the greatest amount of Government service is performed, the commercial business is greatly depressed and but comparatively small revenue derived by the telegraph companies from this source. Yet to meet the wants of the Government, the telegraph companies are compelled to keep up the same organization and equipments as formerly, and without extra compensation. In no place have I found the Government service upon a line large enough to justify Government in assuming the entire running expenses of the same, nor can I recommend such a course as long as the telegraph companies perform the service promptly and faithfully.

Greater efficiency and economy in the Telegraph Department would be secured by the prompt supply of funds for the pay of operators, builders, etc., and purchase of materials. No provision is made for the expenses of operators, except their salary, and as they are subject to frequent and sudden changes, they should be paid promptly. A better class of operators can be engaged, if pay is certain and prompt. The labor and responsibility of the General Manager can be relieved and the public service greatly enhanced by the allowance of one clerk for each military department. The Quarter-master General has not yet made any allowance for clerk hire, although the disbursements of this service extend to the outposts of all the military departments. I respectfully ask the attention of the Secretary of War to the above requests.

It is but just to state, in concluding this report, that the telegraph companies have at all times evinced the most liberal and loyal spirit. They have tendered to the Government all facilities possessed by their lines. With few exceptions, they have transmitted the messages of the Government Telegraph Department regarding purchase and shipment of material, and other urgent correspondence, without charge. At telegraph stations where military lines terminate, they furnish room, light and fuel, and the labor of repeating army business without compensation. No commissions are charged for the material manufactured or supplied the Government.

The assistant managers appointed are all practical telegraphers, of tried integrity and extensive experience. Many of the operators employed rank as the best in the country, and are devoted in their efforts to render the telegraph invaluable to the public service. It is estimated that the number of Government messages transmitted over the lines in the "Department of the Potomac," averages one thousand per day.

Very respectfully, ANSON STAGER, *Capt. and A. Q. M.*,
Supt. U. S. Military Telegraphs.

REPORT OF COLONEL A. STAGER, ASSISTANT QUARTER-MASTER AND SUPERINTENDENT OF THE UNITED STATES MILITARY TELEGRAPH, FOR THE FISCAL YEAR ENDING JUNE 30, 1863.

GENERAL M. C. MEIGS, Q. M. G.,
Washington, D. C.:

GENERAL: I have the honor to acknowledge the receipt of your "General Order, No. 13," dated Washington, D. C., July 22, 1863, and submit the following report of the operations and conditions of the United States Military Telegraph, for the fiscal year 1863, as my reply thereto, respectfully calling your attention to the accompanying reports of the different quarter-masters under my control. Their reports contain details of the operation of the Military Telegraph in the respective Departments to which they are assigned. The report of Major Eckert, Assistant Superintendent of the Department of the Potomac, is marked "A;" Captain Bruch, Assistant Superintendent of the Departments of Ohio, Cumberland and Mississippi "B;" Captain David, Assistant Superintendent Department of Western Virginia, "C;" Captain Wade, Cleveland, Ohio, "D;" Captain Smith, Assistant Superintendent Department Missouri, "E," and that of Captain Bulkley, Assistant Superintendent Department of the Gulf, "F."

Since my appointment as Aide-de-camp, Assistant Quartermaster and Superintendent of the United States Military Telegraph, I have been commanding officer in charge of the Military Telegraph in the United States. My head-quarters, from the commencement of the fiscal year, ending in June, 1863, up to the 1st of April, were at Washington, D. C. In April, I was ordered by the Honorable Secretary of War, to make a personal tour of observation of the Military Telegraph in the West and South-west, and to make my head-quarters, hereafter, in Ohio. I was on special duty at Memphis, Tenn., and Cairo, Ill., by the order of the Honorable Secretary of War.

It is my duty as commanding officer of the Military Telegraph Department, to exercise a general supervision of all its lines; to give such orders and direction to the subordinate officers in this branch of the public service, as may from time to time be necessary for the better conduct of the relative affairs and business therein, and to supervise the purchase of all the material which the wants or exigencies of the various Departments may demand.

The Military Telegraph lines required by the Government have been constructed over an extensive and scattered territory, embracing parts of the States of Pennsylvania, Ohio, Indiana, Illinois, Maryland, Delaware, Virginia, North Carolina, South Carolina, Louisiana, Mississippi, Alabama, Arkansas, Tennessee, Kentucky, Missouri, Kansas and the Indian Territory.

Under the immediate direction of Major Eckert, Assistant

Superintendent Department of Potomac, three hundred miles have been constructed during the fiscal year. Under the direction of Captain Smith, Assistant Superintendent Department of Missouri, five hundred and forty-eight miles have been constructed during the year. Under the direction of Captain Bruch, Assistant Superintendent Departments of Ohio, Cumberland and Mississippi, five hundred and ten miles have been constructed during the year. Under the direction of Captain David, Assistant Superintendent Department of Western Virginia, ninety-seven miles have been constructed during the same period. Under the direction of Captain Bulkley, Assistant Superintendent Department of the Gulf, three hundred miles have been built as above. On the first day of July, 1862, there were three thousand, five hundred and seventy-one miles of land and submarine lines in working order. During the fiscal year, one thousand, seven hundred and fifty-five miles of land and submarine line were constructed, making the total number of miles of land and submarine military telegraph lines in operation during the fiscal year, five thousand, three hundred and twenty-six, being length of line sufficient to girdle more than one-fifth of the circumference of the globe.

It should be borne in mind that a large proportion of this labor has been performed as it were, in the presence of an armed foe. Sometimes the lines have been required to take their course through portions of the revolted States, actually in the possession of the enemy, and upon such occasions, the constructing parties have been obliged to perform their labor during the night time only, to avoid detection and capture by the enemy. It would be difficult to enumerate the hazards, vexations and obstacles incident to the construction of military telegraph lines. The telegraph is ever at the front, occupying the post of danger and of honor. It has been frequently in advance of the army, and it can not be denied but that the result of its enterprising hazard has often proven of much advantage to our forces. But few cases of capture of telegraphers by the enemy have occurred, when these experts have voluntarily assumed advanced and exposed positions.

Follow the army where you will, there you will find the telegraph exercising its vigilance and its protection over the surrounding camps. At the foremost picket posts, in the rifle pits and in the advanced parallels, at any hour of the day or of the night, you can listen to the mysterious, yet intellectual click of the telegraph instrument amidst the strife of battle and the whistling of bullets, its swift, silent messengers pass unseen and unharmed. It is through the medium of the telegraph that the vast amount of supplies of various descriptions required for the daily sustenance of the armies are ordered forward from their depositories. If an advance of the army is to be made, all deficiencies to the comfort and necessities of the troops, or any lack of the material of war, can, by the assistance of the military wires, be immediately ordered and speedily procured. If a retrograde movement is contemplated, all detachments adjacent to the line of march are quickly notified by telegraph, and the whole column is in motion at once. The public mind has but a faint conception of the magnitude of the uses of the army telegraph. Its importance and utility in a military campaign is fully understood only by those who are constantly brought into contact with it as a medium for the daily transaction of their important and extensive business. The Military Telegraph offices are kept open day and night—the lines never being closed to the transaction of business. The War Department and the General-in-Chief at Washington are in constant telegraphic communication with the commanders of the armies both East and West.

As an illustration of the importance and usefulness of the military and commercial telegraph to the Government, I will refer to the fact, that under the first call for three hundred thousand volunteers, and within forty days from the time that recruiting actively commenced, three hundred and twenty-seven thousand men were mustered into service, fifty thousand of whom were armed, equipped and placed in the field, and one hundred and fifty thousand armed, equipped and awaiting marching orders. The orders of the Honorable Secretary of War and the detailed instructions from the various bureaus of his department pertaining to the mustering, clothing, equipping and arming of these troops, together with the correspondence of the War Department with the various State authorities, were transmitted by telegraph. Without the aid of the telegraph, weeks instead of days would have been required to accomplish the work. Take a glance along the military railroads of the country and in quiet company with the long continuous band of iron rail, you can not but observe the air lines of iron wire through which the electric winged messengers of thought flash the orders of our commanders from one section of the country to another. A distant command on some part of the line receives, through the means of this lightning communication, its orders to move forward and create a diversion in favor of the struggle which is going on in some other part of the line and perhaps by destroying the enemy's line of communication or his supply trains, a victory is won. General orders are given—armies are moved—battles are planned and fought, and victories won with the assistance of this simple, yet powerful aid-de-camp, the Military Telegraph. Even the history of this unholy rebellion is being recorded by the electric dottings of the telegraph from day to day as the war progresses, let us hope to its speedy termination and a reunion against all enemies for all future time. The military railroads and telegraphs are the great arteries which warm the soul and keep alive the body of our grand Union army.

By a close estimate it appears that at least *one million, two hundred thousand* telegrams have been sent and received over the military lines in operation during the fiscal year ending June 30, 1863, being at the rate of about *thirty-three hundred* per diem. These messages have varied in length from *ten to one thousand words and upwards*, and generally were of an urgent or most important character. All business of an important or confidential nature has been transacted in cipher, and the contents of the telegrams thus transmitted have been known only to the War Department or general officers with whom they originated, and the sworn cipher operators. This mode of secrecy has been invaluable to the Government and of great advantage to military operations.

I take pleasure in acknowledging the valuable services of my assistant superintendents, Major Eckert and Captains Bruch, Smith, David, Bulkley and Wade. These officers have given their personal and undivided attention to the interests of the Military Telegraph, and to their exertions it is indebted for its uniform promptness, reliability and usefulness.

I would call especial attention to the paper herewith, marked "G," it being a report from Captain Bulkley upon the removal of rebel obstructions in the Bayou Teche, La. Major General Banks having called upon Captain Bulkley to remove the impediments to the navigation of the bayou, he undertook the task, and speedily succeeded in accomplishing the same. Also, to the report of Captain Bulkley, marked "H," on Beardslee's Field Telegraph.

The operators in the service of the United States Military Telegraph, as a general rule, have manifested a spirit of patriotism and devotion to their duty, in the highest degree commendable. They are not bound by any military organization or regulations, yet they have undergone all the exposure, the dangers and privations of camp life, with a degree of endurance and forbearance worthy of mention. They have been on duty night and day, and of all the many important trusts and positions bestowed upon them, there is yet to be recorded the first case of recreancy to the task confided to them. The amount of pay generally received by these persons is not considered a fair remuneration for the service performed. Instances of meritorious conduct on the part of telegraphers in the field, I should be glad to have rewarded by favorable mention of, or presentation of suitable medals to such persons.

July 1, 1862, there was remaining in my hands the sum of $9,829.24. During the fiscal year ending June 30, 1863, I received from the Treasurer of the United States at Washington, D. C., the sum of $418,000. Total amount of funds in my possession during the fiscal year, for application to operating and constructing military telegraph air and water lines within the United States, $427,829.24. I have made a proper rendition of the same to the accounting officers of the Government. From this amount, I have transferred to my subordinate quarter-masters of the Military Telegraph, during the year, $295,265.43, and disbursed on service account $55,711.76, and to the purchase of property $70,795.99.

In September, 1862, I officially transferred to Major Eckert, Captains Smith, Bruch, David and Wade, the military telegraph lines and property in the respective departments of which they were previously in nominal charge.

On the 30th of June, 1863, there was remaining in my possession a balance of $6,056.06, in Government funds, which was deposited in my safe at Cleveland, Ohio.

I take this opportunity to express my obligation to the Secretary of War for the unlimited voluntary interest and assistance extended by himself and his department in and to the United States Military Telegraph. This kind support has contributed to a great extent in making the military telegraph what it is to-day, a reliable, useful and indispensable assistant to the Government in the conduct of military operations, both in and out of the field. It has been a pleasant reflection, both to myself and my assistant superintendents in the discharge of our official duties, to know that our efforts to accomplish all that has been required of us, have been duly appreciated and acknowledged by the War Department, although perhaps our energies may not always have realized our most sanguine expectations. The impression that our exertions have generally received the approval of the Honorable Secretary of War has been an agreeable incentive to renewed and untiring efforts to successfully accomplish the various tasks required of us.

Very respectfully, etc., ANSON STAGER,
Col. A. Q. M. and Supt. U. S. Mil. Tel.

CLEVELAND, O., Nov. 2, 1863.

ANNUAL REPORT OF COLONEL A. STAGER, ASSISTANT QUARTERMASTER AND CHIEF OF UNITED STATES MILITARY TELEGRAPHS FOR THE FISCAL YEAR ENDING JUNE 30, 1864.

BREVET MAJOR GENERAL M. C. MEIGS, Q. M. G., U. S. A.,
Washington, D. C.:

GENERAL:—Accompanying this, my official report of the operations and constructions of United States Military Telegraphs for the fiscal year ending June 30, 1864, I have the honor also to submit the reports of the different officers and Assistant Superintendents connected with this branch of this Government service, and very respectfully solicit your consideration of the detailed operations of the Military Telegraph within the United States embraced in the several reports herewith. * * * * *

My report is designed to present in condensed form the operations of the United States Military Telegraph within the several departments:

The amount of public money in my possession July 1, 1863, was,	$ 6,056.06
Received from the United States Treasurer during the year, -	606,000.00
Amount transferred to officers is the total, - - - - - -	$612,056.06
Add amount received by officers of this Corps from other sources as per their reports, - - - - - - - - - - - -	31,848.02
Shows total amount received of money applicable to the Military Telegraphs during the year, - - - - - - - - -	$643,904.08
Deduct amount of balances in the hands of quarter-masters June 30, 1864, - - - - - - - - - - - - -	34,247.77
Total amount expended during the year for construction and operation of United States Military Telegraphs, - - -	$609,656.31
Of the last mentioned amount there was applied to the purchase of material about, - - - - - - - - - -	218,000.00
For incidental expenses say, - - - - - - - - - -	13,000.00
Leaving applicable to service account, - - - - - -	378,656.31
Total, - - - - - - - - - - - - -	$609,656.31

As the cost of material, labor, subsistence, etc., was rapidly advanced the amount of money which will be required for the operation and construction of United States Military Telegraphs for the fiscal year of 1865, if the war continues so long, will be proportionately increased. Probably ninety thousand to one hundred thousand dollars per month will be required to meet indispensable expenditures of the Corps. A monthly average of one thousand persons have been engaged in the Military Telegraph service within the several departments during the fiscal year of 1864. The number at present in the service is considerably greater and increases as the Federal forces advance or the military operations become more active and extended.

	LAND.	SUBMARINE.
The number of miles of telegraph lines in operation July 1, 1863, was, - - - - - - -	2,969	39¾
No miles constructed during year, - - - -	3,692	15¾
Total number of miles in operation during year, -	6,661	55½
Total number of miles abandoned during year, -	1,536	1½
Total number of miles in operation June 30, 1864, -	5,125	54

The number of miles of Military Telegraph lines in operation June 30, 1864, was distributed as follows:

	LAND.	SUBMARINE.
In the Department of the Gulf, Captain Bulkley, -	639	5½
In the Department of the South, Captain Sheldon, -	55	12
In the Department of the Potomac, Major Eckert, -	874½	32½
In the Department of West Virginia, Captain Lynch,	310	½
In the Departments of Tennessee, Cumberland and Mississippi, Captains Bruch, Gross and VanDuzer,	1732½	1¼
In the Department of Missouri and Kansas, Captain Smith, - - - - - - - - - -	1,303	1¾
In the Department of Arkansas, Captain Clowry, -	211	½
Total, - - - - - - - - -	5,125	54

The estimated number of telegrams transmitted over the Military Telegraph lines during the fiscal year is one million, eight hundred thousand.

Herewith is statement "A" of quarter-masters' property received from all sources, transferred, expended and on hand June 30, 1864, condensed from the property statement accompanying the reports of the different quarter-masters.

I take pleasure in stating that the several assistant quarter-masters and assistant superintendents engaged in the service have invariably responded with willing alacrity to all orders issued from these head-quarters, having in view the execution of such demands as have been made upon me from time to time for increased and extraordinary telegraphic facilities. Your perusal of their respective reports is respectfully solicited. I heartily endorse the compliments tendered by the different assistant superintendents to the employés under their charge. These men generally have faithfully pursued their various duties with commendable reliability and often under circumstances requiring force of character sufficient to undergo many privations, resist stampedes and risk captivity by the enemy, adhering to their posts of duty with remarkable tenacity and approving spirit.

Very respectfully, your obedient servant,

ANSON STAGER, *Col., A. Q. M.*
And Chief U. S. Military Telegraphs.

CLEVELAND, O., October 31, 1864.

ANNUAL REPORT OF COLONEL ANSON STAGER, CHIEF UNITED STATES MILITARY TELEGRAPHS FOR THE FISCAL YEAR ENDING JUNE 30, 1865.

CLEVELAND, O., September 15, 1865.

GENERAL:—In obedience to General Order No. 39, Quarter-master General's Department, Washington, D. C., July 1, 1865, I have the honor to submit my annual report of the operations and condition of the United States Military Telegraphs for the fiscal year ending June 30, 1865, etc. I have the honor to hand you also herewith the reports of such of the officers serving under me as have been received up to this time. That of Captain J. C. VanDuzer marked A, Captain R. C. Clowry B, Captain W. G. Fuller C, Captain W. L. Gross D, Captain J. R. Gilmore E, Captain S. G. Lynch F. The report of Major Thomas T. Eckert has not yet reached me.

My annual report for 1864 was forwarded to your Department in the latter part of October, 1864, accompanied by the reports of the several officers acting as assistant superintendents and assistant quarter-masters, attached to this branch of the public service.

It has been my duty during the past year to act as chief officer of United States Military Telegraphs, having a general superintendence over all the lines, providing the necessary funds, and supervising the purchase of material required for the operation and construction of the same. I have been very ably and energetically assisted in the discharge of my duties as General Superintendent of United States Military Telegraphs by the several officers acting as assistant superintendents in the various military departments, and I desire to accord to them the greater portion of whatever merit attaches to this branch of the public service. I have the honor to call attention to the reports of those officers herewith transmitted, and to ask the consideration of the Department to the merits of the officers respectively.

The demands of the proper military authority for telegraphic communication have always been promptly answered and I believe fully satisfied. The Military Telegraph under the immediate charge of the several assistant superintendents thereof has kept company with our armies wherever they have gone, and upon all important expeditions; and it is a well established fact that the mobility of the army has been greatly accelerated by its usefulness and assistance. The Military Telegraph has been an invaluable assistant in the construction and operation of the various military railroads. Trains have been run and many of the roads operated almost exclusively by telegraph. The military railroads and the Military Telegraph have been great auxiliaries to the gigantic and successful efforts of the Government in suppressing the rebellion. The Military Telegraph has operated frequently in the field in conjunction with the Signal Corps, and has rendered efficient aid in this respect by diffusing information from advanced signal stations simultaneously to the head-quarters of the commanding general and the different corps head-quarters.

The Military Telegraph could be made the means of establishing the Signal Corps of the army for active operations upon a more useful basis than heretofore. The telegraph depends not upon the atmosphere, nor the weather, but flashes its thoughts alike unheeded, through storm or sunshine, darkness or light. I am however of the opinion that the Field Telegraph which we have operated independently as well as in connection with the Signal Corps, could be so perfected and operated as to completely and advantageously supplant the use of a Signal Corps for military operations.

In May, 1865, it was decided by the War Department that all commercial telegraph lines throughout the Southern States, lately in armed resistance to the United States authorities, should be supervised and controlled by the officers of the United States Military Telegraph, subject to specified conditions favorable to the different telegraph companies, which privileges however were to be ignored under certain circumstances. Previous to cessation of hostilities these lines had been operated to a great extent by the rebel authorities, making it necessary therefore on account of the *morale* of the former employés and the peculiar power of the telegraph for public good or evil, for the Government to exercise its prerogative in regard to the operations of the lines referred to. Under the present arrangement the telegraph companies are called upon to repair their lines and put in good working order, furnishing all labor and material therefor, the United States to be at no outlay beyond the expense of maintaining purely military lines and military stations. An account is to be kept of all Government business passing over the lines. The account however is not assumed as an indebtedness by the Government, but is left open for future consideration or settlement.

The amount of supplies which will be required for the future operations of the Military Telegraph, so far as I am at present informed, will be such as is necessary only for maintaining the lines at present in use. It is supposed that the Government will require no more lines constructed, unless, perhaps, in Texas or upon the Western frontier.

The amount of money expended during the year ending June 30, 1865, for the purchase of material and supplies required for the United States Military Telegraph was about $300,000, $130,000 of this amount however was for the purchase of 285 miles of English submarine cable, the greater portion of which is still on hand. Probably not more than from $1,000 to $1,500 per month will hereafter be required for the purchase of necessary material and supplies. So long as the Military Telegraph lines in present use are needed by the Government, it will require about $75,000 per month to supply, maintain and operate the same. As the usefulness of some of these telegraph lines to Government shall cease from time to time, and the lines are taken down or otherwise disposed of, the expenditures for maintaining the United States Military Telegraph will be proportionately reduced. A considerable revenue to Government will probably be derived from the sale of the material now on hand so soon as the same shall be of no further use to the Government. Such of the lines as may be located most advantageously for commercial purposes can undoubtedly be disposed of to the owners of the "telegraph right of patent" within the territory through which the lines pass, and at a reasonable consideration. The less important lines can be taken down and the material sold.

The number of miles of land and submarine United States Military Telegraph lines in operation July 1, 1864, was, after deducting error in former report, as follows:

	LAND.	SUBMARINE.	
July 1, 1864,	4,955$\frac{1}{4}$	52$\frac{3}{4}$	
The number of miles constructed during the year,	3,246$\frac{1}{2}$	68$\frac{3}{4}$	
The number of miles in operation within the year	8,201$\frac{3}{4}$	121$\frac{1}{2}$	
Total,			8,323$\frac{1}{4}$
The number of miles taken down or abandoned during the year,	2,049	46$\frac{1}{4}$	
The number of miles in operation June 30, 1865,	6,152$\frac{3}{4}$	75$\frac{1}{4}$	
Total,			6,228

The 3,315$\frac{1}{4}$ miles of line constructed during the year were built under the supervision of the following officers, and within the departments specified, viz.:

Major Thomas T. Eckert, Assistant Quarter-master Department of the Potomac,	1,217$\frac{3}{4}$
Captain J. R. Gilmore, Assistant Quarter-master Department of the South,	86$\frac{1}{4}$
Captain W. G. Fuller, Assistant Quarter-master Department of the Gulf,	228$\frac{1}{4}$
Captain J. C. VanDuzer, Assistant Quarter-master Department of Kentucky, Tennessee and Mississippi,	1,476
Captain R. C. Clowry, Assistant Quarter-master Department of Missouri, Kansas, and Arkansas,	307
Total,	3,315$\frac{1}{4}$

The lines in operation June 30, 1865, were under charge of the following officers, and located as mentioned, viz.:

	LAND.	SUBMARINE.	
In the Department of the Potomac, Major Thomas T. Eckert,	1,824$\frac{1}{4}$	53$\frac{1}{2}$	
In the Department of the South, Captain J. R. Gilmore,	140	13$\frac{1}{4}$	
In the Department of the Gulf, Captain W. G. Fuller,	56	5	
In the Department of West Virginia, Captain S. G. Lynch,	303	0$\frac{1}{2}$	
In the Department of Kentucky, Tennessee and Mississippi, Captain J. C. VanDuzer,	2,127$\frac{1}{2}$	1$\frac{1}{4}$	
In the Department of Missouri, Kansas and Arkansas, Captain R. C. Clowry,	1,702	1$\frac{3}{4}$	
	6,152$\frac{3}{4}$	75$\frac{1}{4}$	
Total,			6,228

It is estimated from the commencement of the rebellion up to June 30, 1865, there have been constructed and operated about *fifteen thousand miles* of United States Military Telegraph—land, submarine and field lines.

From May 1, 1861, up to December 31, 1862, $22,000 per month sufficed to construct and maintain United States Military Telegraphs. For the year 1863, about $38,500 per month was required for the same purpose. During the year 1864 the Military Telegraph was greatly extended and required about $93,-500 per month. From May 1, 1861, to June 30, 1865, $2,655,-500 have been received by me from the United States Treasury and disbursed or transferred by me for the construction, maintainance and operation of United States Military Telegraphs. Herewith please find statement of public moneys received during the year, and form "A" embracing all the articles of property received by me during the same period. I have no occasion to use forms B, C, CC, D, E, F or G. I have the honor to be, General, very respectfully, your obedient servant,

ANSON STAGER, *Colonel,*
And Chief United States Military Telegraphs.

MAJOR GENERAL M. C. MEIGS,
Quarter-master General U. S. Army,
Washington, D. C.

Statement of public moneys in possession of Colonel A. Stager, Assistant Quarter-master, Chief of United States Military Telegraphs, during the fiscal year ending June 30, 1865:

On hand July 1, 1864,		$
Received from Treasury Department during year,		1,360,000.00
Total received,		$1,360,000.00
Expended during the year,	$ 149,304.72	
Transferred to officers during the year,	1,210,695.28	
Remaining on hand June 30, 1865,		
Total,		$1,360,000.00

Statement of Quarter-master's property received by Colonel A. Stager, Chief Quarter-master United States Military Telegraphs, for the fiscal year ending June 30, 1865: English submarine cable purchased during the year ending June 30, 1865, 284 miles and 1,070 yards. English submarine cable transferred during the year ending June 30, 1865, 284 miles and 1,070 yards.

NOTE.— The final report, *viz.*, that for the fiscal year, ending June 30, 1866, is substantialy shown herein, beginning at page 346.

ROLL OF UNITED STATES MILITARY TELEGRAPH OPERATORS.*

A.

Abrams, H. H.
Adams, Milton.
Alexander, E. P.
Allen, J.
Allen, James M.
Allen, Hudson J.
Allen, J. H.
Allen, John C.
Allen, Thomas.
Allen, W. E.
Allen, W. F.
Allis, Geo. B.
Allyn, John.
Altermeyer, Wm F.
Anderson, Joseph.
Anderson, Wm. C.
Andrews, M. S.
Aplin, A. N.
Applebaugh, W. K.
Armes, W. J.
Armor, Thos.
Armstrong, Ewing L.
Armstrong, S. T.
Ashley, W. N.
Ashley, G. W.
Atwell, Jos. W.
Atwater, E. W.
Atwater, H. H.
Atwater, Wm.
Aughinbaugh, D. C.

B.

Bacon, Duncan T.
Baker, A. L.
Baker, C. H.
Baker, Robert A.
Baker, Wm. H.
Baldwin, A. J.
Baldwin, C. P.
Baldwin, G. Wrisley.
Baldwin, M. C.
Bancroft, Wm. A.
Barnes, Cassius M.
Barrett, Clinton S.
Barron, W. C.
Barry, J. B.
Barth, M. H.
Barth, Samuel.
Bassett, F. H.
Bassett, Milton H.
Bassett, W. F.
Bates, Ed. E.
Bates, D. Homer.
Bauer, W. H.
Baxter, G. W.
Bay, J. W.
Bay, Wm.
Beal R. J.
Beaumont, Henry P.
Bear, Jacob K.
Beck, G. H.
Beckwith, Samuel H.
Bell, Geo. W.
Beman, C. D.
Bender, R. W.
Benedict, C. H.
Benner, F.
Benner, R. M.
Bennett, W. R.
Besanson, Chas. W.
Benson, F. N.
Bentley, Jas. N.
Benton, H. W.
Berry, C.
Berry, D. T.
Berry, H. B
Berryhill, Thos. R.
Bickford, F T.
Biggert, W. L.
Bigger, Samuel S.
Bishop, H. H.
Blackburn, Robert.
Blades, L. J.
Blair, J.
Blawcan, G.
Bliestine, G. W.
Blish, Jr., Joseph.
Bliss, A. H.
Bliss, J. E.
Bliven, C. E.
Bliven, R. H.
Bodell, Wm. J.
Bogardus, Henry A.
Bohle, Rudolph H.
Booth M. K.
Booth, O. H.
Bort, C. F.
Bosquet, A. J.
Bowers, J. W.
Bowers, Philip.
Bowen, James H.
Bowerman, Henry.
Boyle, E. C.
Boyd James W.
Bradnack, Fowler.
Bracken, John H.
Brannon, Clark.
Brenaman, A. T.
Brice, N. D.
Brigham, R. H.
Briggs, Henry G.
Britney, W. H.
Brooks, J. N.
Bromell, W. H.
Bromley, O. B.
Bross, H. L.
Brown, C. I.
Brown, Edgar H.
Brown, E. O.
Brown, Edw. B.
Brown, G. H.
Brown, H. R.
Brown, J. A.
Brown, J. K.
Brown, N. H.
Brown, Samuel M.
Brown, R.
Bruner Philip.
Brush, Samuel T.
Brush George M.
Bruch, Adam.
Bruton, J. G.
Bryant, E. R.
Bryant D. N.
Bryant, James.
Bryan, N. M.
Buck, Chester H.
Buck, Ed. A.
Bucklin, W. B.
Buehler, Richard E.
Buell, Henry C.
Buell, M. V. B.
Bull, H. P.
Bunnell, Jesse H.
Burhans, W. W.
Burnapp, G.
Burnett, Brace M
Burnett, Dug.
Burnett, G. A.
Burnette, L. F.
Burns, Silas C.
Burr, Platt.
Bush, Alonzo.
Bush, Charles.
Bush, Frank.
Bushnell, Daniel W.
Butler, C. D.
Butler, Edwin D.
Butler, Edw. F.
Butler, Emmet J.
Burucker, J. L.
Byington, Dwight.
Byrne, John H.
Byrns, I. Oliver.

C.

Caldwell, A. H.
Caldwell, James C.
Camp, Samuel P.
Camp, George H.
Campbell, W. J.
Campbell, F. H.
Campbell, Wm. L.
Carey, Samuel.
Carhart, C. B.
Carroll, Alfred W.
Carver, John W.
Caruthers, Geo. F.
Case, Hardin.
Cassell, John A.
Chaddock, Wm. H.
Chamberlain, A. R.
Chandler, Albert B.
Chandler, Chas. E.
Chandler, Jos. W.
Chapman, E. T.
Chapman, W. S.
Chapman, Harry.
Chapman, G. H.
Chappell, Scott R.
Chase, C. R.
Cheeney, Oscar F.
Cherry, J. L.
Childs, Albert. F.
Childs, M. J.
Chittenden, John H.
Clapp, R.
Clark, J. B.
Clark, J. E.
Clark, J. R.
Clark, T. B.
Clarke, F. S.
Clarke, H. K.
Clowry, Robt. C.
Cochrane, A. P.
Cochrane, S. D.
Cogan, L.
Congdon, J. DeWitt.
Cogley, E. W.
Cole, Eli.
Cole, Geo.
Collins, W. H.
Collings, Jos. W.
Colstock, Daniel.
Connor, Chas. O.
Conway, Ed.
Cook, F. C.
Cornelison, John R.
Cotton, J. S.
Covell, S. J.
Cowan, H. W.
Cowlan, Geo. B.

* NOTE. This list is as perfect as I have been able to make it, yet it is possible that a few names have been omitted. Those whose names are italicized are reported dead.

Cox, R. Emmet.
Couls, Dwight.
Craig, Archibald.
Craig, Hugh.
Crain, Geo. H.
Crain, Mark D.
Crittenden, J. N.
Cromwell, Geo. E.
Crouse, Jesse W.
Crowe, James N.
Cruise, John D.
Cryan, Thomas.
Cubitt, Jr., Richard S.
Culgan, E. A.
Culbertson, Cam.
Cunningham, R.
Curran, John.
Curtiss, G. H.
Curtiss, J. W.
Curtiss, Leverett E.
Cutter, Henry.

D.

Dameron, Charles B.
Darlington, H. P.
Davin, Thos. A.
Davidson, Thomas.
Davis, C. N.
Davis, C. S.
Davis, F. W.
Davis, Samuel
Davis, Geo. B.
Dealey, Wm. J.
DeBree, N.
Deetz, Geo. M.
DeForrest, C. L.
DeGrove, Wm. M.
Demorest, Ed. H.
Denny, James C.
Dennis, L. B.
DePew, H. S.
DeWitte, Wm. K.
Deslondes, J. M.
Dixon, John R.
Dodge, S. C.
Dolan, Thomas.
Dollard, James.
Dorsey, Benj.
Dorchester, J. C.
Dorrence, O. H.
Dougal, W. H.
Dougherty, Abe D.
Dougherty, Christopher.
Douglass, Charles.
Douglass, James H.
Douglass, Wm. J.
Doughty, Wm. L.
Downey, G. W.
Doyle, P. C.
Drake, Wm. H.
Drew, I. N.
Drummond, Frank.
Duell, John F.
Duesner, Phil.
Duffie, P. A.
Dunn, A. O.
Dunn, Wm. A.
Dunlap, J. R.
Dunning, Jay D.
Durant, Geo. F.
Dwight, J. H.
Dwyer, Chas. E.
Dwyer, Con.
Dyer, Derrick O.
Duxbury J. W.

E.

Eaton, Wm. H.
Eckert, B. F.
Eckman, Wm. H.
Eddy, Charles G.
Edwards, Jno. R.
Edwards, Samuel H.
Egan, John J.
Elphick, David.
Elliott, J. A.
Elliott, Richard H.
Ellison, A.
Ellsworth, Henry L.
Embree, W. N.
Emerick, J. H.
Erringer, T. F.
Evans, T. A.
Everette, Geo. H.

F.

Fancher, Charles I.
Farnham, G. N.
Fawcett, G. E.
Fay, Frank P.
Felton, G. C.
Fenton, Frank A.
Ferris, D. V.
Fish, E. G.
Fish, Harry G.
Fitch, Derrick H.
Fitchett, Hamilton.
Flack, B. W.
Flagg, John A.
Flannagan, J. J.
Flesher, Jr. Thos.
Floersh, Andrew.
Florence, J. C.
Flynn, J. B.
Foley, John.
Foley, Wm.
Follett, R. F.
Folwell, Geo. E.
Fonda, Ten Eyck H.
Forbes, W. W.
Forker, James.
Forrest, J. W.
Forsey, Wm. S.
Forster, Wm.
Foster, J. J.
Fowler, J. J.
Fowler, Peter.
Fowler, Robert S.
Fox, Fred.
Francis, D. T.
Franks, H. D.
Frank John R.
Frank, Wm. H.
Franklin, Ed. A.
Frazey, G. S.
Frazier, P. B.
Freeman, Edwin C.
Freeland, John W.
French, Wm. F.
Frey, J. J. B.
Fuller, Edwin.
Fuller, Joseph A.
Furr, R. A.

G.

Galbraith, Franklin G.
Gallup, George.
Galvin, J. H.
Gamble, J. E.
Gamble, R. E.
Gardner, O. A. A.
Garland, J. G.
Garland, Geo.
Geiger, Jno. M.
Gelpin, W. D.
Gentry, W. D.
Geyer, H. P.
Gibson, Wm.
Gifford, Carl.
Gilbert, Homer W.
Gile, Courtland, H. B.
Gilkerson, Joseph H.
Gilmore, B. F.
Gilmore, James R.
Glasier, Julius M.
Glass, B. B.
Glass, Frank M.
Glasscott, H. H.
Glazier, J. H.
Goalding, Geo. J.
Godfrey, Geo. H.
Gordon, M.
Graham, M. B.
Graham, Richard.
Graham, Wm. N.
Gransby, Wm.
Graves, E. H.
Gregg, Chas. M.
Gregg, Henry L.
Gregg, John C.
Gregg, J. W.
Gregory, G. M. D.
Green, E. C.
Green, C. M.
Griffin, Alonzo D.
Griffin, Robt. B.
Griffin, Stockton L.
Griffith, C. H.
Griswold, H.
Griswold, Martin E.
Grogan, John.
Gross, Wm. L.
Gross, Chas. F.
Gullihur, James K.
Gustin, A. J.
Guthridge, Jules F.

H.

Halin, J. F.
Hall, Edw. A.
Hall, H. H.
Hall, W. C.
Hall, W. H.
Hallum, J. W.
Hambright, C. K.
Hammann, Chas. A.
Hammond, C. D.
Hammond, Chas. W.
Hammond, J.
Hanna, Geo. S.
Hanchett, D.
Hanson, Joseph.
Hanson, Peter.
Hancock, A. G.
Harris, F. C.
Harris, J. B.
Harris, L. M.
Harrison, Jas. S.
Hartman, Wm. H.
Harvey, A. O.
Harvey, T. G.
Hatton, O. C.
Hawkins, A. S.
Hawkins, W. D.
Havens, Paul C.
Hebard, L. I.
Hefferin, John.
Henby, J. E.
Henderson, Geo.
Henderson, H. B.
Henry, C. C.
Hicks, Wm. W.
Hickson, Wm.
Hill, Jacob V.
Hixon, Wm.
Hoge, O. E.
Hodge, Wm.
Hogan, Daniel.
Holbrook, Spencer C.
Holden, A. H.
Holden, H. T.
Holdridge, John C.
Holmes, W. J.
Holloway, Wilbur F.
Holt, Theo.
Holtham, Wm.
Holly, M. Y.
Homan, C. A.
Hoover, R. B.
Horner, John.
Hotchkiss, Z. P.
House, Wm. T.
Howe, K. B.
Howe, G. W.
Howard, Alvin A.
Howard, James A.
Howard, S. D.
Howell, A. W.
Howell, Albert J.
Howell, Richard N.
Howell, Stephen V.
Hruby, Julius M.
Hughes, Alex. M.
Hughes, R. J.
Hull, Henry P.
Hull, A. K. V.
Humphreys, J. M.

Humphreys, R. C.
Hunter, Alex.
Hunter, D. W.
Huntington, Geo. M.
Hutchings, W. S.
Huyck, Maynard A.
Hyne, Joseph W.

I.

Ingram, F. M.
Ingram, Samuel E.
Ingle, John O.
Ives, W. L.

J.

Jacques, C. W.
Jacobs, C. D.
Jameson, Nelson W.
Jenks, Jr., Orrin.
Johns, C. H.
Johnson, Barton H.
Johnson, Benj. F.
Johnson, E. H.
Johnson, C. S.
Johnson, Thos. S.
Johnson, W. F.
Johnson, W. P.
Johnston, C. J.
Jones, A. C.
Jones, Geo. E.
Jones, H. P.
Jones, J.
Jones, James.
Jones, James H.
Jones, L. A.
Jolls, J.
Joyce, J. T.
Joyce, Morris.

K.

Kanode, A. H.
Kay, Geo. T.
Keefer, Chas. A.
Keith, J. S.
Kelsey, Wm. H.
Kelley, Seymour J.
Kendrick, Geo. W.
Kent, Douglass.
Kerner, M. H.
Kettles, Wm. E.
Kimber, S. P.
King, W. A.
Kingsbury, Fred. W.
Kinnaman, J. J.
Kinnaman, O. H.
Kinney, Ed.
Kirkman, M. G.
Kirby, J. S.
Kirtland, D. P.
Klunk, James P.
Knapp, Allen C.
Knapp, Seeley B.
Knight, F. B.
Knittle, Joseph.
Knox, Claude M.
Knox, J. W.
Korty, Louis H.
Kress, Wm. B.
Krum, M. H.
Kuhn, Rufus.
Kunkle, H. B.

L.

LaBonte, J.
Laird, Thos. A.
Lamb, Charles.
Lamb, Frank H.
Lancaster, W. H. H.
Lancy, S. H.
Langhorne, A. T.
Lantz, J.
Lathrop, D. B.
Lathrop, J. P.
Lawler, Edward.
Lawrence, F. A.
Lawrence, G. J.
Lawrence, G. T.
Lavin, Mike C.
Laverty, Robt. C.
Lea, Samuel,
Lehr, Charles,
Lennox, G. P.
Lessell, Ernest.
Lewis, A. C.
Lewis, E. A.
Lewis, Richard T.
Lewis, W. S.
Lindauer, A. C.
Lindley, W. T.
Lines, Henry F.
Lines, Robert B.
Livergood, W. W.
Livingston, E. M.
Livsey, B.
Lithgow, C. H.
Lithgow, Hector.
Lock, John M.
Logue, W. S.
Long, F. C.
Lonergan, John.
Loomis, Richard.
Loomis, Robt. C.
Loucks, T. N.
Lounsberry, Chas. H.
Low, Geo. A.
Low, James.
Ludwig, Danl. J.
Ludwig, John F.
Lynch, S. G.
Lyle, J. S.

M.

MacIntosh, William.
MacKenzie, J. W.
McCaine, R. R.
McCarty, D.
McConnell, C. C.
McClure, James.
McClure, William.
McCleverty, Joseph D.
McCormick, John.
McCoy, Mortimer A.
McCoy, S. G.
McCutcheon, J.
McDill, Edgar B.
McDonald, S.
McDonald, W. J.
McGaughey, D. C.
McGintey, E. H.
McGintey, T. J.
McGintey, W. T.
McGuire, T.
McIlvaine, James F.
McIlvaine, J. P.
McInnes, W. N.
McKelvey, A. T.
McKenna, Edwd.
McKenna, J. A. L.
McLean, Natl.
McMillen, James.
McMichael, Isaac.
McMurray, Alex. K.
McMurtrie, B.
McMullen, John W.
McNairn, Edgar B.
McNeil, James.
McReynolds, C. W.
Maiden, William P.
Magehan, Wm. H.
Maguire, James D.
Magonigle, S. R.
Maize, Isaiah D.
Manley, John C.
Mapes, H. R.
Marean, Anne M.
Marean, Wm. H.
Martin, Philip.
Martin, Robt. W.
Martyn, D. E.
Martin, Henry S.
Mason, D.
Mason, E. W.
Mason, John Q.
Mason, W. S.
Mason, W. T.
Mathews, Chas. B.
Matthews, John
Matlock, H. H.
Matlocks, R. M.
May, Marion.
Maynard, George C.
Maynard, Wm. S.
Meagher, James.
Meehan, John H.
Meehan, W. F. D.
Meldrum, W. W.
Merkley, Wm. H.
Merrick, E. D.
Miller, Frank.
Miller, F. B. G.
Miller, Wm.
Miller, Wm. H.
Minnehan, G. C.
Mixer, C. H.
Moloney, Patrick.
Moltz, John J.
Montanye, C. D.
Montalvo, Wm. W.
Moore, C. W.
Moore, J. F.
Morcalla, Stephen.
Morgan, John.
Morgan, John B.
Morlock, W. H.
Morris, Absolom M.
Morris, Charles.
Morris, James C.
Morrison, Thomas.
Mullarkey M.
Mullarkey, Patrick.
Mulligan, J. J.
Munro, W. H.
Munson, J. A.
Murdock, Charles C.
Murphy, J. J.
Murphy, Robert.
Murray, Daniel.
Murray, James A.
Murray, P. J.

N.

Nail, Geo. W.
Nagle, J. E.
Nappin, A.
Nash, F. A.
Nash, W. H.
Neff, G. W.
Nelson, W. B.
Nesbitt, Joseph G.
Newell, James.
Newton, Ed. C.
Newton, O. K.
Nichols, A. M.
Nichols, J. Hervey.
Nichols, Horace W.
Nicholson, John.
Nohe, A. W.
Nolen, Eugene.
Norman, Alfred T.
Norris, J. B.
Northrup, C. C.
Norton, G. N.
Noxon, S. M.
Nunan, P. H.
Nye, J. M.

O.

Oakes, T. F.
O'Brien, John E.
O'Brien, Richard.
Olmstead, John Q.
O'Neil, Algernon W.
O'*Neil John.*
O'Neil, Wm. C.
O'Neill, John L.
O'Reiley, Luke.
O'Ryan, Michael W.
Orton, A. W.
Orvis, John.

P.

Painter, J. I.
Painter, L. M,
Paisley, S. T,
Palmer, Mrs.
Pardridge, Albert J.
Park, John B.
Parker, Leander D.
Parsons, James K.

Parsons, John B.
Parsons, J. W.
Parsons, Wayne H.
Patterson, Wm.
Paxson, Charles A.
Paxson, Oscar W.
Payne, Charles S.
Payne, W. N.
Peabody, S. P.
Pearson, W. S.
Peck, G. H.
Peck, Rufus L.
Peel, Edwin.
Peebles, Benj. H.
Peeler, Thomas M.
Peirce, Geo. C
Penn, George R.
Perdue, Ford.
Peterson, J. H.
Pettit, James E.
Phelps, H. W.
Phelps R.
Phillips, Chas.
Phillips, C. C.
Pidgeon, G. S.
Pierce, J. B.
Pierson, C. W.
Pitcairn, W. W.
Pitfield, J. W.
Pitton, James.
Platt, G. W.
Platt, John T.
Plum, Henry W.
Plum, Wm. R.
Pond, C. H.
Porter, J. K.
Power, Jr. R.
Powers, James.
Pratt, W. H.
Pritchard, A. P
Pritchard, Charles M.
Printz, Geo. W.
Purcell, Patk. J. A.
Purdon, George.
Purdy, Geo. A.
Purnell, J. W
Pyle, Henry P.

Q.

Quan, P. F.
Quate, James L.
Quinn, T. M.

R.

Rabbeth, J. T.
Railton, Geo.
Rand, D. E.
Ransford, Jr., H.
Rawlins, Thos. E,
Raymond, John H.
Reaser, Robt.
Reeves, James.
Reese, Samuel.
Reddington, Wm. B.
Reid, Douglass.
Reid, W. J.
Reynolds, C. R.
Rice, Geo. H.
Rich, Thos. H.
Richardson, John D.
Riley, J. J. G.
Risdon, Samuel L.
Roby, B. F.
Roberts, Samuel B.
Roberts, C. R.
Roberts, M. E.
Robinson, Byron L.
Robinson, E. H.
Robinson, Heber C.
Robinson, J. H.
Robinson, Steve L.
Roche, Thos.
Rodgers, T. J.
Rockwell, Eugene.
Rockwell, John W.
Root, Albert G.
Rose, L. A.
Rosewater, E.
Rouser, G. W.
Rowe, Chas. O.
Rowe, R. E. D.
Royce, D. P.
Royce, H. P.
Rugg, J. H.
Rugg, Norman H.
Rumsey, S. R.
Runyan, C. A.
Rupley, S. K.
Ryan, C. J.
Ryan, R. H.
Ryland, John M.

S.

Sabin, John F.
Sackett, H. R.
Safford, A. G.
Salmons, J. P.
Sampson, Thos. M.
Sampson, J. W.
Sandborn, F. A. H.
Sargent, Wm. D.
Sargeant, Stephen.
Schermerhorn, Ed.
Schnell, A C.
Schnell, Jr., Joseph.
Schnell, T. M.
Scott, E. Alex.
Scott, S. W.
Sears, Joseph L.
Sellers, D. C.
Sellers, Calvin T.
Sewell, C. H.
Seymour, M. T.
Shape, E. M.
Shaffer, —.
Sheldon, Geo. D.
Sheldon, Lemuel F.
Sheldon, Wm. A.
Sheridan John A.
Sherman, Israel A.
Sherman, Thos. H.
Sherman, C. T.
Shock, G. W.
Shock, W. W.
Sholes, Cass G.
Showacre, H. C.
Showerman, I. C.
Shreffler, W. H.
Shrigley, James A.
Shuman, W. A.
Shurr, S. M.
Shutt, R. A.
Sieberg, H.
Sigler, James H.
Simpson, W. H.
Sisson, Wm. H.
Skinner, Ira G.
Sloat, Henry T.
Slocum, W. P.
Smith A.
Smith, C. B.
Smith, Chas. H.
Smith, Dexter.
Smith, Dewitt Wilmott.
Smith, Day K.
Smith, G. K.
Smith, Hosea.
Smith, H. L.
Smith, Isaac A.
Smith, James A.
Smith, J. W.
Smith, Mary E.
Smith, Richmond.
Smith, R. Hector.
Smith, Samuel H.
Smith, Thos. H.
Smith, T. R.
Smith, Wm. K.
Smithers, H. S.
Snell, Fred. W,
Snow, H. N.
Snyder, A. C.
Snyder, C. L.
Somers, L. A.
Somerville, Wm. B.
Spare, Frank.
Speed, F. M.
Spellman, Chas. H.
Spellman, Geo. E.
Spellman Lewis B.
Spencer, H. B.
Spencer, W. H,
Spencer, J. M.
Spinner, Wm,
Sponagle, J. L.
Spring, Parker.
Starling, C. C.
Standifer, Geo. W.
Stebbins, C, M.
Stephenson, Saml.
Stewart, Danl. N.
Stewart, Frank.
Stewart, J. B.
Steward, Neal,
Stevens, F. H.
Stevens, J. H.
Stidham, P. A.
Stillman, Geo.
Stone, Ellis W.
Stopp, Robt. M.
Stouder, Henry.
Stough, John B.
Street, C. A.
Strong, G. N.
Strouse, David.
Strothers, C. G.
Strubble, Wm. F.
Stull, Thos. G.
Stumm, F. A.
Sturgeon, Geo.
Sullivan, John C.
Sutherland, Norris.
Sweeney, W. B.
Swift, J. A.

T.

Talbot, Robt. M.
Taylor, C. J.
Taylor, Henry H.
Taylor, W. S.
Talmadge, Geo. J.
Teachout, Ed.
Temple, Henry C.
Terry, G. W.
Thayer, Wm. A.
Thomas, Chas. J.
Thomas, John.
Thompson, J. R.
Thorn, Robt.
Thorp, J. B.
Thropp, C. A.
Thurston, J. D.
Tiffany, J. T.
Tingle, E. C.
Tinker, Chas. A.
Tinker, Wm. A.
Tinney, J. D.
Tinney, Wm. E.
Toby, John G.
Todd, A. W.
Todd, Edgar L.
Tomlinson, D. B.
Tompkins, E. P.
Torrance, J. A.
Torrence, J. C.
Town, A. C.
Townsend G.
Townsend, N. S.
Tracey, Richard H.
Tracey, Simon T.
Trimm, William E.
Truax, Jas. D.
Tull, C. D.
Tupper, J. K.
Turner, Geo.
Tyler, F. B.
Tyler, J. D.
Tyrrill, Augustus.

U.

Ulmer, S. S.
Upham, B. H.

V.

Valentine, H. S.
Van Derhoef, Robt. B.
Van Gelder, Jacob H.
Van Gilder, A. C.
Van Gesel, W. T.
Van Valkenburgh, F. S.
Van Zandt, H. E.

Vernetson, W. E.
Vermilion, J. W.
Vincent, H. C.
Vincent, O. B.
Volker, Rudolph C,
Volker, Louisa E.
Voltz, John D.
Von Eye, Edw.

W.

Wagner, John C.
Wagner, Robt.
Wagstaff, John H.
Wait, E. O.
Waite, J. R.
Wallace, Chas.
Walsh, Anthony R.
Walsh, Jas. W.
Ward, Edward T.
Ward, Joseph C.
Ware, W. F.
Warren, Geo.
Warren, J. H.
Warren, Walter.
Warner, D. W.
Warring, Thos. P.
Washburn, Melancthon, E.
Waterhouse, C. D.
Waterhouse, Edwin J.
Waterhouse, T. Q.
Waterhouse, H. F.
Watts, H. M.
Watts, J. C.
Wayne, —.
Ways, Chas. E.
Webb, J. G.
Weber, C.
Weed, W. W.
Weir, L. C.
Wells, A. B.
Wells, V. H.
Weems, J. D.
Weitbrecht, Robt.
Whelpley, C. L.
Whipple, Hiram E.
Whitford, E. P.
Whitney, Charles.
Whitney, C. O.
Whitney, Leonard.
Whittlenife, Chas. S.
Whittlesey, Chas. S.
Whittlesey, J. B.
Wickard, J. W.
Wicker, Frank N.
Wickham, J. J.
Winder, Alfred.
Wilbur, John F.
Wilcox, Stephen T.
Wilkinson, G. D.
Wilkinson, D. W.
Wilson, Ellis J.
Wilson, G. W.
Wilson, John J.
Wilson, J. W.
Wilson, W. B.
Wilson, W. H.
Wilson, W. T.
Williams, D. A,
Williams, F. D.
Williams, Geo. W.
Williams, J. S.
Williams, R. D.
Winker, George.
Winter, Garrett.
Wintrup, John.
Wise, E. P.
Wolf, Charles.
Wood, William B.
Woodring, W. H.
Woods, H. A.
Woodmansy, J. F.
Woodward, B. F.
Woodward, G. D.
Worden, Wm. M.
Worth, Robt. G.
Wortsman, L. W.
Wright, M. E.

Y.

Yeakle, J. B.
York, C. W.
York, Geo. C.
Younkers, Simon T.
Young, Charles.
Young, W. H.

Z.

Zeublin, J. E.

TOTAL, - - - - - 1,079.

INDEX.

A.

Adam, I. L., II, 300, 303, 343.
Adam, J. R., II, 343.
Adams, Taylor, II, 343.
Alexandria captured by Federals, I, 69.
Allatoona saved by the telegraph, II, 232.
Allen, Geo., I, 336.
Allen, Hudson H., I, 317, 323.
Allen, James N., II, 267, 315.
Allen, John, I, 144 147 ; II, 114.
Allen, Wm., II, 120.
Allen, W. F., II, 149.
Alleyn J. T., II, 343.
Allis, Geo. B., I, 336.
Ancient means of communication in war, I, 9.
Anderson, Jos., II, 63, 77, 115, 176, 181, 242-3; 275, 335 ; a close call, 277-8 ; taps a line, 334.
Anderson, W. D. S., II, 120.
Anderson, Wm., II, 209.
Andrews, daring railroad raid in Ga., I, 209.
Andrews, M. S., II, 28, 267, 315.
Angel, Lee, II, 344.
Antietam, battle of, I, 236.
Applebaugh, W. K., II, 260, 266, 315 ; as a pilot, 137 ; first Union operator in Richmond, 320.
Appomattox, surrender of Lee at, II, 326.
Arkansas Post, capture of, I, 272.
Armor, Thos., I, 158, 160, 217 ; under fire, 84.
Armstrong, E. L., I, 348 ; II, 219.
Assassination of President Lincoln, II, 327.
Atlanta, campaign against, II, 173 ; battle of, 183 ; evacuated, 186 ; ruins of, 241.
Atwater, E. W., II, 54 ; captured, 56.
Atwater, H. H., II, 315, 329.
Atwater, Wm., I, 289.
Atwell, J. W., I, 253, 258, II, 80.
Aughinbaugh, D. C., II, 12, 267, 315.
Aveithe, J. B., II, 344.
Averill, Gen., worsted by Gen. J. H. Morgan, II, 143.

B.

Bacon, D. T., I, 114, 117, 185, 214, 242, 250, 253, 283 ; II, 108 ; builds line in S. E. Mo., I, 179.
Baker, C. H., I, 218.
Baker, G. M., II, 342.
Baker, L. C., I, 121 ; II, 341.
Baldwin, G. W., I, 70, 144 ; II, 12, 114, 314, 350.
Baldwin, M. C., I, 107 ; II, 54, 75.
Balloons, their uses in war, I, 15 ; uses in Federal army, 15 ; in siege of Paris, 17.
Balls Bluff disaster, and how it happened, I, 84 ; important telegrams, 86, 87.
Baltimore, riots, I 64 ; threatened, II, 152.
Baltimore and Ohio Railroad, I, 92.
Banks, Gen. N. P., operations in Shenandoah Valley, I, 153 ; in Dept. Gulf, II, 38 ; Port Hudson, 41; Red River, 45 ; Compliments U S. M. T., II, 355.
Barnes, C. F., II, 117.
Barnes, C. M., I, 340.
Barnes, Ed. F., II, 146.
Barnes, L. J., I, 340.
Barr, M. W., II, 120, 344.
Barrett, C. S., I, 345 ; II, 136, 219, 223.
Barron, W. C., II, 123, 267.
Barry, J. B., II, 28.
Barth, M., I, 205, 292 ; II, 54.
Bassett, F. H., II, 106, 209.
Bassett, M. H., II, 106, 209, 211, 343.
Bates, B. R., II,346.
Bates, D. Homer, I, 59, 60, 67, 70, 82, 234 ; II, 123, 350 ; translates enemy's cipher, I, 41.
Bates, Edward E., II, 209.
Bauer, W. H., I, 70.
Baxter, G. W., I, 317, 323; II, 44, 209, 298.
Beard, G. W., II, 120, 344.
Beardslee Magneto-instrument, II, 88 ; described, 90 ; disapproved, 96.
Backwith, Samuel H., I, 35, 56, 119, 250, 253, 317, 324 ; II, 71, 130, 133, 260, 262, 317, 324, 328, 330 ; removed and reinstated, 170.
Bell, Prof. A. Graham, his photophone, I, 32.
Bell, Geo. W., II, 54.
Benner, Frank, I, 260 ; II, 115, 343, 357 ; important service of, 286.
Benson, F. N., I, 169; II, 12, 123.
Bentley, J. N., II, 314.
Benton, H. W., I, 69, 70, 76.
Berry, Curley, II, 149.
Berry, D. T., II, 242, 275.
Berryhill, T. R., I, 253, 349.
Bertrand, C. P., II, 341.
Besanson, C. W., I, 208.
Bickford, F. T., I, 217 , II, 12 114, 136, 260 ; narrow escape of, I, 158
Big Bethel, defeat of Federals, I, 72.
Biggert, W. L., I, 288.
Bishop, H. H., II, 315.
Black, J. H., I, 336.
Blades, L. J., II, 315.
Blair, J., II, 357.
Blestine, G. W., I, 208, 289.
Blish, Jr., Jos., I, 253; II, 80, 188, 219.
Bliss, A. H., I, 254, 358, 361; II, 12, 114, 122.
Bliss, J. E., II, 137, 260.
Bliven, C. E., I, 250, 287.
Bodell, W. J., II, 314.
Bogardus, A. H , I, 117, 144.
Bohanna John, II, 300, 304, 343.
Booth, M. K., I, 317; II, 310 ; death of, I, 323.
Booth, O. H., I, 144.
Bowerman, Henry, II, 209.
Boyd, J. W., II, 315
Boyle, E. C., I, 124-5; II, 235.
Boyle, T. R., I, 47, 124.
Bracken, John H., II, 357.
Bramlette, Governor, opposition by, to enlistment of colored troops, II, 154.
Bragg, Gen., advance into Ky., I, 279.
Brannon, Alfred, II, 343.
Brenaman A. T., II 149.
Brenner, J. A., II, 248, 344.
Brennecke, C. Y., II, 344.
Briggs, H. G., killed in battle, I, 342.
Brigham, R. H., II, 122.
Bromell, W. H., I, 185.
Brooks, J. N., II, 54 ; captured, I, 275.
Brown, C. I., I, 115 ; II, 114.
Brown, Gen. E. B., defends Springfield, Mo., I, 340.
Brown, Edgar H., II, 223.
Brown, E. O., I, 102.
Brown, H. R., II, 315.
Brown, J. K., I, 107.
Brown, J. R., II, 63, 115.
Brown, John W., II, 344.
Brown, N. H., I, 69, 70.
Brown, R., I, 189.
Brown, Samuel M., I, 67.
Brown, T L., II, 344.
Bruch, Adam, I, 288 ; II, 63, 76.
Bruch, Samuel, I, 191, 204, 210, 282, 287, 293, 295 ; II, 104, 111, 112, 115, 158, 205 ; appointed Asst. Manager Telegraphs, in Ky., 1, 125 ; commissioned, 131 ; repairs line to Nashville, 203 ; important service of, II, 77 ; death of, 292 : monument, 293 : biographical, 293.

Bruner, Philip, I, 107, 189, 205, 242, 289, 290, 298.
Brush, Geo. M., I, 253, 274 ; captured, 329 ; death, 330.
Brush, Samuel T., I, 120, 253.
Bryan, T. M., II, 120, 344.
Bryant, D. N., II, 28.
Bryant, E. B., I, 98.
Buck, C. H., I. 152.
Buell, Gen. D. C., takes command in Ky., I, 125 ; advances on Nashville, 201 ; retires army to Ky., 279 ; army-train saved by operator, 209; advances from Louisville, 294.
Buell, H C., captured, I, 158, 163, 358 ; II, 114.
Buell, M. V. B., I, 69, 70, 76 ; II, 114, 315, 350.
Bulkley, C. H., commissioned, I, 131 ; II, 47, 104 ; supt. Gulf dept., II, 38 ; clears Bayou Teche, 39 ; resigns, 47, 208 ; condemns Beardslee system, 98.
Bull, Run., Federals defeated ; the news thereo., I, 75 ; 2d battle of, 226.
Bull, H. P., II, 315.
Bullock, R. B., II, 345.
Bunnell, J. H., I, 69, 70, 71, 142, 232, 239 ; II, 53, 67, 75, 114, 176, 311 ; prevents defeat at Gaines' Mills, I, 149.
Burbridge, Gen., enforces law about enlistments in Ky., II, 154.
Burch, E. M., II, 344.
Burhans, W. W., II, 54.
Burke, Ed., II, 343.
Burnapp, G., I, 185.
Burnett, B. M., I, 253.
Burnett, G. A., I, 47, 185.
Burns, S. C., II, 114, 123, 131, 261.
Burnside, Gen. E. A., on N. C. coast, I. 146 ; supersedes Mc Clellen, 241 ; Fredericksburgh campaign, 351 ; commands Dept. Ohio, II, 54 ; captures Knoxville, 60 ; repulses Longstreet, 75 ; compliments U. S. M. T., 354.
Burr, M. P., II, 228-9.
Burucker, J. L., I, 56.
Bush, Chas., captured, II, 167.
Bush, W. H., II, 344.
Butler, Gen. B. F., takes command S. E. Va. Dept., I, 70 ; operations about Bermuda Hundred, II, 130, 140 ; at Ft. Fisher, 274.
Butler, E. D., I, 253, 318, 329 ; II, 188, 205.
Butler, E. F., I, 253 ; II, 80, 188, 204, 205 ; joins in fight, II, 81.
Byington, Dwight, I, 56, 336, 346 ; II, 215, 217, 292.
Byrne, J. H., I, 117, 180 ; II, 218.
Byrnes, J. O., II, 203.

C.

Cairo, Ills., during the war, I, 326.
Caldwell, A. Harper, I, 58, 142, 149, 232, 241 ; II, 12, 122, 129, 130-6, 141, 267, 315, 317, 327, 340, 345, 350 ; death, 351.
Caldwell, James, II, 126, 130, 315, 317.
Camp, R. O., I, 209.
Campbell, C. T., II, 249.
Campbell, F. H., II, 327.
Campbell, V. F., II, 344.
Canby, Gen., operations of, in Div. of W. Miss., II, 199 ; against Mobile, 297.
Canadian operators, II, 356.
Cape Girardeau, Marmaduke, defeated at, I, 344.
Carey, Samuel, I, 208 ; II, 63.
Carnegie, Andrew, first military telegraph officer, I, 66, 67 ; mention of, 347.
Carnifax, Ferry fight I, 103.
Carrier Pigeons, employment of, in war, I, 13.
Carroll, J. J., II, 315.
Carver, J. W., II, 315.
Case, Hardin, II, 222.
Cassell, J. A., I, 191, 200, 287; II. 54, 205, 285.
Cathcart, W R., I, 135 ; II, 345.
Caton Hon. John D., I, 119.
Cedar Mountain, Battle of, I, 219.
Chamberlain A., II, 28.
Chamberlain, H B., II, 278.
Chancellorsville campaign, I, 359.
Chandler, Albert B., I, 53, 60, 234 ; II, 15, 22, 123 314, 333, 350 ; translates Enemy's cipher, I, 41.
Chandler, J. W., 1, 348.
Chantilly, Battle of, I, 228.
Chapman, E. T., II, 64, 114 ; valuable service of, II, 163.
Chappell, Scott R., II, 209.
Cherry, J. L., I, 102, 217, 220, 232, 368 ; II, 28, 114, 126 ; capture and escape, II, 352 ; first army typo, I, 98.
Chattanooga, semi seige of, II, 69.
Chickamauga campaign, II, 64.
Chickasaw Bluffs, Battle of, I, 272.
Childs, A. F., II, 315.
Chisholm, W. H., II, 343.
Cincinnati threatened, I, 288.
Cipher, see Cryptographs.
Clarke, F. S., II, 278.
Clarke, H. K., II, 278.
Clarke, John B., I, 112, 172; II, 312.
Clark, John E., II, 186.
Clark, J. R., I, 200
Clarke, J. S., II, 117.
Clarke, W. H., II, 117.
Clapp, R., II, 278.
Clowry, Robt. C., I, 119, 251, 344, 346, 349; II, 48, 215, 217, 223, 224, 267, 292, 312, 342, 348; biographical, I, 346.
Cloyd's Mountain battle, II, 144.
Clute, C. C., II, 341.
Coburn, J. R., II, 344.
Cochrane, S. D., II, 81, 84, 188; death of, 309.
Cody, W. H., I, 136; II, 343.
Cold Harbor, battle of, II, 138.
Cole, Geo., II, 63, 77, 115.
Colestock, D., II, 149, 268.
Collier, B. W., II, 120, 343.
Collings, J. W., II, 131.
Columbus, Ky., occupied by Confederates, I, 120.
Congdon, J. D., I, 56.
Conner, Charles O., II, 209, 298.
Conway, Edward, I, 158, 217, 232, 240, 246; II, 38, 45, 114, 356 ; office shelled, I, 84; brings news of battle, 226; narrow escape, II, 47.
Cook, F. C., I, 125, 288; II, 63.
Corinth, Miss., capture of, I, 246; battle of, 255.
Corrigan, Sergeant, killed, II, 137.
Cotton, J. S., I, 205.
Cowan, H. W., I, 246, 254, 368 ; II, 12, 114, 122, 137 262.
Cowlam, Geo B., bravery of, I, 139, 144.
Cox, Gen., advances in W. Va., I, 103, 166 ; defeats Heth, 168; retires, 168.
Craig, Hugh, I, 200, 328; II, 54, 357.
Crane, Mark D., I, 205, 243, 253, 256, 258.
Crittenton, J. N., successes in N. Ala., I, 208, 210, 284; captured, 291.
Cromwell G. E., I, 208, 253; II, 114, 315.
Crook, Gen., advance of to Dublin, Va., II, 143; Cloyd's Mountain battle, 144.
Crouse, Jesse, I, 69, 70.
Cryptographs—war, chapter on, I, 33; Confederate States, 37-9; C. S. key, 38; C. S. mail system, 41; facsimile hieroglyphs, 41; discovery of key, 42; important translations, 42 ; Federal system, its beginning and growth, 44 ; No. 9, complete—see Appendix, Vol. I; interference by Grant, II, 170; facsimile message, 287.
C. S. Telegraphic Club, II, 119.
Cubitt, Jr., Richd., II, 209.
Culbertson, C., I, 328; II, 188.
Culgan, E. W., II, 315.
Cumberland Gap, captured, I, 192; evacuated, 286.
Cunningham, R., II, 114.
Curtis, Gen., at Pea Ridge, I, 179; along White River, Ark., 180.
Curtiss, G. H., I, 169; II, 149.
Curtiss, J. W., II, 115, 159.

D.

Daniel, Arthur, II, 344.
Daniel, F. V., II, 342.
Daniel, Lee S., I, 321 ; II, 342; saves Vicksburg from capture, I, 271.
Darlington, H. P., II, 315.
David, T. B. A., I, 94, 347; II, 104, 148; opens line on B.& O.R.R., I, 95, 97, 100, 101: successes and difficulties in W. Va., 166 ; sent to Mo., 250; biographical, II, 145.

David, N. C., II, 120, 343.
Davidson, Gen. J. W., joins movement on Little Rock, I, 334.
Davis, D. L., II, 341.
Davis, F. W., II, 267. 315.
Davis, Geo. B., II, 315.
Davis, H. C., II, 343.
Davis, Jefferson, quits Richmond, II, 319.
Davis, Wm., II, 342.
Dealey, Wm. J., I, 83, 84. 216, 231, 260-1, 346; captured, I, 65, 233, 236.
DeBree, N., I, 84; II, 149.
DeParis, Comte, compliments U. S. M. T., II, 356.
DePew, C. I., II, 119, 344.
Deetz, Geo. M., I, 94, 96.
DeForrest, C. L., II, 278.
Delgado, Philip, II, 300.
Dennis, L. B., I. 169; II, 149.
Denny, J. C., II. 106.
Deslondes, J., II, 209.
DeWitt, W. K., I, 297, 363; II, 114.
Dixon, J. R., II, 203, 211, 298, 299, 310, 342.
Dodd, D. O., hung as a spy, I, 349.
Dodge, Gen., expedition of, I, 316.
Dodge, E. R., II, 344,
Dodge, S. C., II, 344.
Dolan, Thos., I, 144, 146, 240.
Dorchester, J. C., I, 303.
Doren, Dennis, II, 11, 127, 129, 136, 137, 138, 141, 320, 324, 350.
Dorrence, O. H., I, 217, 357; II, 114, 123, 274.
Dorsey, Benj., II, 300, 310.
Dougherty, A. D., I, 177, 258.
Dougherty, Chris., I, 209; his air loom, 281; as a warrior, 283, captured, 292.
Douglass, J. H., I, 185 253; II, 125, 261.
Drake, Wm. H., I, 189, 194; II, 232, 242-4, 254; deceived by Morgan's operator, I, 194; as a dragoon, II, 293.
Drummond, Frank, I, 84, 154, 356; II, 114, 357; captured, 160; his prison diary, 161.
Duell, J. T., captured, II, 236.
Dugan, G. M., II, 119.
Dulin, W.E., II, 344.
Duncan, J. C., II, 344.
Dunn, W. A., I, 218; II, 315.
Dunlap, J. R., I, 185; II, 28.
DuPont, Commodore, operations along coast, II, 25.
Dwyer, Con, I, 204, 210, 276, 292, 296, 298, 302.
Dwight, H. P., II, 79, 123.
Dwight, J. H. II, 314.
Dyer, D. O., I, 56, 117; II, 236.

E.

Earley, Gen., advances into Md.; threatens Washington, II, 151; defeats in valley, II, 270.
Eastland, Thomas, II, 300, 301, 304.
Eckert, E. A., II, 320
Eckert, Thos. T., I, 47, 60, 91, 146, 220, 246, 357-9, 364; II. 17, 23-8, 103-4, 129, 132, 141, 223, 260, 267, 314, 327-8, 331-2-3-5, 344-8, 350; Biographical, I, 142.
Eckman, Wm. H., I, 67; II, 114.
Edison, T. A., I, 347.
Edwards, C. H., II, 198.
Edwards, Samuel H., I, 363; II, 11, 12, 15, 20, 96, 123, 127, 130, 134, 139, 260; nearly killed, I, 368.
Eddy, Charles G., I, 56; II, 106, 181, 182, 185, 233, 241, 242-4-6, 275.
Egan J. J., I, 253, 328; II, 80, 83, 188, 189, 285, 299, 343.
Elison, A., I, 189, 208, 357.
Ellsworth, Geo., II, 244-6; taps Union lines, I, 193; deceives Union operators, 194, 197; misconduct of, 275; captures Union operator, II, 56.
Elphick, D., II, 38, 209, 304.
Embree, W. N., I, 152, 217, 358; II, 114, 132, 260, 315, 327; captured, 361.
Emerick, J. H., I, 144, 147, 217, 232, 235, 239, 240, 358, 363, 368; II, 11, 12, 114, 127, 130, 133, 136, 315, 316, 318, 321, 345.
Evans, S. P., II, 344.
Evans, W. J., II, 344.
Everett, Geo., II, 236.

F.

Fairchild, S. B., II, 38, 45, 46, 209, 306.
Fair Oaks, battle of, I, 147.
Fall, P. H., prevents surprise of Vicksburg, I, 271.
Fancher, C. I., II, 106, 209.
Farmington, battle of, I, 243.
Farnham, G. M., II, 232.
Farragut, Admiral, captures New Orleans, II, 37; before Vicksburg, 37, 40; in Mobile Bay, 210.
Felton, G. C., II 278.
Fires, their use for signaling by the ancients, I, 18.
Fish, E. G., II, 166.
Fish, H. J., I, 117; II, 85, 166.
Fitchet, Hamilton, I, 70.
Five Forks, battle of, II, 316.
Flack, B. W., II, 123.
Flagg, J. A., I, 217, 223, 225, 228; captured, 357.
Flanery, David, II, 190, 311, 344.
Flannagan, J. J., II, 71.
Flesher, Jr. Thos., I, 70.
Flippen, M., II, 119, 343.
Flippen, W. E., II, 343.
Florence, J. C., II, 149.
Floyd, Gen., in W. Va., I, 103.
Flynn, J. D., II, 11, 122, 259-60, 315.
Foley, Wm., I, 120, 323, 328, 329; II, 44, 45, 46.
Follett, R. F., I, 144.
Fonda, T. H., I, 70, 144; II, 12, 17, 114.
Fort Pillow, Massacre, II, 192.
Forbes, W. W., I, 209, 253, 328; II, 66-7-8; captured, I, 262.
Forker, James, II, 54; captured, 58.
Forrest, Gen., raid in W. Tenn., I, 257, 260, 261; captures Murfreesboro, 273; defeated, 275; captures Streights command, II, 51; in W. Tenn., Dec. 1862, 82; another raid, defeated at Paducah, 191; Ft. Pillow massacre, 182; defeats Sturgis, 198; in Memphis, 203; Last raid in Middle Tenn., 227.
Forsey, W. S., II, 220.
Forster, Wm., II, 28, 357; taps enemy's line, 29; capture and death, 30.
Fowler, Peter, I, 178, 277; II, 188.
Fowler, R. S., II, 80, 84, 188, 202-5.
Fox, Fred., I, 217, 224.
Franco-German war, telegraph in, I, 28.
Frank, H. D., II, 315.
Frank, J. R., II, 85, 188, 205, 298.
Franklin, Tenn., battle of, II, 234.
Franklin, Gen., compliments U. S. M. T., II, 354.
Frazier Farm, battle of, I, 151.
Frazier, P. B., I, 180.
Fredericksburgh, Burnsides, campaign, I, 351.
Freeman, J. K., II, 344.
Fremont, Gen., appoints supt. telegraphs in Mo., I, 114; pursues Price's army, 115; superseded, 116; operations in W. Va., 153, 167.
Frey, J. J. B., II, 28.
Ft. Donelson, operations against and capture of, I, 173; attacked by Woodward, 277; by Forrest and others, II, 52; Ft. Donelson line, 166.
Ft. Fisher captured, II, 274.
Ft. Henry, operations against and capture of, I, 173.
Ft. McAllister, II, 252.
Ft. Monroe, I, 137; line to, 137.
Ft. Sumter attacked, I, 63.
Fulda, E. A., II, 120, 344.
Fuller, Edward, II, 84.
Fuller J. A., I, 208, 283, 289, 298; II, 67, 68, 182; captured, 299; homesick, II, 52; leads colored men, 52.
Fuller, W. G., I, 47, 51, 56, 94, 188, 189, 190, 200, 317, 329; II, 79, 81, 82, 84, 108, 202, 205, 292, 297, 304, 340, 341, 342, 343; opens line on B. & O. R. R., I, 95, 97, 101; brave conduct, 102; more frightened than hurt, 106; commissioned, 131; goes to Ky., 168; builds line to Cumberland Gap, 192; dangerously ill, 193; biographical, 262; repairs Grant's lines, 264, 325.

Furr, R. A., I, 169; II, 149.

G.

Gaines' Mills, battle of, I, 149.
Galbraith, Jr., J. A., II, 120, 343.
Gallup, Geo., I, 328; II, 188.
Galvin, J. H., I, 289.
Gamble, James, I, 119.
Gantt, Robt. S., I, 122.
Gardner, O. A. A., I, 179.
Garland, Geo., II, 114, 122.
Garland, J. G., I, 200, 287, 356.
Gaston, C. A., II, 117, 345; taps Union lines, II, 264.
Gentry, W. G., II, 125.
Gettysburg campaign, II, 9.
Gibson, Wm., II, 85, 188.
Giffard, Carl, II, 278.
Gile, C. H. B., I, 258.
Gilkerson, Jos. H., II, 63, 115.
Gilmore, J. R., I, 69, 70, 71, 82, 362; II, 27, 255, 275, 342, 348; succeeds Strouse as supt., I, 82; builds line Washington to Banks' army, 82; commissioned, 131; in N. C., II, 33; biographical, 35.
Glascott, W. H., II, 272-3-4, 357.
Glass, B. B., II, 38, 236; a perilous ride, 41; a bomb proof, 42; secures a trophy, 42.
Glazier, J. H., I, 217; II, 12, 114, 122, 127, 267, 315.
Goalding, G. J., II, 63, 115.
Gordon, M., I, 169.
Grace, Fred J., II, 38.
Grace, G. H., I, 347.
Graham, M. B., II, 149, 268.
Graham, Richard, I, 158, 217, 224; II, 114, 123, 315.
Grant, Gen. U. S., occupies Paducah, Ky., I, 120; captures Ft. Donelson, 173; at Shiloh, 212; succeeds Halleck, 251; advances against Pemberton, 257; communications cut by Forrest and VanDorn, 260, 261; appoints Riggin Supt. Telghs., 264; operations against Vicksburg, 313; a present to, from operators, 324; commands division of Miss., II, 71; wilderness, 128; Spottsylvania, 135; Cold Harbor, 138; Petersburg, 139; interference with cipherer, 170; siege of Petersburg, 257; Five Forks, 316; surrender of Lee to, 326.
Grasby, Wm., II, 278.
Green, Dr. Norvin, II, 296, 345.
Gregg, John C., I, 56, 164; II, 54; bravery of, 70.
Gregg, J. W., II, 149, 223.
Grierson, Gen., expedition of, I, 316; raid to Vicksburg, II, 206.
Griffin, R. B., I, 178, 258, 330; II, 188, 192.
Griffin, S. L., I, 117, 253; II, 38, 42.
Griffith, C. H., I, 125; II, 54; taps rebel wires, 201, 205; captured, 206.
Grimes, A., I, 321.
Griswold, Martin E., II, 237.
Gross, Wm. L., I, 253, 328; II, 78, 80, 84, 164-5, 175, 340, 341-8; commissioned, 131; biographical, 61; report by, 157.
Gross, C. T., I, 56.
Groveton, battle of, I, 221.
Guerrillas, made drunk, I, 354; capture of a General in camp at night, 360; shoot an operator, II, 201; operator nearly shot by, 308.
Gulick, Palmer, II, 320, 321.
Gulliher, J. K., II, 204, 205.
Guntown, battle of, II, 198.
Guthridge, Jules F., I, 69, 70; II, 25.

H.

Hall, E. A., II, 11, 13, 20, 122, 130, 134, 135, 139, 260.
Hall, W. C., I, 76, 158, 216; II, 114, 122, 126, 135, 139.
Hall, W. H., I, 249.
Halleck, Gen., in command in Mo. and Ky., I, 170; made Gen.-in-Chief, 218, 251; advance on Corinth, 242.
Hallum, J. W., I, 144.
Hambright, C. K., I, 362; II, 139, 259-60, 327.
Hammann, C. A., I, 346; II, 217, 223, 292.
Hammond, C. D., II, 266-7, 315.
Hammond, Chas. W., I, 178, 208.
Hampton, Gen. Wade, raids of, I, 355, 356; captures, 2,486 cattle, II, 265.
Hancock, A. G., II, 315.
Hanna, Geo. S., II, 209, 304.
Hanson, Jos., I, 348; II, 50.
Hardee, Gen., invited by telegraph to Christmas dinner with Union General, II, 247; evacuates Savannah, 255.
Harpers Ferry, surrendered by Miles, I, 233.
Harvey, C. B., II, 343.
Harvey, T. G., II, 63, 115.
Harris, Merritt, I, 182; II, 342.
Hassard, J.R.G., translates campaign ciphers, I, 33.
Havens, P. C., II, 28.
Hawkins, A. S., I, 85, 243, 253, 256; II, 83, 237, 352.
Hawkins, W. D., II, 84, 204.
Hebard, L. J., II, 305.
Hefferin, John, II, 209.
Heiss, W. H., I, 138.
Helena, Prentiss, victory at, I, 333.
Heliograph, Heleotrope, Heliostat, I, 29, 30.
Henderson, Ed., I, 209.
Henderson, Geo., I, 246; II, 135, 139, 343, 346.
Henderson, J. H., II, 119, 344.
Henley, W., II, 343.
Hess, G., II, 343.
Hewitt, W. S., I, 47, 114, 175.
Hill, Jacob V., I, 253, 258, 328, 329; II, 188; death of, 192.
Hindman, Gen., invades Mo., I, 338.
Hine, C. C., I, 119.
Hogshead, E. H., II, 119, 120, 343.
Hoffman, Milliage, II, 343.
Holbrook, S. C., II, 267, 315.
Holden, A. H., II, 274.
Holden, D. L., II, 341.
Holdridge, J. C., I, 253; II, 53, 67; captured, 260.
Holloway, Wilbur F., I, 91, 144, 146, 362; II, 114, 131, 346.
Holly, M. Y., II, 114.
Holly Springs, captured by VanDorn, I, 257.
Holt, S. E., II, 343.
Holt, Theo., I, 56, 180, 185, 336; II, 217, 224, 292.
Homan, C. A., I, 362; II, 132, 261, 315, 346.
Hood, Gen., advance on Nashville, II, 227; defeat, 238, 283.
Hooker, Gen., takes command army Potomac, I, 359; Chancellorsville campaign, 359; invasion of Penn., II, 9; captures Lookout Mountain, 74; on U. S. M. T., 355.
Hoot, R. D., II, 343.
Hoopes, L. R., II, 248, 288.
Hoover, R. B., II, 54, 179.
Horner, John, II, 315.
Hotchkiss, Z. P., I, 185; II, 298.
House, Wm. T., II, 298.
Howard, A. A., II, 209.
Howard, Emmet, I, 120, 181, 182; II, 119, 190, 343.
Howard, J. A., II, 180.
Howard, S. A., II, 344.
Howard, S. D., I, 185.
Howe, K. B., II, 85, 205.
Howell, A. J., I, 208, 258, 289, 321; II, 63, 115.
Howell, A. W., II, 357.
Howell, R. N., II, 221, 222, 357.
Howell, S. V., II, 291.
Hughes, Barney, II, 343.
Hughes, R. J., II, 80, 83, 188.
Hull, A. K. V., II, 149.
Humphreys, Jos., I, 280, 292.
Humphreys, R. C., II, 149.
Hunter, Alex., I, 117.
Hunter, D. W., death of, II, 209.
Hunter, Wm., I, 47.
Hunter, Gen., supersedes Fremont in Mo.; retires the army from Springfield, I, 116; supersedes Sigel in Va., II, 151; advances on Lynchburg, 151.
Huntington, Geo., II, 264, 274.
Huyck, M. A., II, 260, 315.
Huyler, G. B., II, 344.
Hynes, J. W., II, 209.

I.

Incendiary designs, I, 54.
Ingle, J. O., I, 258, 316, 317, 322 ; II, 83, 138, 239, 285.
Ingram, F. M., I, 84, 169.
Irvine, Hugh, II, 352.
Island No. 10, captured, I, 183.
Iuka, Miss., captured by Price, I, 255.
Ives, W. L., II, 315.
Izard, James, II, 79.

J.

Jackson, Gen. T. J., his famous raid down the Shenandoah, I, 154.
Jacques, Charles W., I, 69, 70, 76, 77, 142, 147; II, 54, 59, 60, 63, 75, 76, 77, 114, 175, 177, 181, 185.
Jenneson, D. C., I, 347.
Johns, C. H., I, 168, 217; II, 59, 63, 77.
Johnson, B. H., I, 258, 259.
Johnson, C. S., II, 209, 211.
Johnson, D. H., II, 149.
Johnson, E. H., I, 253, 317, 323; II, 44, 209.
Johnson, Thos. S., I, 70; II, 344.
Johnson, W. R., II, 344.
Johnston, Gen. J. E., wounded at battle of Fair Oaks, I, 147; opposes Sherman in Ga., II, 176; surrenders to Sherman, 335.
Johnston, J. R. W., II, 344
Jolls, J., II, 149, 269.
Jones, A. C., I, 303.
Jones, F. C., captured, II, 189.
Jones, Geo. E., II, 60, 64, 115.
Jones, H. P., I I, 334.
Jones, James, I, 191, 200, 303; II, 184; frightens off Morgan's command, I, 199 ; captured, 287.
Jones, Mervin, captured, II, 189.
Jones, W. P., II, 343.
Jonesboro, Ga., battle of, II, 185.
Joyce, J. T., II, 54, 239.

K.

Kanode, A. H., II, 114, 123.
Kates, J. W., I, 135; II, 344.
Kay, Geo. T., II, 315.
Keefer, C. A., II, 63, 343, 357.
Keith, J. S., I, 169; II, 149.
Kelly, J. M., II, 155.
Kelly, W. O., II, 343.
Kelsey, W. H., I, 208, 292.
Kennan, F. W., II, 344.
Kennesaw, battle of, II, 178.
Kent, D., death of, II, 35.
Kentucky, early affairs in, I, 122 ; opposition to Negro enlistments, II, 154.
Kerner, M. H., I, 83-4, 144, 169; saved by Nancy Hart, 104.
Kettles, W. E., II, 123, 314, 322.
Kilpatrick, Gen., narrow escape, II, 277.
Kimber, S. P., I, 54; II, 209.
King, Geo., II, 344.
King, Thomas, II, 274.
King, W. A., I, 70.
Kingsbury, F. W., II, 28.
Kinneman, O. H., I, 70, 217.
Kinney, Ed., II, 80, 84.
Kirby, J. S., II, 11, 114.
Klunk, J. P., II, 149.
Knapp, S. B., I, 349.
Knight, F. B., II, 131.
Knox, Claude, I, 210, 291.
Knoxville, Burnside's objective, II, 54; captured, 60; Longstreet's move against, 75.
Korty, Louis H., II, 299, 342; a dread ride, 308.
Kress, Wm. B., I, 70.
Kuhn, R., I, 117.
Kunkle, H. B., I, 180.

L.

La Bonte, J., I, 144.
Lacey, C. E., II, 343.
Laird, T. A., II, 314.
Lamb, Frank, I, 115; II, 28, 114; captured, I, 158, 163, 355.
Lanbright, W. H., II, 343.
Lancaster, W. H. H., I, 169; II, 149, 269.
Lathrop, D. B., I, 70, 144; killed, 145.

25

Lathrop, John P., II, 54, 169.
Laverty, R. C., II, 264, 315, 320, 324; nearly killed, II, 262.
Lawrence, F. A., I, 144.
Lawrence, G. J., I, 159, 231; II, 114, 315.
Le Barron, Ch., I, 136.
Lee, N. A., II, 344.
Lee, Gen. R. E., I, 149; invades Md., I, 229; invasion of Pa. thwarted, 235; invades Pa., II,9; retreats from Gettysburg, 21; Wilderness, 128; Spottsylvania, 135; Cold Harbor, 138; defense of Petersburg, 257; surrender, 326.
Leete, G. K., I, 347.
Lehr, Charles, II, 59, 64; repairs line under fire, I, 111; captured, 117; important service of, 289; remarkable success of, II, 60.
LeLoup, Ed., II, 300, 304.
Lenhart, John, I, 124; II, 58.
Lennox, G. P., I, 253; arrested, 258.
Lewis, T. R., I, 253, 256, 258.
Lewis, W. S., II, 188.
Lexington captured by Confeds,, I, 288.
L'Hommedieu, W. A., II, 120.
Lies, *artful*, II, 246.
Ligon, J. R., II, 343.
Lincoln, President, A , his cipher, I, 34; his solicitude, 91, 234; II, 20, 22; proclaims emancipation, I, 239; relieves Hooker, II, 15; fears Lee will escape Meade, 22, 324; assassination of, 327; stories by, 331.
Lindauer, A. C., II, 209.
Lindley, W. T., I, 83, 84, 169.
Lindsey, L. T., II, 120, 204.
Lines, H. F., II, 285-7.
Lines, R. B., II, 284, 343.
Lithgow, C. H., I, 144, 217, 221.
Lithgow, H., I, 208, 289.
Little Rock, capture of, I, 332.
Livergood, W. W., I, 185.
Livingston, E. N., I, 288.
Lock, John M., I, 71.
Logue, W. S., II, 132, 260.
Lonergan, John, II, 54, 66, 242-7, 251-3, 275-7; hid by negroes, 55; narrow escape, 72.
Long, F. C., II, 267.
Longstreet, Gen., before Knoxville, II, 75.
Lookout Mountain, Hooker carries, II, 74.
Loomis, Robt. C., II, 209.
Loucks, T. N., I, 217; II, 114, 123, 261, 267, 315.
Lounsberry, C. H., I, 70, 158, 216, 231; II, 114, 315; under fire, 84; captured, 233.
Low, Geo. A., II, 314.
Lowe, James, II, 73.
Lucas, B. H., II, 343.
Ludwig, D. J., II, 125, 307, 315; captured, I, 233.
Ludwig, John F., II, 203, 228-9; captured, 230, 327, 343.
Lyle, J. S., I, 208, 258, 289; captured, 261; II, 63.
Lynch, S. G., II, 149, 340-8; biographical, II, 146.
Lyon, Gen. Natl., captures enemy at St. Louis, I, 110; routs enemy at Booneville, Mo., 110; goes to Springfield, 110; his danger and death, 113; battle of Wilson Creek, 113.

M.

Macon, Ga., captured by Wilson, II, 288.
Magonigle, S. R., I, 144.
Maiden, W. P., II, 123.
Maize, I. D., I, 169.
Malvern Hill, battle, I, 151.
Manson, Gen., defeated in Ky., I, 288.
Mapes, H. R., II, 54.
Marchant, E. L., I, 321; II, 119, 343.
Marmaduke, Gen., in Mo., I, 339; attacks Springfield, Mo., 340.
Marshall, E. J., II, 344.
Marshall, T. F., II, 120, 344.
Marshall, W. M., II, 343.
Martin, E. O., II, 344.
Martin, Wm., II, 285.
Martyn, D. E., I, 189, 194.
Mason, C. T., II, 344.
Mason, E. W., II, 267, 315, 317.
Mason, John Q., II, 166, 327.
Mason, W. T., I, 56, 328.
Matthews, Col. S., defeats Wheeler, I, 297.

Matlock, H. H., II, 114.
May, Marion, I, 328.
Maynard, Geo. C., II, 314, 346.
Maynard, W. S., II, 315.
McCaine, J. H., II, 315.
McCaine, R. R., I, 83-4, 154-9, 217, 221-2; II, 12, 123, 151, 271, 315, 317, 325.
McCandless, L. D., I, 154, 220-1, 232; II, 17.
McCann, Geo., I, 136.
McCargo, David, I, 347.
McCarty, D., I, 208; II, 114, 123.
McClellan, Gen., relieves Lt-Gen. Scott, I, 78, 128; Telegraph follows his army first, 97; Peninsular campaign, 141-7; Halleck's unkind treatment of, 218; brings army to Pope's aid, 219; resumes command in Dep't, 229; pursues Lee into Md., 232; South Mountain, 234; Antietam, 236; superseded by Burnside, 241; compliments U. S. M. T., II, 355.
McCleskey, Wm., I, 182.
McCleverty, J. D., captured, I, 344.
McConnell, C. C., II, 12.
McCrickett, M. J., I, 224, 225; II, 357; death of, 272-4.
McCutcheon, J. C., II, 335.
McDowell, Gen., at Bull Run, I, 75; operations against Lee and Jackson, 153.
McGaughey, D. C., brave conduct of, II, 34; death, 35.
McGintey, E. H., I, 144; II, 28, 324.
McGintey, Terry, II, 28, 278.
McGintey, W. T., II, 28.
McGovern, Joseph, II, 344.
McGuire, J. D., II, 223.
McIlvaine, J. F., II, 28.
McIlvaine, J. P. I, 208.
McInnes, W. N., I, 83-4.
McIntosh, Wm., II, 136; captured, I, 158.
McKelvey, A. T., II, 149.
McKenna, E. W., II, 267.
McKenna, J. A. L., II, 314, 318.
McLaughlin, Jerry, II, 288.
McMichael, Isaac, I, 117.
McMullen, Frank U., II, 198, 343.
McMullen, J. W., I, 217.
McMurchy, W. A., II, 117, 119.
McMurchy, G. W., II, 343.
McMurray, A. K., II, 28.
McMurtrie, B., I, 208; II, 219.
McNairn, E. B., II, 85, 188, 191; remarkable adventures of, 194.
McPherson, Gen., advances from Chattanooga, II, 174; killed, 183.
McReynolds, C. W., II, 80, 85.
Meade, Gen., supersedes Hooker, II, 15; Gettysburg campaign, 15; operations preceding the Wilderness campaign, 126; Wilderness, 128; Spottsylvania, 135; Cold Harbor, 138; Petersburg, 139; siege of Petersburg, 257.
Meagher, James, I, 189, 200.
Mechanicsville, battle of, I, 149.
Meehan, John H., II, 209.
Memphis, Tenn., Forrest s dash into, II, 202.
Merkley, W. H., I, 328; II, 188, 357.
Merrimac, iron-clad, destroyed, I 146.
Merriwether, C. G., I, 347, 248, 250, 300, 304, 343.
Miller, Frank, I, 96.
Miller, F. B. G., 11, 149.
Miller, John F., captured, II, 276.
Miller, W. H., II, 54, 67, 75.
Mill Springs, battle of, I, 187.
Minnehan, J., II, 315.
Missionary Ridge, carried by Grant's troops, II, 74.
Missouri, affairs early in the War, I, 109; in Winter of 1861-'2, 170; in Summer '62, 338.
Mingle, John K., II, 249, 268, 343.
Mitchell, F. W., II., 344.
Mitchell, Gen. O. M., successes in N. Ala., I, 208; sends raiders into Ga., 209.
Mixer, C. H., II, 267, 315.
Mobile evacuated, II, 303.
Mobile Bay, entrance to, captured, II, 212.
Moffatt, A. H., II, 248.
Molony, P., II, 209, 343.
Monitor and Merrimac fight, I, 139.
Montanye, C. D., II, 239.
Montgomery, H. A., I, 121; II, 341.
Moreland, T. E., I, 144, 220; II, 114, 123.
Morgan, Gen. G. W., captures Cumberland Gap, I, 192; evacuates the Gap, 287.
Morgan, J. B., II, 106.
Morgan, Gen. John H., captures builders, I, 189; raid into Ky., July, 1862, 193; telegraphs to Unionists, 200, 277; captures Gallatin, 273; success at Hartsville, 297; raid in Ky., Dec., 1862, 303; Ohio raid, II, 55; captured, 59; fights Averill, 143; raid of June, 1864, 159-62.
Morley, R. F., I, 67.
Morris, Charles, I, 56.
Morris, Jas., I, 276.
Morris, J. B., I, 136; II, 119, 120, 344.
Morris, Dr. Wm. S., I, 135; II, 118, 286.
Morris, Zeph, I, 288.
Morrison, Thos., I, 144, 147; II, 114, 123, 267, 315.
Morse, Prof. S. F. B., invents telegraph, I, 24; II, 339, 25; in Berlin, 28.
Mosby, Gen., raid of, I, 360.
Mullarkey, P., I, 254, 289, 298; taps enemy's lines, 307, 328; captured, II, 58.
Mulligan J. J., II, 315.
Mundle, S. D. II, 344.
Munfordsville, captured by Bragg, I, 289.
Munro, W. H., II, 45, 46, 343.
Munson, J. A., II, 315.
Murfreesboro, captured by Forrest, I, 273; occupied by Rosecrans, 302.
Murray, Danl., I, 209, 288.
Murray, James, I, 363, 368.
Murray, J. A., II, 123, 129, 267.
Murray, P. J., II, 242.
Murphy, R. J., II, 131.
Myer, A. J., invents signal system, I, 21; Gen. Butler appoints him telegraph manager, 71; inspects the corps, 72; his field telegraph plans, II, 86.
Nagle, J. E., I, 303; II, 63, 115.
Nail, Geo. W., I, 144; II, 114, 123.
Nash, W. H., I, 168, 169.
Nashville, Tenn., entered by Federals, I, 203; besieged, 295; Hood's advance on, II, 227; defeat, 238.
Nesbitt, J. G., I, 96.
Nettles, W. M., II, 248, 286, 343.
New Madrid, besieged and captured, I, 183.
New Orleans, capture of, II, 37.
Newton, O. K., II, 211.
Nichols, A. M., II, 54.
Nichols, Henry E., II, 304.
Nichols, H. W., I, 120, 185, 253, 258, 260 323; II, 44, 45, 166, 209.
Nichols, J. H., I, 144, 217, 221 ; II, 114, 123, 131, 132, 260, 315, 318, 327, 345 ; his remarkable office, I, 151 ; brings news of 2d Bull Run, 226.
Nicholson John, II, 357; death of, 79.
Norris, D. E., II, 123, 248.
Norris, J. B., I, 144,
Northrup, C. C., II, 315.
Northrup, D. B., II, 344.
Norton, G. W., II. 25, 357.
Nunan, P. H., I, 144; II, 114, 267, 315.
Nye, J. M., II, 28, 357.

O.

Oath, operators, form of, I, 96; II, 109.
O'Brien, John E., I, 138, 146; II, 114, 131, 278, 315.
O'Brien, Richard, I, 67, 69, 71, 137, 140, 146; II, 114, 131, 132, 260, 274, 316, 335, 340, 344-5, 350.
Olmstead, Jas. Q., I, 253, II, 83, 188.
O'Meara, T., I, 113.
O'Neil, A. W., I, 120, 253, 256, 328 ; II, 45, 209, 298, 301, 342, 343.
Operators, see Status of, etc.; reviewed, I, 72.
O'Reilly, Henry, I, 132, 133.
O'Reilly Luke, I, 117, 179, 180, 181, 316, 339.
Orton, A. W., I, 217; II, 114, 123.
Osborn, J. M., II, 119.

P.

Painter, L. M., II, 260, 315.
Palmer, Sol., I, 173, 175, 243, 258, 261, 317, 323; II, 85, 176.

Paris, France, The balloon in the siege of, I, 17; Tellegraph, 29.
Parsons, John B., I, 70.
Parsons, Jas. K., I, 185, 253.
Parsons, J. W., II, 114.
Parsons, W. H., I, 208, 212, 213, 242, 253, 317, 322, 328; II, 75, 81; driven from Farmington, I, 243; important service of, II, 73.
Parker, L. D., I, 120, 177, 212, 213, 214.
Patterson, Gen., failure of, in the valley, I, 74.
Patterson, William, I, 103, 289; II, 54, 67, 68, 75.
Patton, W. H., II, 343.
Paxson, C. A., I, 340, 346.
Paxson, O. W., I, 248.
Payne, C. S., I, 117, 180.
Peabody, S. P., I, 204-5, 210, 276, 291.
Pea Ridge, battle of, I, 179.
Pearson, C. W., I, 258; II, 188, 192, 299, 343; captured, 203.
Peck, G. H., I, 340.
Peebles, B. H., I, 243, 253.
Peel, Edwin, I, 253, 328, 357; II, 188, 192.
Peeler, T. M., I, 349.
Pegues, R. H., II, 120.
Peirce, Geo. C., II, 54; captured, 65.
Pemberton, Gen. J. C., supersedes VanDorn, I, 257; surrenders Vicksburg, 319.
Penn, Geo. R., II, 211, 299.
Peninsular, campaign, 1, 141.
Pensacola, evacuated, II, 197.
Perdue, S. Ford, II, 54.
Perryville, battle of, 1, 294.
Petersburg threatened by Butler, II, 131; siege, 257; mine disaster, 257; evacuated, 320.
Peterson, J. H., I, 328.
Pettit, James E., I, 56; II, 203-6; captured, 230, 327.
Phelps, R., II, 28.
Phillippi, battle of, I, 96.
Phillips, C. C., II, 149.
Photophone, I, 31.
Pidgeon, G. S., I, 120, 253.
Pierce, J. B., I, 169, 216, 226, 363; II, 12.
Pierson, W. S., II, 54.
Pitcairn, Robt., I, 347.
Pitcairn, W. W., I, 287.
Pitton, James, II, 84, 88.
Pittsburg Landing—see Shiloh.
Plant, H. B., II, 345.
Plum, H. W., I, 296; II, 54, 75; first military experiences of, I, 277.
Plum, Wm. R., I, 56, 191, 194, 197, 210, 296, 305, 306, 328; II, 54, 87, 166, 169, 180, 185, 232, 238-9, 281, 292, 327.
Polk, Gen., occupies Columbus, Ky., I, 120.
Pond, C. H., II, 63; captured, I, 292.
Pope, Gen., captures New Madrid and Island No. 10, I, 183; commands Army of Va.—his campaign, 216.
Port Hudson—Banks moves against, II, 40; siege and capture, 41.
Porter, Admiral, up the Red River, II, 45; at Ft. Fisher, 274.
Porter, Gen. F. J., his command saved by telegraph, I, 149.
Porter, James D., II, 343.
Porter, J. K., II, 223.
Porterfield, driven out of Phillippi, W. Va., I, 96.
Potts, J. D., II, 344.
Power, Jr., R., I, 144, 359, 360; II, 123, 272-4, 315.
Powers, J. M., II, 120.
Prairie Grove, battle of, I, 345.
Presentiment, remarkable, II, 179.
Pressly, James, II, 190, 343.
Price, Gen., early operations in Mo., I, 170, 179; fight with Rosecrans, 255; attacks Corinth, 255; invasion of Mo., 214, 217.
Prison experience of operators, I, 161, 330; II, 203, 230.
Pritchard, A. P., cures a cold, I, 83, 84.
Printz, Geo. W., I, 107.
Purdon, Geo., I, 209; II, 63, 115.
Purdy, Geo. A., I, 180; captured, 181.
Purnell, J. W., I, 178.

Q.

Quarles, J. T., II, 344.
Quate, J. L., I, 117, 180: captured, 181.

R.

Rabbeth, J. T., II, 54: railroad raid, daring, I, 209; railroads, military, service of, II, 273, 337.
Railton, Geo., II, 54, 357.
Rand, D. E., II, 12, 122, 136, 260, 261, 266, 315, 318.
Ransford, Jr. H., I, 47.
Rawlings, T. E., II, 54; captured, 236.
Reaser, Robt., death of, II, 240.
Record, A. M., II, 119, 344.
Reconstruction of the States, II, 339.
Reddington, W. B., I, 209, 258.
Red River, Bank's campaign up, II, 45.
Reed, Jerry L., II, 288, 343.
Reese, Sam, II, 60, 64.
Reflection, means of signaling, I, 29.
Reid, Douglass, I, 243.
Reid, J. D., II, 146.
Repairers, captured, I, 189, 255, 264, 277, 349; II, 266; hang a guerrilla, II, 49; killed, I, 325; II, 49, 168; as daring couriers, 77; their unrequited service, 336.
Reprieves by telegraph, II, 123.
Richmond evacuated, II, 320.
Rich Mountain battle, I, 97.
Richardson, J. D., I, 288; II, 54.
Richardson, J. F., II, 343.
Richardson, J. W., I, 208.
Riley, J. J. G., I, 70.
Risdon, S. L., I, 201.
Roberts, C. M., I, 144.
Roberts, Chas. R., II, 114, 315.
Roberts, M. E., II, 28.
Roberts, S. B., II, 59, 63, 114.
Robinson, E. N., I, 144.
Robinson, Steve L., I, 253, 262, 318; II, 188, 205-6, 341; wounded, 201.
Roche, Thos., II, 28.
Rodgers, T. J., II, 343.
Romney, W. Va, enemy driven off by Wallace, I, 97.
Rose, L. A., I, 75, 76, 77, 83, 144, 232; II, 11, 12, 17, 122, 126, 139, 263; important service of, 133.
Rosecrans, attacks Price, I, 255; defends Corinth, 255; supersedes Buell, 295; at Stone River, 298; advances from Murfreesboro, II, 53; crosses the Tenn., 64; in Missouri, 214; repels Price, 217.
Rosewater E., I, 169, 217, 228; II, 114.
Rouser, G. W., I, 201.
Rowe, Charles O., I, 168-9; II, 148, 149, 268.
Rowe, R. E. D., II, 268.
Royce, H. P., I, 144; II, 114.
Rugg, J. H., I, 117.
Runners, news carried by, I, 11.
Rupley, S. K., II, 12, 266, 315, 318.
Russell, A. G. McN., II, 120, 343.
Ryan, Crosby J., I, 70, 84; II, 139.
Ryan, R. H., II, 12, 114, 123, 127.

S.

Sabin, Charles M., II, 304.
Sabin, J. J., II, 315.
Sabine Cross Roads, battles of, II, 46.
Sabine John F., II, 264, 267.
Safford A. G., II, 28, 123, 264, 315.
Salaries of operators, II, 106.
Sampson, J. W., I, 217; II, 12, 114.
Sampson, T. M., I, 169, 208, 289; II, 63, 115.
Sanborn, Gen. J. B., compliments the U. S. M. T., II, 353.
Sanborn, F. A., II, 209.
Sanborn, E. S., renders valuable aid, I, 68.
Sandford, Wm., II, 117, 248, 300, 343.
Sargent, Stephen, is babtized, I, 83.
Savage Station, battle, I, 151.
Savannah, campaign, II, 226, 241; capture of, 255.
Saville, Alf, 1, 249.

Saville, E., II, 117, 119; taps Union line, I, 249.
Schermerhorn, Ed., I, 119, 178, 253, 258, 264, 315, 317, 320, 324, 325, 328; II, 267.
Schnell, A. C., I. 144.
Schnell, Jos., I, 144; II, 114.
Schnell, T. M., I, 144.
Schofield, Genl., at battle of Franklin, II, 234. advances from Chattanooga, II, 174; fights battle of Franklin, 234.
Scott, Lt.Gen., practically relieved by McClellan, I, 78.
Scott, S. W., II, 64.
Scott, Col. Thomas A., restores communication to Washington, I, 66; aid of, II, 17.
Sears, Jos. L., I, 348; II, 49, 223, 267, 278.
Sedgwick, Gen., in battle of Fredericksburg, I, 363.
Selma, Ala., captured; II, 286.
Sellers, D. C., I, 125, 201, 205.
Semaphores, their invention and use in war, I, 21, 22.
Seven Pines battle, I, 148.
Seymour, M. T., II 149.
Shape, E. M., II, 236.
Shape, G. H., II, 236.
Sheldon, Geo. D., I, 144. 169, 362; II, 27, 131, 260, 315, 350.
Sheldon, L. F., commissioned, I, 131; II, 27.
Sheldon, W. A., I, 316; hunts guerrillas, II, 42.
Shepard, D. P., II, 341.
Sheppard, J. E., II, 344; arrested, 43; sick 45.
Sheridan,Gen., victories of in Shenandoah Valley, II, 270; at Five Forks, 316; compliments the U.S.M.T., II, 353.
Sheridan, J. A., II, 131, 267.
Sherman, I. A., II, 209, 211, 286, 299, 310.
Sherman, Gen. T. W., operations along the Atlantic coast, II, 25.
Sherman, Gen. W. T., I, 176; takes command in Ky., 124; repulse at Chickasaw Bluffs, 272; goes to Chattanooga, II, 72, at Missionary Ridge, 74; saved by Telegraph, 80; Vicksburg district, 189; advances into Ga., 174; prepares for Savannah campaign, 226; the campaign, 241; S. and N. C. campaign, 275; surrender of Johnston, 335.
Shiloh, battle of, I, 211.
Shock, W. W., II, 315.
Sholes, Cass G., II, 75. 176, 181-2; tastes lead, 177; loses cipher key, 185.
Showerman, I. C., II, 54.
Shreffler, W. H., I, 144 217, 224, 225; II, 123, 127, 263.
Shrigley, J. A., I, 349; II, 224
Shurr, S. M., I, 107, 169; II, 149, 278.
Shutt, R. A., II. 267.
Sigel, Gen., joins Lyon, I, 110, in Shenandoah Valley, II, 143.
Sigler, J. H., I, 208.
Signals, fire, I, 19; semaphoric, 21; flag, translated by Federals, I, 35; news by, in early times, I. 23.
Signal Corps versus the U. S. M. T., II, 86.
Sisson, W. H., II, 63, 115.
Skinner, Ira G., I, 242, 253, 258, 317.
Slavery, Proclamation of Emancipation, I, 239.
Sledge, P. II, 344.
Smith, Gen. A. J., expedition of, I, 316; II, 199, 202.
Smith, A. W., I, 56.
Smith, C. B., II, 310.
Smith, D. K., I, 258,; II, 149
Smith, D. W., II. 28, 99, 328.
Smith, E. T., II, 345.
Smith, George H., I, 47, 144, 145, 217, 225, 242, 265, 333, 336, 346; II, 104; appointed supt. of Fremont's lines, I, 113; his plans for military organization, 114; commissioned, 131, 251; at Savannah, 210; daring and success of, 247; resigns, II, 215; biographical, 215.
Smith, G. K., I, 107, 169; II, 149, 268.
Smith, H. L., I, 70, 71.
Smith, Lt. I. Elliot, supt. Butler's telegh's, II, 38.
Smith, J. W., I, 69.
Smith, Gen. Kirby, moves into Ky., I, 280, 286; defeats Banks in La., II, 45; surrenders, 312.
Smith, Richmond, II, 242-3-6, 251, 252.
Smith, R. Hector, I, 336.
Smith, S. E., II, 343.
Smith, Gen. Sooy, expedition of, II, 189.
Smith, T H., II. 58, 64.
Smith, W. K., II, 151, 153, 270; taps wires, 143.
Snell, Fred W., I, 316; II, 80, 188, 191, 193, 198, 204.
Snow, H. N., II, 131, 318, 327.
Snyder, A. C., I, 69, 70, 83; II, 315.
Snyder, C. L., I, 144, 217; II, 114, 123, 131, 260.
Somerville, Wm. B., I, 253, 256, 258; II, 80, 84, 176-7, 188, 357; arrested, I, 267.
Southern Telegraphic Association, II, 116.
South Mountain, battle of, I, 234.
South-western Telegraphic Association, II, 119.
Speed, F. M., II, 233, 239.
Spellman, C. H., II, 80, 188, 203.
Spellman, G. E., II, 82, 204.
Spellman, L. B., I, 36, 56, 60, 242, 258, 261, 316, II, 80, 84, 176, 188, 191, 341.
Spencer, J. M., II, 63.
Spies hung, I, 305, 349.
Sponagle, J. L., I, 348; II, 278.
Spottsylvania, battle of, II, 135.
Springfield, Mo., attacked I, 340.
Stager, Anson, I, 60, 97, 106, 125, 228, 265, 287, 321, 328, 353, 357; II, 39, 47, 104, 108, 110, 113, 115, 147-8, 193, 215, 292, 296, 337, 340; arranges first cipher key, I, 44; takes charge lines in Ohio, 92; order of assignment in Dept., 93; organizes the corps, 128; biographical, 132; line to Ft. Monroe, 137; disagreement with Halleck, 175; with Grant, 265; and Stanton meet Grant, II, 71; important service of, 77; orders line to Memphis, 82; recommends exclusion of Signal wires in field, 100; prevents conscription of operators, 107; interference by Grant with cipherer, 170; final report of, 346; annual reports of, 359.
Starling, C. C., I, 169, 231, 235; II, 149, 239; captured, 233.
Status of U. S. and C. S. Mil. operators, chapter on, II, 106.
Stebbins, C. M., I, 121; II, 146.
Steele, Gen., capture of Little Rock, I, 332; advance to join Banks, II, 45, 46, unsuccessful, 48.
Stevens, J. F., II, 314.
Stewart, F., II, 314, 346, 350.
Stidham, P. A., I, 96, 169.
Stillman, Geo., I, 185, 253.
St. John, C., II, 119, 343.
Stone, Ellis, I, 258; II, 80, 82, 188.
Stone River, battle of, I, 298.
Stoneman, Gen., raids of, I, 363, 368; II, 162, 281.
Stough, John B., I, 71; II, 25.
Stouder, H., II, 38, 43, 209.
Streight, Col., command captured by Forrest, II, 51.
Strike, operators, II, 114.
Strouse, David, I, 67 75; services and death, 79.
Strubbe, W. T., II, 115.
Stuart, Gen., raids of, I, 356; to Chambersburg, Pa., 240; into Pa., II, 16.
Stumm, F. A., I, 114.
Sturges, Gen., raid of, from Memphis, defeat, II, 198.
Suffolk, siege, of, I. 362.
Sultana disaster; 1,400 killed or drowned, II, 307.
Sullivan, J. C., I, 242, 248. 258, 261, 288, 317; II, 205.
Sweet, E. D. L., I, 119.
Swift, James A., captured, I, 66.

T

Talbot, Robt. M., II, 53.
Talmadge, G. J., I, 180.
Taylor, Gen., surrenders to Canby, II, 304.
Taylor, H. B., II, 343.

Telegrams seized by Govt. throughout North, I, 69.

Telegraph, Electric, its invention, I, 24-5; Congressional aid, 25; its first use in war, its use in India (1857), 26; in Germany, in England, in Spain, in Ashantee war, 27; at Sadowa and in U. S., 28; in siege of Paris, 28.

Telegraphs, origin of the U. S. Military, I, 62; during Bull Run fight, 72; No. miles in Dept. Potomac to Oct. 31, 1861, 91; first to follow army in U.S., 97; erected to Nov. 15, 1861, 125; C. S., 134; before Yorktown, 143; abandoned in W. Va., 168; toward Pea Ridge, 179; for Pope at New Madrid, 183; for Buel and Thomas in Ky., 187; to Nashville, 203; Mitchell's in N. Ala., 208, to Savannah, Tenn., 210; in Pope's eastern campaign, 217; in Antietam campaign, 230 to 241, under Halleck and Grant, 252; with Buell from Louisville, 295, 297; on Stone River campaign, 298; at Vicksburg, 317; in Ark., 336; on Fredericksburg campaign, 351; in Chancellorsville campaign, 359; Gettysburg campaign, II, 9; along Atlantic coast and in N. C., 25; in the Gulf Dept., 37; in siege of Pt. Hudson, 41; into E. Tenn. under Burnside, 59; in Ky. and E. Tenn. under Burnside, 63; in Chickamauga campaign, 64; complimented by Sec. War, 104, 321, 355; with Meade, 126; in Wilderness, 128; about Petersburg, 131, Spottsylvania, 135; in Early's dash into Maryland, 152; about Washington, 153; in Ky. and E. Tenn., 160; on Atlanta campaign, 176; Canby's in Gulf Dept., 208; in Price's Mo. campaign, 214; in Nashville campaign, 236; siege of Petersburg, 257; with Schofield in N. C., 273; in Stoneman's last raid, 282; in Mo. and Ark. in Fall of 1864, 289; in Mobile campaign, 297; summary, 335; in the reconstruction of the States, 339; resume, 349; evidence of its value, 352; annual reports of Gen. Stager, 359.

Terry, Gen., capture of Ft. Fisher, II, 274.

Thayer, Wm. A., I, 253; II, 80, 82, 188.

Thielgard, S. C., II, 343.

Thode, John, II, 236.

Thomas, C. J., I, 169; II, 149, 268, 278.

Thomas, John, II, 99, 274.

Thomas, Dr. John C., II, 343.

Thompson, Frank, II, 342.

Thompson, J. R., I, 178.

Thornton, J. G., I, 136; II, 198, 343.

Tiffany, J. T., I, 120, 175, 243, 253, 259.

Tinker, Charles A., I, 35, 60, 84, 144, 234; II, 123, 314, 323, 350; translates enemy's cipher, I, 41.

Tinker, W. A., II, 42, 209.

Tinney, J. D., I, 220, 243; II, 11, 12, 122, 132, 315, 317, 352.

Tinney, W. E., I, 69, 83.

Tomlinson, D. B., I, 217; wounded, 223.

Topping, N. B., II, 344.

Torrence, J. A., II, 64, 77, 316, 324.

Thomas, Gen. G. H., defeats Zollicoffer, I, 187; see "Stone River," "Chickamauga;" in command Dept., Cumberland, II, 72; Missionary Ridge, 74; Defense of Nashville, 227; defeats Hood, 238, 283; on U. S. M. T., 355.

Thurston, J. D., II, 267, 315.

Tidd, W. L., II, 60; important service of, I, 287, 298; wounded and captured, 299.

Townsend, G., I, 169.

Townsend, N. S., II, 176, 186; brave service, 70.

Trabue, G. W., II, 119, 344.

Tracy, R. H., II, 188.

Tree, J. B., II, 248, 285, 304, 344.

Treuenfeld, B. von F., author "*Krieg's Telegraphie*," I, 27; his book, 28, 341.

Trumpets, their use in war for signaling, I, 17.

Tull, C. D., I, 107, 169; II, 149, 269.

Turner, S. S., II, 120.

Turner, W. H., II, 344.

Tweedy, A., II, 343.

Tyler, F. B., I, 208; II, 54; captured, I, 292; death, II, 240.

Tyler, J. D., I, 159, 216, 231, 236; captured, 233.

Tyrrill, Aug., I, 253; II, 85, captured, 262.

U.

Ulmer, S. S., II, 106, 209.

Uniform of operators, II, 110.

Upham, B. H., II, 209, 211, 299.

V.

Valentine, H. S., II, 209.

VanDerhoef, R. B., II, 278.

VanDorn, Gen., supersedes Price, I, 179; raid of, 257, 260.

VanDuzer, John C., I, 173, 175, 250, 256, 258, 259; II, 69, 70, 78, 158, 159, 175, 177, 181, 182, 239, 242, 250, 293, 340, 342, 348; commissioned 131; relieved by Grant, 265; order countermanded, 268; biographical, 303; reports operations, II, 53, 64, 67; recommends mil. organization, 111.

VanGesel, W. T., II, 315.

VanGilder, A. C., II, 166.

VanHorne, John, I, 135; II, 344.

VanValkenburg, F. S., I, 114, 175, 253; II, 54, 75, 175, 176, 177, 239; arrested, 186; taps enemy's lines, 307.

Vermillion, J. W., I, 169; II, 149.

Vicksburg, siege and capture of, I, 313.

Vincent, H. C., I, 280, 287, 293.

Vincent, O. B., II, 132, 260.

Voice, messages vocally transmitted, I, 12.

Volker, Louisa E., I, 345; II, 218.

Volker, R. C., II, 219.

VonEye, E., II, 38, 209, 306-7.

W.

Wade, J. H., II, 344.

Wade, R. P., I, 250, 269; commissioned, 131.

Wagner, R., I, 200, 208, 287, 296.

Waite, J. R., I, 144.

Wallace, Chs., I, 253; II, 85, 188.

Walsh, A. R., I, 253; II, 85, 188.

Walton, Nat., captured, II, 189, 343.

Ward, P. H., II, 344.

Ware Will F., captured, II, 284.

Warner, D. W., II, 149.

Warren, Geo., II, 267.

Washburn, M. E., II, 219.

Washington saved, I, 63.

Waterhouse, C. D., I, 114.

Waterhouse, E. J., I, 336.

Waterhouse, H. F., death of, II, 35.

Waterhouse, T. Q., I, 217, 224; II, 260-7, 315; nearly captured, I, 157.

Ways, Ch. C., II, 12.

Webb, J. G., I, 178; captured, 277.

Webb, ram, destruction of, II, 305.

Weber, C., II, 136.

Weeden, R. B., II, 343.

Weems, J. D., I, 191, 288.

Weir, L. C., I, 47, 212, 243; delivers important order, I, 116.

Weitbrecht, R. F., I, 358; II, 12, 114, 123-6, 151; captured, I, 360.

Weitzel, Gen., enters Richmond, Va., II, 320.

Wells, V. H., II, 264, 315.

Westervelt, J. D., II, 120.

Wheeler, Gen., defeated, I, 297; driven, II 184.

Whitaker, S. W., II, 344.

White, Gen. Julius, abandons Winchester, I, 231.

Whitford, Edward P., II, 242, 342.

Whitney, C. O., II, 81.

Whitney, C. D., II, 176, 181, 185, 239.

Whitney, Wm., II, 285.

Whitthorne, F. C., II, 119, 246.

Whittlesey, C. S., I, 253.

Wickard, J. W., I, 56, 239.

Wickham, J. J., I, 56, 282; captured, 274.

Wilber, J. T., II, 63, 115, 176.

Wilderness, II, 132.

Wilkie, J. J., II, 38, 166.

Wilkinson, G. W., II, 115.

Wilkinson, G. D., II, 63, 239.

Williams, D. A., opens U. S. Grant's first office, I, 112.

Williams, J. S., II, 149.
Williams, R. D., II. 315.
Wilmington, N. C., captured II, 274.
Willis, Robert, nearly murdered, II, 290.
Wilson Creek, Mo., battle of, I, 113.
Wilson, E. K., II, 344.
Wilson, Ellis J., I, 204-5 276, 296, 305-6; II, 54, 75.
Wilson, G. W., II, 176, 181.
Wilson, Gen. J. H., raid of, in Ala. and Ga., II, 285.
Wilson, J. J. S., I, 47, 111, 173-5-6, 242; appointed supt. in U. S. M. T., 117; biographical, 118; builds line to Paducah, 120.
Wilson, W. B., I, 70. 77, 82.
Wilson, W. H., II, 63, 132, 149, 260-3, 315, 327.
Winder, Alf., I, 169; II, 186; captured, 149.
Windsor, W. B., I, 122.
Winter, G., II, 344.
Wintrup, John, I, 144; II, 114, 315.
Witherspoon, J. K., II, 276.
Wood, W.B., II, 123, 315; sends news of fall of Richmond, 321.
Woodhouse, E. N., II, 120.
Wolf, C., I, 169.
Woodring, W. H., I, 117, 179, 341; II, 49, 106, 223, 267; brave service of, I, 341-3.
Woodward, Col., captures Clarksville, I, 277; defeat at Ft. Donelson, 277.
Woolums, J. W., I, 196.
Worden, J. M., II, 304, 343.
Worth, Robt. G., II, 149.
Wortman, E. F., II, 278.
Wortsman, L. W., II, 28.

Y

Yellow Fever, deaths from, II, 35.
Yonkers, S. T., I, 185, 209.
York, A. W., II, 248.
York, C. W., I, 243; II, 312.
York, Geo. C., I, 117.
Yorktown besieged by McClellan, I, 143; evacuated, 145.
Young, W. H., II, 344.

Z.

Zantzinger, Louis F., II, 300, 304.
Zollicoffer, Gen., killed, I, 187.
Zook, Paul Gen., killed, II, 18.

TELECOMMUNICATIONS

An Arno Press Collection

Abramson, Albert. **Electronic Motion Pictures:** A History of the Television Camera. 1955

[Bell, Alexander Graham]. **The Bell Telephone:** The Deposition of Alexander Graham Bell in the Suit Brought By the United States to Annul the Bell Patents. 1908

Bennett, A. R. **The Telephone Systems of the Continent of Europe** and Webb, Herbert Laws, **The Development of the Telephone in Europe.** 1895/1910. Two vols. in one

Blake, George G. **History of Radio Telegraphy and Telephony.** 1928

Bright, Charles. **Submarine Telegraphs:** Their History, Construction and Working. 1898

Brown, J. Willard. **The Signal Corps U. S. A. in the War of the Rebellion.** With an Introduction by Paul J. Scheips. 1896

Chief Signal Officer, U. S. Signal Corps. **Report of the Chief Signal Officer: 1919.** 1920

Danielian, N[oobar] R. **A. T. & T.:** The Story of Industrial Conquest. 1939

Du Moncel, Count [Theodore A. L.] **The Telephone, the Microphone, and the Phonograph.** 1879

Eckhardt, George H. **Electronic Television.** 1936

Eoyang, Thomas T. **An Economic Study of the Radio Industry in the United States of America.** 1936

Everson, George. **The Story of Television:** The Life of Philo T. Farnsworth. 1949

Eyewitness to Early American Telegraphy. 1974

Fahie, J[ohn] J. **A History of Electric Telegraphy to the Year 1837.** 1884

Federal Communications Commission. **Investigation of the Telephone Industry in the United States.** 1939

Federal Communications Commission. **Public Service Responsibility of Broadcast Licensees.** 1946

Federal Trade Commission. **Report of the Federal Trade Commission on the Radio Industry.** 1924

Fessenden, Helen M. **Fessenden:** Builder of Tomorrows. 1940

Hancock, Harry E. **Wireless at Sea:** The First Fifty Years. 1950

Hawks, Ellison. **Pioneers of Wireless.** 1927

Herring, James M. and Gerald C. Gross. **Telecommunications:** Economics and Regulations. 1936

Lodge, Oliver J. **Signalling Through Space Without Wires:** Being a Description of the Work of Hertz and His Successors. [1900]

McNicol, Donald. **Radio's Conquest of Space:** The Experimental Rise in Radio Communication. 1946

Plum, William R[attle]. **The Military Telegraph During the Civil War in the United States.** With an Introduction by Paul J. Scheips. Two vols. 1882

Prime, Samuel Irenaeus. **The Life of Samuel F. B. Morse,** L.L. D., Inventor of the Electro-Magnetic Recording Telegraph. 1875

The Radio Industry: The Story of Its Development. By Leaders of the Radio Industry. 1928

Reid, James D. **The Telegraph in America:** Its Founders, Promoters and Noted Men. 1879

Rhodes, Frederick Leland. **Beginnings of Telephony.** 1929

Smith, Willoughby. **The Rise and Extension of Submarine Telegraphy.** 1891

Special Reports on American Broadcasting: 1932-1947. 1974

Thompson, Silvanus P., **Philipp Reis:** Inventor of the Telephone; A Biographical Sketch. 1883

Tiltman, Ronald F., **Baird of Television:** The Life Story of John Logie Baird. 1933

Wile, Frederic William. **Emile Berliner:** Maker of the Microphone. 1926

Woods, David L., **A History of Tactical Communication Techniques.** 1965